Design of Steel-Concrete Composite Bridges to Eurocodes

Design of Steel-Concrete Composite Bridges to Eurocodes

Ioannis Vayas and Aristidis Iliopoulos

CRC Press is an imprint of the
Taylor & Francis Group, an **informa** business
A SPON PRESS BOOK

CRC Press
Taylor & Francis Group
6000 Broken Sound Parkway NW, Suite 300
Boca Raton, FL 33487-2742

Printed on acid-free paper
Version Date: 20130725

International Standard Book Number-13: 978-1-4665-5744-4 (Hardback)

Library of Congress Cataloging-in-Publication Data

Vayas, Ioannis.
Design of steel-concrete composite bridges to eurocodes / Ioannis Vayas, Aristidis Iliopoulos.
pages cm
Includes bibliographical references and index.
ISBN 978-1-4665-5744-4 (hardback)
1. Concrete bridges--Standards--Europe. 2. Iron and steel bridgesIron and steel bridges--Standards--Europe. 3. Reinforced concrete. I. Iliopoulos, Aristidis. II. Title.

TG55.V39 2014
725'.98--dc23 2013027554

Visit the Taylor & Francis Web site at
http://www.taylorandfrancis.com

and the CRC Press Web site at
http://www.crcpress.com

To my father Dinos

Ioannis Vayas

To my mother Maria

Aristidis Iliopoulos

Contents

Foreword

Composite structures of steel and concrete have become popular for a number of reasons. One reason is that while concrete is excellent for dealing with compressive forces, steel also can carry large tensile stresses. In some sense, any reinforced concrete beam is a composite structure, with the reinforcement in composite action with the concrete.

Furthermore, it is often necessary when constructing concrete structures to carry the concrete before it has hardened. The steel girders' ability to support the formwork, the rebars, and the wet concrete has indeed contributed to the increasing popularity of composite structures. For the case of a road bridge, every month of earlier opening saves money for the contractor, but also for the road user. This fact is often neglected when evaluating different structural solutions.

I strongly believe that this book could spread knowledge about composite bridges while teaching students and experienced designers the techniques of designing composite bridges according to the Eurocodes.

It has been a long process to write the Eurocodes, but the advantages of having a common set of codes are obvious, making it easier to design and construct bridges all over Europe, which may not only lower prices in individual projects but also provide cost-effective solutions. Moreover, it will also be more effective to implement in European R&D projects, since researchers and designers will be using the same set of codes. Last but not least, it will give engineers an opportunity to work in other countries, which is important in order to attract talented young engineers. In the long run, I also think that the Eurocodes will be spread to countries outside Europe. The picture shows a haunched composite box girder bridge over the river of Ljusnan, Sweden.

Peter Collin
Professor, Luleå University of Technology
Bridge Designer, Ramböll
Luleå, Sweden

Preface

Bridges have a strong symbolism as they connect opposite sides. It is not a coincidence that bridges are illustrated on one side of Euros. For many engineers, bridge design assumes top priority in their practice. In fact, the design of the optimal, technical, economical, and aesthetical solution with existing tools and means is a challenge for any structural engineer. International best practices show that the search for alternative solutions and the choice of the optimal one are essential for the construction of a successful bridge. The criteria for selection may be governed by technical, economical, operational, aesthetical, and environmental considerations and may be the choice of either the owner, the contractor, the designer, or the user. As a result, bridge construction builds a healthy competition as regards structural types, construction materials, construction methods, and other parameters that result in a polymorphy of bridges. In some aspects, bridges reflect the technological development in a period or in a country.

Among the different types of bridges, composite bridges have a significant place, because the combination of the most common construction materials, steel and reinforced concrete, allows the construction of safe, operational, durable, and robust bridges economically. Bridge design is strongly based on prescriptive normative rules regarding loads and their combinations, safety factors, material properties, analysis methods, required verifications, and other issues that are included in the codes. Composite bridges may be designed in accordance with the Eurocodes, which have recently been adopted across the European Union and many other countries worldwide. Eurocode 4, part 2 (EN 1994-2), is exclusively devoted to the design of composite bridges. However, many Eurocodes and their different parts would need to be consulted as the design of bridges includes a variety of constructional issues and due to the fact that two construction materials are involved.

This book presents in 13 chapters the main information needed for the design of composite bridges in accordance with the Eurocodes:

- Chapter 1 introduces the subject and provides a list of symbols used in the book.
- Chapter 2 presents the main types of common composite bridges. It discusses structural forms and structural systems, describes preliminary design aids and erection methods, and delves into the structural details.
- Chapter 3 summarizes the relevant design codes. These refer to actions, combinations of actions, safety factors, material properties, and limit state design.
- Chapter 4 discusses the actions to be considered, including traffic loads for road and railway bridges, temperature, wind, and earthquake, as well as the effects of shrinkage and creep of concrete.
- Chapter 5 introduces the limit states, presents safety factors for actions and resistances and factors for combinations of actions, and discusses durability issues.

- Chapter 6 presents the properties for structural materials used with reference to concrete, structural, reinforcing, and prestressing steel, as well as mechanical connectors, and provides criteria for their selection.
- Chapter 7 is devoted to modeling for global analysis. It presents alternative models for global analysis, shear lag effects, effects of rheological behavior, and cracking of concrete and models for slab analysis.
- Chapter 8 discusses the effects of plate buckling and deals with the critical and post-critical plate behavior and the column-buckling behavior of stiffened plated elements.
- Chapter 9 presents the verifications at ultimate limit states for members and cross sections with reference to various cross-sectional classes and for lateral torsional buckling.
- Chapter 10 refers to verifications at serviceability limit states, which include stress limitations, web breathing, control of cracking of concrete, deflections, and vibrations.
- Chapter 11 lays down the rules for the shear connection between concrete and steel by means of headed studs.
- Chapter 12 presents fatigue analysis and design. It presents fatigue load models, detail categories, and fatigue verifications for structural steel, reinforcement, concrete, and shear connectors.
- Chapter 13 covers structural bearings and dampers, with an emphasis on reinforced elastomeric bearings.

Covering all topics related to composite bridge design is a challenging task. The authors have tried to provide as comprehensive a coverage as possible. Although not all types of bridges, for example, arch, cable-stayed, or suspension bridges, are fully covered and their substructures and foundations not addressed, the basic knowledge on steel structure is dealt with and topics related to analysis and design for the overwhelming majority of composite bridge superstructures are addressed.

The book is didactical and is addressed primarily to structural engineering students. However, it might also be helpful for bridge designers or practicing engineers who are converting from other codes to Eurocodes. For better understanding of the design procedures and the use of code provisions, several design examples are incorporated in the chapters.

Acknowledgments

We would like to thank the following for their kind contributions: Prof. P. Collin from the Luleå University of Technology, Sweden; Mike Banfi, Matt Carter, and Tim Hackett from ARUP, UK; and Daniel Bitar and Theodore Adamakos from Centre Technique Industriel de la Construction Metallique (CTICM), France.

Ioannis Vayas would like to thank all his friends, colleagues, and PhD students for their patience, tolerance, and support.

Aristidis Iliopoulos is grateful to Topi Paananen, Taru Leinonen, and Simo Peltonen from the Peikko Group Corporation, Finland, for their support and their common vision in innovative engineering projects.

Ioannis Vayas
Aristidis Iliopoulos

Authors

Ioannis Vayas is professor and director of the institute of steel structures at the National Technical University of Athens. He graduated in civil engineering at the same university and received his Dr-Ing from the Technical University of Braunschweig, Germany, and welding engineering degree from SLV Hannover, Germany. He has been involved in research, national and European codification, and consultancy on steel structures and bridges for over 30 years.

Aristidis Iliopoulos is a structural engineer and received his Dr-Ing from Ruhr University Bochum, Germany. He has participated in numerous projects such as constructing longspan roofs, steel–concrete composite bridges, and steel buildings in seismic areas. Iliopoulos currently serves as an R&D engineer for the multinational company Peikko Group Corporation. He is also a member of the CEN/TC 250/SC4 Evolution Group for EN 1994-2 for composite bridges.

Chapter 1

Introduction

1.1 GENERAL

Bridges are built to overcome an obstacle, whether a valley, a strait, a river, or an existing road or railway line. They carry traffic from pedestrians, cyclists, vehicles, or trains, being distinguished as footbridges, road, or railway bridges. In some cases, they carry exclusively pipelines (e.g., for gas or water). Bridges are mostly fixed but may be movable if they obstruct sea traffic in harbors or channels or are used for lifting of entire ships.

A bridge is composed of the superstructure, the substructure, and the foundation. Bearings are inserted between super- and substructure or, in some cases, between substructure and foundation, unless for integral bridges where bearings are missing. The superstructure may be composed of plate, box, or truss girders that act alone or are supported by arches, portal frames, stay cables, or suspension cables. The substructure includes the abutments, the piers, and any pylons or towers. Spans vary from a few meters to nowadays 2 km, but short spans, let's say <5 m, are defined as culverts and not bridges. The construction materials for super- and substructures may be stone, timber, steel, and reinforced or prestressed concrete.

Bridge design and construction is one of the most challenging issues for a structural engineer. Choices have to be made in respect to structural systems, construction materials, foundation types, or execution processes that are based on structural performance, construction and maintenance costs, local conditions, and aesthetics. Architects are recently involved, not always with success, in the conceptual design phase. However, structural engineers are the main actors in design that includes the final and the construction stages, the latter being so important that it is said that a bridge design is primarily the design at construction stages.

Composite bridges are composed of steel girders and reinforced concrete decks to combine the benefits of both structural materials. They are suitable for almost all bridge spans, except the very long spans, having the advantages of reduced dead loads, high prefabrication, simpler and quicker erection, and simpler maintenance procedures when compared with reinforced concrete ones.

Following a decision by the European Commission to develop harmonized technical specifications in the field of construction, and after an effort of almost 40 years, the *Eurocodes* were prepared to be used as design codes in replacement of national standards. By the correct use of the Eurocodes and all underlying standards, it is demonstrated that construction works including bridges are sufficient safe, fit for purpose, and robust.

This book intends to cover the design of composite bridges, primarily of girder-type bridges, following the provisions of the Eurocodes as they are understood by the authors. It provides information for their background and guidance on their application. Design examples are introduced in each chapter for better understanding of the design methods.

1.2 LIST OF SYMBOLS

Effort was made for the symbols to be in accordance with those used by the Eurocodes. A single symbol is used for those quantities where different symbols are used by different Eurocodes.

General Symbols for Geometric Properties

b	Width
d	Depth
h	Height
t	Thickness
L, l	Length

General Symbols for Mechanical Properties

A	Area
I	Second moment of area (moment of inertia)
S	First moment of area (static moment)
W	Cross-sectional modulus

General Symbols for Internal Forces and Moments

M	Bending moment
M_T	Torsional moment
N	Axial force
V	Shear forces

General Symbols for Stresses

σ	Direct (normal) stress
τ	Shear stress

Indexes

a	Structural steel
b	Beam
bear	Bearing
c	Concrete, compression
d	Design value, diagonal
dur	Durability
eq	Equivalent
eff	Effective

(continued)

el	Elastic
f	Flange, effect due to fatigue loading
inf	Lower value
k	Characteristic value
l	Lightweight concrete
long	Longitudinal
max	Maximum value
min	Minimum value
nom	Nominal values
o	Top
p	Steel sheeting
pl	Plate, plastic
s	Reinforcement
ser	Serviceability
sup	Upper value
sur	Surface
sa	Steel + reinforcement
t	Tension
tot	Total
x	Longitudinal x-axis of member, longitudinal direction of bridge
y	Yield, cross-sectional major axis of bending, transverse direction of bridge
z	Cross-sectional minor axis of bending, vertical direction of bridge
u	Ultimate
w	Web, warping
x	Longitudinal x-axis of member, longitudinal direction of bridge
y	Yield, cross-sectional major axis of bending, transverse direction of bridge
z	Cross-sectional minor axis of bending, vertical direction of bridge
u	Limit value, bottom
E	Action effect, equivalent
H	Horizontal
L	Longitudinal, long term
P	Permanent
PT	Temporary permanent
R	Resistance
S	Shrinkage
T	Torsional
V	Vertical
0	Initial
1	Uncracked
2	Fully cracked
I	First order
II	Second order

Axes

x	Longitudinal axis of bridge or member
y	Transverse axis of bridge, major principal axis of cross section
z	Vertical axis of bridge, minor principal axis of cross section

Operators

Δ	Difference

Latin Small Letters

a	Length, distance between rail supports
a_g	Peak ground acceleration
b	Width
b_o	Half distance between webs
b^s_{eff}	Effective width due to shear lag
b_{fo}	Width of top flange of steel girder
b_{fu}	Width of bottom flange of steel girder
c	Outstand flange width, concrete cover of reinforcement, smeared spring constant
c_e	Exposure factor
c_f	Wind coefficient
c_{min}	Minimum value of concrete cover
c_{nom}	Nominal value of concrete cover
c_φ	Stiffness of rotational spring
d	Differential, diameter, shank diameter of shear connector, length of diagonal
$d_{head,sc}$	Head diameter of shear connector
d_{ref}	Reference height of bridge
d_{tot}	Total height of bridge
e	Eccentricity, distance of rail from girder flange, center distance between stiffeners
e_D	Edge distance of shear connectors from steel flange
e_h	Horizontal distance of tendon to steel web
e_L	Spacing of shear connectors in longitudinal direction
e_T	Spacing of shear connectors in transverse direction
e_V	Edge distance of shear connectors from concrete haunch, vertical distance between tendon and plane of shear connection
e_0	Imperfection
f	Reduction factor
f_{yd}	Design strength of structural steel
f_{yk}	Characteristic yield stress of structural steel
$f_{y,red}$	Reduced yield stress due to simultaneous shear
f_{cd}	Compression strength of concrete, design value
f_{ck}	Compression strength of concrete, characteristic value
f_{cm}	Compression strength of concrete, mean value
$f_{ct,eff}$	Tensile strength of concrete, mean value

(continued)

f_{ctm}	Tension strength of concrete, mean value
$f_{ctk\ 0.05}$	Tension strength of concrete, 5% fractile
$f_{ctk\ 0.95}$	Tension strength of concrete, 95% fractile
f_{clm}	Compression strength of lightweight concrete, mean value
f_{clk}	Compression strength of lightweight concrete, characteristic value
f_{cltm}	Tension strength of lightweight concrete, mean value
$f_{cltk\ 0.05}$	Tension strength of lightweight concrete, 5% fractile
$f_{cltk\ 0.95}$	Tension strength of lightweight concrete, 95% fractile
f_{pk}	Tensile strength of prestressing steel
$f_{p0,1k}$	Proof strength of prestressing steel
f_Q	Coefficient for shear area
f_{sd}	Design strength of reinforcement
f_{sk}	Characteristic yield strength of reinforcement
f_{tk}	Characteristic tensile strength of reinforcement
f_y	Yield stress of structural steel
f_u	Tensile strength of structural steel
g	Permanent load, acceleration of gravity
g_a	Self-weight of steel girder
g_c	Self-weight of concrete
h	Height, depth of concrete
h_c	Height of concrete slab
h_w	Height of web
h_0	Notional size
i	Index
k	Spring constant
k_1, k_2	Reduction coefficients for concrete strength
k_σ	Plate buckling coefficient
k_τ	Shear buckling coefficient
l	Length
l_k	Buckling length
l_y	Effective loaded length
n	Number, modular ratio of concrete, number of shear connectors
n_0	Modular ratio for short-term loading, fundamental natural frequency
n_i	Applied number of cycles of constant amplitude
n_L	Modular ratio depending on the type of loading
m	Mass, distributed moment, slope of fatigue curve
p	Uniformly distributed load
q	Uniformly distributed load, behavior factor
q_{fk}	Uniformly distributed load on footways
$q_{fk,comb}$	Combination value of uniformly distributed load on footways
q_{il}	Uniformly distributed traffic load on lane i
q_{vk}	Uniformly distributed traffic load on railway bridges

(continued)

(continued)

r	Radius, transverse distance between rails
s	Reinforcement bar spacing, coefficient for cement
s_f	Spacing of transverse reinforcement in longitudinal direction
s_s	Length of stiff bearings
t	Thickness, time, age
t_f	Thickness of flange
t_{fo}	Thickness of top flange of steel girder
t_{fu}	Thickness of bottom flange of steel girder
t_i	Thickness of elastomeric layers
t_{Ld}	Design life
t_s	Thickness of steel plates of elastomeric bearings
t_w	Thickness of web
u	Perimeter, deformation along in x direction
v	Loading speed, deformation in y direction
v_b	Basic wind velocity
$v_{c,Rd}$	Crushing design resistance of struts
v_{Mt}	Shear flow due to torsional moments
v_L	Longitudinal shear flow
$v_{L,Ed}$	Longitudinal shear flow, design value
$v_{L,Rd}$	Longitudinal shear flow, design resistance
$v_{x\ or\ y}$	Deformation of bearings in x, y directions
w	Width, deformation in z direction, width of carriageway
w_k	Crack width
w_i	Width of lane i
x	Longitudinal axis
x_{pl}	Depth of plastic neutral axis

Greek Small Letters

α	Aspect ratio of panel, modification factor for railway loads, imperfection factor
α_{crit}	Critical load multiplier
α_{LT}	Imperfection factor for lateral torsional buckling
α_{ult}	Yield load multiplier
α_q	Weight factor for uniform traffic loads
α_Q	Weight factor for axle loads
α_t	Coefficient of thermal expansion
β	Reduction factor for shear lag, reduction factor for plastic sagging moment
β_c	Coefficient for the development of creep
γ	Safety factor, specific weight
γ_A	Partial safety factor of accidental actions
γ_{AE}	Partial safety factor of seismic actions
γ_c	Partial safety factor for concrete

(continued)

γ_f, γ_F	Partial safety factors for actions
$\gamma_{F,f}$	Partial safety factor for fatigue actions
γ_G	Partial safety factor of permanent actions
γ_M	Partial safety factor for a material property
γ_{Mf}	Partial safety factor for fatigue strength
γ_P	Partial safety factor for prestress
γ_Q	Partial safety factor of variable actions
γ_{Rd}	Partial safety factor for resistance
γ_s	Partial safety factor for reinforcement
γ_v	Partial safety factors for shear connectors
γ_I	Importance factor
γ_{M0}	Partial safety factor for yield
γ_{M1}	Partial safety factor for stability
γ_{M2}	Partial safety factor for fracture and connections
γ_{M3}	Partial safety factor for slip
δ	Deflection
ε	Strain
ε_{ca}	Autogenous shrinkage strain
ε_{cd}	Drying shrinkage strain
ε_{cs}	Shrinkage strain
f_{yd}	Design strength of structural steel
θ	Rotation, twisting angle, angle of inclination of strut
κ	Curvature
$\bar{\lambda}$	Nondimensional slenderness for flexural buckling
$\bar{\lambda}_{LT}$	Nondimensional slenderness for lateral torsional buckling
$\bar{\lambda}_p$	Nondimensional slenderness for plate buckling
$\bar{\lambda}_w$	Nondimensional slenderness for web, for shear
μ	Friction coefficient
ν	Poisson ratio
ξ	Interpolation factor for column-like behavior, damping ratio
ρ	Density, reduction factor for plate buckling, reduction factor for presence of shear
ρ_s	Reinforcement ratio
σ	Direct stress
σ_a	Stress of structural steel
σ_c	Stress in concrete
$\sigma_{cr,p}$	Critical stress for plate buckling
$\sigma_{cr,c}$	Critical stress for column buckling
σ_s	Stress in reinforcement
σ_w	Stress in web
τ	Shear stress
τ_{cr}	Critical shear buckling stress
τ_{sm}	Mean value of bond stress
φ	Creep coefficient
φ_0	Notional creep coefficient

(continued)

(continued)

χ	Buckling reduction factor, relaxation factor
χ_c	Reduction factor for column buckling
χ_w	Reduction factor for shear buckling
χ_{LT}	Reduction factor for lateral torsional buckling
ψ	Stress ratio
ψ_L	Creep multiplier
ψ_0	Basic value of combination factor
ψ_1	Frequent value of combination factor
ψ_2	Quasi permanent value of combination factor
ω	Warping function
ω_0	Natural circular frequency

Capital Letters

A	Cross-sectional area, plan area of bearings, accidental action
A_a	Cross-sectional area of structural steel
A_b	Area of bottom reinforcement of slab
A_{bh}	Area of bottom transverse haunch reinforcement
A_c	Cross-sectional area of concrete
$A_{c,eff}$	Effective area of compression flange
A_{cp}	Partial area of concrete
$A_{c,tot}$	Total area of concrete flange
A_{ct}	Area of tension zone before cracking
A_{eff}	Effective cross-sectional area due to plate buckling
A_E	Seismic action
A_{net}	Net section area at holes
A_p	Gross area of plate
A_r	Reduced area of elastomeric bearings
A_{ref}	Reference area for wind force
A_s	Area of reinforcement
A_{sl}	Gross area of longitudinal stiffener
A_{sp}	Partial area of reinforcement
$A_{s,tot}$	Total area of reinforcement
A_{sf}	Area of transverse reinforcement cutting a section
$A_{s,min}$	Minimum reinforcement area
$A_{sl,eff}$	Effective area of longitudinal stiffeners
A_t	Area of top reinforcement of slab
A_v	Shear area
A_1	Area of steel plates of elastomeric bearings
C	Concrete, creep of concrete, wind load factor, spring constant
D_c	Action due to replacement of bearings
Δl	Elongation, contraction
$\Delta\sigma_{E2}$	Equivalent direct stress range for $2 \cdot 10^6$ cycles
$\Delta\tau_{E2}$	Equivalent shear stress range for $2 \cdot 10^6$ cycles
$\Delta\sigma_R$	Fatigue resistance to direct stresses
$\Delta\tau_R$	Fatigue resistance to shear stresses

(continued)

ΔT	Temperature difference
ΔT_M	Linear temperature difference
ΔT_N	Uniform temperature difference
E	Modulus of elasticity
E_a	Modulus of elasticity of structural steel
E_b	Compression modulus of elastomer
E_c	Modulus of elasticity of concrete
$E_{c,28}$	Modulus of elasticity of concrete at 28 days
E_{cm}	Modulus of elasticity of concrete—mean value
E_d	Design value of the effects of actions
$E_{d,dst}$	Design value of the effects of destabilizing actions
$E_{d,stb}$	Design value of the effects of stabilizing actions
E_D	Absorbed hysteretic energy
E_{lcm}	Modulus of elasticity of lightweight concrete—mean value
E_s	Modulus of elasticity of reinforcement
F	Force
F_{cr}	Critical concentrated load
F_{Rd}	Design buckling resistance to concentrated transverse forces
F_W	Wind force
G	Weight, shear modulus, permanent action
G_a	Shear modulus of structural steel
G_c	Shear modulus of concrete
G_1	Self-weight
G_2	Superimposed dead weight
G_{set}	Permanent action due to settlement permanent action due to settlement
H	Horizontal force, lateral force
I	Second moment of area (moment of inertia), length, influence length
I_{net}	Second moment of area of net section
I_p	Second moment of area of plate, polar second moment of area of a stiffener
I_{sl}	Second moment of area of stiffened plate
I_T	Torsional constant of cross section
I_W	Warping constant
I_1	Second moment of area of uncracked section
$I_{1,0}$	Second moment of area of uncracked section for short-term loading
I_2	Second moment of area of fully cracked section
$I_{2,sa}$	Second moment of area of fully cracked section (structural steel + reinforcement)
J	Creep function, impact energy, torsional constant
K	Spring stiffness of bearings, stiffness of system
K_{eff}	Effective stiffness
L	Length, span
L_e	Distance between zero moments
L_{eff}	Effective length for resistance to concentrated forces

(continued)

(continued)

L_f	Influence length
LM	Load model
L_Φ	Determinant length
M	Bending moment, mass
$M_{a,el,Rd}$	Elastic design moment resistance of steel girder
M_{cr}	Bending moment at cracking of concrete
M_{el}	Elastic moment resistance
$M_{el,Rd}$	Elastic design moment resistance
M_{Ed}	Design moment
$M_{f,Rd}$	Design bending resistance of cross section consisting of the flanges only
$M_{max,f}$	Maximum moments due to fatigue loading
$M_{max,f,Ed}$	Maximum moments in the fatigue combination
$M_{min,f}$	Minimum moments due to fatigue loading
$M_{min,f,Ed}$	Minimum moments in the fatigue combination
$M_{N,pl,Rd}$	Design bending resistance of cross section allowing for axial forces
M_{perm}	Moments due to all actions in combination except fatigue traffic loads
M_{pl}	Plastic moment
$M_{pl,Rd}$	Design plastic bending resistance
$M_{pl,V,Rd}$	Design plastic bending resistance allowing for shear forces
M_{Rd}	Design bending resistance
M_{sh}	Primary shrinkage moment
$M_{T,Ed}$	Design torsional moment
M_x	Torsional moment
M_{xp}	Uniform torsional moment
M_{xs}	Nonuniform torsional moment
M_I	Moment from first-order theory
M_{II}	Moment from second-order theory
M_w	Bimoment
M_1	Bending moment acting on noncracked composite section (state 1)
M_2	Bending moment acting on fully cracked composite section (state 2)
N	Axial force, number of cycles
$N_{b,Rd}$	Design buckling resistance
N_c	Axial force in concrete
$N_{c,el}$	Force in concrete at elastic resistance of steel girder
$N_{c,f}$	Force in concrete for full shear connection
N_{cr}	Euler buckling load, axial force at cracking of concrete
$N_{c,Rd}$	Design resistance to compression
N_{Ed}	Design axial force
N_{obs}	Number of lorries per year in the slow lane
N_{Ed}	Design axial force
$N_{t,Rd}$	Design resistance to tension
$N_{u,Rd}$	Design resistance to tension for sections with holes
$N_{pl,Rd}$	Plastic design resistance force

(continued)

N_s	Axial force in reinforcement
N_{sh}	Primary shrinkage axial force
P	Load, force, prestressing, permanent
PT	Secondary effects of creep and shrinkage
P_{Rd}	Shear resistance of shear connectors at ultimate limit state (ULS)
$P_{Rd,ser}$	Shear resistance of shear connectors at serviceability limit state (SLS)
Q	Variable action, traffic load
Q_{ik}	Axle load
Q_{lk}	Braking force
Q_{ml}	Average gross weight of lorries in slow lane
Q_{tk}	Centrifugal force
Q_{vk}	Concentrated vertical force for rail traffic
R	Resistance, relaxation factor
R_d	Design resistance
RH	Relative humidity
S	Shrinkage of concrete, soil factor, static moment (first moment of area), shape factor of bearings
$S_{a,d}$	Design response spectrum, design spectral acceleration
S_e	Elastic response spectrum, elastic spectral acceleration
S_w	Sectorial area
T	Temperature, vibration period
T_b	Total thickness of elastomeric bearings
T_e	Total nominal thickness of elastomeric layers
T_q	Total thickness of elastomeric layers including upper and lower layers
TS	Tandem system
UDL	Uniformly distributed load
V	Shear force, vertical load, train speed
$V_{b,Rd}$	Design shear resistance
$V_{bf,Rd}$	Design shear resistance contribution of the flange
$V_{bw,Rd}$	Design shear resistance contribution of the web
V_L	Force due to longitudinal shear
$V_{max,f}$	Maximum shear forces due to fatigue loading
$V_{min,f}$	Minimum shear forces due to fatigue loading
$V_{bw,Rd}$	Shear buckling resistance
V_{Ed}	Design shear force
$V_{pl,Rd}$	Plastic shear resistance
V_{Rd}	Design shear resistance
W	Section modulus, wind load
W_{eff}	Elastic section modulus of effective cross section
W_{el}	Elastic section modulus
W_{pl}	Plastic section modulus
Φ	Diameter of bars
Φ_s^*	Maximum bar size for crack control
Φ_2	Dynamic factor
Φ_3	Dynamic factor
X	Material property
Z	Through-thickness property

Examples

$M_{I,Ed,0}$	Design (Ed) bending moment (M) acting on noncracked (I) composite section due to short-term (0) actions
$M_{el,Rd,\infty}$	Long-term (∞) elastic (el) design (d) bending moment (M) resistance (R)
$\sigma_{c,Ed,\infty}$	Design (Ed) concrete (c) normal stress (σ) due to long-term (∞) actions

Structural Analysis Programs Used

- RSTAB (www.dlubal.com)
- SOFISTIK (www.sofistik.com)

Chapter 2

Types of steel–concrete composite bridges

2.1 GENERAL

Reinforced concrete slabs rigidly connected with steel girders have been used to form the basic superstructure of large numbers of deck bridges for many decades in Europe. This is due to the fact that the composite construction method offers the bridge engineers a great variety of solutions for different types of problems. An illustration of such problems, arising during the conceptual planning of a bridge, is given in Figure 2.1.

It is therefore obvious that finding the optimum solution is a difficult exercise for designers. Knowledge and experience from different fields of civil engineering are required, and therefore, the success of the final choice highly depends on teamwork. The aforementioned complexity in combination with the notorious difficulty of assessing costs keeps inexperienced designers closer to conventional concrete solutions. Furthermore, for small and medium spans, many contractors prefer concrete bridges due to the fact that they tend to maintain the same building technique and materials for the entire structure (concrete foundations, piers, abutments, and superstructure).

The percentage of composite bridges in the European market mainly depends on the cost level of labor and the price of structural steel. In high labor-cost countries (central and north Europe), fast-track solutions with prefabricated elements are undoubtedly the most cost-effective ones. Low weather temperatures also boost this trend. The prefabrication techniques are continuously under development, and composite bridges with partially or fully prefabricated slabs have already a relative big share in the domestic markets. In south countries, concrete bridges dominate the markets due to cheap labor. However, many interesting composite bridges with more cast-in-place elements can be found.

In literature, composite bridges are considered to be competitive for spans larger than 35 m. In practice, there is no proof that composite bridges are less competitive than the concrete ones even for smaller spans; see [2.17] and [2.18]. As mentioned earlier, many contractors neglect fundamental advantages of steel–concrete composite bridges in their evaluations and maintain their opinion about “unforeseeable price increases in the steel market.”

Nowadays, innovations in welding technology, high-strength steel qualities, and new types of precast slabs, specially fabricated for cooperating with steel beams, prepare the ground for a successful comeback for composite bridges. Table 2.1 provides an overall insight on their advantages mainly connected with safety (S), economy (E), constructional simplicity (CS), functionality (F), and aesthetics (A).

The main disadvantage of structural steel in bridge construction, its susceptibility to corrosion, is being increasingly overcome by improved protective coatings. Weathering steel is also used with great success giving the ability of avoiding painting (or future repainting), thus keeping maintenance costs low.

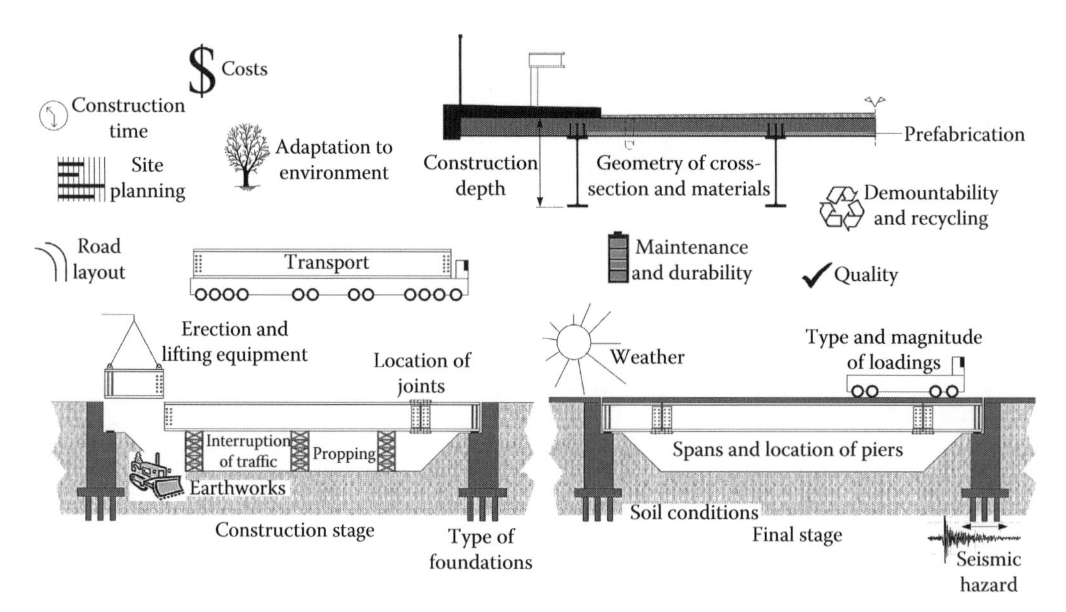

Figure 2.1 Conceptual bridge design considerations.

Table 2.1 Advantages of steel–concrete composite bridges

Low self-weight of superstructure	→	• Cheaper foundations and bearings (E) • Lower seismic forces (E, S) • Cheaper reconstruction and retrofitting (E)
Assembly capability on site	→	• Lower transport and lifting costs (E) • Flexible site planning (F, E)
No propping during construction	→	• No traffic interruption (E, F) • Elimination of formworks (CS)
Big spans and low construction depth	→	• Slender appearance (A) • Fewer piers (F)
Maximum prefabrication	→	• High quality (S) • Fewer cast-in-place activities (CS) • High speed of construction (E) • Low labor costs (E)

2.2 COMPOSITE BRIDGES: THE CONCEPT

A typical composite cross section of a highway bridge is shown in Figure 2.2. A series of parallel steel girders are rigidly connected with a reinforced concrete slab through shear connectors. The shear connectors installed are mostly welded studs allowing the use of the deck as part of the top flange (deck-plate girders). The longitudinal bending of the composite T girders, at positive bending areas, results in tension in steel and compression in concrete. The simultaneous operation of both of these materials generates the composite action that is the most important feature for the formation of stiff and high-strength cross sections. In areas of negative moments, concrete is considered to be fully cracked. Despite the contribution of the steel reinforcement to the hogging moment of resistance strengthening locally, the steel cross section, by adding cover plates or concreting, can sometimes be necessary.

Direct loads from the wheels are distributed by the bending action of the reinforced concrete deck slab to the longitudinal composite girders. In addition, deck slab acts as a

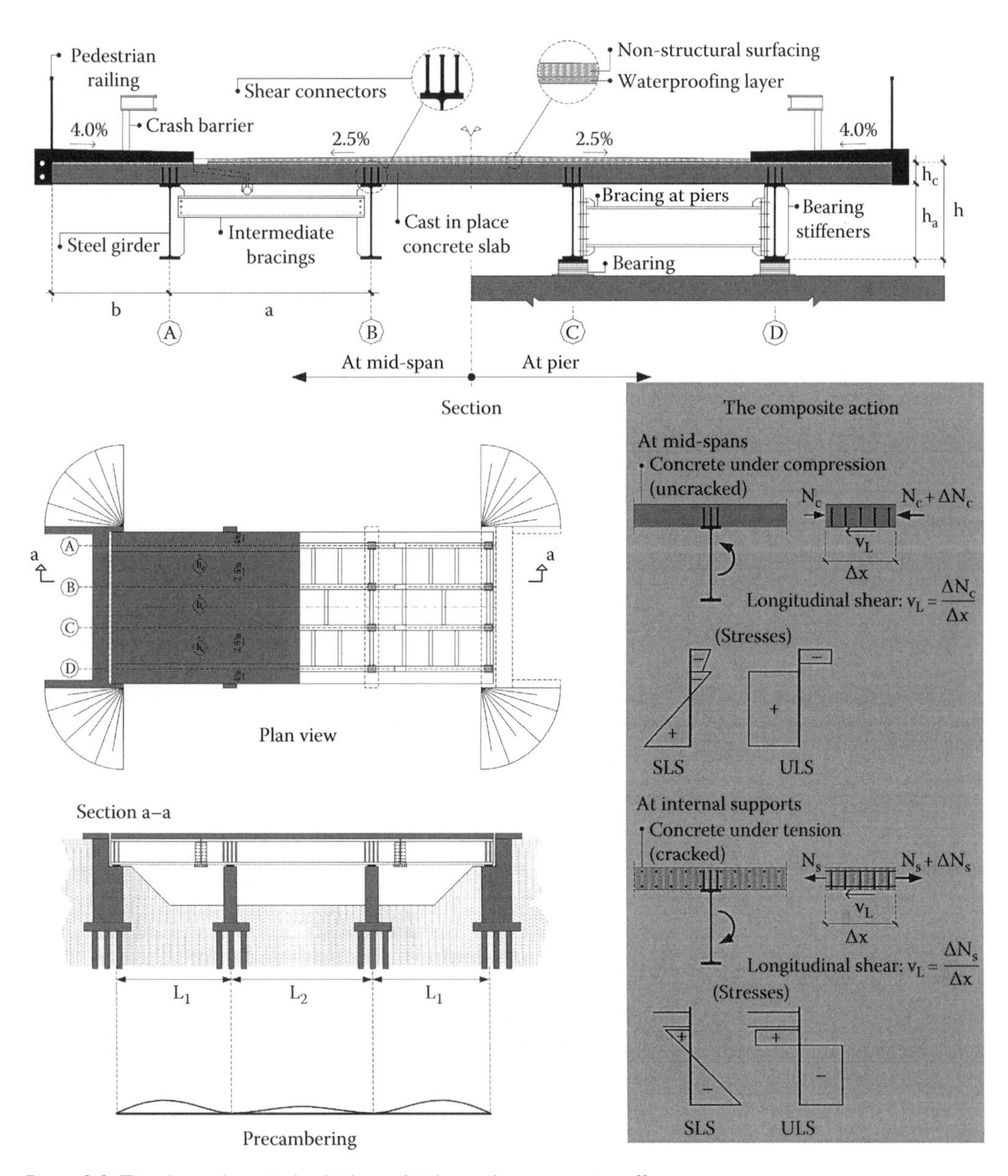

Figure 2.2 Two-lane plate-girder highway bridge—the composite effect.

diaphragm in cases of seismic loadings, braking forces, etc. Therefore, the slab's thickness h_c must be high enough to guarantee both an adequate out-of-plane and in-plane stiffness. In correspondence to the girder's spacing a, which usually varies from 2.5 to 4.0 m, the slab's thickness varies from 25 to 30 cm. It is advantageous to choose a girder's spacing not larger than the effective width, calculated according to EN 1994-2 provisions, so that the entire concrete slab contributes to the structural performance of the superstructure.

Cantilevers at the edge of the deck slab should normally be less than 1.6 m. Cast-in-place activities for the cantilevers are always a difficult part of the construction process, and subsequently, minimizing the edge length b should always be in the designer's mind. Bridge decks with wide sidewalks can however lead to increased cantilever lengths sometimes even greater than 3 m.

An even number of steel girders achieves better material optimization and allows bracing at piers. In cases of medium spans, the designer will have to choose between twin- or multi-girder bridges. Both of these options offer different advantages, which will be discussed in the following paragraphs.

At piers, steel girders rest normally on bearings. Bearings are structural assemblies installed to secure the safe transfer of all reactions from the superstructure to the substructure without generating any harmful restraining forces. They must be able to spread the reactions over adequate areas of the substructure, adapt thermal and other deformations of the superstructure, and isolate the superstructure from seismic excitations coming from the substructure. Transverse bearing stiffeners are required to transfer support reactions from the web into the bearings and to introduce concentrated loads into the webs.

For short and medium spans, steel girders are often preassembled in pairs through intermediate bracings and then lifted into final position by mobile cranes enabling rapid erection. Intermediate bracings offer increased stability during deck concreting, and therefore, temporary supports can be avoided. This is of great importance especially when traffic disruption under the bridge must be kept at a minimum level.

	Short	Up to 35 m
Span lengths are described as	**Medium**	35–80 m
	Long	Greater than 80 m

Transverse bracing at supports is also needed to transfer horizontal loads from wind, earthquake, and centrifugal forces to the bearings. Since the bearing's design life is less than that of the bridge itself, bracing at piers should be also able to resist high jacking loads in case of replacement of bearings.

For simply supported bridges with spans up to 25 m and for continuous bridges with spans up to 30 m, steel girders can be of rolled sections HEB or HEM type. For longer spans, nonsymmetrical welded plate girders are the most commonly used cross sections. With depths to main span ratios h/L varying from 1/20 to 1/30, an appealing design can be reached. More slender bridge sections can be in some cases feasible, but considerations of deflections or vibrations may be critical for the design. In order to reach the highest possible slenderness, the number of the girders can be increased so that the bay width a becomes even lower than 2.0 m; see [2.4].

In Figure 2.2, one can also see the precambering's shape of the steel girders (superelevation). Precambering of the steel girders is necessary to compensate for deflections under permanent loads, temperature effects, and creep and shrinkage of concrete, so as to avoid any appearance of excessive downward deflection. Curved in elevation, rolled sections can only be fabricated by few companies with heavy rolling equipment. Fabricating welded plate girders with an upward deflection is much easier as webs are cut based on the precambering's geometry and then welded with the flanges.

The previous offered a brief description of a common plate-girder composite bridge. In the following paragraphs, more types of composite bridges are presented and commented on.

2.3 HIGHWAY BRIDGES

2.3.1 Plate-girder bridges with in situ concrete deck slab

The main geometric and structural aspects of a deck-plate-girder bridge have already been demonstrated in the previous paragraph. The most commonly used structural steel quality is S355, but in some European countries, S420, S460, and even S690 have already been

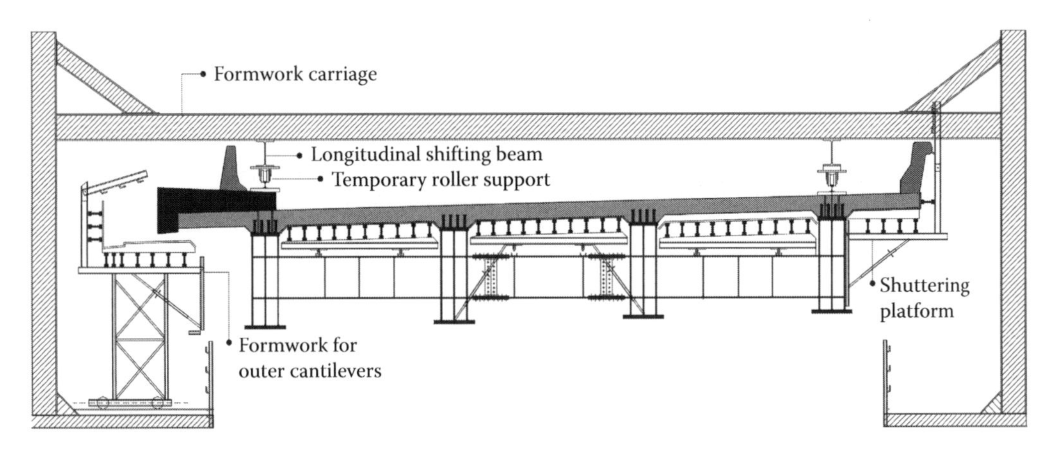

Figure 2.3 Typical mobile formwork superstructure.

implemented; see [2.2]. For the in situ parts of the deck concrete qualities, C30/37 and C35/40 are the most appropriate ones.

The use of full in situ concrete deck slabs is by far the most popular building method. If the steel girders are not supported during concreting, a mobile formwork runs along the steel beams concreting sections with a maximum length of 25 m; see Figure 2.3. Since cast-in-place activities with temperatures below 5°C can be problematic, warm weather is an important requirement. The mobile formwork technique is offered by many construction companies, and therefore, low prices can be expected due to high competition. The main disadvantages of this method are long execution time, high shrinkage forces, and use of large amount of structural steel due to the noncomposite action during concreting.

Concreting of the deck slab on temporary soffit formworks, usually made of timber, in conjunction with supporting towers (props), is in many cases the most cost-effective solution. Experienced contractors always come up with tailor-made formwork ideas, which are less expensive than mobile formworks. Furthermore, support of the deck during concreting leads to lighter cross sections for the steel girders since composite action is generated for both permanent and traffic loads. Precambering values are also significantly reduced. It is particularly worth mentioning that propping during concreting results in considerably less amount of intermediate bracings since stability problems are eliminated. This has a positive effect on both the economy and the appearance of the bridge.

Releasing the props in continuous bridges after concrete hardening can produce cracks due to the imposed rotations at internal supports. This can be avoided if the construction sequence of Figure 2.4 is followed.

Single-span steel girders are placed on bearings, and thereafter, concrete is poured on timber soffits supported by steel towers. After concrete hardening, props are released, and the simply supported composite girders deflect due to the bridge's self-weight. Finally, concreting of the internal support transforms the former isostatic system into a stiffer continuous one. Hogging bending moments due to permanent loading, creep, and shrinkage are significantly reduced; the bridge maintains its static indeterminacy, and cracking of concrete is easier controlled.

In situ deck concreting allows different slabs' shapes. Typical geometries are shown in Figure 2.5. Slabs with uniform thickness (case A) are, from a constructional point of view, the easiest ones, and they are mainly chosen for short-span straight bridges with decks narrower than 7 or 8 m. Thicknesses from 220 to 250 mm are commonly preferred. Slabs with

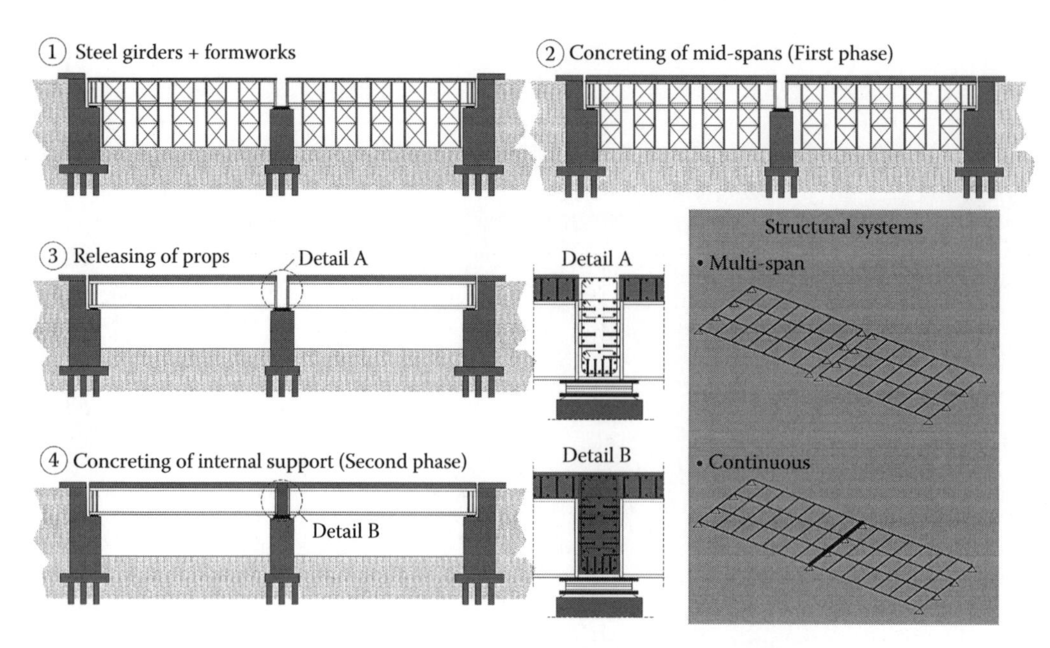

Figure 2.4 Construction sequence for controlling cracking in a continuous bridge.

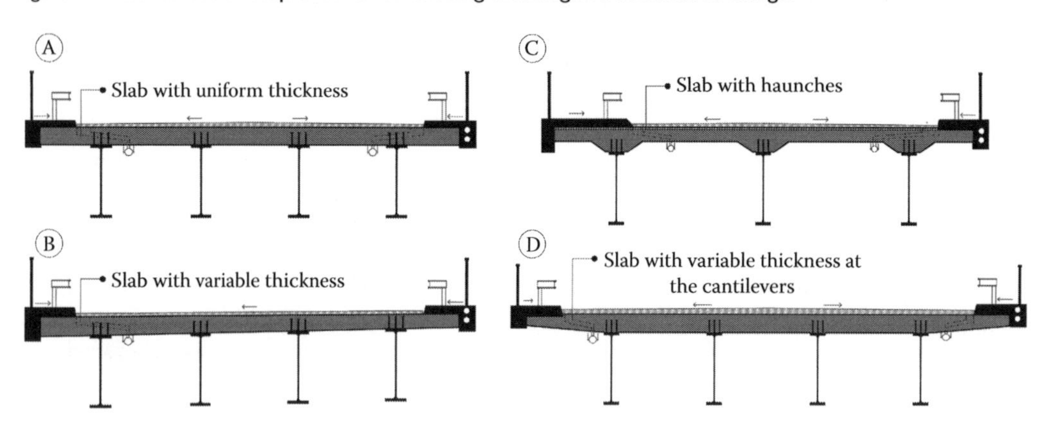

Figure 2.5 Bridges with different slabs' geometries. (A) With uniform slab thickness, (B) with variable slab thickness, (C) with haunches, and (D) with variable thickness at cantilevers.

variable thickness (case B) provide the desired inclination without increasing the thickness of surfacing. The slab's top surface follows the road's geometry, while the soffit is aligned with the girders' upper flanges. In cases of decks with significant inclination, steel girders are usually at different levels. For wide decks, a cost-optimized steel and concrete consumption can be achieved by designing haunched slabs in conjunction with increased girder spacing (case C). Geometries with a bay width 6 m, a central slab thickness 35 cm, and a haunch of 45 cm are feasible without transverse prestressing.

2.3.2 Plate-girder bridges with semiprecast concrete deck slab

In Figure 2.6, a simply supported composite bridge with lightweight precast concrete elements supplemented with in situ concrete is demonstrated (semi-precast deck slab). The prefabricated slabs are used as permanent formwork for the topping, and they are supported

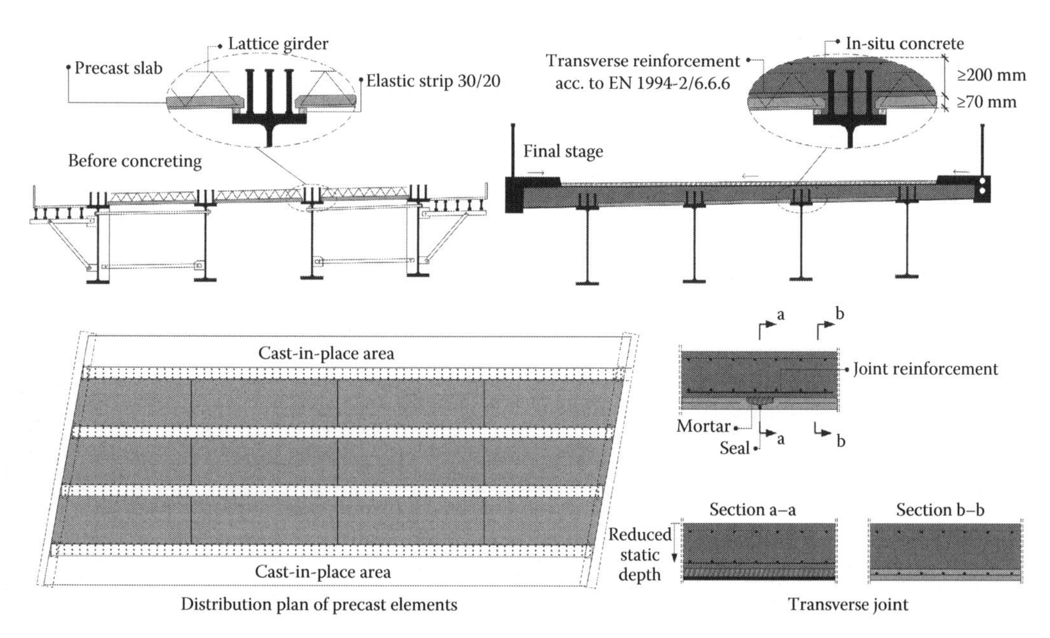

Figure 2.6 Composite bridge with a partially prefabricated deck slab.

statically determined between the steel girders. The outer cantilevers are casted in situ through the use of conventional formwork, with supports attached to the edge beams. This type of bridges is generally preferred when passing over existent rail- and highways without any restriction to traffic during erection. The main advantages of this method are short construction time, easy erection, and reduced creep and shrinkage effects.

The thickness of the precast slabs depends mainly on the spacing of the girders and is typically 70–100 mm. The concrete quality for the prefabricated elements is preferred to be C40/50 or greater and for the in situ concrete topping C30/37. Lattice girders embedded with their lower chord in the precast slabs bar ensure the composite action, so that the longitudinal reinforcement of the precast elements can be taken into account in the capacity design of the entire deck slab. Increasing the height of lattice girders results in greater permissible spans in bridges with large girder spacing. It should be noted that the structural design of lattice girder precast slabs has to be regulated by official approvals based on static and dynamic tests with special reference to bridges. This design issue is not covered in detail by Eurocode 2.

In Figure 2.6, one can also see that steel girders are delivered on site with elastic strips on top. The elastic strips are usually made of elastomeric material in order to obtain a compensation for tolerances and a waterproof joint between steel and concrete. Especially in cases of bridge sections with steel girders at different levels (see section B in Figure 2.6) or high precambering values, elastic supporting strips can be described, from a constructional point of view, as mandatory. In many bridges, shims have been used, instead of elastic strips, due to lower cost. The contact area of the shims with the top of the girders has proved to be sensitive to corrosion, and therefore, shims should be avoided.

In bridges with semi-precast deck slabs, the laying plan of the prefabricated elements is of primary importance. Indeed transverse joints between adjacent precast slabs are considered to be the weak points of the deck slab, and minimizing their number must always be in the designer's mind. Typical lengths for precast slabs are 8–10 m, and for multi-girder bridges, a width lower or equal to 2.5 m is convenient to select due to easy transportation. Furthermore, the detailing of the joints is equally important. Fixing of the joints with nonshrinkable mortar before concreting gives the superstructure the adequate horizontal stiffness during concreting so that horizontal braces can be reduced and even eliminated. It should also be noted that in locations of transverse joints, the longitudinal reinforcement of the precast slabs is interrupted and therefore, additional rebars (joint reinforcement) are necessary. Comparing section a–a with b–b in Figure 2.6 shows that the required amount of joint reinforcement should be increased due to its reduced effective depth. It is also considerably higher than the longitudinal reinforcement of the precast slabs.

The prefabrication rate can be considerably increased by using precast deck elements that cover the cantilever parts together with the adjacent internal bays; see Figure 2.7. Precast elements of this type are specially designed with box-shaped pockets so that an arrangement of the shear connectors in groups is possible. Transverse reinforcement fully anchored in the precast units between the adjacent rows of the connectors should be sufficient enough in order to prevent premature local failure of the precast or the in situ concrete. Obviously, the pocket's reinforcement has to be perfectly suited to shear connectors' distribution since the available tolerances are limited. It has to be noted that EN 1994-2 allows the arrangement of shear connectors in groups but special precautions should be taken into account from the designers considering fatigue of connectors and concentrated forces due to nonuniform longitudinal shear.

In order to prevent overturning of the outer precast planks during concreting of the deck, the direction of concreting has to be from the inside to the outside. In cases of decks with large sidewalks, the outer elements must be anchored to the internal girders. This requires special detailing and extra costs.

It is obvious from Figure 2.7 that steel girders during concreting carry their self-weight and also the weight of the precast slabs and fresh concrete. Additional construction loads may need to be taken into account. Excessive deformations due to this loading may result in increasing the dimensions or the precambering of the steel girders. Moreover, buckling phenomena of the girders may lead to an increased number of bracings and crossbeams.

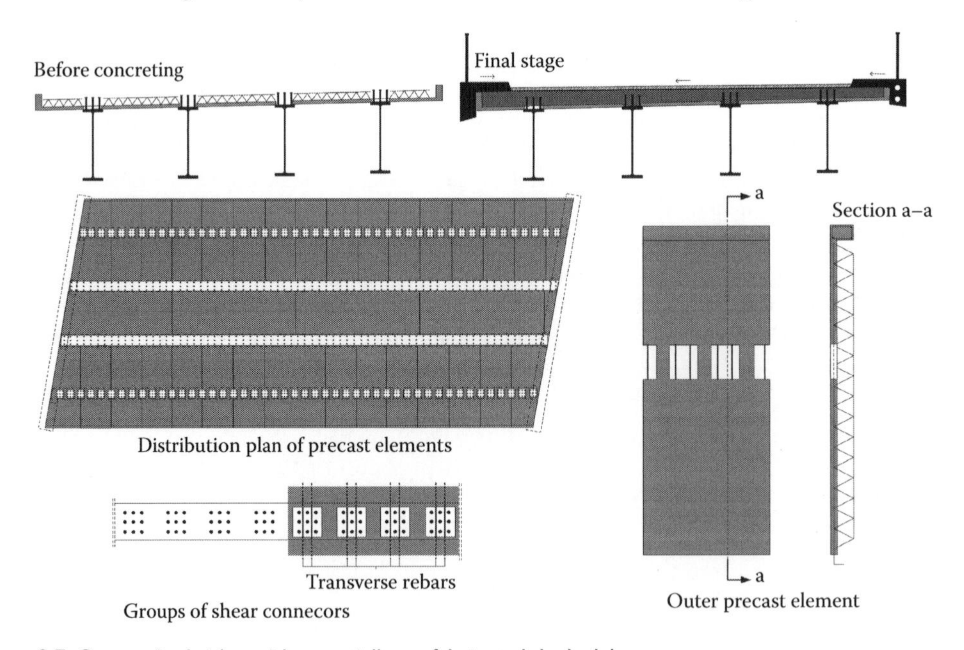

Figure 2.7 Composite bridge with a partially prefabricated deck slab.

Figure 2.9 shows a simple solution that activates shear connection of the girders with the precast slabs prior to casting the in situ concrete topping (partial composite action). The latter is easily achieved by using short studs welded together—*double-headed studs*; see Figure 2.8.

After placing the precast planks, grouting of the lower studs with mortar follows. Therefore, a composite T section of high stiffness is formed that ensures limitation of deflections and the desired stability during in situ deck concreting. Therefore, a buckling analysis is conducted for an ultimate limit state (ULS) load combination that only includes the dead loads of the steel girders, the prefabricated elements, and the mortar, including possible construction loads.

Due to the low magnitude of the previous loads, the stability check during erection can be of minor importance even for medium-span bridges. The partial composite action method may offer bridges free of intermediate braces, thus combining aesthetics with economy.

From a structural point of view, it is worth mentioning that precast slabs of Figure 2.9 must be equipped with transverse hooked rebars that surround the shear studs. The hooked rebars transmit the longitudinal shear forces from the headed studs' area to the precast slabs ensuring the cooperation of the entire precast concrete flange with the steel girder (dowel action). From a constructional point of view, it has to be noted that the designer should provide the necessary detailing so that a conflict between the hooked rebars and the studs during erection is avoided. This is important since bending of the hooks on site is impossible due to their short length.

In conclusion, the use of double-headed studs represents a very interesting solution for composite bridge engineering. Unfortunately, there are not many bridges constructed with this type of shear connectors, and consequently, experience is limited. EN 1994-2 does not exclude this type of connectors and can be treated similar to an equivalent single one. A soft pad placed under the intermediate head in order to prevent a mechanical interlocking is also advised to be used; see Figure 2.8.

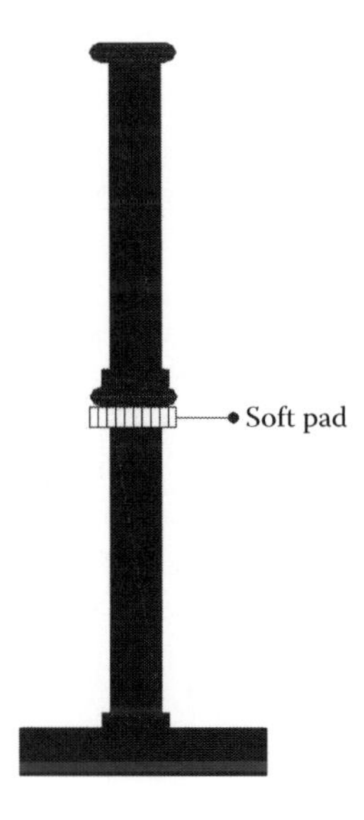

Figure 2.8 Double-headed stud.

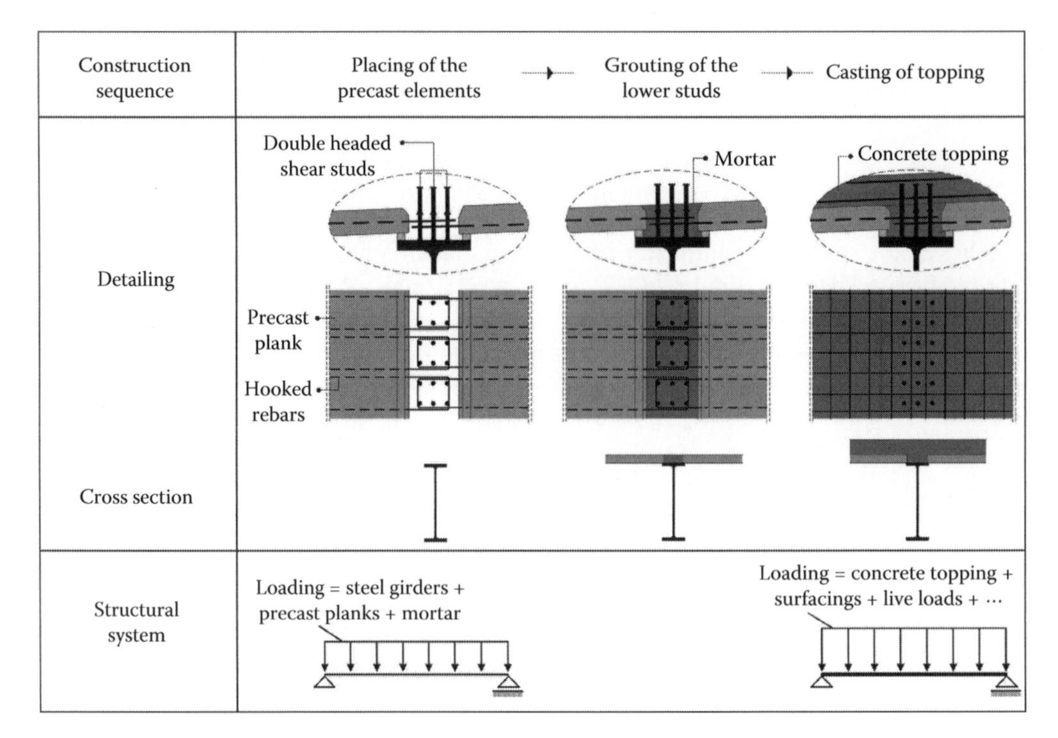

Figure 2.9 Partial composite action concept.

2.3.3 Plate-girder bridges with fully precast concrete deck slab

In cases of temperatures below 0°C, cast-in-place activities are avoided, and fully prefabricated deck slabs are preferred. Nowadays, innovative fully precast concrete elements are used offering multiple advantages as rapid construction, high quality, and reduced labor costs and construction during freezing periods; see [2.4]. In Figure 2.10, a composite bridge with a deck slab as full precast unit is shown. The precast elements are equipped with overlapping tongues so that dry joints can be achieved. The tongues should be specially designed in order to prevent loss of contact between the prefabricated slabs during vehicles passing over the joints. Above the steel girders, void channels are filled with in situ concrete through grout holes. Transverse reinforcement in the slab with spacing equal to this of the shear connectors ensures safety against longitudinal splitting and damage of the console during erection.

The transverse joints between the precast units should always be under compression in order to prevent cracking and water leakage under service loads. Therefore, the dry joint technique is recommended only in cases of simply supported bridges. Continuous bridges with fully precast deck slabs can be designed when the deck is prestressed by tendons before mounting of the shear connectors (pretension). Posttension can result in overloading of the studs and compression of the steel girders reducing their rotational capacity.

Despite the advantages of the fully prefabricated deck slab method, the following disadvantages have to be taken into account. From both design and constructional point of view, a high know-how is required. In addition, not many precast companies can deliver full precast deck slabs, and therefore, high prices due to limited competition should be expected.

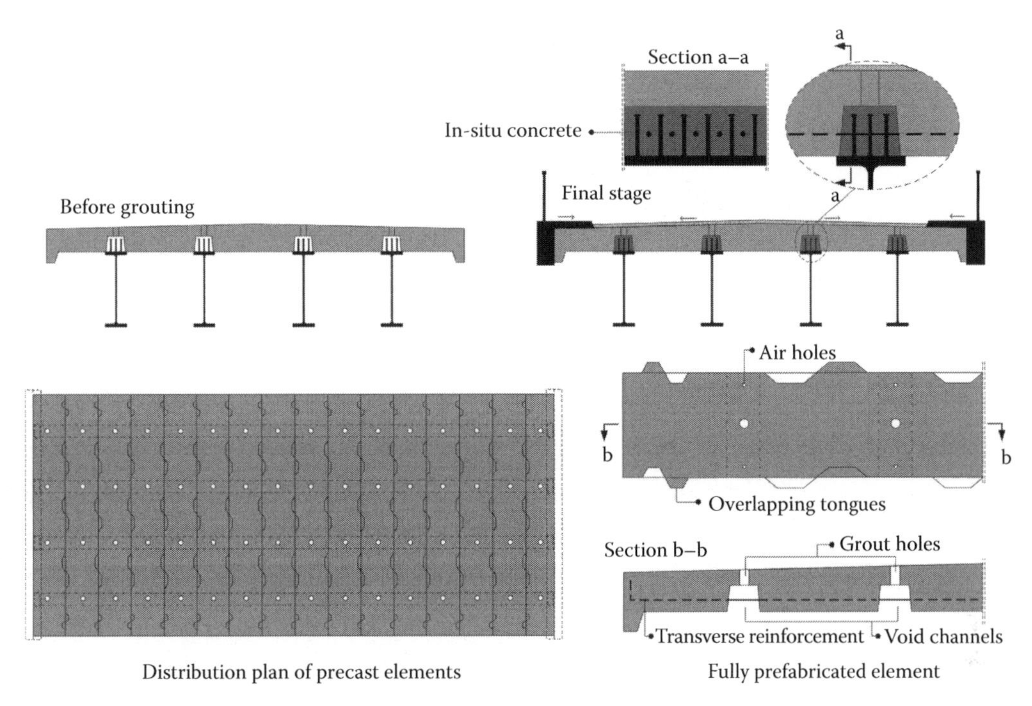

Figure 2.10 Composite bridge with a fully prefabricated deck slab.

2.3.4 Plate-girder bridges with composite slab deck with profile steel sheeting

Instead of using prefabricated elements as permanent formwork, cold-formed steel sheets can be used as well; see Figure 2.11. This is an interesting fast-track construction method for multi-girder bridges with spacing not greater than 3 m. Profile steel sheets are much lighter than precast planks, and they act both as a platform for construction and as shuttering for the wet concrete; elastic strips are not needed. Moreover, profile steel panels can be easily field cut using a grinder or nibbler. Yet, field cut should be kept to a minimum.

Steel panels should be placed simply supported since through deck welding of the studs is not allowed. Furthermore, panels should be fixed with shot-fired fastenings to the underlying girders to prevent movements prior or during concreting. It is important to note that the deck slab must be designed as a reinforced concrete slab with a total height equal to the depth of concrete above the ribs. Unfortunately, EN 1994-2 does not provide any regulations for the application of steel decking composite slabs in bridges.

The concrete's depth above the ribs should be at least 25 cm. This is due to the fact that overloading of an underlying girder is in cases of slim slabs possible. Normally, deck slabs with profile steel panels have a total thickness with a minimum value of 30 cm. In addition, the slab's main reinforcement should be located above the panels' ribs. Increasing reinforcement should not be followed for raising the slab's stiffness.

There is a huge variety of profile steel panels in the domestic markets. Use of open-through profile sheets with a minimum thickness of 1.50 mm is recommended. It has to be noted that profile steel sheets must be dimensioned against buckling in the wet-concrete condition. The selection of a suitable decking is usually made using manufacturers' design charts or tables.

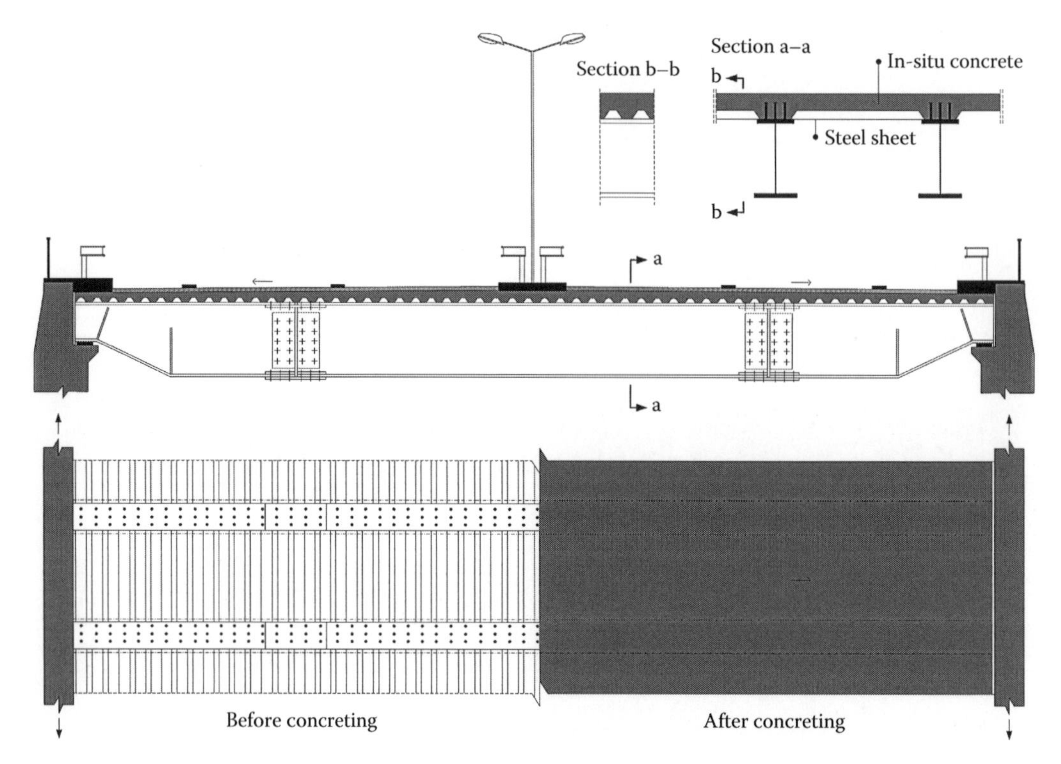

Figure 2.11 Composite bridge with profile steel sheets as shuttering.

In verification for steel sheeting, the ponding effect has to be taken into account (increased weight of wet concrete due to deflection of sheeting).

Temporary supports for the steel panels during concreting may be in some cases necessary; see Figure 2.12. Trusses or telescopic beams attached on the main girders offer the ability of bridging long bays with steel decking of reduced height. However, this is not always a cost-effective solution since connecting and removing the support elements demand numerous man-hours.

An easier method to set up a composite bridge with a steel decking on long distanced girders can be followed by casting the deck slab in two phases; see Figure 2.13. The first concrete layer is poured on high-bond steel sheets. The layer must be as thin as possible so that lateral torsional buckling of the steel girders during concreting is prevented and plan bracing for the upper flanges is reduced and even eliminated. Limited deflections for the sheeting and a negligible ponding effect can also be achieved.

After concrete hardening, the second layer is then poured on a reinforced concrete slab. The hardened layer offers a stiff construction platform and acts as diaphragm against wind or earthquake loadings during second-phase concreting. Attention must be paid to the reinforcing of the layers' interface against slipping. Bent-up bars or lattice girders can be used. A reinforcing detail is shown in Figure 2.13.

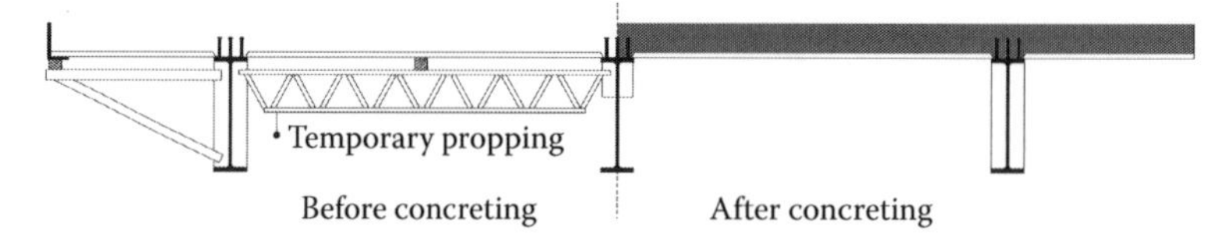

Figure 2.12 Bridge with temporary propping during concreting.

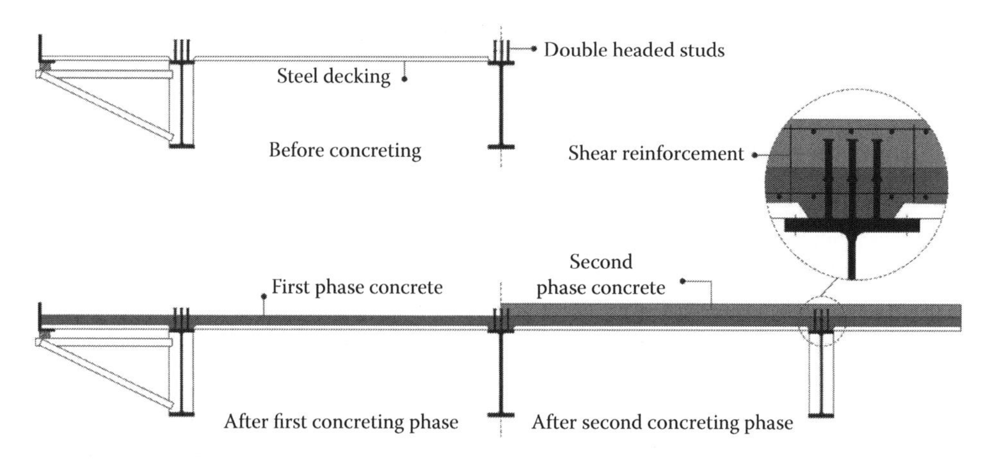

Figure 2.13 Bridge with two concreting phases.

In Figure 2.13, one can also see that the solution of concreting the deck slab in two phases can be effectively combined with double-headed studs. A partial shear connection during construction can be consequently achieved for both transverse and longitudinal direction. This solution may be convenient to follow in cases of self-supported bridges with spans larger than 25 m.

In cases of non-compact thin webs, the configuration of Figure 2.14 is an interesting one. Steel panels are placed below the top flange of the steel girders on angle cleats welded on the webs. With this solution, thick slabs do not affect the slenderness of the bridge; this can be important when the available construction depth is highly restricted.

Excessive flexural stresses can limit the web's resistance and lead to local buckling. A situation like this can be critical for non-stiffened thin webs both during construction and in the final stage. Angle cleats act as longitudinal stiffeners enabling the web to reach an adequate level of strength. An optimum design should obviously ensure full plasticity, but this is not always feasible. One can also see that angle cleats are welded in an asymmetrical way so that a greater part of the web is protected.

Due to encasement in concrete, solution A provides upper flanges of small dimensions (economical steel consumption). Additionally, no cover is provided for the top of the connectors

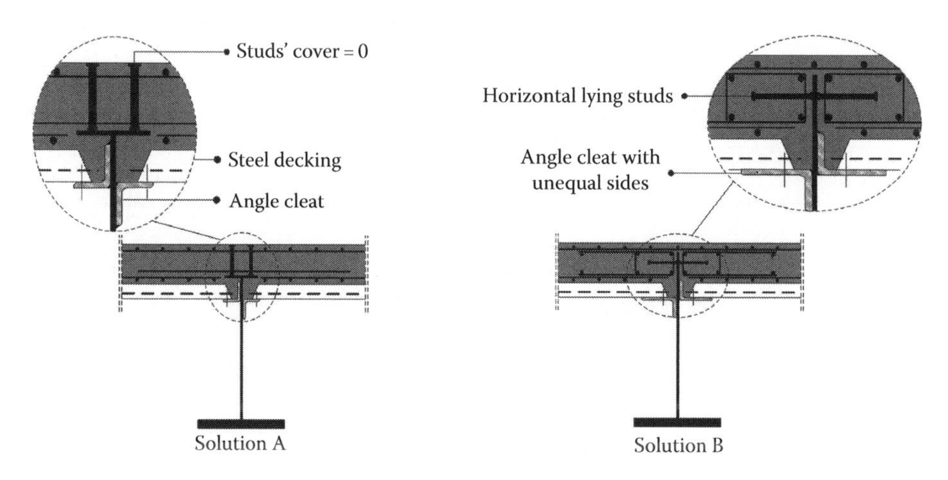

Figure 2.14 Shallow composite sections with angle cleats and steel decking.

allowing a shallower construction. A zero cover provision for the studs can be found in EN 1994-2 and is recommended for nonaggressive environments.

Solution B may be more attractive for small-span bridges. EN 1994-2 provides calculation formulas for the shear resistance of horizontal lying connector studs taking into account the surrounding reinforcement. This gives the designer the ability to form cross sections without upper flange leaving this role for the angle cleats. Looking Figure 2.14 closer, one can also see that angle cleats with unequal members can sometimes be preferable. The longer members give the steel girders a greater bending capacity around the weak axis, thus increasing their lateral torsional buckling resistance in wet-concrete condition.

2.3.5 Plate-girder bridges with partially prefabricated composite beams

For reasons discussed before, it is often preferable, where possible, to install shear connection prior to concreting. Middle-span bridges with lengths larger than 35 m can be erected in an economic way by the use of partially prefabricated composite girders; see Figure 2.15. Steel girders rigidly connected through shear studs with thin precast slabs 100–120 mm can be fabricated in factory. Obviously, the connected precast slabs serve as formworks for the in situ concrete supplement and stabilize the steel girders, thus eliminating the use of intermediate bracings.

Prefabricated girders with a maximum length of 60 m can be transported by road and lifted into place by using mobile cranes. Girders with a length of 80 m can be transported by water.

During the last 15 years, partially prefabricated composite girders are implemented in Germany with great success. This solution is known as the VFT construction method and leads to numerous advantages for both the owner and the users; see [2.43], [2.44], and [2.52]. These advantages are short construction time, low steel consumption, and high sustainability. In cases of bridges rigidly connected to piers/abutments (integral bridges) with

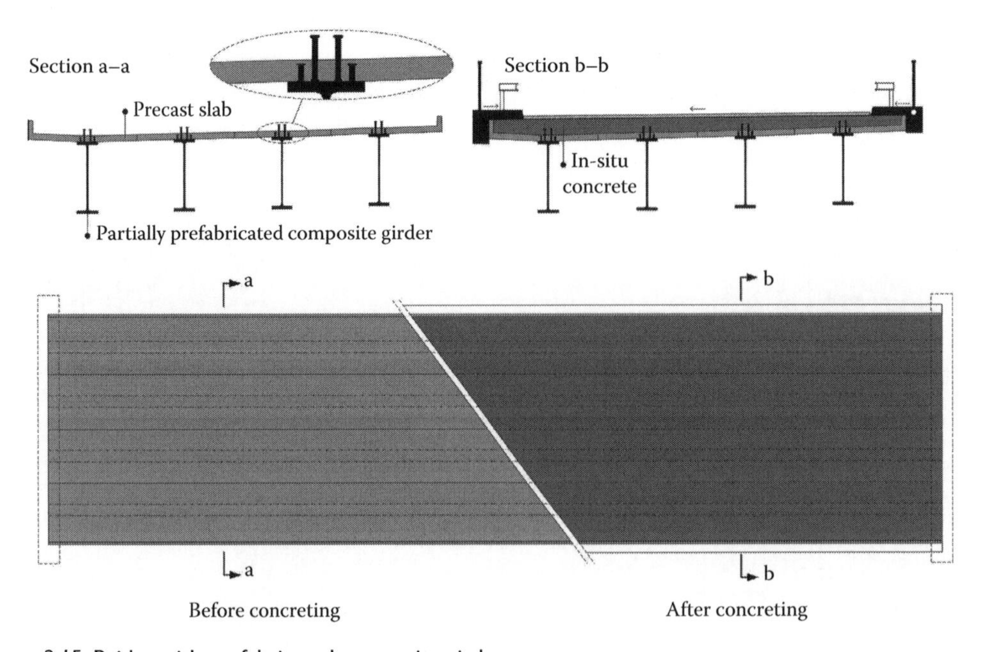

Figure 2.15 Bridge with prefabricated composite girders.

variable construction depth, impressive slenderness ratios are feasible, h/L = 1/45 for highway bridges and 1/35 for railroad ones.

2.3.6 Double-girder bridges

For the curved bridge of Figure 2.16, two different section types are shown. Cross section A is a double-girder bridge with main girders of unequal heights. In this way, a severe inclination of the deck slab can be achieved. Cross section B is equipped with identical steel girders that are positioned at different levels. This results in a lower steel consumption but can be problematic when launching is selected as the bridge's construction method.

Double-girder bridges are mainly suitable for medium spans. The main advantage is that there are fewer girders to erect and fewer piers to construct. The main negative feature of a double-girder bridge is its zero redundancy. Indeed, if a girder is damaged, internal forces cannot be redistributed to an adjacent beam and collapse is highly possible.

Bridges with symmetrical cross sections are also called twin-girder bridges; see Table 2.2.

In Figure 2.16, one can also see the variable depth of the deck slab in transverse direction. The slab's depth is usually 25–30 cm at deck center and 35–45 cm at main girders. Deck slabs with spans less than 15 m can be easily constructed as reinforced concrete ones. For larger widths, a transverse prestress is unavoidable.

Main girders are fabricated by shop welding into I sections. In cases of curved bridges, main girders are usually accommodated in I-beam construction by connecting a series of straight sections. For the flanges, plates wider than 600 mm and thicker than 60 mm are not rare. Designers should always have in mind possible steel's strength reduction in cases of thicker plates.

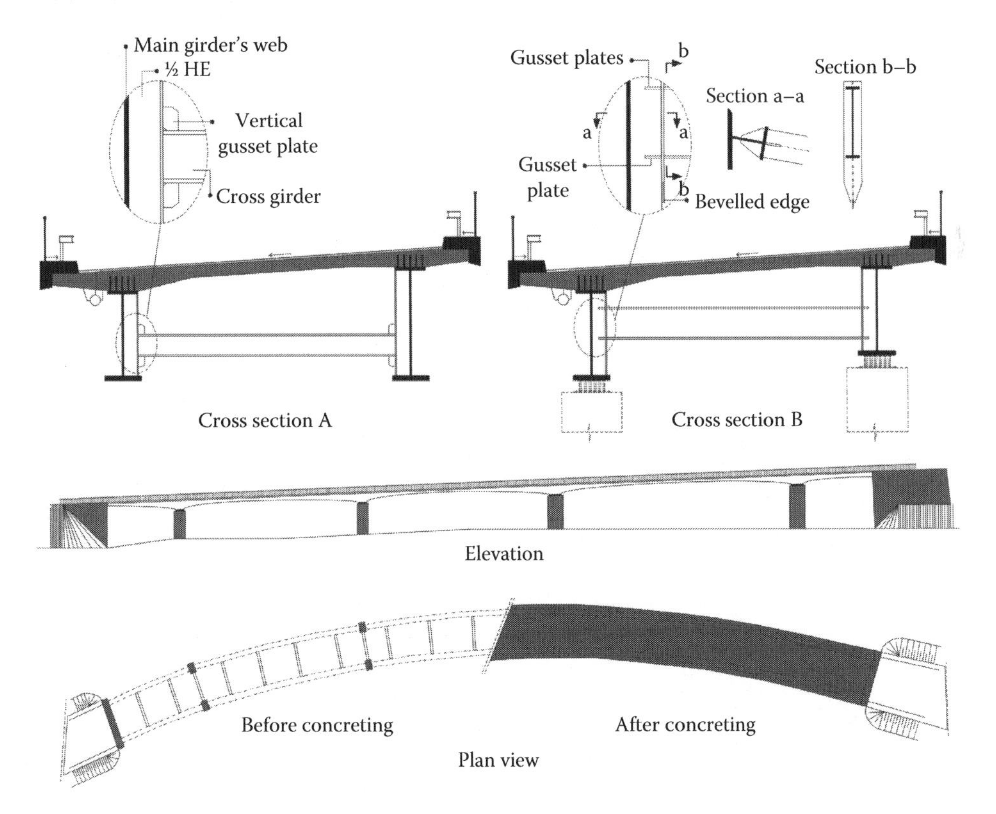

Figure 2.16 Double-girder bridge.

Table 2.2 Preliminary design for twin-girder bridges

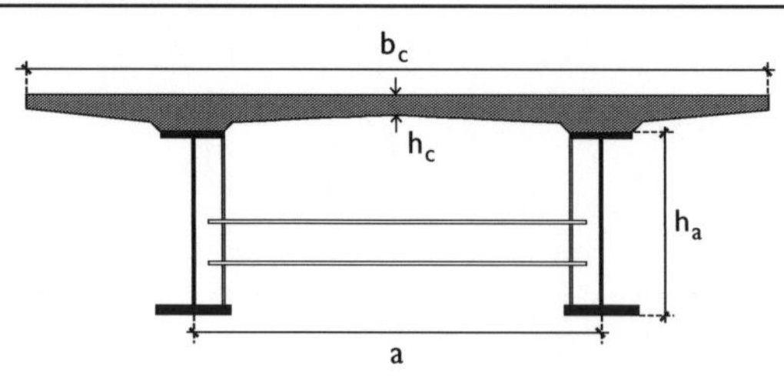

Length L_{eq}	For isostatic spans $L_{eq} = 1.4 \cdot L$ For internal spans in continuous systems $L_{eq} = (2 \cdot L_i + L_{i+1})/3,\ L_i > L_{i+1}$ End spans should be multiplied with 1.25. L_i, L_{i+1} are the two longest consecutive spans.
Steel consumption	$G = 63 + 0.9 \cdot L_{eq}^{1.2} \cdot \left(1.34 - \frac{b_c}{40}\right) + \frac{L_{eq}}{4} \left[kg/m^2\right]$
Slab reinforcement ratio	$\approx 250\ kg/m^3$
Main girder depth (for a constant depth deck)	$h_a = \max\left[\frac{L_{eq}}{28} \cdot \left(\frac{b_c}{12}\right)^{0.45}, 0.40 + \frac{L_{eq}}{35}\right]$, L_{eq} and b_c in m
Main girder depth (for a variable depth deck with more than two spans)	$h_a = \frac{L_{eq}}{24}$ at pier, $h_a = \frac{L_{eq}}{36}$ at midspan
Slab thickness (m)	$h_c = \frac{b_c - a}{26} + 0.13$ at main girders $h_c = \frac{a}{50} + 0.12$ at deck center
Main girder c/c distance	$a = \text{approx.}\ 0.55 \cdot b_c$
Bottom flange width (m)	$b_{fu} = \left(0.25 + \frac{b_c}{40} + \frac{L_{eq}}{125}\right) \cdot \left(0.92 + \frac{b_c}{150}\right)$
Top flange width (m)	$b_{fo} = b_{fu} - 0.1$ for a two-lane deck $b_{fo} = b_{fu} - 0.2$ for a four-lane deck
Cross girder	IPE 500 to IPE 700 at spans

Example
A three-span bridge with a deck slab width: $b_c = 12.3$ m

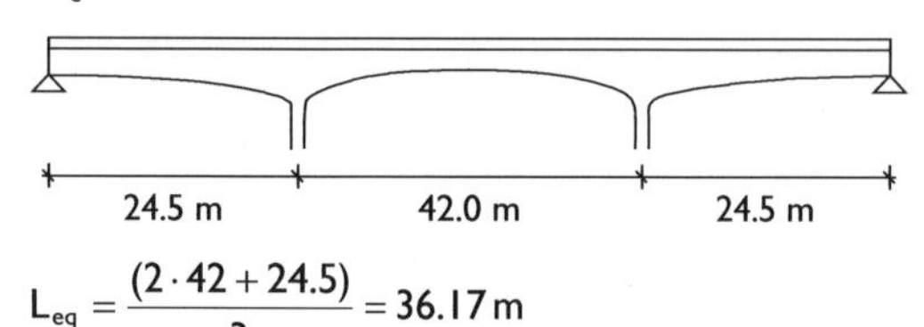

$$L_{eq} = \frac{(2 \cdot 42 + 24.5)}{3} = 36.17\,m$$

Girder spacing (a)	8.42 m
Main girder depth (h_a)	1.51 m at pier
	1.51 m at midspan
Slab thickness (h_c)	0.39 m at main girders
	0.29 m at deck center
Bottom flange width (b_{fu})	0.94 m
Top flange width (b_{fo})	0.84 m
Steel consumption	135.9 kg/m²

For continuous systems, steel girders with variable depth are preferred in order to achieve a better aesthetic and an improved clearance. At internal supports, steel cross sections are always heavier than those at spans due to the interaction of strong negative bending moments and shear forces. The web thickness mainly depends on the steel girder's height and at spans rarely exceeds 20 mm. In multilane bridges, the web thickness at hogging moment areas can reach 35 mm.

Cross girders at spans are 500–700 mm high and are placed at a constant distance of no more than 8 m. Through this, a steel frame with an adequate stability during erection is formed. The connection of the main girders with the transverse ones must be classified as rigid, and therefore, thick gusset plates are used. Bolted connections are usually avoided. A T-section link welded to the cross girders' web (see section a–a in Figure 2.16) ensures the members' continuity. For small deck widths, standard half T sections are preferred; for larger widths, welded T sections are more suitable. Cross girders at piers are stiffer than those at spans with depth ranging from 600 to 1600 mm. In continuous bridges, high values of hogging moments at piers can result in lateral torsional buckling of bottom flanges. In such cases, the cross girders spacing is reduced. Table 2.2 summarizes a preliminary design for twin-girder bridges with a deck slab of variable depth; see [2.32].

2.3.6.1 Ladder deck bridges

Twin-girder bridges with cross girders rigidly connected with the deck slab are generally known as ladder deck bridges; see Figure 2.17. Due to its simplicity, the ladder deck configuration has been successfully implemented in many European countries and especially in the United Kingdom. Experience shows that this arrangement is convenient for bridges with a dual two-lane carriage way.

The girders' spacing fits to the width of the slab. Main girders are mostly welded cross sections with wide and thick plates. Cross girders are normally spaced every 3.0–3.5 m so that a depth of 25 cm for the deck slab can be suitable. The main and cross girders effectively act as supports for the deck slab allowing an economical reinforcing against global and local bending. Cross girders transfer the vertical loads from the deck slab to main girders. Torsional overloading of the main girders must be avoided, and therefore, cross girders are designed as simply supported beams.

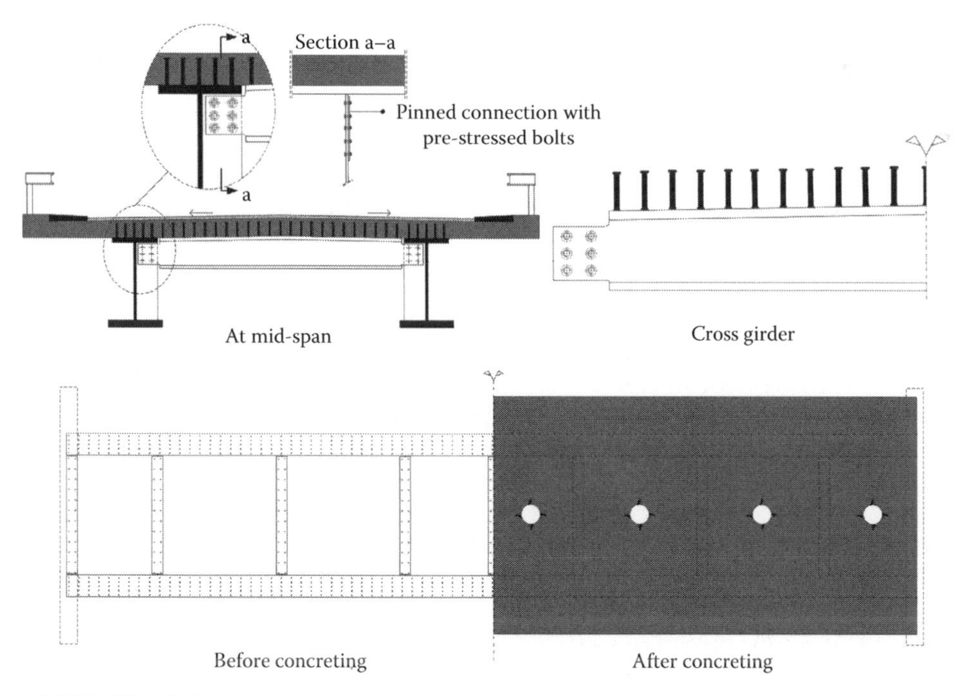

Figure 2.17 Ladder deck bridge.

A bolted connection of the cross girder's web with the stiffener of the main girder is shown in Figure 2.17. The connection must be designed as slip resistant at ULS according to EN 1993-2 so that cross girders restrain the main girders against lateral torsional buckling. In addition, tensile stresses in concrete due to hogging moments at the centroid of the bolt group should be carefully calculated, and slab cracking must be avoided. EN 1994-2

Table 2.3 Preliminary design for ladder deck bridges

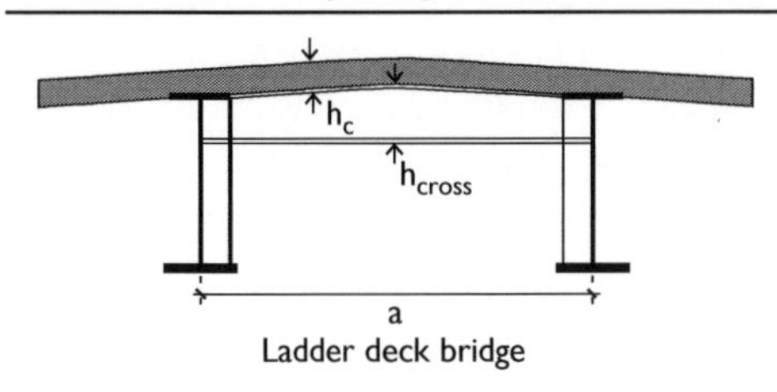

Ladder deck bridge

Cantilever girder

h_c

Ladder deck bridge with cantilever girders

Length L_{eq}	See Table 2.2.
Steel consumption	$G = 65 + 0.9 \cdot L_{eq}^{1.2} \cdot \left(1.43 - \frac{b_c}{30}\right) + 0.22 \cdot L_{eq} + 2 \cdot b_c \left[kg/m^2\right]$
Slab reinforcement ratio	≈275 kg/m³
Main girder depth (for a constant depth deck)	$h_a = \max\left[\frac{L_{eq}}{28} \cdot \left(\frac{b_c}{12}\right)^{\frac{1}{3}}, 0.40 + \frac{L_{eq}}{35}\right]$, L_{eq}, b_c in m
Main girder depth (for a variable depth deck with more than two spans)	$h_a = \frac{L_{eq}}{24}$ at pier, $h_a = \frac{L_{eq}}{36}$ at midspan
Slab thickness	24–26 cm
Main girder c/c distance	$a = \text{approx.}\, 0.55 \cdot b_c$ for decks without cantilever girders $a = b_c - 4$ for decks with cantilever girders
Bottom flange width (m)	$b_{fu} = \left(0.25 + \frac{b_c}{40} + \frac{L_{eq}}{125}\right)$
Top flange width (m)	$b_{fo} = b_{fu} - 0.1$ for a two-lane deck $b_{fo} = b_{fu} - 0.2$ for a four-lane deck
Cross girder (m)	$h_{cross} = \text{approx.}\, \frac{a}{11} \geq 0.3\,\text{m}$

Example

A three-span bridge a deck slab width: $b_c = 12.3$ m

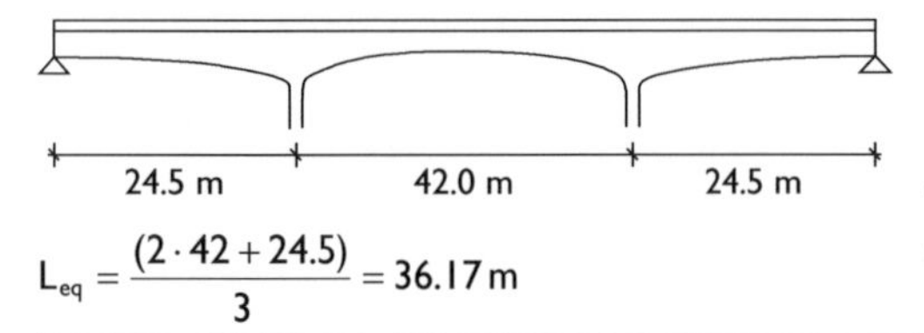

$$L_{eq} = \frac{(2 \cdot 42 + 24.5)}{3} = 36.17\,\text{m}$$

Girder spacing (a)	8.42 m 8.3 m with cantilever girders
Main girder depth (h_a)	1.51 m at pier 1.51 m at midspan
Cross girder (h_{cross})	750 mm
Slab thickness (h_c)	24–26 cm
Bottom flange width (b_{fu})	0.92 m
Top flange width (b_{fo})	0.82 m
Steel consumption	164.9 kg/m²

does not provide information on the calculation procedure of these stresses, thus forcing designers to adopt conservative assumptions; see [2.16].

In Table 2.3, one can find preliminary design equations for ladder deck bridge cross sections; see [2.32].

In some countries, main and cross girders are welded together on site. The top flanges of the crossbeams are welded straight onto the main girders' flanges; see figures in Table 2.3. These welds are hidden under the deck slab after concreting making future inspection impossible. Fracture of these welds due to fatigue is at these regions highly possible, and therefore, welded joints are not recommended. Nowadays, cantilever girders are implemented in the majority of the ladder deck bridges. This helps considerably the construction procedure and ensures a proper load transfer between the bottom flanges of both the cross girders and the cantilever girders.

2.3.7 Bridges with closed box girders

For spans larger than 60 m, plate girders can be uneconomical due to excessive flange sizes, complicated bracing systems, and/or temporary supports. In such cases, the use of box girders can result in less steel consumption and to a more aesthetic appearance; see Figure 2.18. Box girders are closed cross sections with a high value of torsional rigidity leading to an improved stability during both the erection and the final stage. Therefore, intermediate bracings can be avoided, and cross girders are located mainly at pier areas. Box-girder sections may have an orthogonal or a trapezoidal shape and are very effective in cases of curved road layouts where torsional loadings are critical.

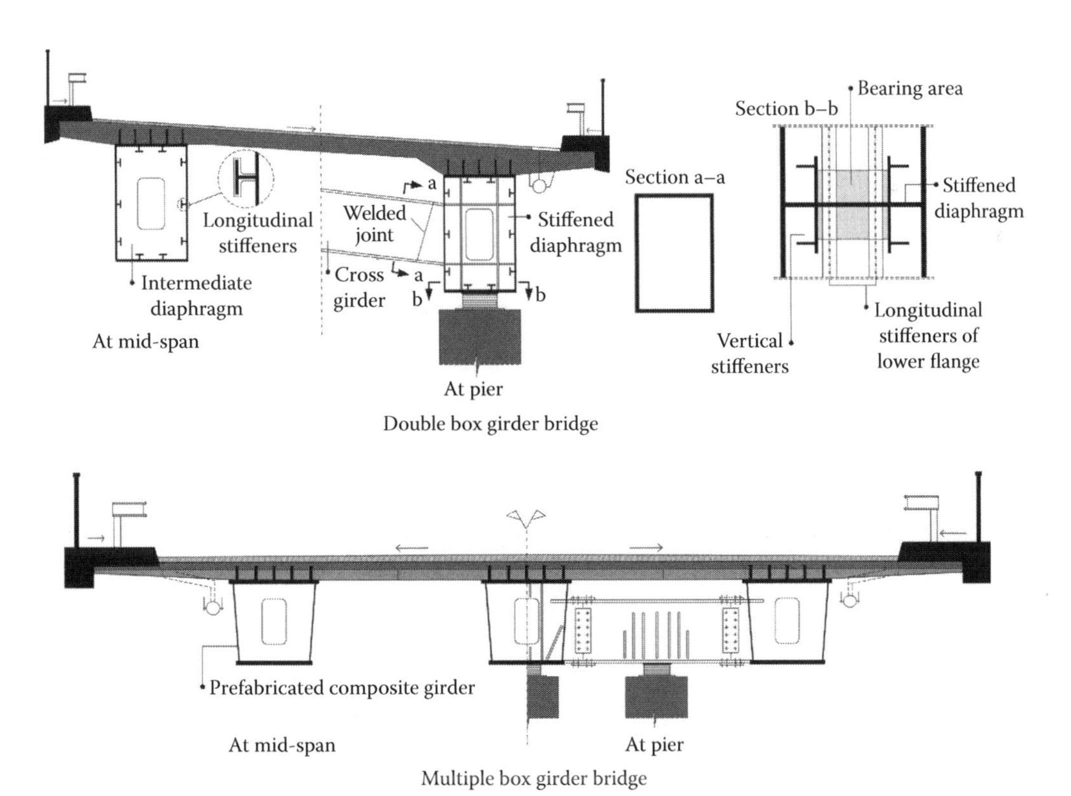

Figure 2.18 Closed box-girder bridges.

Bridges with two separate box girders are successfully implemented in many European countries mainly for carrying minor roads with two lanes. The steel beams are fabricated in the shop and then are transferred on site. Where possible long steel girders are easily erected by cranes, otherwise launching can be chosen as the most convenient construction method. A depth to span ratio equal to 1/30 is for continuous systems feasible.

Due to the high slenderness of the plates, longitudinal and transverse stiffeners for the webs and/or for the flanges may be unavoidable (plates of class 4). The stiffeners secure the plates against buckling, but complicated welding procedures and a great amount of man-hours are required. Furthermore, the designer must allow through manholes sufficient ventilation for internal inspection and openings for access in emergency situations. The previous lead to a demanding detailing that considerably raises the structure's cost. It is also worth mentioning that a dense net of welds is associated with a high failure risk due to fatigue. EN 1993-1-9 does not cover all the detail categories that may arise during a fatigue design procedure of such bridges. Therefore, a multi-box-girder configuration with narrower steel cross sections may sometimes be seen as a simpler solution.

With multiple narrow box girders, the need for longitudinal stiffeners can be eliminated and a slenderness h/L equal to 1/45 is achievable. Prefabricated composite girders can also be erected offering a high prefabrication rate and a protection for the upper flange during concreting or launching; see [2.38]. Multiple boxes are usually preferred for wide roads with more than two lanes and for spans up to 50 m.

It is important to note that closed cross sections exhibit a different structural behavior compared to typical I girders. Due to the high torsional rigidity of the closed girders, strong torsional moments will appear. Torsion is followed by an out-of-plane deformation of the section (distortion) that is associated with an interaction between shear and normal stresses. The previous stress situation is known as warping and can result in unexpected failure modes during erection and the final stage; such failure modes can be plate buckling at serviceability limit state (SLS), weld ruptures due to fatigue at welds, and local yielding of structural steel at unexpected positions. Thus, a detailed investigation for appropriate internal stiffening that prevents excessive distortional effects and high warping stresses should be conducted.

Figure 2.19 shows a comparison of a closed cross-sectional beam with and without intermediate plated diaphragms. Intermediate diaphragms ensure that the shape of the cross

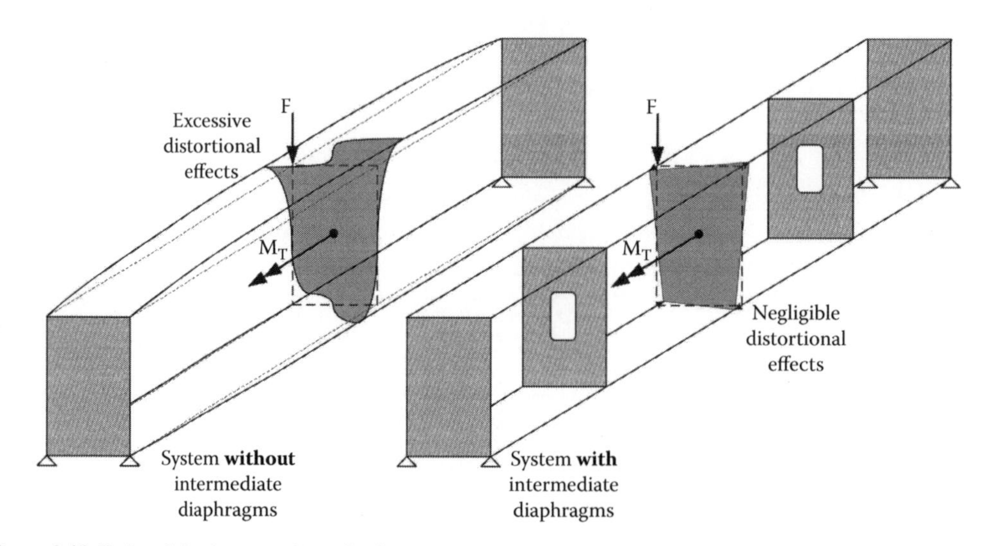

Figure 2.19 Role of the intermediate diaphragms.

section remains unchanged and that torsion is distributed as a shear flow along the box's perimeter. Moreover, they act as rotational springs "captivating" the distortional effects between adjacent diaphragms. Plated diaphragms must be connected along all four edges of the box for a better fatigue performance.

The efficiency of the internal stiffening depends on the flexibility and the spacing of the diaphragms. Flexible diaphragms result in excessive out-of-plane deformations that may considerably increase the instability risk. On the other hand, very stiff diaphragms may lead to high values of warping stresses especially at positions where webs and flanges are met. It is obvious from the previous that intermediate diaphragms should not be considered as secondary structural elements and neglected during structural analysis. The finite element method is preferred as the most accurate analysis method for closed girder bridges due to the calculation complexity of the distortional effects.

At supports, plated diaphragms transfer strong shear forces from the webs of the box girders to the bearings; see Figure 2.18. Support diaphragms are much thicker than the intermediate ones and are equipped with vertical stiffeners; see section b–b. The stiffeners protect the diaphragm against buckling and are connected to the top flange to avoid fatigue problems. In other words, a stiff and high-capacity internal column connecting the lower with the upper flange at supports ensures a safe loading transmission from the upper structure to the piers.

Alternative stiffening solutions to diaphragms can be triangular cross frames or ring frames. The selection of the appropriate type mainly depends on the cost-effectiveness and the fabrication simplicity. Unfortunately, there are no specifications in the Eurocodes for structural elements such as cross girders and diaphragms. Guidance notes for best practice based on experience can be found from country to country; see, for example, [2.23].

Attention must be paid at the final stage since the torsional rigidity of the main girders may lead to an overloading of the adjacent elements such as cross girders or the deck slab.

The torsional restraint offered to the slab by the main girders results in high-tension forces on the shear studs at the corner of the box; see Figure 2.20. Without the appropriate reinforcing, a pullout concrete failure is possible; see Figure 12.5.

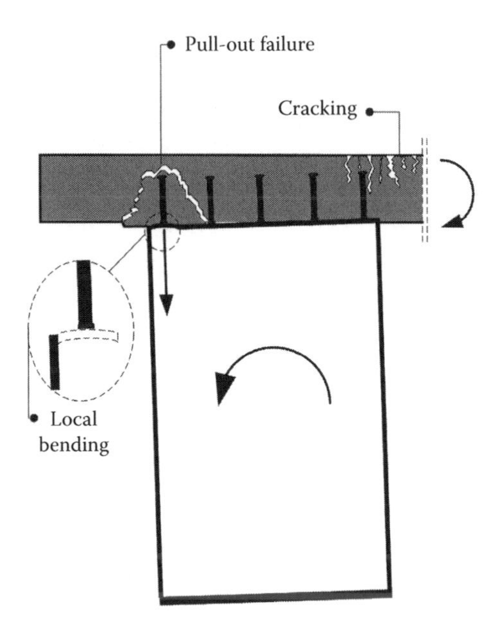

Figure 2.20 Local failures due to torsional rigidity.

Care should also be taken to avoid local bending of the top flange due to tension of the studs. Such failure can take place when the flange is very thin.

A non-negligible part of the torsional moment is expected to be transferred as bending moment to the deck slab. As a consequence, cracking in the vicinity of the webs at SLS is possible. This is more critical in cases of long distance girders.

Closed box girders have wide flanges, and due to the shear lag effect, a part of them may be considered as ineffective during design. Indeed, according to EN 1993-1-5 for flanges with a width b_f greater than $L_e/25$, where L_e is the length of the equivalent span (distance between zero bending moments), a reduced effective width must be applied. The designer must be aware of this limitation and if possible avoid flanges with excessive widths. Shear lag can significantly affect the magnitude of the peak stresses during fatigue design.

It can be concluded that closed box-girder bridges are offered in certain cases as an attractive solution. The designer should be experienced with this type of bridges since their structural behavior differs significantly from that of the I-girder ones. Usually designers contact the steel fabricator for consulting prior to the design process. Pre-dimensioning equations is almost impossible to offer, and therefore, a trial and error procedure must be followed to reach an optimal solution.

2.3.8 Open-box bridges

For continuous bridges with a maximum span length greater than 50 m, a solution with a single open-box girder is in many cases the most cost-effective and aesthetical one. A typical cross section of an open-box-girder bridge that carries a symmetrical bidirectional banked road is demonstrated in Figure 2.21. The box girder has a trapezoidal shape and consists of a wide bottom flange equipped with longitudinal stiffeners; the stiffeners protect the flange from plate buckling due to high compression stresses that may arise during the final and/or the erection stage.

The webs are slightly inclined with the angle θ being equal to a value between 15° and 25°. This gives the cross section a more attractive appearance. Furthermore, the webs' inclination is important for the following reasons:

- The reduced width of the bottom flange leads to a better structural performance because a smaller part of the flange will be ineffective due to the shear lag effect. It makes also the plate buckling verification easier and achievable with fewer stiffeners.
- The distance between the bearings is smaller, and thus, bending of the transverse frames at supports becomes easier to control.
- Small-sized abutments can be designed allowing the construction of a slender substructure.

The steel girders are opened cross sections and without internal stiffening are vulnerable to torsional and distortional effects. For this reason, transverse frames of T sections are welded to the webs and the bottom flange and are placed every 4–7 m at spans; see also sections a–a and b–b in Figure 2.21. At piers, the transverse frames must form a fork-bearing support capable of transferring high torsional support reactions from the upper structure to the abutments. Therefore, the transverse frames at supports are closer with each other and much stiffer than those at spans.

It therefore becomes obvious that the bending rigidity of the transverse frames determines the torsional rigidity of the entire bridge. In most cases, the torsional rigidity due to the transverse frames is insufficient, and additional stiffening is necessary. This is easily achieved by supporting the frames with internal trusses consisting of diagonals and tension members.

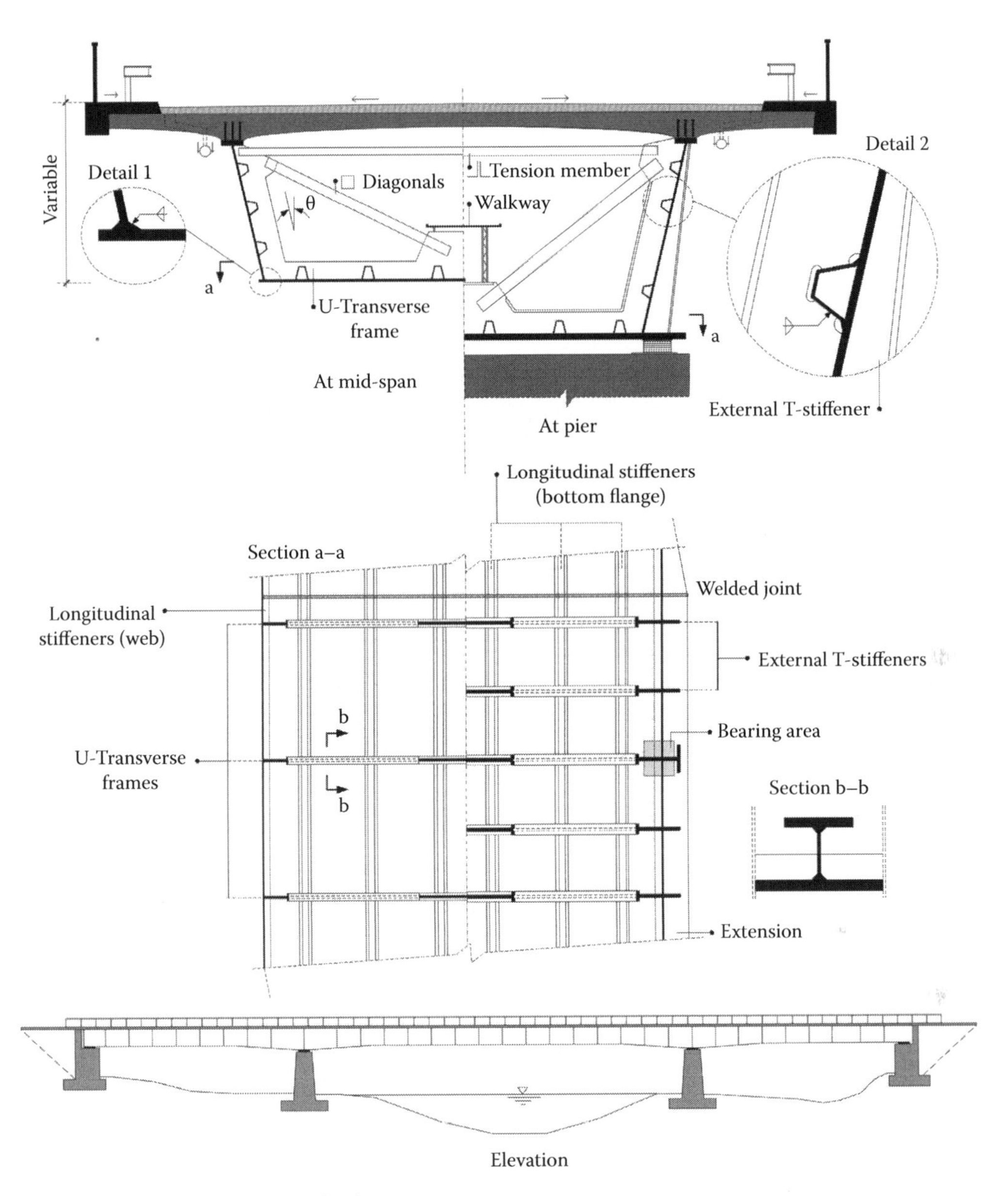

Figure 2.21 Single-box composite bridge.

Diagonals are usually closed cross sections ensuring an adequate buckling resistance. Double L or C sections are used as tension members. The designer must be sure that no reversed loading will take place during erection in these members; this can happen due to wind or earthquake during the erection phase. In such cases, closed cross sections must be used for all internal members.

Taking a closer look in Figure 2.21, one can also observe that the plates of the box girder at supports are thicker than those at spans. This is mainly due to the strong support reactions and the zero tension capacity of concrete in hogging moment areas. The thickness of the bottom flange usually varies longitudinally between 25 and 35 mm within the spans and between 60 and 80 mm at supports. At piers, the bottom flange is also slightly wider than the standard width allowing a connection with an external stiffener above the bearing area; see detail 1. The width of the bottom flange should not exceed the maximum width determined by the fabrication. This is usually 5 m for bridges with straight road layouts. For bridges with curved deck, the width to be chosen must be less than 5 m, since the bottom flange will be cut from rectangular plates with a maximum width of 5 m.

The thickness of the webs varies also longitudinally between 14 and 18 mm at spans and 20–25 mm at supports. The previous values are feasible only through the application of longitudinal stiffeners capable of resisting high normal and shear stresses; see detail 2.

The thickness of the top flanges ranges between 20 and 40 mm at spans and at supports can exceed 100 mm. Accordingly, the flanges' width at spans varies longitudinally from 600 to 800 mm and at supports can reach 1200 mm.

In many bridges, the overall width of the bottom flange is slightly greater than the distance between the webs; see detail 1. This gives enough space for the outer welding and is advantageous during launching because launching devices can be placed directly under the webs. However, it has the disadvantage of water or dirt accumulation in the corner.

In the previous, a typical box-girder bridge and its main structural elements were described. Different cross-sectional geometries can also be implemented. The choice of the optimum solution depends on several factors that were already commented in Section 1.1. Different cross-sectional cases are shown in Figure 2.22.

Case A depicts a box-girder bridge in which the top beams of the transverse frames are rigidly connected with a semi-precast deck slab through shear studs. The combination of directly supporting beams together with cantilever girders and precast slabs offers a high prefabrication rate. Therefore, complex concreting operations with mobile formworks are avoided; see Figure 2.3. It is also easy to observe that internal trusses are missing. This is due to the fact that the rigidity of the transverse frame is considered as sufficient. Indeed, the transverse frame consists of double T sections (see section b–b in Figure 2.22) rigidly connected with the composite crossbeam and the bottom flange. Such frames can offer an adequate stiffness in cases of narrow boxes. A load path through additional plates between the crossbeams and the cantilever beams ensures a secure transmission of internal forces through the webs. At piers, instead of transverse frames, plated diaphragms have been chosen. Diaphragms are preferred in cases of very large torsional effects, for example, due to high horizontal curvature. The diaphragms have a thickness of 30–60 mm and a top flange that is connected to the slab. They are heavily stiffened in order to possess sufficient resistance and in-plane stiffness. The stiffening of the diaphragms is not a straightforward procedure and is conducted through complicated plate buckling analysis. Manholes necessary for inspection complicate the analysis, and finite element models are used.

In case B, the bottom flange has an inclination corresponding to the road banking. This geometric configuration is followed in cases of heavily banked roads (>2.5%) so that webs of equal depth can be designed. Banked bottom flanges can be problematic when launching is chosen as the bridge's erection method. One can also see the existence of a reinforced concrete slab rigidly connected through headed studs with the bottom flange. This is known as *double composite action* and is followed when a great amount of longitudinal stiffeners or an excessive thickness increase is necessary for ensuring the stability of the bottom flange at supports. The designer must pay attention to the fact that secondary internal forces will develop due to the rheological behavior of the bottom composite slab. This needs a detailed time-dependent analysis in which cracking of the deck slab and creep and shrinkage of the bottom slab are taken into account. Unfortunately, EN 1994-2 does not offer a design procedure for cases of double composite action.

Case C shows a cross section with a heavily banked road, a usual case for curved bridges with high values of horizontal curvature. The weight difference at webs' heights causes

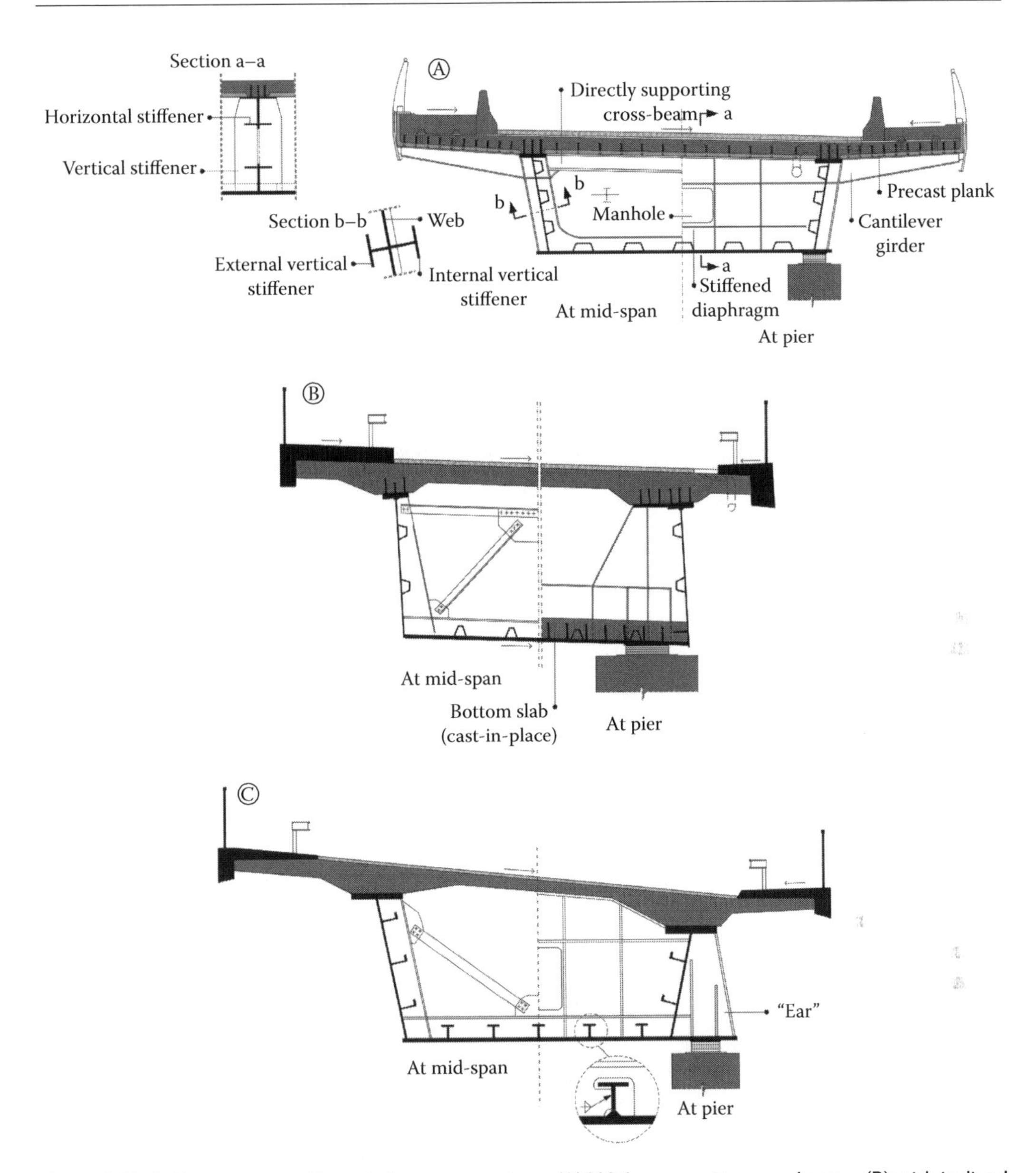

Figure 2.22 Different types of box-girder cross sections. (A) With composite cross-beams, (B) with inclined bottom flange and haunches, and (C) with heavily banked road.

distortional forces during both the erection and the final stage. These effects must be carefully considered during analysis. One can also see that the distance between the support bearings is larger than that of the webs. This can be the case when a large lever arm for the bearings is needed due to high torsional support reactions. This problem is solved through the lateral extension of the stiffened diaphragm; the extended part is generally known as ear. The diaphragm's extension is an interesting solution that allows also having a wider bottom flange at piers. It may also be observed that open longitudinal stiffeners were used for both the webs and the bottom flange. Instead of trapezoidal stiffeners, T, C, and Γ sections are also used in composite bridges. The choice of the appropriate stiffener is discussed in Chapter 8.

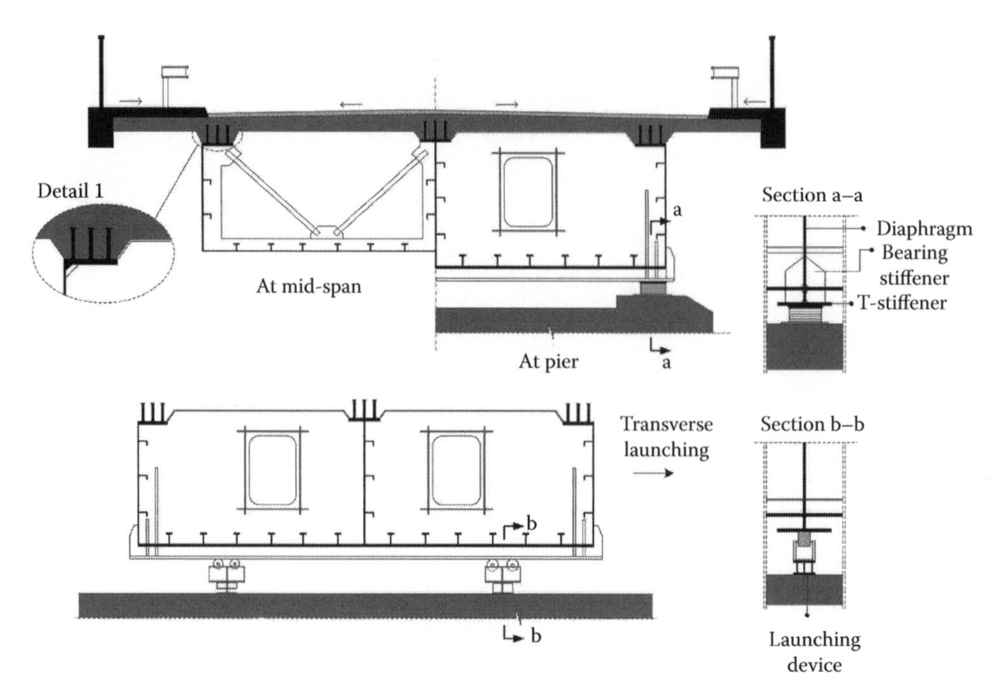

Figure 2.23 Open-box-girder bridge with three webs.

With Figure 2.22, cross-sectional configurations, depth to main span ratios h/L varying from 1/15 to 1/25, are usually feasible. More appealing slenderness values can be achieved with multicell box girders; see Figure 2.23. Due to the web in the middle, a greater part of the bottom flange becomes effective, and therefore, both resistance and stiffness are increased. This is important in cases of bottom flanges with widths larger than 5 m. Furthermore, shear and torsional capacity of the cross section is obviously higher.

Sections a–a and b–b depict an interesting detailing that can be advantageous when a transverse launch is employed to move the steel girder to its permanent bearing location. At piers, an external transverse frame is formed by welding a T stiffener to the bottom flange and the webs. This increases considerably the bending resistance of the cross frame. Moreover, concentrated forces induced from the launching devices to the diaphragm are distributed in a more uniform manner.

In addition, the multi-supported deck slab requires less reinforcement, and transverse prestress is avoided. The middle flange is also slightly higher than the adjacent ones, and the deck slab is easily built with the desired inclination. This leads to a reduced volume of surfacing materials that is important for bridges with wide decks.

There is also an aesthetical feature in the bridge of Figure 2.23 that has to be noticed. The outermost flanges project only on the inner side of the web. This is believed to improve the appearance of the bridge although many designers argue that it is not noticed by most people.

Bridges carrying roads with more than two traffic lanes are in general difficult to construct, and therefore, twin bridges with separate deck slabs and piers are preferred by many designers as the simplest solution; see Figure 2.24. The main advantage of the twin bridges configuration is that the full traffic can be diverted from the one bridge on the other during maintenance period. Disadvantageous is that two bridges must be built instead of one.

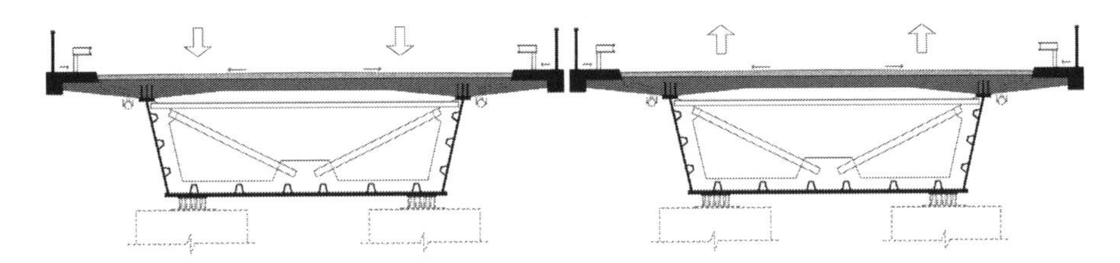

Figure 2.24 Twin bridges for different traffic directions.

When for both bridges the road layout is exactly the same, only one structural system should be analyzed. It is important to notice that in seismic regions the joint between the decks must be large enough in order to avoid an earthquake-induced pounding.

Although twin bridges are convenient solutions, designers sometimes prefer more difficult alternatives such as the *super bridges* of Figure 2.25. This can be due to high aesthetical and/or environmental demands. Single-box bridges with very wide decks (>20 m) carrying multilane roads are very attractive and slender in constructions. Case A is a single-box girder that carries two roads of different directions. The left road has one lane more than the right one. Therefore, the cross section is constantly under torsion. A system of internal and external diagonals forms a transverse truss that supports the deck slab and resists the traffic loads. The external diagonals act also as props for the cantilevers whose length may in certain cases be greater than 10 m.

Case B shows also a wide deck bridge. Interesting is that the top part of the transverse truss is not a "heavy" crossbeam but a flexible steel chord; see section a–a. The cord consists of a steel plate with shear studs welded on it. It becomes after concreting part of a composite member that is subjected to different types of loading; see section b–b. These are bending due to local traffic loads and obviously tension. The designer must be aware of the difficulties that may arise during the fatigue verification of the connection of the chords with the longitudinal elements. A special detailing is therefore necessary. Different types of transverse chords with comments on advantages and drawbacks can be found in [2.12]. Numerous applications of bridges with transverse composite tension members are mainly found in Germany; see also [2.31].

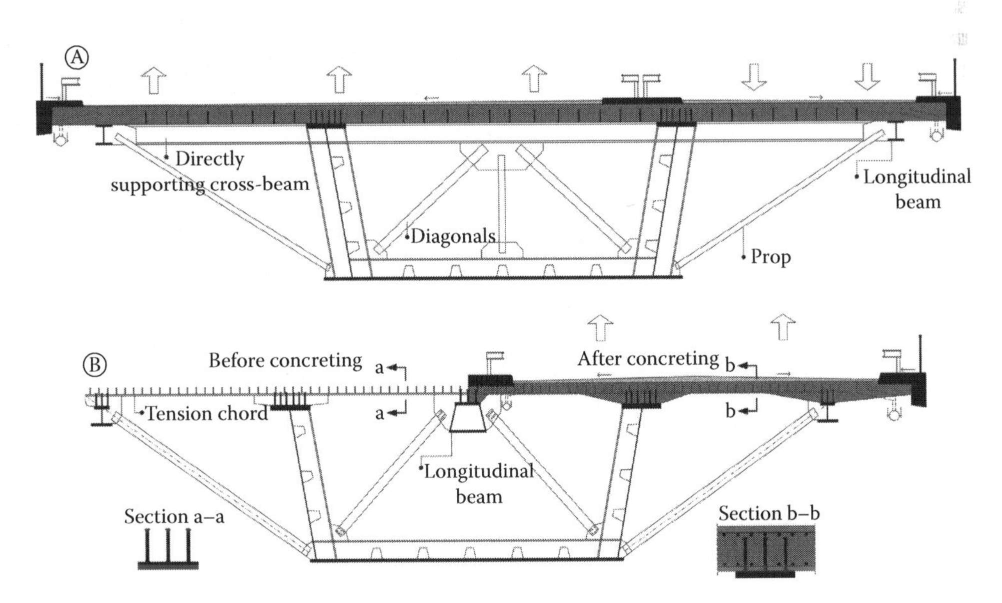

Figure 2.25 Bridges with wide decks.

2.3.9 Arch bridges

Typical arch bridges consist of two or more parallel arches that carry the bridge's deck through hangers. The hangers are connected with a plane grillage system of transverse and longitudinal beams that supports the deck slab; see Figure 2.27. This is a very elegant configuration that is preferred in cases of rivers and canals because it gives enough under clearance. Arch bridges are usually simply supported systems that are used for span lengths ranging from 50 to 180 m. The maximum value of the total construction depth is in most cases between 1/5 and 1/6 of the span length. More demanding slenderness values can be asked due to architectural reasons and not structural ones.

The structural performance of arched bridges is schematically demonstrated in Figure 2.26.

The dead weight of the deck together with the traffic loads is transferred from the transverse beams to the edge girders that are also known as stiffening girders. The hangers behave as intermediate supports for the stiffening girders and transmit the vertical loads to the upper part of the structure, the arches. Obviously, the arches are under compression, and therefore, buckling is highly possible. Due to this reason, arches are made of reinforced concrete or of steel closed cross sections. In pure arched bridges (Figure 2.26a), the stiffening girders are connected to the arch only by the hangers. The arch thrust is transferred to the foundation and the soil. When the stiffening girders and the arches are rigidly connected with each other at the bridge ends (Figures 2.26b, c and 2.27), the thrust is transferred to the stiffening girders that behave as tension ties. These systems are known as *bowstring arches* [2.13] or arch-and-tie bridges.

The arches have usually parabolic form that follows the bending moment diagram of a simply supported beam with a uniform loading. This ensures that arches are uniformly compressed under self-weight and bending moments remain low. Biaxial bending due to wind or earthquake can take place, and therefore, symmetrical cross sections of closed geometry are chosen. It has to be stated that a precise buckling analysis must be conducted. This has to be done by calculating the critical buckling factors for every load combination and the corresponding buckling modes. This is a demanding design procedure that severely affects the morphology and the cross-sectional geometry of arches. EN 1993-2 provides buckling factors and imperfection values for a second-order theory analysis of arched structures.

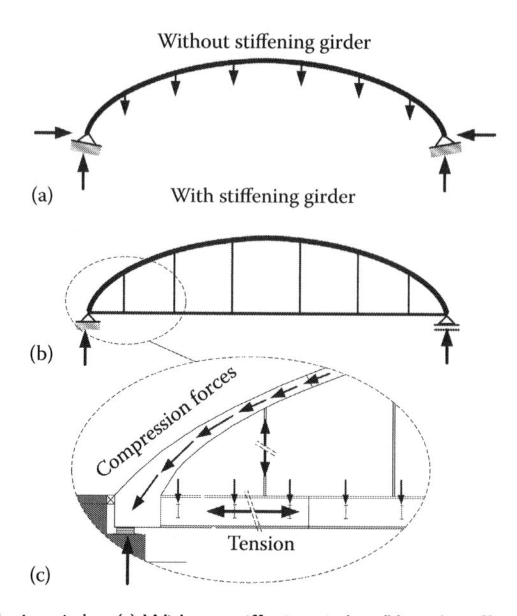

Figure 2.26 Role of the stiffening girder. (a) Without stiffening girder, (b) with stiffening girder, and (c) load path.

A validation of the code's proposed buckling factors with the results of a buckling analysis of a 3D model is in any case to recommend.

Stiffening girders behave as continuous systems with equally spaced concentrated forces; these are the support reactions of the transverse beams. As mentioned earlier, they also act as tension ties due to their "cooperation" with the arches. Thus, an interaction between bending and tension must be carefully taken into account in different positions along the bridge's length. The most critical verification points are usually those of the hangers. For this reason, designers place the hangers at positions that are coincident to those of the transverse beams. This obviously allows a more direct and economical force transfer.

Hangers are compact tension rods with the commonly used diameters ranging from 50 to 140 mm. They are connected with the arches and the stiffening girders through gusset plates; see details 1, 2, and 3 in Figure 2.27. In details 1 and 2, one can see that the

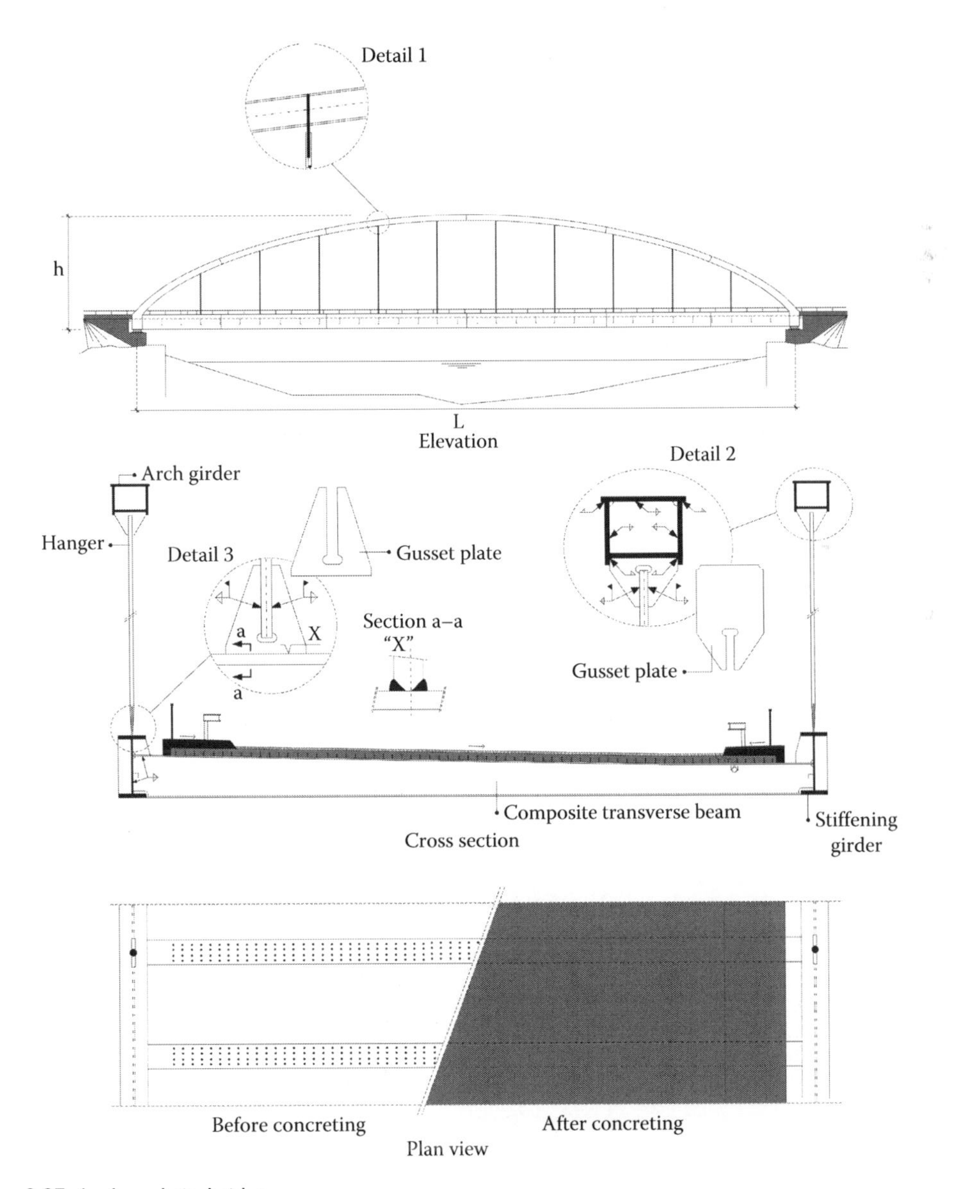

Figure 2.27 Arch-and-tie-bridge.

hollow-box cross section of the arc is penetrated by a thick plate. The external part of the gusset plate transfers the tension load from the hanger to the internal part that works as a diaphragm. It is welded with the hanger through double-sided fillet welds; different types of welds are also possible. It is important to note that hangers constantly vibrate under the action of wind and therefore, a careful fatigue detailing is necessary. The dynamic behavior of the hangers needs therefore to be investigated since measurements have shown that their damping capacity is very low [2.51]. In case of resonance, the stress variations at the edge connections will be maximized and may be the reason of an unexpected brittle failure. Central hangers are usually more sensitive against wind-induced vibrations due to higher slenderness. Designers usually calculate the natural frequencies of the hangers (as isolated elements) by taking into account the effects of the axial forces due to the imposed loads and the shrinkage stresses of the welds and by considering different types of supports at the edges; the natural frequencies should be greater than a minimum value (recommended is 7 Hz); otherwise, dampers are installed. EN 1993-1-11 covers issues of tension members but does not offer adequate guidance on the vibration control of hangers. Finally, hangers should be replaceable since they cannot endure the total design working life of the bridge.

Arch bridges with inclined hangers are shown in Figure 2.28. The main advantage of such structures is mainly aesthetical since the inclination of the hangers gives to the structure an aerodynamic shape. Due to their increased length, inclined hangers are more sensitive to vibrations and construction difficulties arise especially at the connections with the arches and the stiffening girders. In order to minimize the effects of vibrations, additional horizontal stabilizing elements connecting adjacent hangers may be used, but this will have a negative effect on the appearance of the bridge.

In Figure 2.29, one can also see that the arch girders may be connected through top bracings. These braces are necessary for increasing the stability of the arch girders due to compression but also for enhancing the lateral stiffness in case of wind or seismic actions. Designers can choose different shapes for the bracing systems such as X, K, and Λ-diagonals or rigid frames; see also the network arched bridges in 2.4.7 that are preferably chosen for railroad applications. The decision of applying or not a top-bracing system is a difficult one since heavier arches may be finally more cost-effective and elegant alternatives.

The concrete slab may be separated from the stiffening girders in which case concrete is allowed to shrink or expand without affecting the steel elements. This may be seen advantageous, but the positive effects of the composite action are lost. However, the structure loses its redundancy that is a necessary characteristic for modern bridges. In other arch bridges, the concrete slab is connected with the steel elements and in bowstring arches additionally prestressed by tendons in the longitudinal direction; this is due to the fact that concrete is part of the tie and its cracking may not be taken into account in design realistically. Figure 2.29 shows a bowstring arch where a horizontal end bracing system is provided to allow for the participation of the entire bridge deck in the transfer of the arch thrust. The deck is a cracked composite tension member with effective cross-sectional properties, as provided by EN 1994-2. This leads to more slender structural members and raises the competitiveness of the "arched solution."

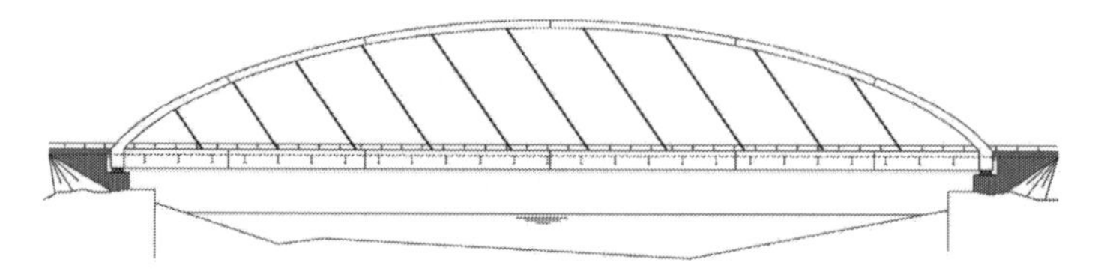

Figure 2.28 Arch bridge with inclined hangers.

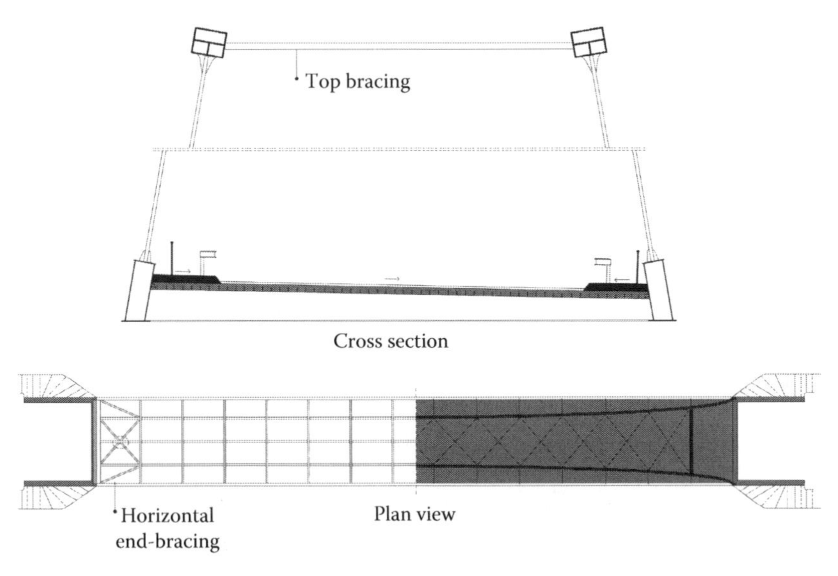

Figure 2.29 Composite bowstring arches.

A sensitive point in design of a concrete deck under tension is its shear resistance against the point loads of the wheels. When the crack width is excessive, the shear transfer is achieved by the dowel action of the reinforcement. The aggregate interlock becomes negligible, and the shear resistance decreases. In such a case, the reinforcing bars are constantly under cyclic bending, and fatigue failure is highly possible. For this reason, the maximum crack width should be limited to 0.1 mm for normal forces and to 0.2 mm for combined local bending and normal forces [2.13]. These values are smaller than those of EN 1992-2 for durability.

The erection of pure arch bridges starts with the construction of the arch that is done either on temporary falsework, if possible in swallow valleys, or by cantilevering in steep valleys or over water. In cantilevering, construction starts from the two springing points and continues with the position of new segments until the two halves are joined at midspan. During the progress of works, the parts of the arch in place must be temporarily tied back by cables. The prefabricated deck modules are then lifted and hung from the deck hangers that are connected to the arch.

Arch-and-tie bridges have the advantage of been external statically determinate. All steel parts including arches, hangers, and stiffening girders may be erected near the site and the entire bridge moved in place by barges. This is a common solution for medium-span river bridges. At the end, the deck is concreted. In that manner, the weight of the bridge during its put in place is reduced. Additional braces to ensure diaphragm action of the deck and temporary compression elements between the arch and the deck during moving operations are needed since the hangers are not in tension and therefore not effective.

2.3.10 Cable-stayed bridges

Cable-stayed bridges offer an economical and elegant solution for spans larger than 150 m and can be competitive for spans as long as 600 m; see *Rion-Antirion* multi-span cable bridge (www.gefyra.gr). The static behavior of a cable-stayed bridge is considerably different from that of an arch bridge. A typical case is shown in Figure 2.30. The inclined cables that are connected with a central pylon work as stable, however, elastic supports for the deck. The vertical loads (permanent and traffic) are transferred through tension forces from

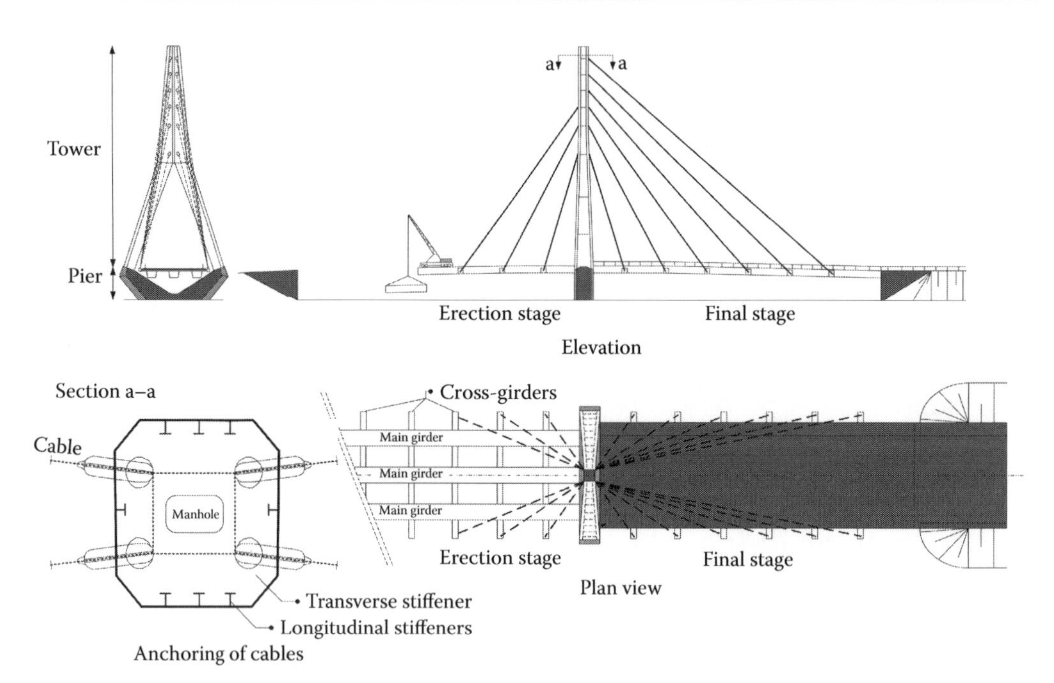

Figure 2.30 Cable-stayed bridge.

the cables to the tower and finally to the concrete pier. In contrast to arch bridges, the main girders are in most cases under compression, and therefore, closed cross sections such as composite boxes or steel boxes with orthotropic decks are preferred. The girders are in most cases shallow, and depths usually range from 1/60 to 1/80 the main span.

The towers may have different shapes and be made of structural steel or prestressed concrete. In general, the height of a pylon is about 1/6–1/8 the span. For reasons of strength and cost-efficiency, steel pylons are composed of segments with different dimensions that follow the normal force distribution. In Figure 2.30, one can see that the tower's cross section is a closed one, equipped with longitudinal and transverse stiffeners. Such cross sections "suffer" not only from high normal stresses but from high shear stresses due to shear and torsion. The complicated stress situation, the residual stresses due to the large number of welds, and the existence of holes in the plated elements make the design significantly difficult. In Chapter 8, one will read that EN 1993-1-5 for the dimensioning of plated elements does not refer to plates with holes. Therefore, nonlinear finite element analysis and an experimental verification may be necessary. Attention should be paid at the anchoring positions of the cables where stress concentrations are expected.

Modern cables are composed of groups of strands that are formed from steel galvanized or stainless wires. A commonly used type of cables is shown in Figure 2.31a, *bundle of parallel strands (BPSs)*. A number of strands are closed around a core strand. Wires have a diameter 3–7 mm and are made of high-strength steel with a tensile strength f_u ranging from 1300 to 1800 Mpa. After installation of the cables, additional corrosion protection is applied that usually consists of a polyethylene tube. Such tubes have discontinuous spiral on their surface in order to combat combined effects of wind and rain. The space between the rope and the tube is filled with a suitable hydrophobic material such as cement grout so that circulation of water and heat is avoided. An alternative solution is the *parallel wire strand (PWS)* shown in Figure 2.31b in which wires are bundled in parallel. PWS cables may have different shapes, usually a circular or a hexagonal one, and are usually chosen in cases of smaller loads.

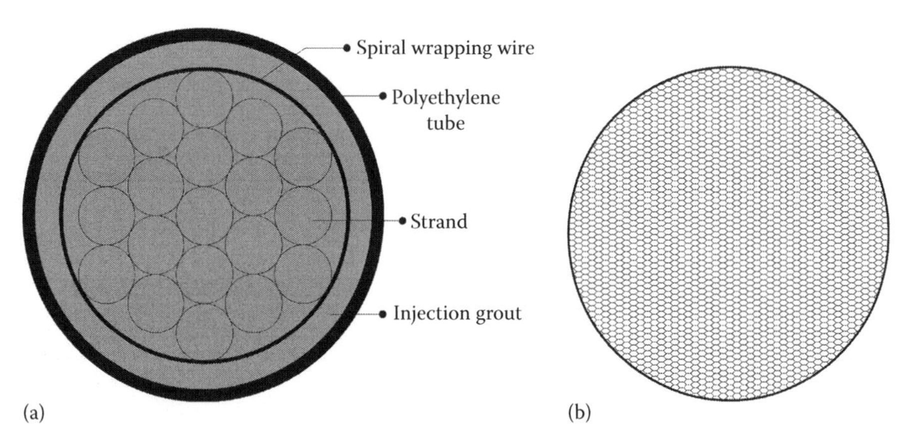

Figure 2.31 Parallel strand ropes. (a) Bundle of parallel strands and (b) parallel wire strand.

Another type of cables is shown in Figure 2.32 and is known as *locked coil rope (LC)*. One can see that deformed wires are used for the outside layer of a spiral rope. The main advantage of the LC cables is that the special corrosion protection that is necessary for the BPS and the PWS cables can be avoided; therefore, inspection and replacement are much easier. A disadvantage is the limited size that can be fabricated.

In Figure 2.32, the tension force–strain diagram of a LC is also shown [2.1]. One can observe that the initial value of the modulus of elasticity is the same as for structural steel. With increased stresses, the elastic modulus is decreased; for serviceability conditions, this decrease may be up to 50%. The behavior of cables is considerably nonlinear, and this has to be taken into account by conducting a third-order or a large displacement global analysis at both SLS and ULS. EN 1993-1-11 offers a stress-adjusted modulus of elasticity $E = f(\sigma)$ so that nonlinear deformation effects can be estimated. It is important to note that in cable-stayed bridges, composite members may be severely compressed. Therefore, the nonlinear behavior of the cables should be combined with the reduced cross-sectional properties of the composite members due to creep of concrete; see Section 6.1.2.

Cables are preloaded elements. This means that during and/or after erection, cables are prestressed so that the structure adopts the required geometric profile and stress distribution.

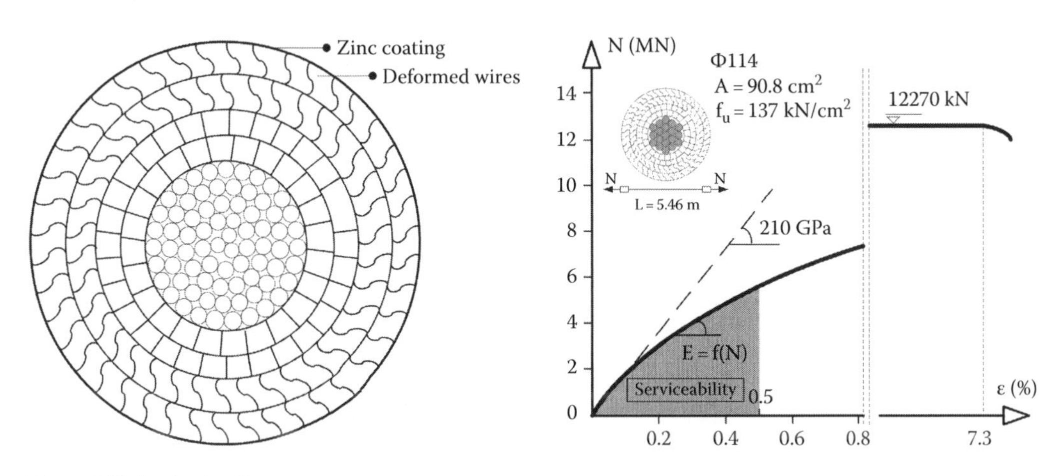

Figure 2.32 Locked coil rope.

Finally, the BPS, PWS, and LC cables are mainly used as final-stage elements. During erection, other types of cables are used such as strand or spiral strand ropes; see EN 1993-1-11.

The erection of cable-stayed bridges starts with the construction of the pylons. It is followed by successive cantilevering out of the pylons of prefabricated deck units suspended from the stay cables. This constitutes the main advantage of cable-stayed bridges over deep valleys and over waters.

2.3.11 Suspension bridges

Suspension bridges are generally preferred for spans over 600 m. A typical case of cable-suspended bridge is depicted in Figure 2.33. The stiffening girder is suspended through hangers from flexible main cables. The main cables are curved and are hung from towers. At the edges, main cables are anchored in massive concrete blocks capable of resisting strong tension forces. However, there are some bridges where the cables are connected to the main girders. This avoids the heavy anchorage blocks but introduces high compression forces in the girders and needs the completion of the deck before starting the cable erection. Main cables and hangers may be BPS, PWS, or LC ropes.

Stiffening girders distribute concentrated loads, act as chords for the lateral system, and secure the aerodynamic stability of the structure. They may be single-box cross sections with orthotropic deck. An orthotropic deck is a continuous flat steel plate with stiffeners welded to its underside. Single-box girders with orthotropic deck allow the design of different bridge types with considerable slenderness. The surfacing material is usually asphaltic, but due to the elastic and thermal properties of the steel plate, problems may arise, especially due to frost. Therefore, many designers are fond of stiffening trusses with concrete deck plate. Stiffening trusses are less slender than boxes with orthotropic deck, but they are advantageous due to their smaller air resistance.

A suspended bridge is obviously a flexible system, generally more flexible than cable-stayed bridges. The long span associated with the narrow deck makes the aerodynamic stability for the major bridges the most significant issue. Fluctuating wind loads may lead to an unexpected buffeting response; see Figure 2.34a. This is a dangerous resonance phenomenon in which certain wind velocities favor vibration modes with higher frequency, usually the torsional ones. For this reason, the cross section of the stiffening girder has slanted walls and rounded corners. In certain cases, stiffening girders are equipped with *vortex spoilers* so that the performance of the bridge against wind gusts and vortices is further improved; see Figure 2.33. It is also worth mentioning that the cross-sectional shape of the deck is optimized by wind tunnel measurements in which the structure is tested at different angles

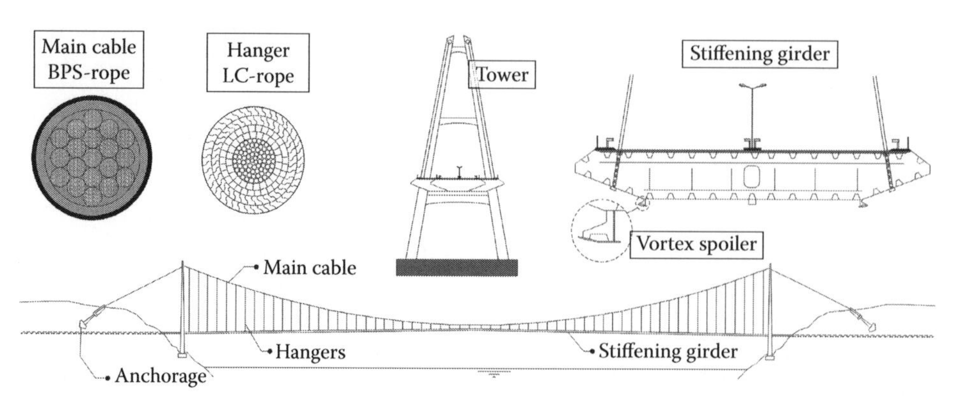

Figure 2.33 Cable-suspended bridge with orthotropic deck.

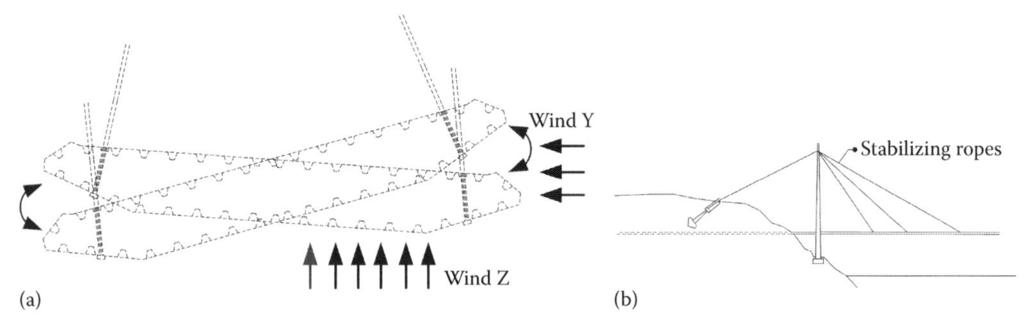

Figure 2.34 (a) Torsional vibration due to wind and (b) temporary stabilization during erection.

of wind attacks. The effects of vortex-excited vibrations can be estimated according to the regulations of EN 1991-2-4.

Wind effects may be more severe during erection. The most efficient way to stabilize the structure against buffeting is to increase its fundamental frequency. This is achieved by *temporary ropes* (or *stabilizing ropes*); see Figure 2.34b. Use of stabilizing ropes can also help reduce the unbalanced bending moment in the tower during construction. After placing the permanent cables, the bridge takes its final form and stabilizing ropes are disconnected (Figure 2.33).

The erection of suspension bridges starts with the pylons. Subsequently, the main cables are erected. The most usual method for cable erection is the *air-spinning method* in which small, for example, 5 mm, wires are spun one by one. The wire is reeled onto an unreeling/reeling winch then it is pulled out from this winch making round trips over the catwalk. Each wire is placed around the strand shoe and fixed to both anchorages. The cable is made by uniting the individual wires together. The stiffening girders are assembled in unit blocks that are transported to place, lifted from the main cables to position, connected to the hangers, and field welded.

2.4 RAILWAY BRIDGES

2.4.1 General

A typical cross section of a railway bridge is shown in Figure 2.35. One can see the railway infrastructure that comprises the permanent way (track), the access ways beside the track, and the associated plant equipment that allows the proper function of the railway. The *track* carries the railway traffic and consists of the *rails*, the *sleepers*, and the *ballast*. When the track is on curves, it is superelevated to compensate for the effects of centrifugal forces.

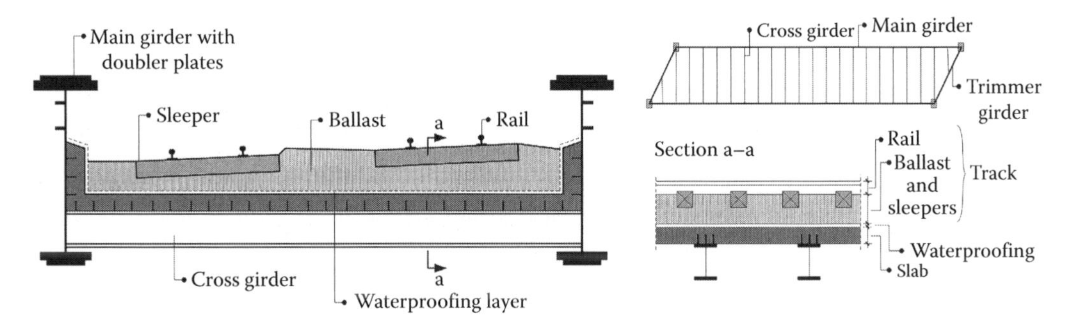

Figure 2.35 Half-through plate-girder bridge.

The rails are mainly steel bars laid on the sleepers, and their weight ranges from 55 to 80 kg/m. In older railways, rails used to be bolted on the sleepers, but nowadays, continuous welded rails are preferred. The rails are pretensioned in order to remain stress-free and avoid buckling due to compression from braking–acceleration forces and thermal actions.

Sleepers pass under the rails and hold them at the right spacing. In high-speed routes, sleepers are usually of prestressed concrete, but when construction depth is very limited, shallow timber or steel alternatives are also used.

Ballast works as a resilient bed for the sleepers since it distributes the wheel loads onto the deck plate and allows for drainage. It consists of coarse stone, slag, or clinker about 50–65 mm with a density of 20–22 kN/m^3 and is mechanically compacted. The usual ballast depth under the sleeper ranges from 250 to 300 mm depending on the requirements of the local authorities. These values ensure a satisfactory load distribution and drainage. When construction depth is severely restricted, lower ballast depths may be adopted. In some bridges, ballast and sleepers were omitted by fastening the rails directly to the bridge deck, *direct fastening*. This method seems to solve the problem of the restricted construction depth, but special detailing and experience are necessary. Moreover, track maintenance problems may arise when a rail is damaged. In general, a reduced ballast depth is the simplest solution and often the most preferable one.

Railway bridges are very demanding structures. The magnitude of both permanent and live loads may be several times larger than that of a typical highway bridge. As already noted, the construction depth is more difficult to optimize due to the additional depths of the ballast and the rails. Therefore, serviceability verifications (mainly deformation and vibration controls) are in many cases onerous, especially in multitrack railways. Designers should ensure that the track geometry remains and that the contact between the rails and the wheels is not lost.

Another important design aspect is the fatigue verification due to the dynamic nature of live loads. A railway bridge must be able to endure repeated actions whose magnitude depends on the annual tonnage of the traffic on every lane. Connections and especially the welded ones should be carefully designed and located where inspection, blast cleaning, and repainting is possible. It is true that in most railway bridges the inspection costs throughout the design life are comparable with the total cost of the structure. For these reasons, railway engineers are more concerned with the life cycle costs than with initial construction costs. From the design point of view, this is an important difference between rail- and roadway bridges.

Most of the cross sections described previously for the roadway bridges are also applied for railway applications. Modifications and special issues are discussed subsequently.

2.4.2 Half-through bridges

A popular solution is the bridge that is depicted in Figure 2.35. Two main girders are connected together through crossbeams. The crossbeams consist of composite cross sections and carry the tracks. The main girders are welded double T sections, and the connection between them and the crossbeams is rigid so that a stiff frame is formed. If the plate thicknesses of the steel flanges are not sufficient, then *doubler plates* are welded. This is a practical solution but may lead to corrosion problems at the interface of the two plates, and therefore, noncontinuous welds are not allowed. If possible, doubler plates should be avoided and thicker plates should be chosen.

Rolled steel beams have better fatigue resistance than welded ones and are recommended to be used as crossbeams. One can also observe that longitudinal stiffeners at the web are

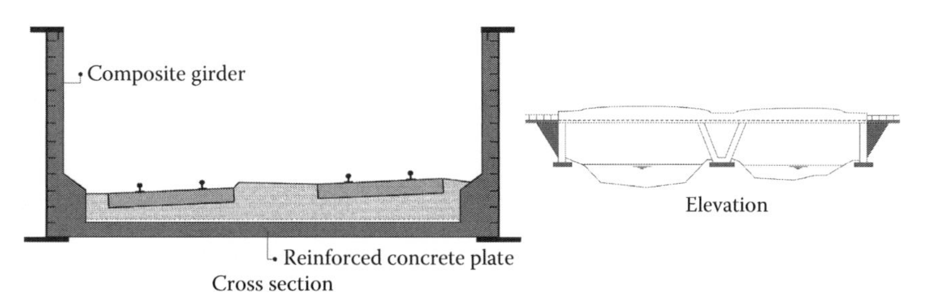

Figure 2.36 Half-through plate-girder bridge with composite main girders.

placed outside the bridge. This is done in order to avoid damage during track maintenance activities (relaying and reballasting operations).

Half-through bridges are usually preferred for railway underline bridges. The underline clearance is in such cases limited, and the half-through construction offers the most suitable alternative. The spans are generally simply supported; this simplifies construction and replacement activities during traffic conditions. For spans less than 17 m, the top of the main girders need not be more than 100 mm above the rail level. This minimizes the construction depth that is the vertical dimension between the tops of the rails and the bridge soffit as much as possible. However, the main girder cannot be considered to provide a safe support (*robust kerb*) against derailment loads; additional parapets acting as robust kerbs are necessary. A main girder is generally considered as a robust kerb when the top of the main girder extends at least 300 mm above the rail level [2.22].

Half-through plate-girder bridges with a twin-track railway may be competitive for spans up to 50 m. For larger spans, half-through box-girder bridges are preferred. In the literature, one can find a variety of half-through bridges. One is shown in Figure 2.36. For large spans, the longitudinal steel girders suffer from lateral torsional buckling. During construction, additional bracing systems can be used but not at the final stage. For this reason, main girders are casted so that composite action is activated. This leads to an elegant structure, and excessive steel consumption is avoided. In cases of twin-track continuous bridges, slenderness values 1/10–1/15 are feasible [2.43].

2.4.3 Plate-girder bridges

Plate-girder bridges are chosen for railway applications when the construction depth is not critical. The longitudinal girders consist of welded cross sections, and the concrete deck's geometry is similar to that employed for roadway bridges. Figure 2.37 shows some common cases. Deck plates with precast planks and in situ topping offer a fast and simple construction method (case A). Fully precast concrete deck slabs with dry joints are the best solutions for reconstruction activities, but experience on durability issues for railway bridges is limited; see Figure 2.10. In situ concrete deck slabs (cases B and C) is applied in more complex geometries or when transverse prestressing with tendons is necessary. In case B, the deck plate is designed to provide a robust kerb; thus, additional support parapets are avoided. A modern solution is depicted in case D. Prefabricated composite girders arrive on site, and the final deck concreting follows, *VFT construction method* [2.43]. Such cross sections seem to be very effective since it makes the limitation of deformations less laborious, especially in continuous and integral bridges.

Composite bridges with multiple girders can be applied for spans up to 50 m. For larger spans, twin-girder bridges are more cost-effective and easier to construct. Twin-girder bridges are less redundant structures than those with multiple girders, and therefore, fatigue

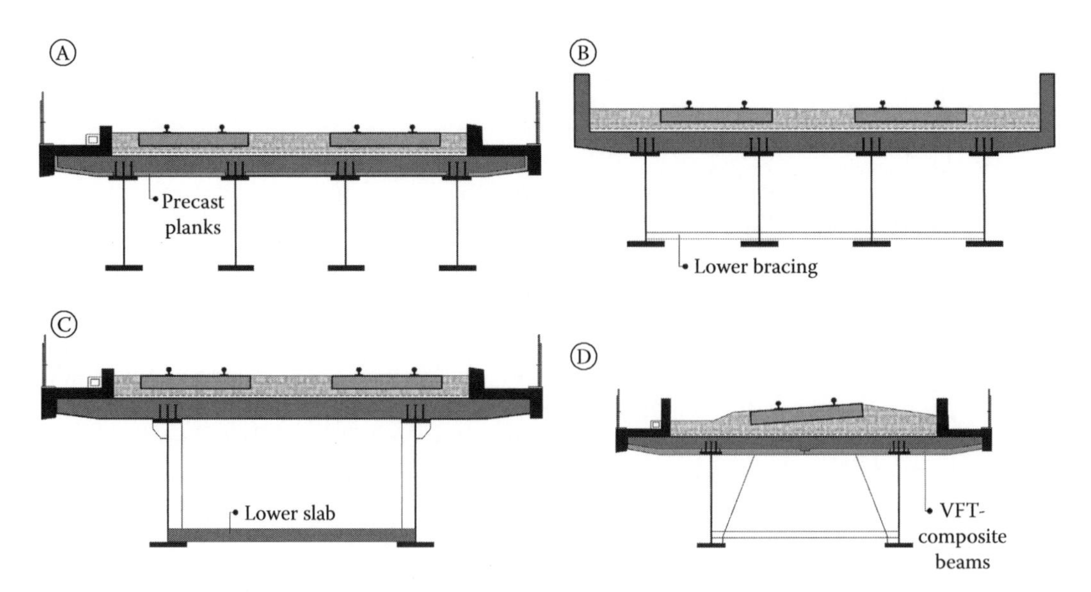

Figure 2.37 Plate-girder railway bridges. (A) With precast planks, (B) with lower steel bracing, (C) with lower reinforced concrete slab, and (D) with prefabricated composite beams.

resistance seems to be their weak point. However, inspection and repair are less complicated, and with a rigorous maintenance policy, this risk is minimized.

In Figure 2.37, one can see that some plate-girder bridges are equipped with lower bracing systems. These may be trusses or stiff frames. Lower braces increase the redundancy of the structure and its torsional stiffness. Indeed the cross section behaves as a closed one, and expensive solutions with steel boxes are avoided. In case C, a concrete slab has been chosen to serve as lower bracing. This activates also a double composite action that is advantageous in continuous systems.

2.4.4 Box-girder bridges

Closed or opened box girders are used in a similar way as for the highway bridges. Due to their increased flexural and torsional strength, they are mainly preferred for long spans. For small and medium spans, they are less competitive, and simpler solutions are obviously chosen. One of the main disadvantages of the boxed sections is the large areas that have to be repainted due to corrosion; this significantly increases maintenance costs. Moreover, repainting is time consuming, and this can be very dangerous under load conditions. In coastal regions, corrosion is considerably accelerated by airborne salt, and this has been detected as the main reason for many damages. A solution to this can be *ship-bottom-shaped cross sections*; see Figure 2.38. This configuration helps rainwater to wash away the airborne salt and

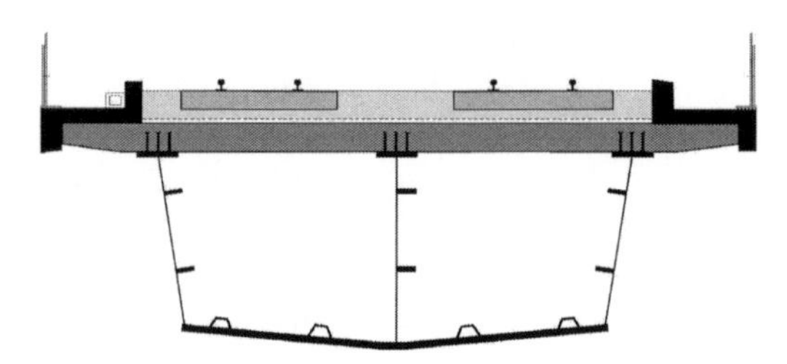

Figure 2.38 Ship-bottom-shaped cross section.

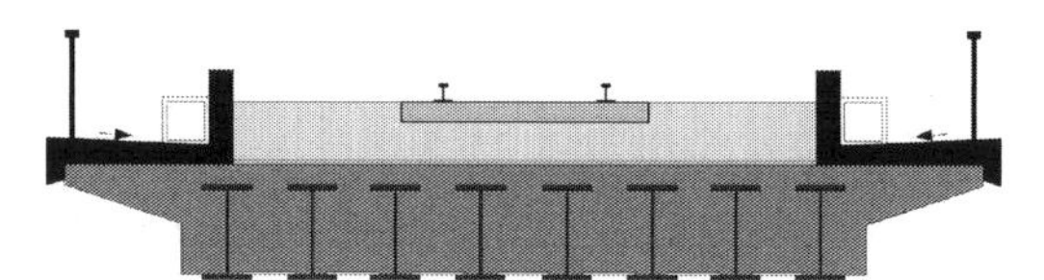

Figure 2.39 Filler-beam bridge.

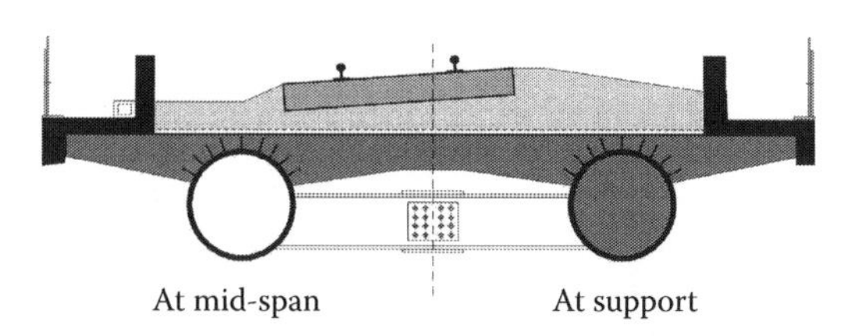

Figure 2.40 Pipe beam bridge.

decreases the repainting frequency. Ship-bottom-shaped cross sections in combination with special types of weathering steel can enhance the salt corrosion resistance considerably [2.19].

2.4.5 Filler-beam bridges

Encasing steel beams in concrete is beneficial from many points of view. Steel beams do not require coating, a composite action can be achieved without the use of headed studs, local buckling of steel plates is avoided, and a better stiffness with less steel is achieved. In Figure 2.39, a *filler-beam bridge* is shown. These are small-span bridges and are used both as simply supported (max. span ≈ 15 m) and continuous systems (max. span ≈ 30 m). Transverse reinforcement passes through holes at the webs of the steel beams; thus, shear connection is ensured. EN 1994-2 offers detailed guidance on the design of filler-beam deck bridges.

2.4.6 Pipe-girder bridges

For small and medium spans, the cross-sectional configuration of Figure 2.40 may be seen in certain cases as an attractive solution. Steel pipes do not suffer from lateral torsional buckling, and therefore, horizontal bracing systems during concreting are omitted. This has a positive effect on both the speed of construction and the appearance of the bridge. Steel pipes can be filled with concrete at hogging moment areas so that a double composite action is achieved. At midspans, pipes can be filled with low-density mortar so that the noise level is improved [2.34]. Finally, the tubular shape of the pipes increases the corrosion resistance since it minimizes the accumulation of airborne salt especially in coastal regions. A pipe-girder cross section has been successfully designed by the authors for a 25 m simply supported one-track railway bridge.

2.4.7 Arch bridges

In arch bridges, nonsymmetrical vertical loads result in strong bending moments both to the arch and the stiffening girders. Therefore, in cases of railway applications, cross sections of excessive size and weight may arise. An alternative solution is *network arches* in which inclined hangers form internal trusses (see Figure 2.41). This develops a diaphragmatic action between the arch and the stiffening girders, a characteristic that is missing from a common arch bridge. The hangers in the networks are positioned in such a way so that

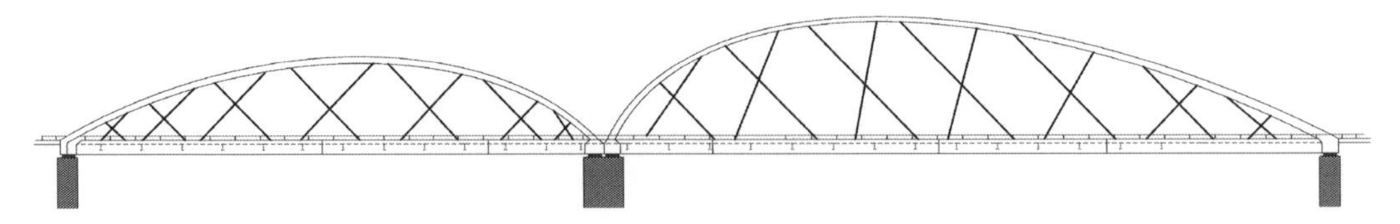

Figure 2.41 Network arched bridges.

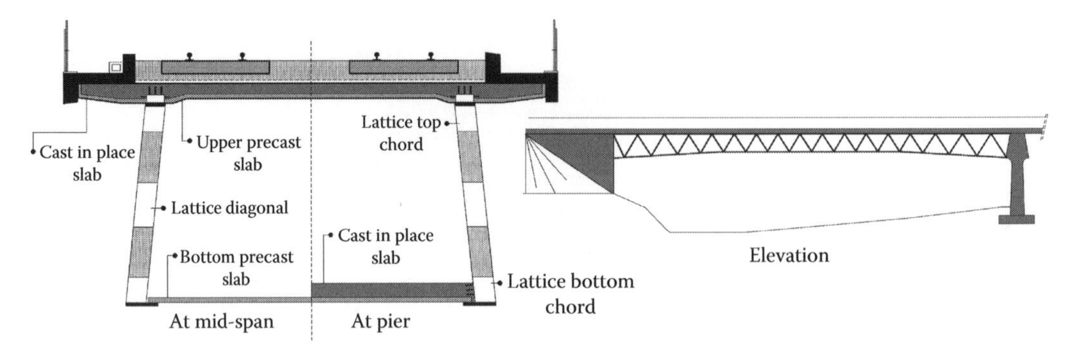

Figure 2.42 Haunch lattice girder bridge.

bending moments in arch and stiffening girders are practically negligible and only normal forces arise. Network arches are stiffer than regular ones and with more slender structural elements [2.9]. Therefore, deflections and vibrations are easier to control.

2.4.8 Lattice girder bridges

Composite lattice girder bridges may be a viable option for long spans with no underclearance limitations (Figure 2.42). The steel structural elements of the girder are usually closed cross sections with adequate resistance against buckling. The top chord is rigidly connected with the deck slab. In Figure 2.42, this is achieved through an additional boxed beam encased in concrete so that longitudinal shear forces are closer to the gravity center of composite upper chord. In this way, local forces and bending moments in the connection are minimized [2.33].

Lattice girder bridges may be equipped with lower bracing systems or concrete slabs in order to develop closed box behavior. At hogging moment areas, a thicker concrete slab increases the flexural resistance and stabilizes the bottom chords laterally in plane. It is worth mentioning that regulations for composite lattice girders are not given in EN 1994-2. Attention should be paid to the rheological and the temperature effects. In order to neutralize members' shortening due to shrinkage of concrete, longer segments may need to be erected. The use of prefabricated elements has reduced creep- and shrinkage-induced deformations and is in such cases to recommend.

2.5 CONSTRUCTION FORMS

2.5.1 General

Choosing the most appropriate structural system for a bridge project is not an easy task. The final decision is determined by many different parameters such as the available construction depth, the soil quality, the seismicity of the area, future reconstruction activities, and maintenance. A discussion on the advantages and the disadvantages of each system follows.

2.5.2 Simply supported bridges

Simply supported bridges (Figure 2.2) and bridges consisting of isostatic spans can be described as the simplest structural forms. Thermal effects, creep, and shrinkage do not cause any additional internal forces, and cracking of concrete is avoided since the deck plate is constantly under compression. Due to compression in the deck, many designers favor the implementation of an isostatic solution when prefabricated elements are to be used. Tension in the deck plate in hogging moment areas of continuous bridges may cause cracking in concrete or fatigue damages in reinforcement due to the movement of the joints between the precast planks during passage of vehicles.

Bridges with multiple isostatic spans facilitate erection and increase the prefabrication rate. Indeed main girders have smaller lengths and weights and can be placed on the piers with cranes of lower capacity. For larger spans, temporary falseworks must be used at intermediate positions. In some cases, isostatic systems are used during construction and at final stage are constructed as continuous ones. An example has been given in Figure 2.4.

Isostatic systems are also preferable when strong support movements are expected due to weak or compressible soils. Moreover, such bridges can be easier reconstructed or replaced, for example, after an earthquake. This is important in the case of urban areas where such activities must be conducted as quick as possible and with the less traffic disturbance.

2.5.3 Continuous bridges

Simply supported bridges can be cost-effective solutions mainly for small and medium spans. However, for longer spans, serviceability verifications are onerous and steel consumption becomes excessive. Therefore, stiffer systems are chosen. Continuous composite bridges (Figure 2.16) are associated with a limited deformability, increased redundancy, and redistribution capabilities. These are important structural characteristics that modern bridges should possess.

The structural topology of a continuous bridge often depends on the available positions for placing the piers. Piers are constructed in positions of "healthy soil conditions," and the trend is to reduce their number as much as possible. For the superstructure, the size of end spans should be equal to 80% of the internal spans so that bending moments at internal supports are approximately of the same magnitude. Obviously, this is not always feasible. Excessive hogging moments arising from the decompensation of long next to small spans can be reduced by lowering the structure at intermediate supports after concrete hardening. This is an indirect way of prestressing concrete without longitudinal tendons. However, designers have additional means for reducing bending moments at supports. An example is depicted in Figure 2.36 in which a central V-shaped pier reduces the lengths of the main spans and contributes to the bridge's performance both structurally and aesthetically.

Continuous bridges may be erected by *launching*. This is a sequential construction quite appropriate for box girders that starts from one end of the bridge. A new segment is added, and the whole deck including diaphragms and lateral bracing that rests on a system of guided rollers is pushed by hydraulic rams. A temporary launching nose is attached to the front of the first span to limit the weight and reach the next pier. The bridge may be launched downhill if a small inclination is present. Plate buckling of the girder webs must be checked since they are subjected to concentrated compression forces from the rollers during launching operations; see Figure 8.31.

2.5.4 Frame bridges

In many bridges, longitudinal movements due to temperature, shrinkage, and any kind of horizontal support movement are supposed to be absorbed by *expansion joints* (Figure 2.43). These elements need to be replaced due to leakage of the joints. Moreover, expansion joints lead to discomfort for the drivers during passage of vehicles. Bearings should also be replaced during the design life of a bridge. Therefore, many designers tend to eliminate all movement joints and, if possible, all bearings as well. This is achieved through frame and integral construction in which super- and substructure act together in response to loading and imposed deformations.

In framed bridges, piers are rigidly connected with the main girders. This type of construction is suitable for long spans and in general where deformation, resonance, and fatigue requirements are difficult to fulfill. In the railway bridge of Figure 2.42, the lattice deck and the piers form a stiff composite frame. Through the trial and error method, bridge engineers try to optimize the stiffness ratio between horizontal and vertical components. For example, deck rotations should be avoided, and for this reason, the deck is more flexible than the piers. However, in seismic regions, an earthquake may cause plastic hinges in the deck, and this is not acceptable. Therefore, an adequate number of weaker piers should be placed capable of developing high-ductile plastic hinges (Figure 2.44).

Piers at side spans experience higher deformations and loadings due to temperature effects, creep, and shrinkage. This is because of their larger distance from the *neutral displacement point* of the deck. This is defined as the stiffness center of the deck in which horizontal displacements are equal to zero. In Figure 2.45, one can see the estimation procedure of this point whose coordinate depends on the stiffnesses of the piers and the bearings if any. Moreover, the maximum values of deformations and forces arise at the abutments. For this reason, most of the designers avoid the rigid connection of the deck with the abutments by the use of bearings and expansion joints.

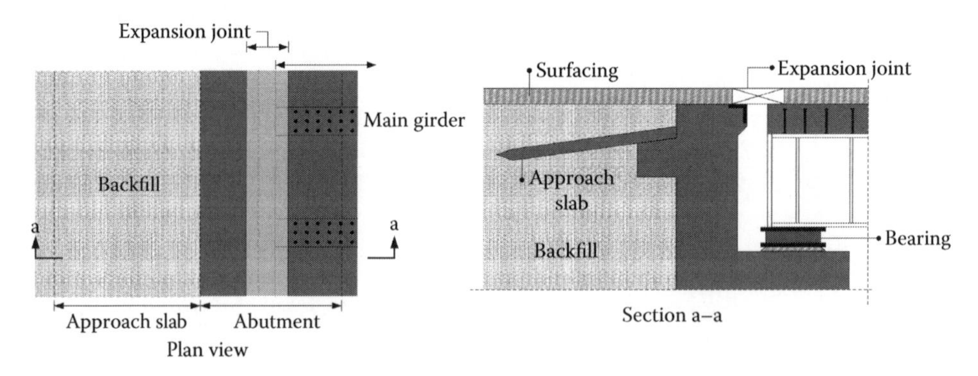

Figure 2.43 Simple support detailing with expansion joint.

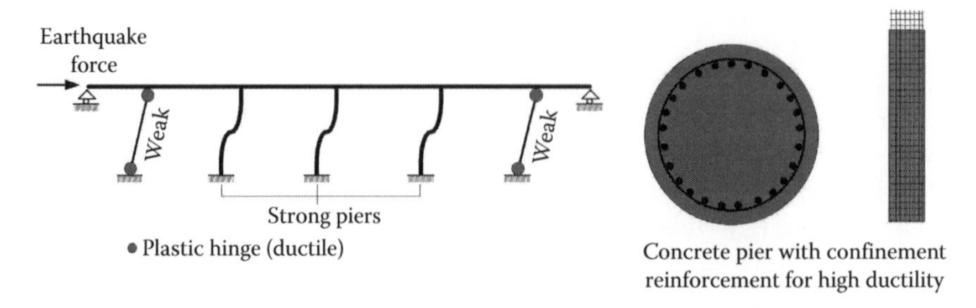

Figure 2.44 Frame bridge with weak and ductile piers at side spans.

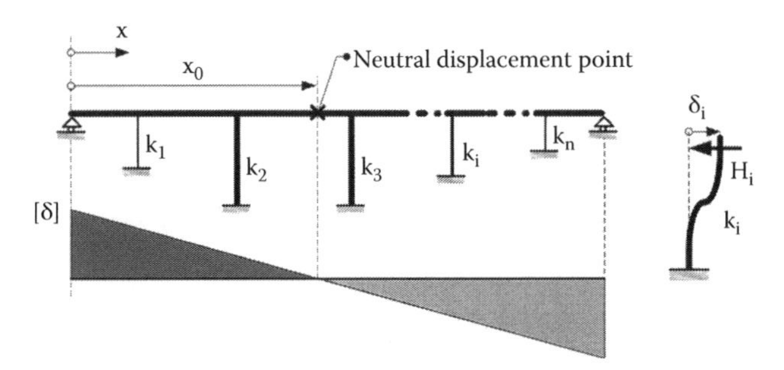

Figure 2.45 Calculation of neutral displacement point.

Horizontal deformation of pier i:

$$\delta_i = \lambda \cdot (x_i - x_0) = \frac{H_i}{K_i}$$

$$\Rightarrow H_i = \lambda \cdot K_i \cdot (x_i - x_0)$$

It is

$$\sum H_i = 0 \Rightarrow x_0 = \frac{\sum K_i \cdot x_i}{\sum K_i} \tag{2.1}$$

Piers in bridges crossing over valleys may be extremely high. Piers with a total height larger than 200 m are not peculiarities. Obviously, such slender structural elements are associated with an increased buckling risk. It is worth noting that architects favor "very thin" piers that do not attract the observer's attention from the surrounding environment. Framed bridges ensure for the piers a reduced buckling length due to their rigid connection with the deck. This is an additional advantage that designers always consider.

2.5.5 Integral and semi-integral bridges

Bearings and expansion joints can be omitted by connecting the main girders rigidly with the abutments. The elements of the superstructure act together with those of the substructure, and thus, a rigid frame is formed. Integral bridges are very robust structures that can reach spans up to 100 m with impressive slenderness values (min $h/L \approx 1/50$). This made them very popular both for high- and railway applications in many European countries.

One can find a variety of end connections for integral bridges. A typical one is shown in Figure 2.46. Steel or concrete piles are connected with each other through a reinforced concrete crossbeam (pile cap). On the upper side of the pile cap, the main girders are placed on the pile cap through the use of temporary bearings, for example, steel plates. Thereafter, the endscreen wall is completed after the deck steelwork has been erected and the deck slab cast. In this way, bending moment continuity is achieved. Moreover, the encased part of the steel girders has to be well anchored. Therefore, girders are equipped with shear connectors, hoops, and/or tie beams. Holes in the webs are also needed for the reinforcement continuity of the endscreen wall.

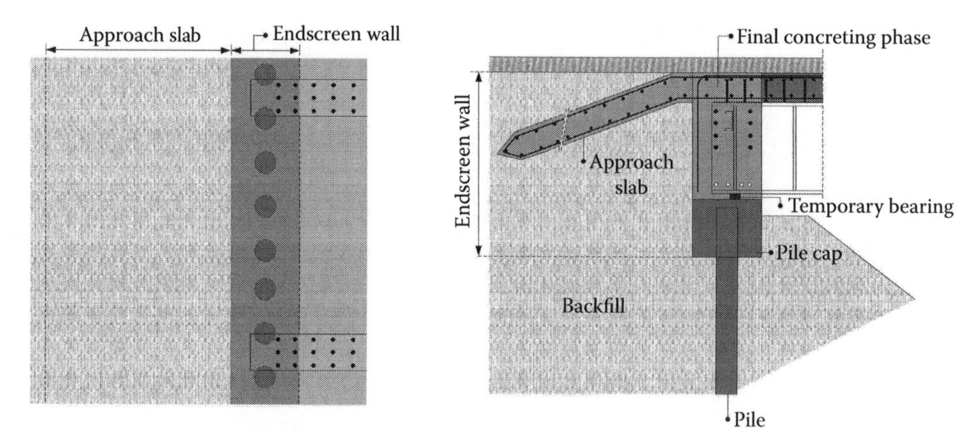

Figure 2.46 Fully integral bridge on piled foundation.

The piles have to be flexible enough in order to absorb the horizontal deformations of the deck due to thermal expansion and contraction, concrete shrinkage, etc. Therefore, they are positioned in a straight line. The soil pressures due to the aforementioned displacements should be carefully calculated in order to achieve an economic and safe design. This is a difficult task because soils are often inhomogeneous. To further complicate matters, second-order theory deformations due to buckling of the piles may arise.

When the endscreen wall does not provide support to the main girders, the bridge is defined as *semi-integral*. In such cases, horizontal deformations are accommodated by conventional bearings that are placed on footings or piles; design is easier and uncertainties due to soil conditions do not need detailed consideration. An end connection of a semi-integral bridge is depicted in Figure 2.47. The main girders are placed on the bearings and grouting of the joint between them, and the retaining wall with nonshrinkable mortar follows. The moment continuity is ensured by means of high-strength anchor bolts that connect the upper flanges with the backfill and the lower flanges with the retaining wall.

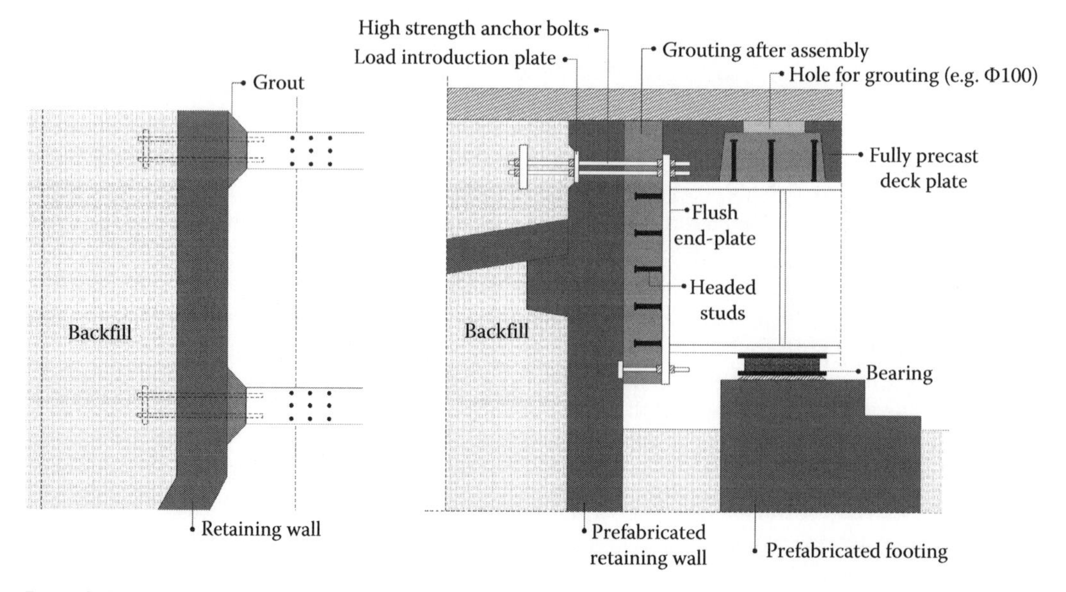

Figure 2.47 Semi-integral bridge.

Prefabricated footings and back walls are used when old bridges need to be replaced and traffic disturbance must be minimized. Erection time can be further reduced with fully precast elements for the deck plate.

It is worth mentioning that in cases of skew bridges, soil pressure tends to cause plan rotation of the deck. For this reason, integral construction should be considered for skews up to 30°.

In seismic regions, abutments have a major contribution to the seismic resistance, and the bridge should be designed to remain elastic during the earthquake event. EN 1998-2 provides limit values for the displacements at the abutments so that the soil or the embankments behind them are not severely damaged.

Interesting information and experiences from integral and semi-integral bridges are given in numerous papers of [2.3].

2.6 ERECTION METHODS

2.6.1 General

The final choice of an appropriate erection procedure is of great importance for achieving the desired geometry, construction speed, and cost-efficiency. The erection method defines the loading history of the bridge and has a primary influence on the evolution of stresses and deformations. For the majority of medium- and long-span bridges, the cross sections, the bracing, and the strengthening members mainly depend on the internal forces that emerge during erection. This means that the static design for the transient situations of erection may be more crucial than for the final-stage design; usually it is more laborious as well. Experienced designers discuss all the possible alternatives together with steel fabricators and contractors. The risk assessment of the proposed alternatives is a demanding task, and detailed safety plans need to be prepared. The erection of bridges is a complicated issue and cannot be covered in few paragraphs. However, a brief introduction follows. Interesting information on erection techniques is found in [2.48].

2.6.2 Lifting by cranes

Lifting steel members with cranes is the most economic method for the erection of small- and medium-span bridges. Mobile cranes can lift 50 tones at 50 m radius or 100 tones at 28 m radius. For lifting heavier elements, cranes can be used in tandem, but this may increase the costs considerably. Cranes are used in tandem when erection has to be carried out during a limited period of time and I girders are lifted in pairs. Similarly, lifting of complete composite members may be preferable. This may be the case for the VFT beams described in Section 2.3.5.

The most common type of cranes is the road-mobile ones; see Figures 2.30 and 4.26. Other types are the rail-mounted cranes or those on a floating vessel. In all cases, the advice of a specialist contractor is necessary so that important factors are considered, for example, overhead electrification equipment, access to site, and exposure to wind.

2.6.3 Launching

This method is based on the concept of assembling the bridge's steel part and launching it forward on rollers or sliders to its final position; see Figure 8.31. A tapered *launching nose* is connected to the main girders so that stressing and cantilever deflections can be kept to a minimum. The launching nose is a light truss construction with a length approximately equal to 30% of the length to be crossed. Rollers and sliders are placed on the piers and on temporary piers in the case of very long spans. The slide path generally runs parallel to the lines of the substructure, and temporary piers are located where stable solid exists. It is important to note that during launching laborious stability investigations have to be conducted for all the possible launching phases since plate buckling due to shear, bending, or concentrated support reactions may occur; see Figure 8.1. Moreover, the system needs to be robust against pulling forces that are produced from the launching devices. These devices consist of a system of hydraulic jacks for heavy structures or winches for the lighter ones.

Rollers are usually ball bearings constrained in a channel (Figure 2.23) and are used for light superstructures. Sliders are devices composed of different materials with sliding interfaces. These may be phosphor/bronze or a PTFE sledge on stainless steel. Sliders during launching may exhibit a friction coefficient up to 8% and are used for heavier structures.

Launching operations are usually conducted for the pure steel girders; thereafter, reinforcement is placed and concreting follows. Launching the steel frame with its slab reinforcement on it has been applied in few bridges where steel cage handling had to be avoided. However, this alternative should be treated as a nonstandard solution. In cases of bridging very busy roads or railways, launching steel–concrete composite frames has been seen by some engineers as an attractive option. Launching the completed composite structure is sometimes preferred when casting with mobile formwork (Figure 2.3) is considered as expensive, more time consuming, or causing under clearance problems. Obviously such an erection method leads to an increased steel consumption since it is more difficult to ensure stability. Moreover, stronger launching devices are needed. The picture in Figure 2.48 shows the world's longest composite bridge ever constructed with composite launching.

Figure 2.48 Box-girder bridge of the Montreal Autoroute A30 in Canada during construction (ARUP©).

The launching method, known also as *incremental launching*, is usually chosen for multiple continuous span bridges with constant height girders and when lifting is seen as impossible, for example, bridges crossing deep valleys. It is a complicated procedure and more expensive than crane lifting and carries higher risks. An additional drawback is the need of an assembly area behind the abutments. However, launching is associated with a high erection speed that ranges from 30 to 50 cm/h.

2.6.4 Shifting

The steel structure is fully constructed on temporary supports alongside its final position. Then it is moved transversely trough rollers and jacked down onto its permanent bearings (Figure 2.23). The main advantages of this method are the brief traffic interruption and the reduced need of stability strengthening since the structure remains always in its final structural configuration. Many bridge engineers are fond of combining shifting with launching. In urban areas, it may be difficult to find a sufficiently wide area next to the final position.

2.6.5 Hoisting

Lifting devices attached to the cantilever parts of the bridge hoist up vertically central parts to their final level (Figure 4.26). This is a rarely used method that is mainly appropriate for bridges crossing waterways. The hoisting operations are performed by cables drawn by launching jacks or winches with pulley blocks. Heavy and very large beams can be erected in few hours with this method. The wind speed during erection must be very low.

2.6.6 Segmental construction

This is the standard construction method for box-girder frame bridges crossing deep valleys, also known as the *cantilever method*. Steel segments are transported and hoisted up to their final level. The next segments are suspended in place by cranes and on-site welding follows; see Figure 2.49. The construction is repeated until the span is completed. The segmental construction may begin from different starting points, usually central piers. The length of the steel segments usually varies from 3 to 6 m. For equilibrium reasons, segments at supports are obviously rigidly connected with the piers.

2.7 CONCRETING SEQUENCE

Simple span bridges with a length up to 25 m are concreted in one stage. The concrete of single-span bridges of middle to long spans may be cast in several stages in order that the weight of the wet concrete is not resisted by the pure steel girders. The concrete may be cast

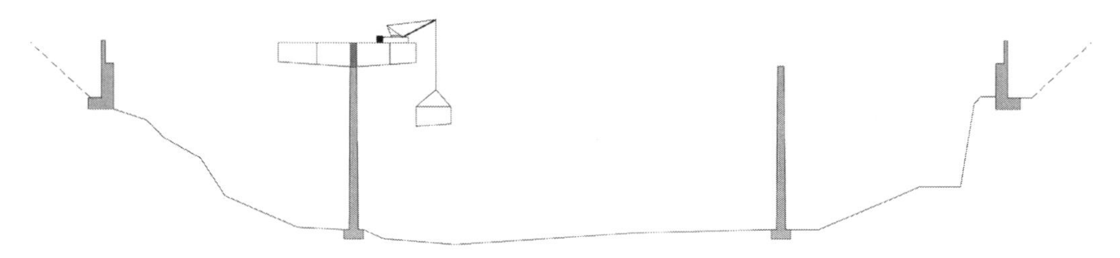

Figure 2.49 Cantilever method.

first in the span region and then near the ends so that the most stressed middle region acts as a composite section when the ends are concreted.

The concrete of continuous bridges is usually cast also in stages in order to limit the negative moments on the composite section and consequently the tension stresses of concrete. In small-span bridges, hogging moments is not an issue, and continuous concreting may be followed; see method 1 in Figure 2.50. This is the simplest method that contractors prefer. For longer spans, hogging moments need to be reduced as much as possible. This may be achieved by concreting first the span regions and then the regions at internal supports; the length of the casted parts varies between 15 and 25 m. In practice, two sequences of concrete casting are used, one where the sequence of casting follows the direction of concreting (method 2) and one where the sequence of casting is opposite to this direction (method 3). One can see that with the methods 2 and 3, positive bending moments at internal supports arise and negative at spans [2.11]. Combining the aforementioned methods with imposed settlements at supports makes control of cracking at final stage feasible. However, due to cement hydration (see Section 6.1.5) and shrinkage (see Section 6.1.3), cracks may emerge after the 1st days of concreting in unexpected areas, that is, spans. For this reason, the

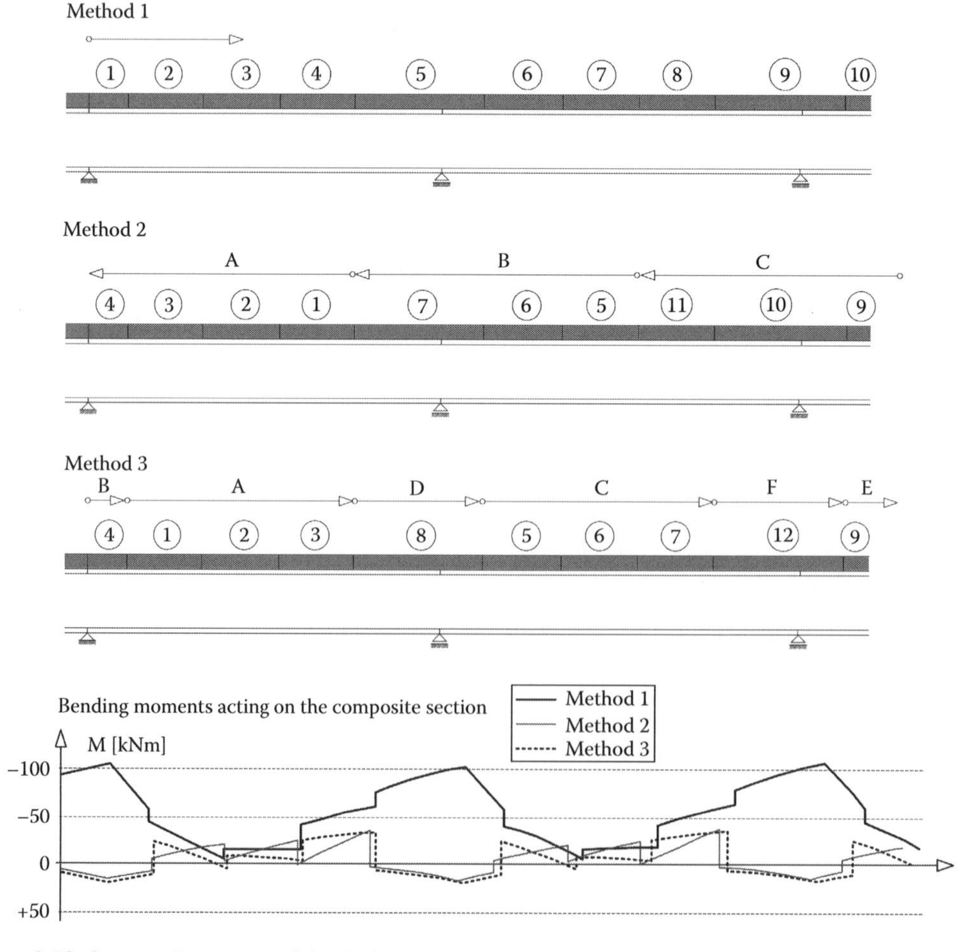

Figure 2.50 Casting of concrete of the deck slab in stages.

cement type should be carefully selected and cracking reinforcement should be placed in all parts of the structure. If possible, concreting operations during summertime should take place during evening.

Finally, it is important to note that in methods 2 and 3 the mobile formwork needs to travel backward a distance equal to the span length before starting casting again. This is time consuming and onerous for the concrete contractors. In long bridges, two or more mobile formworks are used.

2.8 EXECUTION

Execution covers all activities necessary for the realization of a bridge project. This includes procurement of material, fabrication, surface treatment, transportation, erection, inspection, and documentation. Involved in the execution is the owner who decides, supervises, checks, and pays; the designer who is the technical authority for the implementation of his or her design; the general contractor who is responsible for the delivery of the completed bridge; and the steelwork contractor who fabricates and possibly erects the steelwork as well as any subcontractors.

EN 1090-2 is the European document on execution of steel structures including bridges. This document covers specification and documentation, specifications for constituent products (structural steels, steel castings, welding consumables, mechanical fasteners, studs and shear connectors, grouting materials, expansion joints, cables, structural bearings), preparation and assembly, welding, mechanical fastening, erection, surface treatment, geometric tolerances, inspection, testing, and correction.

Crucial for execution is the definition of an execution class that is given in the design phase to ensure consistency between design assumptions and requirements for execution of the work. Bridges belong to service category 2 (SC2) since they are subject to fatigue loading and contain welded components from steel grade not smaller than S355, so they are assigned a production category 2 (PC2). The combination of CS2 and PC2 gives normally an execution class 3 (EXC3) and in exceptional cases of extreme consequences of structural failure EXC4.

Although many items are based on a common execution class that is given for the whole of works, other items demand selection of the execution class on the basis of a component, or connection detail, that is, some component or connections (e.g., bolted connection), may have a different execution class. For instance, items of generic application refer to quality documentation, identification, traceability and marking of constituent products, thickness tolerances, surface conditions and special properties of structural steels, inspection certificates of finished products, thermal cutting, flame straightening execution of holing during preparation and assembly, qualification of welding procedures, welders and operators, joint preparation, acceptance criteria for welds, handling and storage on site, fit up and alignment, inspection after welding, inspection of preloaded bolted connections, and geometric survey of connection nodes. On the other side, differentiations are possible for specific components or connection details. For instance, although quality level B in accordance with EN ISO 5817 [2.7] is assigned to EXC3 for acceptance of weld imperfections (misalignment, weld shape, porosity, lack of fusion or penetration, etc.), for welding of secondary members, the lower quality level C could be sufficient.

A quality documentation is required that includes, among others, the allocation of tasks and authority during execution, procedures, methods and work instructions to be applied, an inspection plan, and procedures to handling modifications and nonconformities.

It should be emphasized that matters related to execution require considerable expertise and experience and are not within the scope of this book.

2.9 INNOVATION IN COMPOSITE BRIDGE ENGINEERING

In the last 20 years, European road freight transport has increased by 40%, and this strong growth is forecasted to continue. At the same time, architectural and environmental requirements become more demanding. Bridge engineers will have to cope with new problems that will be difficult to solve with today's design codes. Some of the research trends that may lead to innovative types of composite bridges are the following:

- Steel plates of variable thickness (*LP plates*) have been successfully used in long-span bridges in order to avoid welded joints. This offers improved fatigue behavior and better cost optimization (Figure 2.51a). Unfortunately, few mills can produce such plates.
- Steel girders with corrugated webs (Figure 2.51b) and no top flange are associated with higher values of transverse bending stiffness [2.28]. Therefore, cross frames and braces for the erection stage are significantly reduced. In addition, corrugated webs form an own shear connector. The combination with horizontally lying studs leads to an increased longitudinal shear resistance.
- Rolled sections can be cut into halved ones and connected with the deck slab without the use of headed studs (Figure 2.51c). With the appropriate reinforcing, detailing *dowel action* is activated and shear transmission is ensured. Light prefabricated composite elements can be fabricated and erected in very short time. Bridges with impressive slenderness values are feasible.
- As already noticed, welded joints should be kept to a minimum. Lattice girder composite bridges with casted joints lead to a reduced number of welding operations and an increased fatigue resistance [2.28]. Moreover, higher flexibility concerning geometry is achieved.

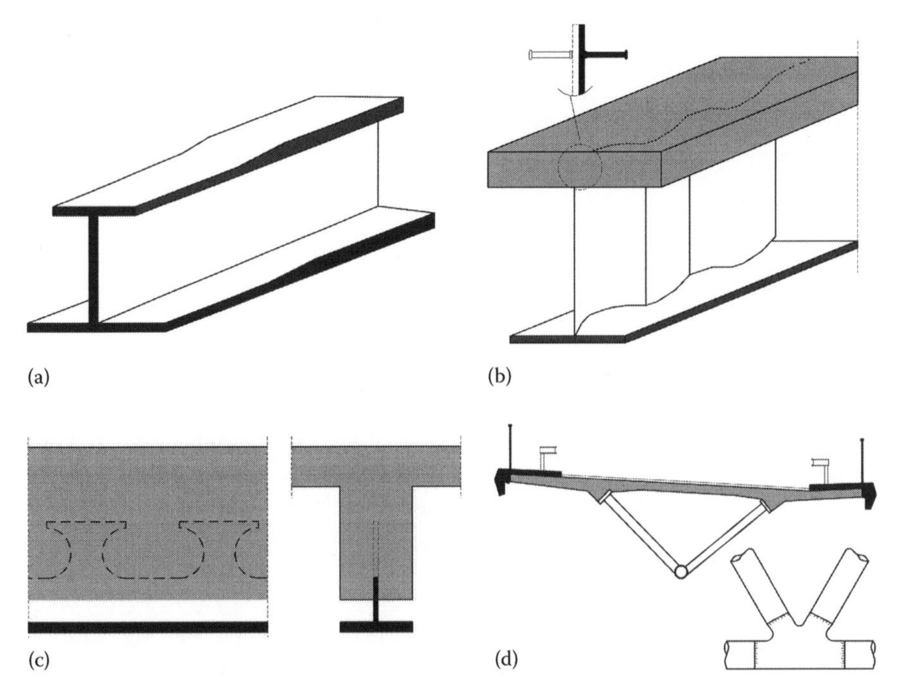

Figure 2.51 Innovative developments in steel–concrete composite bridges. (a) Steel girders with LP plates, (b) steel girders with corrugated webs, (c) halved rolled section with clothoidal composite dowels, and (d) road bridge with hollow sections and casted joints.

REFERENCES

[2.1] Albrecht, G.: Seile. Stahlinformationszentrum. Stahlbau-Lehrprogramm, 2003.

[2.2] Collin, P., Lundmark, T.: Competitive Swedish composite bridges. *IABSE Symposium*, Melbourne, Victoria, Australia, 2002.

[2.3] Collin, P., Veljkovic, M., Pe'tursson, H.: *International Workshop on the Bridges with Integral Abutments*. Topics of relevance for the INTAB project. Luleå University of Technology Department of Civil and Environmental Engineering Division of Structural Engineering, Sweden, 2006.

[2.4] *Composite Bridge Design for Small and Medium Spans. Design Guide*. ECSC Steel Programme, 2002.

[2.5] Daniels, B. J., Brekelmans, W. P. M., Stark, W. B.: State of the art for composite bridge research. *Journal of Constructional Steel Research* 27, 123–141, 1993.

[2.6] EN 1090, CEN (European Committee for Standardization): Execution of steel structures and aluminium structures—Part 2: Technical requirements for steel structures, 2011.

[2.7] EN ISO 5817, CEN (European Committee for Standardization): Welding—Fusion-welded Joints in steel, nickel, titanium and their alloys—Quality levels for imperfections, 2005.

[2.8] Fischer, J. W., Merz, D. R.: Hundreds of bridges—Thousand of cracks. *Civil Engineering*, 55(4), 64–67, April 1985.

[2.9] Gauthier, P., Krontal, L.: *Experiences with Network Arches for Railway Bridges*. Stahlbau 79, Heft 3, pp. 199–208, Ernst und Sohn, Berlin, Germany, 2010.

[2.10] Haensel, J., Kina, J., Schaumann, P.: *An Extension to the Field of Application of Composite Girders*. Stahlbau 63, Heft 9, Ernst und Sohn, Berlin, Germany, 1994.

[2.11] Hanswille, G.: Ausbildung und Herstellung der Fahrbahnplatten von einteiligen Verbundquerschnitten. Bergische Universität Wuppertal. Lehrstuhl für Stahlbau und Verbundkonstruktionen, 2004.

[2.12] Hanswille, G.: Composite bridges in Germany. State of the Art. *International Workshop on Eurocode 4-2—Composite Bridges*, Stockholm, Sweden, March 2011.

[2.13] Hanswille, G.: Composite bridges recently built in Germany. *Composite Bridges, State of the Art in Technology and Analysis*, Madrid, Spain, 2001.

[2.14] Hayward, A. C. G.: *Composite Steel Highway Bridges*. Corus Construction and Industrial, Glasgow, Scotland, 2005.

[2.15] Hayward, A., Sadler, N., Tordoff, D.: *Steel Bridges. A Practical Approach to Design for Efficient Fabrication and Construction*. The British Constructional Steelwork Association Ltd., London, U.K., December 2002.

[2.16] Hendy, C.: UK experience with EN 1994-2. *International Workshop on Eurocode 4-2—Composite Bridges*, Luleå, Sweden, March 2011.

[2.17] Hoorpah, W., Hever, M.: Innovative design of short span steel concrete composite bridges with rolled beams. Some examples from France and Germany. *Seventh International Conference on Short and Medium Span Bridges*, Montreal, Quebec, Canada, 2006.

[2.18] Hoorpah, W.: *Innovative Design of Short Span Steel Bridges with Rolled Beams*. Seoul, South Korea, 2007.

[2.19] Hosaka, T., Tanaka, M., Homma, K., Kihira, H.: Application of new advanced weathering steel to unpainted viaducts in coastal regions. *Steelbridge, OTUA International Symposium*, Millau, France, June 2004.

[2.20] Iles, D. C.: *Design Guide for Composite Highway Bridges*. The Steel Construction Institute, Spon Press, London, U.K., 2001.

[2.21] Iles, D. C.: *Design Guide for Simply Supported Composite Bridges*. The Steel Construction Institute, Ascot, U.K., 1991.

[2.22] Iles, D. C.: *Design Guide for Steel Railway Bridges*. The Steel Construction Institute, Ascot, U.K., March 2006.

[2.23] Iles, D. C., Hendy, C. R.: Steel Bridge Group. Guidance notes on best practice in steel bridge construction. Steel Construction Institute, Fifth Issue, 2010.

[2.24] Järvenpää, E.: *Innovative Composite Bridges in Finland*. WSP Finland Ltd., Stockholm, Sweden, 2010.

[2.25] Kim, H-Y., Jeong Y-J.: Steel–concrete composite bridge deck slab with profiled sheeting. *Journal of Constructional Steel Research* 65, 1751–1762, 2009.

[2.26] Kindmann, R., Krahwinkel, M.: *Stahl-und Verbundkonstruktionen*. B. G. Teubner, Stuttgart, Leipzig, 1991.

[2.27] Koch, E.: *Railway Bridges of Deutsche Bahn AG. New Technical and Strategical Approaches*. Stahlbau 75, Heft 10, pp. 786–790, Ernst und Sohn, Berlin, Germany, 2006.

[2.28] Kuhlmann, U.: Innovative developments for composite bridges. *International Workshop on Eurocode 4-2—Composite Bridges*, March 2011.

[2.29] Larena, J. B., Larena, A. B.: *On the Development of Sections in Composite Bridges*. Stahlbau 80, Heft 3, pp. 185–197, Ernst und Sohn, Berlin, Germany, 2011.

[2.30] Lerchner, C., Schimeta, G.: *Concrete Frame with an Integrated Rolled Steel in Concrete Structure*. Beton-und Stahlbetonbau 97, Heft 10, pp. 530–535, Ernst und Sohn, Berlin, Germany, 2002.

[2.31] Marzahn, G., Hamme, M., Prehn, W., Swadlo, J.: Wupper river valley bridge: A state of the art composite bridge, *Structural Engineering International*, 17(1), 35–39, 2007.

[2.32] Matteis, D., Chauvel, G., Cordier, N., Corfdir, P., Leconte, R., Faucheur, D., Léglise, R.: *Steel—Concrete Composite Bridges. Sustainable Design Guide*. Service d'études sur les transports, les routes et leurs aménagements, May 2010.

[2.33] Mato, F. M., Rubio, L. M., Cornejo, M. O., Agromayor, D. M., Bujalance, B. E.: Development of steel and composite solutions for outstanding viaducts on the Spanish H.S.R lines. VII Congresso de Construção Metálica e Mista.

[2.34] Nakamura, S., Momiyama, Y., Hosaka, T., Homma, K.: New technologies of steel-concrete composite bridges. *Journal of Constructional Steel Research* 58, 99–130, 2002.

[2.35] Naumann, J.: *Current Development of Road Bridges in Germany*. Stahlbau 10, pp. 779–785, Ernst und Sohn, Berlin, Germany, 2006.

[2.36] Olipitz, M.: Verbundbrücken mit Teilfertigteilen im kleineren und mittleren Spannweitenbereich. Beispiele in Kärnten. Betonzement 2005.

[2.37] Pauser, A.: *Konstruktions-und Gestaltungskonzepte im Brückenbau*. Beton Kalender, Ernst und Sohn, Berlin, Germany, 2004.

[2.38] Schäpertöns, B.: A new modular concept for composite steel bridges. *Steelbridge, OTUA International Symposium*, Millau, France, June 2004.

[2.39] Schaumann, P., Upmeyer, J.: *Composite Bridges with Precast Concrete Slabs*. University of Hanover, Institute for Steel Construction, Hanover, Germany.

[2.40] Schmackpfefer, H.: *Standardized Design of Composite Bridges in Middle Span Area*. Stahlbau 68, Heft 4, pp. 264–276, Ernst und Sohn, Berlin, Germany, 1999.

[2.41] Schmitt, R.: *Die Schalungstechnik., Systeme, Einsatz und Logistik*. Ernst und Sohn, Berlin, Germany, 2001.

[2.42] Schmitt, V.: *Eisenbahnbrücken kleiner und mittlerer Stützweiten in Stahlverbund*. 11 Dresdner Brückenbausymposium, Dresden, Germany, März 2001.

[2.43] Schmitt, V., Seidl, G.: *Railway Bridges Using Steel-concrete Composite Structures*. Stahlbau 79, Heft 3, pp. 159–166, Ernst und Sohn, Berlin, Germany, 2010.

[2.44] Schmitt, V.: *Verbundbrücken in der Praxis*. Beton Kalender, pp. 273–335, Ernst und Sohn, Berlin, Germany, 2002.

[2.45] Schmitt, V.: Verbundbrücken mit kleinen Spannweiten. Brücken-und Ingenieurbau. Dokumentation 658, Vortragsreihe II, Deutsher Stahlbautag, S. 9-12, Bauen mit Stahl, 2002.

[2.46] Seidl, G., Braun, A.: *VFT-WIB Viaduct in Vigaun—A Composite Bridge with External Reinforcement*. Stahlbau 78, Heft 2, pp. 86–93, Ernst und Sohn, Berlin, Germany, 2009.

[2.47] Sobrino, J. A.: *Two Steel Bridges for the High Speed Railway Life in Spain*. Stahlbau 79, Heft 3, pp. 181–187, Ernst und Sohn, Berlin, Germany, 2010.

[2.48] Structural Engineering International: Bridge erection techniques and construction equipment. *International Association for Bridge and Structural Engineering*, SEI 21(4), November 2011.

[2.49] Taner, P., Bellod, J. L.: Salto del Carnero railway bridge. Concept and design. *Steelbridge, OTUA International Symposium*, Millau, France, June 2004.

[2.50] Vayas, I., Iliopoulos, A.: *Steel-Concrete Composite Bridges—Design Guide According to the DIN-Fachberichte and the Eurocodes*, Klidarithmos Publications, Athens, Greece, 2006 (in Greek).
[2.51] Verwiebe, C., Sedlaceck, G.: *Frequenz-und Dämpungsmessungen an Hängern von Stabbogenbrücken*, Forschung Straßenbau und Staßenverkehrstechnik, Heft 77, 1999.
[2.52] Weizenegger, M.: Hybrid frame bridge, River Saale, Merseburg, Germany. *Structural Engineering International* 13(3), 179–181, August 2003.
[2.53] Yandzio, E., Iles, D. C.: *Precast Concrete Decks for Composite Highway Bridges*. Steel Construction Institute, Ascot, U.K., Publication 316.

Chapter 3

Design codes

3.1 EUROCODES

3.1.1 General

Civil engineering design is generally guided by design codes. In some construction works, the conditions are not fully known in advance so that the design assumptions and consequently the design have to adjust to the new conditions that appear during the progress of works. An example is tunneling, where during construction differences between initial design assumptions concerning soil/rock properties, existence of underground water veins, and the real site conditions may occur so that design and construction have to be appropriately modified during construction works. In contrast, bridge design is subject to extensive regulations concerning loads, materials, required verifications, etc., provided by codes. How a bridge is to be built and how it is designed are strongly linked. The design then provides construction and erection methods and their sequences and includes the relevant verifications for situations during construction. Modifications of the design assumptions during construction should generally be avoided as they may result in increase in time and costs. This book makes reference to the Eurocodes [3.21] as design standards, which will be briefly presented and discussed in the following.

The structural Eurocode program comprises standards shown in Table 3.1 generally consisting of a number of parts.

The first two Eurocodes are of general application for design, EN 1990 describing the requirements for safety, serviceability, and durability of structures and EN 1991 providing the actions ("loads") on structures. All other Eurocodes related to superstructures (i.e., all except EN 1997) concern specific construction materials (concrete, steel, composite steel and concrete, etc.). A special case is EN 1998 that includes both a generic part on seismic actions and specific parts related to types of structures with relevant construction materials.

Despite the vertical distinction, there exist horizontal connections between the material-related Eurocodes. EN 199X-1 (part 1) of each Eurocode gives generic design rules intended to be used with the other parts and also gives supplementary rules applicable to buildings. The horizontal connection between Eurocodes in relation to bridges is created by EN 199X-2 (part 2). Consequently, the leading document for the design of composite bridges is Eurocode 4, part 2 (EN 1994-2). However, since composite construction combines the use of both structural steel and reinforced concrete, EN 1994 calls, besides the generic Eurocodes, both relevant material Eurocodes, EN 1992 and EN 1993. Figure 3.1 offers a schematic navigation to the regulations necessary for bridge design.

In composite bridge construction, structural steel is of primary importance. Indeed, the strength and stability verifications in the construction phase refer to the steel girders alone, but also most checks during service concern the steel parts. Consequently, the provisions of

Table 3.1 The Eurocodes

EN 1990	Eurocode 0	Basis of structural design
EN 1991	Eurocode 1	Actions on structures
EN 1992	Eurocode 2	Design of concrete structures
EN 1993	Eurocode 3	Design of steel structures
EN 1994	Eurocode 4	Design of composite steel and concrete structures
EN 1995	Eurocode 5	Design of timber structures
EN 1996	Eurocode 6	Design of masonry structures
EN 1997	Eurocode 7	Geotechnical design
EN 1998	Eurocode 8	Design of structures for earthquake resistance
EN 1999	Eurocode 9	Design of aluminium structures

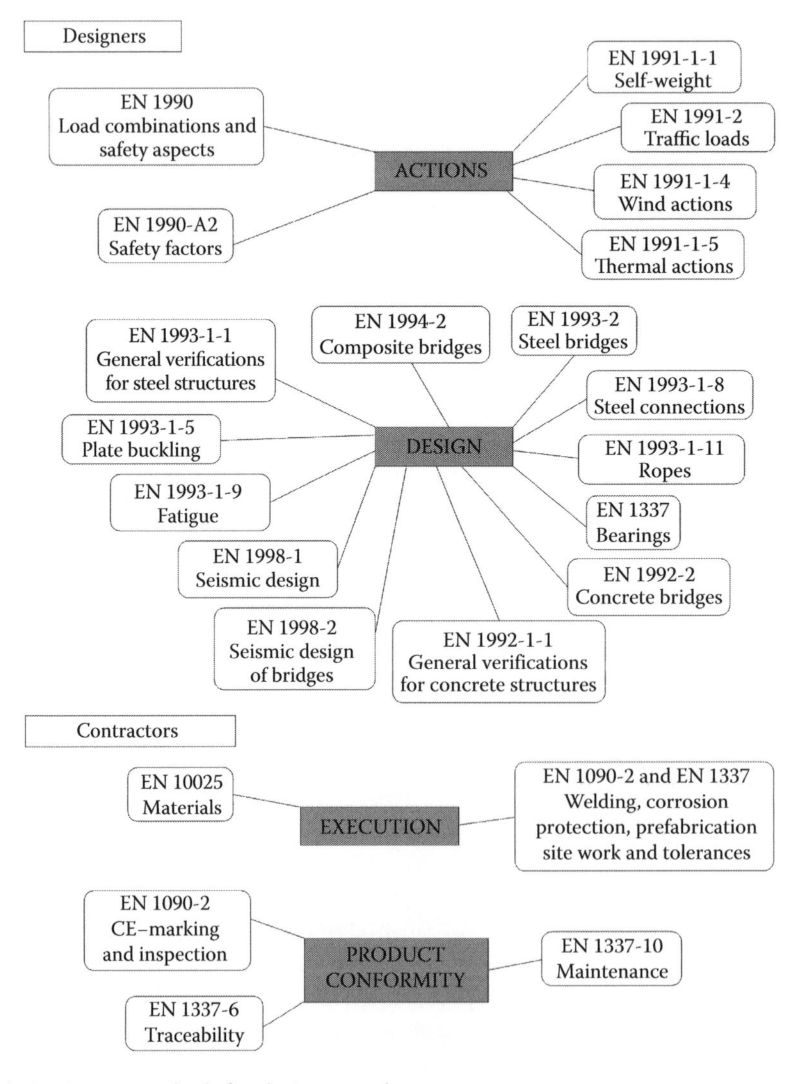

Figure 3.1 Navigation to standards for designers and contractors.

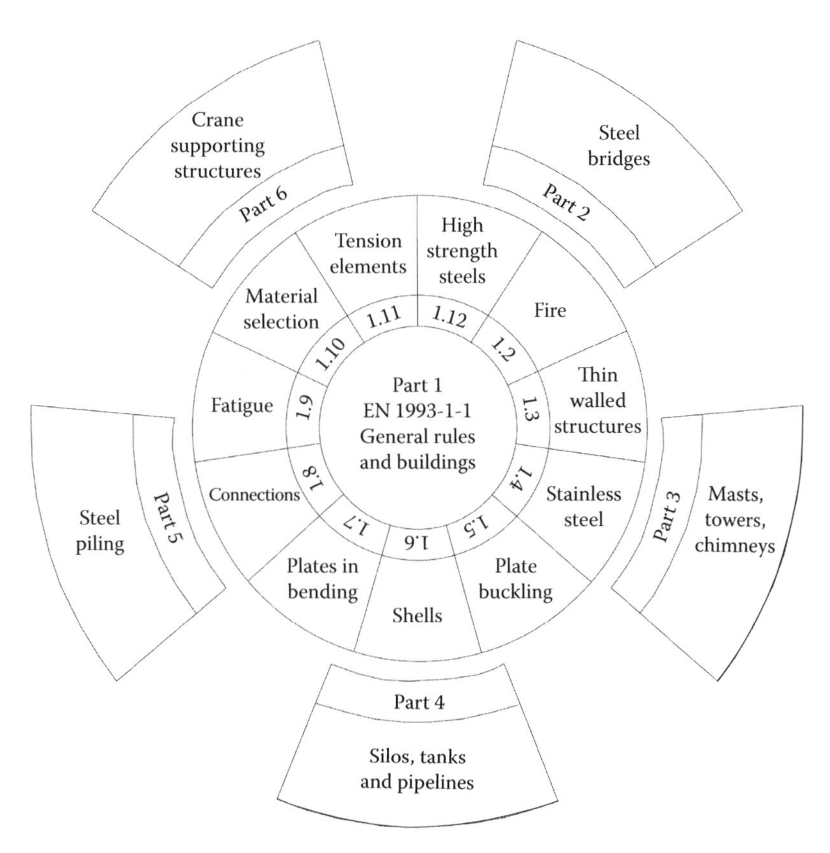

Figure 3.2 The Eurocode 3—family.

EN 1993 are of crucial importance for the design of composite bridges. This Eurocode is the most extensive from all others and is structured in a special way. EN 1993 consists of a "core" composed of 12 parts (EN 1993-1-1 to EN 1993-1-12) and a "periphery" composed of five parts (EN 1993-2 to EN 1993-6). The "core" is for general application for all types of steel construction, while the "periphery" covers specific steel constructions, like bridges (EN 1993-2), towers, masts, and chimneys. The periphery makes reference to the core whenever special verifications, like plate buckling and fatigue, or design issues are needed. Figure 3.2 gives a survey on all parts relevant to Eurocode 3.

In the following, the Eurocodes that should be mainly consulted in composite bridge design are given and shortly described.

3.1.2 EN 1990: Basis of structural design

Basis of Design [3.1]
This code defines the basic requirements, introduces the limit state design, provides the classification for actions, gives design values for actions and resistances for various limit states and design situations, and defines combinations for actions. Bridges are given the design working life category of 5, with a design working life of 100 years.

Annex 2 [3.2]
This annex gives rules and methods for establishing combinations of actions for bridges. It also gives the recommended design values and combination factors for actions during construction and at service to be used in the design for bridges.

3.1.3 EN 1991: Actions on structures

EN 199-1-1: Actions: General Actions [3.3]
This part provides specific weights for materials, self-weights, and imposed loads for buildings.

EN 199-1-4: Wind Actions [3.4]
Chapter 8 of this part provides wind actions for bridges.

EN 199-1-5: Thermal Actions [3.5]
Chapter 6 of this part provides temperature changes in bridges.

EN 199-1-6: Actions during Construction [3.6]
Annex A2 of this part provides rules for bridges and gives design values for settlements in the longitudinal and transverse directions.

EN 1991-2: Traffic Loads on Bridges [3.7]
This part defines models of traffic loads for the design of road bridges, footbridges, and railway bridges.

3.1.4 EN 1998: Design of structures for earthquake resistance

EN 1998-1: Seismic Design: General Rules and Seismic Actions [3.8]
This part refers to the design and construction of buildings and civil engineering works in seismic regions.

EN 1998-2: Seismic Design: Bridges [3.9]
This part is applicable to the design of bridges in seismic areas. It includes provisions for analysis, verifications, detailing, and seismic isolation.

3.1.5 EN 1994: Design of composite steel and concrete structures

EN 1994-2: Composite Structures: Rules for Bridges [3.10]
This part gives basic design rules for composite bridges or members of bridges. It does not fully cover the design of cable-stayed bridges.

3.1.6 EN 1993: Design of steel structures

EN 1993-1-1: Design of Steel Structures [3.11]
This part provides general rules and rules for buildings made of steel.

EN 1993-1-5: Plated Structural Elements [3.12]
This part includes plate buckling rules and provides design requirements for unstiffened and stiffened plates.

EN 1993-1-8: Design of Joints [3.13]
This part gives methods for analysis and design of bolted and welded joints.

EN 1993-1-9: Fatigue [3.14]
This part gives methods for the assessment of the fatigue resistance of members and joints.

EN 1993-1-10: Selection of Materials for Fracture Toughness [3.15]
This part contains design rules for selecting steel materials to provide resistance to brittle fracture and lamellar tearing.

EN 1993-1-11: Design of Structures with Tension Components [3.16]
This part gives design rules for structures with replaceable wire ropes.

EN 1993-2: Steel Bridges [3.17]
This part gives a general basis for the structural design of steel bridges and steel parts of composite bridges covering issues related to resistance, serviceability, and durability.

3.1.7 EN 1992: Design of concrete structures

EN 1992-1-1: Design of Concrete Structures [3.18]
This part applies to the design of buildings and civil engineering works in concrete.

EN 1992-2: Concrete Bridges [3.19]
This part gives the basis for the design of reinforced and prestressed concrete bridges.

3.2 NATIONAL ANNEXES

It is noted that the national standards implement the various Eurocodes by means of a relevant *National Annex*. This annex contains country-specific data, for example, for climatic actions. It also provides design values of some parameters where the Eurocodes give recommended values. All these data are referred as nationally determined parameters. This book uses the recommended values of those parameters.

Finally, it should be mentioned that Eurocode 3 is accompanied by a very important document concerning construction. This is EN 1090-2 "Execution of steel structures and aluminium structures" [3.20], which specifies requirements for any type and shape of steel structure including bridges.

REFERENCES

[3.1] EN 1990, CEN (European Committee for Standardization): Basis of structural design, 2002.

[3.2] EN 1990, CEN (European Committee for Standardization): Basis of structural design. Annex 2: Application on bridges (normative), 2004.

[3.3] EN 1991-1-1, CEN (European Committee for Standardization): Actions on structures—Part 1–1: General actions—Densities, self-weight, imposed loads for buildings, 2002.

[3.4] EN 1991-1-4, CEN (European Committee for Standardization): Actions on structures—Part 1–4: General actions—Wind actions, 2005.

[3.5] EN 1991-1-5, CEN (European Committee for Standardization): Actions on structures—Part 1–5: General actions—Thermal actions, 2003.

[3.6] EN 1991-1-6, CEN (European Committee for Standardization): Actions on structures—Part 1–6: General actions—Actions during execution, 2005.

[3.7] EN 1991-2, CEN (European Committee for Standardization): Actions on structures—Part 2: Traffic loads on bridges, 2003.

[3.8] EN 1998-1, CEN (European Committee for Standardization): Design of structures for earthquake resistance—Part 1: General rules, seismic actions and rules for buildings, 2004.

[3.9] EN 1998-2, CEN (European Committee for Standardization): Design of structures for earthquake resistance—Part 2: Bridges, 2005.

[3.10] EN 1994-2, CEN (European Committee for Standardization): Design of composite steel and concrete structures—Part 2: General rules and rules for bridges, 2005.

[3.11] EN 1993-1-1, CEN (European Committee for Standardization): Design of steel structures, Part 1–1: General rules and rules for buildings, 2005.

[3.12] EN 1993-1-5, CEN (European Committee for Standardization): Design of steel structures, Part 1–5: Plated structural elements, 2006.
[3.13] EN 1993-1-8, CEN (European Committee for Standardization): Design of steel structures, Part 1–8: Design of joints, 2005.
[3.14] EN 1993-1-9, CEN (European Committee for Standardization): Design of steel structures, Part 1–9: Fatigue, 2005.
[3.15] EN 1993-1-10, CEN (European Committee for Standardization): Design of steel structures—Part 1–10: Material toughness and through-thickness properties, 2005.
[3.16] EN 1993-1-11, CEN (European Committee for Standardization): Design of steel structures—Part 1–11: Design of structures with tension components, 2005.
[3.17] EN 1993-2, CEN (European Committee for Standardization): Design of steel structures—Part 2: Steel bridges, 2006.
[3.18] EN 1992-1-1, CEN (European Committee for Standardization): Design of concrete structures, Part 1–1: General rules and rules for buildings, 2004.
[3.19] EN 1992-2, CEN (European Committee for Standardization): Design of concrete structures—Part 2: Concrete bridges—Design and detailing rules, 2005.
[3.20] EN 1090, CEN (European Committee for Standardization): Execution of steel structures and aluminium structures—Part 2: Technical requirements for steel structures, 2011.
[3.21] R. P. Johnson: Eurocodes, 1970–2010: Why 40 years? *Proceedings of the Institution of Civil Engineers, Structures and Buildings* 162, 371–379, December 2009.

Chapter 4

Actions

4.1 CLASSIFICATION OF ACTIONS

Actions are classified according to EN 1990 [4.1] in relation to their duration, magnitude, and probability of occurrence as

- *Permanent (G)*, which has a small variation in magnitude during a reference period, or this variation is monotonic until it reaches a certain limit value.
- *Variable (Q)*, for which the variation in magnitude is neither negligible nor monotonic.
- *Accidental (A)*, actions of short duration and significant magnitude that have a small probability to occur during the design life.
- *Seismic (AE)*, which develops during an earthquake ground motion.

Further on, the actions are distinguished as

- *Direct*, which are forces (loads) applied to the structure, for example, self-weight, traffic loads, and wind.
- *Indirect*, which develop due to imposed deformations or accelerations, for example, during temperature changes, uneven settlements, creep, shrinkage, or earthquake.

The actions taken under consideration for the superstructure of composite bridges are given in the following with reference to their duration.

4.1.1 Permanent actions

Types of permanent actions are listed in Table 4.1. Longitudinal prestressing in composite bridges is in many cases provided by controlled imposed deformations by means of settlements at internal supports. Such a prestressing causes positive bending at supports and therefore compression in the concrete slab. Prestressing by tendons refers either to the slab in the transverse direction, in case of large span (distance between steel girders and large cantilever), or to prestress by external tendons, usually during upgrading or retrofitting works. Prestressing by tendons is a direct action.

4.1.2 Variable actions

Types of variable actions are listed in Table 4.2.

4.1.3 Accidental actions

Accidental actions are due to collision from vehicles or derailment of railway vehicles. These accidental actions are direct and are given in EN 1991-2 [4.4] by relevant forces. Other accidental actions (e.g., collision of ships) are examined in individual cases.

Table 4.1 Permanent actions

Symbol	*Description*	*Type*
G_1	Self-weight of the structure (structural and reinforcing steel, concrete and shear connectors)	Direct
G_2	Superimposed deadweight, for example, road surfacing, rails, fixed equipment like crashing barriers, pedestrian railing, and parapets	Direct
G_{set}	Differential settlements between supports	Indirect
P	Prestressing by controlled imposed deformations	Indirect

Table 4.2 Variable actions

Symbol	*Description*	*Magnitude*	*Type*
Q	Traffic loads and loads on footways	Vertical–horizontal Dependent on the type of bridge (road, railway, footbridge) Load models dependent on the considered limit state (ultimate, fatigue)	Direct
T	Temperature	Uniform temperature component ΔT_N either contraction or expansion Vertical linear temperature component ΔT_M is as follows: $\Delta T_{M,heat}$ when top is warmer than the bottom $\Delta T_{M,cool}$ when bottom is warmer than the top Combination of ΔT_N and ΔT_M	Direct and indirect
W	Wind	For loaded bridges For unloaded bridges	Direct
D_c	Replacement of bearings	Imposed deformation (jacking) of supports to replace a bearing	Indirect

4.1.4 Seismic actions

Seismic actions are described in the relevant Eurocodes EN 1998-1 [4.5] and EN 1998-2 [4.6].

4.1.5 Specific permanent actions and effects in composite bridges

Composite bridges have some peculiarities compared to pure steel or pure concrete bridges. This is due to the fact that they are composed of two different structural materials: steel (elastic) and concrete (viscoelastic). These peculiarities are summarized as follows:

a. The concrete part of the cross section is subjected to time-dependent deformations due to concrete's rheological behavior (creep and shrinkage), while the steel part is not. This leads to reduced cross-sectional values of the composite section and redistribution of the internal forces and moments from the concrete slab to the steel girder.
b. The steel girder carries the fresh concrete during the construction phase. The composite action becomes effective after the hardening of the concrete. Therefore, apart from the time-dependent property changes of the composite section, the sections change during construction from pure steel to composite. This change does not happen at the same time for the entire structure due to the fact that concreting usually takes place at distinct phases, known also as concreting stages.
c. In statically indeterminate bridges, additional bending moments at supports (secondary effects) develop due to creep and shrinkage, which have influence at both serviceability limit states (SLS) and ultimate limit states (ULS).

Table 4.3 Special permanent actions

Symbol	*Description*	*Type*
C	Creep of concrete	Indirect
S	Shrinkage of concrete	Indirect

4.1.6 Creep and shrinkage

The effects of creep and shrinkage of concrete on the structure change with time. However, these actions are considered as temporarily permanent, because unlike the variable actions, which may appear or not, creep and shrinkage are always present (Table 4.3). It should be noted that concrete creeps only due to permanent actions and shrinkage and differently for each type of action. *Concrete does not creep due to variable actions* because of their short duration.

The effects of creep and shrinkage are discussed in Chapters 6 and 7.

4.1.7 Actions during construction

As mentioned in Chapter 2, concreting of the slab usually takes place in phases. The weight of the fresh concrete, the formwork, and the possible mechanical or other equipment are supported from the steel girder alone. The composite section resists lighter construction loads due to hardening of the concrete and removal of the formwork. The fresh reinforced concrete according to EN 1991-1-1 Table A1 has a density of 26 kN/m^2, while the hardened concrete 25 kN/m^2. The described actions are direct (loads).

4.2 TRAFFIC LOADS ON ROAD BRIDGES

4.2.1 Division of the carriageway into notional lanes

The width of the carriageway (w) is measured between the inner limits of vehicle restraint systems and does not include neither the distance between fixed vehicle restraint systems nor kerbs of a central reservation nor the widths of these vehicle restraint systems. This width is divided into notional lanes (Figure 4.1), whose width (w_l) and number (n_l) are given in Table 4.4.

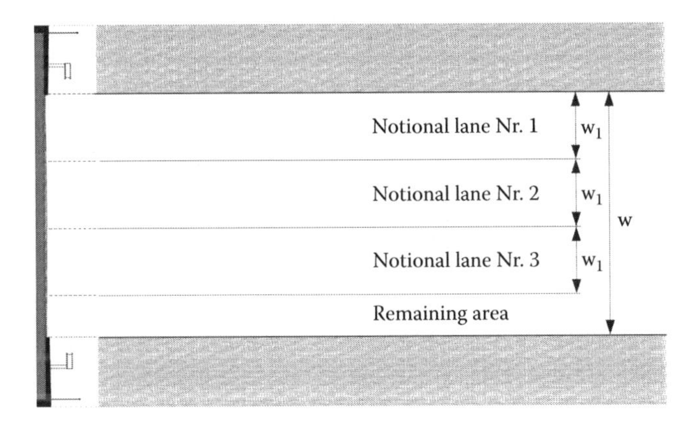

Figure 4.1 Example of lane numbering.

Table 4.4 Number and width of the notional lanes

Width of carriageway (w)	*Number of notional lanes*	*Width of a notional lane w_l*	*Width of the remaining area*
$w<5.4$ m	$n_l=1$	3 m	$w-3$ m
$5.4\ m\leq w<6$ m	$n_l=2$	$\frac{w}{2}$	0
$w\geq 6m$	$n_l=Int\left(\frac{w}{3}\right)$	3 m	$w-3\cdot n_l$

For example, the number of notional lanes will be

- $n_l=1$ if $w<5.4$ m
- $n_l=2$ if $5.4\ m\leq w<9$ m
- $n_l=3$ if $9\ m\leq w<12$ m

Where the carriageway on a bridge deck is divided into two parts separated by a central restraint system, then

a. Each part should be separately divided in notional lanes if this system is permanent.
b. The whole carriageway, central part included, should be divided in notional lanes if this system is temporary.

Where the carriageway consists of two separate parts on two independent decks, each part is considered as a carriageway and separate numbering is used for the design of each deck.

w is the carriageway width
w_l is the notional lane width

The lane giving the most unfavorable effects is numbered lane 1, followed by lane 2, etc. As traffic loads are variable actions, they are placed in such a way that the most adverse effects are obtained. That means that they are placed longitudinally and transversely only in regions where the influence line of the relevant effect (internal moment, shear, support reaction, deformation, etc.) is either positive or negative. Their position is not necessarily as indicated in Figure 4.1 (1–2–3) but may be different (e.g., 1–3–2). In fact, the position and numbering of the lanes may be different for each member (e.g., main beam and transverse beam).

4.2.2 Vertical loads on the carriageway

The following load models apply for loaded lengths less than 200 m. For greater lengths, the load model may be defined in the National Annex or for the individual project.

4.2.2.1 Load model 1 (LM1)

Current traffic situations in European roads are covered by means of load models. EN 1991-2 [4.4] gives four load models, out of which load model 1 (LM1) should be used for global and local verifications. It covers most of the effects of the traffic of lorries and cars.

Table 4.5 Characteristic values of LM1

	TS	*UDL system*
Location	*Axle loads Q_{ik} [kN]*	*q_{ik} [kN/m^2]*
Lane number 1	300	9
Lane number 2	200	2.5
Lane number 3	100	2.5
Other lanes	0	2.5
Remaining area	0	2.5
Recommended values[a]	$\alpha_{Qi} = 1.0$	$\alpha_{qi} = 1.0$

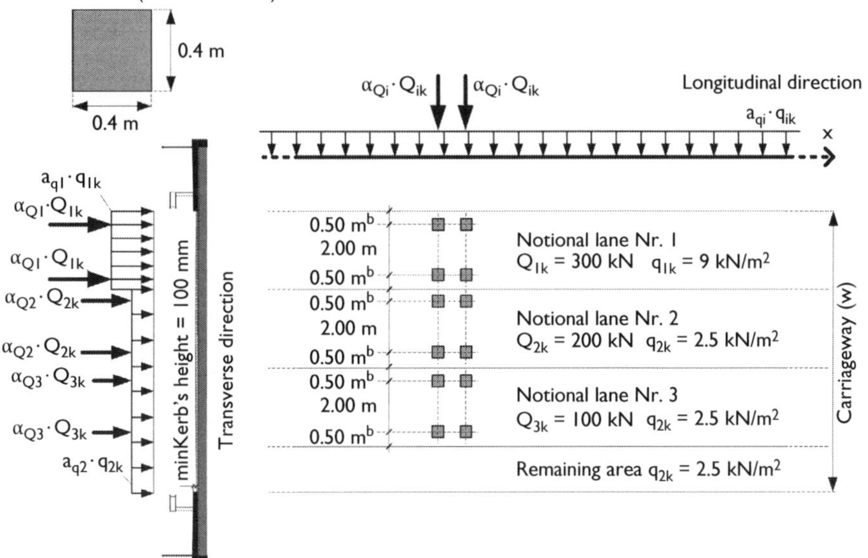

[a] The values for the adjustment factors are given in the National Annex.
[b] For $w_l = 3$ m.

LM1 consists of two partial systems (see Table 4.5):

- Double axle concentrated loads (*tandem system (TS)*) with weight $\alpha_{Qi} \cdot Q_k$ per axle
- Uniformly distributed loads (UDL), with weight $\alpha_{qi} \cdot q_{ik}$

It is noted that

- No more than one TS should be taken into account per notional lane.
- One complete TS should be taken into account, meaning that all four wheels should be loaded, even if some of them produce a favorable effect when they load the influence line in the opposite direction.
- Each TS travels centrally along the notional lane.
- The axle load is equally divided into the two wheels, each one being loaded by $0.5 \cdot \alpha_{Qi} \cdot Q_{ik}$.
- Amplifications due to dynamic effects are included in the characteristic values of Q_{ik} and q_{ik}, and thus no further magnifications are required.
- The contact surface of each wheel is taken as square with a surface of 0.4×0.4 m^2.

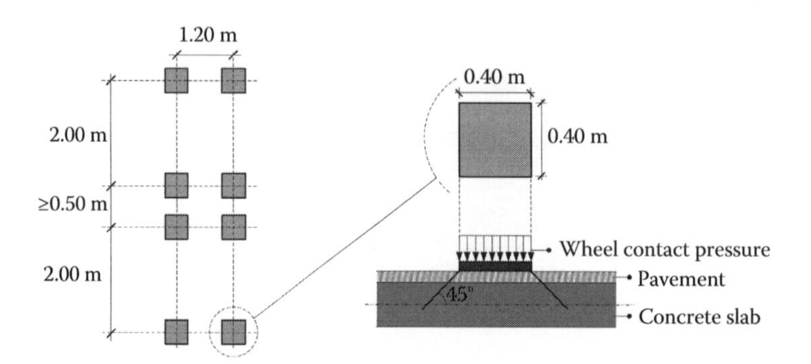

Figure 4.2 Application of the TS for local verifications.

The adjustment factors represent the return period for traffic on the main roads in Europe. An adjustment factor a_q equal to 1 is equivalent to a return period of 1000 years for the characteristic values and corresponds to a heavy industrial international traffic; fatigue is excluded. This is also equal to a probability of exceedance in 50 years equal to 5%. For lighter traffic conditions, such as cases of highways and motorways, reduced adjustment factors may be found in the National Annexes.

For local verifications, for example, for the design of the slab, the wheel loads may be considered as uniformly distributed taking into account the contact area of the wheel and the dispersal of the load through the pavement and the concrete slab (Figure 4.2).

4.2.2.2 *Load model 2 (LM2)*

Load model 2 (LM2) is a single axle model, which is applied when a local verification for short structural elements is necessary. Such elements can be crossbeams, upper flange stiffeners of orthotropic decks, or deck panels of composite slabs with profile steel sheeting. The internal forces due to LM2 may then be more critical compared to those of LM1. This can also happen at the vicinity of the expansion joints as discussed subsequently.

LM2 is applied at any location of the carriageway and consists of a single axle load with a magnitude equal to $\beta_Q \cdot Q_{ak}$. β_Q is an adjustment factor whose value may be defined in the National Annex. According to EN 1991-2, it is recommended that $\beta_Q\ Q_{ak} = \alpha_{Q1} \cdot Q_{ak}$ is equal to 400 kN.

The contact surface of the wheels is of rectangular shape with dimensions $0.35 \times 0.60\ m^2$ corresponding to twin tires. Therefore, LM2 wheels result in different stress distributions in the deck slab than the LM1 ones.

The adjustment factor β_Q includes dynamic amplifications. Attention must be paid in the vicinity of expansion joints (critical area in Figure 4.3) where β_Q must be multiplied with the following amplification factor:

$$\Delta\varphi_{fat} = 1.30 \cdot \left(1 - \frac{D}{26}\right) \geq 1.0 \tag{4.1}$$

where D is the distance of the cross section verified from the expansion joint in m.

Despite the fact that expression (4.1) is recommended by the code, a more simplified and conservative approach is to adopt an amplification factor equal to 1.3. As mentioned earlier,

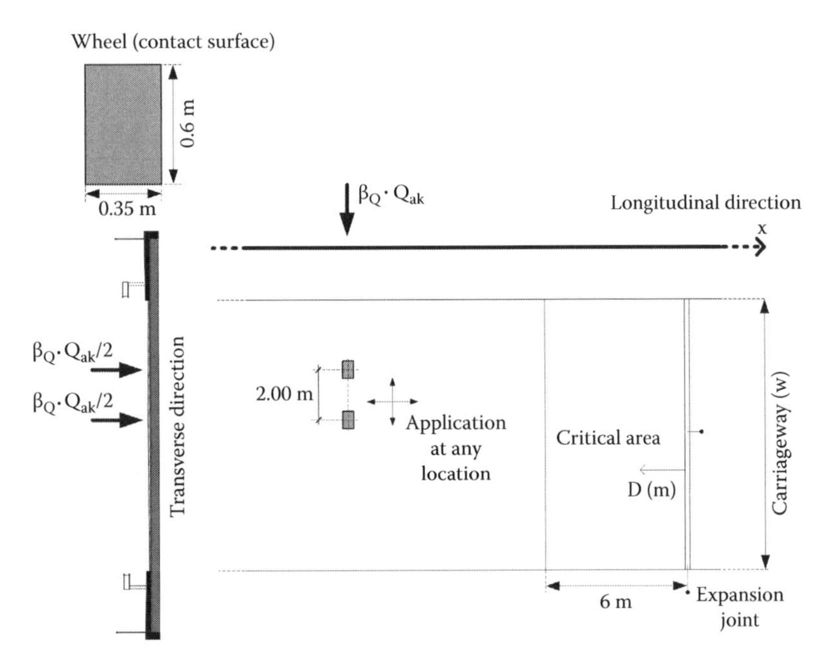

Figure 4.3 Load model 2.

this has to be done for any cross section within the critical area: 6 m from the expansion joint. The increase of the adjustment factor at the vicinity of the expansion joints is due to impact loading, which takes place during vehicle overpassing.

4.2.2.3 Load model 3 (LM3)

Some road bridges must be designed against special traffic loads. This is the case of bridges that may experience a military use during their lifetime (see [4.11]). Annex A of EN 1991-2 defines standardized models of special vehicles whose total weight ranges from 600 to 2400 kN. For vehicles that are expected to move at speeds greater than 70 km/h, the following dynamic amplification must be taken into account:

$$\varphi = 1.40 - \frac{L}{500} \geq 1.0 \tag{4.2}$$

where L is the influence length in m.

4.2.2.4 Load model 4 (LM4)

Load model 4 is called *crowd loading* and is represented by a UDL equal to 5 kN/m^2. It includes dynamic amplification and in certain cases can be more critical than LM1. In [4.12], comparative analyses have shown that crowd loading may be more critical in

- Bridges with span lengths $L \geq 47.8$ m and with a carriageway width $w = 14.5$ m
- Bridges with span lengths $L \geq 145$ m and with a carriageway width $w = 10.0$ m

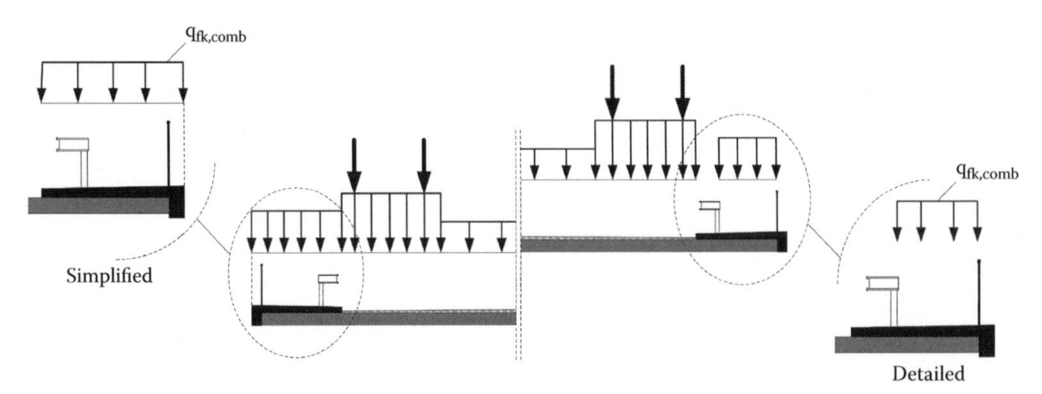

Figure 4.4 Vertical loading on footways.

Similar bridge configurations are usually found in urban areas. Therefore, designers should not underestimate the effects of the crowd loading by comparing its magnitude with that of LM1 because this can be misleading.

LM4 can be applied at every location of the bridge's deck and should be associated only with a transient design situation.

4.2.3 Vertical loads on footways and cycle tracks

Vertical loads on footways and cycle tracks include a UDL $q_{fk} = 5$ kN/m^2 (Figure 4.4) that acts on the unfavorable parts of the influence line in longitudinal and transverse directions. For local verifications, the previously mentioned load may be replaced by a concentrated load $Q_{fwk} = 10$ kN acting on a surface of sides 0.10×0.10 m^2. For road bridges, a vertical load $q_{fk,comb} = 3$ kN/m^2 in combination with the traffic loads is taken into account.

Figure 4.4 demonstrates two modeling approaches for a pedestrian loading. In the detailed one, the load is only applied between the railing and the crash barrier. Many designers though prefer the simplified modeling in which the pedestrian loading covers the entire width from the railing up to the kerbs.

4.2.4 Horizontal forces

4.2.4.1 Braking and acceleration forces

The braking force is a longitudinal force that acts at the surfacing level of the carriageway. It is transferred to the expansion joints, the bearings, and the substructure.

The characteristic value of the braking force Q_{lk} for the total width of the carriageway is equal to

$$Q_{lk} = 0.6 \cdot \alpha_{Q1} \cdot (2 \cdot Q_{1k}) + 0.10 \cdot \alpha_{q1} \cdot q_{1k} \cdot w_1 \cdot L \tag{4.3}$$

with

$$180 \cdot \alpha_{Q1}\ (\text{kN}) \leq Q_{lk} \leq 900\ \text{kN}$$

where L is length of the deck or of part of it under consideration.

This force should be taken into account as acting along the axis of any lane. However, if the eccentricity effects are not significant, this force may be transformed to a uniform line load Q_{lk}/L over the length L acting along the carriageway axis. In Figure 4.5, a calculation example for a simply supported bridge is shown.

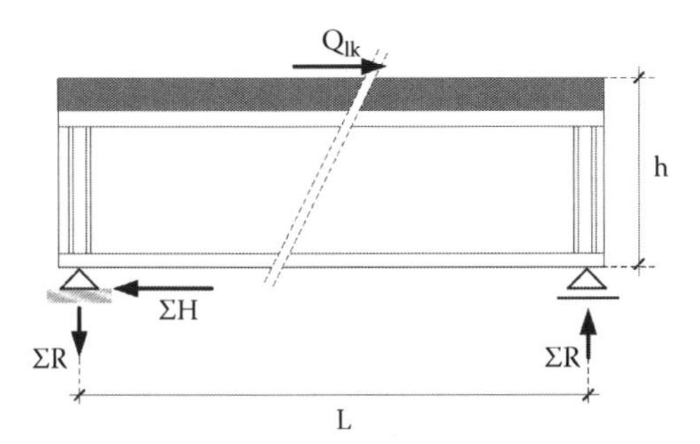

Figure 4.5 Braking force calculation example.

EXAMPLE 4.1

- Bridge: h = 1.2 m, L = 30 m
- Traffic loads and notional lane 1

Q_{lk} = 300 kN, q_{lk} = 9 kN/m², w_l = 3 m

- Adjustment factors: $\alpha_{Q1} = \alpha_{q1} = 1.0$
- Braking force: Q_{lk} = 441 kN
- Support reactions: H = Q_{lk} = 441 kN
  R = 441 · 1.2/30 = 17.64 kN

From the earlier example, one can see that the horizontal support reactions ΣH are of considerable magnitude. Moreover, braking forces cause vertical support reactions ΣR as well. Their magnitude is very small though and cannot result in uplifting of the bearings.

In bridges with expansion joints, the horizontal force that is transmitted by them may be found in the National Annex. Otherwise, the following expression is recommended to be used:

$$Q_{lk,exp} = 0.6 \cdot \alpha_{Q1} \cdot Q_{lk} \tag{4.4}$$

Acceleration forces are of the same magnitude as the braking forces but act in opposite direction, meaning that both types of forces are to be considered as $\pm Q_{lk}$ (Figure 4.6).

In curved bridges, lateral forces due to skew braking or skidding should be taken into account. The transverse force Q_{tk} is equal to 25% of the longitudinal force Q_{lk}. Both forces act simultaneously and at the finished carriage level. Forces due to earthquake, wind, or collision on Kerbs are in most cases more critical.

4.2.4.2 Centrifugal forces

The centrifugal force is a transverse force that acts at the level of the finished carriageway level and radially to the carriageway axis. It is acting as a point load Q_{tk} at any point of the carriageway. Its value is given in Table 4.6.

From the diagram of Figure 4.7, the centrifugal forces for the most common cases of loaded notional lanes 1 + 2 and 1 + 2 + 3 can be estimated.

Bridges with curved layouts may have a nonconstant horizontal radius along the axis of the bridge. Therefore, the centrifugal force should be calculated at different locations.

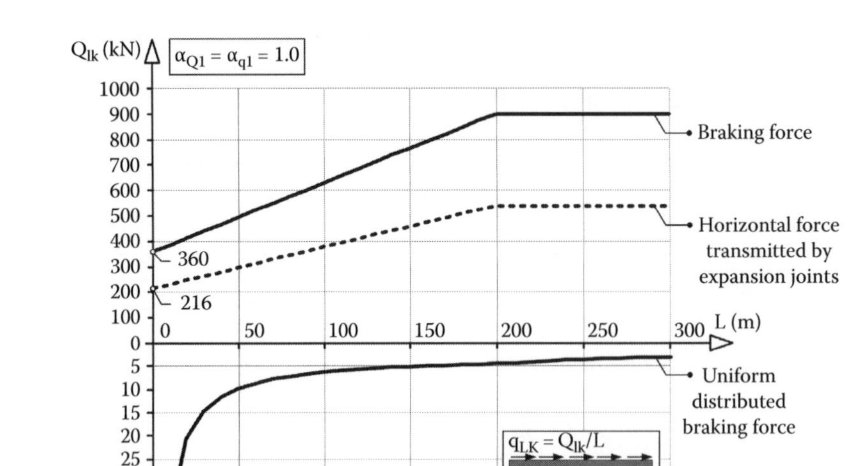

Figure 4.6 Calculation diagram for braking forces.

Table 4.6 Characteristic values of centrifugal forces

$Q_{tk} = 0.2 \cdot Q_v$ (kN)	If $r < 200$ m
$Q_{tk} = 40 \cdot \dfrac{Q_v}{r}$ (kN)	If $200 \leq r \leq 1500$ m
$Q_{tk} = 0$	If $r > 1500$ m

Notes:

r, the horizontal radius of the carriageway centerline in m.

$Q_v = \sum_i \alpha_{Qi} \cdot (2 \cdot Q_{ik})$ = weight of vertical TS forces of the LM 1.

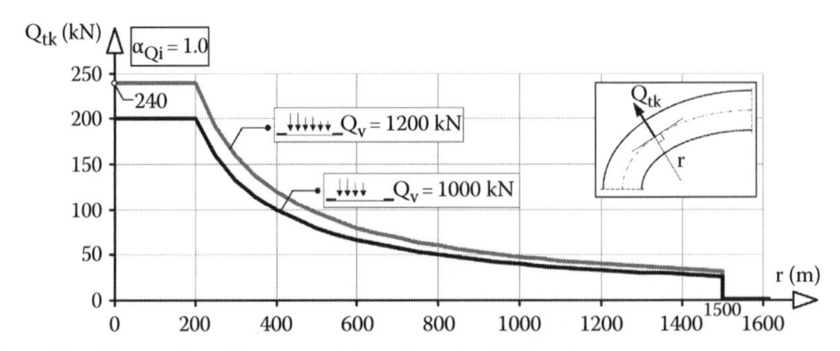

Figure 4.7 Centrifugal forces for different weights of vertical TS loads.

4.2.5 Groups of traffic loads on road bridges

As shown before, the traffic loads include vertical and horizontal forces on the carriageway and on footways. The probability that those loads appear simultaneously with their characteristic values is small. Accordingly, groups of loads are considered, where one type is taken with its characteristic value, while the others with their combination values. Each of these groups of loads is considered as a single characteristic traffic action to be combined with other nontraffic loads. Table 4.7 gives the groups of loads for LM1, marking the primary loads in the group. It is noted that there are additional groups that refer to the other load models. Group gr1a is used for global verifications of the structural elements both at ULS and SLS. Group gr1b is used for local verifications of the deck slab such as punching resistance. Group gr2 mainly consists of horizontal loadings and is critical for the design of bearings and expansion joints. Group gr3 consists of a UDL with a recommended value $q_{fk} = 5$ kN/m^2, which represents loads due to pedestrians or cyclists on footways or cycle tracks. Gr4 (dense crowd load) and Gr5 (abnormal vehicles) are applied in cases of specific design specifications imposed by the responsible authorities. If group gr4 is specified, then this substitutes group gr3.

4.3 ACTIONS FOR ACCIDENTAL DESIGN SITUATIONS

The accidental design situations include the following:

4.3.1 Collision forces from vehicles moving under the bridge

4.3.1.1 Collision of vehicles with the soffit of the bridge, for example, when tracks are higher than the clear height of the bridge

Such accidental situations may be critical in cases of bridges with very light decks, for example, footbridges without a concrete deck slab. They may lead to inelastic horizontal deformations, and therefore appropriate protection measures should be taken. A high clearance is in most cases the most common and cost-effective solution.

A bridge is recommended to be classified as sensitive to collision when the vertical support reaction per bearing is smaller than 250 kN.

4.3.1.2 Collision of vehicles on piers

Impact loads on piers or other structural elements, which support the bridge, should be taken into account according to the regulations of the National Annex. A concentrated force of 1000 kN acting in the direction of vehicle travel and 500 kN perpendicular to that are recommended by EN 1991-2 as minimum values. These forces act 1.25 m higher than the ground level.

In general, reinforced concrete piers with a minimum width are not susceptible to buckling during collision damage and do not endanger the superstructure. Investigations have shown that a minimum width of 0.9 m for piers at motorways can be described as safe (see [4.12]).

Impact loads on piers should be combined with the frequent values of the traffic loads so that an interaction of vertical and horizontal forces is considered.

4.3.2 Actions from vehicles moving on the bridge

4.3.2.1 Vehicles on footways or cycle tracks up to the position of the safety barriers

An accidental axle load with a magnitude of $\alpha_{Q2} \cdot Q_{2k}$ (see Table 4.5) should be placed on the parts of the footways or the cycle tracks, which are considered as unprotected. The part of the footway that is going to be classified as unprotected depends on the stiffness and the location

Table 4.7 Groups of the multicomponent traffic actions

		Carriageway				*Footways and cycle tracks*		
Load type		*Vertical forces*				*Vertical forces*	*Horizontal forces*	
Load system		*LM1*	*LM2*	*LM3*	*LM4*	*UDL*	*Braking forces and acceleration*	*Centrifugal and transverse forces*
Groups of loads	gr1a	**Characteristic values (see Table 4.5)**	—	—	—	**Combination value (see NA, Section 4.2.3)**	(see NA)	(see NA)
	gr1b	—	**Characteristic values (see Figure 4.3)**	—	—	—	—	—
	gr2	**Frequent values (see NA; 3 kN/m² is recommended)**	—	—	—	—	**Characteristic values (see Equation 4.3 and Figure 4.6)**	**Characteristic values (see Table 4.6)**
	gr3[b]	—	—	—	—	**Combination value[a] (see NA, Section 4.2.3)**	—	—
	Gr4	—	—	—	**Characteristic value**	—	—	—
	Gr5	**Characteristic value**	—	**Characteristic value**	—	—	—	—

Source: EN 1991-2: Eurocode 1: Actions on structures—Part 2: Traffic loads on bridges, 2003.

Notes:

NA means National Annex.

Dominant component actions are shown in bold face.

[a] One footway should only be considered to be loaded if the effect is more unfavorable than that of two loaded footways.
[b] gr3 is irrelevant if Gr4 is considered.

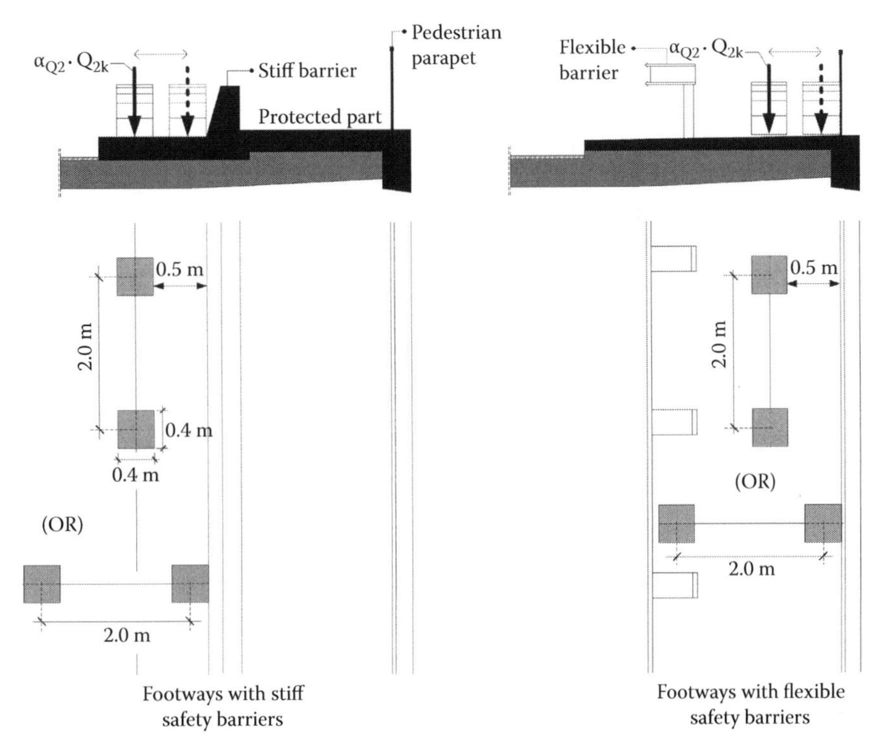

Figure 4.8 Load arrangement for footways with safety barriers of different stiffness.

of the crash barrier. In Figure 4.8, one can see two different cases of load arrangement. In the first case, a part of the footway is protected by a stiff concrete barrier. The axle load is placed adjacent to the outer side of the barrier so that the most adverse effects can be investigated. This is done with two different orientations: a longitudinal and a transverse one.

In many bridges, flexible safety barriers are used. It is obvious that in such cases high-speed vehicles may crash on the barrier. Flexible safety barriers can be found at any location of the footway. The traffic arrangement is different now, and the axle loads are placed near the pedestrian parapet. If the designer is not sure of the type of barrier to be used, then both load arrangements of Figure 4.8 should be carefully examined. Further information may be found in the National Annex.

The accidental loads on footways or cycle tracks should not be combined with other variable loads. They are only applied for local verifications.

4.3.2.2 Collision forces on kerbs

Kerbs are high-quality concrete elements, usually prefabricated, which are used for bordering the carriageway and limiting the footways. The actions taken into account for the capacity design of the kerbs and the footways are shown in Figure 4.9.

The collision force has a magnitude of 100 kN and acts in the horizontal direction 5 cm below the top of the kerb. A base line of 50 cm long can be assumed for the transmission of the collision force by the kerbs into the supporting elements. The angle of dispersal is considered equal to 45° for rigid elements. In cases of flexible footways, different values of dispersal angles may need to be investigated.

In Figure 4.9, one can also see that a vertical force equal to $0.75 \cdot \alpha_{Q1} \cdot Q_{1k}$ acts simultaneously with the collision force. Q_v is only applied when this has an unfavorable effect. For α_{Q1}

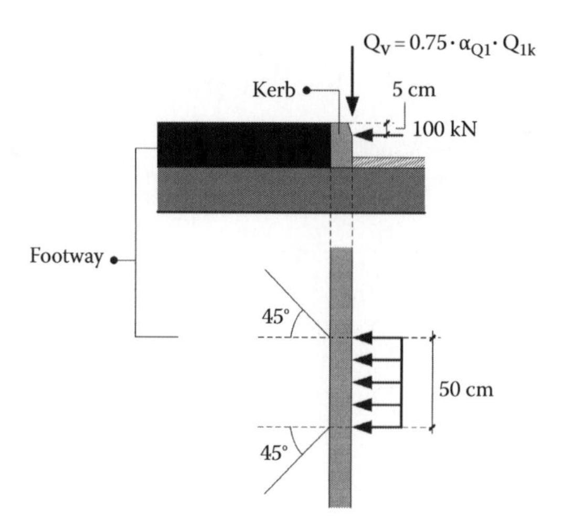

Figure 4.9 Vehicle collision forces on kerbs.

and Q_{1k}, see Table 4.5. For the most conservative case with an adjustment factor α_{Q1} equal to 1.0, the vertical force Q_v becomes equal to 225 kN.

4.3.2.3 Collision forces on safety barriers

The magnitude of the collision forces on barriers and other vehicle restraint systems may be given in the National Annex. EN 1991-2 recommends the following classification depending on the stiffness of the barrier.

The recommended values of Table 4.8 are based on experimental tests for barriers, which are commonly used for bridges. It has to be stated that the stiffness of a barrier highly depends on the stiffness of its connection with the footway. For example, a stiff reinforced concrete barrier with a weak connection should be classified as a flexible one.

In Figure 4.10, the arrangement of the collision forces is schematically demonstrated. The horizontal force H, given in Table 4.8, may be applied for the lowest value of the following distances:

- 100 mm below the top of the barrier
- 1.0 m above the carriageway or footway

Furthermore, force H acts along a base line of 50 cm and is distributed in the bridge deck in a similar way with the collision force of Figure 4.9. When unfavorable, force H should act

Table 4.8 Horizontal force transferred by safety barriers

Recommended class	*Horizontal force (kN)*	*Stiffness of barrier*
A	100	Very flexible
B	200	—
C	400	—
D	600	Very stiff

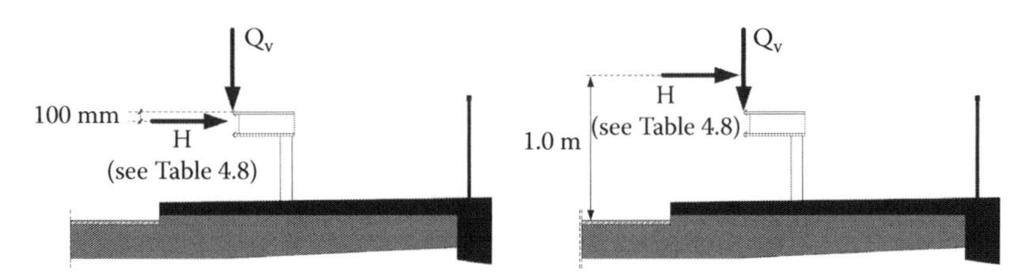

Figure 4.10 Loading cases for collision forces on safety barriers. (From EN 1991-2: Eurocode 1: Actions on structures—Part 2: Traffic loads on bridges, 2003.)

simultaneously with a vertical force Q_v. The magnitude of Q_v may be found in the National Annex. EN 1991-2 recommends a value equal to $0.75 \cdot \alpha_{Q1} \cdot Q_{1k}$.

4.3.2.4 Collision forces on unprotected structural members

Unprotected structural elements above or beside the carriageway may be designed according to the prescriptions of the National Annex or in agreement with the bridge owner. EN 1991-2 recommends collision forces, which are the same with those at piers. These forces should not be in combination with other variable loads.

4.4 ACTIONS ON PEDESTRIAN PARAPETS AND RAILINGS

Actions transferred by the pedestrian parapets to the bridge deck should be considered as variable loads. EN 1317-6 gives loading classes for pedestrian parapets and railings. Class C should be chosen as the recommended minimum class for bridges.

A line force of 1.0 kN/m, which acts horizontally or vertically on the top of the parapet, is described in EN 1991-2 as the recommended loading type in case of footways. For service side paths, a reduced value of 0.8 kN/m is suggested. As already mentioned, these are variable loads in which accidental effects are not taken into account. More information may be found in the National Annex.

The line force should be combined with a vertical surface load q_{fk} acting on the supporting footway or cycle track (see Figure 4.11). This must be the case during the capacity design of the supporting footway or cycle track and only when the pedestrian parapet is considered adequately protected. The recommended value of q_{fk} according to EN 1991-2 is 5 kN/m^2.

If the pedestrian parapet is not protected, then the supporting structure should be designed against an additional accidental load H_A equal to 1.25 times the characteristic resistance of the parapet or its base connection. The variable load q_{fk} is then not taken into account.

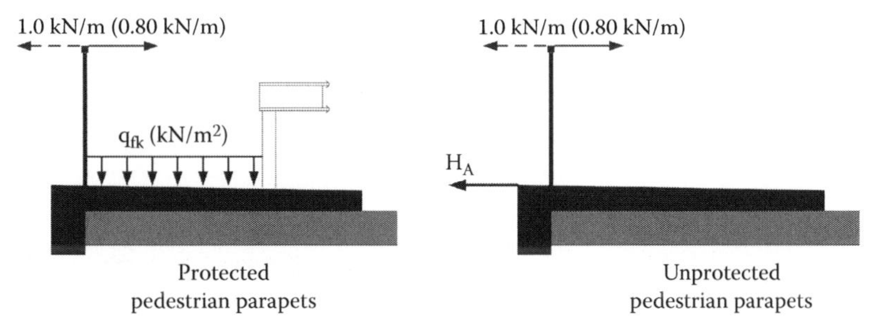

Figure 4.11 Actions on pedestrian parapets.

4.5 LOAD MODELS FOR ABUTMENTS AND WALLS IN CONTACT WITH EARTH

4.5.1 Vertical loads

A simplified load model representing the LM1 for the capacity design of the carriageway located behind abutments, wing walls, and side walls is recommended by EN 1991-2. The TS is allowed to be replaced by an equivalent uniformly distributed surface load q_{eq}. This load is spread over a rectangular surface depending on the dispersal of the loads through the backfill. A simplified model is proposed in Figure 4.12. The load surface for the TS is assumed to be $3 \times L$ m². L is calculated based on the dispersal of the axle loads. For the notional lane 1, the equivalent surface load is then

$$q_{eq,i} = \alpha_{Qi} \cdot \frac{2 \cdot Q_{ik}}{3 \cdot L}.$$

In EN 1997, one can find detailed information on the dispersal of loads through the backfills. If the backfill is properly consolidated, EN 1991-2 recommends a dispersal angle value equal to 30°; L becomes then equal to 2.2 m. With a load surface of 3×2.2 m², the load q_{eq} is 90.91 kN/m² for the notional lane 1 and 60.61 kN/m² for the notional lane 2 (for $a_{Qi} = 1$).

4.5.2 Horizontal loads

At the surfacing level of the carriageway, no horizontal loadings should be considered over the backfill.

The upstand walls of the abutments should be designed against the forces shown in Figure 4.13; $0.6 \cdot \alpha_{Q1} \cdot Q_{1k}$ represents a longitudinal braking or accelerating force. When unfavorable, this force should act simultaneously with the axle load $\alpha_{Q1} \cdot Q_{1k}$ from the LM1 and with the earth pressure from the backfill.

It should be noted that the vertical load alone, without the horizontal force, should be also investigated as a different loading case. The opposite need not to be examined since there can be no braking force without the vertical force of the axle loads.

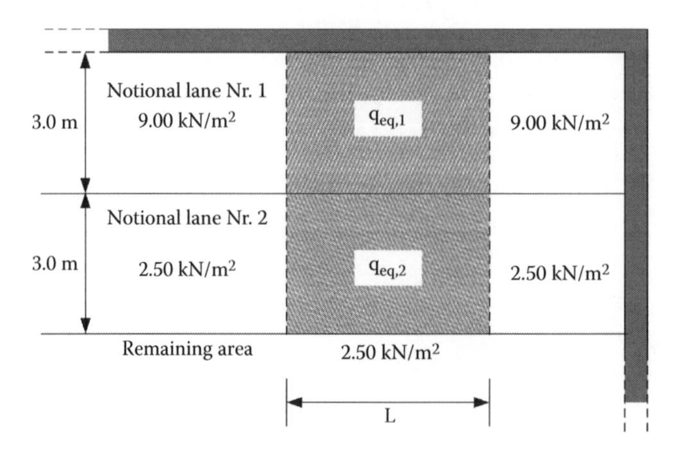

Figure 4.12 Simplified load model for the carriageway behind abutments.

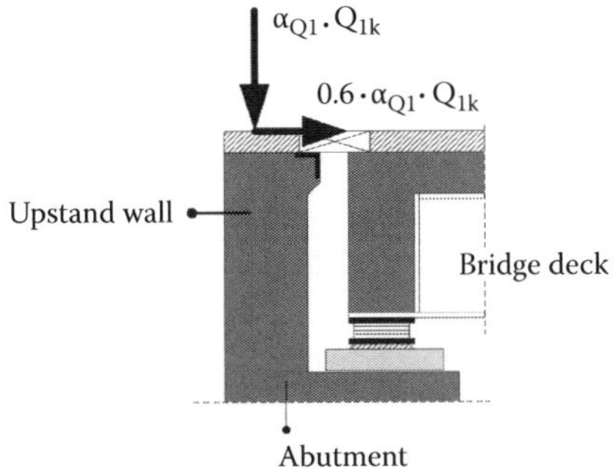

$\alpha_{Q1} \cdot Q_{1k}$

$0.6 \cdot \alpha_{Q1} \cdot Q_{1k}$

Upstand wall

Bridge deck

Abutment

Figure 4.13 Loads on upstand walls.

4.6 TRAFFIC LOADS ON RAILWAY BRIDGES

4.6.1 General

The loads due to railway traffic on the European mainline network, excluding narrow-gauge railways and tramways, are defined by various load models. EN1991-2 [4.4] gives five load models denoted as load model 71 (LM71), SW/0, SW2, high-speed load model (HSLM), and "unloaded train," which represent different traffic conditions. In the following, the first three, which are the most usual, are presented. For the other two, HSLM representing traffic at speeds higher than 200 km/h and the effects of an unloaded train, reference is made to the code. It is noted that the load models do not describe actual loads but have been selected so that their effects, with due dynamic enhancement, represent the effects of service traffic.

4.6.2 Vertical loads

4.6.2.1 Load model 71

This represents the static effect of normal rail traffic (see Figure 4.14).

During global analysis, the local effects due to the concentrated loads can be neglected. Therefore, an equivalent UDL $q_{Q,vk}$ can be applied (see Figure 4.15). The magnitude of this load is obviously equal to $250/1.6 = 156.25$ kN/m.

The earlier characteristic values are multiplied by a factor α on lines carrying rail traffic, which is heavier or lighter than normal rail traffic. When multiplied with this factor, these loads are called *classified vertical loads.*

Factor α takes the following values:

0.75 – 0.83 – 0.91 – 1.00 – 1.21 – 1.33 – 1.46

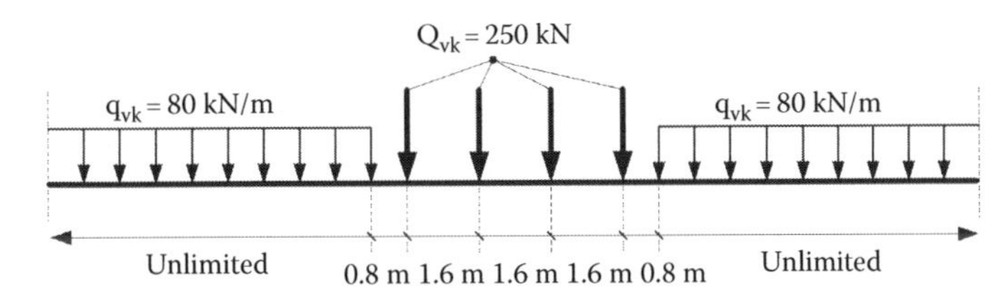

Figure 4.14 Characteristic values of vertical loads for load model 71. (From EN 1991-2: Eurocode 1: Actions on structures—Part 2: Traffic loads on bridges, 2003.)

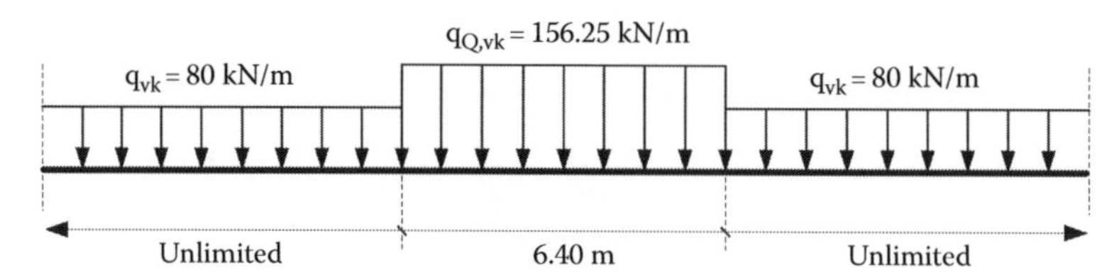

Figure 4.15 Simplified load model 71 for global analysis.

Factors $\alpha > 1.0$ are usually applied in cases of bridges, which carry international lines. The value to be used for nonnormal rail traffic may be found in the National Annex and has to be verified by both the track fabricator and the authorities.

When $\alpha \neq 1.0$, then the following actions shall be multiplied with the same factor α:

- Vertical loads for earthworks and earth pressure effects
- Load model SW/0 for continuous span bridges
- Accidental actions due to derailment
- Nosing forces (only if $\alpha > 1$)
- Centrifugal forces
- Traction and braking forces
- Combined response of structure and track to variable actions

Load model 71 is mostly applied in cases of **simply supported bridges.**

4.6.2.2 *Load models SW/0 and SW/2*

Load model SW/0 represents the static effect of normal rail traffic on **continuous systems** and is multiplied by factor α. For continuous bridges, it is considered as an alternative to LM 71. This means that continuous bridges are loaded separately with LM 71 and SW/0 and the most critical internal forces are chosen from the envelope of the two. Usually, LM 71 provides the largest effects at the spans, while SW/0 at the support regions. Load model SW/2 represents heavy rail traffic on lines that are specifically designated as carrying heavy traffic. **Load model SW/2 is not multiplied by factor α.**

The earlier load models are described in Figure 4.16 and in Table 4.9.

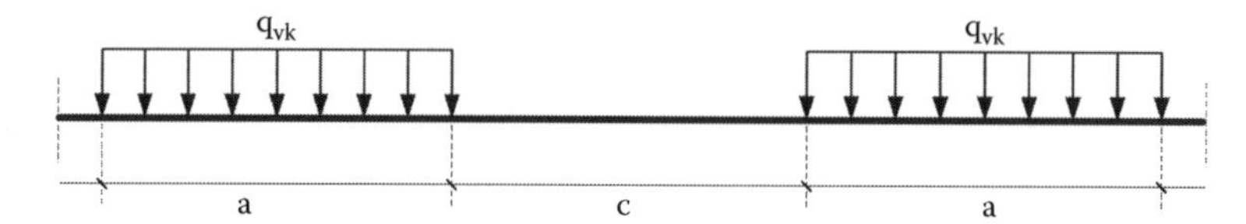

Figure 4.16 Load models SW/0 and SW/2.

Table 4.9 Characteristic values for vertical loads of SW/0 and SW/2

Load model	q_{vk} *[kN/m]*	*a [m]*	*c [m]*
SW/0	133	15	5.3
SW/2	150	25	7.0

4.6.2.3 Load model "unloaded train"

A special type of load model is the "unloaded train." It is represented by a UDL equal to 10 kN/m. Different values may be given by the National Annex. In certain cases, the unloaded train may result in a critical loading situation. This can arise during the stability analysis of a railway bridge, which carries an "unloaded train" in combination with strong wind forces (verification of adequate torsional rigidity; see also load group gr15 in Table 4.13).

4.6.2.4 Eccentricity of vertical loads (load models 71 and SW/0)

The uniform and concentrated loads of LM 71 and SW/0 are not distributed equally to the two rails but with the ratio 1.25:1.0 in order to take into account possible eccentricities as shown in Figure 4.17. The maximum resulting eccentricity is equal to $e=r/18$, where r is the distance between rails (usually $r=1.435$ m). This eccentricity may be neglected in the fatigue verifications.

4.6.2.5 Longitudinal distribution of concentrated loads by the rail and longitudinal and transverse distribution by the sleepers and ballast

For the design of local deck elements (concrete slabs, cross girders, etc.), the concentrated loads of LM 71 are considered to be distributed over three rail support points (Figure 4.18a). These loads are further distributed beneath sleepers up to the upper surface of the deck, in the longitudinal direction by an angle 4:1 (Figure 4.18b) and in the transverse direction, so that the point loads are transformed to distributed loads. Accordingly, local effects for the capacity design of the deck slab can be taken into account.

4.6.2.6 Transverse distribution of actions by the sleepers and ballast

The distribution of actions in transverse direction for bridges with ballast depends on the existence of cant (u) (see Figure 4.19b). Cant is the relative vertical distance between the uppermost surfaces of the two rails at a particular location along the track.

In Figure 4.19a, one can see the vertical axle load Q_v (see Figure 4.17) and the horizontal force Q_h. Q_h can be due to wind and centrifugal actions (curved tracks). Vertical loads are distributed through the ballast by an angle 4:1.

It should be noted that eccentricities between the mass center of the train's section and the axis of the bridge and of the track should be carefully taken into account. Equivalent horizontal forces Q_h can also be applied for considering additional loadings due to the previously mentioned eccentricities. A simple example is shown in Figure 4.20. A composite crossbeam,

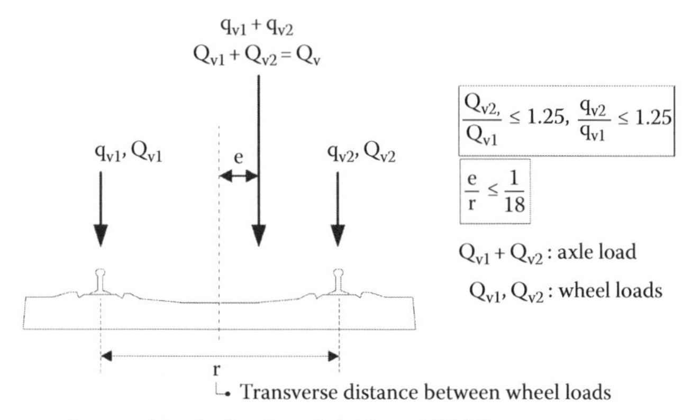

Figure 4.17 Eccentricity of vertical loads (load models 71 and SW/0).

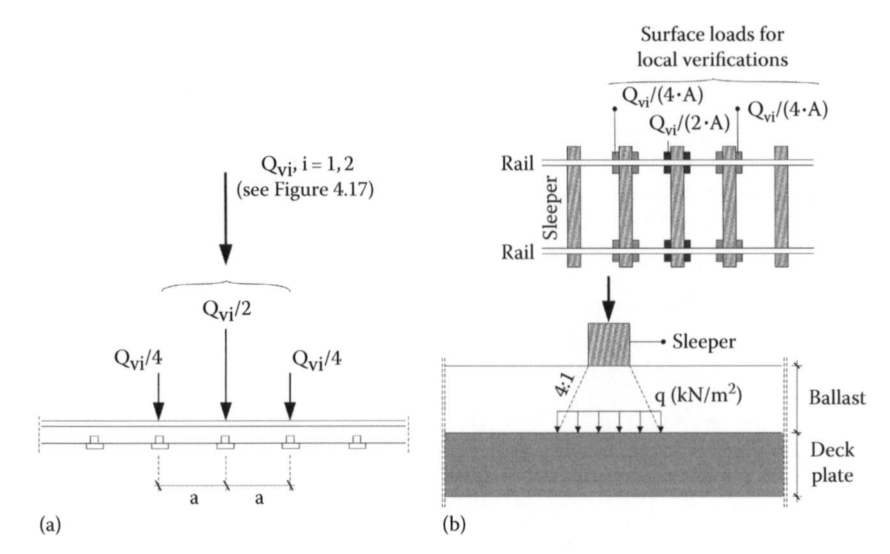

Figure 4.18 Longitudinal distribution of concentrated loads (a) by the rail and (b) the sleepers and ballast.

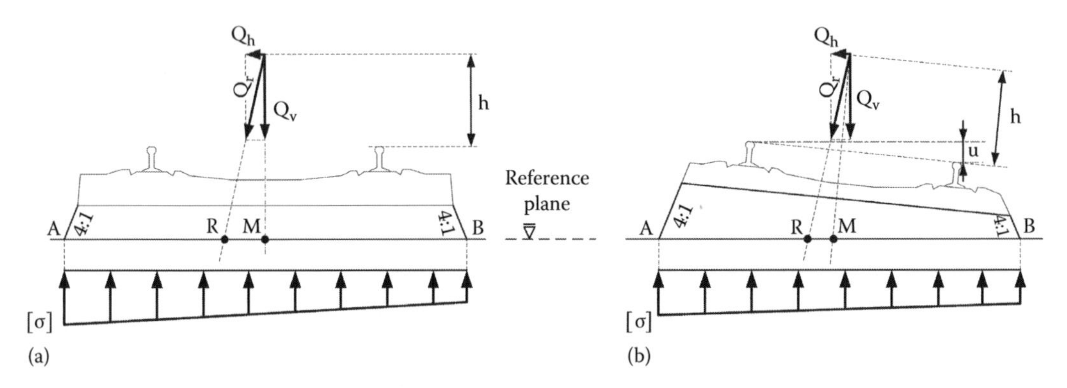

Figure 4.19 Transverse distribution of loads by the sleepers and ballast (a) without cant and (b) with cant.

which is considered as simply supported, carries the traffic load Q_v that acts at the gravity center of the train's section. This vertical load is distributed through the ballast and is transformed into a UDL q. Thus, the internal forces of the crossbeam due to Q_v can be calculated. The eccentricity e of Figure 4.17 must also be taken into account. This can be done with a fictitious horizontal force $Q_{h,v}$, which results in an additional loading on the crossbeam equal to Δq. When unfavorable, the internal forces due to Δq should be added to those of q (Q_v).

4.6.3 Dynamic effects (including resonance)

Since the trains are moving at high speeds over the bridge, the actual loading is obviously dynamic in nature, which may be enhanced due to possible resonance effects and inevitable track or train imperfections. Accordingly, a dynamic analysis including these effects should be conducted. However, for simplification reasons, static analysis in accordance with the load models described earlier may be performed and their effects subsequently being multiplied by a *dynamic factor* Φ. Such an analysis is allowed (a) for continuous bridges with train speeds V ≤ 200 km/h, (b) for simply supported bridges with train speeds V ≤ 200 km/h and fundamental bending natural frequency within the limits of Table 4.10, and (c) for

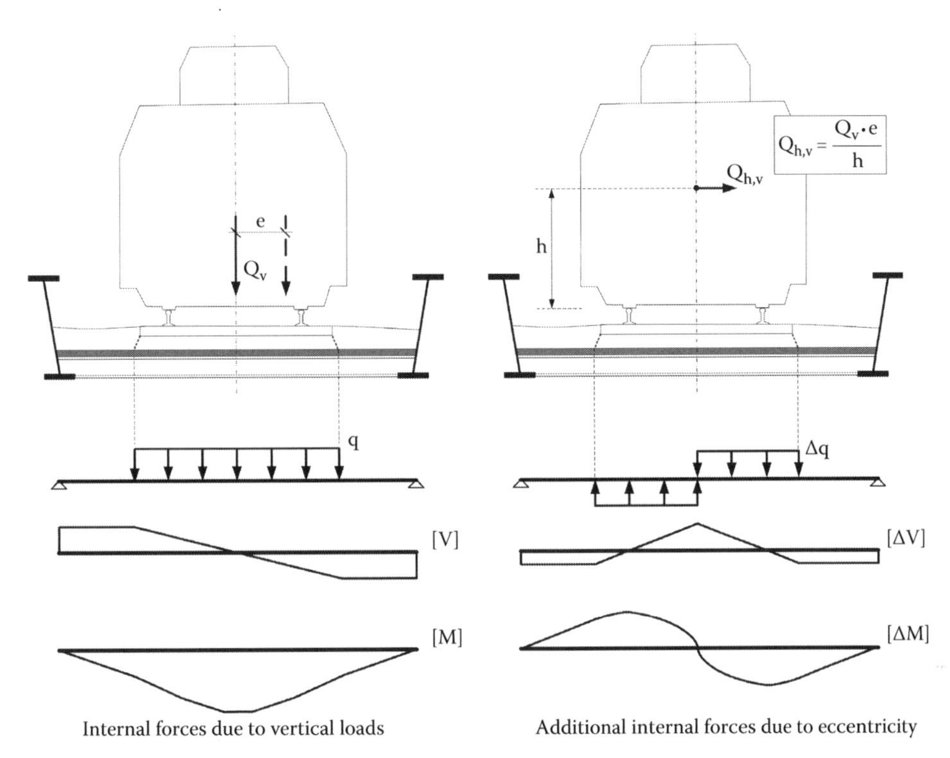

Figure 4.20 Internal forces in a crossbeam due to horizontal forces and eccentricities.

Table 4.10 Limits of bridge natural frequency to avoid dynamic analysis

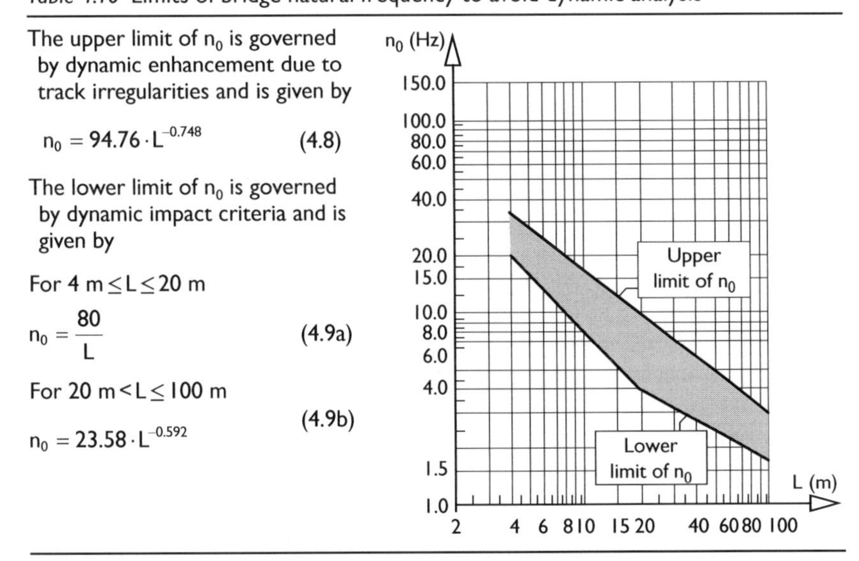

The upper limit of n_0 is governed by dynamic enhancement due to track irregularities and is given by

$$n_0 = 94.76 \cdot L^{-0.748} \tag{4.8}$$

The lower limit of n_0 is governed by dynamic impact criteria and is given by

For 4 m ≤ L ≤ 20 m

$$n_0 = \frac{80}{L} \tag{4.9a}$$

For 20 m < L ≤ 100 m

$$n_0 = 23.58 \cdot L^{-0.592} \tag{4.9b}$$

Source: EN 1991-2: Eurocode 1: Actions on structures—Part 2: Traffic loads on bridges, 2003.

Note: n_0, fundamental bridge frequency in bending taking account of mass due to permanent loads; L, span length for simply supported bridges or L_φ for other types of bridges.

simply supported bridges with train speeds V>200 km/h, span ≥40 m, and fundamental bending natural frequency within the limits of Table 4.10. In all other cases, a dynamic analysis is required, reference for which is made to EN 1991-2.

The fundamental natural frequency in bending for a *simply supported* bridge may be estimated by the formula

$$n_0 = \frac{17.75}{\sqrt{\delta_0}} \tag{4.5}$$

where δ_0 is deflection at mid-span [mm] due to permanent loads using the short-term modular ratio from Table 6.4.

REMARK 4.1

Equation 4.5 is defined as follows:

The deflection at mid-span of a simply supported beam is given by

$$\delta_0 = \frac{5}{384} \cdot \frac{(m \cdot g) \cdot L^4}{E \cdot I_{1,0}} \tag{R4.1}$$

where

m is the mass of the structure (permanent loads)

g is 9.81 m/s²

$E \cdot I_{1,0}$ is the short-term stiffness of the composite cross section

L is the span length of the beam

The natural circular frequency of a simply supported beam is

$$\omega_0 = \pi^2 \cdot \sqrt{\frac{E \cdot I_{1,0}}{m \cdot L^4}} = \pi^2 \cdot \sqrt{\frac{5 \cdot g}{\delta_0 \cdot 384}} \tag{R4.2}$$

The first natural frequency is calculated as follows:

$$n_0 = \frac{\omega_0}{2 \cdot \pi} = \frac{\pi}{2} \cdot \sqrt{\frac{5 \cdot g}{\delta_0 \cdot 384}} \tag{R4.3}$$

With δ_0 [mm] and g [m/s²], Equation R4.3 gives the following expression:

$$n_0 = \frac{\pi}{2} \cdot \sqrt{\frac{5 \cdot 9.81 \cdot 10^4}{384}} \cdot \frac{1}{\sqrt{\delta_0}} \Rightarrow n_0 = \frac{17.75}{\sqrt{\delta_0}} \text{ [Hz]}$$

The dynamic factor Φ, which enhances the static load effects for load models 71, SW/0, and SW/2, is either Φ_2 or Φ_3 depending on the track maintenance as follows:

For a carefully maintained track,

$$\Phi_2 = \frac{1.44}{\sqrt{L_\Phi} - 0.2} + 0.82 \quad \text{with } 1.00 \le \Phi_2 \le 1.67 \tag{4.6}$$

For a track with regular maintenance,

$$\Phi_3 = \frac{2.16}{\sqrt{L_\Phi} - 0.2} + 0.73 \quad \text{with } 1.00 \le \Phi_3 \le 2.00 \tag{4.7}$$

where L_Φ is "determinant" length given in Table 4.11.

EXAMPLE 4.2

Simply supported railway bridge
Span length L = 25 m
Deflection at mid-span due to permanent loads $\delta_0 = 37$ mm
Equation 4.5 gives the fundamental natural frequency:

$$n_0 = \frac{17.75}{\sqrt{37}} = 2.92\ \text{Hz}$$

Upper limit of n_0 is taken from Equation 4.8 (Table 4.10):

$$\max n_0 = 94.76 \cdot 25^{-0.748} = 8.53\ \text{Hz}$$

Lower limit of n_0 is taken from Equation 4.9b (Table 4.10):

$$\min n_0 = 23.58 \cdot 25^{-0.592} = 3.51\ \text{Hz} > n_0$$

The bridge is not stiff enough!

4.6.4 Horizontal forces

4.6.4.1 Centrifugal forces

In regions where the track on the bridge is curved, a centrifugal force acting horizontally at height 1.80 m above the running surface is considered. This force acts simultaneously with the vertical forces. Its characteristic value is equal to

$$Q_{tk} = \frac{V^2}{127 \cdot r} \cdot (f \cdot Q_{vk}) \tag{4.10a}$$

Table 4.11 Determinant length L_Φ

Case	*Structural element*	*Determinant length L_Φ*
Main girders		
1.1	Simply supported girders and slabs including filler beam decks	Main girder span
1.2	Continuous girders and slabs with average length $L_m = \frac{1}{n} \cdot (L_1 + L_2 + \cdots + L_n)$ where n is number of spans	$L_\Phi = k \cdot L_m \geq \max L_i,\ i = 1, 2, \ldots, n$ where n = 2, 3, 4, ≥5; k = 1.2, 1.3, 1.4, 1.5
1.3	Portal frames or boxes	
	Single span	As case 1.2 with k = 1.3
	Multi-spans	As case 1.2 with k = 1.5
1.4	Single arch, stiffened girders of bowstrings	Half span
1.5	Suspension bars (in conjunction with stiffened girders)	Four times the spacing of the suspension bars
Concrete deck slab for local and transverse design (with ballast)		
2.1	Deck slab as part of box girder or upper flange of main beams spanning	
	Transversely to the main girders	Three times span of deck plate
	In the longitudinal direction	Three times span of deck plate
	On cross girders	Twice the length of the cross girder
	Transverse cantilevers supporting railway loading	e ≤ 0.5 m: three times the distance between the webs e > 0.5 m: requires special study
2.2	Deck slab continuous over cross girders	Twice the cross girder spacing
2.3	Deck slab for half through and through bridges spanning	
	Perpendicular to the main girders	Twice the span of deck slab + 3 m
	In the longitudinal direction	Twice the span of deck slab
2.4	Deck slabs in filler beam decks	Twice L_Φ in longitudinal direction
2.5	Longitudinal cantilevers of deck slab	e ≤ 0.5 m: 3.6 m e > 0.5 m: requires special study
2.6	End cross girders or trimmer beams	3.6 m
Structural supports		
3.1	Columns, trestles, bearings, tension anchors, and for the calculation of contact pressures under bearings	L_Φ of the supported members

Source: EN 1991-2: Eurocode 1: Actions on structures—Part 2: Traffic loads on bridges, 2003.

$$q_{tk} = \frac{V^2}{127 \cdot r} \cdot (f \cdot q_{vk}) \tag{4.10b}$$

where

V is the maximum train speed [km/h]
r is the radius of curvature [m]
Q_{vk}, q_{vk} are the vertical forces of the corresponding load models
f is the reduction factor for LM 71 and SW/0 (see Equation 4.11 or Table 4.12)

The reduction factor is calculated by the following equation:

$$f = \left[1 - \frac{V-120}{1000} \cdot \left(\frac{814}{V} + 1.75\right) \cdot \left(1 - \sqrt{\frac{2.88}{L_f}}\right)\right] \geq 0.35 \tag{4.11}$$

$f = 1$ if $V \leq 120$ km/h or $L_f \leq 2.88$ m.

In the earlier relation, L_f [m] is the influence length of the part of the curved track, which is loaded. The part of the track that is loaded must be the one that results in the most unfavorable effects.

Expression (4.11) can be avoided through the following simplified method. The centrifugal forces are rewritten as follows:

$$Q_{tk} = \gamma_{fV} \cdot \frac{Q_{vk}}{r} \tag{4.12a}$$

$$q_{tk} = \gamma_{fV} \cdot \frac{q_{vk}}{r} \tag{4.12b}$$

where

$$\gamma_{fV} = \frac{V^2 \cdot f}{127} \tag{4.12c}$$

γ_{fv} is in [m] and its value can be taken from Table 4.12. Then, through multiplication with the known ratios Q_{vk}/r and q_{vk}/r, the centrifugal forces are calculated.

EXAMPLE 4.3

$V = 160$ km/h, $r = 560$ m, $L_f = 40$ m
From Table 4.12: $\gamma_{fV} = 161.26$
Centrifugal forces are calculated as follows:
$Q_{tk} = 161.26 \cdot (250/560) = 71.99$ kN
$q_{tk} = 161.26 \cdot (80/560) = 23.04$ kN/m

For curved bridges, the loading case without the effects of the centrifugal forces shall also be considered.

When the maximum train speed V is greater than 120 km/h, the following two loading cases should be considered:

a. The load model 71 (or SW/0 if required) multiplied by its dynamic factor Φ_2 or Φ_3 (see Section 4.6.3) and the centrifugal forces calculated with $\gamma_{fV} = 113.39$ ($V = 120$ km/h).
b. A "reduced" load model 71 (or SW/0 if required) without the dynamic factor Φ_2 or Φ_3 and with axle loads equal to $f \cdot Q_{vk}$ and $f \cdot q_{vk}$. The centrifugal forces are calculated with γ_{fV} for the maximum speed V specified.

Table 4.12 Simplified method for the calculation of centrifugal forces

Factor: $\gamma_{fV} = \frac{V^2 \cdot f}{127}$ [m]

L_f [m]	Maximum train speed V [km/h] ≤120	160	200	250	≥300
≤2.88	113.39	201.57	314.96	492.13	708.66
3	113.39	199.56	311.81	487.20	694.49
4	113.39	193.51	292.91	442.91	623.62
5	113.39	187.46	280.31	413.39	574.02
6	113.39	185.45	270.87	393.70	531.50
7	113.39	181.42	261.42	378.94	503.15
8	113.39	179.40	255.12	364.17	481.89
9	113.39	177.39	251.97	354.33	460.63
10	113.39	175.37	245.67	344.49	446.46
12	113.39	173.35	239.37	329.72	418.11
15	113.39	171.34	233.07	310.04	389.76
20	113.39	167.31	223.62	295.28	354.33
30	113.39	163.28	214.17	270.67	318.90
40	113.39	161.26	207.87	255.91	290.55
50	113.39	159.24	204.72	246.06	276.38
60	113.39	159.24	201.57	241.14	262.20
70	113.39	157.23	198.43	236.22	255.12
80	113.39	157.23	195.28	231.30	248.03
90	113.39	157.23	195.28	231.30	248.03
100	113.39	155.21	192.13	226.38	248.03
≥150	113.39	153.20	188.98	216.54	248.03

Calculation diagram for f and γ_{fV}

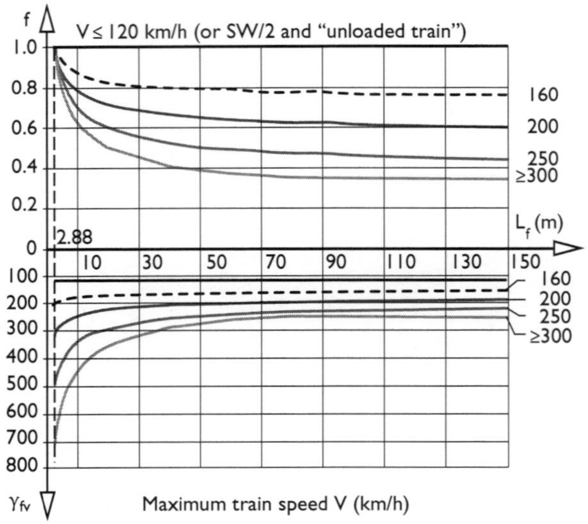

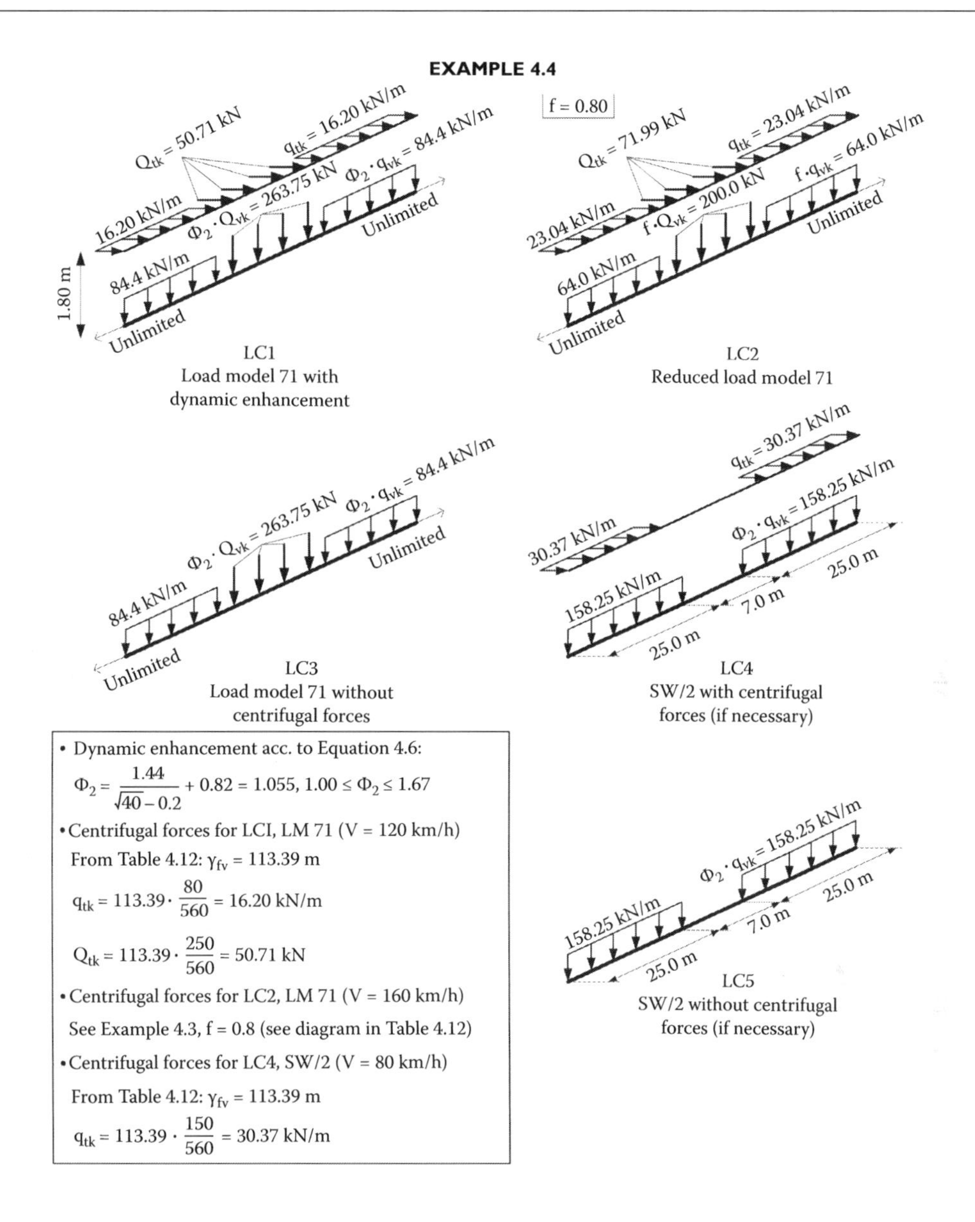

Figure 4.21 Load cases (traffic loads + centrifugal forces) for a simply supported curved bridge with L = 40 m and r = 560 m.

It is important to note that centrifugal forces are not multiplied by the dynamic factor Φ of Section 4.6.3.

In Figure 4.21, five different load cases are demonstrated for a simply supported curved bridge with a span of 40 m and a radius of curvature equal to 560 m. Load cases LC4 and LC5 refer to load model SW/2, which in most cases may not need to be taken into account.

4.6.4.2 Nosing force

This is a horizontal concentrated force $Q_{sk} = 100$ kN, transverse to the track axis applied at the top of the rails in combination with the vertical traffic loads. It is not multiplied with Φ or f but with α if this is larger than 1 (see Section 4.6.2).

4.6.4.3 Actions due to traction or braking

These are forces in the longitudinal direction acting at the top of the rails in combination with the vertical traffic loads. They are taken as uniformly distributed line loads over the corresponding influence length $L_{a,b}$ [m] for the structural element considered. Their direction corresponds to the permitted direction of travel on each track. The characteristic values of these forces are the following:

Traction force for LM 71, SW/0, SW/2, and HSLM

$$Q_{lak} = 33 \cdot L_{a,b} \leq 1000 \text{ kN} \tag{4.13}$$

Braking force for LM 71, SW/0, and HSLM

$$Q_{lbk} = 20 \cdot L_{a,b} \leq 6000 \text{ kN} \tag{4.14a}$$

Braking force for SW/2

$$Q_{lbk} = 35 \cdot L_{a,b} \tag{4.14b}$$

These forces are not multiplied with Φ or f but with α for LM 71 and SW/0.

For loaded lengths $L_{a,b} > 300$ m, additional requirements may be found in the National Annex.

Different values than the aforementioned may need to be taken into account in cases of lines carrying special traffic. EN 1991-2 [4.4] accounts for 25% of the sum of the axle loads of the real train, which act on the loaded length. Requirements may depend on the type of the project though.

When the track is continuous at one or both ends of the bridge, only one part of the earlier forces is transferred through the deck to the bearings. The other part is transmitted through the track and resisted behind the abutments. EN 1991-2 contains detailed provisions on how to calculate the part of the longitudinal forces due to traction/braking, thermal effects, and temperature variations that are transferred through the deck.

When comparing railway with roadway bridges, one can easily observe that braking/traction forces of the second ones are much more critical for the design of the entire structure (sub- and superstructure). Moreover, it should be underlined that braking forces in railway bridges have different magnitudes than the traction forces.

4.6.5 Consideration of the structural interaction between track and superstructure

In most bridges, rails are continuous over discontinuous locations such as the support joints. Obviously, a part of the longitudinal forces due to thermal effects, braking or traction, and other secondary effects is transferred through the rails to the embankment and the bearings. Therefore, the structural model should take into account both the superstructure and the track system. EN 1991-2 suggests a model in which the track and the superstructure are connected with nonlinear springs, which represent the longitudinal stiffness of the rails (see Figure 4.22a).

The spring that is defined as K_1 in Figure 4.22 may have a stiffness whose value is defined in the National Annex or by the fabricator of the rails. A typical load–displacement diagram is shown in Figure 4.22b. One can observe that the longitudinal behavior of the track–superstructure system depends on the type of the rail and on the loading condition. Indeed,

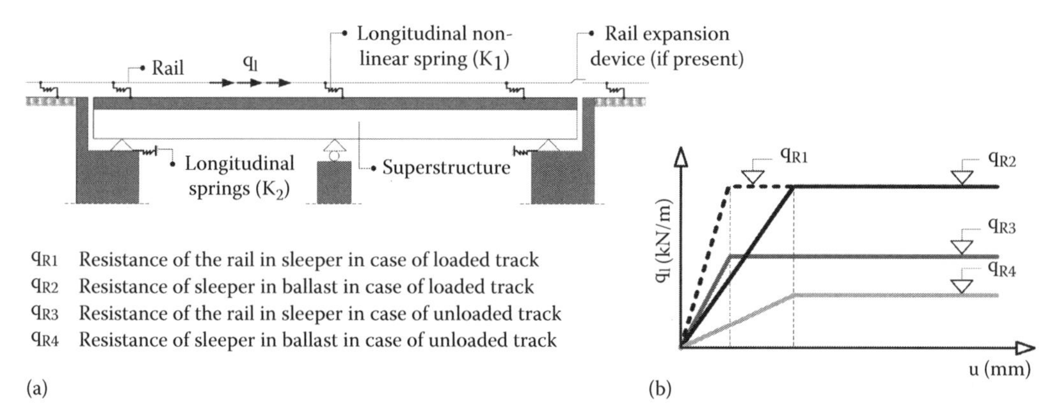

Figure 4.22 (a) Modeling example of a track–structure system. (b) Relation of longitudinal shear flow (q_l) with relative longitudinal displacement (u) for different loading cases.

if the track is loaded, then, due to high friction forces, both the longitudinal resistance and the stiffness become higher. Therefore, designers should take into account a variation of the longitudinal shear flow q_l.

For the capacity design of the rails, a longitudinal support reaction F_L should be calculated as follows:

$$F_L = \sum \psi_{0i} \cdot F_{li} \tag{4.15}$$

where

F_{li} is the individual longitudinal support reaction that corresponds to action i

ψ_{0i} is combination factor that is taken as equal to 1.0 for the calculation of rail stresses or from EN 1990 A2 for the calculation of load effects in the superstructures, bearings, and substructures

The stresses on rails should be limited to 72 N/mm^2 for compression and to 92 N/mm^2 for tension. The previous values are valid for tracks complying with the following criteria:

- UIC 60 rails with a tensile strength ≥900 N/mm^2
- Track with a straight layout or a radius ≥1500 m
- For ballasted tracks with a consolidated ballast with a depth of at least 30 cm
- For ballasted tracks with stiff concrete sleepers with a center-to-center spacing ≤65 cm

In cases that differentiate considerably from the preceding text, experimental tests should be conducted.

EN 1991-2 imposes additional limitations for the deformability of the rails. Some of the limitations are the following:

- Traction and braking
  The relative longitudinal displacement between the deck's end and the adjacent abutment shall not exceed 30 mm in the case of tracks with continuous ballast and with rail expansion devices at both ends of the deck. If a ballast with a movement gap over the supports and rail expansion devices is provided, then values greater than 30 mm are allowed.

- Vertical traffic actions (load models LM 71 and SW/0)
  The longitudinal displacement of the upper surface of the deck at its ends shall not exceed 8 mm when the structural interaction between the track and the superstructure is taken into account. For simple models in which the structural interactions are neglected, the maximum allowable value is 10 mm.
- Variable actions
  The relative vertical displacement between the upper surface of the deck and an adjacent construction shall not be greater than 3 mm for a maximum line speed up to 160 km/h. For higher speed values, the displacement limit is 2 mm.

4.6.6 Other actions and design situations

Other actions include aerodynamic actions from passing trains that develop in noise barriers, overhead protective structures, or platform canopies. Accidental design situations include derailment of rail traffic running on the bridge. EN 1991-2 gives detailed provisions for their calculations.

Longitudinal forces due to breakage of rails, actions from catenaries and other overhead line equipment attached to the structure, and other railway infrastructure and equipment are also referred to without specific provisions in the earlier code.

4.6.7 Groups of loads

Horizontal and vertical traffic loads act simultaneously on the bridge. This is taken into account considering groups of loads that define a single variable characteristic action to be combined with nontraffic action. The load groups are denoted as $gr_{i,j}$, where i is the number of tracks on the bridge (i = 1, 2, 3 ...) and j the current group number. The groups of loads for bridges with one track (i = 1) are given in Table 4.13. For more tracks, reference is made to EN 1991-2 [4.4].

Table 4.13 Group loads for rail traffics for one track according to EN 1991-2

	Vertical forces			*Horizontal forces*			
Load group	*LM 71, SW/0, HSLM*	*SW/2*	*Unloaded train*	*Traction, braking*	*Centrifugal force*	*Nosing force*	*Comments*
gr11	1	—	—	1[b]	0.5[b]	0.5[b]	c
gr12	1	—	—	0.5[b]	1[b]	1[b]	d
gr13	1[a]	—	—	1	0.5[b]	0.5[b]	e
gr14	1[a]	—	—	0.5[b]	1	1	f
gr15	—	—	1	—	1[b]	1[b]	g
gr16	—	1	—	1[b]	0.5[b]	0.5[b]	h
gr17	—	1	—	0.5[b]	1[b]	1[b]	i

Source: EN 1991-2: Eurocode 1: Actions on structures—Part 2: Traffic loads on bridges, 2003.

[a] It should be reduced to 0.5 in case of favorable effects.
[b] When favorable, these values shall be taken as equal to zero.
[c] Unfavorable vertical and longitudinal effects.
[d] Unfavorable vertical and transverse effects.
[e] Unfavorable effects in the longitudinal direction.
[f] Unfavorable lateral effects.
[g] Verification of global stability.
[h] Unfavorable vertical and longitudinal effects with SW/2.
[i] Unfavorable vertical and transverse effects with SW/2.

The vertical forces have to be considered with the eccentricities shown in Figure 4.17. In cases of nonnormal traffic loads, “classified vertical loads” must be applied (see factors α in Section 4.6.2.1).

The combination of vertical and horizontal loads should be performed as in Example 4.4 (see Figure 4.21). Attention must also be paid to whether dynamic amplifications (Φ_2 and Φ_3) are taken into consideration or not.

4.7 TEMPERATURE

4.7.1 General

A detailed investigation of the temperature effects is necessary during both the erection and the final stages. Provisions are described in EN1991-1-5 (see Section 6 of the code) [4.3], but additional guidelines may be found in the National Annex. Figure 4.23 demonstrates the division of a “real” temperature profile [ΔT_{REAL}] into four independent components. The first component [ΔT_N] is uniformly distributed along the height of the cross section causing longitudinal deformations. Components [ΔT_{MY}] and [ΔT_{MZ}] result in rotations around the strong and the weak axis, respectively. One can see a fourth component [ΔT_E], which represents the nonlinear part of the temperature’s profile. This distribution may cause out-of-plane deformations, which may be critical in certain types of bridges such as box girder bridges.

The real distribution is nonlinear and obviously time dependent. Therefore, [ΔT_{REAL}] is described by a combination of the previously mentioned temperature components together with the enhancement factors ω(t). The magnitude of these factors is mainly dependent on the

- Geometry of the cross section
- Time of the day and the season
- Thermal conductivity and the density of the materials
- Orientation of the bridge
- Thickness and color of the surfacing
- Variation of air temperature
- Wind speed
- Humidity

From the preceding text, it can be concluded that none of the components can be considered as more important than the others. The consideration of [ΔT_N] and [ΔT_{MY}] can be described for the majority of the plate-girder bridges as adequate. The first component leads to longitudinal deformations Δu, which are associated with forces and shear drifts in bearings

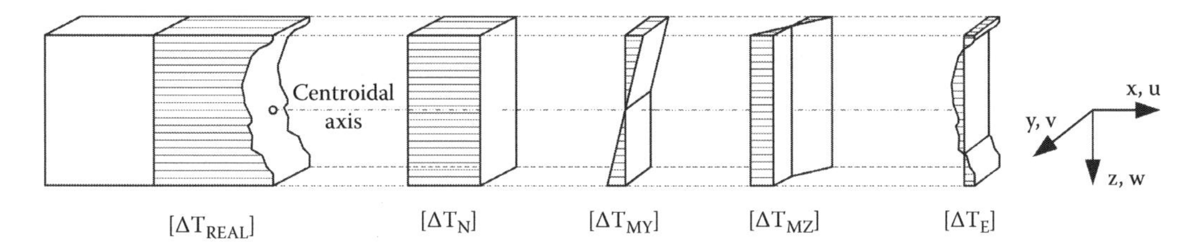

$[\Delta T_{REAL}] = \omega_N(t) \cdot [\Delta T_N] + \omega_{MY}(t) \cdot [\Delta T_{MY}] + \omega_{MZ}(t) \cdot [\Delta T_{MZ}] + \omega_E(t) \cdot [\Delta T_E]$

$\omega_N(t)$, $\omega_{MY}(t)$, $\omega_{MZ}(t)$, $\omega_E(t)$: time-dependent factors

Figure 4.23 Division of temperature profile in four parts.

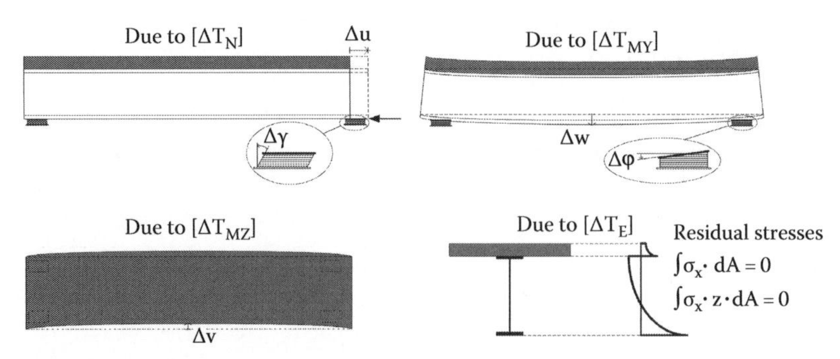

Figure 4.24 Temperature effects on a simply supported composite bridge.

(see Figure 4.24). When the temperature varies linearly over the cross section, additional deflections Δw and bearings' rotations $\Delta\varphi$ should be taken into account.

In cases of bridges with flexible deck slabs, for example, footbridges, bending of the bridge due to $[\Delta T_{MZ}]$ along the weak axis may occur; $\Delta v \neq 0$. Nonlinear temperature variations produce self-equilibrating stresses, which are also known as *eigen-* or *residual stresses.* This is because any fiber, being attached to other fibers, cannot perform free temperature expansion. More information about the consideration of nonlinear temperature effects is given in Section 4.7.5.

Both for global analysis and stress estimations, the coefficient for thermal expansion α_T is needed. For the composite sections, a value equal to 10^{-5}/°C both for concrete and for steel can be used. For steel cross sections during the erection stage, an increased thermal expansion factor equal to 1.2×10^{-5}/°C must be applied.

When the deformations and rotations due to temperature are restrained, additional internal forces are developed (also see Section 4.7.5). This is always the case in indeterminate systems such as continuous, frame, or integral bridges. Stresses that are caused by these forces are also known as *continuity stresses.*

The main provisions given in EN 1991-1-5 are presented in the following paragraphs. A discussion about the effects of temperature during the erection phase can be found in Section 4.7.6.

4.7.2 Uniform temperature component ΔT_N

ΔT_N expresses a global increase or decrease in temperature of the structure due to the corresponding temperature changes in the environment. If the minimum and maximum shade air temperatures are T_{min} and T_{max}, respectively the corresponding minimum and maximum temperatures of the bridge are $T_{e,min}$ and $T_{e,max}$ respectively. The former are nationally determined parameters provided in the National Annex of EN 1991-1-5. For group 2 decks (composite decks), the bridge temperatures are approximately 5°C above the air temperatures. The initial temperature T_0 is the temperature at which the structure is finished or when the bearings are placed. If unknown, it may be taken as the mean temperature during the construction period with a recommended value of $T_0 = 10$°C.

The characteristic value of the maximum contraction range is given by

$$\Delta T_{N,con} = T_0 - T_{e,min} \tag{4.16}$$

The characteristic value of the maximum expansion range is given by

$$\Delta T_{N,exp} = T_{e,max} - T_0 \tag{4.17}$$

For the design of bearings and expansion joints, it is recommended to increase the aforementioned values by 20°C in general or by 10°C if the temperatures at which the bearings or joints are set and specified. For example, for values $T_{e,min}=-15°C$, $T_{e,max}=+45°C$, and $T_0=10°C$, it is

$$\Delta T_{N,con} = 10-(-15) = 25°C \quad \text{and} \quad \Delta T_{N,exp} = +45-10 = 35°C$$

For bearings and expansion joints with unspecified temperatures at placement, the corresponding values are

$$\Delta T_{N,con} = 25+20 = 45°C \quad \text{and} \quad \Delta T_{N,exp} = +35+20 = 55°C$$

Unfavorable effects due to the aforementioned temperature differences may obviously arise in cases of longitudinal restraints during both the erection and the final stages. An expansion temperature of 35°C may result in excessive compression forces causing buckling phenomena, lateral deformations, local failures, etc. Due to creep, the magnitude of these forces is considerably reduced in concrete bridges or filler beam decks. Unfortunately, this relief does not take place in steel and composite bridges. Therefore, greater attention must be paid.

4.7.3 Temperature difference component ΔT_M

This expresses the fact that not all parts of the bridge change temperature at the same rate. It includes a linear varying temperature component along the vertical axis, a linear varying temperature component along the horizontal axis, and a nonlinear temperature component that produces self-equilibrating stresses. Out of the three, the first component only, denoted as Approach 1 in EN1991-1-5 [4.3], is usually considered in bridge design. According to Approach 1, ΔT_M is a temperature difference between the top and the bottom of the bridge deck. Two values are considered, $\Delta T_{M,heat}$ when the top is warmer than the bottom and $\Delta T_{M,cool}$ when the bottom is warmer than the top.

The recommended values for composite decks with a surfacing of 50 mm are

$$\Delta T_{M,heat} = 15°C \quad \text{and} \quad \Delta T_{M,cool} = 18°C$$

For different depths of surfacing, the aforementioned values should be multiplied with the factor k_{sur} of Table 4.14. These values represent upper bound values for surfacing of dark color.

Table 4.14 k_{sur} values according to EN 1991-1-5

Temperature	*Unsurfaced*	*Water proofed*	*50 mm*	*100 mm*	*150 mm*	*Ballast 750 mm*
$T_{top} > T_{bottom}$ 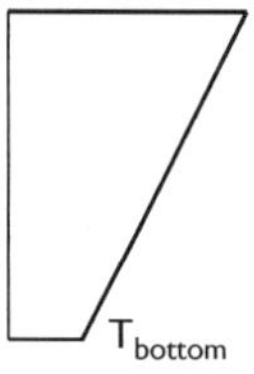 (T_{bottom})	0.9	1.1	1.0	1.0	1.0	0.8
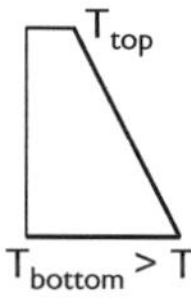 (T_{top}) $T_{bottom} > T_{top}$	1.0	0.9	1.0	1.0	1.0	1.2

Source: EN 1991-1-5: Eurocode 1: Actions on structures—Part 1–5: General actions—Thermal actions, 2003.

4.7.4 Combination between ΔT_N and ΔT_M

In some cases (e.g., in integral or frame bridges), the temperature effects ΔT_N and ΔT_M should be combined and regarded as single actions. The relevant combination rule is

$$\Delta T_{M,heat}(\text{or } \Delta T_{M,cool}) + \omega_N \cdot \Delta T_{N,exp}(\text{or } \Delta T_{N,con}) \tag{4.18a}$$

or

$$\omega_M \cdot \Delta T_{M,heat}(\text{or } \Delta T_{M,cool}) + \Delta T_{N,exp}(\text{or } \Delta T_{N,con}) \tag{4.18b}$$

The aforementioned are considered as a single action, and the most adverse effect should be chosen. The recommended values for the combination factors are $\omega_N = 0.35$ and $\omega_M = 0.75$.

The temperature combinations (4.18) are highly consistent with the $\omega(t)$ factors shown in Figure 4.23. For reasons of simplicity, the code avoids time-dependent factors by offering the earlier combinations. In (4.18a), the linear temperature component [ΔT_M] is the dominant one. In contrast, (4.18b) covers the case of a dominant uniform distribution [ΔT_N].

4.7.5 Nonuniform temperature component ΔT_E

Nonlinear temperature variations may be in certain cases much more critical than the linear ones. As already mentioned, residual stresses are developed and can accelerate cracking and yielding procedures. In compressed areas, these stresses may also play a negative role and increase the buckling risk. For compact cross sections (classes 1 and 2), nonlinear variations can be neglected. For noncompact cross sections, a more detailed analysis with [ΔT_E] may be necessary. EN 1991-1-5 suggests that nonlinear temperature variations should be considered without any reference to cross sections, materials, or structural systems.

In Figure 4.25, one can find the temperature differences over composite cross sections with a surfacing of 100 mm. EN 1991-1-5 offers two kinds of procedures: the normal and the simplified one. For different depths of surfacing, recommended values for ΔT_1 and ΔT_2 are given in Annex B of EN 1991-1-5.

It must be pointed out that [ΔT] incorporates [ΔT_M] and [ΔT_E] together with a small part of [ΔT_N]. When designers wish to consider all the temperature components and include nonlinear effects, then in combinations (4.18a) and (4.18b), [ΔT_M] should be replaced by [ΔT].

Generally, the computation of the residual stresses and the corresponding internal forces due to nonlinear effects is laborious. Interesting information on this issue is also given in [4.7] and [4.10].

4.7.6 Temperature effects during erection

Steel cross sections in bridges consist of thin plates, which may easily deform due to thermal effects, thus leading to erection difficulties. Figure 4.26 shows the erection procedure of a composite bridge with an open box steel cross section over a river bank.

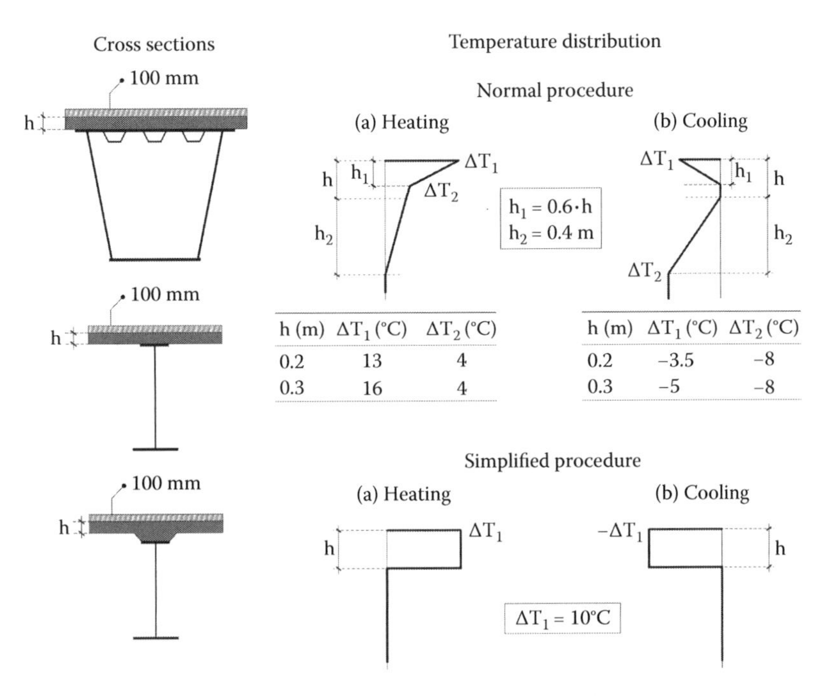

Figure 4.25 Temperature variations for composite bridge decks. (From EN 1991-1-5: Eurocode 1: Actions on structures—Part 1–5: General actions—Thermal actions, 2003.)

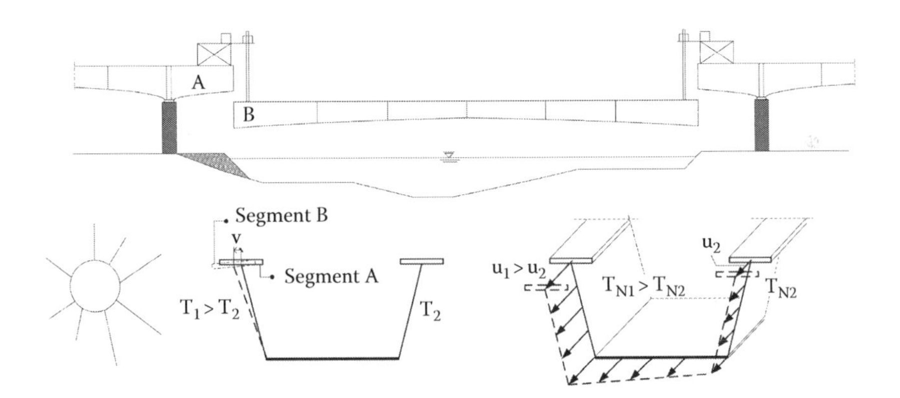

Figure 4.26 Deformations due to thermal effects during the erection stage.

The solar radiation on one side of the steel girder leads to lateral deformations v. Usually, segment A is equipped with thicker plates than those of segment B. Therefore, the relative lateral displacements between A and B may cause considerable assembly problems.

Additional difficulties will arise due to different temperatures between the webs ($T_{N1} > T_{N2}$). This results in an out-of-plane deformation of the cross section whose unfavorable effects may only be discovered on site.

Experienced designers are aware of the "tricky" situations that may emerge due to temperature effects. Unfortunately, the codes do not provide the necessary guidelines, and assumptions are unavoidable. Explicit calculations with sophisticated FE-structural models are usually used, and the following parameters are considered:

- Different temperature variations.
- The torsional flexibility of the girders. A 2nd-order theory analysis is recommended.
- Additional temperature effects due to welding procedures.
- Changes of the structural system during construction.
- Fabrication tolerances according to EN1990 [4.1].

4.8 WIND

4.8.1 General

Guidance on the determination of wind actions on structures is given by EN1991-1-4 [4.2]. This includes the whole structure, parts of it, and elements attached to the structure. Section 4.8 of this code gives the wind loads for plate and single or multiple box-girder bridges of constant depth. It does not cover arch, suspension, or cable-stayed bridges. In the usual coordination system for bridges, x-axis is along the longitudinal direction of the bridge, y-axis is along the transverse direction, and z-axis is the vertical axis, while in EN1991-1-4, the longitudinal axis is denoted with y and the transverse axis with x (see Figure 4.28). Wind along the longitudinal direction x (y according to EN1991-1-4) is usually negligible. Wind in the vertical direction z is taken into account for large span bridges, when aerodynamic stability is to be examined.

The wind force is given as a function of

- The basic wind velocity v_b
- The exposure factor c_e
- The force coefficient c_f
- The reference area A_{ref}

4.8.2 Wind force in bridge transverse direction y

When aerodynamic effects are not relevant, the wind force may be determined from

$$F_{Wyk} = \frac{1}{2} \cdot \rho \cdot v_b^2 \cdot C \cdot A_{ref,y} \, [kN] \tag{4.19a}$$

where

v_b is basic wind velocity [m/sec] (see Section 4.8.3)
ρ is density of air
C is wind load factor, $C = c_e \cdot c_{f,y}$; for c_e (see Section 4.8.4)
$c_{f,y}$ is force coefficient in transverse direction; $c_{f,y} = c_{fy,0}$ (see Section 4.8.5)
$A_{ref,y}$ is reference area in transverse direction [m^2] (see Section 4.8.6)

The density of air is taken as equal to 1.25 kg/m³, and therefore Equation 4.19a is rewritten as follows:

$$F_{Wyk} = \frac{1}{1600} \cdot v_b^2 \cdot C \cdot A_{ref,y}\ [kN] \tag{4.19b}$$

4.8.3 Basic wind velocity

The basic wind velocity is given by

$$v_b = c_{dir} \cdot c_{season} \cdot v_{b0} \tag{4.20}$$

Since the recommended values for the directional and seasonal factors c_{dir} and c_{season}, respectively, are equal to 1.0, the basic wind velocity is usually equal to the fundamental value of the basic wind velocity v_{b0}. This is determined statistically as the characteristic 10 min mean wind velocity 10 m above the ground level in open country terrain with low vegetation which is given in the National Annex of each country.

4.8.4 Exposure factor

The exposure factor c_e is a function of

- The terrain category. Five terrain categories, denoted as 0, I, II, III, or IV, are distinguished so as to express the terrain roughness.
- The distance z of the axis of the structure, that is, the middle height of the bridge, from the ground.

Values of the exposure factor as a function of the height z above terrain and the terrain category are illustrated in Figure 4.27.

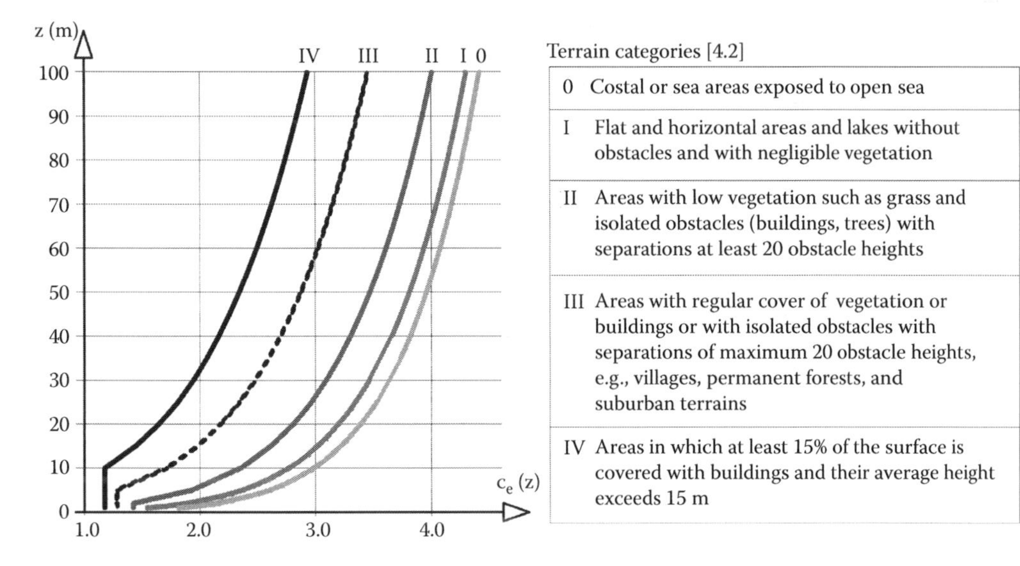

Figure 4.27 Exposure factor c_e.

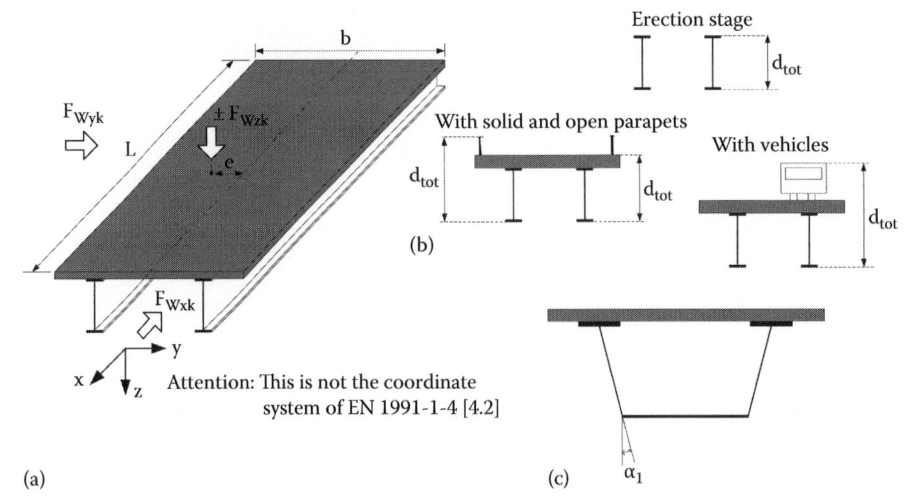

Figure 4.28 (a) Wind forces. (b) Determination of d_{tot} for different cases. (c) Cross section with an inclined windward face.

4.8.5 Force coefficient $c_{fy,0}$

For normal bridges, it may be taken as $c_{fy,0}=1.3$. This value is valid for $b/d_{tot}\geq 4$, where b = width of the deck and d_{tot} = total height of deck, including vehicles, parapets, and noise barriers (see Figure 4.28b). For $b/d_{tot}=0$, it is $c_{fy,0}=2.4$, while for intermediate values, a linear interpolation applies.

Where the windward face is inclined to the vertical (e.g., in box girders), $c_{fy,0}$ may be reduced by 0.5% per degree of inclination from the vertical, up to 30% (see Figure 4.28c).

Where a bridge deck is sloped transversely, $c_{fy,0}$ should be increased by 3% per degree of inclination but no more than 25%.

Where two similar decks are at the same level and separated transversally by a gap smaller than 1 m, the wind force may be calculated, by increasing the width b, as if it were a single structure.

4.8.6 Reference area $A_{ref,y}$

The reference area is given by

$$A_{ref,y} = L \cdot d_{ref} \tag{4.21}$$

where

L is total length of the bridge (see Figure 4.28a)
d_{ref} is reference height of the bridge, dependent on whether traffic is present or not

The reference height for wind without traffic is given in Table 4.15.

The reference height *in presence of traffic* is taken, independent of the position of the vehicles, as equal to

- d + 2 m for road bridges
- d + 4 m for railway bridges

Table 4.15 Reference height d_{ref} for wind without traffic

Road restraint system	*On one side*	*On both sides*
Open parapet or open safety barrier	$d + 0.3$ m	$d + 0.6$ m
Solid parapet or solid safety barrier	$d + d_l$	$d + 2 \cdot d_l$
Open parapet and open safety barrier	$d + 0.6$ m	$d + 1.2$ m

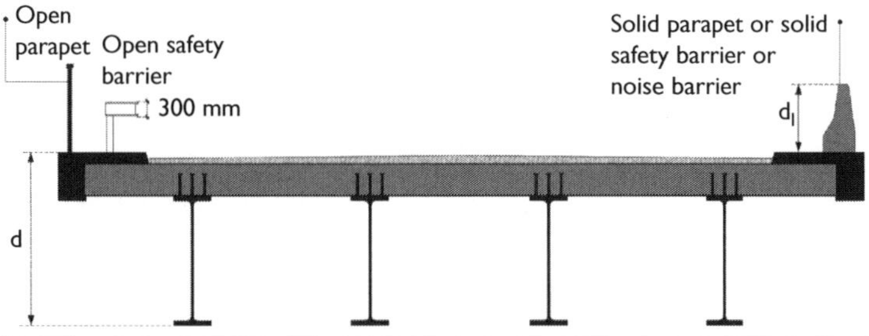

Notes:

For decks with truss girders, d is equal to the sum of the solid face area (one cornice, footway, or ballasted track) and the solid parts of the truss girders above or below that area in normal projected elevation divided by L.

For decks with several main girders during construction prior to concreting of the slab; d includes the height of two main girders.

Obviously, if this value is smaller than for the bridge without traffic, for example, due to the presence of high solid parapets, the larger value should be considered.

On a bridge with traffic, wind is combined with traffic loads so that the wind force on the bridge is considered with the combination value $\psi_0 \cdot F_{Wk}$. This force should be limited to

- $\psi_0 \cdot F_{Wk} \leq F_W^*$ for road bridges
- $\psi_0 \cdot F_{Wk} \leq F_W^{**}$ for railway bridges

The forces F_W^* and F_W^{**} are determined as aforementioned, with a reduced wind velocity with recommended values:

- $v_{b0}^* = 23$ m/s for road bridges
- $v_{b0}^* = 25$ m/s for railway bridges

4.8.7 Wind force in bridge vertical direction z

Wind force may be determined from Equation 4.19 with a force coefficient $c_{f_z} = 0.9$. This force acts both downward and upward. The reference area is equal to

$$A_{ref,z} = L \cdot b \tag{4.22}$$

This force is eccentric to the axis of the bridge, the relevant eccentricity being $e = b/4$ (see Figure 4.28). This force should only be considered if it is of the same order as the dead weight of the bridge.

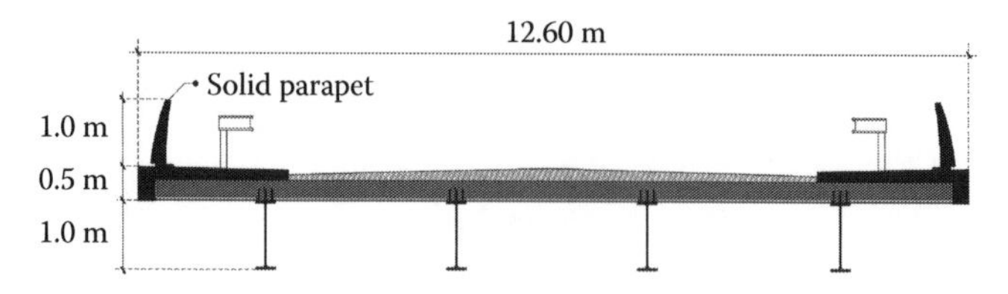

Figure 4.29 Cross section of the road bridge of Example 4.5.

EXAMPLE 4.5

The wind force in the transverse direction on a simply supported bridge of span L=25 m and a cross section shown in the succeeding text is to be determined. The fundamental value of the basic wind velocity is 33 m/s, the distance of the deck from the ground is 6 m, and the area is flat and horizontal with low vegetation (Figure 4.29).

Equation 4.20, basic wind velocity: $v_b = v_{b0} = 33$ m/s

Middle height of the bridge from the ground: $z = 6 - (1.5/3) = 5.25$ m

The area is classified as terrain category II.

Figure 4.27, exposure factor: $c_e = 1.7$

Width of deck: $b = 12.6$ m

Bridge without traffic

Total height: $d_{tot} = 1.50 + 1.0 = 2.5$ m

Force coefficient:

$$\frac{b}{d_{tot}} = \frac{12.6}{2.5} = 5.04 > 4.0 \Rightarrow c_{fy,0} = 1.3$$

Table 4.15, reference height: $d_{ref} = 1.5 + 2 \cdot 1.0 = 3.5$ m (solid parapets on both sides)

Equation 4.21, reference area:

$$A_{ref,y} = 25 \cdot 3.5 = 87.5\ \text{m}^2$$

Wind load factor: $C = 1.7 \cdot 1.3 = 2.21$

Equation 4.19b, wind force:

$$F_{Wyk} = \frac{1}{1600} \cdot 33^2 \cdot 2.21 \cdot 87.5 = 131.6\ \text{kN}$$

Bridge with traffic

Total height: $d_{tot} = 1.50 + 2.0 = 3.5$ m

Force coefficient:

$$\frac{b}{d_{tot}} = \frac{12.6}{3.5} = 3.6 < 4.0 \Rightarrow c_{fy,0} = 1.3 + \frac{2.4 - 1.3}{4} \cdot (4 - 3.6) = 1.41$$

In presence of traffic, the reference height is $d_{ref} = 1.5 + 2.0 = 3.5$ m.

Equation 4.21, reference area:

$$A_{ref,y} = 25 \cdot 3.5 = 87.5\ m^2$$

Wind load factor: $C = 1.7 \cdot 1.41 = 2.4$

Equation 4.19b, wind force:

$$F_{Wyk} = \frac{1}{1600} \cdot 33^2 \cdot 2.4 \cdot 87.5 = 142.9\ kN$$

The combination value with traffic loads is

$$\psi_0 \cdot F_{Wyk} = 0.6 \cdot 142.9 = 85.74\ kN$$

The maximum value in the combination is

$$F_W^* = \frac{1}{1600} \cdot 23^2 \cdot 2.4 \cdot 87.5 = 69.4\ kN < 85.74\ kN$$

Accordingly, the combination value is 69.4 kN.

Notes

a. The vertical component of the wind force of the bridge without traffic is calculated as follows:

Force coefficient: $c_{fz} = 0.9$

Wind load factor: $C = 0.9 \cdot 1.7 = 1.53$

Reference area, Equation 4.22:

$$A_{ref,z} = 25 \cdot 12.6 = 315\ m^2$$

$$F_{Wzk} = \frac{1}{1600} \cdot 33^2 \cdot 1.53 \cdot 315 = 328.02\ kN$$

b. The transverse wind load in the construction period (concreting of the slab), where the area of two girders is considered, is equal to

Total height: $d_{tot} = 1.25$ m (0.25 m slab's depth)

Force coefficient:

$$\frac{b}{d_{tot}} = \frac{12.6}{1.25} = 10.1 > 4.0 \Rightarrow c_{fy,0} = 1.3$$

Reference height: $d_{ref} = 0.25 + 2 \cdot 1.0 = 2.25$ m

Reference area:

$$A_{ref,y} = 25 \cdot 2.25 = 56.25\ m^2$$

Wind load factor: $C = 1.7 \cdot 1.3 = 2.21$

Construction phases constitute transient design situations with return periods smaller than 50 years that are the basis of the definition of the basic wind velocity. For example, the return period for wind and other climatic actions is 5 years if the construction period is up to 3 months but more than 3 days. The recommended value for the basic wind velocity is then 20 m/s:

$$F_{Wyk} = \frac{1}{1600} \cdot 20^2 \cdot 2.21 \cdot 56.25 = 31.08\ \text{kN}$$

4.9 EARTHQUAKE

Seismic actions are described in EN 1998-1 [4.5]. The seismic action has two horizontal and one vertical component. In usual bridges, the earthquake forces are determined by spectrum analysis in which the seismic motion is described by means of a *response spectrum*. In general, like for bridges on low damping elastomeric bearings (see Chapter 13), the elastic response spectrum should be used. For the horizontal components of the seismic action, the elastic response spectrum is defined by following expressions:

$$0 \le T \le T_B : S_e(T) = a_g \cdot S \cdot \left[1 + \frac{T}{T_B} \cdot (2.5 \cdot \eta - 1)\right] \tag{4.23a}$$

$$T_B \le T \le T_C : S_e(T) = a_g \cdot S \cdot \eta \cdot 2.5 \tag{4.23b}$$

$$T_C \le T \le T_D : S_e(T) = a_g \cdot S \cdot \eta \cdot 2.5 \cdot \frac{T_C}{T} \tag{4.23c}$$

$$T_D \le T : S_e(T) = a_g \cdot S \cdot 2.5 \cdot \eta \cdot \frac{T_C \cdot T_D}{T^2} \tag{4.23d}$$

where

$a_g = \gamma_I \cdot a_{gR}$ is the peak ground acceleration (PGA) on type A ground; ground of type A refers to all rock-like geological formations
γ_I is the importance factor (see Table 4.16)
a_{gR} is the reference value of the peak ground acceleration
S is the soil factor (see [4.5] or National Annex)
$\eta = \sqrt{\frac{10}{5+\xi}} \ge 0.55$ is the damping correction factor
ξ is the viscous damping ratio in %
T is the fundamental vibration period of the structure in sec
T_B, T_C, T_D are the characteristic periods of the spectrum as a function of the soil in sec

The values of the parameters of the spectrum are nationally determined parameters described in the National Annex of EN 1998-2. For the vertical component, the values of PGA and the periods T_B, T_C, and T_D are different. However, the vertical component is usually small and may be, like vertical wind loading, disregarded.

Table 4.16 Importance categories and importance factors

Importance category	*Description*	*Importance factor* (γ_I)
I	Bridges not critical for communication with a design life of 50 years	0.85
II	Road and railway bridges generally	1.0
III	Bridges of vital importance for retaining communications after the seismic event, major bridges with larger design life, or bridges the failure of which would result in a large number of fatalities	1.3

The application of the elastic response spectrum implies elastic structural behavior during the seismic event. However, in regions of moderate to high seismicity, ductile design could be envisaged, mainly by allowing the formation of plastic hinges at piers rigidly connected to the superstructure while the bridge deck remains elastic. Seismic forces are then reduced compared to elastic behavior. This force reduction is taken into account by consideration of a design response spectrum where a global reduction factor, called *behavior factor* q, is introduced. The design response spectrum for the horizontal seismic components is given by the following equations:

$$0 \le T \le T_B : S_{a,d}(T) = a_g \cdot S \cdot \left[\frac{1}{1.5} + \frac{T}{T_B} \cdot \left(\frac{2.5}{q} - \frac{1}{1.5}\right)\right] \quad (4.24a)$$

$$T_B \le T \le T_C : S_{a,d}(T) = a_g \cdot S \cdot \frac{2.5}{q} \quad (4.24b)$$

$$T_C \le T \le T_D : S_{a,d}(T) = a_g \cdot S \cdot \frac{2.5}{q} \cdot \frac{T_C}{T} \ge \beta \cdot a_g \quad (4.24c)$$

$$T_D \le T : S_{a,d}(T) = a_g \cdot S \cdot \frac{2.5}{q} \cdot \frac{T_C \cdot T_D}{T^2} \ge \beta \cdot a_g \quad (4.24d)$$

where

β is the lower bound factor of the horizontal design spectrum; the recommended value is equal to 0.2

q is the behavior factor

The maximum values of the behavior factor are summarized in Table 4.17. Different values of q may be used in the two horizontal directions. Upon the decision of the owner/designer, smaller values of q than those suggested by Table 4.17, implying less ductility demands and less potential damage, may be employed.

For higher seismic forces, high damping elastomeric bearings, dampers, or combinations of dampers with bearings may be used. For long bridges with lengths between 300 and 600 m, dependent on the soil conditions, or if the soil conditions vary considerably along the bridge, the spatial variability shall be considered in order to take into account asynchronous soil motions [4.6]. Modified response spectrums apply (see Chapter 13).

EN 1998-2 [4.6] allows the implementation of several analysis methods for the seismic design of bridges such as linear and nonlinear methods, equivalent static force methods, and pushover or time-series analysis. The method to be chosen mainly depends on the complexity level of each case, for example, the geometry of the bridge and the seismicity of the region. For complicated cases, designers usually compare different analysis methods and validate the final results.

For "normal bridges," the *fundamental mode method* can be described as the most convenient. In this method, the earthquake excitations are represented by equivalent

Table 4.17 Maximum values of the behavior factor q for non-isolated bridges

Members	*Limited ductility*	*High ductility*
Reinforced concrete piers		
Vertical piers in bending	1.5	$3.5 \cdot \lambda(\alpha_s)$
Inclined struts in bending	1.2	$2.1 \cdot \lambda(\alpha_s)$
Steel piers		
Vertical piers in bending	1.5	3.5
Inclined struts in bending	1.2	2.0
Piers with normal bracing	1.5	2.5
Piers with eccentric bracing	—	3.5
Deck rigidly connected with the abutments		
In general	1.5	1.5
Locked-in structures ($T_{horizontal} \leq 0.03$)	1.0	1.0
Arches	1.2	2.0

Source: EN 1998-2: Eurocode 8: Design of structures for earthquake resistance—Part 2: Bridges, 2005.

Notes:

$\alpha_s = \frac{L_s}{h}$ is the shear span ratio of the pier.

L_s is the distance from the plastic hinge to the point of zero-bending moment.

h is the depth of pier's cross section in the direction of flexure of the plastic hinge.

If $\alpha_s \geq 3$, then $\lambda(\alpha_s) = 1.0$.

If $3 > \alpha_s \geq 1.0$, then $\lambda(\alpha_s) = \sqrt{\frac{\alpha_s}{3}}$.

In cases of piers with rectangular shape, the minimum of the α_s corresponding to the two sides of the cross section should be used.

The q-factors for high ductility are valid only when special detailing for the structural members is followed (see [4.6]). If not, the low ductility q-factors should be chosen.

For reinforced concrete members of high ductility, the normalized axial force $n_k = N_{Ed}/(f_{ck}/A_c)$ should not exceed 0.3. If $0.30 < n_k \leq 0.60$, the q-factors should be reduced as follows:

$$q_r = q - \frac{n_k - 0.3}{0.3} \cdot (q - 1) \geq 1.0.$$

If $n_k > 0.6$, then $q_r = 1.0$ (elastic response).

N_{Ed} is the compressive force at the plastic hinge for the seismic design situation (compression with a positive sign).

If the locations of the plastic hinges are difficult to inspect, then q-factors should be multiplied with 0.6. Final values should be at least equal to 1.0. However, in the cases where plastic hinges are located in piles, the final q-values need not to be less than 2.1 for vertical piles and 1.5 for inclined piles. This remark is valid only for the case of high ductility.

static forces acting at the mass center of the superstructure. A normal bridge is the one that fulfills the following criteria:

- The bridge is straight with a continuous deck.
- The theoretical eccentricity between the mass center of the deck and the stiffness center of the supporting members does not exceed 5% of the deck's length.
- The mass of the piers is less than 20% of the tributary mass of the deck, the piers carry simply supported spans, and no significant interaction between piers is expected.

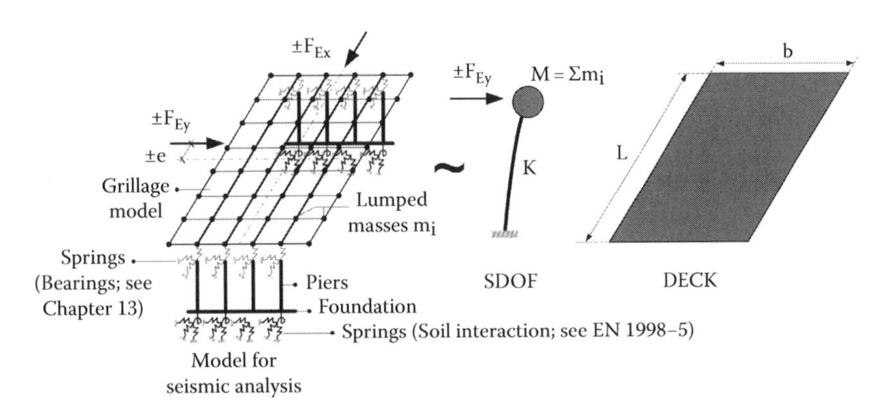

Figure 4.30 Model of a simply supported bridge under seismic loads.

When the earlier conditions are satisfied, then the system can be represented by a single dynamic degree of freedom model (SDOF). This is briefly demonstrated in Figure 4.30.

From the aforementioned, it is obvious that all the structural elements that contribute to the mass and the stiffness of the system should be included in the dynamic model. The reason for this is the following. The calculated seismic forces are equal to the effective mass of the structure M multiplied by the spectral acceleration $S_d(T)$. The magnitude of the spectral acceleration is highly dependent on the natural period T, thus the mass M, and the stiffness K.

The mass M should be taken as a combination of the mean values of all the permanent loads and the quasi-permanent values of the masses corresponding to the variable actions, $\psi_{2,1} \cdot Q_{k,1}$, where $Q_{k,1}$ is the characteristic value of the traffic load. Table 4.18 provides the recommended values according to [4.6] for the combination factor $\psi_{2,1}$.

Different approaches for the *fundamental mode method* are found in EN 1998-2: the *rigid deck*, the *flexible deck*, and the *individual pier models*. The rigid deck model is the most common one because in most cases road- and railway bridges consist of reinforced concrete deck slabs, which can be considered as rigid diaphragms. Bridges of elongated geometries or without diaphragms should be analyzed with the flexible deck model. When the seismic action is mainly resisted by piers, then the individual pier model is applied. Table 4.19 offers an overview of the previously mentioned approaches.

In the case of the rigid deck model, the seismic force in the transverse direction should be distributed along the deck proportionally to the distribution of the effective mass.

Table 4.18 Recommended values for $\psi_{2,1}$

$M = \sum M_{G,k} + \psi_{2,1} \cdot Q_{k,1}$	
For normal traffic and foot bridges:	$\psi_{2,1} = 0$ (see also Annex A2 of [4.1])
For severe traffic:	
Road bridges	$\psi_{2,1} = 0.2$
Railway bridges	$\psi_{2,1} = 0.3$

Notes:

Road bridges with severe traffic carry motorways and other roads of national importance.

Railway bridges with severe traffic carry intercity rail links and high-speed railways.

For the traffic loads $Q_{k,1}$, the adjustment factors a_Q and a_q should be taken into account.

Table 4.19 Approaches of the fundamental mode method

Rigid deck model	*Flexible deck model*	*Individual pier model*
Applied when (a) Longitudinal direction Always for straight bridges with continuous deck (b) Transverse direction and $L/b \leq 4.0$ (see Figure 4.30) Or $\frac{\Delta_d}{d_a} \leq 0.20$ where L is the total bridge length in continuous bridges b is the width of the deck Δ_d and d_a are respectively the maximum difference and the average of the displacements in the transverse direction of all pier tops under F_{Ey}.	**Applied when** The limitations for rigid deck model are not fulfilled.	**Applied when** When the seismic action in the transverse direction is mainly resisted by piers and there is no significant interaction between adjacent piers and $0.90 \leq \frac{T_{pier,i}}{T_{pier,i+1}} \leq 1.10$
Fundamental period $T_{dir} = 2 \cdot \pi \cdot \sqrt{\frac{M_{dir}}{K_{dir}}}, dir = x, y$ where M_{dir} is the effective mass of the structure in the direction i K_{dir} is the total stiffness of the system in the direction i	**Fundamental period** $T = 2 \cdot \pi \cdot \sqrt{\frac{\sum_i M_i \cdot d_i^2}{g \cdot \sum_i M_i \cdot d_i}}$ where M_i is the mass at the ith nodal point d_i is the displacement of the ith nodal point in an approximation of the shape of the 1st mode g gravity acceleration	**Fundamental period** $T_{pier} = 2 \cdot \pi \cdot \sqrt{\frac{M_{pier}}{K_{pier}}}$ where T_{pier} is the fundamental period of the same pier, considered *independently* of the rest of the bridge M_{pier} and K_{pier} are the mass and the stiffness, respectively, attributed to each pier
Seismic forces $F_{Edir} = M_{dir} \cdot S_{a,d}(T_{dir})$	**Nodal seismic forces** $F_{Ed,i} = \frac{4 \cdot \pi^2 \cdot d_i}{g \cdot T^2} \cdot S_{a,d}(T) \cdot M_i$	**Seismic forces** $F_{pier,y} = M_{pier} \cdot S_{a,d}(T_{pier,y})$

For skewed bridges, an eccentricity in mass distribution shall apply. In case the rigid deck model is used, this eccentricity shall be equal to

$$e = e_a + e_d \tag{4.25}$$

where

e_a is $0.03 \cdot L$ or 0.03b is the accidental eccentricity of the mass

e_d is $0.05 \cdot L$ or 0.05b is an additional eccentricity taking into account the simultaneous presence to translational and torsional modes

L is the total bridge length for single span or continuous bridges

b is the deck width

REMARK 4.2

When traffic loads are dominating and $\psi_{2,1} \neq 0$ (see Table 4.18), it is recommended to investigate different arrangements of the LM1 on the carriageway. This should be done because seismic forces may be applied with eccentricities much greater than e. This usually happens in small span bridges. A seismic analysis may also be necessary during the erection stage.

REFERENCES

[4.1] EN 1990: CEN (European Committee for Standardization). Eurocode basis of structural design, 2002.
[4.2] EN 1991-1-4: CEN (European Committee for Standardization): Actions on structures—Part 1-4: General actions—Wind actions, 2005.
[4.3] EN 1991-1-5: CEN (European Committee for Standardization): Actions on structures—Part 1-5: General actions—Thermal actions, 2003.
[4.4] EN 1991-2: CEN (European Committee for Standardization): Actions on structures—Part 2: Traffic loads on bridges, 2003.
[4.5] EN 1998-1: CEN (European Committee for Standardization): Design of structures for earthquake resistance—Part 1: General rules, seismic actions and rules for buildings, 2004.
[4.6] EN 1998-2: CEN (European Committee for Standardization): Design of structures for earthquake resistance—Part 2: Bridges, 2005.
[4.7] Ghali, A., Neville, A. M., Brown, T. G.: *Structural Analysis. A Unified Classical and Matrix Approach*, 5th edn., pp. 177–184, Spon Press, Taylor & Francis Group.
[4.8] I. Vayas: *Verbundkonstruktionen auf der Grundlage von Eurocode 4*, Ernst & Sohn, Berlin, Germany, 1999.
[4.9] Mangerig, I., Lichte, U., Beucher, S.: *Validation of the Safety Coefficients for Thermal Load on Bridges*. Stahlbau 79, pp. 167–180, Ernst & Sohn, Berlin, Germany, 2010.
[4.10] Priestley, M. J. N.: Design of thermal gradients for concrete bridges. New Zealand Engineering 31(9), 213–219, September 1976.
[4.11] STANAG 2021. Military STANDardization AGreements.
[4.12] Timm, G., Großmann, F.: Einwirkungen auf Brücken. In: *Beton Kalender*, Ernst & Sohn, Berlin, Germany, 2004.

Chapter 5

Basis of design

5.1 GENERAL

The design life of bridges is conventionally set to 100 years [5.1]. During this life, the bridge has to comply with certain basic requirements that refer to structural resistance, serviceability, and durability and are met by appropriate design, production, execution, and use. Concerning the design, this is based on consideration of ultimate and serviceability limit states (SLSs) that have to be verified for *persistent*, *transient*, and *accidental* design situations.

Ultimate limit states (*ULSs*) are associated with the safety of people and of the structure and for composite bridges refer to

- *EQU*: Loss of static equilibrium of the structure or parts of it, regarding them as a rigid body, design of hold-down anchors or verification of uplift of bearings in continuous bridges
- *STR*: Failure by collapse or excessive deformation of the superstructure or its members and more specifically to
  - Resistance of cross sections and connections
  - Stability of members
  - Resistance of shear connection
- *FAT*: Failure caused by fatigue
- *GEO*: Failure or excessive deformation of the foundation and the ground

REMARK 5.1

The loss of static equilibrium (EQU) for the majority of bridges during persistent situations is quite impossible. In contrast, the risk of losing equilibrium can be considerably high during erection, transient situations. Typical examples are the cantilever method (Figure 2.49) in which destabilizing effects may lead to collapse; see Equation 5.9.

SLSs concern the functioning of the structure under normal use, the comfort of people, and the structural appearance and are associated with

- Deformations
- Cracking of concrete
- Vibrations

5.2 LIMIT STATE DESIGN

In limit states verification, design values are considered. The design values of actions are defined as

$$F_d = \gamma_f \cdot \psi \cdot F_k \tag{5.1}$$

where
- F_d is the design value of the action
- F_k is the characteristic value of the action
- γ_f is the partial safety factor of the action that takes into account the possibility of unfavorable deviations of the action values from the characteristic values
- ψ is the combination value of the action with other action and is either 1.0 or ψ_0, ψ_1, or ψ_2

However, verifications are not made in practice by direct comparison between design values and limit values of actions. Actions result in internal forces and moments, deformations, and vibrations in bridges that are characterized as *action effects* and are evaluated by appropriate structural analysis. The design values of the effects of one action are given by

$$E_d = \gamma_{Sd} \cdot E[\gamma_f \cdot \psi \cdot F_k] \tag{5.2}$$

where γ_{Sd} is the partial safety factor that takes into account uncertainties in modeling the actions and in modeling the bridge structure in analysis.

Usually, factors γ_{Sd} and γ_f are merged together to a single partial safety factor:

$$\gamma_F = \gamma_{Sd} \cdot \gamma_f \tag{5.3}$$

The effects of actions are then determined from

$$E_d = E[\gamma_F \cdot \psi \cdot F_k] \tag{5.4}$$

Following the classification of actions presented in 4.1 in respect to their duration, partial safety factors are distinguished in

- γ_G for permanent actions
- γ_Q for variable actions
- γ_A for accidental actions
- γ_{AE} for seismic actions

Furthermore, two values of the safety factors for permanent actions $\gamma_{F,inf}$ and $\gamma_{F,sup}$ are used, depending on whether they produce favorable or unfavorable effects. An example is the EQU verification for a cable-stayed bridge that is erected by cantilevering from the pylon, where the permanent loads are multiplied by $\gamma_{F,inf}$ on one side and $\gamma_{F,sup}$ on the other side of the pylon to produce the largest overturning moment on the pylon; see Figure 5.1b.

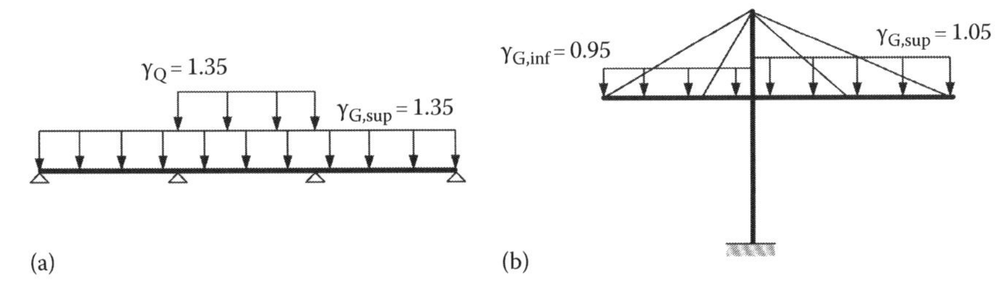

Figure 5.1 (a) Safety factors for the span moment of a continuous bridge and (b) the overturning moment at the pylon foot during construction of a cable-stayed bridge.

For a linear structural response, the analysis may be performed with the characteristic action values and the design values of the action effects determined by multiplication with the safety factors and the combination values, that is, the action effects are given by

$$E_d = \gamma_F \cdot \psi \cdot E[F_k] \tag{5.5}$$

This has an important implication in design when regarding combinations of actions. For linear structural response, analysis is made for each individual action separately, and the combination refers to the resulting actions. **However, for nonlinear response, the design values of the actions are combined, and analysis is made for each combination.**

The design resistances are similarly determined from

$$R_d = \frac{1}{\gamma_{Rd}} \cdot R\left[\frac{X_k}{\gamma_m}\right] \tag{5.6}$$

where

X_k is the characteristic value of a material property
γ_{Rd} is the partial safety factor covering uncertainties in modeling the resistances

As for the actions, factors γ_{Rd} and γ_m of resistances are usually merged together to a single partial safety factor:

$$\gamma_M = \gamma_{Rd} \cdot \gamma_m \tag{5.7}$$

The design resistances are then determined from the relevant characteristic values:

$$R_d = \frac{R_k}{\gamma_M} \tag{5.8}$$

5.3 ULTIMATE LIMIT STATE (ULS)

5.3.1 Design formats

The design format for the limit state of static equilibrium (EQU) may be written as

$$E_{d,dst} \leq E_{d,stb} \tag{5.9}$$

where

$E_{d,dst}$ is the design value of the effects of destabilizing actions
$E_{d,stb}$ is the design value of the effects of stabilizing actions

The design format for the limit state of collapse or excessive deformation (STR and GEO) may be written as

$$E_d \leq R_d \tag{5.10}$$

where

E_d is the design value of the effects of actions, like internal forces or moments
R_d is the design value of the corresponding resistances

5.3.2 Combination of actions

The effects of individual actions are combined to form load cases to take into account their simultaneous presence. Three combination types are distinguished: *basic*, *accidental*, and *seismic*. In the basic combinations, one variable action is considered as *leading action*, the others being *accompanying actions*. The leading action in the accidental combination is the accidental action itself, while variable actions are introduced with their combination values and multiplied by the relevant factors ψ. The combinations at ULS other than fatigue are presented in Tables 5.1 and 5.2.

Table 5.1 Combinations of actions at ULS other than fatigue

Basic combinations

$$\sum_{j\geq 1} \gamma_{Gj,sup} \cdot G_{kj,sup} + \sum_{j\geq 1} \gamma_{Gj,inf} \cdot G_{kj,inf} + \gamma_P \cdot P_k + \gamma_{Q1} \cdot Q_{k1} + \sum_{i>1} \gamma_{Qi} \cdot \psi_{0i} \cdot Q_{ki} \tag{5.11a}$$

$$\sum_{j\geq 1} \gamma_{Gj,sup} \cdot G_{kj,sup} + \sum_{j\geq 1} \gamma_{Gj,inf} \cdot G_{kj,inf} + \gamma_P \cdot P_k + \gamma_{Q1} \cdot \psi_{01} \cdot Q_{k1} + \sum_{i>1} \gamma_{Qi} \cdot \psi_{0i} \cdot Q_{ki} \tag{5.11b}$$

$$\sum_{j\geq 1} \xi \cdot \gamma_{Gj,sup} \cdot G_{kj,sup} + \sum_{j\geq 1} \gamma_{Gj,inf} \cdot G_{kj,inf} + \gamma_P \cdot P_k + \gamma_{Q1} \cdot Q_{k1} + \sum_{i>1} \gamma_{Qi} \cdot \psi_{0i} \cdot Q_{ki} \tag{5.11c}$$

Accidental A

$$\sum_{j\geq 1} G_{kj,sup} + \sum_{j\geq 1} G_{kj,inf} + A_d + P_k + (\psi_{1,1} \text{ or } \psi_{2,1}) \cdot Q_{k1} + \sum_{i>1} \psi_{2,i} \cdot Q_{ki} \tag{5.12}$$

Seismic E (see also Table 5.11)

$$\sum_{j\geq 1} G_{kj} + P_k + \gamma_1 \cdot A_{Ed} + \sum_{i\geq 1} \psi_{2,1} \cdot Q_{ki} \tag{5.13}$$

Notes:

+ does not mean summation but "combination with."

Σ means "the combined effect of."

G_{sup} are the permanent actions with unfavorable effects.

G_{inf} are the permanent actions with favorable effects.

P are the prestressing actions.

Q_1 is the leading variable action.

Q_{ki} are the accompanying variable actions.

A_d is the leading accidental action.

A_{Ed} is the seismic action.

γ_1 is the importance factor from Table 4.16.

Table 5.2 Combinations of actions at construction stages ULS

Basic EQU/STR/GEO

$$\sum_{j\geq 1} \gamma_{Gj,sup} \cdot G_{kj,sup} + \sum_{j\geq 1} \gamma_{Gj,inf} \cdot G_{kj,inf} + \gamma_P \cdot P_k + \gamma_Q \cdot Q_{c,k} \quad (5.14)$$

Accidental A

$$\sum_{j\geq 1} G_{kj,sup} + \sum_{j\geq 1} G_{kj,inf} + P_k + A_d + \psi_2 \cdot Q_{c,k} \quad (5.15a)$$

Seismic E (see also Table 5.11)

$$\sum_{j\geq 1} G_{kj,sup} + \sum_{j\geq 1} G_{kj,inf} + P_k + \gamma_I \cdot A_{Ed} + \psi_2 \cdot Q_{c,k} \quad (5.15b)$$

Notes:

$Q_{c,k}$ are construction loads.

The importance factor γ_I takes into account the lower return period during construction.

It should be pointed out that for the basic combinations at ULS, EN 1990 offers three different expressions that can be applied for both persistent and transient design situations; see Equations 5.11a through c in Table 5.1. For EQU limit states, Equation 5.11a applies with the safety and combination factors given in Tables 5.3 through 5.5. For the design of structural elements (STR limit states) **without the involvement of geotechnical actions,** the choice between 5.11a, b, and c will be found in the National Annex. Again, safety and combination factors are given in Tables 5.3 through 5.5. For the verification of structural elements (STR and GEO) **involving geotechnical actions** (footings, side walls, wing walls, piers, piles, retention walls, etc.), three different approaches are envisaged that are not presented in this book.

For construction three combinations are similarly examined, the basic, accidental, and seismic as illustrated in Table 5.2.

REMARK 5.2

Permanent loads that result from a single source, for example, self-weight of the structure, the road surface, rails, and fixed equipment, are multiplied by either $\gamma_{G,sup}$ or $\gamma_{G,inf}$ depending on whether the resulting action effect is unfavorable or favorable. Normally, for such actions, a unique value of the partial factor (γ_G) is applied. Nevertheless, when the permanent actions are sensitive to variations, then the upper and lower limits $\gamma_{G,sup}$ and $\gamma_{G,inf}$ should be considered. A typical example of a permanent action with a "variable" magnitude is the fresh concrete, concreting stages. This issue is further discussed in Chapter 7.

5.3.3 Safety factors and combination values

For the basic combinations, values of safety factors γ_F and combination factors ψ_0 for road bridges are given in Table 5.3, for railway bridges in Table 5.4, and for footbridges in Table 5.5. It is noted that shrinkage of concrete is not a separate action. However, it is included separately in the tables since it is treated as a loading by imposing equivalent temperature gradients as outlined in Section 7.4.3. Geotechnical designs often do not work with

Table 5.3 Safety and combination factors of road bridges for limit states EQU and STR/GEO without geotechnical actions

Action situation	*Symbol*		*Unfavorable effect*	*Favorable effect*	*Factor ψ_0*
Permanent for persistent and transient situations (EQU)	G	γ_G	1.05	0.95 or 0.8 when self-weight is not well defined	—
Permanent for persistent and transient situations (STR/GEO)	G	γ_G	1.35 (ξ=0.85)	1.0	—
Uneven settlements	G_{set}	$\gamma_{G,set}$	1.2 (for linear elastic analysis) 1.35 (for nonlinear analysis) 0[a]	0	—
Secondary effect of shrinkage	S_{sec}	γ_S	0[a] or 1.0	0[a] or 1.0	—
Prestress by imposed deformations at internal supports	P	γ_P	1.0	1.0	—
Stability for external prestress	P	γ_P	1.35	1.0	—
Traffic loads gr1a (LM1 + loads on footways and cycle tracks q_{fk}^*)	Q	γ_Q	1.35	0	TS: 0.75 UDL: 0.40 Footways and cycle tracks: 0.40
Traffic loads gr1b, gr2, gr3, gr4, gr5	Q	γ_Q	1.35	0	0
Wind	W	γ_Q	1.50	0	Persistent 0.6 Execution 0.8 For F_W^* 1.0
Snow[c] during construction	S	γ_Q	1.50	0	0.8
Thermal	T	γ_Q	0[a] or 1.5[b]	0	0[a] or 0.6[b]
Construction loads (EQU)	Q_c	γ_Q	1.35	0	1.0

[a] For bridges where all cross sections are of class 1 or 2.

[b] For bridges with cross sections of class 3 or 4.

[c] Snow is considered to be removed during service so that it is only accounted for during construction. This is implied in EN 1990-A2 where ψ_0-values only during construction are given for snow. Cases where snow is the leading action, where ψ_0-values are not required, are obviously less unfavorable than cases where traffic is the leading action.

Table 5.4 Safety and combination factors of railway bridges for limit states EQU and STR/GEO without geotechnical actions

Action situation	*Symbol*		*Unfavorable effect*	*Favorable effect*	*Factor ψ_0*	
Traffic loads. Individual components[a]	Q	γ_Q	1.45	0	LM 71, SW/0	0.8
					SW/2	0
					Unloaded train	1.0
					HSLM	1.0
					Real trains	1.0
					Nosing force	1.0
					Aerodynamic effects	0.8
Traffic loads Groups of loads gr11-15, gr21-25, gr31, and gr26, 27 associated with LM1, SW/0 (STR/GEO)	Q	γ_Q	1.45	0	0.80	
Traffic loads Groups of loads gr16,17, SW/2, and gr26,27 associated with SW/2 (STR/GEO)	Q	γ_Q	1.20	0	0.80	
Traffic loads gr1b, 2,3,4,5 (STR/GEO)	Q	γ_Q	1.35	0	0.80	
Wind	W	γ_Q	1.50	0	0.75[b]	
Wind during construction	W	γ_Q	1.50	0	0.80	
Snow during construction	S	γ_Q	1.50	0	0.80	
Thermal	T	γ_Q	0[c] or 1.5[d]	0	0[c] or 0.6[d]	
Construction loads (EQU)	Q_c	γ_Q	1.35	0	1.0	
Construction loads (STR/GEO)	Q_c	γ_Q	1.50	0	1.0	

Notes:

Permanent for persistent and transient situations (EQU), permanent for persistent and transient situations (STR), uneven settlements, secondary effect of shrinkage, and prestress by imposed deformations at internal supports and stability for external prestress as in Table 5.3.

For EQU limit states for all the rail traffic loads, $\gamma_Q = 1.45$ (0 where unfavorable). For road and pedestrian traffic actions, $\gamma_Q = 1.35$.

[a] Individual components of traffic actions (i.e., traction and braking, centrifugal forces, and interaction forces due to deformation under vertical traffic loads) in design situations where the traffic loads are not considered as groups of loads should use the same values of ψ_0 as those for the associated vertical loads.

[b] For wind forces F_W^{**}, it is $\psi_0 = 1.0$.

[c] For bridges where all cross sections are of class 1 or 2.

[d] For bridges with cross sections of class 3 or 4.

Table 5.5 Safety and combination factors of footbridges for limit states EQU and STR/GEO without geotechnical actions

Action situation	*Symbol*		*Effect*		*Factor* ψ_0	
			Unfavorable	*Favorable*		
Permanent (STR/GEO)	G	γ_G	1.35	1.0	—	
Permanent (EQU)	G	γ_G	1.05	0.95	—	
Secondary effect of shrinkage	S_{sec}	γ_S	0[a] or 1.0[b]	0[a] or 1.0[b]	—	
Prestress by imposed deformations at supports	P	γ_P	1.0	1.0	—	
Stability for external prestress	P	γ_P	1.35	1.0	—	
Traffic loads gr1, Q_{fwk}, gr2	Q	γ_Q	1.35	0	gr1	0.40
					Q_{fwk}	0
					gr2	0
Thermal	T	γ_Q	0[a] or 1.5[b]	0	0[a] or 0.6[b]	
Wind	W	γ_Q	1.5	0	0.3	
Snow during construction	S_c	γ_Q	1.5	0	0.8	
Construction loads	Q_c	γ_Q	1.35	0	1.0	

[a] For bridges where all cross sections are of class 1 or 2.
[b] For bridges with cross sections of class 3 or 4.

factored loads and include the safety factors in the right side of Equation 5.10. In such cases, other γ-values for GEO are employed.

REMARK 5.3

- In the previous tables, one can observe that effects due to shrinkage and thermal differences are neglected during the ULS verifications **when all** cross sections are class 1 or 2. This is due to the adequate rotation capacity of the compact sections that allows the imposed deformations to be released. This is only allowed when lateral torsional buckling failure at hogging moment areas is excluded.
- It is also worth mentioning that both for favorable and unfavorable effects, the partial safety factor γ_F due to imposed deformations at ULS is taken equal to 1.0. EN 1994-2 allows this simplification only in cases of controlled deformations. However, EN 1993-2 permits the effects of imposed deformations to be ignored when all cross sections are class 1. In such a case, designers should follow the recommendation of EN 1994-2.
- The safety factor γ_G for the differential settlements depends on the type of analysis. This can be explained by the fact that a linear elastic analysis offers more conservative results than a nonlinear one for cases of time-dependent settlements since the ability of redistributions is omitted. Therefore, the code proposes a reduced safety factor equal to 1.20 for linear global analysis.

Figure 5.1 gives simple examples for appropriate selection of partial safety factors. For the span moment of the continuous bridge for an STR limit state, $\gamma_{G,sup} = 1.35$ applies to all permanent loads, while variable loads are considered only in the relevant span and are multiplied by $\gamma_Q = 1.35$ (Figure 5.1a). On the other side, the overturning moment for an EQU limit state at the pylon foot during segmental erection of a cable-stayed bridge that is required to verify the stability of the structure as a rigid body to overturning is determined by the application of $\gamma_{G,sup} = 1.05$ on the longer side and $\gamma_{G,inf} = 0.95$ on the shorter side. This overturning moment is critical for checking the foundation. However, the resistance of the pylon section at its base is verified by application of $\gamma_{G,sup} = 1.35$ to the self-weight on both sides of the pylon, since this belongs to an STR/GEO situation.

REMARK 5.4

It can be said that the main difference between STR and EQU limit states is that for the latter, the partial factor γ_G for the permanent actions is not uniform over the whole structure; it gets higher values in regions with destabilizing actions. This book mainly covers STR limit states since the EQU ones are mainly associated with the design of foundations (see EN 1997).

5.3.4 Basic combinations

The effects of temperature, creep, and shrinkage may be neglected at ULSs for bridges in which **all cross sections are of class 1 or 2** (see Remark 5.3) **and not susceptible to lateral torsional buckling** (see Section 9.13). Where structural systems like continuous bridges or frame bridges are sensitive to differential settlements, such settlement shall be taken into account. Such settlements are considered as permanent actions for which G_{set}, as given in Table 4.1 must be specified. G_{set} may be represented by a set of values that correspond to the calculated settlement $d_{set,i}$ due to permanent loads of individual foundations or groups of foundations. The predicted values of $d_{set,i}$ are in accordance with the requirements found in EN 1997. In addition, two individual foundations or groups of foundations are considered to settle at a value $d_{set,i} + \Delta_{dset,i}$, where the latter takes into account uncertainties connected to the estimation of the settlement. Possible value of $\Delta_{dset,i}$ is 10 mm. It is worth mentioning that box-girder bridges are very sensitive to differential settlements in both longitudinal and transverse bearing lines and therefore, designers should be very careful with the design.

For **road bridges,** the following rules apply:

- Wind needs only to be combined with gr1a of Table 4.7.
- Snow is generally not combined with traffic loads, except in cases of roofed bridges; see notes in Table 5.3.
- Wind and temperature are not considered to act simultaneously.
- Load group gr1b is combined with no other variable nontraffic action; see Table 4.7.

The most usual ULS basic combinations for road bridges are summarized in Table 5.6 and for bridges where all cross sections are of class 1 or 2 in Table 5.7. The box of the leading action is shown in bold face. Combinations with traffic loading as the leading action are critical for the superstructure, while those with temperature and wind may be critical for piers and bearings. **Variable actions are obviously considered if they act unfavorably.**

Table 5.6 Basic combinations according to Equation 5.11a at ULS for road bridges generally (STR/GEO)

	$G+C_{sec}$	S_{sec}	Q	T	W	G_{set}
No.	*Permanent and secondary effects of creep*	*Shrinkage secondary effects*	*Traffic loads*	*Temperature*	*Wind*	*Differential settlements (Table 5.3)*
1	1.35	1.0	*gr1a · 1.35*	0	1.5 · 0.6 or 1.5 · F_W^*	1.2[a] or 1.35[b]
2	1.35	1.0	*(gr1b,2,3,4,5) · 1.35*	1.5 · 0.6	0	1.2 or 1.35
3	1.35	1.0	1.35 · (TS · 0.75 + UDL · 0.4 + q_{fk}^* · 0.4)	*1.5*	0	1.2 or 1.35
4	1.35	1.0	1.35 · (TS · 0.75 + UDL · 0.4 + q_{fk}^* · 0.4)	0	*1.5 loaded bridge*	1.2 or 1.35
5	1.35	1.0	0	0	*1.5 unloaded bridge*	1.2 or 1.35

[a] For linear elastic analysis.
[b] For nonlinear analysis.

Table 5.7 Basic combinations according to Equation 5.11a at ULS for road bridges with all cross sections of class 1 or 2 (STR/GEO)

	G	Q	W
No.	*Permanent neglecting effects of creep*	*Traffic loads*	*Wind*
1	1.35	*gr1a · 1.35*	1.5 · 0.6 or 1.5 · F_W^*
2	1.35	*(gr1b,2,3,4,5) · 1.35*	0
3	1.35	1.35 · (TS · 0.75 + UDL · 0.4 + q_{fk}^* · 0.4)	*1.5 loaded bridge*
4	1.35	0	*1.5 unloaded bridge*

For **railway bridges,** the following rules apply:

- Wind is not combined with gr13, gr16, gr17, gr23, gr26, and gr27 and individual load model SW/2.
- Snow is generally not considered.
- Aerodynamic actions of rail traffic should be combined with wind actions. Each action should be taken into account individually as a leading variable action.

The most usual ULS basic combinations for railway bridges are summarized in Table 5.8 and for bridges where all cross sections are of class 1 or 2 in Table 5.9. The box of the leading action is shown in bold face. Combinations with traffic loading as the leading action are critical for the superstructure, while those with temperature and wind may be critical for piers and bearings. Variable actions are obviously considered if they act unfavorably.

Table 5.8 Basic combinations according to Equation 5.11a at ULS for railway bridges generally (STR/GEO)

No.	$G+C_{sec}$ Permanent and secondary effects of creep	S_{sec} Shrinkage secondary effects	Q Traffic loads	T Temperature	W Wind	G_{set} Differential settlements
1	1.35	1.0	*(gr11, gr12, gr14, gr15, gr21, gr22, gr24, gr31) · 1.45*	1.5 · 0.6	1.5 · 0.75 or 1.5 · F_W^{**}	1.2[a] or 1.35[b]
2	1.35	1.0	*(gr13, gr23) · 1.45*	1.5 · 0.6	0	1.2 or 1.35
3	1.35	1.0	*(gr16, gr17, gr26, gr27, SW/2) · 1.20*	1.5 · 0.6	0	1.2 or 1.35
4	1.35	1.0	(gr11, gr12, gr14, gr15, gr21, gr22, gr24, gr31) · 1.45 · 0.8	*1.5*	1.5 · 0.75 or 1.5 · F_W^{**}	1.2 or 1.35
5	1.35	1.0	(gr13, gr23) · 1.45 · 0.8	*1.5*	0	1.2 or 1.35
6	1.35	1.0	(gr16, gr17, gr26, gr27, SW/2) · 1.20 · 0.8	*1.5*	0	1.2 or 1.35
7	1.35	1.0	(gr11, gr12, gr14, gr15, gr21, gr22, gr24, gr31) · 1.45 · 0.8	1.5 · 0.6	*1.5 loaded bridge*	1.2 or 1.35
8	1.35	1.0	0	1.5 · 0.6	*1.5 unloaded bridge*	1.2 or 1.35

[a] For linear elastic analysis.
[b] For nonlinear analysis.

Table 5.9 Basic combinations according to Equation 5.11a at ULS for railway bridges with all cross sections of class 1 or 2 (STR/GEO)

No.	G Permanent neglecting effects of creep	Q Traffic loads	W Wind
1	1.35	*(gr11, gr12, gr14, gr15, gr21, gr22, gr24, gr31) · 1.45*	1.5 · 0.75 or 1.5 · F_W^{**}
2	1.35	*(gr13, gr23) 1.45*	0
3	1.35	*(gr16, gr17, gr26, gr27, SW/2) · 1.20*	0
4	1.35	(gr11, gr12, gr14, gr15, gr21, gr22, gr24, gr31) · 1.45 · 0.8	*1.5 loaded bridge*
5	1.35	0	*1.5 unloaded bridge*

5.3.5 Accidental combinations

Accidental combinations include all permanent actions, prestress only due to tendons but not due to imposed support deformations, and the accidental action itself. Traffic loads need not be included, unless otherwise specified. In this case, traffic loads are reduced by the relevant combination factors ψ_1 of Table 5.13. However, for accidental combinations at con-

Table 5.10 Accidental combinations

During service

$$\sum_{j\geq 1} G_{kj,sup} + \sum_{j\geq 1} G_{kj,inf} + P_k + A_d + (\psi_{1,1} \text{ or } \psi_{2,1}) \cdot Q_{k1} + \sum_{i>1} \psi_{2,i} \cdot Q_{ki} \quad (5.16)$$

At construction stages

$$\sum_{j\geq 1} G_{kj,sup} + \sum_{j\geq 1} G_{kj,inf} + P_k + A_d + \psi_2 \cdot Q_{c,k} \quad (5.17)$$

Notes:

For the combination factors ψ_1 and ψ_2, see Table 5.13.

The main variable action may be taken into account in combination (5.16) with its frequent or its quasipermanent value according to the recommendation of the National Annex.

In combination (5.17), $Q_{c,k}$ is the characteristic value of construction loads as defined in EN 1991-1-6.

struction stages, the full construction loads are accounted for. The accidental combinations during service and at construction stages are given in Table 5.10.

5.3.6 Seismic combinations

Seismic combinations include all permanent actions, prestress only due to tendons but not due to imposed support deformations, and earthquake as the leading action. The only variable loads are traffic loads and need to be considered only for major bridges of important

Table 5.11 Seismic combinations of actions

Road and railway bridges of importance categories I and II—footbridges

$$\sum_{j\geq 1} G_{kj} + P_k + \gamma_I \cdot A_{Ed} \quad (5.18)$$

Road bridges of importance category III

$$\sum_{j\geq 1} G_{kj} + P_k + \gamma_I \cdot A_{Ed} + 0.2 \cdot Q_{k1} \quad (5.19)$$

Railway bridges of importance category III

$$\sum_{j\geq 1} G_{kj} + P_k + \gamma_I \cdot A_{Ed} + 0.3 \cdot Q_{k1} \quad (5.20)$$

Q_{k1} are traffic loads where only one track is loaded and SW/2 is neglected.

Construction stages

$$\sum_{j\geq 1} G_{kj} + P_k + \gamma_I \cdot A_{Ed} + Q_{c,k} \quad (5.21)$$

$Q_{c,k}$ are construction loads.

Note: γ_I is importance factor from Table 4.16.

category III (see Table 4.16) with relevant combination factors. However, the masses of traffic loading to be considered in the seismic situation as given in Table 4.18 do not depend on the importance category but on the type of traffic (normal or severe). This means that, for example, for a motorway with severe traffic but of importance category II in the sense of Table 4.16, the **masses** to determine the structural periods and the basic shear include 20% of the mass due to traffic but the **loads** due to traffic are not considered in the seismic combination; see Equation 5.18.

At construction stages, the full construction loads are taken into account. However, the value of the importance factor γ_I is reduced due to the smaller return period of the seismic event during construction. The seismic combinations are given in Table 5.11.

5.4 SERVICEABILITY LIMIT STATE (SLS)

Design formats: The design format for SLSs may be written as

$$E_d \leq C_d \tag{5.22}$$

where

E_d is the design value of the effects of actions, like stresses, deflections, frequencies, and crack widths in concrete

C_d is the corresponding limiting design value

Three combinations of actions associated to different verifications are considered at SLS. Table 5.12 gives a summary of them with the related verifications. It should be noted that in the SLSs, both primary and secondary effects of creep and shrinkage of concrete are to be taken into account. In addition, snow loads are not examined for SLSs.

Table 5.13 gives the combination factors ψ_1 and ψ_2 for road-, rail-, and footway bridges.

The resulting combinations at SLSs for road- and railway bridges under consideration of the aforementioned ψ-values are summarized in Tables 5.14 through 5.18.

Table 5.12 Combinations of actions at SLS

Combinations		*Application*
Characteristic		
$\sum_{j\geq 1} G_{kj} + P_k + Q_{k1} + \sum_{i>1} \psi_{0i} \cdot Q_{ki}$	(5.23)	Stress limitation for structural steel and reinforcement Stress limitation for concrete for exposure classes XS, XF and XD (see Remark 5.5) Resonance control for railway bridges Cracked regions for placing min. reinforcement (see Figure 7.19)
Frequent		
$\sum_{j\geq 1} G_{kj} + P_k + \psi_{1,1} \cdot Q_{k1} + \sum_{i>1} \psi_{2i} \cdot Q_{ki}$	(5.24)	Web breathing Deformations and vibrations for road bridges
Quasi-permanent		
$\sum_{j\geq 1} G_{kj} + P_k + \sum_{i\geq 1} \psi_{2i} \cdot Q_{ki}$	(5.25)	Stress limitation for concrete Limitation of crack width for non-prestressed members

Table 5.13 Combination factors ψ_1 and ψ_2

Road bridges		ψ_1		ψ_2
Traffic loads	gr1a			
	TS	0.75		0
	UDL	0.4		0
	Foot, cycle tracks	0.4		0
	gr1b	0.75		0
	gr2	0		0
	gr3	0		0
	gr4	0.75		0
	gr5	0		0
Wind	Service condition	0.2		0
	Construction stage	—		0
	F_W^*	—		—
Temperature	T	0.6		0.5
Construction loads	Q_c	—		1.0
Snow (during construction)	S_c	—		—
Railway bridges		ψ_1		ψ_2
Traffic loads Single components	LM71 and SW/0	0.8	If one track is loaded	0[a]
		0.7	If two tracks are loaded	
		0.6	If three or more tracks are loaded	
	SW/2	1.0		0
	Traction or breaking	The same values as for vertical loads if traffic loads are considered as a single leading action		
	Centrifugal forces			
	Nosing force	0.8		0
	Loads in footways	0.5		0
Traffic loads	gr11 to gr17	0.8		0
Groups of loads	gr21 to gr27	0.7		0
	gr31	0.6		0
Wind	Service condition	0.5		0
	F_W^{**}	0		0
Thermal	T	0.6		0.5
Construction loads	Q_c	—		1.0
Footbridges		ψ_1		ψ_2
Traffic loads	gr1	0.4		0
Wind	W	0.2		0
Thermal	T	0.6		0.5
Construction loads	Q_c	—		1.0

[a] If deformations are taken into account for persistent and transient design situations, then $\psi_2 = 1.0$ for rail traffic loads.

Table 5.14 Characteristic SLS combinations for road bridges

	G	*S*	*Q*	*T*	*W*
No.	*Permanent and creep (primary and secondary effects)*	*Shrinkage (primary and secondary effects)*	*Traffic loads*	*Temperature*	*Wind*
1	1.0	1.0	gr1a · 1.0	0.6	0.6[a]
2	1.0	1.0	(gr1b,2,3,4,5) · 1.0	0.6	0
3	1.0	1.0	(TS · 0.75 + UDL · 0.4 + q_{fk}^{*} 0.4)	1.0	0
4	1.0	1.0	(TS · 0.75 + UDL · 0.4 + q_{fk}^{*} · 0.4)	0	1.0
5	1.0	1.0	0	0	1.0

[a] For wind forces F_W^{*}, it is $\psi_0 = 1.0$.

Table 5.15 Characteristic SLS combinations for railway bridges

	G	*S*	*Q*	*T*	*W*
No.	*Permanent and creep (primary and secondary effects)*	*Shrinkage (primary and secondary effects)*	*Traffic loads Individual components or groups, except gr1b and gr2 to 5*	*Temperature*	*Wind*
1	1.0	1.0	1	0.6	0.75[a]
2	1.0	1.0	1.0 · 0.8	1	0.75
3	1.0	1.0	1.0 · 0.8	0.6	1

[a] For wind forces F_W^{**}, it is $\psi_0 = 1.0$.

Table 5.16 Frequent SLS combinations for road bridges

	G	*S*	*Q*	*T*	*W*
No.	*Permanent and creep (primary and secondary effects)*	*Shrinkage (primary and secondary effects)*	*Traffic loads*	*Temperature*	*Wind*
1	1.0	1.0	(TS · 0.75 + UDL · 0.4 + q_{fk}^{*} · 0.4)	0.5	0
2	1.0	1.0	gr4 · 0.75	0.5	0
3	1.0	1.0	0	0.6	0
4	1.0	1.0	0	0	0.2

Table 5.17 Frequent SLS combinations for railway bridges

	G	S	Q	T	W
No.	*Permanent and creep (primary and secondary effects)*	*Shrinkage (primary and secondary effects)*	*Traffic loads Individual components or groups, except gr13 and gr2 to5*	*Temperature*	*Wind*
1	1.0	1.0	(LM1, SW/0) · 0.8 to 0.6 depending on the number of the loaded tracks + Footways · 0.5	0.5	0
2	1.0	1.0	Nosing force · 0.8	0.5	0
3	1.0	1.0	(gr11 to gr 17) · 0.8	0.5	0
4	1.0	1.0	(gr21 to gr27) · 0.7	0.5	0
5	1.0	1.0	gr31 · 0.6	0.5	0
6	1.0	1.0	0	0.6	0
7	1.0	1.0	0	0	0.50

Table 5.18 Quasi-permanent SLS combinations for road- and railway bridges and footbridges

	G	S	Q	T
No.	*Permanent and creep (primary and secondary effects)*	*Shrinkage (primary and secondary effects)*	*Traffic loads*	*Temperature*
1	1.0	1.0	0[a]	0.5

[a] If deformations are taken into account for persistent and transient design situations, then $\psi_2 = 1.0$ for rail traffic actions.

EXAMPLE 5.1

A two-span continuous composite bridge with 25 m + 25 m spans is casted at one stage. The internal forces at midspan and at internal support are given in Table 5.19. At sagging moment areas, the composite cross sections are class 1 and at the hogging ones class 3. The design forces will be calculated for ULS and SLS verifications (persistent situation).

It will be shown in Chapters 6, 7, 9 and 10 that internal forces and cross-sectional properties of composite members are time dependent due to the rheological behavior of concrete (creep and shrinkage). This means that the effects of several actions need to be calculated and combined at different times (short- and long-term effects). For the majority of the composite bridges, design combinations are calculated before ($t_0 = 0$ days) and after ($t_\infty = 30{,}000$ days) the development of creep and shrinkage. However, in part of the literature, t_0 is the time of traffic opening. In such a case, creep and shrinkage are active, and they are taken into account both for t_0 and t_∞. In this book, the first approach is adopted; short term means without creep and shrinkage.

In this bridge, the cross section at the internal support is class 3, and therefore, differential temperatures $\Delta T_{M,cool}$ and $\Delta T_{M,heat}$ are considered. For this reason, secondary effects due to creep and shrinkage are also taken into account. One can observe the high value of the hogging moment at mid-support due to shrinkage (−1901.12 kN-m). These effects are discussed in detail in Chapters 6 and 7.

Table 5.19 Bending moments and shear forces for a two-span composite bridge

	Span (class 1)	*Internal support (class 3)*	
Actions	*maxM_{Ed} (kN-m)*	*V_{Ed} (kN)*	*minM_{Ed} (kN-m)*
G1—steel elements	155.24	62.47	−346.32
G2—concrete slab	1029.95	386.51	−2099.53
G3—surfacing	93.99	26.98	−148.23
G4—concrete caps on footways	558.57	188.86	−916.97
G5—parapets, cornices	31.30	10.66	−51.60
$\mathbf{S_{sec}}$—secondary effects due to shrinkage	−668.52	80.38	−1901.12
$\mathbf{C_{sec}}$—secondary effects due to creep	−324.89	45.87	−650.45
$\Delta \mathbf{T_{M,heat}}$—positive differential temperature	373.58	−41.56	1034.79
$\Delta \mathbf{T_{M,cool}}$—negative differential temperature	−448.30	49.88	−1241.74
UDL—uniformly distributed load	1306.79	329.74	−1620.31
TS—tandem system (two axles)	921.74	154.26	−426.28
$\mathbf{q_{fk}^*}$—crowd loading at footpaths	155.24	62.47	−346.32
D(t=0)—imposed deformation of 10 mm at internal support (short-term effects)	119.68	81.97	321.4
D(t=∞)—imposed deformation of 10 mm at internal support (long-term effects)	123.07	62.36	245.44
$\mathbf{G_{set}}$**(t=0)**—uneven settlement (short-term effects)	−45.08	29.37	−99.76
$\mathbf{G_{set}}$**(t=∞)**—uneven settlement (long-term effects)	−38.09	12.34	−76.90

Basic Combination at ULS according to Equation 5.11a for Resistance Verifications

Span

Short-term design

Leading variable gr1a:

$$\max M_{Ed,0} = 1.35 \cdot \sum_{1}^{5} G_i + 1.35 \cdot (TS + UDL + q_{fk}^*) + 1.5 \cdot 0.6 \cdot \Delta T_{M,heat} + 1.0 \cdot D(t=0)$$

$$= 6197.21\,\text{kN-m}$$

Leading variable $\Delta T_{M,heat}$:

$$\max M_{Ed,0} = 1.35 \cdot \sum_{1}^{5} G_i + 1.5 \cdot \Delta T_{M,heat} + 1.0 \cdot D(t=0)$$

$$+1.35 \cdot (0.75 \cdot TS + 0.4 \cdot UDL + 0.4 \cdot q_{fk}^*) = 4871.93\,\text{kN-m}$$

Long-term design

Leading variable gr1a:

$$\max M_{Ed,\infty} = 1.35 \cdot \sum_{1}^{5} G_i + 1.35 \cdot C_{sec} + 1.0 \cdot S_{sec} + 1.35 \cdot (TS + UDL + q_{fk}^*)$$

$$+1.5 \cdot 0.6 \cdot \Delta T_{M,heat} + 1.0 \cdot D(t=\infty) = 5093.48\,\text{kN-m}$$

Leading variable $\Delta T_{M,heat}$:

$$\max M_{Ed,\infty} = 1.35 \cdot \sum_{1}^{5} G_i + 1.35 \cdot C_{sec} + 1.0 \cdot S_{sec} + 1.5 \cdot \Delta T_{M,heat} + 1.0 \cdot D(t = \infty)$$

$$+1.35 \cdot (0.75 \cdot TS + 0.4 \cdot UDL + 0.4 \cdot q_{fk}^{*}) = 4004.22 \text{ kN-m}$$

Shear forces at midspan are low, and therefore, they are not presented.

Note: Secondary effects C_{sec} due to self-weights were multiplied with 1.35 despite having a favorable effect. EN 1990:2002 imposes that all actions originating from the same source should be multiplied with the same safety factor.

Internal support

Short-term design

Leading variable gr1a:

$$V_{Ed,0} = 1.35 \cdot \sum_{1}^{5} G_i + 1.20 \cdot G_{set}(t = 0) + 1.35 \cdot (TS + UDL + q_{fk}^{*})$$

$$+1.5 \cdot 0.6 \cdot \Delta T_{M,cool} + 1.0 \cdot D(t = 0) = 1811.74 \text{ kN}$$

$$\min M_{Ed,0} = 1.35 \cdot \sum_{1}^{5} G_i + 1.20 \cdot G_{set}(t=0) + 1.35 \cdot (TS + UDL + q_{fk}^{*})$$

$$+1.5 \cdot 0.6 \cdot \Delta T_{M,cool} + 1.0 \cdot D(t = 0) = -8955.9 \text{ kN-m}$$

Leading variable $\Delta T_{M,cool}$:

$$V_{Ed,0} = 1.35 \cdot \sum_{1}^{5} G_i + 1.20 \cdot G_{set}(t = 0) + 1.5 \cdot \Delta T_{M,cool} + 1.0 \cdot D\,(t = 0)$$

$$+1.35 \cdot (0.75 \cdot TS + 0.4 \cdot UDL + 0.4 \cdot q_{fk}^{*}) = 1471.91 \text{ kN}$$

$$\min M_{Ed,0} = 1.35 \cdot \sum_{1}^{5} G_i + 1.20 \cdot G_{set}(t = 0) + 1.5 \cdot \Delta T_{M,cool} + 1.0 \cdot D\,(t = 0)$$

$$+1.35 \cdot (0.75 \cdot TS + 0.4 \cdot UDL + 0.4 \cdot q_{fk}^{*}) = -7964.09 \text{ kN-m}$$

Long-term design

Leading variable gr1a:

$$V_{Ed,\infty} = 1.35 \cdot \sum_{1}^{5} G_i + 1.20 \cdot G_{set}(t = \infty) + 1.35 \cdot (TS + UDL + q_{fk}^{*}) + 1.5 \cdot 0.6 \cdot \Delta T_{M,cool}$$

$$+1.0 \times D(t = \infty) + 1.0 \times S_{sec} + 1.35 \times C_{sec} = 1914 \text{ kN}$$

$$\min M_{Ed,\infty} = 1.35 \cdot \sum_{1}^{5} G_i + 1.20 \cdot G_{set}(t = \infty) + 1.35 \cdot (TS + UDL + q_{fk}^{*}) + 1.5 \cdot 0.6 \cdot \Delta T_{M,cool}$$

$$+1.0 \cdot D(t = \infty) + 1.0 \times S_{sec} + 1.35 \times C_{sec} = -11783.6 \text{ kN-m}$$

Leading variable $\Delta T_{M,cool}$:

$$V_{Ed,\infty} = 1.35 \cdot \sum_{1}^{5} G_i + 1.20 \cdot G_{set}(t = \infty) + 1.5 \cdot \Delta T_{M,cool} + 1.0 \cdot D\,(t = \infty)$$

$$+ 1.35 \cdot (0.75 \cdot TS + 0.4 \cdot DL + 0.4 \cdot q_{fk}^{*}) + 1.0 \cdot S_{sec} + 1.35 \cdot C_{sec} = 1574.17\ \text{kN}$$

$$\min M_{Ed,\infty} = 1.35 \cdot \sum_{1}^{5} G_i + 1.20 \cdot G_{set}(t = \infty) + 1.5 \cdot \Delta T_{M,cool} + 1.0 \cdot D\,(t = \infty)$$

$$+ 1.35 \cdot (0.75 \cdot TS + 0.4 \cdot UDL + 0.4 \cdot q_{fk}^{*})$$

$$+ 1.0 \cdot S_{sec} + 1.35 \cdot C_{sec} = -10791.8\ \text{kN-m}$$

Due to secondary effects, the bending moment at support was increased by 31.5%. By imposing a deformation at internal support greater than 10 mm, this unfavorable increase could be lower.

In the following, only the more critical combinations for SLS are demonstrated.

Characteristic Combination at SLS according to Equation 5.23 for Stress Limitations of Structural Steel and Reinforcement

Span

Short-term design

Leading variable gr1a:

$$\max M_{Ed,ser,0} = \sum_{1}^{5} G_i + gr1a + 0.6 \cdot \Delta T_{M,heat} + 1.0 \cdot D(t = 0) = 4596.65\ \text{kN-m}$$

Internal support

Long-term design

Leading variable gr1a:

$$V_{Ed,ser,\infty} = \sum_{1}^{5} G_i + G_{set}(t = \infty) + gr1a + 0.6 \cdot \Delta T_{M,cool} + 1.0 \cdot D(t = \infty)$$

$$+ 1.0 \cdot C_{sec} + 1.0 \cdot S_{sec} = 1452.83\ \text{kN}$$

$$\min M_{Ed,ser,\infty} = \sum_{1}^{5} G_i + G_{set}(t = \infty) + gr1a + 0.6 \cdot \Delta T_{M,cool} + 1.0 \cdot D(t = \infty)$$

$$+ 1.0 \cdot C_{sec} + 1.0 \cdot S_{sec} = -9083.63\ \text{kN-m}$$

Quasi-permanent Combination at SLS according to Equation 5.25 for Stress Limitations of Concrete at Span

Span

Short-term design

$$\max M_{Ed,ser,0} = \sum_{1}^{5} G_i + G_{set}(t = 0) + 0.5 \cdot \Delta T_{M,heat} + 1.0 \cdot D(t = 0) = 2175.52\ \text{kN-m}$$

5.5 SAFETY FACTORS OF RESISTANCES γ_M

The design values for resistances are determined by provision of appropriate safety factors γ_M as described by Equation 5.8. These safety factors depend on the limit state, the type of material, and the failure mode under consideration. Table 5.20 gives a summary of the γ_M-factors.

The safety factors for connections are as follows:

Resistance of bolts, pins, welds, and plates in bending: $\gamma_{M2} = 1.25$
Slip resistance: $\gamma_{M3} = 1.15$
Resistance of joints in hollow sections: $\gamma_{M5} = 1.10$
Resistance of pins at SLS: $\gamma_{M6,ser} = 1.10$
Preload of high-strength bolts: $\gamma_{M7} = 1.10$

5.6 DURABILITY

Durability is a structural property that ensures compliance with the basic requirements of safety and serviceability throughout the intended design life. The basic factor affecting durability is corrosion of steel, which is highly influenced by the environmental conditions. Steel reinforcement and shear connectors are protected by concrete while structural steel by appropriate coating systems. Accordingly, the rules ensuring durability refer to the concrete cover of reinforcement and the protection of steel as a function of the environmental conditions.

5.6.1 Concrete cover

Reinforcement

EN 1992-1-1 [5.2] defines 18 exposure classes related to environmental conditions ranging from X0 (no risk of corrosion) to XA3 (highly aggressive chemical environment). The relevant exposure classes for the concrete elements of composite bridges are as follows:

- XC3 for bridges away from the sea. Moreover, this is the recommended exposure class for deck slabs protected by waterproofing according to EN 1992-2.
- XS1 for bridges near the coast.
- XF4 bridges exposed to deicing agents.

Table 5.20 Safety factors for resistances γ_M except fatigue

Limit state	*Combination*	*Structural steel (resistance of members)*		*Reinforcing steel*	*Concrete*	*Shear connectors*
				γ_s	γ_c	γ_v
ULS	Basic and seismic	Yield	$\gamma_{M0} = 1.0$	1.15	1.5	
		Stability	$\gamma_{M1} = 1.1$			
		Fracture	$\gamma_{M2} = 1.25$			1.25
SLS	Accidental		1.0	1.0	1.3	1.0
			1.0	1.0	1.0	1.25

The nominal concrete cover may be determined from

$$c_{nom} = c_{min} + \Delta c_{dev} \tag{5.26}$$

where

c_{min} is the minimum value of concrete cover

Δc_{dev} is an extra safety element to allow for deviations with a recommended value of 10 mm. For precast units where accurate measurements of cover are feasible, a reduced value may be applied

Minimum concrete cover, c_{min}, is provided for the following:

- The safe transmission of bond stresses between steel and concrete
- The protection of the steel against corrosion (durability)

The minimum value is the largest of three values:

$$c_{min} = \max(c_{min,b}; c_{min,dur}; 10\,\text{mm}) \tag{5.27}$$

where

$c_{min,b}$ is the minimum value that assures good bond with the concrete. It is equal to the diameter of the reinforcement for aggregate sizes up to 32 mm

$c_{min,dur}$ is the minimum value due to environmental conditions and is given in Table 5.21 as function of the exposure and the structural class

The structural class for bridges with a service life of 100 years is 6 in general. The structural class may be reduced by 1 if the strength class of concrete is ≥35/45 for exposure class XC3 and ≥40/50 for exposure class XS1. If the position of reinforcement is not affected by the construction process then the construction class can be reduced by 1. Finally, if special quality control measures are taken, the structural class may be further reduced by 1.

For exposure class XF4, special attention must be given to the concrete composition, so that class XS1 applies.

Table 5.21 Minimum concrete cover $c_{min,dur}$ [mm]

Type of steel	*Reinforcement steel*		*Prestressing steel*	
exposure class	*XC3*	*XS1*	*XC3*	*XS1*
Structural class 3	20	30	30	40
Structural class 4	25	35	35	45
Structural class 5	30	40	40	50

REMARK 5.5

It should be noted that XS1 and XF4 exposure classes affect the stress limit of concrete under the characteristic combination; the maximum compressive stress should be lower than $k_1 \cdot f_{ck}$ with recommended $k_1 = 0.6$; see EN 1992-2. Therefore, the choice of a durable concrete may in some cases lead to a concrete quality higher than what is required for structural design.

Shear connectors

The concrete cover over the shear connectors should be as specified earlier for reinforcing steel less 5 mm but ≥20 mm.

EXAMPLE 5.2

The concrete cover for the longitudinal-slab reinforcement of a composite bridge deck shall be determined. The bridge is situated 2 km from the sea. The strength class of concrete is C 35/45 and the diameter of transverse reinforcement 20 mm.

The distance from the sea is considered low, so that an exposure class XS1 is selected.

The calculation of the structural class is based on Table 4.3N, of EN 1992-1-1.

The initial structural class is 6. Since the concrete quality is not greater than C40/50, the structural class is not reduced. Moreover, the position of reinforcement is not affected by construction process, and therefore, the construction class is reduced to 5. A further reduction is applied due to special quality control; the final structural class is 4.

Table 5.21: $c_{min,dur} = 35$ mm

$c_{min,b} = 20$ mm

Equation 5.27: $c_{min} = \max\{20; 35; 10\} = 35$ mm

Equation 5.26: $c_{nom} = 35 + 10 = 45$ mm

The required concrete cover of the longitudinal reinforcement is accordingly 45 + 20 = 65 mm.

The concrete cover for the shear connectors is then equal to 65 − 5 = 60 mm, which is larger than 20 mm.

5.6.2 Structural steel

The steel of steel girders must be protected by appropriate coating systems in accordance with the environmental conditions. Steel that is in contact with concrete is protected by the concrete and need no coating. However, steel surfaces that are in contact with concrete including the underside of baseplates shall be coated for a minimum of the first 50 mm of the embedded length. Uncoated parts of the surface must be blasted or wire brushed to remove loose mill scale and cleaned to remove dust, oil, and grease.

When precast slabs are used that rest directly on the girder, the top flange of the steel girder must be protected over the entire width as the rest of the steelwork except the top coating provided after erection.

REFERENCES

[5.1] EN 1990: 2002, CEN (European Committee for Standardization). Eurocode—Basis of structural design.
[5.2] EN1992-1-1: 2004, CEN (European Committee for Standardization). Design of concrete structures—Part 1-1: General rules and rules for buildings.
[5.3] EN 1992-2: 2004, CEN (European Committee for Standardization). Design of concrete structures. Part 2: Concrete bridges: Design and detailing rules.
[5.4] EN 1993-2: 2003, CEN (European Committee for Standardization). Design of steel structures—Part 2: Steel Bridges.
[5.5] Calgaro, J.-A., Tschumi, M., Gulvanessian, H.: *Designer's Guide to Eurodode 1: Actions on Bridges*, Thomas Telford, London, U.K., 2010.

Chapter 6

Structural materials

6.1 CONCRETE

6.1.1 Strength classes

6.1.1.1 Normal concrete

Normal concrete is characterized by the letter C followed by two figures that express the characteristic (5%) cylinder strength f_{ck} and the cube strength $f_{ck,cube}$ at 28 days. Accordingly, C35/45 is a concrete with $f_{ck}=35$ MPa and $f_{ck,cube}=45$ MPa. In composite bridges designed by Eurocode 4-2, concrete of strength classes between C20/25 and C60/75 should be used. However, the most usual strength class of the concrete slab is C35/45. The properties of normal concrete are given in Table 6.1, where the aforementioned strength class is shown in italic face [6.3].

Other properties of concrete are as follows:

Specific weight	$\gamma_c = 25$ kN/m^3
Specific weight of wet concrete	$\gamma_{c,wet} = 26$ kN/m^3
Poisson ratio for uncracked concrete	$\nu_c = 0.2$
Poisson ratio for cracked concrete	$\nu_c = 0$
Coefficient of thermal expansion	$a_t = 10 \cdot 10^{-6}$ [per °C]

The design value of the compressive stress of concrete is defined as

$$f_{cd} = \alpha_{cc} \cdot \frac{f_{ck}}{\gamma_c} \tag{6.1}$$

where

f_{ck} is the characteristic value of the compressive stress (see Table 6.1)
γ_c is the relevant safety factor (see Table 5.20)
α_{cc} is a reduction factor that takes into account the long-term effects on the compressive strength

The recommended values for α_{cc} are 0.85 for unconfined concrete and 1.0 for confined one. Different values may be found in the National Annex.

For the capacity design of steel–concrete composite cross sections, the stress–strain relations of Figure 6.1 can be used. The parabola–rectangle diagram describes the "exact"

Table 6.1 Properties of concrete (units in MPa)

Grade	f_{ck}	f_{cm}	f_{ctm}	$f_{ctk,0.05}$	$f_{ctk,0.95}$	E_{cm} ($\times 10^3$)
C20/25	20	28	2.2	1.5	2.9	30
C25/30	25	33	2.6	1.8	3.3	31
C30/37	30	38	2.9	2.0	3.8	33
C35/45	*35*	*43*	*3.2*	2.2	4.2	*34*
C40/50	40	48	3.5	2.5	4.6	35
C45/55	45	53	3.8	2.7	4.9	36
C50/60	50	58	4.1	2.9	5.3	37
C55/67	55	63	4.2	3.0	5.5	38
C60/75	60	68	4.4	3.1	5.7	39

Notes:

f_{cm} is the mean compressive strength at the age of 28 days.
f_{ctm} is the mean tensile strength.
$f_{ctk0.05}$ = 5% fractile of tensile strength.
$f_{ctk0.95}$ = 95% fractile of tensile strength.
E_{cm} is the mean value of modulus of elasticity.

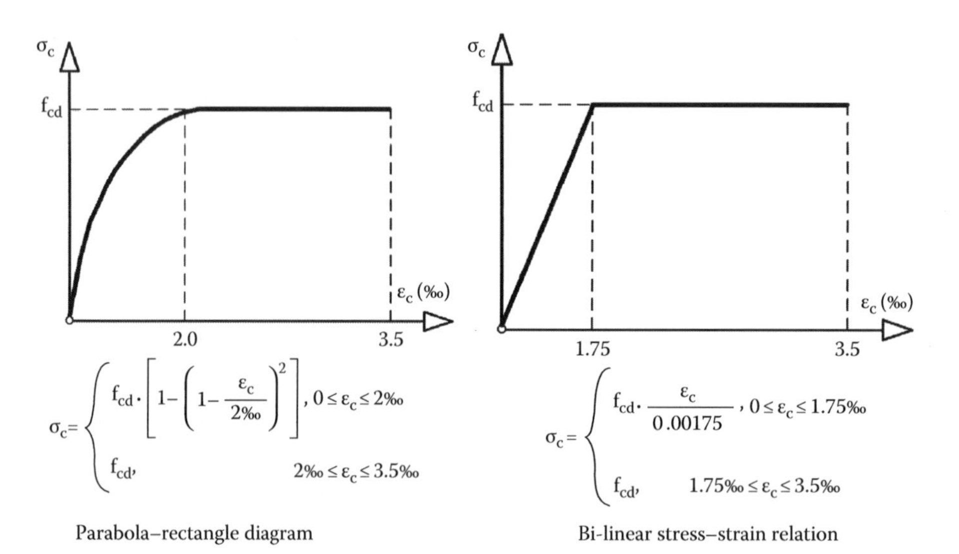

Figure 6.1 Stress–strain relations for the capacity design of cross sections for C20/25 till C50/60 (concrete under compression).

behavior of the compressed concrete but it obviously makes the calculations more onerous. The bilinear diagram offers a more simplified approach.

The stress–strain diagrams are used when a nonlinear computation of the bending resistance is mandatory. This can be the case for composite girders with excessive compression.

6.1.1.2 Lightweight concrete

Lightweight concrete is denoted as LC followed by the two figures of cylinder strength and the cube strength. The strength and stiffness of lightweight concrete depend on its density. In Table 6.2, one can find the density classification according to [6.18].

Table 6.3 provides the main mechanical characteristics for lightweight concrete.

Table 6.2 Design densities ρ (kg/m³) for lightweight aggregate concrete for density classes according to EN 206-1

Density class	*1.0*	*1.2*	*1.4*	*1.6*	*1.8*	*2.0*
max ρ	1000	1200	1400	1600	1800	2000
min ρ	801	1001	1201	1401	1601	1801
Plain concrete	1050	1250	1450	1650	1850	2050
Reinforced concrete	1150	1350	1550	1750	1950	2150

Table 6.3 Properties of lightweight concrete (units in MPa)

Quality	f_{lck}	f_{lcm}	f_{lctm}	$f_{lctk,0.05}$	$f_{lctk,0.95}$	E_{lcm} ($\times 10^3$)
LC20/25	20	28				
LC25/30	25	33				
LC30/37	30	38				
LC35/45	35	43				
LC40/50	40	48	$= f_{ctm} \cdot n_l$	$= f_{ctk,0.05} \cdot n_l$	$= f_{ctk,0.95} \cdot n_l$	$= E_{cm} \cdot n_E$
LC45/55	45	53				
LC50/60	50	58				
LC55/67	55	63				
LC60/75	60	68				

Notes:

$n_E = (\rho/2200)^2$, coefficient for the determination of the secant modulus E_{lcm}.
$n_l = 0.40 + 0.6 \cdot (\rho/2200)$, coefficient for the determination of the tensile strength.
ρ is the upper limit of the density for the relevant class according to Table 6.2.
For f_{ctm}, $f_{ctk,0.05}$, and E_{cm}, see Table 6.1 for normal concrete.

It should be noted that the density ρ used in the design calculations should be verified on site by appropriate measurements. More information on lightweight concrete are provided in Section 11 of EN 1992-1-1 [6.3].

6.1.2 Time-dependent deformations due to creep

6.1.2.1 General

Concrete is subjected to time-dependent deformations. If a compression stress σ_{c0} is applied, concrete is subjected initially to elastic deformations $\varepsilon_{c0} = \sigma_{c0}/E_{cm}$ that are followed by time-dependent deformations so that the final deformations at the time t are equal to

$$\varepsilon_c(t, t_0) = \varepsilon_{c0} \cdot \left[1 + \varphi(t, t_0)\right] \tag{6.2}$$

The time-dependent deformations under compression stresses are called *creep deformations*, while the coefficient $\varphi(t, t_0)$ *creep coefficient*. Creep is mainly due to movement of unbounded water molecules from regions of low to regions of high pressure and depends on

a. The age of concrete at time of load application (t_0)
   The time of load application is essential for creep, since the movability of the water molecules decreases as the time increases due to the fact that more water is bounded to cement.

b. The density of concrete
   The movability of the water molecules decreases with increasing density and consequently the creep deformations.
c. The humidity of the environment
   An increase in humidity reduces the pressure difference between the inside of the concrete and the outside environment resulting in a reduction of the creep deformations.
d. The temperature of the environment
   In environments with high temperatures, the unbounded part of water evaporates faster. Therefore, creep deformations are larger than in cold environments.
e. The dimensions of the concrete element
   Thick concrete elements sustain greater part of the humidity; creep is reduced.

6.1.2.1.1 Creep due to permanent loads (P)

The time-dependent behavior of concrete is influenced also by the **type of loading**. Figure 6.2 depicts the strain development in a concrete cylinder subjected to a constant compressive force N at time t_0; as already mentioned, t_0 is the age of concrete at loading. One can observe that the additional deformations due to creep $\varepsilon_{c\varphi}$ can be 2–3 times greater than the elastic ones. Indeed, from Equation 6.2, the creep strains are calculated:

$$\varepsilon_{c0} + \varepsilon_{c\varphi} = \varepsilon_{c0} \cdot [1 + \varphi(t, t_0)] \Rightarrow \varepsilon_{c\varphi} = \varphi(t, t_0) \cdot \varepsilon_{c0} \tag{6.3}$$

Taken into account that the creep coefficient is in most cases between 2 and 3, one can easily understand the importance of considering creep in calculations of stresses and deformations. Creep due to permanent loads, for example, self weights, will be notated with the letter P, referring to *permanent*.

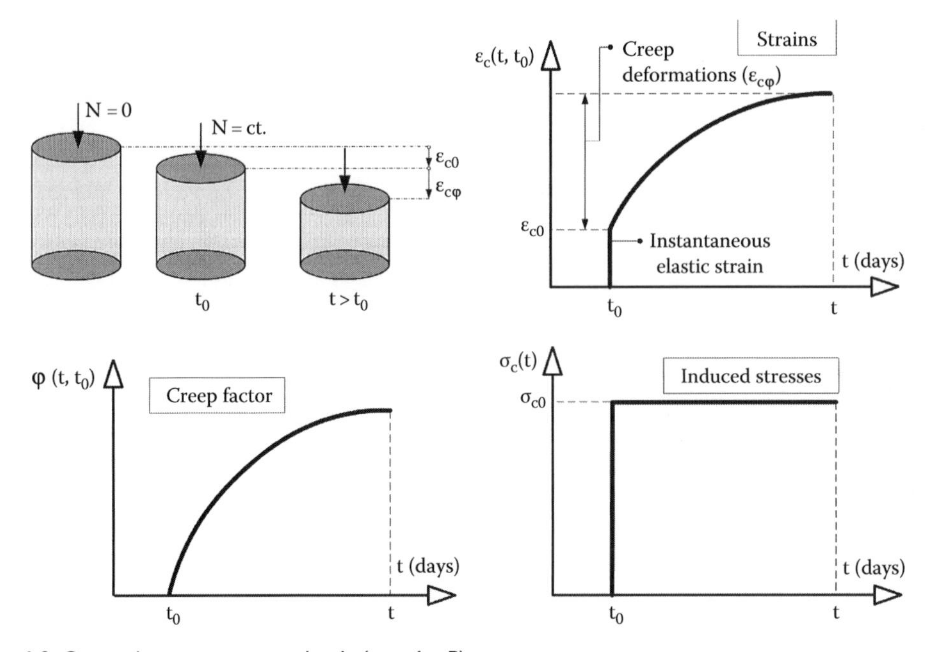

Figure 6.2 Creep due to permanent loads (type L = P).

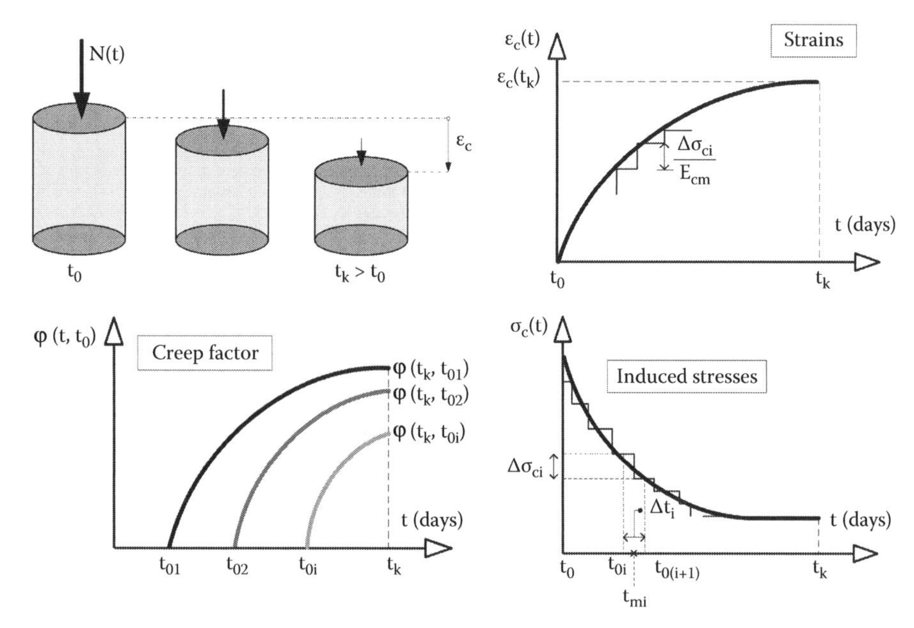

Figure 6.3 Creep due to temporary permanent loads (type L = PT).

6.1.2.1.2 Creep due to temporarily permanent loads (PT)

In bridges, there is also an important type of loading that refers to permanent loads whose magnitude changes constantly with time. They are not described as permanent because of their time-dependent magnitude; therefore, they are called *temporarily permanent* actions and are notated with PT. These may be stresses due to secondary internal forces that are developed in statically indeterminate structures or due to longitudinal prestressing. An example is shown in Figure 6.3. The axial force N(t) diminishes with time and the stress variation $\sigma_c(t)$ is changed into a series of stress increments $\Delta\sigma_i$. Time is also divided into n intervals Δt_i, in the middle of which $\Delta\sigma_{ci}$ is introduced. Obviously, the total strain at time t_k due to $\Delta\sigma_{ci}$ is estimated with Equation 6.2 and thus is equal to $(\Delta\sigma_{ci}/E_{cm})\cdot[1+\varphi(t_k,t_{mi})]$.

According to the *theory of viscoelasticity*, the superposition of strains caused by stress increments is allowed. The final strain at time t_k is represented by the following summation:

$$\varepsilon_c(t_k)=\sum_{i=1}^{k}\Delta\sigma_{ci}\cdot\left[\frac{1}{E_{cm}}+\frac{\varphi(t_k,t_{mi})}{E_{cm}}\right]=\sum_{i=1}^{k}\Delta\sigma_{ci}\cdot J(t_k,t_{mi}) \tag{6.4}$$

$J(t, t_0)$ is known as *creep function* and expresses the total strain at time t due to a unit stress introduced at age t_0.

EN 1994-2 [6.7] offers a more simplified approach than the aforementioned incremental procedure. This is achieved with the *creep factor* ψ_L explained subsequently.

The aforementioned numerical method is also used for considering the effects of creep during construction stages, for example, segmental construction.

6.1.2.1.3 Creep due to imposed deformations (D)

Imposed deformations in bridges may be due to support settlements. These displacements may be sudden or time varying. Sudden support settlements are introduced to the

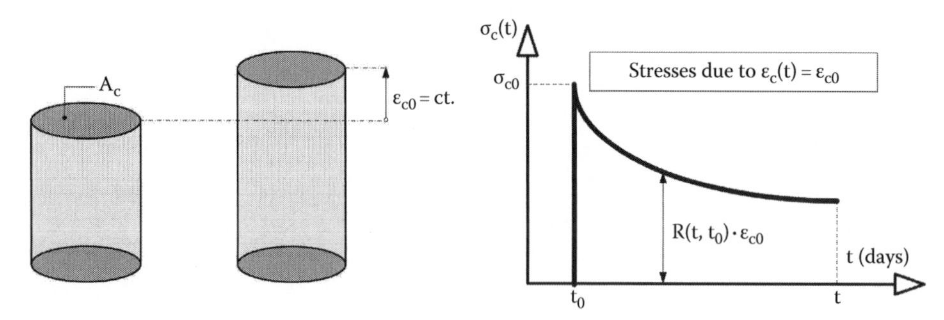

Figure 6.4 Relaxation of concrete (type D).

intermediate supports of continuous composite bridges to limit cracking. This is an alternative solution to longitudinal prestressing. Time-varying support movements may arise due to soil consolidation. In both cases, strains are introduced in concrete that are developed under the influence of creep.

In Figure 6.4, the reduction of stresses after an instantaneous induced strain is illustrated. The resulting stresses decrease gradually due to creep.

Time-dependent stresses can be easily calculated through Equation 6.4, which is rewritten as follows:

$$\varepsilon_c(t_k) = \Delta\sigma_{ck} \cdot J(t_k, t_{mk}) + \sum_{i=1}^{k-1} \Delta\sigma_{ci} \cdot J(t_k, t_{mi}) \tag{6.5}$$

The strain $\varepsilon_c(t_k)$ is constant and equal to the initially induced strain ε_{c0}. The stress increment $\Delta\sigma_{ck}$ is therefore given by Equation 6.6:

$$\varepsilon_c(t_k) = \varepsilon_{c0} \Rightarrow \Delta\sigma_{ck} = \frac{\varepsilon_{c0} - \sum_{i=1}^{k-1} \Delta\sigma_{ci} \cdot J(t_k, t_{mi})}{J(t_k, t_{mk})} \tag{6.6}$$

The concrete stress at time $t_n > t_k$ is calculated after the summation of the stress increments $\Delta\sigma_{ck}$:

$$\sigma_c(t_n) = \sum_{k=1}^{n} \Delta\sigma_{ck} = \sum_{k=1}^{n} \frac{\varepsilon_{c0} - \sum_{i=1}^{k-1} \Delta\sigma_{ci} \cdot J(t_k, t_{mi})}{J(t_k, t_{mk})} \tag{6.7}$$

Therefore, the stress evolution due to creep after a successive application of Equation 6.7 can be calculated.

The phenomenon of stress reduction due to imposed strains is also known as *relaxation* of concrete. In EN 1992-1-1, a *relaxation function* $R(t, t_0)$ is found. Multiplying $R(t, t_0)$ with the imposed strain ε_{c0} offers the concrete stress at time t; thus,

$$\sigma_c(t_n) = R(t_n, t_0) \cdot \varepsilon_{c0} \tag{6.8}$$

Comparing Equations 6.7 and 6.8 enables the numerical evaluation of the relaxation function R.

The influence of relaxation of concrete on continuous composite bridges is discussed in Chapter 7. Creep due to imposed deformations is notated with D.

6.1.2.1.4 *The creep coefficient $\varphi(t, t_0)$*

According to EN 1992-1-1, Annex B [6.3], the creep coefficient in the time interval (t, t_0) may be determined from

$$\varphi(t, t_0) = \beta_c(t, t_0) \cdot \varphi_0 \tag{6.9}$$

where

$\beta_c(t, t_0)$ is the coefficient to describe the development of creep with time after loading
φ_0 is the notional creep coefficient
t_0 is the age of concrete at loading in [days], valid for normal hardening cements (class N). This age is corrected according to Equation 6.17 for other types of cement.

The concrete of the slab is connected with the girders by the shear connectors and is stressed in the longitudinal direction from the first day of casting if the girders are unsupported. Therefore, the age t_0 varies between individual segments for bridges cast in several stages. According to EN 1994-2, one mean value of time t_0 may be used for all segments. The final age of concrete may be taken as $t = t_\infty = 30.000$ days that corresponds to a service bridge life of 100 years.

The two factors of Equation 6.9 may be determined according to the following relations.

6.1.2.1.4.1 NOTIONAL CREEP COEFFICIENT

$$\varphi_0 = \varphi_{RH} \cdot \beta(f_{cm}) \cdot \beta(t_0) \tag{6.10}$$

where

φ_{RH} is a factor to account for the influence of relative humidity
$\beta(f_{cm})$ is a factor to account for the influence of concrete strength
$\beta(t_0)$ is a factor to account for the effect of concrete age at loading

The aforementioned three factors may be determined from the following expressions:

$$\varphi_{RH} = 1 + \frac{1 - RH/100}{0.1 \cdot \sqrt[3]{h_0}} \quad \text{for } f_{cm} \leq 35 \text{ MPa} \tag{6.11a}$$

$$\varphi_{RH} = \left(1 + \frac{1 - RH/100}{0.1 \cdot \sqrt[3]{h_0}} \cdot \alpha_1\right) \cdot \alpha_2 \quad \text{for } f_{cm} > 35 \text{ MPa} \tag{6.11b}$$

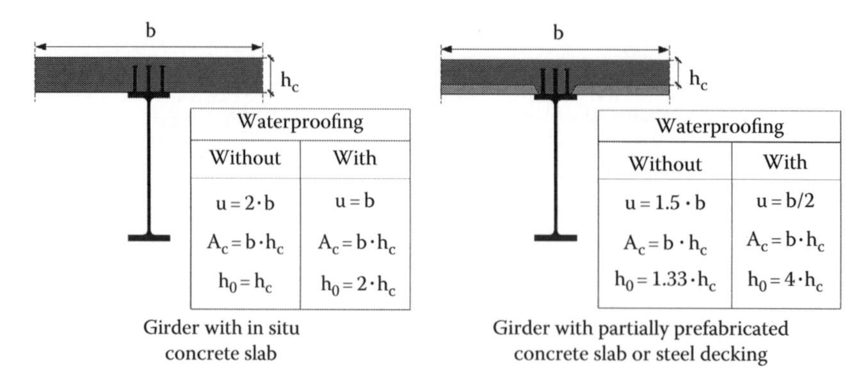

Figure 6.5 Notional size h_0 for deck slabs of constant thickness with and without waterproofing layer.

$$\beta(f_{cm}) = \frac{16.8}{\sqrt{f_{cm}}} \tag{6.12}$$

$$\beta(t_0) = \frac{1}{0.1 + t_0^{0.2}} \tag{6.13}$$

where

RH is the relative humidity of the ambient environment in %

$h_0 = \dfrac{2 \cdot A_c}{u}$ is the notional size of the member in [mm]

A_c is the cross-sectional area

u is the perimeter of the member in contact with the atmosphere

f_{cm} is the mean compressive strength of concrete from Table 6.1 in [MPa]

Figure 6.5 shows recommended values of h_0 for in situ and partially prefabricated slabs or slabs with steel decking. It is obvious that for bridges with prefabricated planks and water proofing layers, lower values of $\varphi(t, t_0)$ are expected; *reduced creep.*

It has to be mentioned that b is the geometrical and **not** the effective width.

6.1.2.1.4.2 COEFFICIENT FOR THE DEVELOPMENT OF CREEP WITH TIME

$$\beta_c(t, t_0) = \left(\frac{t - t_0}{\beta_H + t - t_0}\right)^{0.3} \tag{6.14}$$

$$\beta_H = 1.5 \cdot \left[1 + (0.012 \cdot RH)^{18}\right] \cdot h_0 + 250 \le 1500 \quad \text{for } f_{cm} \le 35 \text{ MPa} \tag{6.15a}$$

$$\beta_H = 1.5 \cdot \left[1 + (0.012 \cdot RH)^{18}\right] \cdot h_0 + 250 \cdot \alpha_3 \le 1500 \cdot \alpha_3 \quad \text{for } f_{cm} > 35 \text{ MPa} \tag{6.15b}$$

$$\alpha_1 = \left(\frac{35}{f_{cm}}\right)^{0.7}, \quad \alpha_2 = \left(\frac{35}{f_{cm}}\right)^{0.2}, \quad \alpha_3 = \left(\frac{35}{f_{cm}}\right)^{0.5}, \quad f_{cm} \text{ in MPa} \tag{6.16}$$

The aforementioned equations are valid for normal cements of type 32.5R and 42.5. For other types of cements, time t_0 has to be modified to

$$t_{0,eff} = t_{0,T} \cdot \left(\frac{9}{2 + t_0^{1.2}} + 1 \right)^{\alpha} \geq 0.5 \quad (6.17)$$

where

$\alpha = -1$ for slowly hardening cements (class S)
$\alpha = 1$ for rapid hardening cements (class R)
$t_{0,T}$ is the modified value of t_0 according to the following expression:

$$t_{0,T} = \sum_{i=1}^{n} \exp\left[13.65 - \frac{4000}{273 + T(\Delta t_i)} \right] \cdot \Delta t_i \quad (6.18)$$

where

Δt_i is the time interval in which temperature $T(\Delta t_i)$ prevails
$T(\Delta t_i)$ is the temperature in °C during ΔT_i

Expression 6.17 takes into account the effects of the temperature variation on the maturity of concrete within the range of 0°C–80°C.

For lightweight concrete, the creep coefficient $\varphi(t, t_0)$ is equal to the value of normal density concrete multiplied by a factor $(\rho/2200)^2$.

REMARK 6.1

The creep factor $\varphi(t, t_0)$ calculated earlier is valid for compressive stresses not exceeding $0.45 \cdot f_{ck}(t_0)$, where $f_{ck}(t_0)$ is the compressive strength of concrete at age t_0.

- For $3 < t < 28$ days,

$$f_{ck}(t_0) = \beta_{cc}(t_0) \cdot f_{cm} - 8 \text{ (MPa)} \quad (R6.1)$$

where

$$\beta_{cc}(t_0) = \exp\left[s \cdot \left(1 - \sqrt{\frac{28}{t_0}} \right) \right], \text{ is a time function with} \quad (R6.2)$$

$s = 0.38$, for cements 32.5N
$s = 0.25$, for cements 32.5R and 42.5N
$s = 0.20$, for cements 42.5R, 52.5N, and 52.5R

- For $t \geq 28$ days, $f_{ck}(t_0) = f_{cm}$
When $\sigma_c \leq 0.45 \cdot f_{ck}(t_0)$, the creep strains ε_{cc} are proportional to $\varphi(t, t_0)$ and Equation 6.2 is valid. The strains' time-dependent development is known as *linear creep*.

Nonlinear creep occurs for higher stresses and the creep factor $\varphi(t, t_0)$ is replaced by $\varphi_k(t, t_0)$ as follows:

$$\varphi_k(\infty, t_0) = \varphi(\infty, t_0) \cdot \exp\left[1.5 \cdot \left(\frac{\sigma_c(t_0)}{f_{cm}(t_0)} - 0.45\right)\right] \quad \text{(R6.3)}$$

Nonlinear creep **should be avoided** and concrete stresses should be limited accordingly.

6.1.2.1.5 Creep in composite girders

In composite bridges with one flange composite, the effects of creep are taken into account by introduction of the modular ratio of concrete, which is the ratio between the modulus of elasticity of steel to the modulus of elasticity of concrete. For short-term loading, for example, for traffic loads, wind, temperature, and earthquake, the modular ratio is given by (Table 6.4)

$$n_0 = \frac{E_a}{E_{cm}} \quad (6.19)$$

where
- E_a is the modulus of elasticity of steel
- E_{cm} is the mean value of modulus of elasticity of concrete

For long-term loading, the modular ratio may be determined from

$$n_L = n_0 \cdot [1 + \psi_L \cdot \varphi(t, t_0)] \quad (6.20)$$

where ψ_L is the creep multiplier depending on the type of loading in accordance with Table 6.5.

Table 6.5 shows that creep is affected mostly by imposed deformations and less by secondary effects (see Section 7.4.3). Consequently, prestress by imposed deformations becomes less effective in time (see Figure 7.48), since a large part of it is lost due to relaxation of concrete.

Table 6.4 Short-term values for n_0 for different qualities of normal concrete

	C20	*C25*	*C30*	*C35*	*C40*	*C45*	*C50*	*C55*	*C60*
n_0	7.00	6.77	6.56	6.18	6.00	5.83	5.68	5.53	5.38

Table 6.5 Creep multiplier ψ_L

Type of action	*Description*	ψ_L
Permanent (P)	Permanent actions invariant in time (e.g., self weights)	1.10
Secondary effects (PT)	Secondary effects of creep and shrinkage	0.55
Imposed deformations (D)	Prestressing by imposed deformation (e.g., support settlement)	1.50

Source: EN 1994-2, Design of composite steel and concrete structures, Part 2: Rules for bridges, 2005.

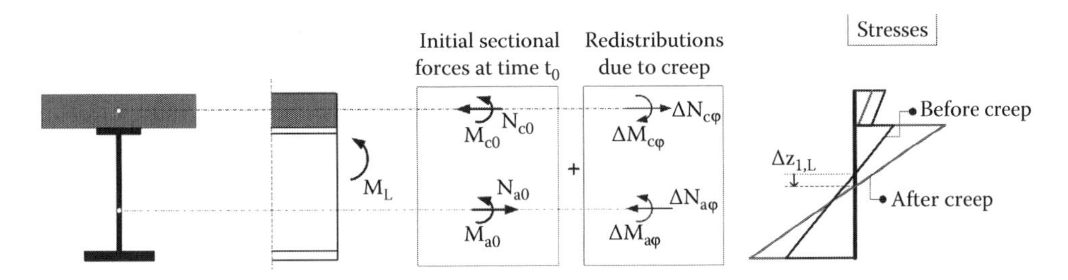

Figure 6.6 The effect of creep on stresses of a composite girder.

A clearer picture on the effects of creep on composite girders at sagging moment areas is given in Figure 6.6. One can see that the deck slab is at time t_0 under compression. This is the time that loading M_L is imposed. Due to creep, time-dependent cross-sectional forces are developed that redistribute tension from concrete to steel; thus, concrete stresses become lower and steel stresses higher. Indeed, the cross-sectional properties of the concrete slab are reduced through the long-term modular ratio n_L of Equation 6.20. In contrast, structural steel keeps its stiffness and as a result, time-dependent redistributions arise.

The redistributions are not only time- but loading dependent as well; the magnitude of the redistribution and the final results depend on the type of loading. If the bending moment M_L is constant and due to permanent loads, then creep of type P will be developed (see Figure 6.2). If the bending moment acting on the girder is due to an imposed deformation, then in this case, creep of type D should be considered (see Figure 6.4).

In statically determinate structures, additional displacements and rotations due to creep are developed freely and they are called *primary effects*. In the case of indeterminate structures, primary deformations are restrained so that additional internal forces arise; these forces are known as *secondary effects*. The secondary internal forces are developed parallel with creep. This means that they are permanent actions whose magnitude constantly changes. Hence, secondary effects are associated with creep of type PT (see Figure 6.3).

Figure 6.7 illustrates the effects of two different types of creep on a two-span bridge. The support settlement δ acts on a continuous system with a short-term stiffness determined by the modular ratio n_0 (Figure 6.7a). Due to creep, the modular ratio n_0 is changed into n_D and the system's stiffness is reduced. Obviously, the bending moment diagram

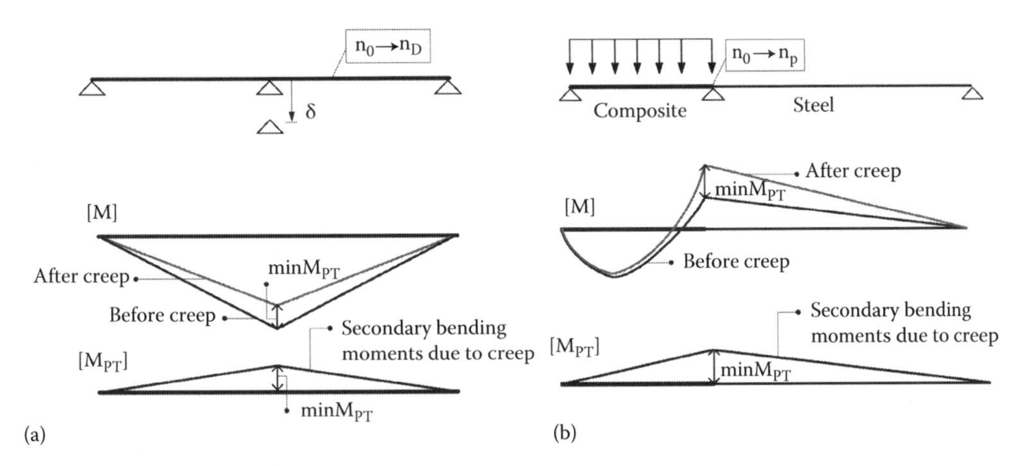

Figure 6.7 The effect of creep on continuous systems. (a) Due to support settlement and (b) due to partial concrete loading.

changes as well. In the second case, one part of the bridge is composite and the other pure steel (Figure 6.7b). Due to the rheological behavior of the composite part ($n_0 \rightarrow n_P$), secondary bending moments (M_{PT}) are developed. Secondary internal forces are developed in the cracked areas of composite bridges; in mixed bridges, for example, steel–concrete composite; in prestressed composite bridges; and during construction stages.

Computation methods for the previously described primary and secondary effects are found in [6.21] and [6.23]. A more detailed discussion on the effects of creep and shrinkage on composite bridges is found in Section 7.4.

EXAMPLE 6.1

A bridge deck is casted in five segments at different times as shown in Figure 6.8. The modular ratios n_L shall be determined

a. For short-time loading
b. At the time of traffic opening t = 100 days
c. At time $t_\infty = 3 \cdot 10^4$ days

The slab thickness is 250 mm.

Waterproofing is provided at the top of the concrete slab.

The relative humidity is supposed to be RH = 80%.

The concrete grade is C35/45.

Conservatively, it is assumed that the age of concrete t_0 at time of loading L is for all segments equal to 1 day. The duration of casting is for all segments 1 day.

From Table 6.1, $f_{cm} = 43$ MPa and $E_{cm} = 34$ GPa

Equation 6.16: $\alpha_1 = \left(\frac{35}{43}\right)^{0.7} = 0.866, \quad \alpha_2 = \left(\frac{35}{43}\right)^{0.2} = 0.960, \quad \alpha_3 = \left(\frac{35}{43}\right)^{0.5} = 0.902$

Equation 6.12: $\beta(f_{cm}) = \frac{16.8}{\sqrt{43}} = 2.562$

The notional size of the slab, for simplification not extracting the width of the girder top flange that is in contact with the slab, is equal to $h_0 = 2 \cdot 250 = 500$ mm (see Figure 6.5).

Equation 6.11b: $\varphi_{RH} = \left(1 + \frac{1 - 80/100}{0.1 \cdot \sqrt[3]{500}} \cdot 0.866\right) \cdot 0.960 = 1.169$

Equation 6.15b: $\beta_H = 1.5 \cdot \left[1 + (0.012 \cdot 80)^{18}\right] \cdot 500 + 250 \cdot 0.902 = 1335.203 \leq 1500 \cdot 0.902 = 1353$

Modular ratio for short-time loading (Table 6.4) $n_0 = 6.18$

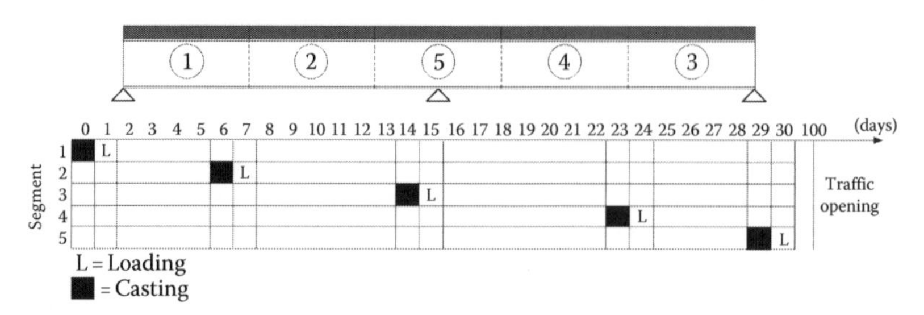

Figure 6.8 Numbering of segments and time of loading L.

Segment 1

Modular ratios at traffic opening ($t = 100$ days, $t_0 = 1$ day)

Equation 6.13: $\beta(t_0 = 1) = \dfrac{1}{0.1 + 1^{0.2}} = 0.909$

Equation 6.10: $\varphi_0 = 1.169 \cdot 2.562 \cdot 0.909 = 2.722$

Equation 6.14: $\beta_c(100,1) = \left(\dfrac{100-1}{1335.203 + 100 - 1}\right)^{0.3} = 0.448$

Equation 6.9: $\varphi(100, 1) = 0.448 \cdot 2.722 = 1.219$

The long-term modular ratios are calculated through Equation 6.20:

$$n_P = 6.18 \cdot (1 + 1.1 \cdot 1.219) = 14.47$$

$$n_{PT} = 6.18 \cdot (1 + 0.55 \cdot 1.219) = 10.32$$

$$n_D = 6.18 \cdot (1 + 1.5 \cdot 1.219) = 17.48$$

Modular ratios at $t_\infty = 30 \cdot 10^4$ days

The repetition of the calculations gives the following modular ratios:

$$n_P = 24.45, \quad n_{PT} = 15.32, \quad n_D = 31.09$$

Figure 6.9 shows the development of the modular ratios due to creep for segment 1. It may be seen that creep is well developed after 3000 days ($\varphi_t/\varphi_\infty > 90\%$) and is largely influenced by the type of loading, whether permanent loading, secondary effects, or imposed deformations.

Segments 2–5

Modular ratios at traffic opening ($t = (100$-casting time$)$, $t_0 = 1$ day)

At traffic opening, the calculations must be repeated for a time at loading $t_0 = 1$ day for all segments, but for different times t for each segment corresponding to the time difference between the 100 days and the time at concrete casting. The results are summarized in Table 6.6.

Modular ratios at $t_\infty = 30 \cdot 10^4$—casting time

The modular ratios for $t = t_\infty$ casting time may not be repeated, since the influence of casting time on the creep factor is very low. Therefore, for all segments at t_∞,

$$n_P = 24.45, \quad n_{PT} = 15.32, \quad n_D = 31.09$$

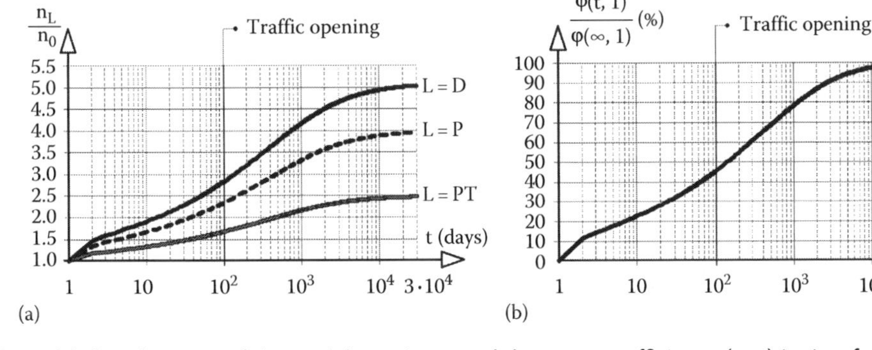

Figure 6.9 Development of the modular ratios n_L and the creep coefficient $\varphi(t, t_0)$ in time for segment 1.

Table 6.6 Modular ratios at traffic opening

Segment	t_0 *(days)*	*t (days)*	$\varphi(t, t_0)$	n_P	n_{PT}	n_D
2	1	93	1.20	14.34	10.26	17.30
3	1	85	1.17	14.13	10.16	17.03
4	1	76	1.13	13.86	10.02	16.66
5	1	70	1.10	13.66	9.92	16.38

REMARK 6.2

According to EN 1994-2 [6.7], for bridges cast in several stages, one mean value of t_0 may be used for all segments. In Example 6.1, this was conservatively assumed equal to 1 day. However, both the creep factor $\varphi(t, t_0)$ and the modular ratios n_L are sensitive to deviations in t_0. This is illustrated in Table 6.7.

The Code proposes this simplification because its implication on the overall cross-sectional properties and later on the distribution of internal forces and moments is assumed to be relatively small. However, this is not always true and more conservative values for t_0 need to be considered.

For box-girder bridges, torsional redistributions may also occur. The effect of creep may be taken into account by integrating in the torsional stiffness of the composite cross section the following modular ratio:

$$n_{LG} = n_{0G} \cdot [1 + \psi_L \cdot \varphi(t, t_0)] \tag{6.21}$$

where

n_{0G} is the short-term torsional modular ratio equal to G_a/G_c; given in Table 6.8
G_a is the elastic shear modulus of structural steel
G_c is the elastic shear modulus of concrete
ψ_L is the creep multiplier in Table 6.5

For cross sections with double composite action, the modular ratios described earlier should not be used. The effects of creep can only be estimated by applying the incremental method based on Equation 6.4.

Table 6.7 Sensitivity of creep factor and long-term modular ratios against t_0 (RH = 80%, h_0 = 500 mm, C35/45)

t_0 *(days)*	*1*	*7*	*14*	*21*	*28*	*56*
$\varphi(30 \cdot 10^4, t_0)$	2.69	1.88	1.65	1.53	1.44	1.26
n_P/n_0	3.96	3.07	2.82	2.68	2.58	2.39
n_{PT}/n_0	2.48	2.03	1.91	1.84	1.79	1.69
n_D/n_0	5.04	3.82	3.48	3.30	3.16	2.89

Table 6.8 Short-term values for n_{0G} for different qualities of normal concrete

If concrete is considered as **uncracked** ($\nu_c = 0.2$)									
	C20	**C25**	**C30**	**C35**	**C40**	**C45**	**C50**	**C55**	**C60**
n_{0G}	6.46	6.25	6.06	5.70	5.54	5.38	5.24	5.10	4.97
If concrete is considered as **cracked** ($\nu_c = 0$)									
	C20	**C25**	**C30**	**C35**	**C40**	**C45**	**C50**	**C55**	**C60**
n_{0G}	5.38	5.21	5.05	4.75	4.62	4.49	4.37	4.25	4.14

Note: EN 1992-1-1 gives Poisson's ratio v_c as 0.2 and zero, depending on whether the concrete is uncracked or cracked. This obviously has influence on the value of the short-term ratio n_{0G}.

6.1.3 Time-dependent deformations due to shrinkage

6.1.3.1 General

Time-dependent deformations develop also if the concrete is not under stress. These deformations are due to the gradual migration of water through the hardened concrete, so that the crystals of the vehicular come closer together. This contraction of the concrete is denoted as **shrinkage.**

Shrinkage increases

a. With development of high temperatures and low humidity in the beginning of the hardening
b. With the increase of cement content or the reduction of the aggregates content, since only the cement jelly contracts
c. With increase in water
d. With inappropriate composition of the aggregates

Shrinkage has two components, the *drying shrinkage* and the *autogenous shrinkage*. The former develops slowly, the latter during hardening of concrete (see Figure 6.10). In composite

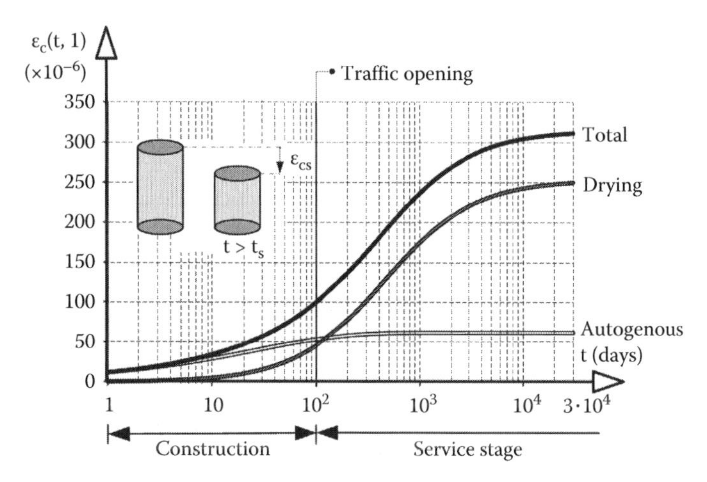

Figure 6.10 Development of shrinkage strains.

bridges, only *drying shrinkage* is considered directly for the calculation of stresses and deformations, while *autogenous shrinkage* should be taken into account together with concrete hydration (see Section 6.1.5) during concreting.

6.1.3.1.1 Drying shrinkage

The drying shrinkage strain at time t may be determined from

$$\varepsilon_{cd}(t) = \beta_{ds}(t, t_s) \cdot k_h \cdot \varepsilon_{cd,0} \tag{6.22a}$$

where

- t is the age of concrete at the considered time in days
- t_s is the age of concrete at the beginning of drying shrinkage that is taken as $t_s = 1$ day
- k_h is a coefficient depending on the notional thickness h_0 of the slab being taken equal to 1.0 for $h_0 = 100$ mm, 0.85 for $h_0 = 200$ mm, 0.75 for $h_0 = 300$ mm, and 0.7 for $h_0 \geq 500$ mm

The function describing the time-dependent development of the drying shrinkage is

$$\beta_{ds}(t, t_s) = \frac{t - t_s}{t - t_s + 0.04 \cdot \sqrt{h_0^3}} \tag{6.22b}$$

The basic drying shrinkage strain $\varepsilon_{cd,0}$ is given by the following expression:

$$\varepsilon_{cd,0} = 0.85 \cdot \left[(220 + 110 \cdot \alpha_{ds1}) \cdot \exp\left(-\alpha_{ds2} \cdot \frac{f_{cm}}{10} \right) \right] \cdot 10^{-6} \cdot \beta_{RH} \tag{6.22c}$$

where

- f_{cm} is the mean compressive strength of concrete (see Table 6.1)
- $a_{ds1} = 3$ for class S cements, 4 for class N cements, and 6 for class R cements
- $a_{ds2} = 0.13$ for class S cements, 0.12 for class N cements, and 0.11 for class R cements

$$\beta_{RH} = 1.55 \cdot \left[1 - \left(\frac{RH}{100\%} \right)^3 \right], \text{ RH relative humidity } (\%) \tag{6.22d}$$

6.1.3.1.2 Autogenous shrinkage

As noted before, the effects of *autogenous shrinkage* may be neglected. However, during construction stages, tension stresses may develop in some cross sections at serviceability limit state that result in cracking of concrete. The corresponding shrinkage strain is equal to

$$\varepsilon_{ca}(t) = \beta_{as}(t) \cdot \varepsilon_{ca}(\infty) \tag{6.23a}$$

where

$$\varepsilon_{ca}(\infty) = 2.5 \cdot (f_{ck} - 10) \cdot 10^{-6} \tag{6.23b}$$

$$\beta_{as}(t) = 1 - \exp(-0.2 \cdot t^{0.5}) \tag{6.23c}$$

6.1.3.1.3 Total shrinkage

As already mentioned, the total shrinkage strain during construction stages may be determined by adding both *drying* and *autogenous shrinkage* strains so that it is equal to

$$\varepsilon_{cs} = \varepsilon_{ca} + \varepsilon_{cd} \tag{6.24}$$

The creep multiplier for ψ_S, for calculating the long-term modular ratio for the primary effects due to shrinkage, is equal to 0.55. Moreover, shrinkage in cracked areas is of negligible magnitude and is not taken into account.

EXAMPLE 6.2

Determine the shrinkage strain for the bridge deck of Example 6.1 at the time of traffic opening t = 100 days. Waterproofing is provided at the top of the slab.

The basic drying shrinkage strain $\varepsilon_{cd,0}$ is calculated as follows:

Equation 6.22d: $\beta_{RH} = 1.55 \cdot \left[1 - \left(\frac{80\%}{100\%}\right)^3\right] = 0.756$

Cement class N: $\alpha_{ds1} = 4, \alpha_{ds2} = 0.12$

Equation 6.22c: $\varepsilon_{cd,0} = 0.85 \cdot \left[(220 + 110 \cdot 4) \cdot \exp\left(-0.12 \cdot \frac{43}{10}\right)\right] \cdot 10^{-6} \cdot 0.756 = 253.16 \cdot 10^{-6}$

The notional size of the slab, for simplification not extracting the width of the girder top flanges that are in contact with the slab, is equal to $h_0 = 2 \cdot 250 = 500$ mm. Therefore, $k_h = 0.7$.

The time that shrinkage starts is $t_s = 1$ day.

The coefficient for the development of shrinkage in time is equal to

Equation 6.22b: $\beta_{ds}(t, t_s) = \frac{100 - 1}{100 - 1 + 0.04 \cdot \sqrt{500^3}} = 0.18$

The drying shrinkage at 100 days is then

Equation 6.22a: $\varepsilon_{cd}(t) = 0.18 \cdot 0.7 \cdot 253.16 \cdot 10^{-6} = 31.9 \cdot 10^{-6}$

If autogenous shrinkage has to be taken into account, the relevant shrinkage strain is

Equation 6.23b: $\varepsilon_{ca}(\infty) = 2.5 \cdot (35 - 10) \cdot 10^{-6} = 62.5 \cdot 10^{-6}$

Equation 6.23c: $\beta_{as}(t) = 1 - \exp(-0.2 \cdot 100^{0.5}) = 0.86$

Equation 6.23a: $\varepsilon_{ca}(t) = 0.86 \cdot 62.5 \cdot 10^{-6} = 53.75 \cdot 10^{-6}$

The total shrinkage strain may be obtained from

Equation 6.24: $\varepsilon_{cs} = (31.9 + 53.75) \cdot 10^{-6} = 85.65 \cdot 10^{-6}$

Note: The development of shrinkage over time may be seen in Figure 6.10. It may be observed that the long-term values of autogenous shrinkage are considerably lower compared to those of drying shrinkage. In contrast, during the initial stages of evolution, autogenous shrinkage is the dominant one. For this reason, autogenous shrinkage should be taken into account during casting.

6.1.4 Time-dependent deformations due to time-dependent development of the modulus of elasticity of concrete

For reasons of simplicity, the elastic modulus of concrete was considered constant and equal to E_{cm}. For more detailed calculations, the modulus of elasticity of concrete should be calculated considering time as follows:

$$E_c(t) = \beta_E(t) \cdot E_{c,28} \tag{6.25a}$$

where $E_{c,28}$ is the modulus of elasticity at 28 days, approximately equal to E_{cm}. For a more detailed estimation, see [6.3].

$$\beta_E(t) = \sqrt{\beta_{cc}(t)} = \sqrt{\exp\left[s \cdot \left(1 - \sqrt{\frac{28}{t}}\right)\right]} \text{ is a time function with} \tag{6.25b}$$

$s = 0.38$, for cements 32.5N
$s = 0.25$, for cements 32.5R and 42.5N
$s = 0.20$, for cements 42.5R, 52.5N, and 52.5R

Figure 6.11 gives β_E values for different cement types. One can observe that for $t < 28$ days, the modulus of elasticity is highly dependent on time.

t	s = 0.38	s = 0.25	s = 0.2
1	0.44	0.58	0.65
3	0.68	0.77	0.81
7	0.83	0.88	0.90
14	0.93	0.95	0.96
28	1.00	1.00	1.00
50	1.05	1.03	1.02
75	1.08	1.05	1.04
100	1.09	1.06	1.05
200	1.13	1.08	1.06
365	1.15	1.09	1.07

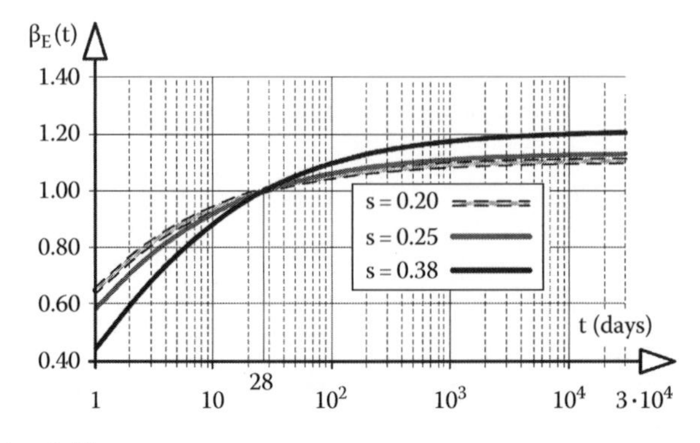

Figure 6.11 Time function $\beta_E(t)$.

Table 6.9 Recommended creep multipliers ψ_L

t_0 (days)	1	3	7	14	28	50	100	365
ψ_P	—	—	1.35	1.25	1.20	1.15	1.10	1.05
ψ_{PT}	—	—	0.75	0.80	0.80	0.85	0.90	0.95
ψ_S	0.15	0.30	0.40	0.45	0.50	0.55	0.60	0.70
ψ_D	—	—	1.60	1.40	1.25	1.20	1.10	1.05

Source: Hanswille, G., Zur Behandlung der Einflüsse aus dem Kriechen und Schwinden des Betons, Forschungsvorhaben: Eurocode 4 Teil 2—Verbundbrücken, Wuppertal, Germany, 1998.

Notes:
Long-term modular ratio: $n_L = n_0 \cdot \left[1 + \beta_{E0} \cdot \psi_L \cdot \varphi(t, t_0)\right]$.
Short-term modular ratio: $n_0 = E_a / E_c(t_0)$; for $E_c(t_0)$ and β_{E0} (see Equation 6.25).

The creep coefficient $\varphi(t, t_0)$ in Equation 6.9 is related to the elastic deformation at 28 days. For loadings imposed at different ages than 28 days, the creep coefficient is recommended to be multiplied with the factor $\beta_E(t = t_0)$. The creep function is thus given by

$$J(t, t_0) = \frac{1}{E_c(t_0)} + \beta_E(t_0) \cdot \varphi(t, t_0) \tag{6.26}$$

With the aforementioned modification of the creep coefficient, a more refined approach is achieved.

REMARK 6.3

There is an inconsistency between EN 1994-2 and EN 1992-1-1 concerning the use of the creep factor $\varphi(t, t_0)$. EN 1992-1-1 relates $\varphi(t, t_0)$ to an elastic deformation at 28 days. For this reason, when $t_0 \neq 28$ days, the creep factor should be replaced by the quantity $\beta_{E0} \cdot \varphi(t, t_0)$, as mentioned earlier. EN 1994-2 gives Equation 6.20 for the long-term modular ratios in which the influence of the time-dependent behavior of the elastic modulus of concrete is missing.

Equation 6.20 and the creep multipliers in EN 1994-2 are based on the creep functions of the old German Codes for prestressed concrete, DIN 4227 [6.1], and the dissertation of *Haensel* in [6.20]. Thus, it can be stated that the creep factor $\varphi(t, t_0)$ according to EN 1992-1-1 and the creep multipliers ψ_L of EN 1994-2 are incompatible.

Numerical investigations on new creep multipliers in [6.21] and [6.23] that are based on the creep function of EN 1992-1-1 led to improved values for ψ_L. With Table 6.9, the influence of the time-dependent behavior of E_c is considered. One can also observe that the creep multipliers depend on the age of concrete t_0 at the time of loading.

6.1.5 Time-dependent deformations due to hydration of cement

The chemical reaction of hydration of cement generates heat over the curing period. As a result, time-dependent temperature variations in concrete lead to induced stresses that in certain cases may cause cracking. The procedure is schematically described in Figure 6.12. After concrete casting, temperature gradually starts to rise; this usually takes place in a week. Due to restraints in deformations, compressive stresses arise. These stresses are generally not very high due to

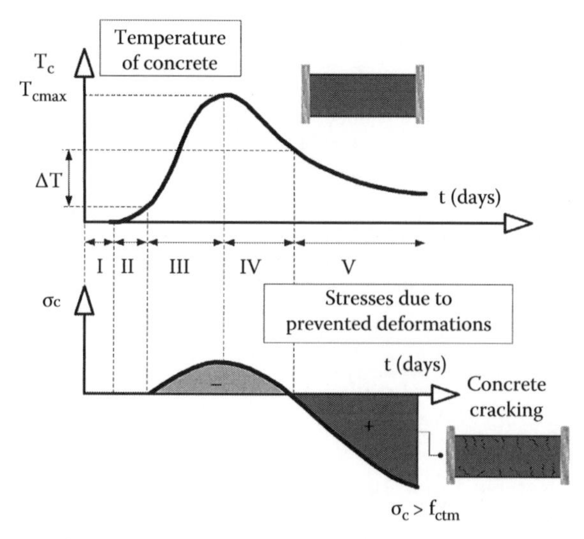

Figure 6.12 Induced stresses due to hydration of cement. (I) Concrete casting and compacting; (II) development of temperature; (III) due to increase of the modulus of elasticity of concrete, compressive stresses are developed; (IV) cooling phase leads to reduction of compressive stresses; (V) due to further cooling, tensile stresses are developed.

interaction with creep. After the temperature's peak, the cooling phase starts and temperature vanishes slowly over a much longer period. When temperature reaches its lowest value, high tensile stresses have been already developed. As mentioned earlier, cracking may be the result.

Due to the rigid connection of the deck slab with the upper flange of the steel cross section, excessive tensile stresses due to hydration of cement are developed at the bottom surface of the slab. This usually occurs when casting takes place in the summer period. It has to be mentioned that the phenomenon of cement's hydration can be strong enough to also influence the precambering of simply supported composite bridges erected by propped construction. According to EN 1994-2, for simplification, a temperature difference $\Delta T = 20°C$ between steel and concrete (concrete cooler) may be assumed.

6.1.6 Cracking of concrete

6.1.6.1 General

Tensile stresses higher than the concrete's tensile strength lead to cracking. Knowing the background of the cracking mechanism is of great importance and its main features are illustrated in Figure 6.13. A reinforced concrete section is subjected to tension and for an axial force N smaller than N_{cr1}, the section remains uncracked; N_{cr1} is the force producing the first crack. The noncracked state is also known as *state 1*. When the first crack is produced, several cracks (1st, 2nd,…, ith) follow corresponding to tensile forces $N_{cr1} < N_{cr2} < \cdots < N_{cri}$. This stage is the *crack formation stage* during which the axial stiffness of the section continuously drops. Thereafter, cracks stop to develop and the axial stiffness stabilizes. When the force N continues to rise the width, but not the number of cracks, increases too. For high tensile forces, yielding of the reinforcement results in an additional stiffness decrease.

6.1.6.1.1 Minimum reinforcement to avoid steel yielding

The axial force N_{cr1} occurs when concrete stresses reach a tensile strength f_{ct1} for which the corresponding strain is approximately equal to 0.001; this obviously takes place at the

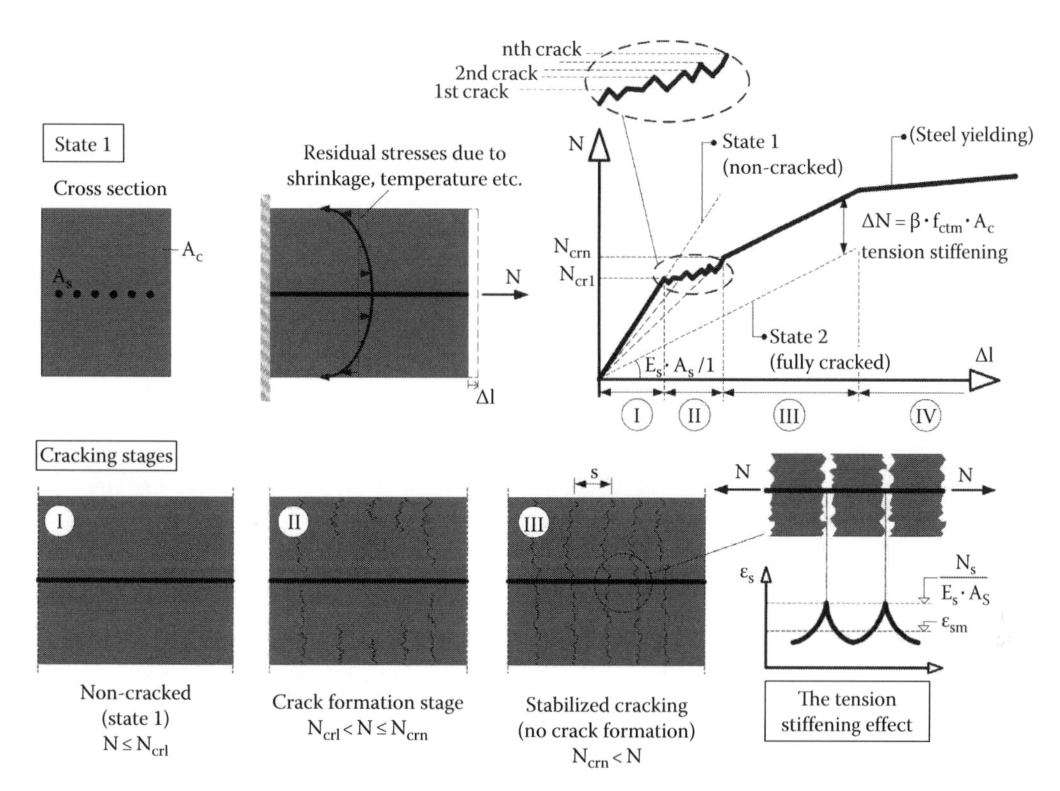

Figure 6.13 Reinforced concrete section subjected to tension.

weakest section. The second crack develops at the second weakest section for a tensile strength slightly greater than f_{ct2} and so on. In general, the difference among the values of the cracking forces N_{cri} is relatively small and employing a mean value N_{cr} is considered as acceptable.

Obviously, N_{cr} depends on the tensile strength of concrete f_{ctm}. As mentioned earlier, self-equilibrating stresses due to shrinkage, humidity, or temperature already exist in non-cracked members and considerably accelerate the cracking procedure. N_{cr} can therefore be written as follows:

$$N_{cr} = \left(f_{ctm} - \sigma_{eq}\right) \cdot A_{i1} \tag{6.27}$$

where

f_{ctm} is the strength of concrete in tension
σ_{eq} is a tensile stress expressing the effect of the self-equilibrating stresses
A_{i1} is the area of transformed section in state 1 at time t given by Equation 6.28:

$$A_{i1} = A_c + \frac{E_s \cdot A_s}{E_c(t)} \tag{6.28}$$

and

E_s is the elastic modulus of reinforcing steel
$E_c(t)$ is the elastic modulus of concrete at time of loading t
A_s is the reinforcement area
A_c is the concrete area

Immediately after cracking, the stress in the reinforcement rebar is $\sigma_s = N_{cr}/A_s$. Setting $\sigma_s = f_{sk}$ and $A_s = A_{s,min}$ gives the minimum reinforcement area required that ensures non-yielding at cracking:

$$A_{s,min} = A_c \cdot \frac{1}{\dfrac{f_{sk}}{f_{ctm} - \sigma_{eq}} - \dfrac{E_s}{E_c}} \tag{6.29a}$$

The difference between the tensile strength f_{ctm} and the mean value of the residual stress σ_{eq} is equal to $k \cdot k_c \cdot f_{ctm}$ where k and k_c are factors that consider the effects of the nonuniform self-equilibrating stresses; $k \le 1.0$ and $k_c \le 1.0$ (see [6.3]).

Equation 6.29a can be rewritten as follows:

$$A_{s,min} = A_c \cdot \frac{1}{\dfrac{f_{sk}}{k \cdot k_c \cdot f_{ctm}} - \dfrac{E_s}{E_c}} = \frac{A_c}{\dfrac{f_{sk}}{f_{ctm}} \cdot \left[\dfrac{1}{k \cdot k_c} - 1 + \left(1 - \dfrac{E_s}{E_c} \cdot \dfrac{f_{ctm}}{f_{sk}}\right)\right]} \tag{6.29b}$$

The term $1 - (E_s/E_c) \cdot (f_{ctm}/f_{sk})$ is approximately equal to unity. Thus, the minimum reinforcement is written in the simpler form:

$$A_{s,min} = \frac{f_{ctm} \cdot k \cdot k_c \cdot A_c}{f_{sk}} \tag{6.29c}$$

The aforementioned expression can be found in EN 1992-1-1 for reinforced concrete members in tension. A similar expression is also offered by EN 1994-2 for the concrete flange of composite girders at hogging areas (see Equation 10.6).

6.1.6.1.2 The effect of tension stiffening

After cracking is stabilized, the sections situated between two cracks are noncracked, state 1. The bond between concrete and the rebar restrains the elongation of the steel, and thus, a part of the tensile force in the reinforcement at a crack is transmitted to the concrete situated between cracks. Therefore, stresses and strains in the rebar vary longitudinally with the maximum values arising in cracks, in locations of state 2. In an opposite way, the axial stiffness of the concrete section in Figure 6.13 reaches its maximum value in the noncracked sections. The contribution of the noncracked concrete in the rigidity of the cracked members is referred in the Codes as *tension stiffening*. Ignoring the effect of tension stiffening in composite bridges results in the following:

- Normal stresses in the reinforcement at hogging areas are underestimated.
- Normal stresses in structural steel are overestimated.
- Deflections of filler-beam decks are overestimated.
- Crack widths are overestimated.
- The shear flow at the interface between structural steel and concrete is underestimated.

As already mentioned, the strain along the length of the member varies and therefore is convenient to adopt a mean value ε_{sm}. This strain is smaller than the maximum strain at cracks $\varepsilon_{s2} = N/(E_s \cdot A_s)$ by a quantity $\Delta\varepsilon_s$ that represents the participation of concrete in carrying tensile stress between the cracks; thus, the tension stiffening effect. The mean strain is given by

$$\varepsilon_{sm} = \varepsilon_{s2} - \Delta\varepsilon_s = \varepsilon_{s2} - \frac{\beta \cdot f_{ctm}}{\rho_s \cdot E_s} \tag{6.30}$$

where

β is a factor that expresses the mean value of the crack spacing (=0.4 for long-term or repeated loading and 0.6 for short-term loading; for fatigue in composite girders $\beta = 0.2$, see Remark 11.9)

ρ_s is the reinforcement ratio ($=A_s/A_c$)

Multiplying both sides of Equation 6.30 with $E_s \cdot A_s$ gives the additional tension force in the reinforcement due to the tension stiffening; $\Delta N_s = \beta \cdot f_{ctm} \cdot A_c$. One can see that the enhancement due to tension stiffening is equal to 40%–60% of the tensile resistance of the concrete cross section; thus, it should not be neglected. Redistributions and stress concentrations due to tension stiffening should be carefully investigated in composite tension members such as chords or ties in bowstring members.

Based on experimental tests, EN 1992-1-1 adopts a hyperbolic variation of $\Delta\varepsilon_s$ with σ_{s2} as follows:

$$\Delta\varepsilon_s = \Delta\varepsilon_{smax} \cdot \frac{\sigma_{sr}}{\sigma_{s2}} \tag{6.31a}$$

where

σ_{sr} is the stress in reinforcing steel at first crack

$\Delta\varepsilon_{smax}$ is the maximum value of the strain enhancement due to the tension stiffening effect

σ_{s2} is the stress in reinforcing steel at state 2

From the geometry of Figure 6.14a,

$$\Delta\varepsilon_{smax} = (\varepsilon_{s2} - \varepsilon_{s1}) \cdot \frac{\sigma_{sr}}{\sigma_{s2}} \tag{6.31b}$$

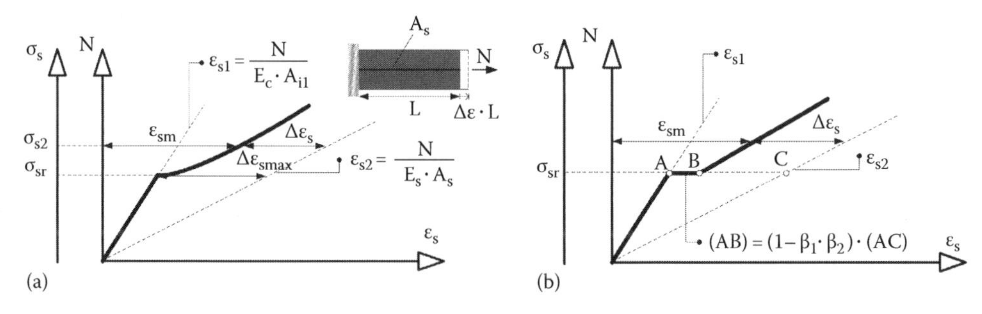

Figure 6.14 Axial stress versus mean strain: (a) exact and (b) simplified.

Comparing Equations 6.31, the mean strain ε_{sm} is calculated:

$$\varepsilon_{sm} = (1-\zeta)\cdot\varepsilon_{s1} + \zeta\cdot\varepsilon_{s2} \tag{6.32a}$$

where ζ is a dimensionless coefficient ranging between 0 (state 1) and 1 (state 2) and is calculated by

$$\zeta = 1 - \left(\frac{N_{cr}}{N}\right)^2 \quad \text{if } N \geq N_{cr} \tag{6.32b}$$

N_{cr} is calculated according to Equation 6.27.

With the aforementioned expressions, the mean strain ε_{sm} can be calculated by taking into account the tension stiffening. EN 1992-1-1 goes a step further and replaces Equation 6.32b with the following:

$$\zeta = 1 - \bar{\beta}\cdot\left(\frac{N_{cr}}{N}\right)^2 \quad \text{if } N \geq N_{cr} \tag{6.33}$$

where

$\bar{\beta}$ is a coefficient taking account of the influence of the duration of the loading
=1.0 for a single short-term loading
=0.5 for sustained loads or many cycles of repeated loading

EN 1992-1-1 extends the use of Equation 6.32a in concrete sections subjected to bending. The curvature is estimated as follows:

$$\kappa = (1-\zeta)\cdot\kappa_1 + \zeta\cdot\kappa_2 \tag{6.34a}$$

with

$$\zeta = 1 - \bar{\beta}\cdot\left(\frac{M_{cr}}{M}\right)^2 \quad \text{if } M \geq M_{cr} \tag{6.34b}$$

where M and M_{cr} are the bending moment acting on the cross section and the bending moment that produces tensile stress f_{ctm} at the extreme fiber.

By integrating the curvatures κ along the length of the concrete beams, deflections can be estimated. This method is used in Section 10.5 for the estimation of deflections in filler-beam decks.

6.1.6.1.3 Control of cracking due to direct loading: Verification by limiting bar diameter or bar spacing

Where at least the minimum reinforcement is provided, the limitation of crack width for direct loading may generally be achieved by limiting the spacing or the diameters of

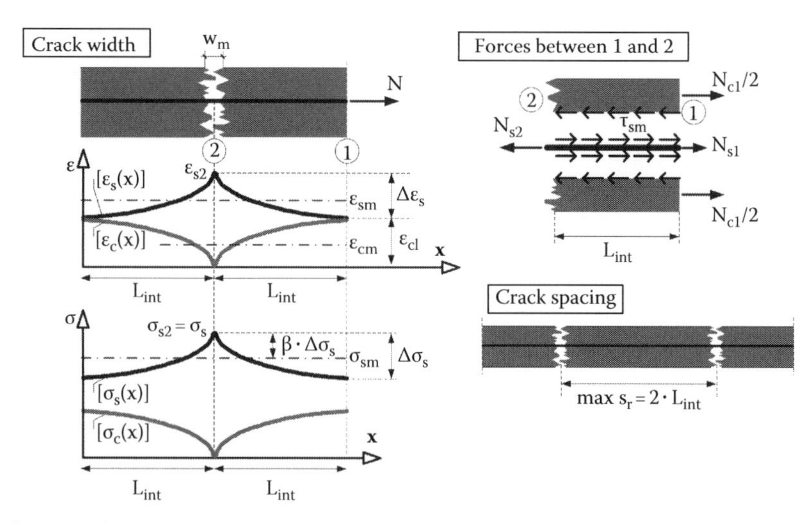

Figure 6.15 Stress and strain variation at cracked positions.

the rebars. The theoretical background of the expressions offered in the Codes is briefly explained in this section.

As shown in Figure 6.15, stresses and strains in steel reach their maximum values in cracked locations (state 2) (position 2 in Figure 6.15). In contrast, concrete stresses and strains reach their maximum values at a distance L_{int} from position 2 where the bond between the two materials is well recovered. After cracking formation in position 2, the axial force in position 1 is $N_{c1} = N_{cr}$ (see Equation 6.27).

The distance L_{int} is calculated from the strain compatibility in position 1 and the equilibrium of forces in steel between positions 1 and 2:

$$\varepsilon_{s1} = \varepsilon_{c1} \Rightarrow \frac{N_{s1}}{E_s \cdot A_s} = \frac{N_{c1}}{E_c \cdot A_{ct,eff}} \tag{6.35a}$$

$$N_{s2} = N_{c1} + N_{s1} \Rightarrow \sigma_s \cdot A_s = N_{c1} + N_{s1} \tag{6.35b}$$

Substitution of Equation 6.35a in Equation 6.35b gives

$$\sigma_s \cdot A_s = N_{c1} \cdot \left(1 + \frac{E_s \cdot A_s}{E_c \cdot A_{ct,eff}}\right) \tag{6.36}$$

Assuming τ_{sm} is an average value of the bond stress, it can be written that

$$N_{c1} = \tau_{sm} \cdot L_{int} \cdot (\pi \cdot \Phi) \tag{6.37}$$

Therefore, Equation 6.36 leads to

$$\sigma_s \cdot \left(\frac{\pi \cdot \Phi^2}{4}\right) = \left[\tau_{sm} \cdot L_{int} \cdot (\pi \cdot \Phi)\right] \cdot \left(1 + \frac{E_s \cdot A_s}{E_c \cdot A_{ct,eff}}\right)$$

$$\Rightarrow L_{int} = \frac{\sigma_s \cdot \Phi}{4 \cdot \tau_{sm}} \cdot \frac{1}{1 + (E_s \cdot A_s / E_c \cdot A_{ct,eff})} \approx \frac{\sigma_s \cdot \Phi}{4 \cdot \tau_{sm}} \tag{6.38}$$

where

Φ is the diameter of the rebar

τ_{sm} is the mean value of the bond stress; for ribbed bars, $\tau_{sm} \approx 1.8 \cdot f_{ct,eff}$

$f_{ct,eff}$ is the mean value of the tensile strength of the concrete effective at the time when the crack may first be expected to occur

$f_{ct,eff} = f_{ctm}$ or lower if cracking is expected earlier than 28 days

$A_{ct,eff}$ is the effective area of concrete in state 2 (see EN 1992-1-1)

σ_s is the stress in rebar in position 2

The maximum spacing between two adjacent cracks is

$$\max s_r = 2 \cdot L_{int} = \frac{\sigma_s \cdot \Phi}{2 \cdot \tau_{sm}} \tag{6.39}$$

The width of crack can be derived by multiplying the spacing max s_r with the difference between the mean values of the strains for reinforcing steel and concrete. Thus,

$$w_k = \max s_r \cdot (\varepsilon_{sm} - \varepsilon_{cm}) \tag{6.40a}$$

with

$$\varepsilon_{sm} = \varepsilon_{s2} - \beta \cdot \Delta\varepsilon_s = \varepsilon_{s2} \cdot (1-\beta) + \beta \cdot \varepsilon_{c1} (\varepsilon_{c1} = \varepsilon_{s1}) \tag{6.40b}$$

$$\varepsilon_{cm} = \beta \cdot \varepsilon_{c1} \tag{6.40c}$$

The crack width can be derived by substituting Equations 6.39 in Equation 6.40a:

$$w_k = \frac{\sigma_s \cdot \Phi}{2 \cdot \tau_{sm}} \cdot \varepsilon_{s2} \cdot (1-\beta) \Rightarrow w_k = \frac{\sigma_s^2 \cdot \Phi}{2 \cdot \tau_{sm} \cdot E_s} \cdot (1-\beta) \tag{6.41}$$

It has to be mentioned that Equation 6.41 gives the most conservative values for the crack width w_k because it is based on the maximum spacing max s_r. Moreover, the width of cracks mainly depends on the stress σ_s in reinforcing steel after cracking. The maximum bar diameter for a required crack width is given by

$$\Phi = \frac{w_k \cdot 2 \cdot \tau_{sm} \cdot E_s}{(1-\beta) \cdot \sigma_s^2} \tag{6.42}$$

For a reference value for the tensile strength of concrete $f_{ct,eff}$ = 2.9 MPa and β = 0.4 for long-term or repeated loading, Equation 6.42 gives the maximum bar size Φ_s^* for crack control:

$$\Phi_s^* \approx \frac{6 \cdot w_k \cdot f_{ct,eff} \cdot E_s}{\sigma_s^2} \tag{6.43}$$

Table 6.10 Maximum bar diameters Φ_s^* for crack control

Steel stress	*Maximum bar diameter* Φ_s^* *(mm)*		
σ_s *[N/mm²]*	$w_k = 0.4$ *mm*	$w_k = 0.3$ *mm*	$w_k = 0.2$ *mm*
160	40	32	25
200	32	25	16
240	20	16	12
280	16	12	8
320	12	10	6
360	10	8	5
400	8	6	4
450	6	5	—

Source: EN 1992-1-1, Design of concrete structures, Part 1-1: General rules and rules for buildings, 2004.

where the minimum reinforcement given by Equation 6.29c is provided; crack widths are unlikely to be excessive if the bar sizes calculated by Equation 6.43 are not exceeded. As a simplification, EN 1992-1-1 offers Table 6.10.

According to EN 1992-1-1, the maximum bar diameter Φ_s^* should be modified.

$$\Phi_s \approx \Phi_s^* \cdot \frac{f_{ct,eff}}{2.9} \cdot \frac{h_{cr}}{8 \cdot (h - d)} \tag{6.44}$$

where

h is the overall depth of the concrete section
h_{cr} is the depth of the zone in tension immediately prior to cracking
d is the effective depth to the centroid of the outer layer of reinforcement
σ_s is the stress in rebar in position 2

Table 6.10 will also be used for the crack control of the concrete flanges of composite girders at hogging moment areas (see Chapter 10). However, the maximum bar sizes Φ_s^* are differently modified.

Instead of limiting the diameter of the bars, EN 1992-1-1 gives an alternative method by limiting the bar spacing according to Table 6.11.

The limiting calculated crack width w_k chosen for using Tables 6.10 or 6.11 depends on the exposure class of the structural members. For reinforced concrete sections and prestressed sections with unbonded tendons, a crack width of 0.3 mm is generally satisfactory

Table 6.11 Maximum bar spacing s for crack control

Steel stress	*Maximum bar spacing s (mm) for width* w_k		
σ_s *[N/mm²]*	$w_k = 0.4$ *mm*	$w_k = 0.3$ *mm*	$w_k = 0.2$ *mm*
160	300	300	200
200	300	250	150
240	250	200	100
280	200	150	50
320	150	100	—
360	100	50	—

Source: EN 1992-1-1, Design of concrete structures, Part 1-1: General rules and rules for buildings, 2004.

in respect of durability. For prestressed members with bonded tendons, 0.2 mm should be applied. However, different recommendations may be found in the National Annex.

6.2 STRUCTURAL STEEL

6.2.1 Steel grades

The European Standard that defines grades and properties of structural steel is EN 10025 [6.2]. According to the system used in this standard, structural steel is designated by the letter S, followed by a number providing its yield strength at thickness $t \leq 16$ mm in [MPa] and one or two symbols specifying the material toughness as indicated in Table 6.12. The toughness is expressed by the minimum *Charpy V-notch impact energy.* The most usual steel grade for bridges is S355 because its cost-to-strength ratio is lower than for other grades. It may be non-alloy, normalized or thermomechanically treated, grades that are described in Parts 2–4 of EN 10025 correspondingly as shown in Table 6.13. It should be noted that the prescribed value of the yield strength of structural steel is, unlike the concrete or the reinforcing steel strength, not a characteristic value but a minimum guaranteed value, allowing thus the application of smaller material safety factors as indicated in Table 5.20.

The mechanical properties of structural steels are mainly characterized by the yield and tensile strengths that are designated in Eurocodes 3 and 4 as f_y and f_u correspondingly. The yield strength for thicknesses up to 16 mm is the number followed by the letter S. The steel grades, the yield strength for higher thicknesses, and the tensile strength are shown in Table 6.14. Table 6.14 includes only steels that are covered by the design rules of the Eurocode (EN 1994-2). Grades higher than S460 need special permission and compliance with additional rules. It should be noted that the yield and tensile strength are differently

Table 6.12 Designation of steel grades according to EN 10025

Letter	*Number*	*Symbol 1 (see Table 6.13)*	*Symbol 2—optional (see Table 6.16)*
S	Yield strength [MPa] at thickness $t \leq 16$ mm	*Charpy* V-notch impact energy in [J] at temperature T	Improved through-thickness properties against lamellar tearing

Source: EN 10025, Hot rolled products of structural steels, 2004.

Table 6.13 Material toughness for symbol 1 of steel designation

		Longitudinal direction	
EN 10025 [6.2]	*Symbol 1*	*Temperature T [°C]*	*Charpy V-notch impact energy [J]*
Part 2	JR	20	27
Non-alloy structural steels	J0	0	27
	J2	−20	27
	K2	−20	40
Part 3			
Normalized/normalized rolled weldable fine-grain structural steels	N	−20	40
	NL	−50	27
Part 4			
Thermomechanically rolled weldable fine-grain structural steels	M	−20	40
	ML	−50	27

Table 6.14 Mechanical properties of structural steels for thicknesses t> 16 mm

	Yield strength f_y in MPa					*Tensile strength f_u in MPa*	
Steel grade	$16<t\leq40$	$40<t\leq63$	$63<t\leq80$	$80<t\leq100$	$100<t\leq150$	$3\leq t\leq100$	$100<t\leq150$
S235JR, J0, J2	225	215	215	215	195	360–510	350–500
S275JR, J0, J2	265	255	245	235	225	410–560	400–540
S355JR, J0, J2, K2	345	335	325	315	295	470–630	450–600
	$16<t\leq40$	$40<t\leq63$	$63<t\leq80$	$80<t\leq100$	$100<t\leq150$	$t\leq100$	$100<t\leq200$
S275N, NL	265	255	245	235	225	370–510	350–480
S355N, NL	345	335	325	315	295	470–630	450–600
S420N, NL	400	390	370	360	340	520–680	500–650
S460N, NL	440	430	410	400	380	550–720	530–700
	$16<t\leq40$	$40<t\leq63$	$63<t\leq80$	$80<t\leq100$	$100<t\leq120$	$63<t\leq80$	$t\leq120$
S275M, ML	265	255	245	245	240	350–510	350–510
S355M, ML	345	335	325	325	320	440–600	430–590
S420M, ML	400	390	380	370	365	470–640	460–630
S460M, ML	440	430	400	390	385	500–670	490–660

Source: EN 10025, Hot rolled products of structural steels, 2004.

Table 6.15 Mechanical properties of structural steels produced to EN 10025, in accordance with EN 1993-1-1

	Nominal thickness of the element t in mm			
	$t \leq 40$ *mm*		*40 mm* $< t \leq 80$ *mm*	
Steel grade to EN 10025	f_y *in MPa*	f_u *in MPa*	f_y *in MPa*	f_u *in MPa*
S 235	235	360	215	360
S 275	275	430	255	410
S 355	355	510	335	470
S 275 N/NL	275	390	255	370
S 355 N/NL	355	490	335	470
S 420 N/NL	420	520	390	520
S 460 N/NL	460	540	430	540
S 275 M/ML	275	370	255	360
S 355 M/ML	355	470	335	450
S 420 M/ML	420	520	390	500
S 460 M/ML	460	540	430	530

designated in EN 10025 compared to the Eurocodes. In EN 10025, the symbol used for the yield strength is R_{eH} while for the tensile strength R_m.

For data related to other properties of structural steels, for example, the chemical composition, as well as for other types of steel, reference is made to EN 10025. Eurocode 3, EN 1993-1-1, allows a simplification regarding the mechanical properties of structural steel produced to EN 10025. These values are presented in Table 6.15 and may be used instead of the values of Table 6.14.

6.2.2 Fracture toughness and through thickness properties

The material thickness in modern bridge construction may be very large. Thicknesses of flange plates up to 150 mm are not unusual in I girders and the top flange of box girders. Thick plates have the advantage of avoiding welding operations to strengthen the flanges by additional plates therefore reducing the manual labor costs and the residual welding stresses. However, thick plates have lower yield strength as discussed before and must have improved properties in respect to their toughness.

6.2.2.1 *Material toughness*

Material toughness is a property of structural steel that indicates its tendency to brittle fracture. In simple words, if steel is insufficiently tough and subjected to tensile stresses, internal cracks propagate rapidly and a non-ductile failure may result. Toughness is measured by the absorbed energy during a *V-notched Charpy* impact test (see Figure 6.16). The apparatus consist of a pendulum axe swinging at a notched specimen. The energy transferred to the specimen can be estimated by comparing the difference in the height of the hammer before and after the fracture. Standard methods can be found in [6.16] and [6.24]. Generally, test specimens should exhibit an impact energy equal or higher than 27J at a specified test temperature T (see Table 6.13).

As shown in Figure 6.16, the toughness depends on the test temperature. Three regions of material toughness may be distinguished: *the upper shelf region, the lower shelf region*, and the *transition region*. In the upper shelf region, steel specimens exhibit elastic–plastic behavior with ductile modes of failure irrespective of the presence of small flaws and welding

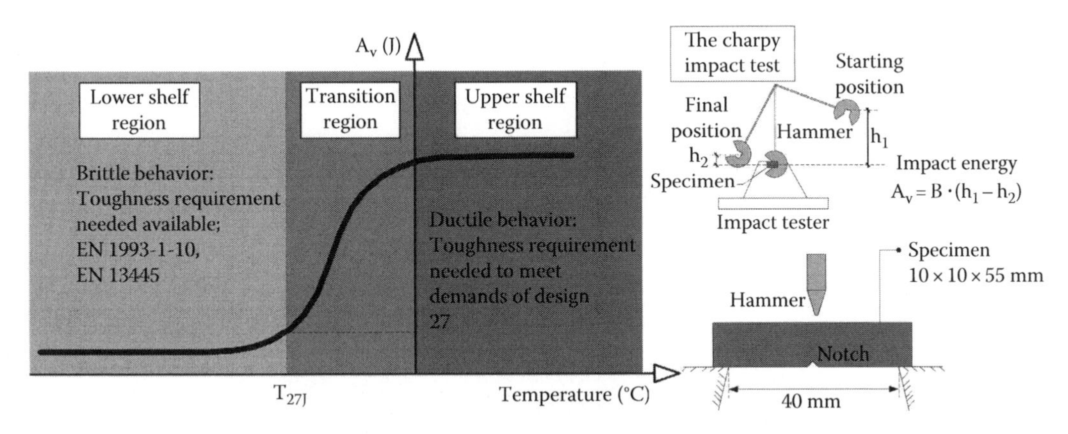

Figure 6.16 Toughness–temperature diagram. (From EN 1993-1-10, Design of steel structures, Material toughness and through thickness properties.)

discontinuities from fabrication. In the lower shelf region, brittle failure due to very low temperature occurs. The region of the toughness–temperature diagram in which the material toughness decreases and the failure mode changes from ductile to brittle is called transition region. The temperature values T_{27J} required are located in the lower part of this region.

Designers and fabricators should ensure sufficient toughness of steel to prevent brittle fracture at lowest service temperature, to provide sufficient ductility at welded details in the upper shelf region, and to assure sufficient through thickness properties.

Important parameters for the required material toughness include

- The lowest service temperature
- The element thickness
- The type of loading (static or dynamic)
- The intensity of the applied stresses due to external loading
- The intensity of the residual stresses due to restraint or fabrication
- The construction detail in reference to stress concentrations and weld details

The aforementioned complexities can be avoided through a simplified method offered by EN 1993-1-10 [6.8]. Tabulated values for the maximum permissible plate thickness allow the selection of appropriate steel grades for the parent material as a function of the reference temperature and the level of applied stresses σ_{Ed}. The stresses σ_{Ed} are determined from the accidental design combination:

$$G_k + \psi_1 \cdot Q_k \tag{6.45}$$

where

Q_k is the characteristic value of the traffic loads
ψ_1 the relevant combination factor according to Table 5.13

It is obviously from the aforementioned that the applied stresses σ_{Ed} refer to the final stage (composite system). This covers the majority of bridges, which are subjected to only moderate tensile stresses or to less severe minimum temperatures during the construction stage. If this is not the case, designers should select an appropriate toughness based on the tension stresses emerging during the construction stage.

Figure 6.17 shows the relevant charts at stress levels equal to ¼, ½, and ¾ of the yield strength for the usual steel grade for bridges S355. In recognition that the minimum bridge

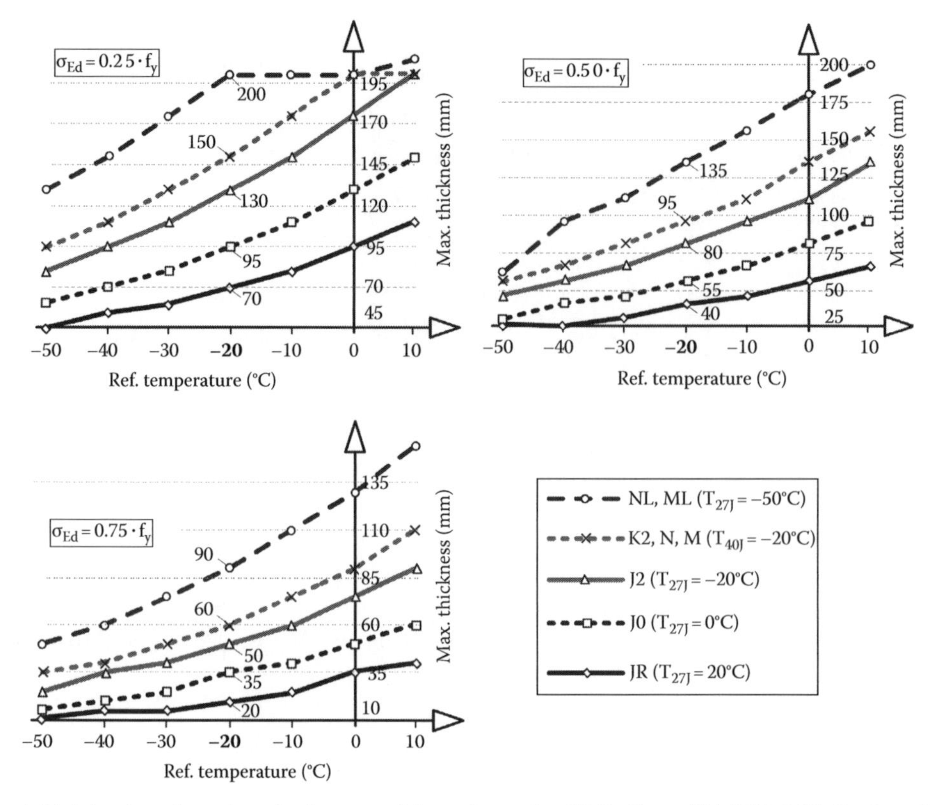

Figure 6.17 Selection of steel grades for material toughness for S355. (From EN 1993-1-10, Design of steel structures, Material toughness and through thickness properties.)

service temperature for the majority of the cases does not follow below –20°C, the corresponding maximum permissible plate thicknesses are notated.

Residual stresses are globally taken into account as equal to 100 MPa. The stress level is obviously well below the yield strength since it is determined from a less severe combination. The level of the applied stress σ_{Ed} may be estimated during preliminary design and checked at final design. Brittle fracture refers mainly to tension stresses. However, for plates in compression, $\sigma_{Ed} = 0.25 \cdot f_y$ may be considered.

EXAMPLE 6.3

The girder of a continuous bridge is made of steel S355. The steel grades shall be determined for flange and web thicknesses as shown in Figure 6.18.

Reference temperature $T_{Ed} = -20°C$.

The yield strength distribution throughout the bridge is **assumed** to be as in Figure 6.18. Steel will be chosen according to EN 10025 (see Table 6.14).

Webs at spans: The stress intensity for the accidental combination of Equation 6.45 is estimated as $\sigma_{Ed} = 0.5 \cdot f_y = 0.5 \cdot 355 = 177.5$ N/mm². Figure 6.17 suggests for JR a maximal allowed thickness of 40 mm for this stress level and a temperature −20°C that is above the thickness used. Therefore, for the webs at spans, steel grade S355JR is selected.

For S355JR with $t < 16$ mm $\rightarrow f_y = 355$ MPa (correct yield strength assumption).

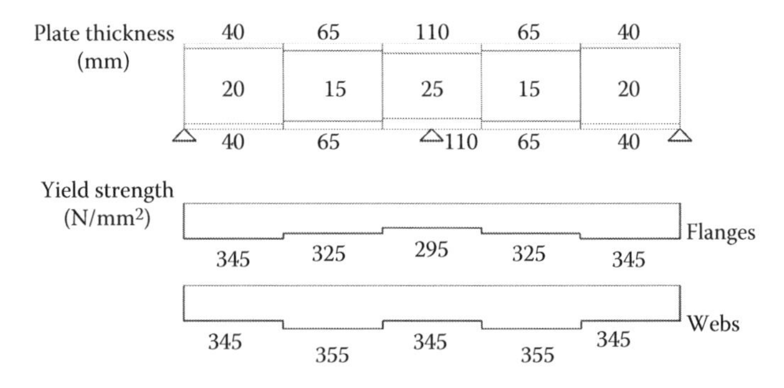

Figure 6.18 Thicknesses of flanges and web of the steel girder of a two-span composite bridge.

Webs at edge and internal supports: The stress intensity for the accidental combination of Equation 6.45 is estimated as $\sigma_{Ed}=0.5 \cdot f_y=0.5 \cdot 345=172.5$ N/mm^2. Figure 6.17 suggests for JR a maximal allowed thickness of 40 mm for this stress level and a temperature −20°C that is above the thickness used. Therefore, for the webs at edge and internal supports, steel grade S355JR is selected.

For S355JR with 16 mm$<t\leq$40 mm$\rightarrow f_y=345$ MPa (correct yield strength assumption).

Flanges at internal support: The bottom flange is in compression at service conditions. Therefore, $\sigma_{Ed}=0.25 \cdot f_y=0.25 \cdot 295=73.75$ N/mm^2 is considered. The required grade for 110 mm thickness is J2. Steel grade S355J2 is selected.

For S355J2 with t = 110 mm$\rightarrow f_y=295$ MPa (correct yield strength assumption).

The top flange is in tension with $\sigma_{Ed}=0.50 \cdot f_y=0.5 \cdot 295=147.5$ N/mm^2. The required grade for 110 mm thickness is NL or ML. Steel grade S355NL is selected.

For S355NL with t = 110 mm$\rightarrow f_y=295$ MPa (correct yield strength assumption).

Flanges at end supports: The end supports are subjected to low bending moments. Therefore, $\sigma_{Ed}=0.25 \cdot f_y=0.25 \cdot 345=86.25$ N/mm^2 is considered for both flanges. The required grade for 40 mm thickness is JR. Steel grade S355JR is selected.

For S355JR with t = 40 mm$\rightarrow f_y=345$ MPa (correct yield strength assumption).

Flanges at span: The top flange is in compression at service conditions. Therefore, $\sigma_{Ed}=0.25 \cdot f_y=0.25 \cdot 325=81.25$ N/mm^2 is considered. The required grade for 65 mm plate thickness is JR. Steel grade S355JR is selected.

For S355JR with t = 65 mm$\rightarrow f_y=325$ MPa (correct yield strength assumption).

The bottom flange is in tension at service conditions. $\sigma_{Ed}=0.50 \cdot f_y=0.5 \cdot 325=162.5$ N/mm^2 is considered. The required grade for 65 mm thickness is J2. Steel grade S355J2 is selected.

For S355J2 with t = 65 mm$\rightarrow f_y=325$ MPa (correct yield strength assumption).

The steel qualities are shown schematically in Figure 6.19.

As already mentioned, the estimated stress intensity should be checked at final design.

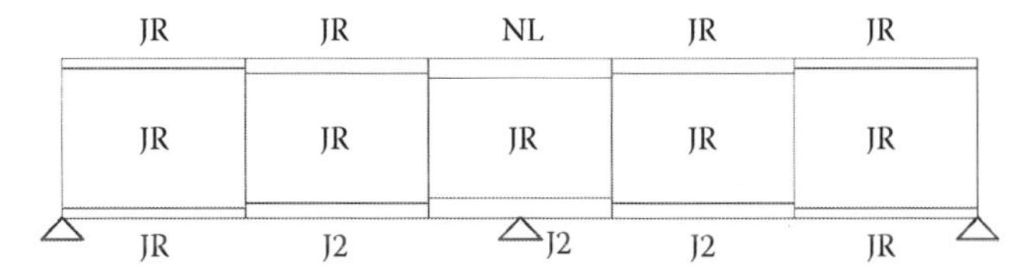

Figure 6.19 S355 subgrade distribution.

6.2.2.2 Lamellar tearing

When plates are subjected to transverse tension, *lamellar tearing* (i.e., separation into leaves) may occur. Lamellar tearing (also called *delamination*) is associated with internal cracks that are caused due to very high out-of-plane stresses in steel. Such cracks can only be inspected by performing *ultrasonic testing*.

Delamination can also occur due the weld shrinkage. Welds shrink during fabrication and cracking in the transverse direction of the connected elements is highly possible. This is especially the case for cruciform, T- and corner joints, and for full penetration welds.

Intermediate webs and intermediate cross girder post flanges in multi-girder bridges are endangered by delamination. Additional examples are shown in Figure 6.20. One can see that the top flange of the main girder is endangered by its welded connection with the gusset plate of the hanger (see section a–a). Internal cracks may also arise in the hanger (see section b-b).

In compliance with EN 1993-1-10 [6.8], designers can minimize the delamination risk by satisfying the following condition:

$$Z_{Ed} \leq Z_{Rd} \tag{6.46}$$

where

Z_{Ed} is the required design Z-value resulting from the magnitude of strains from restrained metal shrinkage under the weld beads

Z_{Rd} is the available design Z-value for the material according to EN 10164 [6.11]

A method for determining the required Z-grade according to EN 1993-1-10 is summarized in Table 6.16. It has to be mentioned that the view of many welding experts is that the method contained in [6.8] is for the most cases very conservative (see [6.22]). Therefore, different methods or values may be found in National Annex.

From Table 6.16, one can see that the Z_{Ed}-value depends on many factors as the type and the size of the weld, the thickness of the material, and the level of restraint and preheating. The through thickness resistance Z_{Rd} is solely dependent upon the delamination risk level. For low and medium risk levels, Z15 and Z25 qualities should be specified accordingly. For high risk situations, Z35 steel should be used [6.17]. Designers should be able to identify by themselves the risk level of the welded connections. Usually T- and cruciform joints with butt and deep penetration welds for plates with s>25 mm are classified as high risk cases. More information may be found in National Annexes or Best Practice Guidance Notes.

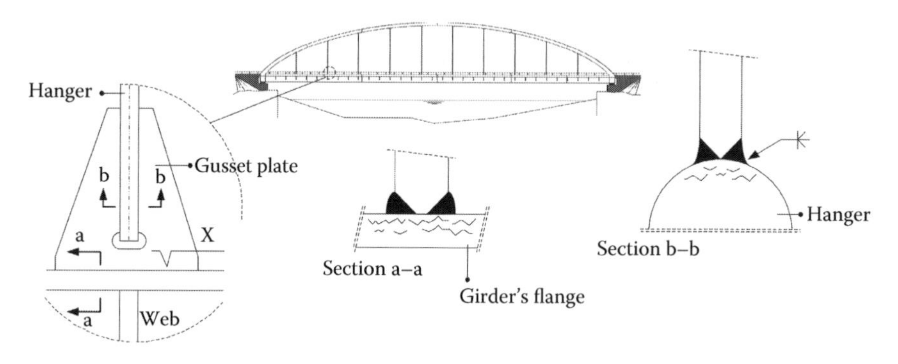

Figure 6.20 Example of potential lamellar tearing due to through thickness tension.

Table 6.16 Design procedure for improving through thickness properties

Design Z-value	$Z_{Ed} = Z_a + Z_b + Z_c + Z_d + Z$	*Available design Z-value*	Z15 if $10 < Z_{Ed} \leq 20$ Z25 if $20 < Z_{Ed} \leq 30$ Z35 if $Z_{Ed} > 30$

Weld depth relevant for straining from metal shrinkage			Z_a
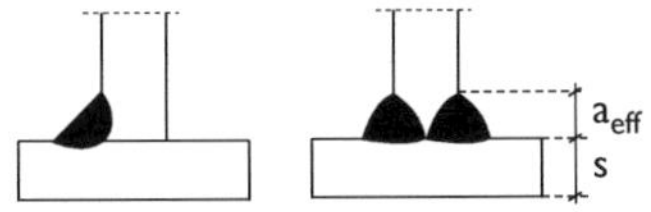Effective weld depth a_{eff} for shrinkage a for fillet welds	$a_{eff} \leq 7$ mm	a = 5 mm	0
	7 mm < $a_{eff} \leq 10$ mm	a = 7 mm	3
	10 mm < $a_{eff} \leq 20$ mm	a = 14 mm	6
	20 mm < $a_{eff} \leq 30$ mm	a = 21 mm	9
	30 mm < $a_{eff} \leq 40$ mm	a = 28 mm	12
	40 mm < $a_{eff} \leq 50$ mm	a = 35 mm	15
	$a_{eff} < 50$ mm	a > 35 mm	15

Shape and positions of welds		Z_b
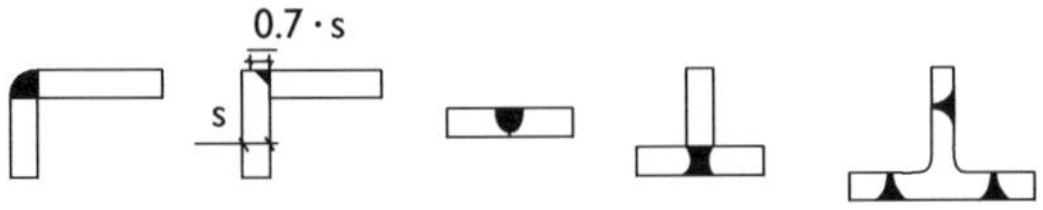		−25
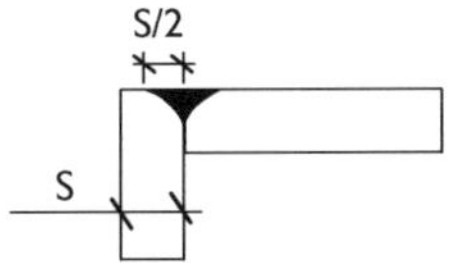	Corner joints	−10
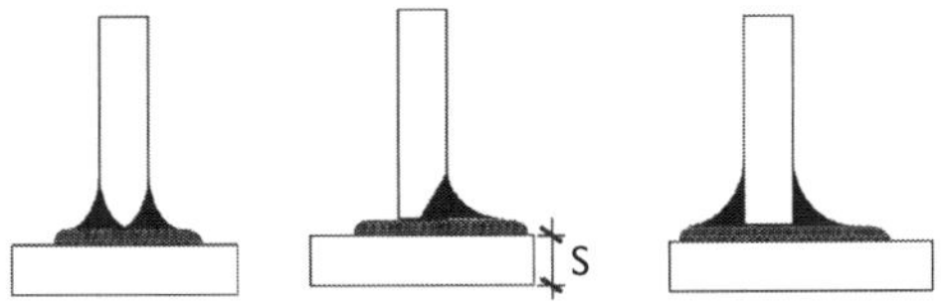	Single-run fillet welds with $a_{eff} \leq 7$mm or fillet welds with $a_{eff} > 7$mm with buttering with weld material of low strength	−5
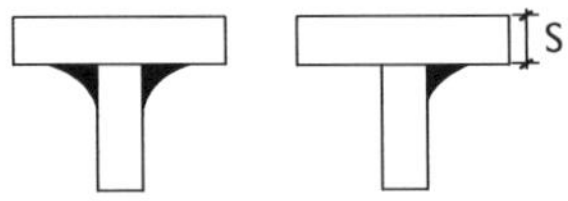	Multi-run fillet welds	0
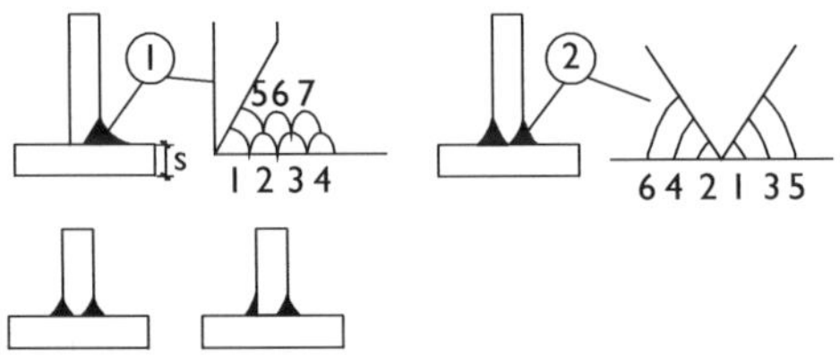	Partial and full penetration welds with appropriate welding sequence to avoid weld shrinkage	3
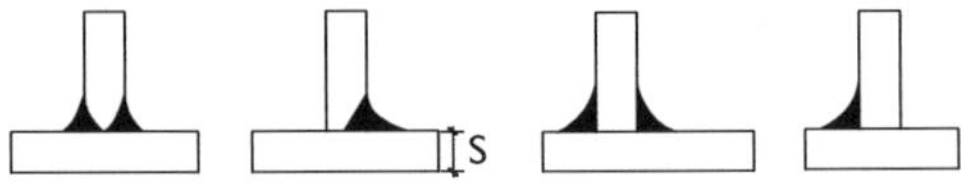	Partial and full penetration welds	5
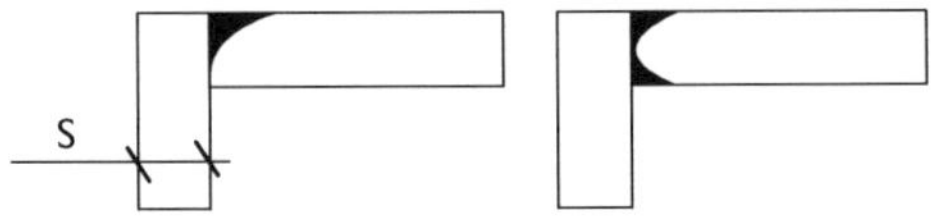	Corner joints	8

(*continued*)

Table 6.16 (continued) Design procedure for improving through thickness properties

Effect of material thickness s	Z_c	**Effect of shrinkage restraint**	Z_d
$s \leq 10$ mm	2	Free shrinkage possible	0
10 mm $< s \leq 20$ mm	4	Free shrinkage restricted	3
20 mm $< s \leq 30$ mm	6	High restraint	5
30 mm $< s \leq 40$ mm	8		
40 mm $< s \leq 50$ mm	10	**Influence of preheating**	Z_e
50 mm $< s \leq 60$ mm	12	Without preheating	0
60 mm $< s$	15	With preheating (≥100°C)	−8

Source: EN 1993-1-10, Design of steel structures, Material toughness and through thickness properties.

6.2.3 Other material properties for structural steel

In composite structures, structural steel is designated by the index a in order to differentiate it from reinforcing steel. Other properties of structural steel are as follows:

Specific weight	$\gamma_a = 78.5$ kN/m^3
Modulus of elasticity	$E_a = 210$ GPa
Poisson ratio	$\nu_a = 0.3$
Shear modulus	$G_a = \dfrac{E_a}{2 \cdot (1 + \nu_a)} = 81$ GPa
Coefficient of thermal expansion	$a_t = 10 \cdot 10^{-6}$ [per °C]

The aforementioned value of a_t is taken as equal to the corresponding one for concrete in order to simplify the design calculations. However, in order to determine the elongation/contraction of the bridge due to uniform temperature changes, the improved value of $a_t = 12 \cdot 10^{-6}$ is used.

6.3 REINFORCING STEEL

In composite structures, reinforcing steel is designated by the index s in order to differentiate it from structural steel. EN 10080 [6.13] describes the requirements for reinforcing steel but does not specify steel grades. The grades are characterized by two letters and a figure in between, as for example, B500B. The first letter B symbolizes weldable high-bond bars that are mainly used in composite bridges. The figure 500 indicates the characteristic value (5% fractile) of the yield strength f_{sk}. In this example, $f_{sk} = 500$ MPa. For composite bridges, the yield strength of reinforcing bars should be between 400 and 600 MPa. The third letter indicates the ratio between tensile and yield strength $k = f_{tk}/f_{yk}$. It is

A for $k \geq 1.05$
B for $1.08 < k < 1.15$
C for $1.15 \leq k < 1.35$

The recommended classes are B or C, that is, $1.08 < k < 1.35$, in order to ensure sufficient ductility. The modulus of elasticity of reinforcing steel may be taken as $E_s = 210$ GPa. This is in contrast to EN 1992-1-1 for concrete structures where E_s is taken equal to 200 GPa. The reason for the adoption of 210 GPa in composite structures is that no differentiation is made between the moduli of elasticity for structural and reinforcing steel. The other properties are those for structural steel.

6.4 PRESTRESSING STEEL

In composite bridges, prestressing refers mainly to the concrete slab in transverse direction. The properties of prestressing tendons are given in EN 10138 [6.14], Parts 2–4. They refer to the 0.1% proof stress $f_{p0,1k}$, the value of the ratio between tensile and proof strength $k = f_{pk}/f_{p0,1k}$ like for reinforcing steel, the class indicating the relaxation behavior, and the size and the surface characteristics. Prestressing steel should have $k \geq 1.1$. Prestressing devices include anchorages that transmit the forces of tendons in the concrete and couple for the connection between tendons. Such devices must be in accordance with relevant Technical Approval.

Cables are used as external tendons or as hangers in cable stayed, arch, or suspended bridges. Three classes, A, B, and C, are distinguished. Class A includes tension rod systems, class B ropes, and class C bundles of parallel wires or strands. The modulus of elasticity is that of steel for class A but varies from 80 to 210 GPa for classes B and C. Specific information for cables is given in EN 1993-1-11 [6.9].

6.5 BOLTS

Bolt classes are characterized by the yield strength f_{yb} and the tensile strength f_{ub}. Table 6.17 gives the classes with reference to EN 1993-1-8 [6.5]. Preloaded bolts for slip-resistant connections must be of classes 8.8 or 10.9. Bolts assemblies consisting of bolts, nuts, and washers (if needed) must be from the same producer.

6.6 STUD SHEAR CONNECTORS

Shear stud connectors are prescribed in EN ISO 13918 [6.15] and [6.19]. The steel grades correspond to mild steel in accordance to either structural steel or steel for bolts. Usual steel grades are S235J2 or 4.8. Depending on the supplier, shaft diameters d are specified in millimeters or inches. The height of the connector must comply with the requirements for sufficient concrete cover as presented in Section 5.6.1, with the additional condition $h \geq 3d$.

Table 6.17 Bolt classes to EN 1993-1-8

Bolt class	*4.6*	*5.6*	*6.8*	*8.8*	*10.9*
f_{yb} [MPa]	240	300	480	640	900
f_{ub} [MPa]	400	500	600	800	1000

Source: EN 1993-1-8, Design of steel structures, Joints, 2008.

REFERENCES

[6.1] DIN 4227: Spannbeton, Teil 1, 1979.

[6.2] EN 10025, CEN (European Committee for Standardization): Hot rolled products of structural steels, 2004; Part 1: General technical delivery conditions; Part 2: Technical delivery conditions for non-alloy structural steels; Part 3: Technical delivery conditions for normalized rolled weldable fine grain structural steels; Part 4: Technical delivery conditions thermomechanical rolled weldable fine grain structural steels. Part 5: Technical delivery conditions for structural steels with improved atmospheric corrosions resistance; Part 6: Technical delivery conditions for flat products of high yield strength structural steels in the quenched and tempered condition.

[6.3] EN 1992-1-1, CEN (European Committee for Standardization): Design of concrete structures, Part 1-1: General rules and rules for buildings, 2004.

[6.4] EN 1992-2, CEN (European Committee for Standardization): Design of concrete structures—Part 2: Concrete bridges—Design and detailing rules, 2005.

[6.5] EN1993-1-8, CEN (European Committee for Standardization): Design of steel structures. Joints, 2008.

[6.6] EN1993-2, CEN (European Committee for Standardization): Design of steel structures. Steel bridges, 2006.

[6.7] EN1994-2, CEN (European Committee for Standardization): Design of composite steel and concrete structures. Part 2: Rules for bridges, 2005.

[6.8] EN 1993-1-10, CEN (European Committee for Standardization): Design of steel structures. Material toughness and through thickness properties.

[6.9] EN 1993-1-11, CEN (European Committee for Standardization): Design of steel structures. Design of structures with tension components, 2006.

[6.10] EN 1090-2, CEN (European Committee for Standardization): Execution of steel structures and aluminium structures. Part 2: Technical requirements for steel structures, 2008.

[6.11] EN 10164, CEN (European Committee for Standardization): Steel products with improved deformation properties perpendicular to the surface of the product. Technical delivery conditions, 2004.

[6.12] EN 1011-2, CEN (European Committee for Standardization): Welding. Recommendations for welding of metallic materials. Arc welding of ferritic steels, 2001.

[6.13] EN 10080, CEN (European Committee for Standardization): Steel for the reinforcement of concrete, weldable, ribbed reinforcing steel, 2006.

[6.14] EN 10138, CEN (European Committee for Standardization): Pre-stressing steel, 2005.

[6.15] EN ISO 13918, CEN (European Committee for Standardization): Welding—Studs and ceramic ferrules for arc stud welding, 2008.

[6.16] EN 10045-1, CEN (European Committee for Standardization): Charpy impact test on metallic materials. Test method (V- and U- notches), 1990.

[6.17] EN 10164, CEN (European Committee for Standardization): Steel products with improved deformation properties perpendicular to the surface of the product, 2004.

[6.18] EN 206-1, CEN (European Committee for Standardization): Concrete—Part 1: Specification, performance, production and conformity, 2000.

[6.19] EN ISO 13918, CEN (European Committee for Standardization): Welding—Studs and ceramic ferrules for arc stud welding, 2008.

[6.20] Haensel, J.: Effects of creep and shrinkage in composite construction. Institute for structural engineering, Ruhr—Universität Bochum, Report 75-12, Bochum, Germany, 1975.

[6.21] Hanswille, G.: Zur Behandlung der Einflüsse aus dem Kriechen und Schwinden des Betons. Forschungsvorhaben: Eurocode 4 Teil 2—Verbundbrücken, Wuppertal, Germany, 1998.

[6.22] Hendy, C. R., Iles, D. C.: Steel bridge group: Guidance notes on best practice in steel bridge construction. Guidance note 3.02, Through thickness properties. The Steel Construction Institute, 2010.

[6.23] Iliopoulos, A.: Zur rechnerischen Berücksichtigung des Kriechens and Schwindens des Betons bei Verbundträgern. Dissertation, Shaker Verlag, Achen, Germany, 2005.

[6.24] ISO 148-1, CEN (European Committee for Standardization): Metallic materials—Charpy pendulum impact test. Part 1: Test method, 2009.

Chapter 7

Modeling and methods for global analysis

7.1 GLOBAL ANALYSIS MODELS

7.1.1 Introduction

Modeling for analysis is required in order to determine the internal bending moments and forces, the deformations, and the vibrations of bridge decks. A bridge analysis model should be based on the following criteria:

- It should reflect the structural response in terms of deformation, strength, and local and global stability.
- It should include as many as possible structural elements and parts (cross frames, stiffeners, bearings, etc.), and their possible eccentric connections.
- It should cover all construction stages and loading cases.
- Loads should be easily introduced.
- It should allow the performance of dynamic analysis and include the most important modes of vibration.
- It should be easily implemented.
- The resulting output should be such that it enables easily the execution of the Code-prescribed verifications.
- It should be supported by commercial analysis and design software.

The most general bridge representation is by means of finite elements (FEM). However, its implementation is not easy if all structural parts, including stiffeners and other construction details, should be represented; it requires large computer time and delivers stresses rather than internal forces and moments that are required for most code-based verifications. Therefore, other structural representations, mostly based on beam elements, are commonly used for bridge global analysis. FEM models may be applied to study local effects for certain construction details, for slab analysis in transverse direction in the presence of concentrated vehicle loads, or for calibration purposes. In the following, some of the current analyses and designs of computer-based models for composite bridges will be presented.

7.1.2 Beam models

7.1.2.1 Bridges with two main girders

For composite bridges with two main girders, the system in transverse direction is statically determinate. Vertical loads may be distributed between the two main girders according to a linear influence line as indicated in Figure 7.1, if the torsional rigidities are neglected and the system deck slab–cross girders are assumed as infinite rigid. Accordingly, in the first step,

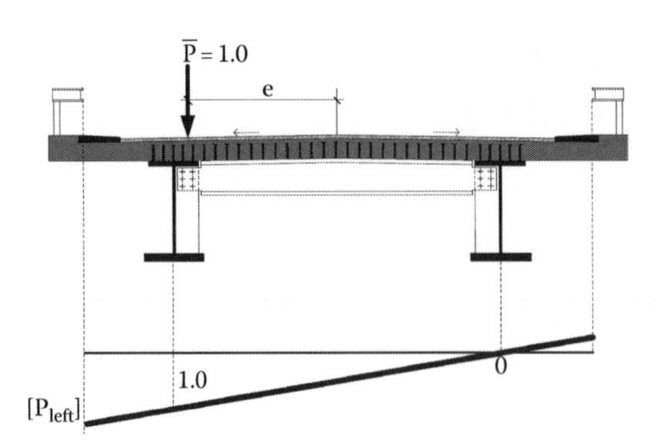

Figure 7.1 Influence line of support reaction for the left main girder.

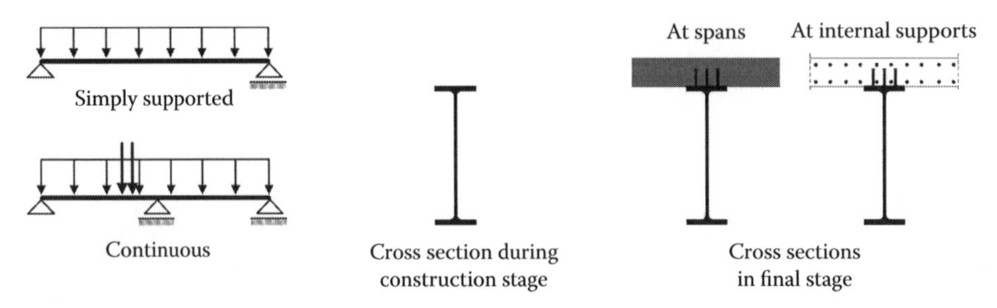

Figure 7.2 Single beam model for bridges with two main girders.

vertical loading due to permanent and traffic actions is distributed transversely to the main girders, while in the second step, the main girders in the longitudinal direction are modeled as beams as shown in Figure 7.2. Global analysis in the longitudinal direction is therefore performed on a single beam.

Depending on the progress of construction, the cross sections of the beam vary; see Figure 7.2. During construction, they are of pure steel section, while after hardening of concrete of composite cross section, composed of the steel section and the concrete slab within the effective width (see Section 7.2). The beam may be simply supported or continuous, depending on the relevant support conditions of the bridge. The vertical loads on the beam due to dead loads and traffic are determined from an analysis in the transverse direction.

Additional vertical loading results in from eccentric wind (Figure 7.3). However, this loading is generally of minor importance for the superstructure of road bridges (not for the bearings) and might be neglected. In cases of railway bridges with traffic, the additional internal forces due to eccentricities should be taken into account.

Single beam models have the advantage that they are supported by commercial analysis and design software that account for the construction phase, the influence of temperature, creep, and shrinkage, and performs automatically all required code verifications at ULS, SLS, fatigue, shear connection, etc.; see [7.11]. However, analysis and design with this model refer to the main beams for vertical actions. It does not cover other structural elements like cross girders, cross bracings, and lateral stability of girders that have to be analyzed and designed separately. It is also noted that this model is inaccurate for skewed or curved bridges.

7.1.2.2 Bridges with multiple main girders and stiff cross girders

In the presence of stiff, closely spaced cross girders, the transverse influence line may also be assumed to be linear, provided that the deck's stiffness in the transverse direction is much

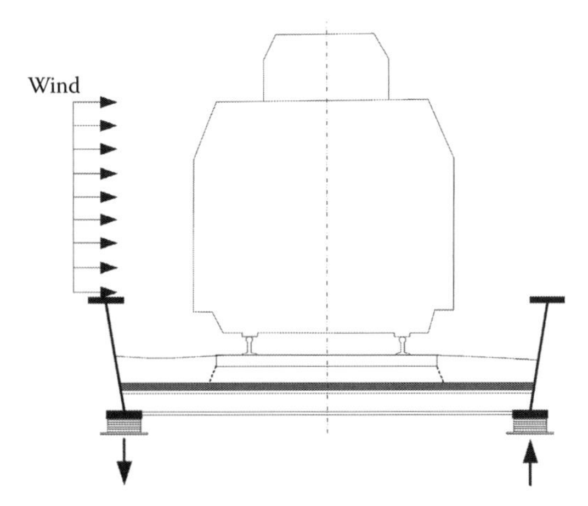

Figure 7.3 Vertical loads due to eccentricity of wind.

higher than the stiffness in the longitudinal one. Thus, the transverse profile of the deck maintains a straight geometry at loading; see Figure 7.4.

Under these conditions, the analysis is performed in two steps:

Step 1: Transverse distribution of vertical loads to the main girders
Step 2: Representation of the main girders as beams

The transverse distribution of vertical loads is performed by the *Courbon method* [7.9] which assumes that

a. The girders are either flexible in torsion or connected to the slab by means of hinges.
b. The girders have the same length so that their stiffness may be represented by the stiffness of their cross sections EI.

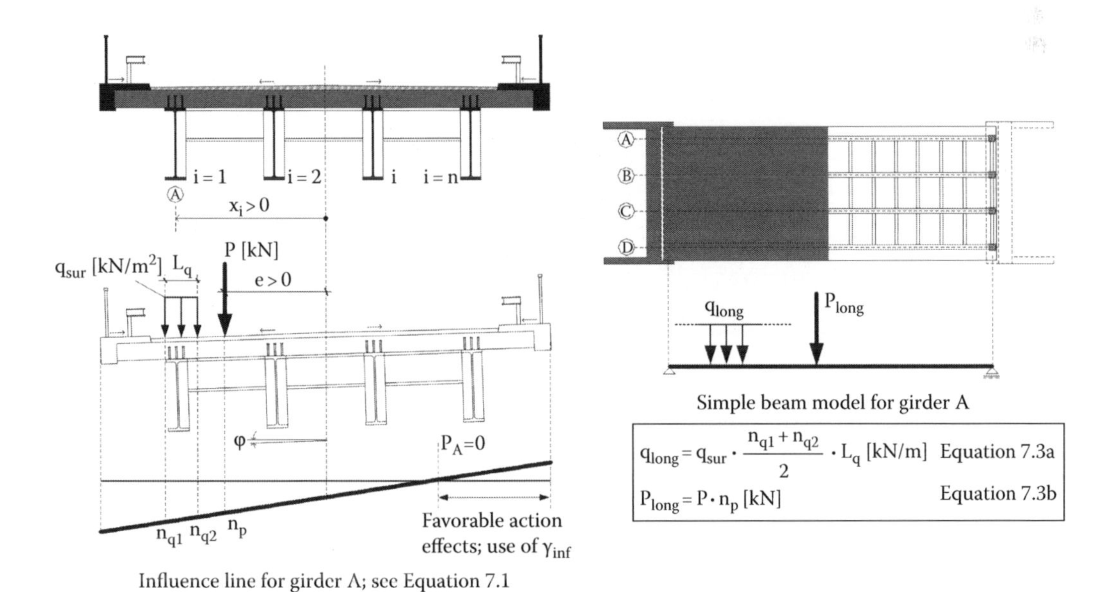

Figure 7.4 Load distribution according to the *Courbon* method.

Considering the transverse distribution of loading, the total reaction of the girders to a unit load P = 1 at distance e (Figure 7.4) is given by Equation 7.1. The first term of this equation refers to the deck translation and the second term to the deck rotation:

$$P_i = \frac{(EI)_i}{\sum (EI)_i} + \frac{(EI)_i \cdot x_i}{\sum (EI)_i \cdot x_i^2} \cdot e \tag{7.1}$$

For main beams with equal stiffness, Equation 7.1 is written as

$$P_i = \frac{1}{n} + \frac{x_i}{\sum x_i^2} \cdot e \tag{7.2}$$

where
- n is the total number of girders i
- x_i is the distance of girder i from the center of gravity of the girders
- e is the eccentricity of the loading

Equations 7.1 and 7.2 refer to an influence line for the support reaction of the girder i. This is shown in detail in Figure 7.4 in which the *Courbon method* for the main girder A is schematically described. The surface load q_{sur} and the point load P are transformed through the influence line into q_{long} and P_{long} for an analysis in the longitudinal direction. The loads can be placed in many different positions both in transverse and longitudinal directions and favorable (and unfavorable) action effects can be easily investigated.

It is noted that this method is inaccurate for skewed or curved bridges since condition (b) is not fulfilled. Actually, in both cases, the main girders have different lengths so that their stiffness may not be represented by the bending stiffness EI of their cross sections alone.

The *Courbon method* can be easily implemented with an Excel sheet. It generally gives conservative results and is especially recommended for pre-dimensioning. An example follows.

EXAMPLE 7.1

Determine the maximum internal forces for the outer girder A of the simply supported bridge in Figure 7.5 (L = 25 m) with the *Courbon method* due to the

a. Self-weight of the concrete slab
b. Load Model 1

The influence line of the total reaction P_A for the main girder A can be calculated with Equation 7.2:

$$P_i = \frac{1}{4} + \frac{4.35}{2 \cdot (4.35^2 + 1.45^2)} \cdot e = 0.25 + 0.1 \cdot e \tag{R7.1}$$

Loads that are imposed on locations with $P_A < 0$ result in favorable action effects and when combined should be multiplied with γ_{inf}.

a. Internal forces due to the self-weight of the concrete slab

The concrete slab's depth is 25 cm and the characteristic value of the permanent surface load is 25 kN/m³ · 0.25 m = 6.25 kN/m².

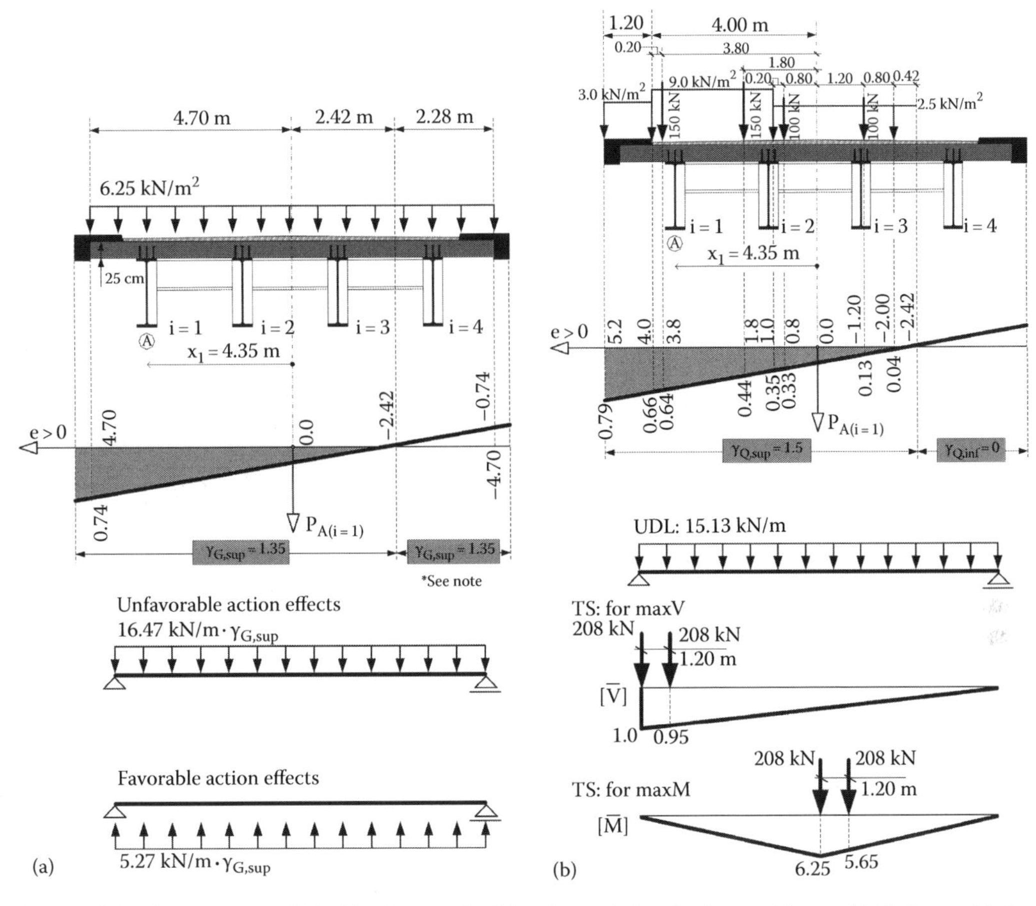

Figure 7.5 Implementation of the *Courbon* method for determining the internal forces V, M due to (a) the self-weight of the concrete slab and (b) Load Model I.

The part of the deck with the unfavorable action effects has a width of 7.12 m. Therefore, the load per unit length with the unfavorable effects is equal to $6.25 \cdot ((1/2) \cdot 7.12 \cdot 0.74) = 16.47$ kN/m

The load per unit length with the favorable effects is equal to $6.25 \cdot [(l/2) \cdot 2.28 \cdot (-0.74)] = -5.27$ kN/m

The design value of the load per unit length at ULS due to the slab's self-weight is

$$g_{slab,Ed} = \gamma_{G,sup} \cdot 16.47 - \gamma_{G,sup} \cdot 5.27 = 1.35 \cdot (16.47 - 5.27) = 15.12 \text{ kN/m}$$

Maximum shear force: $\max V_{c,Ed} = 15.12 \cdot 25/2 = 189$ kN

Maximum bending moment: $\max M_{c,Ed} = 15.12 \cdot 25^2/8 = 1181.25$ kN-m

Note: Theoretically, the loads in the part of the deck with favorable action effects (P < 0 in the influence line of Figure 7.5a) should be multiplied with $\gamma_{G,inf} = 1.0$. However, in EN 1990:2002, it is stated that **all actions originated from the same source should be multiplied with the same safety factor.**

b. Internal forces due to Load Model 1

The surface loads of the Load Model 1 (UDL) can be easily transformed into loads per unit length through the influence line of Equation 7.2. Thus,

$$q_{UDL,k} = 9 \cdot \left(\frac{0.66 + 0.35}{2}\right) \cdot 3 + 3 \cdot \left(\frac{0.79 + 0.66}{2}\right) \cdot 1.20 + 2.5 \cdot \frac{1}{2} \cdot 0.35 \cdot 3.42$$

$$= 17.74 \text{ kN/m}$$

The concentrated loads due to the Tandem System (TS) are:

$$P_{TS,k} = 150 \cdot (0.64 + 0.44) + 100 \cdot (0.33 + 0.13) = 208 \text{ kN}$$

The value of the shear force at supports due to UDL is $17.74 \cdot 25/2 = 221.75$ kN.

The bending moment at mid-span due to UDL is $17.74 \cdot 25^2/8 = 1385.94$ kN-m.

The maximum value of the shear force due to TS is estimated by imposing the concentrated forces at supports as shown in Figure 7.5b. Through the influence line $\bar{V}$ in the longitudinal directions, max $V_{TS,k}$ is calculated as follows:

$$\max V_{TS,k} = 208 \cdot (1.00 + 0.95) = 405.6 \text{ kN}$$

Therefore, the maximum shear force due to Load Model 1 acting on girder A is $221.75 + 405.6 = 627.35$ kN.

The maximum bending moment due to TS is estimated by imposing the concentrated loads at mid-span. Through the influence line $\bar{M}$, the maximum bending moment is equal to $208 \cdot (6.25 + 5.65) = 2475.20$ kN-m.

Therefore, the maximum bending moment acting due to Load Model 1 on girder A is $1385.94 + 2475.20 = 3861.14$ kN-m.

At ULS, the internal forces due to LM1 should be multiplied with the safety factor $\gamma_{G,sup} = 1.35$.

7.1.2.3 Box-girder bridges

Box girders have a large torsional stiffness so that they resist eccentric loading differently than I-girder decks. As indicated in Figure 7.6, an eccentric loading P may be split into a centric loading that results in bending in the cross section and a nonsymmetrical loading that results in torsion. Torsion is resisted generally by *St. Venant torsion* and *warping torsion*. In box girders, warping torsion is usually small due to the high torsional rigidity of the box section. However, the shape of the box-girder cross section tends to distort due to torsional forces that are not distributed in the cross section walls in proportion to the St. Venant torsional stresses. Accordingly, in box girders, three effects due to torsion have to be accounted for: *St. Venant torsion*, *warping torsion*, and *cross section distortion*. Both warping torsion and distortion tend to change the shape of the cross section. Deformations due to warping torsion are out of plane for the cross section but in plane for its walls, while for distortion in plane for the cross section and out of plane for the walls. The resulting bending stresses are added to the stresses due to global bending. Stresses due to warping torsion, illustrated in Figure 7.6, are in the plane of the walls and like the stresses due to global bending are in the longitudinal direction of the bridge. On the other side, stresses due to distortion are through-thickness bending stresses.

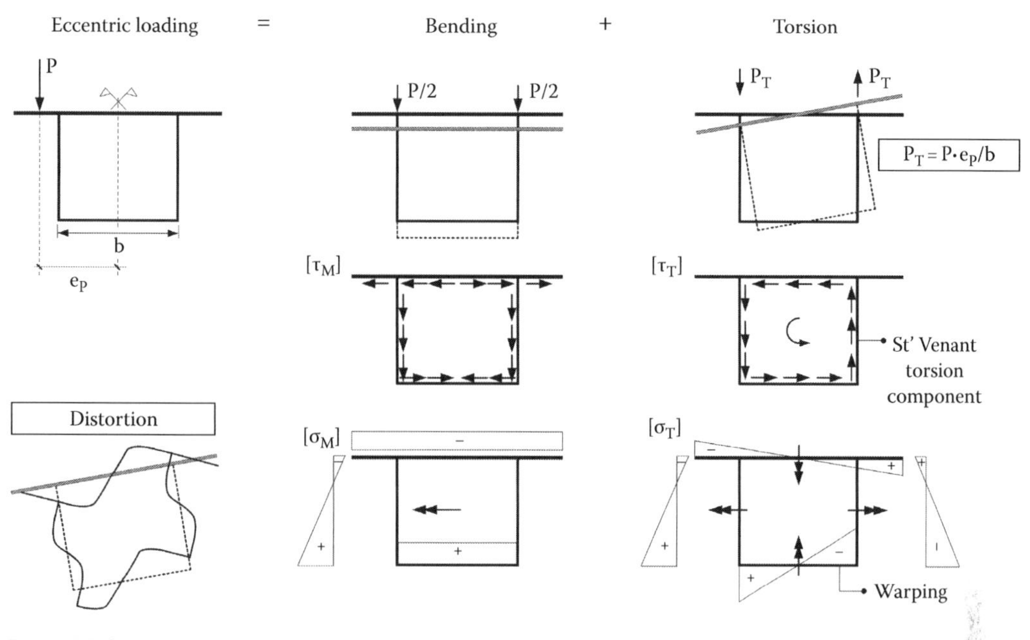

Figure 7.6 Split of eccentric loading in bending and torsion.

Distortion is controlled by diaphragms, internal cross frames, or cross bracings that are spaced along the length of the girder. More information on the treatment of torsion in bridges is given subsequently.

Box-girder bridges with one box may be modeled by a single beam, provided that the walls of the box are sufficiently stiffened by closely spaced diaphragms, cross frames, or cross bracings to prevent cross-sectional distortion; see Figure 7.7. This condition is usually fulfilled when the spacing of transverse elements is between 3.5 and 5 m, depending on the dimensions of the box. Accordingly, the model of a single box-girder bridge without distortion of the box section is a beam consisting of the entire cross section, where the top concrete flange and the bottom steel flange enter with their effective widths. The beam is loaded by the sum of the

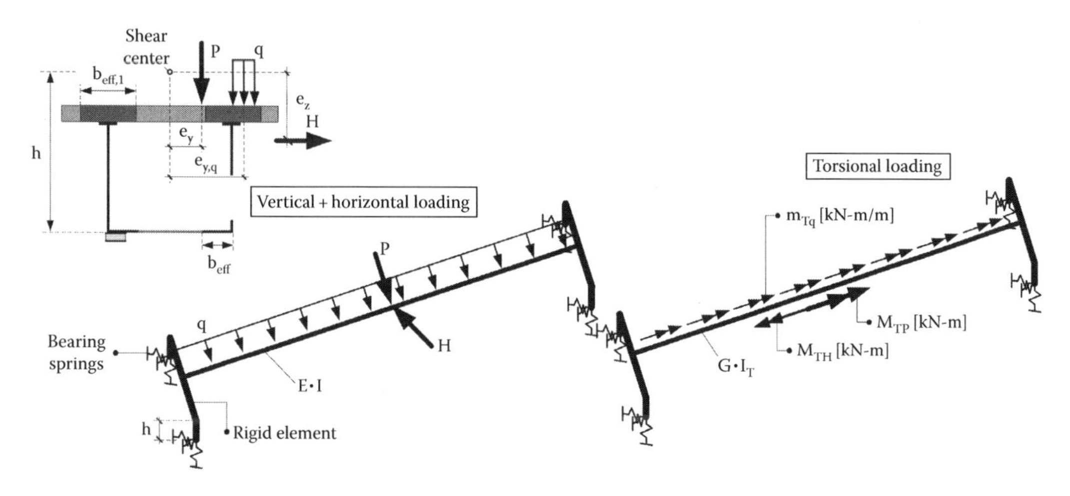

Figure 7.7 Beam model for single box girders without distortion.

vertical loads without the need for distribution in transverse direction as for the bridges with two main girders. This model produces for eccentric loading equal bending stresses in the two webs when the usual beam elements of commercial software are employed in which warping is neglected; see Figure 7.6. In case of large eccentricities and high torsional moments, a grillage model presented in Section 7.1.3 should be preferably used; see Figure 7.23 and design Example 9.4. Another option, if supported by the software, is to use special 7-DOF beam elements that are described subsequently; see Figure 7.10.

Care shall be given to the support conditions of the beam in order to correctly determine the support reactions. Since the bearings are usually positioned near or exactly below the two webs, two support reactions exist at each support section. This may be modeled by providing a rigid transverse element at the supports with a length equal to the distance of the bearings from the shear center. The supports are placed at the ends of the transverse beams so that the torsion in the main beam results in unequal support reactions.

As already mentioned, the warping part of torsion leads to additional out-of-plane deformations of the cross section and the main assumption of the *Bernoulli Theory* (plain sections remain plain) is violated; see Figure 2.19. If restrained, shear and normal stresses are developed, which when integrated over the cross section result in additional internal forces. This is shown in Figure 7.8 and is briefly discussed in the following paragraphs.

The total applied torsion M_x consists of two components; the St. Venant torsion M_{xp} and the warping torsion M_{xs}.

St. Venant torsion leads to circular shear flows τ_{xp} (primary shear stresses) inside the steel plates of the cross section. For open cross sections, for example, box girders in the construction phase where the section has not been closed by the slab, the maximum shear stress in the ith plate is calculated as follows:

$$\max \tau_{xp,i} = \frac{M_{xp}}{I_T} \cdot t_i \tag{7.4}$$

where

t_i is the thickness of the plate i
I_T is the torsional constant of the cross section

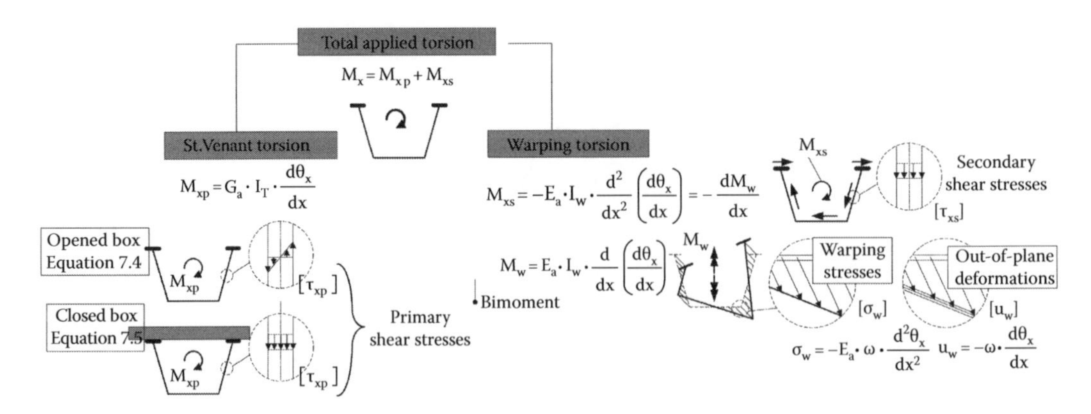

Figure 7.8 Division of torsion into St. Venant and warping torsion.

For closed cross sections, the maximum shear stress is calculated in accordance with the first formula of *Bredt* by Equation 7.5 and is considered constant across the plate's thickness; see Figure 7.8:

$$\max \tau_{xp,i} = \frac{M_{xp}}{2 \cdot A_m} \cdot \frac{1}{\min t_i} \tag{7.5}$$

where
A_m is the enclosed cross-sectional area
$\min t_i$ is the minimum wall thickness

Warping torsion M_{xs} causes direct stresses σ_w that lead to out-of-plane deformations:

$$\sigma_w = \frac{M_w}{I_w} \cdot \omega \tag{7.6}$$

where
ω is the warping function in m^2
M_w is the bimoment in kN-m^2; $\left(M_w = \int \sigma_w \cdot dA\right)$

In addition, secondary shear stresses develop that are calculated from

$$\tau_{xs,i} = \frac{M_{xs}}{t_i \cdot I_w} \cdot S_w \tag{7.7}$$

where
I_w is the warping constant
S_w is the warping sectoral area; $\left(S_w = \int \omega \cdot dA\right)$
t_i is the wall thickness

The *bimoment* M_w is a secondary internal force that causes out-of-plane cross section deformations u_w and as shown in Figure 7.8 is a function of the warping rigidity $E_a \cdot I_w$ and the second derivative of the twisting angle θ_x. The warping function ω expresses the longitudinal deformation shape of the cross section due to a unit bimoment $M_w = 1$ and depends on the location of the shear center. It determines the warping constant according to the following equation:

$$I_w = \int_A \omega^2 \cdot dA \tag{7.8}$$

The variation of the warping function for a closed box section is shown in Figure 7.9. The calculation of $\omega = \omega(y, z)$ is out of the scope of this book. Detailed information on this issue may be found in [7.18], [7.37], and [7.17].

Analysis software with special beam elements with seven degrees of freedom is usually used for investigating the warping behavior of box girders. The seventh DOF is the first derivative of the twisting angle $\theta'_x = d\theta_x/dx$ and its implementation is summarized in Figure 7.10. It can be observed that the DOFs u, θ_y, and θ_z refer to the gravity center of

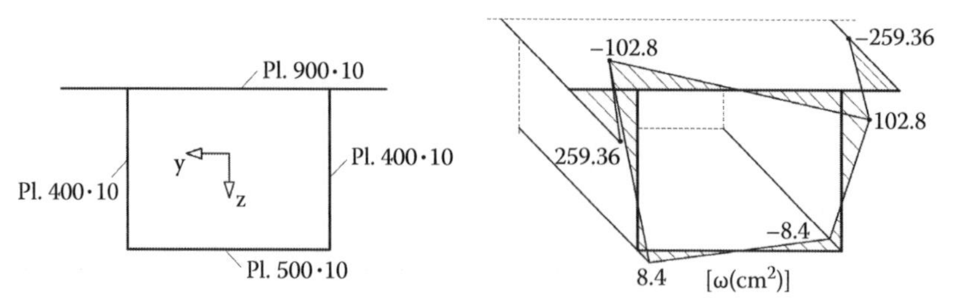

Figure 7.9 Warping function of a closed box section.

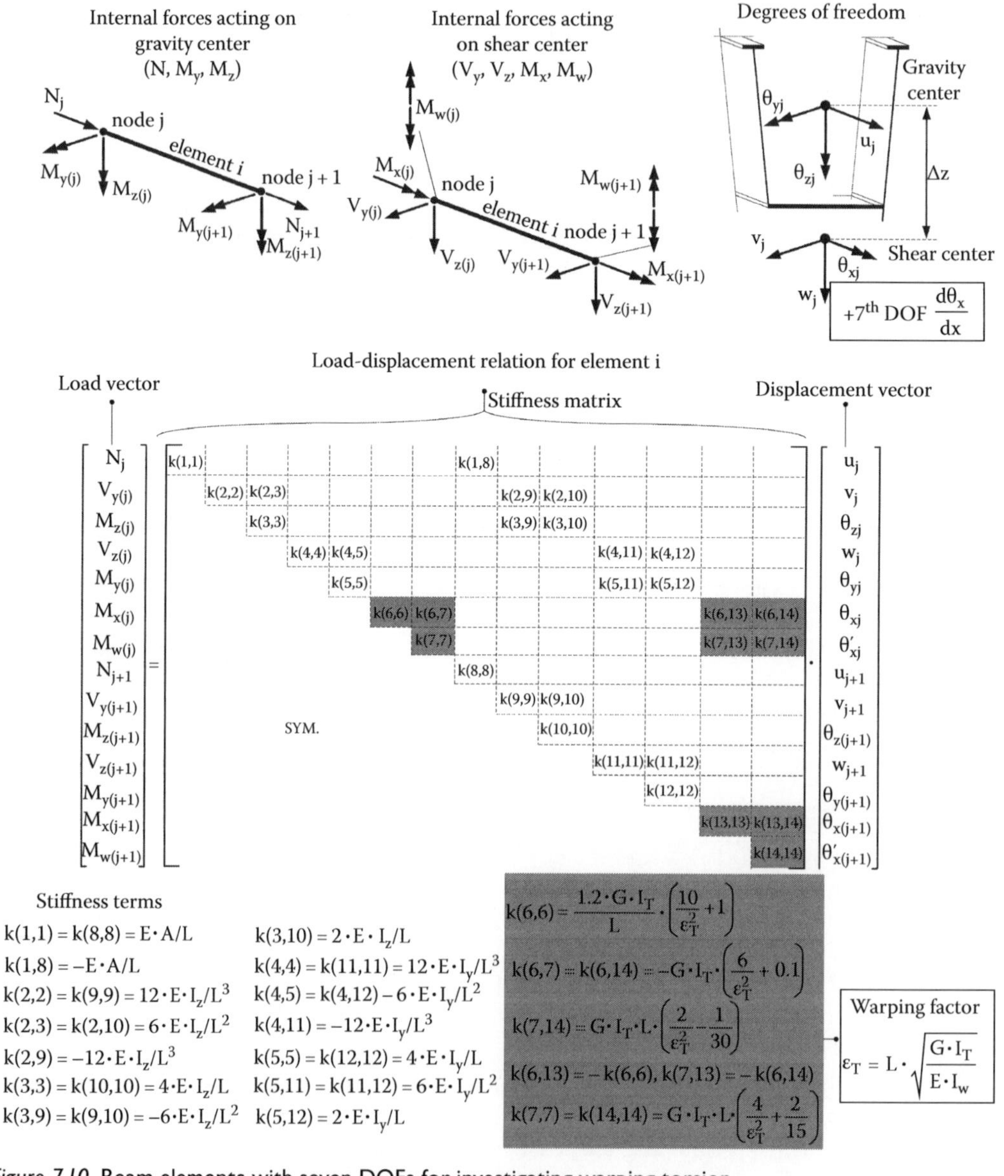

Figure 7.10 Beam elements with seven DOFs for investigating warping torsion.

the cross section and correspond to N, M_y, and M_z. The shear and torsional DOFs θ_x, v, w, and θ'_x correspond to M_x, V_y, V_z, and M_w, respectively and refer to the shear center of the cross section. Therefore, **the distance Δz between the gravity and the shear center is of great importance**. The greater this distance is, the higher the magnitude of the torsional internal forces becomes.

In Figure 7.10, the stiffness matrix of a 7-DOF beam element is demonstrated. One can also see that the torsional stiffness terms (colored with gray in the stiffness matrix) are dependent on the *warping factor* ε_T. Cross sections with high torsional rigidity $G \cdot I_T$ have a high warping factor. In such cases, warping torsion is of negligible magnitude ($M_{xs} \approx 0$, $M_w \approx 0$) and the usual 6-DOF implementation can be considered as acceptable. Cross sections with high values of ε_T are closed box girders or open box girders after concreting. Open box girders may need a detailed warping investigation during the erection stage because the torsional rigidity is low and the distance between the gravity and shear the center is high. After casting the concrete deck, the torsional properties of the cross section are greatly enhanced and also the distance between the gravity and the shear center becomes smaller.

REMARK 7.1

In most cases, shear deformations are quite small because bending moments are much higher than the shear forces acting on the walls of the boxes. The stiffness matrix demonstrated in Figure 7.10 does not include any complementary terms for the shear deformations and can be described for the majority of the bridges as accurate. For small-span bridges, the influence of shear deformations should be taken into account. An improved stiffness matrix for such bridges is found in [7.25].

By taking a closer look at the torsional terms (colored with gray) of the 14×14 stiffness matrix, one can observe that the general expression of k(m, n) is

$$k(m,n) = T(G \cdot I_T, L) \cdot \left[W(\varepsilon_T) + \beta\right], \quad m \text{ (or n)} = 6,7,13,14 \tag{7.9}$$

where

$T(G \cdot I_T, L)$ is a function that expresses the St. Venant torsion

$W(\varepsilon_T)$ is a function that expresses the influence of warping; $W(\varepsilon_T) = \alpha/\varepsilon_T^2$

Obviously, if $\varepsilon_T \to \infty$, then $W(\varepsilon_T) \to 0$ and warping-related forces can be neglected. In practice, systems with $\varepsilon_T > 10$ can be considered as "free of warping." Warping-free cross sections are those with $I_w \to 0$ (i.e., square boxes with flanges and webs of the same thickness). Unfortunately, such sections are rarely used in bridges.

For the majority of the box-girder bridges, the cross-sectional properties vary longitudinally and the expression for ε_T indicated in Figure 7.10 cannot be applied. To account for these variations, the following weighted ε_T calculated for each span can be used:

$$\varepsilon_T = \int_L \sqrt{\frac{G \cdot I_T}{E \cdot I_w}} \cdot dL \approx \sum_i \left[\sqrt{\left(\frac{G \cdot I_T}{E \cdot I_w}\right)} \cdot L\right]_i, \quad i = 1,2,3\ldots\text{segment} \tag{7.10}$$

EXAMPLE 7.2

A stress calculation example of an open box steel cross section with 7-DOF beam elements is given in Figure 7.11. This is the cross section of a straight simply supported bridge with a span of 60 m. The distributed load q = 20 kN/m is a construction load (i.e., wet concrete), and an accidental eccentricity with a peak at mid-span equal to max v_0 = L/240 has been considered. A second-order theory has been applied so that the increase of the torsional flexibility of the cross section due to M_x is taken into account.

At supports, the warping torsion M_{xs} is dominant and therefore, M_{xp} is not shown. However, torsional shear stresses are much smaller than those due to V_z. A bending moment around the weak axis is developed as well but due to its low magnitude is also not illustrated. The most interesting part of the example refers to the normal stresses due to the bimoment M_w at mid-span. One can see that min σ_w = −2.32 kN/cm², almost 50% of normal stresses (−4.89 kN/cm²) due to the bending moment M_y. It may be seen that warping direct stresses add in the web closer to the load and subtract in the web far from the load. Therefore, neglecting direct warping stresses would be unsafe for the most loaded web. This conclusion is justified also from the warping value ε_T that is equal to $0.93 \ll 10$. Moreover, the distance between the shear center and the gravity one is 269.2 cm.

A commonly used method to avoid an analysis with 7-DOF beam elements and to include the warping torsion effects is the *Tension Element Analogy* (TEA) method. It is summarized in Table 7.1. The TEA method is based on the similarity of the equilibrium equations of a deflected element under bending and tension (*second-order theory analysis*) and an element under torsion. Therefore, using a commercial software, the warping components M_{xs} and M_w can be estimated.

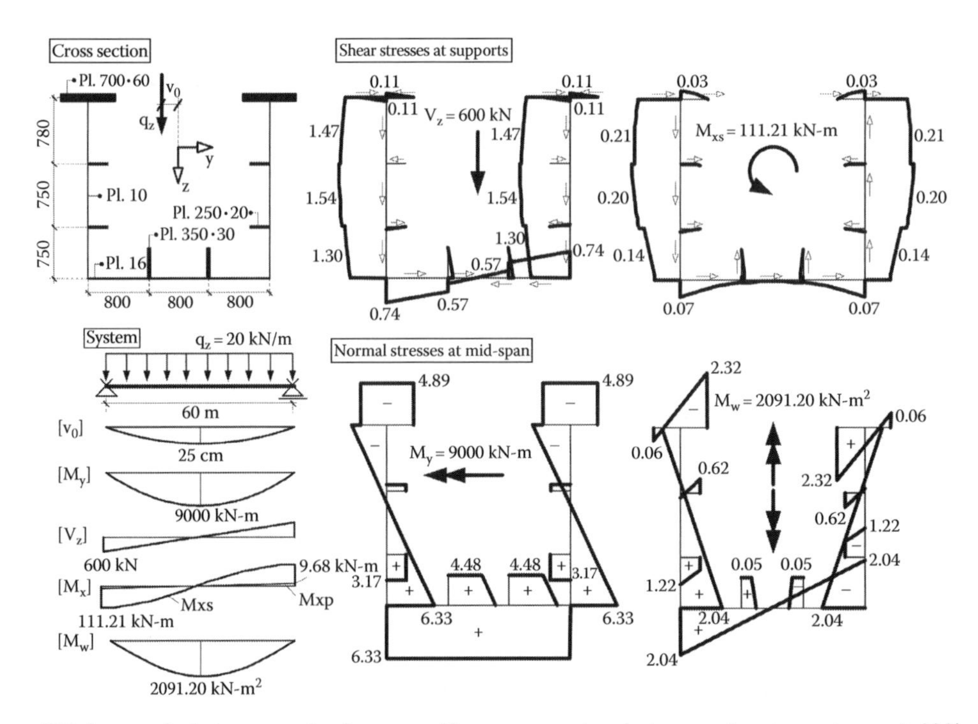

Figure 7.11 Stress calculation example of an opened box cross section during erection stage; stresses in kN/cm².

Table 7.1 Analysis of warping torsion with the *tension element analogy* method

Warping torsion		*Bending with tension force and second-order theory*
$G_a \cdot I_T$, $E_a \cdot I_w$; M_x, m_x; L	→	$E_a \cdot I_y$; P_z; q_z; N^*; L
Equilibrium equations		
$E_a \cdot I_w \cdot \theta'''' - (\mathbf{G_a} \cdot \mathbf{I_T}) \cdot \theta'' = m_x$	→	$E_a \cdot I_y \cdot w'''' - N^* \cdot w'' = q_z$
Twisting angle θ	→	Displacement w
Torsional rigidity $G_a \cdot I_T$	→	Tension force N^*
Warping rigidity $E_a \cdot I_w$	→	Bending stiffness $E_a \cdot I_y$
Concentrated torque M_x	→	Concentrated force P_z
Uniformly distributed torsional moment m_x	→	Uniformly distributed load q_z
Internal forces		
Bimoment M_w [kN-m²]	→	Bending moment M_y [kN-m]
Total torsional moment M_x [kN-m]	→	Transverse forces V_z^* [kN]
Warping torsion M_{xs}	→	Shear forces V_z [kN]
St.Venant torsion $M_{xp} = M_x - M_{xs}$ [kN-m]	→	$\Delta V = V_z^* - V_z$
Support boundary conditions		
Simple torsional support ($M_{w,support} = 0$) $\theta = 0$, $\theta'' = 0$	→	Pinned support ($M_{y,support} = 0$) $w = 0$, $w'' = 0$
Fixed torsional support ($M_{w,support} \neq 0$) $\theta = 0$, $\theta' = 0$	→	Fixed support ($M_{y,support} \neq 0$) $w = 0$, $w' = 0$

Source: Roik, K., *Vorlesungen über Stahlbau—Grundlagen*, Ernst & Sohn, Berlin, Germany, 1983.

Note: Internal forces with (*) refer to the **non-deformed** beam axis.

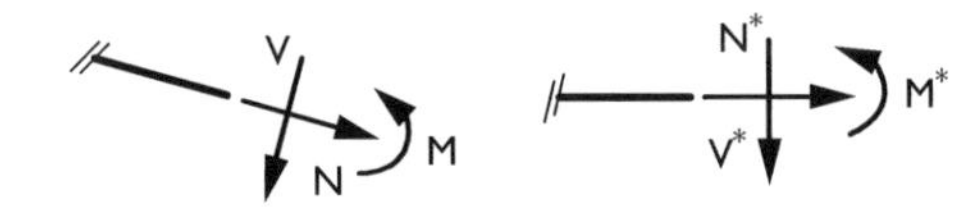

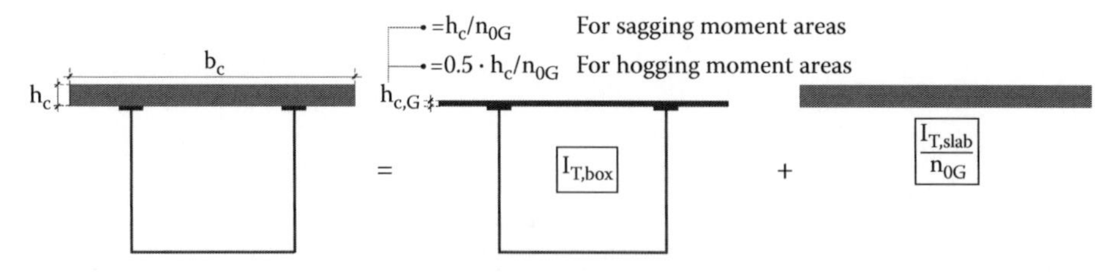

Figure 7.12 Calculation of the torsional constant I_T of composite box girder.

As already mentioned, after concreting, a composite closed cross section with high torsional rigidity is formed; see Figure 7.12. The total torsional constant of the composite box-girder $I_{T,tot}$ section is calculated through the use of the modular ratio n_{0G} in Table 6.8 as follows:

$$G_a \cdot I_{T,tot} = G_a \cdot I_{T,box} + G_c \cdot I_{T,slab} \Rightarrow I_{T,tot} = I_{T,box} + \frac{I_{T,slab}}{G_a/G_c}$$

$$\Rightarrow I_{T,tot} = I_{T,box} + \frac{I_{T,slab}}{n_{0G}} \tag{7.11}$$

where

$I_{T,box}$ is the torsional constant of an equivalent closed cross section composed by the steel cross section and the concrete slab with a thickness equal to h_c/n_{0G}

$I_{T,slab}$ is the torsional constant of the concrete slab $\left(=\frac{1}{3} \cdot b_c \cdot h_c^3\right)$

In case of creep, n_{0G} can be replaced by n_{LG} of Equation 6.21. Obviously, due to creep, torsional redistributions from concrete to structural steel will take place. In continuous bridges, secondary torsional loadings will be developed.

7.1.2.4 Bridges with two main girders and horizontal bracing between the lower flanges

Bridges with two main girders and continuous horizontal truss bracing between the lower flanges (Figure 7.13) may be modeled in a similar way as box-girder bridges by replacing the horizontal bracing with an equivalent continuous steel sheet that closes the section. Then, the torsional constant I_T of the cross section can be calculated. The analysis provides the shear flow in the box. The shear flow in the bottom wall of the box must then be transformed to forces in the bracing members.

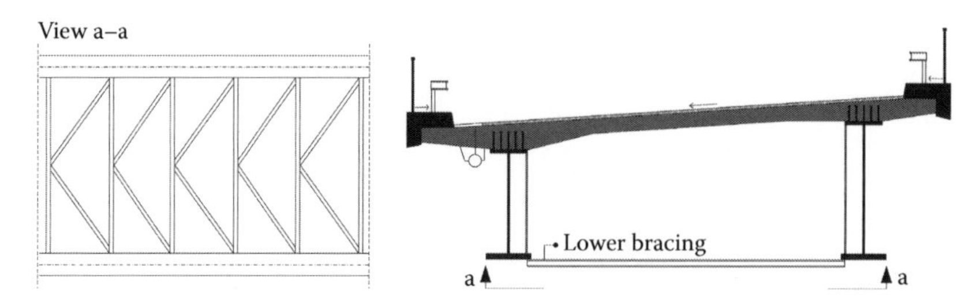

Figure 7.13 Bridge with two main girders and horizontal bracing at the lower flange.

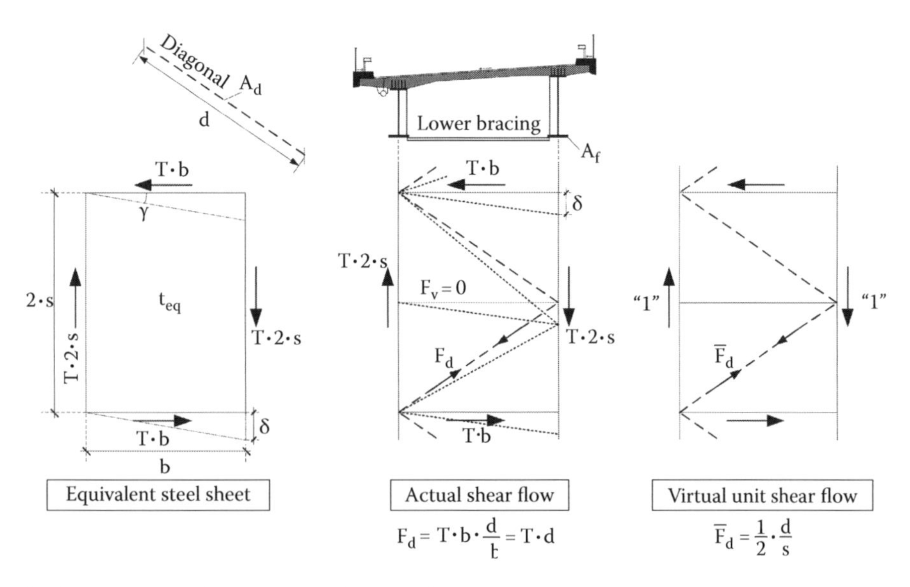

Figure 7.14 Application of the virtual work principle.

The thickness of the sheet is determined by comparing its shear stiffness with the corresponding shear stiffness of the bracing through the use of the *virtual work principle*. It has to be mentioned that cross sections with lower bracings may be more cost-effective than box girders both for cases of roads with straight or curved layouts.

The cross section of Figure 7.14 is equipped with a lower bracing that consists of diagonals A_d and post beams with a center-to-center distance equal to s. Due to the shear flow T in the steel sheet, the displacement δ is

$$\delta = \gamma \cdot b = \frac{T \cdot b}{G_a \cdot t_{eq}} \tag{7.12a}$$

According to the principle of virtual work, the displacement δ can be

$$"1" \cdot \delta = \sum \frac{F_d \cdot \bar{F}_d}{E_a \cdot A_d} \cdot d_i = \frac{F_d \cdot \bar{F}_d}{E_a \cdot A_d} \cdot 2 \cdot d$$

$$\Rightarrow \delta = \frac{T \cdot d^3}{E_a \cdot A_d \cdot s} \tag{7.12b}$$

The equivalent sheet thickness can be derived now from Equations 7.12:

$$t_{eq} = \frac{E_a \cdot A_d \cdot s \cdot b}{G_a \cdot d^3} \tag{7.13}$$

Equation 7.13 is valid only when $A_f \gg A_d$ that is anyway the most usual case in bridge engineering applications.

It is important to note that the "fictitious sheet" with the equivalent thickness t_{eq} participates only in torsion and **not in bending**. The cross sections are therefore different in bending and torsion. This may be taken into account in practical analysis by changing the modulus of elasticity of steel to $E_a = 0$ for the lower sheet but keeping the value of its shear modulus G_a.

More exact values for t_{eq} and for various bracing systems are given in Table 7.2. Forces in the diagonals and the post beams are also offered.

Table 7.2 Equivalent thickness of bracing systems for shear

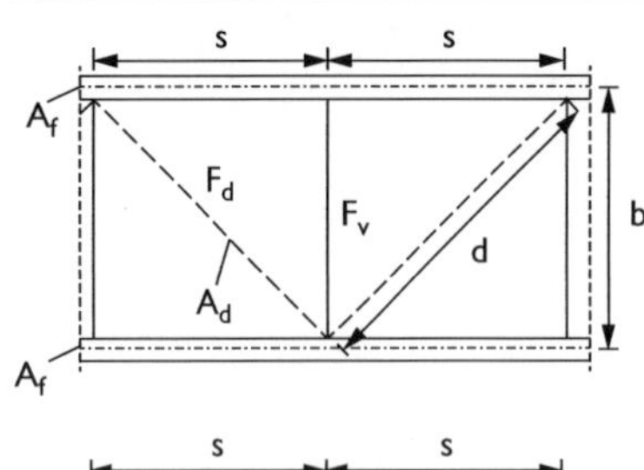

Equivalent thickness of a sheet:

$$t_{eq} = \frac{E_a}{G_a} \cdot \frac{s \cdot b}{(d^3/A_d) + (s^3/6 \cdot A_f)}$$

Forces: $F_v = 0, F_d = T \cdot d$

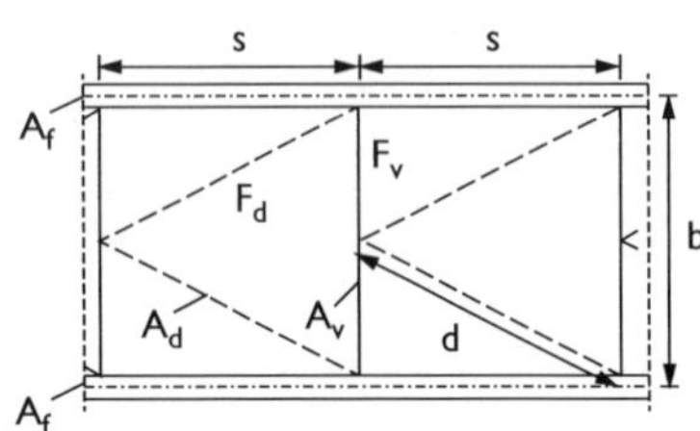

Equivalent thickness of a sheet:

$$t_{eq} = \frac{E_a}{G_a} \cdot \frac{s \cdot b}{(2 \cdot d^3/A_d) + (b^3/4 \cdot A_v) + (s^3/24 \cdot A_f)}$$

Forces: $F_v = T \cdot \frac{b}{2}, F_d = T \cdot d$

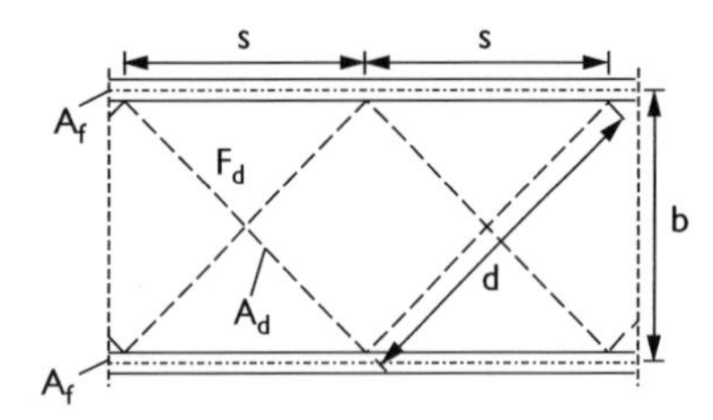

Equivalent thickness of a sheet:

$$t_{eq} = \frac{E_a}{G_a} \cdot \frac{s \cdot b}{(d^3/2 \cdot A_d) + (s^3/24 \cdot A_f)}$$

Forces: $F_d = T \cdot \frac{d}{2}$

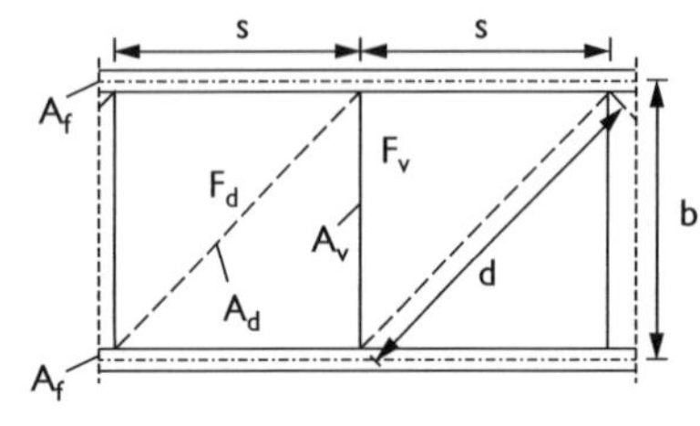

Equivalent thickness of a sheet:

$$t_{eq} = \frac{E_a}{G_a} \cdot \frac{s \cdot b}{(d^3/A_d) + (b^3/A_v) + (s^3/24 \cdot A_f)}$$

Forces: $F_v = T \cdot b, F_d = T \cdot d$

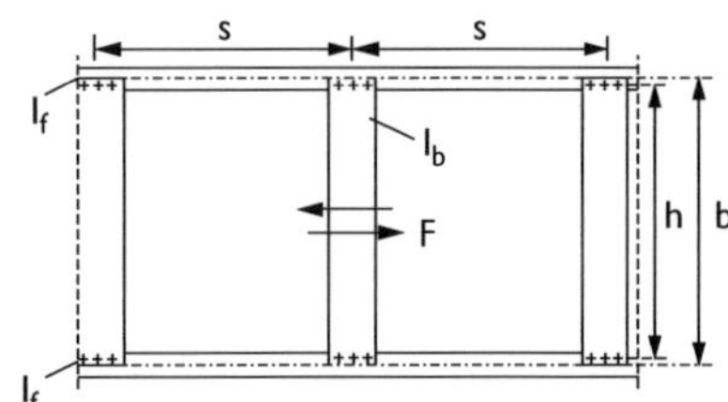

Equivalent thickness of a sheet:

$$t_{eq} = \frac{E_a}{G_a} \cdot \frac{24 \cdot I_f \cdot (h/b)}{h \cdot s^2 \cdot \left(1 + 2 \cdot \frac{h}{s} \cdot \frac{I_f}{I_b}\right)}$$

Forces: $F = T \cdot s$

Source: Roik, K., *Vorlesungen über Stahlbau—Grundlagen,* Ernst & Sohn, Berlin, Germany, 1983.

Note: A_f, cross-sectional area of the flanges; A_d, cross-sectional area of the diagonals; I_f, moment of inertia of the flanges; I_b, moment of inertia of the post beams; b, distance between flanges; s, distance between post beams; F_d, force in diagonal; F_v, force in post beam; and T, shear flow [kN/m].

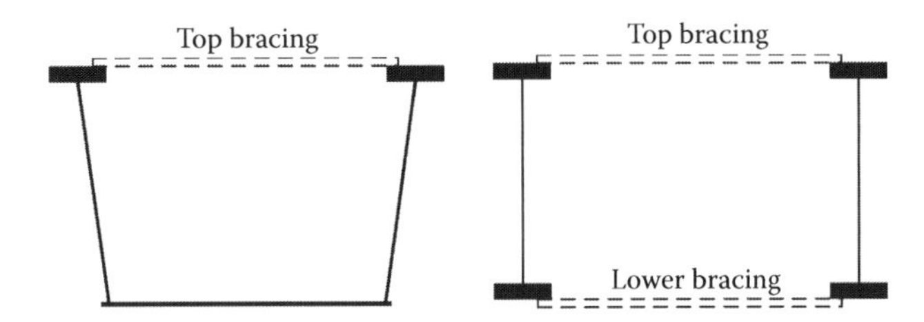

Figure 7.15 Quasi-closed cross sections with bracings.

REMARK 7.2

The equations in Table 7.2 can also be used for determining the torsional sectional properties of opened box girders with a top bracing or plate girders with top and lower bracings during erection, see Figure 7.15. Thus, opened cross sections can be treated as closed ones. As already discussed in Chapter 2, bracings connected with the flanges are provided to ensure stability.

By taking a closer look at the equations for the equivalent thickness in Table 7.2, one can observe that t_{eq} is highly dependent on the distance s between the post beams and the cross-sectional area A_d of the diagonals. The most appropriate values for these two parameters to choose are the ones for which the warping factor ε_T (Figure 7.10) becomes greater than 10. This is a good strategy to avoid warping effects and the laborious analysis with 7-DOF beam elements.

7.1.3 Grillage models

7.1.3.1 General

The most popular computer-aided modeling method for the analysis of composite bridges with multiple main girders is simulation by means of a plane grillage system [7.14], [7.21], [7.27–7.29], [7.34]. This is due to the fact that this system is easy to apply and comprehend as well as that it has been proved to be sufficiently accurate for a wide variety of bridge decks. This refers to both analysis and design of the bridge for the most common design situations as well as the construction stages. In this model, the structure is idealized by means of a series of longitudinal and transverse beam elements rigidly interconnected at nodes. Each element is given an equivalent bending and torsion inertia to represent the relevant portion of the deck.

7.1.3.2 Simply supported plate-girder bridges

Figure 7.16 illustrates a grillage representation of a simply supported composite bridge with four main girders. Longitudinal grillage members are arranged to represent the main girders with the inertia properties of the composite section (steel section with a part of the slab corresponding to the effective width). Transverse members represent the deck slab with thickness h_c equal to the thickness of the slab and width b equal to the distance of the transverse beams. A noncracked flexural rigidity for the slab elements is usually applied. Many designers though consider cracking of the slab by inputting an average rigidity $E_c \cdot I_c = 0.5 \cdot (E_c \cdot I_1 + E_c \cdot I_2)$, where $E_c \cdot I_1$ is the "uncracked" and $E_c \cdot I_2$ the "cracked" flexural rigidity. For the spacing of the slab elements, it is convenient to choose a width b equal to the distance of the axle loads (i.e., for Load Model 1, b = 1.20 m). This leads to dense arrangement of the transverse elements that is due to profound reasons advantageous. The torsion constant per unit width of slab is given by [7.14] and is equal to $I_T = h_c^3/6$. For cracked concrete, *Poisson's ratio* is equal to zero and the torsional rigidity of the slab elements becomes equal

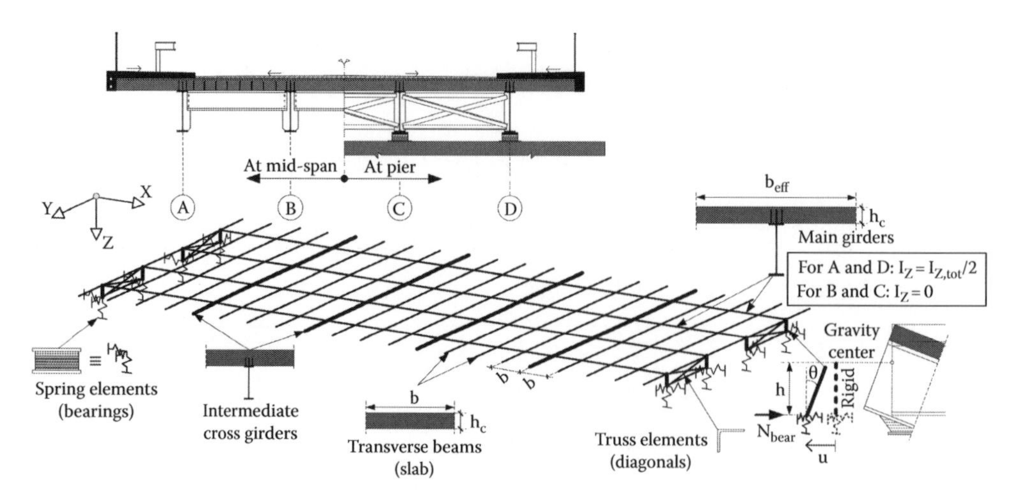

Figure 7.16 Grillage model of a simply supported composite bridge.

to $G_c(\nu_c = 0) \cdot I_T = E_{cm} \cdot b \cdot h_c^3/12$. However, a torsionless approach for the slab elements for both directions may also be followed ($I_T = 0$) so that discontinuities in bending moments are avoided. This has little effect on the final results.

The total in-plane second moment of area of the slab is equally shared between the two extreme main girders (A and D), while the intermediate girders (B and C) are given $I_z = 0$. This is because wind loads mainly act on the edge girders of the bridge.

The bearings are represented by three axial springs of equivalent stiffness corresponding to the relevant stiffness properties in horizontal and vertical directions; the calculation of the bearings' stiffness is presented in Section 13.2.2. The axes of the main beams coincide with the center of gravity of their cross sections. However, the bearings are positioned beneath the lower flange. Accordingly, rotations of the main girders result in horizontal deformations u of the bearings and additional support reactions N_{bear}. The support nodes are therefore put at a lower level from the grillage members and are connected to the longitudinal beams by rigid vertical bars whose height h is equal to the distance between the center of gravity of the main composite beams and the bottom flange; for better accuracy, the shear center of the cross section should be used which is assumed to be the "real" center of rotation. In case of intermediate cross girders whose stiffness may influence the transverse distribution of the vertical loads, these girders are taken into account with beam elements of appropriate stiffness.

At piers, truss elements are used for the representation of the cross braces; see Figure 7.16. Due to the height h of the rigid elements, the geometry of the bracings in the model may not follow the exact geometry of the bracings in real structure. A height adjustment for the rigid elements may then be necessary. This should be done only for the purpose of estimating the forces of the bracing members due to horizontal loadings, that is, wind or earthquake. It has to be stated that in most bridges, the gravity center of the composite cross sections is located near the top flange. For such cases, a height adjustment of "few centimeters" will not cause any considerable difference in the results.

The vertical loads act directly on the slab elements and for many different arrangements that cover all the possible unfavorable situations. A convenient way to avoid the analysis of noncritical loading cases is to locate the areas of the carriageway that lead to unfavorable action effects. This can be done with the influence lines produced by the *Courbon* method. In Figure 7.17, one can see the loaded areas for favorable and unfavorable effects for the extreme girder A.

It is important to note that the slab elements of the grillage model are mainly used for the transverse distribution of the vertical loads on the main girders and for modeling the in- and

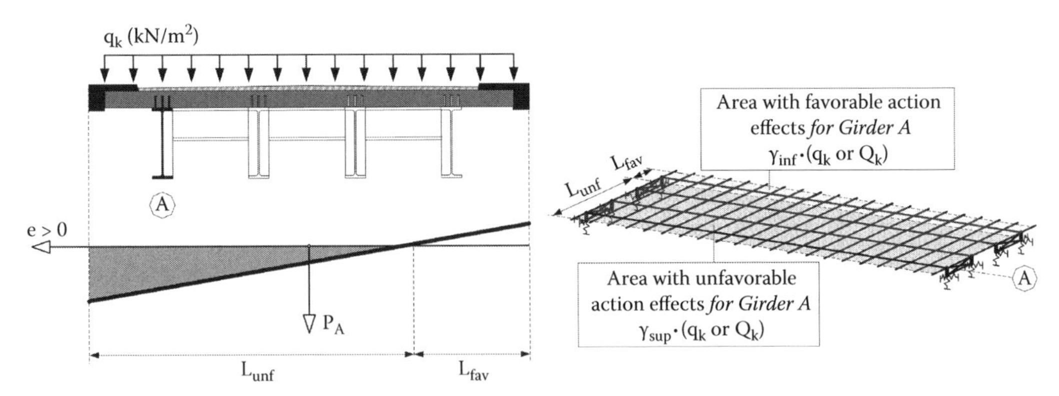

Figure 7.17 Determination of areas with favorable and unfavorable action effects.

out-of-plane stiffness of the deck slab. The internal forces in the slab elements should not be used as final values for the design of the deck slab. This is due to the fact that the slab's monolithic behavior is inaccurately described by transverse slab elements, especially in the case of concentrated wheel loads. Therefore, local analysis of the deck slab should be treated separately from the global analysis preferably with FEM. In case of widely spaced main girders, nominal longitudinal slab elements can be placed in between so that additional bending moments and torques in the slab can be calculated. This also offers the advantage of an easier load application.

7.1.3.3 Continuous plate-girder bridges

A two-span plate-girder bridge is shown in Figure 7.18. In support regions, where the concrete is considered as fully cracked, the properties of the longitudinal elements are determined from the properties of the steel section and the longitudinal reinforcement

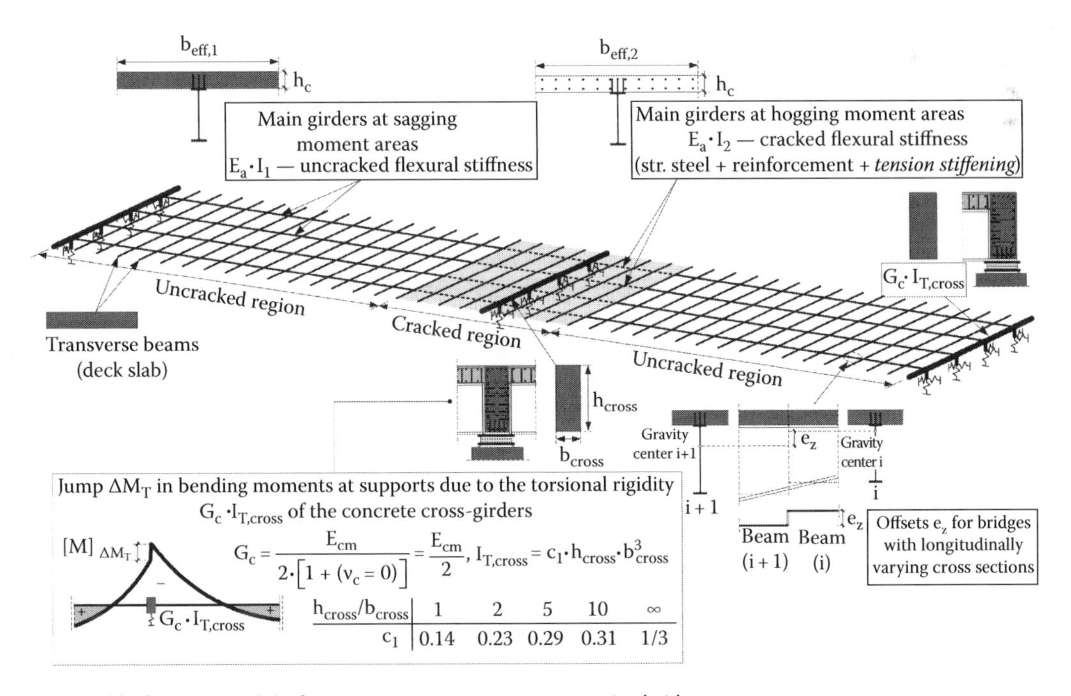

h_{cross}/b_{cross}	1	2	5	10	∞
c_1	0.14	0.23	0.29	0.31	1/3

Figure 7.18 Grillage model of a two-span continuous composite bridge.

(cracked flexural rigidity $E_a \cdot I_2$). The tension-stiffening effect can also be considered in the cross-sectional properties of the main girders; this is also discussed in Remark 7.5. At sagging moment areas, the uncracked flexural rigidity $E_a \cdot I_1$ is applied, exactly as in the case of the simply supported bridge of Figure 7.16.

Concrete cross girders possess a high torsional rigidity ($G_a \cdot I_{T,cross}$ in Figure 7.18) that leads in additional negative bending moments ΔM_T at supports especially for nonsymmetrical loadings, for example, concentrated wheel loads TS. Concrete is considered as cracked and the shear modulus G_c is equal to $E_{cm}/2$. The torsion constant of an orthogonal cross girder $b_{cross} \cdot h_{cross}$ is $c_1 \cdot h_{cross} \cdot b^3{}_{cross}$, where c_1 is a constant that depends on the ratio h_{cross}/b_{cross}.

Attention must be paid during the modeling of composite bridges whose total depth varies longitudinally. A dense discretization of the main girders is necessary so that the longitudinal variation of the cross-sectional properties is adequately taken into account. The eccentricity e_z between the centroids of adjacent cross sections should also be carefully considered because additional bending moments due to axial forces are developed, for example, due to shrinkage and temperature differences. Many software packages do not account for these eccentricities and designers should implement them manually in their models as offsets. Moreover, the eccentricities e_z change with time due to creep and therefore, secondary internal forces will emerge.

In continuous bridges, the difference in stiffness between the cracked cross sections in hogging moment areas and the noncracked ones in sagging moment areas determines the results given by the global analysis. An overestimation of the length of the cracked zones at internal supports will result in an underestimation of the negative bending moments at supports. Excessive deformations will also be calculated and this will have a negative effect on the determination of the precambering values. EN 1994-2 [7.11] offers two methods for calculating the "cracked lengths" based on an elastic global analysis.

The first method, which is also known as the *general method,* is illustrated in Figure 7.19. The first step is to determine the extent of cracking in the longitudinal beams by conducting an uncracked global analysis. The flexural stiffness $E_a \cdot I_1$ is assumed throughout, and an envelope of bending moments for the characteristic SLS combination is calculated. The calculation of the envelope should include the casting sequence. Thereafter, the extreme-fiber tensile stresses max σ_c in concrete are estimated (end of second step). Sections in which max σ_c exceeds **twice** the mean value of the axial tensile strength f_{ctm} given in EN 1992-1-1 or in Table 6.1 should be considered for the next analysis as cracked (third step). The procedure continues till convergence is achieved; see also Section 10.3.4 and Example 10.1.

The length of the cracked regions depends on many parameters such as the casting sequence, the intensity of concrete shrinkage, and the ratio of adjacent spans. In Figure 7.19, one will observe the discontinuities of the tensile stresses in concrete at the end of each slab segment. This is because the cross-sectional properties of the main girders vary longitudinally. It should not be a surprise the fact that in zero-bending moment points, the concrete tensile stresses are not zero. This may be due to shrinkage and/or thermal actions. Further observations have to do with the discontinuity and the symmetry of the cracked zones. Indeed, isolated and/or nonsymmetrical cracked regions may also emerge in cases of symmetrical systems. Both are associated with the casting sequence. It is also important to notice that the calculation accuracy of the length of the cracked regions is linked to the adopted meshing for the beam elements, the offsets e_z discussed previously, and other modeling parameters.

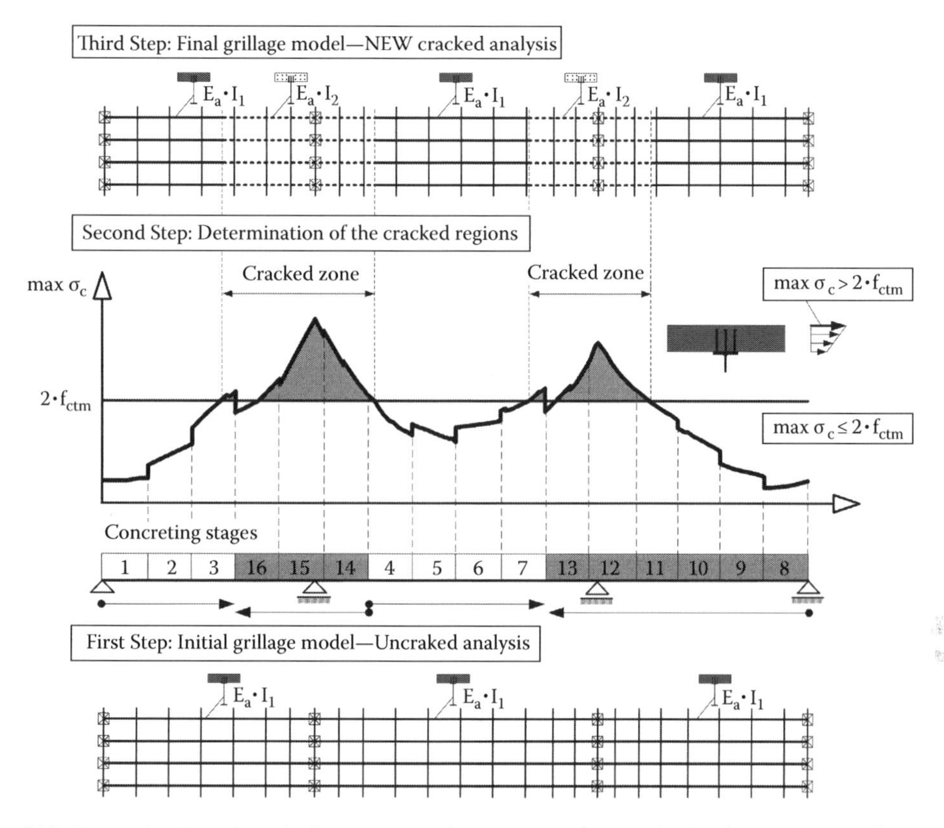

Figure 7.19 Determination of cracked regions in a three-span composite bridge (general method).

REMARK 7.3

It has been stated that the sections in which max σ_c exceeds **twice** the mean value of the axial tensile strength f_{ctm} should be considered as cracked. The main reasons for this enhancement are the following:

- Tension stiffening effects lead to an increased strength against tensile stresses (especially during initial cracking).
- Concrete's real strength is likely to be higher than the mean strength f_{ctm}.
- The general method is based on an envelope of bending moments. Therefore, the cracked regions will not be as extensive as determined by the analysis.

For finding the cracked regions of the longitudinal members, it can be assumed that the short-term cross-sectional values are the critical ones and that creep and shrinkage are neglected. This does not mean that creep and shrinkage are not present, but that their effect on the length of the cracked zones is negligible. Indeed, creep reduces the tensile stresses in concrete due to shrinkage; see Equation 6.20. Therefore, only two characteristic combinations can be applied with traffic and with temperature as leading actions. It is important to note that temperature effects due to hydration of cement are taken into account according to Section 6.1.5. This is further discussed in Section 10.3.4.

The second method for including the effects of cracking in the global analysis is known as the *simplified method*. It is a noniterative method in which the cracked flexural stiffness $E_a \cdot I_2$ is used over 15% of the span on each side of each internal support and the uncracked values $E_a \cdot I_1$ elsewhere. The basic requirement is that all the ratios of the length of adjacent continuous spans (shorter/longer) between supports are at least 0.6. This method is applicable only to some situations as conventional continuous bridges and framed bridges. The simplified method offers good results when it is known that the casting sequence chosen leads to limited cracked zones. In any other case, cracked zones may be two or even three times larger than the 15%—length of the simplified method. Then the general method should be applied.

In case of *uncracked analysis*, the bending moments at internal supports should be reduced by 10% so that redistributions due to cracking are taken into account. For each load case, the internal forces and moments after redistribution should be in equilibrium with the loads. Uncracked analysis is rarely used in composite bridges.

7.1.3.4 Skew bridges

In skew bridges, the support abutments or piers are placed at angles other than 90° (in plain view) from the longitudinal centerlines of the girders. There are different ways of defining the skew angle. Usually, it is defined as the angle between the longitudinal axis of the bridge and a line square to the supports, although a different convention may be used. Figure 7.20 shows different cases of skewed bridges. The presence of skew affects the geometry and the behavior of the structure. Special phenomena, like twisting and out-of-plane rotation of the main girders during concreting, uplifting forces at bearings, and fatigue problems due to out-of-plane web distortion, makes the analysis and design of skewed bridges intricate. The transverse elements representing the slab are usually oriented perpendicular to the main girders (orthogonal mesh); this is the most usual grillage model used by the designers. Alternatively, the transverse members can be placed parallel to the line of supports (skewed mesh). Generally, the skewed mesh is convenient for low skew angles ($\theta < 20°$) or when the intermediate bracing is not arranged square to the main girders.

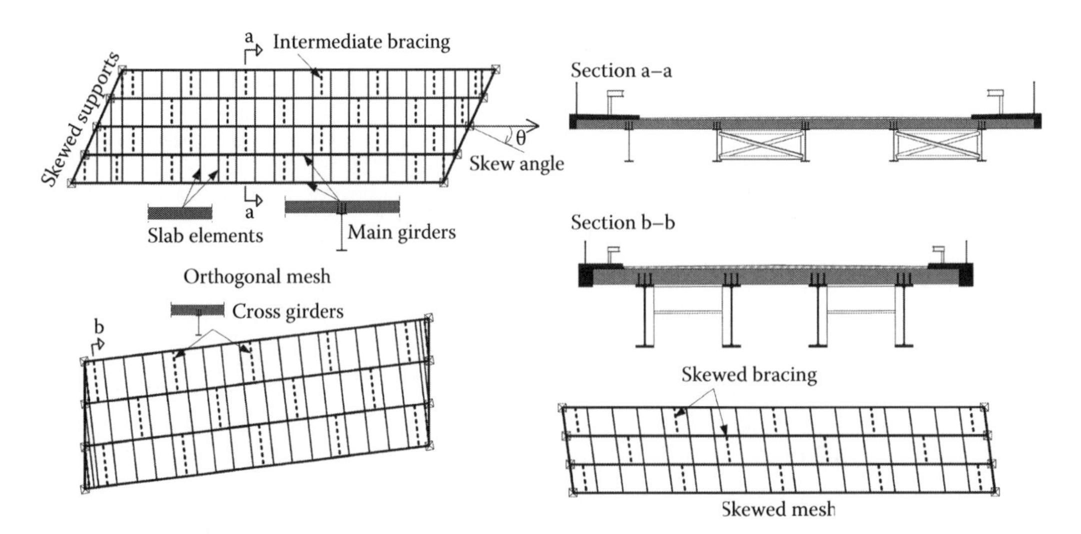

Figure 7.20 Grillage models for skew bridges.

7.1.3.5 Curved bridges

Curved composite bridges display unique behavioral characteristics, some of which are not immediately obvious. The presence of curvature affects the geometry and as a consequence, the behavior of the structure. Figure 7.21 shows the internal forces developed in a deck element of a curved twin-girder bridge due to external loadings q_x, q_y, q_z, and m_x. Due to the plan curvature R and the distance between the shear and the gravity center z_{SC}, longitudinal bending moments M_y and torques M_x are coupled. This leads to a significant enhancement of the bending moments M_y that are usually much greater than the moments developed in straight bridges. Torsional moments m_{xI} are due to loadings that act eccentrically to the shear center of the cross sections. One can see that additional torsional moments m_{xII} are developed due to the curvature R and the distance z_{sc}.

Curved decks pose no particular problem for grillage modeling. A curved bridge deck can be represented by a grillage of curved members or of straight members. Some computer programs support curved members but others do not. Generally, a grillage of straight beams with a very fine mesh is *for small values* of curvature sufficiently accurate. For highly curved bridges, 3D models (see Section 7.1.4) or FE models should be used for comparison.

It has to be pointed out that for the majority of the small- and medium-span bridges with open steel cross sections, the concrete slab attracts a non-negligible part of the torsional loading. This is due to the fact that the torsional rigidity $G_c \cdot I_{Tc}$ of the deck slab may be considerably higher than the torsional rigidity $G_a \cdot I_{Ta}$ of the steel girders. Therefore, the rigidity $G_c \cdot I_{Tc}$ should be included in the cross-sectional properties of the main girders; see Figure 7.22.

7.1.3.6 Box-girder bridges

In a previous paragraph, a one-beam model with 7-DOF beam elements appropriate for the analysis of opened box girders has been discussed; see Figures 7.8 and 7.10. However,

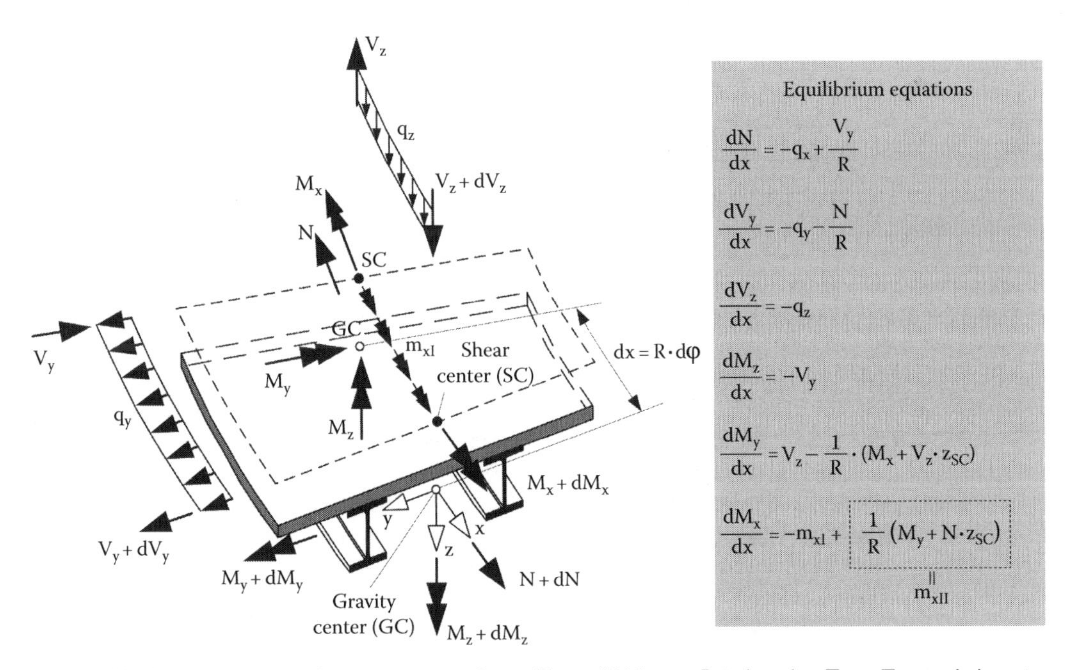

Figure 7.21 Forces on curved composite girders. (From Wehnert-Brigdar, A., Zum Tragverhalten im Grundriss gekrümmter Verbundträger, Dissertation, Fakultät für Bauingenieurwesen, Ruhr – Universität, Bochum, Germany, 2009.)

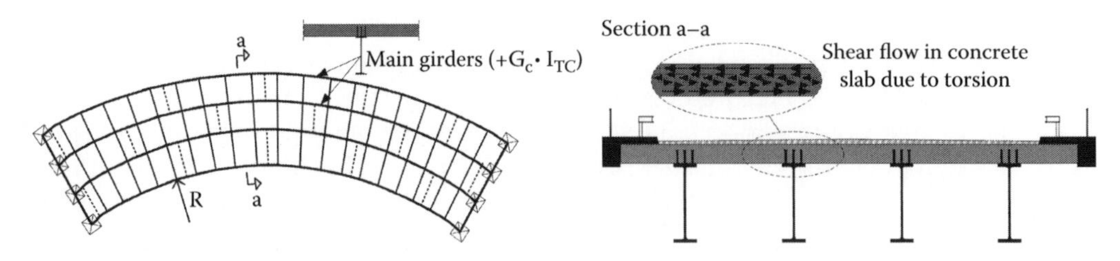

Figure 7.22 Grillage model of a simply supported curved bridge.

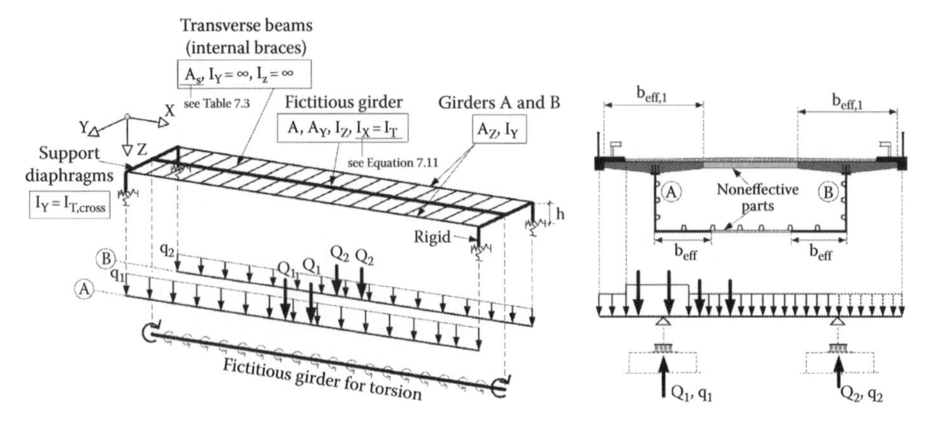

Figure 7.23 Grillage model of a simply supported box-girder bridge.

conventional software packages make use of 6-DOF beam elements. Then the simplified grillage model of Figure 7.23 can be implemented. One can see that the box girder is divided into two opened composite cross sections in which the shear lag effect on the deck slab and the lower flange is considered through the effective widths given in EN 1994-2 and EN 1993-2. The grillage is thus composed of two main composite girders A and B transversely connected with beams representing the internal braces or diaphragms, not the slab. The torsional rigidity of the composite box girder calculated with the Equation 7.11 is represented in the model by a fictitious girder located between the main composite girders. The central girder comprises also the whole bending ($I_{Z,tot}$)—and shear stiffness of the deck slab ($A_Y \approx$ slab area).

The distance h in the model in Figure 7.23 should be equal to the distance between the upper surface of the bearing and the shear center of the cross section.

After the transverse distribution of the vertical loads through the use of a simply supported beam, the reaction forces Q_i and q_i ($i = 1, 2$) on the main girders are calculated; see Figure 7.23. A structural analysis of the grillage system will give shear forces, bending moments, and displacements for the main girders A and B. The fictitious girder obviously gives axial forces, torsional moments, and twisting angles.

The shear flexibility of the internal braces is of great importance for the structural behavior of the bridge since it significantly affects the twisting of the deck due to torsion and the stresses in the cross sections and the horizontal bearing forces. Internal braces and diaphragms are included in the model with transverse members with a shear area A_s given in Table 7.3. Further modeling adjustments are given in [7.27–7.29].

At final stage, the grillage model of Figure 7.23 is considered as an acceptable one because warping moments are of negligible magnitude usually. The grillage model with the fictitious girder for torsion is appropriate for twin-girder bridges and can also be applied in skew and curved structural systems. However, during erection, parts of the bridge may be composed

Table 7.3 Shear area for the equivalent beam element in the global analysis and determination of shear forces and axial forces in the diagonals

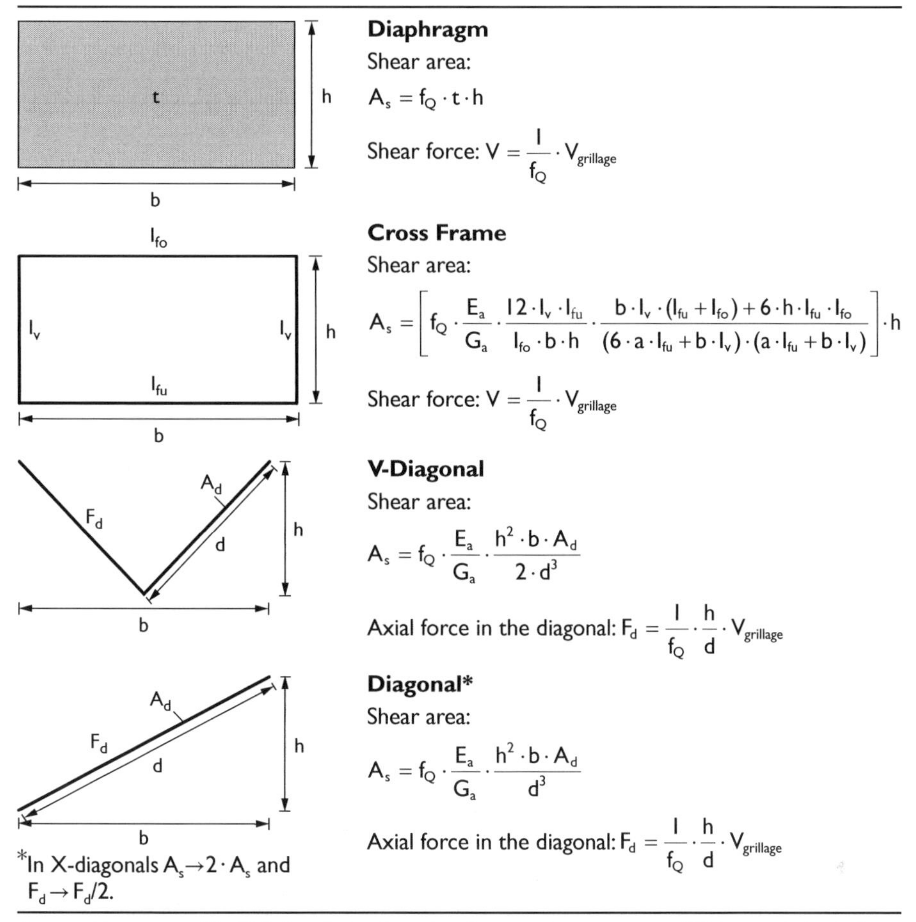

	Diaphragm Shear area: $A_s = f_Q \cdot t \cdot h$ Shear force: $V = \frac{1}{f_Q} \cdot V_{grillage}$
	Cross Frame Shear area: $A_s = \left[f_Q \cdot \frac{E_a}{G_a} \cdot \frac{12 \cdot I_v \cdot I_{fu}}{I_{fo} \cdot b \cdot h} \cdot \frac{b \cdot I_v \cdot (I_{fu} + I_{fo}) + 6 \cdot h \cdot I_{fu} \cdot I_{fo}}{(6 \cdot a \cdot I_{fu} + b \cdot I_v) \cdot (a \cdot I_{fu} + b \cdot I_v)} \right] \cdot h$ Shear force: $V = \frac{1}{f_Q} \cdot V_{grillage}$
	V-Diagonal Shear area: $A_s = f_Q \cdot \frac{E_a}{G_a} \cdot \frac{h^2 \cdot b \cdot A_d}{2 \cdot d^3}$ Axial force in the diagonal: $F_d = \frac{1}{f_Q} \cdot \frac{h}{d} \cdot V_{grillage}$
 *In X-diagonals $A_s \rightarrow 2 \cdot A_s$ and $F_d \rightarrow F_d/2$.	**Diagonal*** Shear area: $A_s = f_Q \cdot \frac{E_a}{G_a} \cdot \frac{h^2 \cdot b \cdot A_d}{d^3}$ Axial force in the diagonal: $F_d = \frac{1}{f_Q} \cdot \frac{h}{d} \cdot V_{grillage}$

Source: Unterweger, H., Global analysis of steel and composite structures—Efficiency of simple beam models. Habilitations—thesis, Technical University Graz, Graz, Austria, 2001 (in German).

Note: A_d, cross-sectional area of the diagonals; t, thickness of the diaphragm; I_{fo}, I_{fu}, moment of inertia of the top and lower flanges of the cross frame; b, h, width and height of cross-bracing; F_d, force in diagonal; a, length of cantilevers; $f_Q = 1$ for open cross sections; and $f_Q = (b_{ao} + b_{au})/b_{ao}$, for closed box cross sections.

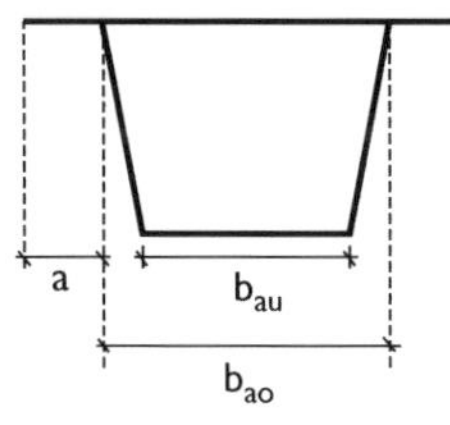

by open cross section that may suffer from high normal stresses and out-of-plane deformations due to warping; see Example 7.2. Then a more complicated analysis with 7-DOF beam elements or FEM is unfortunately unavoidable.

A multiple box-girder bridges can be analyzed with a grillage based on the same principles as the model illustrated in Figure 7.23. The outer beams of the adjacent box girders will be connected with transverse elements that represent the deck slab.

7.1.4 3D models

7.1.4.1 General

The structural representation of bridge decks with truss girders or I-shaped plate girders may be done by means of 3D models as proposed in [7.1–7.2] and [7.30–7.32]. Truss girders are represented by their chord and bracing members, while plate girders are transformed to equivalent trusses. Such models have been proven to be advantageous for modeling orthogonal, skewed, and curved bridges. Unlike grillage models, they are able to consider:

- Eccentricities among the structural elements of a bridge and therefore additional internal forces and possible load distributions.
- The transversal variation in the level of the neutral axis.
- Torsion and distortional warping effects.
- The dispersed structural behavior of the deck slab, in which bending takes place in two directions.
- Buckling phenomena of the steel girders during erection stages.
- Diaphragms, bracing systems, and stiffeners; possible overload or fatigue effects are taken into account.

In the following, the development of such models for representing plate-girder bridge decks by equivalent trusses will be shown. Evidently, the representation of truss girder decks is more straightforward since the trusses are introduced by the properties of the chord and the bracing members.

7.1.4.2 Representation of steel and composite I girders

Steel and composite I cross sections are modeled by a "hybrid" truss as shown in Figure 7.24. For the steel girder, the flanges of the truss are beam elements with a cross section composed of the flange and part of the web of the steel girder. Comparative analyses showed that 1/3 of the web height may be associated with the flange. Therefore, the flanges of the truss are T sections consisting of the flange of the steel girder and 1/3 of the web and are positioned at the center of gravity of the T section. The webs are represented by diagonal truss elements

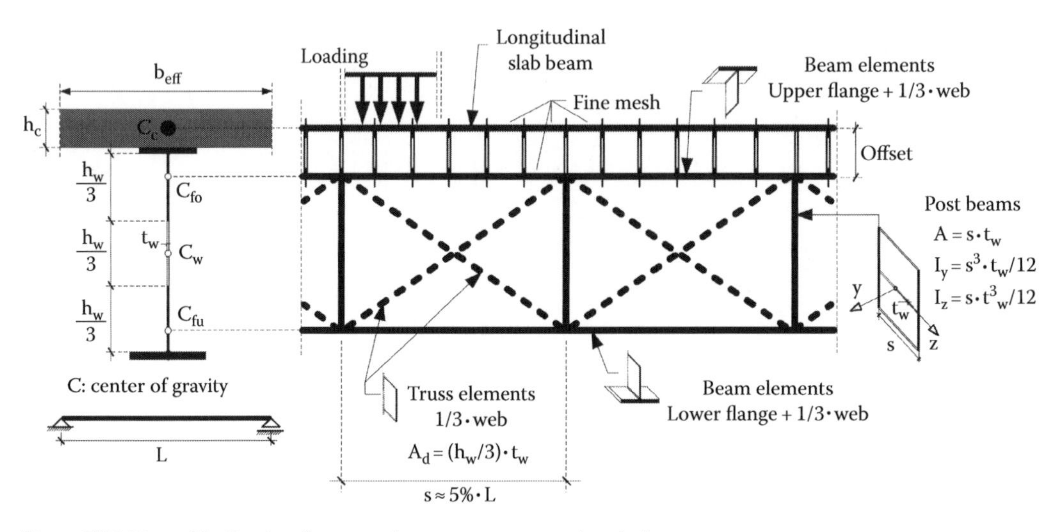

Figure 7.24 Truss idealization for a steel–concrete composite girder.

with width equal to 1/3 of the web height and thickness equal to the web thickness. It has been also shown that the cross-sectional area $A_d = h_w \cdot t_w/3$ for the diagonals adequately corresponds to the shear stiffness of the web.

The post beams are located at a spacing s = 5% of the span of the bridge. This distance is generally acceptable for small and medium-span bridges because the angle between the diagonals and the flange elements usually remains between 35° and 45°. Post beams represent both the in- and out-of-plane stiffness of the web.

For a composite section, the same procedure is followed, with the concrete slab represented by another beam element connected with the upper flange of the truss through the appropriate offset; offset = distance between the centroids C_c and C_{fo}. The nodes of the elements that represent the slab are the same nodes of those representing the upper flange of the truss. It is recommended that a fine mesh is used for the beam elements of the concrete slab and the top flange of the steel girder so that a full shear connection is achieved. Without a fine mesh, the beam elements of the slab may deflect differently than those of the top flange.

In Figure 7.25, one can see the deflected shape of a simply supported equivalent truss, the partial internal forces of the girder, and the stresses in steel and concrete due to a uniformly distributed load. Axial forces and bending moments are developed in the beam elements and therefore, normal stresses can be easily calculated. The N–M forces are discontinuous due to the concentrated forces imposed by the diagonals. The true value of the internal forces can be estimated through an average diagram. The shear forces can be easily calculated from the equilibrium of the normal forces at each node.

The aforementioned modeling for steel and composite girders clearly constitutes a practical approximation. However, comparative results with FEM on a large number of bridge sections indicated its appropriateness; see [7.32]. Indicatively, Table 7.4 shows results for simply supported girders subjected to uniform loading in terms of deflections, stresses, and the fundamental frequencies of vibrations. In addition, the critical load factor for lateral torsional buckling, which is the factor by which the applied load must be multiplied in order to reach the fundamental buckling mode, is determined. It can be seen that the results of the proposed model are in very good agreement with those of the finite element analysis.

The suggested model is also used for the analysis of continuous systems. This is briefly demonstrated with the Table 7.5. A two-span continuous beam is investigated both with pure steel and composite cross sections. For the steel–concrete composite systems, the cracked area is assumed to have a length equal to 15% of the span according to the simplified method of EN 1994-2. In this region, concrete is considered as fully cracked and the

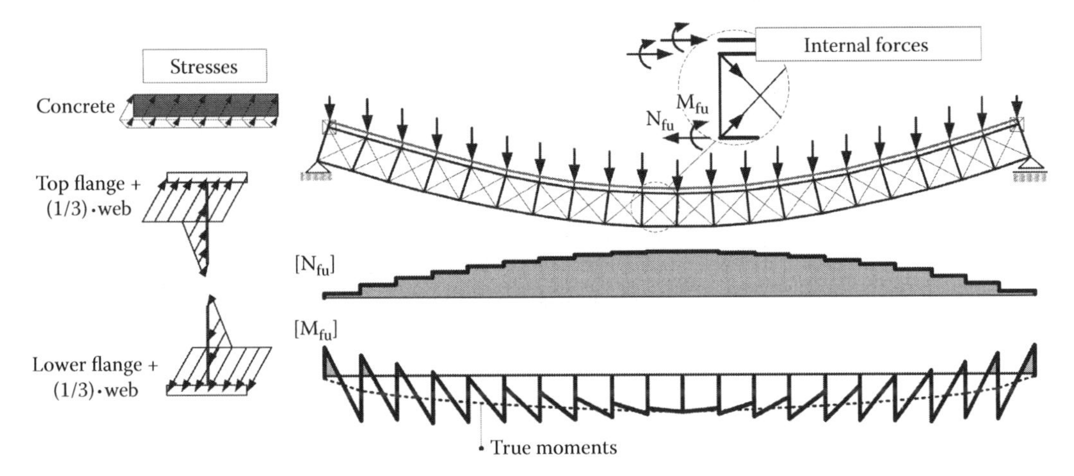

Figure 7.25 Partial internal forces and stresses.

Table 7.4 Comparison of the proposed model [3D], a single line girder model [1D], and a FE model [FEM] for a simply supported beam with L = 25 m and q = 15 kN/m

	Steel girder 1			*Steel girder 2*			*Composite girder 1*			*Composite girder 2*		
	300 · 30 1000 · 12 400 · 40			300 · 30 1500 · 12 300 · 30			3000 200 300 · 30 1000 · 12 400 · 40			2000 350 300 · 30 1500 · 12 300 · 30		
	3D	*1D*	*FEM*	*3D*	*1D*	*FEM*	*3D*	*1D*	*FEM*	*3D*	*1D*	*FEM*
$\mathbf{w}$[a]	*51.0*	49.4	*53.7*	*27.1*	26.1	*27.7*	*17.6*	16.6	*19.6*	*10.1*	9.1	*10.5*
σ_{co}[b]	—	—	—	—	—	—	−2.6	−2.6	−2.8	−1.9	−1.9	−2.0
σ_{ao}[c]	−*101.6*	−100.4	−*99.0*	−*66.4*	−65.7	−*66.5*	−*5.9*	−5.5	−*6.4*	−*1.4*	−1.7	−*1.7*
σ_{au}[d]	*70.7*	70.2	*72.1*	*66.6*	65.7	*66.5*	*52.2*	51.8	*54.3*	*46.8*	44.1	*45.6*
α_{crit}[e]	*0.72*	0.64	*0.68*	*0.51*	0.43	*0.52*	—	—	—	—	—	—
$\mathbf{f}_{dyn}$[f]	*2.51*	2.59	*2.43*	*3.46*	3.57	*3.36*	*4.28*	4.48	*4.00*	*5.69*	6.04	*5.37*

Notes:

All dimensions are given in mm.
For the critical load factors of the 1D model, beam elements with 7-DOF were used.
Modulus of elasticity for concrete $E_c = 33{,}500$ N/mm² and for structural steel $E_a = 210{,}000$ N/mm².

[a] Max. deflection in mm.
[b] Min. stress at the top of the concrete slab in N/mm².
[c] Min. stress at the top of the upper flange in N/mm².
[d] Max. stress at the bottom of the lower flange in N/mm².
[e] Critical load factor for lateral torsional buckling.
[f] Eigenfrequency for vertical bending in Hz.

Table 7.5 Comparison of the proposed model [3D], a single line girder model [1D], and a FE model [FEM] for a two-span continuous beam

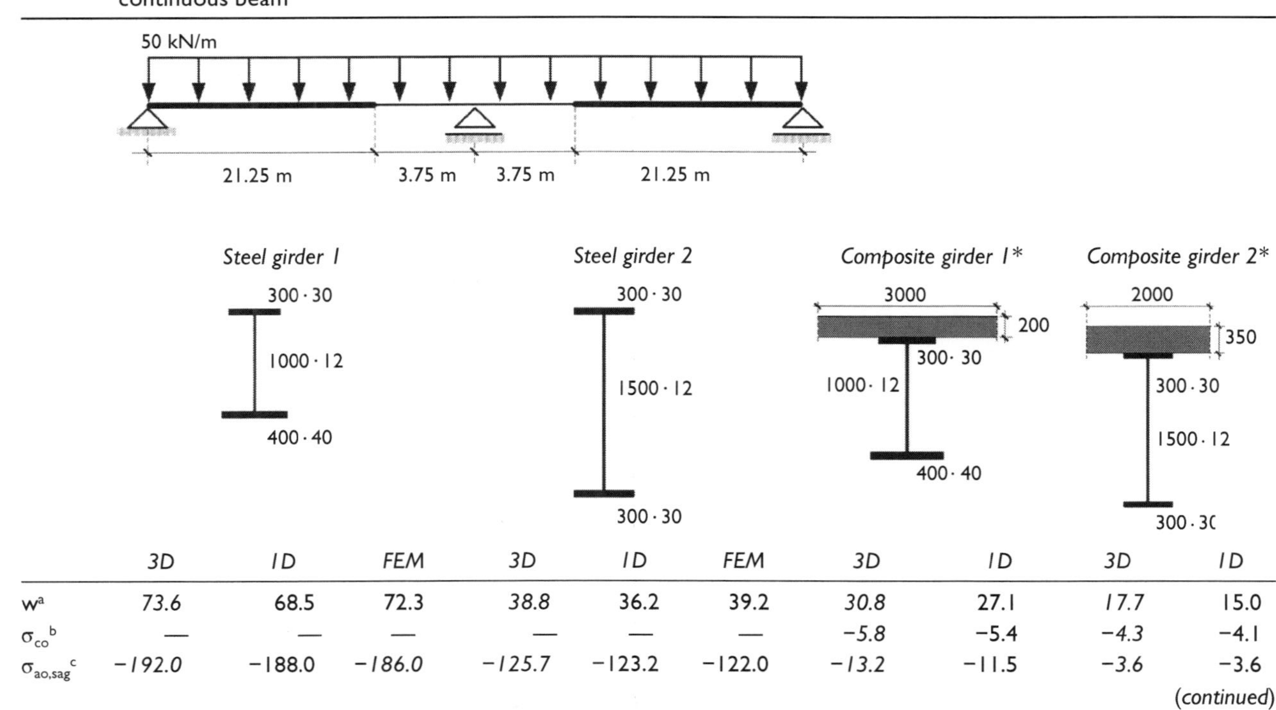

	3D	1D	FEM	3D	1D	FEM	3D	1D	3D	1D
w[a]	73.6	68.5	72.3	38.8	36.2	39.2	30.8	27.1	17.7	15.0
σ_{co}[b]	—	—	—	—	—	—	−5.8	−5.4	−4.3	−4.1
$\sigma_{ao,sag}$[c]	−192.0	−188.0	−186.0	−125.7	−123.2	−122.0	−13.2	−11.5	−3.6	−3.6

(continued)

Table 7.5 (continued) Comparison of the proposed model [3D], a single line girder model [1D], and a FE model [FEM] for a two-span continuous beam

	3D	*1D*	*FEM*	*3D*	*1D*	*FEM*	*3D*	*1D*	*3D*	*1D*
$\sigma_{au,sag}$ [d]	*133.8*	131.5	131.0	*125.7*	123.2	122.0	*109.5*	107.3	*99.4*	93.5
$\sigma_{ao,hog}$ [e]	*346.7*	334.6	321.0	*223.3*	219.0	211.2	*149.1*	111.0	*106.0*	89.3
$\sigma_{au,hog}$ [f]	*−259.0*	−234.1	*−241.0*	*−246.8*	−219.0	*−246.0*	*−186.3*	−165.8	*−181.1*	−150.0
σ_s [g]	—	—	—	—	—	—	*116.5*	136.8	*100.4*	116.1
α_{crit} [h]	*0.33*	0.32	0.31	*0.26*	0.20	0.22	—	—	—	—
f_{dyn} [i]	*1.38*	1.39	1.39	*1.89*	1.95	1.90	*2.32*	2.42	*3.08*	3.28

Notes:

All dimensions are given in mm.
For the critical load factors of the 1D model, beam elements with 7-DOF were used.
Modulus of elasticity for concrete $E_c = 33{,}500$ N/mm² and for structural steel $E_a = 210{,}000$ N/mm².
*At supports $A_c = 0$ and $A_{s,tot} = 46.79$ cm².

[a] Max. deflection in mm.
[b] Min. stress at the top of the concrete slab in N/mm².
[c] Min. stress at the top of the upper flange at span in N/mm².
[d] Max. stress at the bottom of the lower flange at span in N/mm².
[e] Max. stress at the top of the upper flange at support in N/mm².
[f] Min. stress at the bottom of the lower flange at support in N/mm².
[g] Max. stress at reinforcement at support in N/mm².
[h] Critical load factor for lateral torsional buckling.
[i] Eigenfrequency for vertical bending in Hz.

cross section consists of the steel section and the reinforcement that is placed in the centroid of the slab. It should be mentioned that no FEM analysis was carried out for the composite systems due to the difficulty of being able to simulate the cracking of the concrete accurately. One can see the resultant values for the 3D model correlate very well with those obtained from the two other methods. Deflections, eigenvalues for buckling and dynamic analysis, and stresses at sagging moment areas do not show any significant difference between the compared methods. As expected at hogging moment areas, deviations between the 3D and the 1D models become larger.

The 3D model recommended by the authors is based on representing lateral torsional buckling modes of steel girders. Designers will find the model convenient for the lateral torsional buckling investigations during the concreting stages and for the half-through bridges (Figure 2.35). This is discussed in detail in the next section.

7.1.4.3 Slab representation

Slabs are structurally continuous in both directions x and y, and they resist applied loads by shear forces, bending moments, and torques that are coupled with each other. For this reason, it was previously mentioned that the transverse slab elements of the grillage models should not be used for the final design of the slab. A brief description of an isotropic solid slab is given in Figure 7.26.

A slab element with plane dimensions $dx \cdot dy$ is extracted from a deck slab with thickness h_c. The applied load dQ is carried by the vertical shear forces v_x and v_y, the bending moment m_x and m_y, and the torques $m_{yx} = m_{xy}$; all internal forces are all per unit width. After forming the equilibrium equations, it is easy to observe that they differ significantly

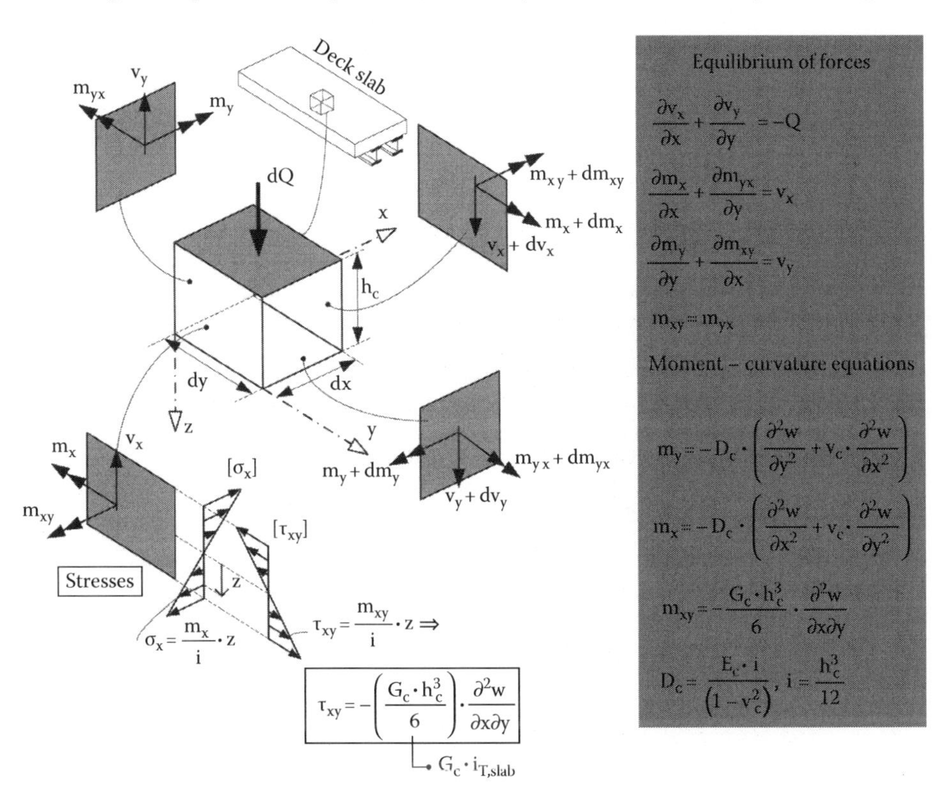

Figure 7.26 Equilibrium and compatibility equations of an isotropic slab element.

from those of a single beam. Due to the slab's infinite indeterminacy, more complicated load paths are developed and internal redistributions take place. One can also see that the shear force is not a simple differential of the bending moment and therefore, it is not the slope of the bending moment diagram. Moreover, bending of the slab is associated with transverse torques in both directions; torques result in cyclic shear flows and stresses τ_{xy}. The torsion shear stresses τ_{xy} have a linear distribution and they are proportional to distance z from the slab's neutral axis. Taking a closer look at Figure 7.26, it is obvious that the magnitude of the shear stresses τ_{xy} are dependent on the torsional rigidity per unit $G_c \cdot i_{T,slab} = G_c \cdot h_c^3/6$. This constant has been multiplied with the width b of the transverse slab elements in the grillage model of Figure 7.16.

It is also worth mentioning that the structural behavior of concrete slabs highly depends on the value of Poisson's ratio. It is reminded that for cracked concrete, $\nu_c = 0$.

The equations shown in Figure 7.26 are for isotropic solid slabs; these are slabs with similar stiffnesses in longitudinal and transverse directions. In orthotropic slabs, Poisson's ratio will be different in the two directions. These are, for example, slabs with significantly different amounts of reinforcement in x and y directions and slabs with profile steel sheeting or filler beam decks.

A grillage model that considers the dispersed bending and torsion stiffness of a solid slab is illustrated in Figure 7.27. The grillage mesh should be sufficiently fine so that the grillage deflects in a smooth surface in a similar way as a real slab. A smooth deflected surface is equivalent to the requirement that the twist $\partial^2 w/\partial x \partial y$ is the same in orthogonal directions and that $m_{yx} = m_{xy}$. The spacing of the beams should not be less than 2.5 times the slab depth. If the local dispersion of concentrated loads has to be considered or in regions of sudden change, then smaller values have to be adopted. Transverse beams should have spacing similar to that of the longitudinal beams. It is also recommended that the row of longitudinal beams at each edge of the grillage should be located at a distance of $0.3 \cdot h_c$ from the edge of the slab, where h_c is the slab depth. This is where the resultant of the shear flows is located; see [7.14]. The width of the edge member for the calculation of I_T should be therefore reduced to $b - 0.3 \cdot h_c$.

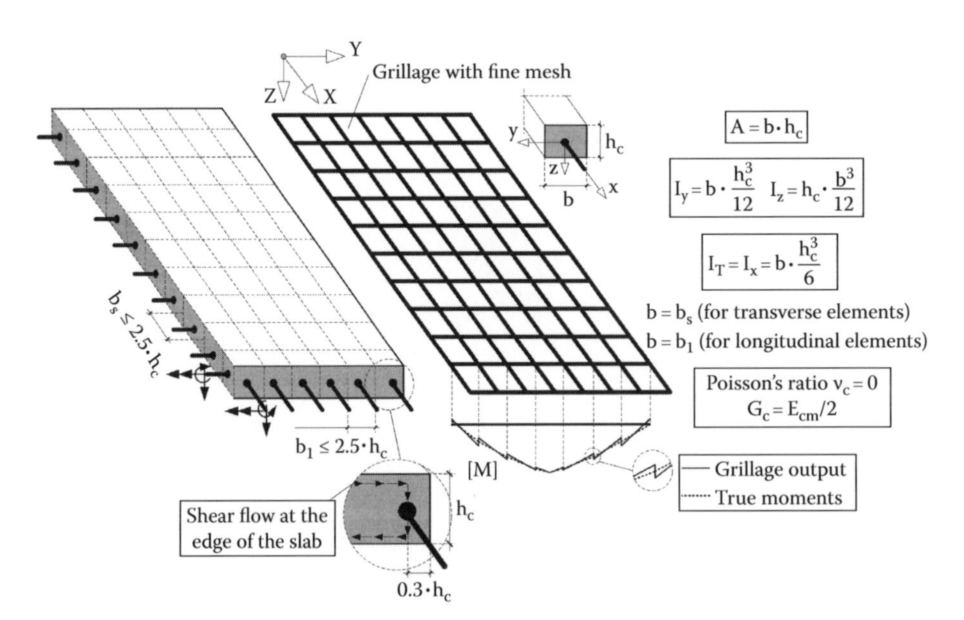

Figure 7.27 Grillage model for an isotropic solid slab.

REMARK 7.4

According to EN 1991-2, for local verification of the deck slab, the wheel loads may be considered as uniformly distributed taking into account the contact area of the wheel and the dispersal of the load through the pavement and the concrete slab. According to Figure 4.2, the effective loaded length is equal to

$$b_{lf} = 0.4 + 2 \cdot h_{pav} + h_c \text{ [m]},$$

where

h_{pav} is the thickness of the pavement
h_c is the slab's depth

If the beam spacing b_l and b_s (see Figure 7.27) are smaller or equal to b_{lf}, then the wheel load can be assumed to be sufficiently dispersed for the grillage to reproduce the moments' distribution throughout the slab.

The slab reinforcement can be calculated from the bending moment diagrams both of the transverse and the longitudinal beams. Due to the torque moments m_{xy}, bending diagrams are usually discontinuous. This is due to the fact that the final moment diagram [M] represents a superposition of a sawtooth moment diagram due to torsion on a continuous moment diagram due to pure bending; see Figure 7.27. However, in a real slab, the bending diagram is not discontinuous and an average diagram can be taken into account.

Reinforced concrete deck slabs used in composite bridges often have similar stiffness in longitudinal and transverse directions and they can be assumed as isotropic. The earlier described grillage model reproduces the behavior of isotropic solid slabs reasonably accurate and it will be used subsequently for the representation of the deck slab of a 3D model.

7.1.4.4 3D model implementation

The grillage model for the slab's representation in Figure 7.27 can be combined with the truss model that is shown in Figure 7.24. Figure 7.28 illustrates a 3D model that is recommended for the structural analysis both of simple and continuous composite bridges. Attention must be paid so that the grillage has its longitudinal members coincident with the center lines of the steel sections. At sagging moment areas, longitudinal slab elements are used with their uncracked properties. At hogging moment areas, concrete is considered as fully cracked and the total reinforcement is considered due to simplicity at the center of the slab. Transverse slab elements can be considered with their uncracked properties.

One can see that the model can be set up in a detailed way by taking into account all the necessary structural elements, that is, cross bracings and bearings. Imperfections, precambering, and girders with variable cross sections can also be implemented in the model. Therefore, structural phenomena that may be difficult or impossible to investigate with plane grillages are included in the outputs of the 3D model, for example, arch effects in integral bridges with longitudinally variable cross sections.

Another interesting issue depicted in Figure 7.28 is the representation of **well-anchored** concrete parapets on the structural performance of the bridge. Solid parapets at the edges of the deck slab usually possess inertia properties comparable to those of the main composite girders. Such stiff parapets attract a significant part of the normal stresses and have a

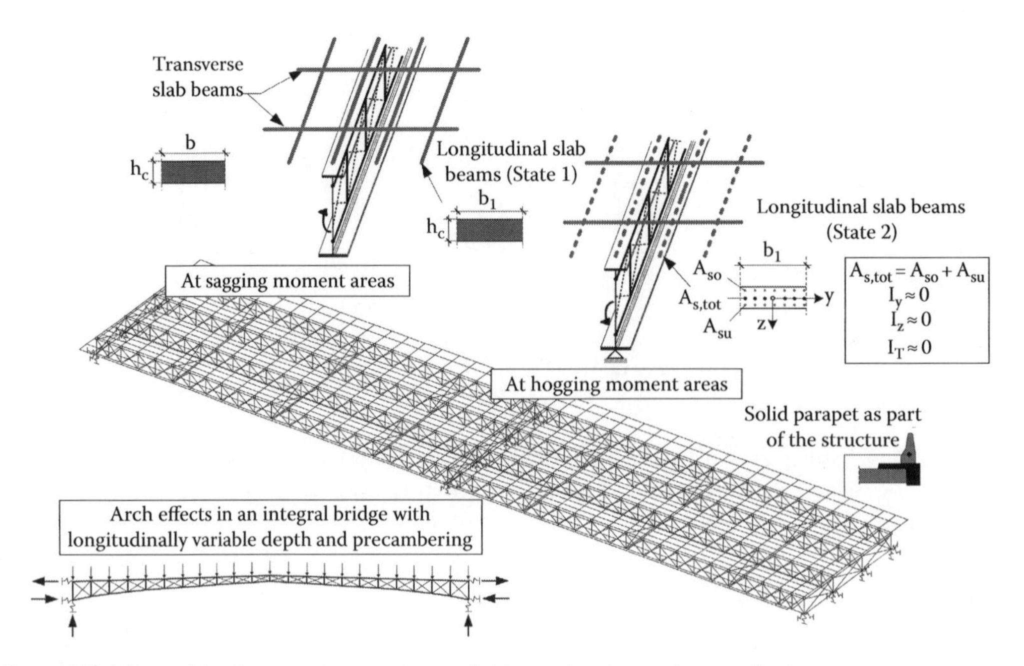

Figure 7.28 3D models of a two-span continuous bridge and an integral one at final stage.

non-negligible effect on the transverse distribution of the vertical loads. With the 3D model, parapets can be included in the model as longitudinal concrete beams.

When the longitudinal slab elements are very close to each other, compression forces away from the main girder decrease and the *shear lag effect* is indirectly taken into account. This is the case of Figure 7.28. As already mentioned, a very fine mesh for the slab beams is used when local effects due to the distribution of the wheel loads in the slab have to be considered in the model. Nevertheless, when the 3D model is not used for the design of the deck slab, then the configuration of Figure 7.29 can be chosen as a simpler solution because it leads to a less fine mesh. Longitudinal concrete beams at sagging moment areas with a width $b_{l,i}$ represent the effective part of the slab. Obviously, the summation of the widths $b_{l,i}$ should be equal to the effective width $b_{eff,1}$ according to EN 1994-2. At hogging moment areas, the effective width of the slab has a different value, and the cross-sectional area of the longitudinal beams is equal to the total reinforcement amount, which can be

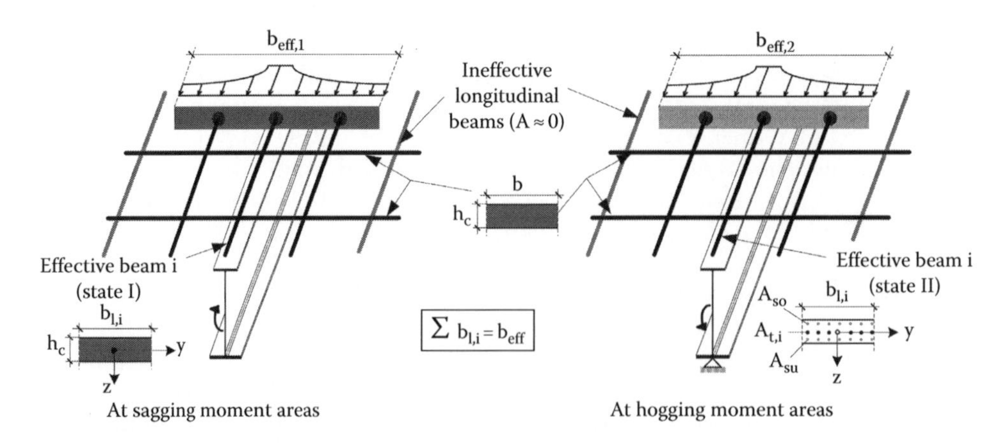

Figure 7.29 Alternative representation of the concrete elements.

assumed at the center of the slab. The *tension stiffening* effect can be considered by using an increased area for the effective beam i as follows; see [7.16]:

$$A_{t,i} = \frac{A_{s,i}}{1-(0.5 \cdot f_{ctm})/(\rho_{s,tot} \cdot f_{sk})} \tag{7.14}$$

where

$A_{s,i}$ is the total amount of reinforcement in the slab element i
$\rho_{s,tot}$ is the total reinforcement ratio
f_{ctm} is the mean tensile strength of concrete; see Table 6.1
f_{sk} is the characteristic yield strength of reinforcement steel

REMARK 7.5

With the Equation 7.14, *tension stiffening* is taken into account at hogging moment areas through a "semi-cracked" concrete cross section A_t. Indeed, according to Equation 6.30, the mean strain in reinforcement (A_s) is due to *tension stiffening* equal to

$$\varepsilon_{sm} = \frac{N_s}{E_s \cdot A_s} - \frac{0.4 \cdot f_{ctm}}{\rho_s \cdot E_s}$$

For an equivalent "semi-cracked" concrete section with $A=A_t$ under an axial tension N_s, the mean strain ε_{sm} is $N_s/(E_s \cdot A_t)$. From the previous expressions, A_t is calculated as a function of N_s:

$$\varepsilon_{sm} = \frac{N_s}{E_s \cdot A_s} - \frac{0.4 \cdot f_{ctm}}{\rho_s \cdot E_s} = \frac{N_s}{E_s \cdot A_t} \Rightarrow A_t = \frac{N_s}{(N_s/A_s)-(0.4 \cdot f_{ctm}/\rho_s)}$$

The semi-cracked concrete section enters state II when $N_s \approx 0.8 \cdot A_s \cdot f_{sk}$; therefore, the equivalent cross-sectional area A_t becomes

$$A_t = \frac{A_s}{1-(0.5 \cdot f_{ctm})/(\rho_{s,tot} \cdot f_{sk})}$$

The longitudinal beams, which are located outside the effective width b_{eff}, do not participate in the distribution of the normal stresses and therefore, their cross-sectional area is set equal to zero.

The long-term behavior of concrete can be taken into account by using the modular ratios n_L for the effective longitudinal beam at sagging moment areas; see Equation 6.20. For the transverse slab elements, the following age-adjusted modulus of elasticity can be used:

$$E_{c,eff} = \frac{E_{cm}}{1+\chi \cdot \varphi(t,t_0)} \tag{7.15}$$

where

E_{cm} is the mean value of modulus of elasticity; see Table 6.1
$\varphi(t,t_0)$ is the creep coefficient according to EN 1992-1-1; see Section 6.1.2
χ is the relaxation factor according to EN 1992-2 (=0.8 for nonprestressed members)

7.1.4.5 Analysis during the concreting stages

3D models allow the investigation of the lateral stability of the steel girders during concreting. At this construction stage, lateral bracings that connect the compressed top flanges are required. The load intensity is equal for all girders since the concrete is not hardened yet and the load of wet concrete between girders and at the cantilever is transferred by simple statics to the girders.

The 3D model is a very important tool for the design of composite bridges because bracings require a lot of man-hours and they have an effect on the appearance of the bridge. Through the eigenforms of the structure, designers are able to make safe conclusions for the buckling type, that is, lateral or lateral torsional buckling, and for the areas that need to be strengthened. Comparative analysis between the 3D and FE models led to very small deviations, lower than 3%; see [7.32]. The proposed model deemed accurate for the stability investigations of the following cases:

- Statically determinate and indeterminate systems
- Systems with longitudinally variable cross sections
- Mixed systems with parts that are composite or pure steel

The 3D model is not used for plate buckling analysis.

Figure 7.30 illustrates an example of a buckling analysis of a simply supported bridge that is casted in one stage. Hot-rolled steel girders that are placed every 2.90 m on elastomeric bearings carry the weight of the wet concrete. The spacing of the post beams is 4% of the span, thus equal to 1.0 m. Bracing members are represented by beam elements with pinned ends so that members' instabilities can be analyzed.

One can see that the first eigenmode of the nonbraced steel structure depicts a typical lateral torsional buckling failure. The corresponding load factor α_{crit} is less than one, which means that the level of the applied loading is higher than the level of elastic stability. Obviously, this cannot be accepted and the stiffness of the structure must be increased through a "cost-effective" combination of plane and vertical braces. This is done with a step-by-step strengthening procedure that starts from the most deflected area; in most cases, this will be the mid-span. In the second analysis, the structure exhibits a lateral buckling mode; rotations of the main girders are zero, but the load factor remains low and therefore, further strengthening is required. A third analysis with more bracing members is performed and then the load factor becomes significantly higher. Additionally, the buckling length of the main girders becomes less than 50% of the span's length. Finally, central bracings are placed near supports so that the load factor α_{crit} becomes higher than 10. According to the clause 5.2.1(4) of EN 1993-2, bridges and components of bridges may be checked with first-order theory if $\alpha_{crit} \geq 10$. Hence, a laborious second-order global analysis can be avoided.

If needed, bracings can be arranged so that the load factor is smaller than 10 but a second-order analysis is then mandatory. The designer must be experienced enough to interpret correctly the outputs of the analysis **by looking closely to the buckling modes**. Then according to the buckling shapes, imperfections should be applied on the structural elements; this is a time-consuming procedure. In some cases, different buckling modes may exhibit the same values of α_{crit}; this makes it difficult to recognize which buckling mode is the critical one. Moreover, in second-order analysis, the principle of superposition is not valid and all loads must be applied to the structure with all the respective combination factors. EN 1993-2 recognizes the previous drawbacks and includes a simplified procedure according to which, when the stability of the structure is mainly governed by the first buckling mode,

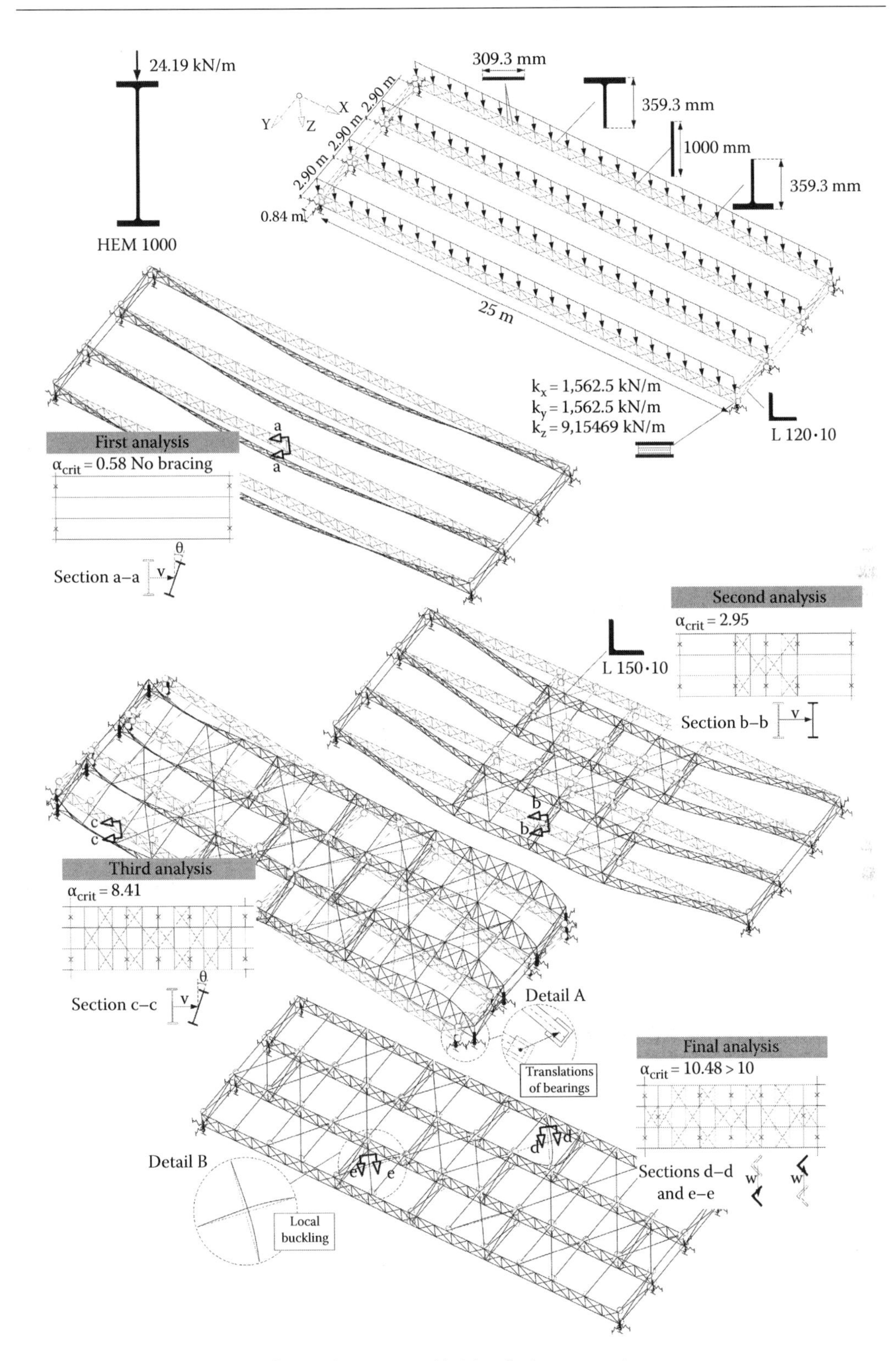

Figure 7.30 Buckling analysis of a simply supported bridge during concreting.

second-order moments M_{II} may be calculated by the application of a magnification factor to the moments from first-order analysis M_I:

$$M_{II} = M_I \cdot \frac{1}{1-(1/\alpha_{crit})} \tag{7.16}$$

From Equation 7.16, one can see that the load factor is

$$\alpha_{crit} = \frac{M_{II}}{M_{II} - M_I} = \frac{M_{II}}{\Delta M} \tag{7.17}$$

The requirement $\alpha_{crit} \geq 10$ for avoiding a second-order analysis becomes then

$$\frac{\Delta M}{M_I} \leq 11\% \tag{7.18}$$

Equation 7.18 implies that the magnification of the bending moments ΔM due to second-order effects is less than 11% of the bending moments calculated with first-order analysis and therefore can be neglected. However, the fulfillment of the criterion $\alpha_{crit} \geq 10$ is essential for the structural performance of a bridge as a whole and it is not just a design facilitation. Experience shows that systems with load factors greater than 10 are laterally stiff (non-sway), and they exhibit an improved load distribution during concreting that is important for the straightness of the deck.

It is worth mentioning that a load factor that is greater than 10 for the loading case of concreting may be much smaller for another loading case as wind or an earthquake. For very wide decks, accidental eccentricities during construction (see Example 7.2) may also lead to different types of buckling modes and load factor values. Peculiarities during a second-order analysis of curved bridges may also arise. For the previous cases, keeping α_{crit} greater than 10 is also recommended.

Looking back at the third analysis in Figure 7.30, one can observe translations in the bearings. Therefore, bearings represent a source of flexibility in the structure that has a non-negligible effect on the buckling shapes and the values of the load factors α_{crit}. In this analysis, bracings are mainly concentrated at the center part of the bridge and therefore, deformations emerged in the most flexible region; near the bearings' area. After an appropriate strengthening, the final analysis offered a load factor greater than 10 that corresponds to a buckling mode of local instability; see detail B. Indeed, in the finalized structure, one observes buckling of two angle cleats connected with the top flanges of the main girders. Generally, bracings should be arranged in such a way that buckling modes with local instabilities become the critical ones. Hence, the possibility of a sway behavior can be excluded; unfortunately, this is not always feasible.

REMARK 7.6

- The magnitude of the flexural torsional buckling load M_{crit} given in Eurocode 3 depends on the height of the load relative to the height of the shear center. In the 3D model, loads apply on the gravity center of the top T section. The destabilizing effect that can be caused due to this inconsistency is however negligible, and this has been proven through numerous comparative analyses with FEM.

- In Eurocode 3, one will find many expressions for the calculation of the elastic critical loads N_{crit} and M_{crit}. The analytical expressions for these loads are usually derived for simple systems that may be appropriate for buildings but not for bridges. This is because various parameters such as the stiffness of the bracings, the bearings, and the geometry of the bridge are not considered in the Codes' equations. Due to the doubts raised, designers often calculate the buckling loads with finite element models. This is an accurate method but time consuming and complicated as well since countless plate buckling modes usually emerge. With the 3D model, unnecessary plate buckling modes are avoided.
- The 3D model is compatible with the simplified method found in EN 1993-2, clause 6.3.4.2(2), in which the compression part of a steel beam subjected to bending is replaced by a compression chord with an effective area $A_{eff} = A_f + A_{wc}/3$ where A_f is the area of the flange and A_{wc} the area of the web under compression. Lateral torsional buckling of the main girders can be represented by lateral buckling of the compression flanges.
- For a buckling analysis, bracing members are recommended to be modeled as beam elements with bending releases at their ends. In contrast to truss elements, the previous configuration allows the investigation of members' instabilities.

As already mentioned, during concreting, the girders deflect under their own weight and the weight of the fresh concrete. On straight bridges, the deflections across any section of the bridge due to the deck weight are almost identical. The point of maximum deflections for each steel girder will be at mid-span of each girder. On a square bridge, these points align across the width of the bridge. By contrast, on a skewed bridge, the deflections are not the same across the width of the bridge, since the girders are longitudinally offset from each other by the skew. On curved bridges, the deflections are not the same across the width of the bridge because the girders do not have the same length, with the outer girders being longer than the inner ones.

Figure 7.31 shows the analysis of a curved bridge that is assumed to be casted at one stage. The uniformly distributed load represents the weight of the fresh concrete. The steel girders are connected with intermediate cross bracings that consist of L sections. One can see the significant difference among the values of the vertical deflections w of the main girders A, B, and C. Moreover, lateral and horizontal deformations (v and u) can be estimated with the 3D model with an adequate accuracy.

Generally, the webs of the I girders are not stiff enough, resulting in web distortion associated with the flange lateral bending between the cross-frame locations. At construction stages, where there is no slab, both the bottom and the upper flanges are subjected to lateral bending between the cross-frame locations. Figure 7.31 also shows the stresses on the upper and lower flanges of the steel section, for the 3D model and the FE model, under the weight of concreting. The results are almost identical. It is worth mentioning that the points of maximum stresses are the positions of the transverse bracing. The differential deflections that occur at these points (which would be much higher without the presence of transverse bracings) are restrained by the transversal bracing, and lateral stresses are developed on the upper and lower flanges. The aforementioned stress situation cannot be investigated with the use of a conventional grillage model.

7.1.4.6 Analysis at final stage

At the final stage, concrete is hardened and the steel girders behave compositely with the slab. The slab restrains laterally the upper flanges so that cross bracings are not required. The 3D model implementation at the final stage has been explained in the

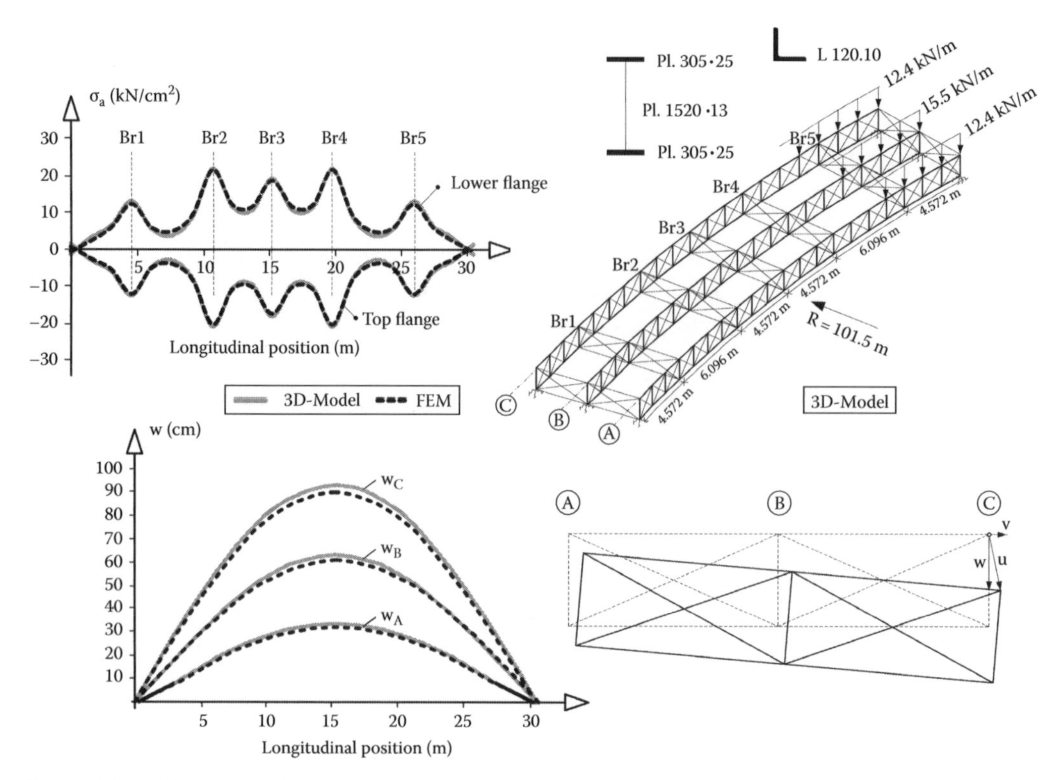

Figure 7.31 Deflections and stresses of a curved bridge during concreting.

previous section and is shown in Figure 7.28. The following example demonstrates the importance of including bracing members in the model and its influence on the distribution of the vertical loads.

The simply supported railway bridge of Figure 7.32 has three main steel girders of 3353 mm total height, connected with a concrete slab of 28 cm thickness. The total length of the bridge is 70 m. Diagonal cross frames are placed every 7 m connecting the main girders between each other. Two different analyses take place for the composite bridge. In the first case, there is no lateral bracing system between the bottom flanges of the steel girders but the diagonal bracing. In the second case, the bottom flanges are connected using the lateral bracing system. Both the 3D and the FE models are illustrated. For the FE model, shell elements have been used for the representation of the steel girders, volume for the slab, and truss elements for the diagonals.

Two different load cases are applied on the composite structure: an eccentric linear load of 50 kN/m, on the middle and outer girder, and the Load Model 71 of Eurocode 1, representing possible rail traffic on bridge. The results for the deflections and the stresses are summarized in Table 7.6. One can see that for both the unbraced and the braced structures, the results of the finite element analysis and the 3D representation correlate very well with each other. Especially, for the second load case of the eccentric train LM71, the reduction in stresses and deformations is significant for the bridge with the bottom bracing. The stress of the outer girder is reduced by 25% (according to FEM results). As far as the vertical deformations are concerned, there is a reduction of 33% for the maximum value of w_A. For the unbraced structure, it seems that the outer girder carries the most part of the load. With the placement of the bottom bracing, the total torsional constant of the whole bridge increases

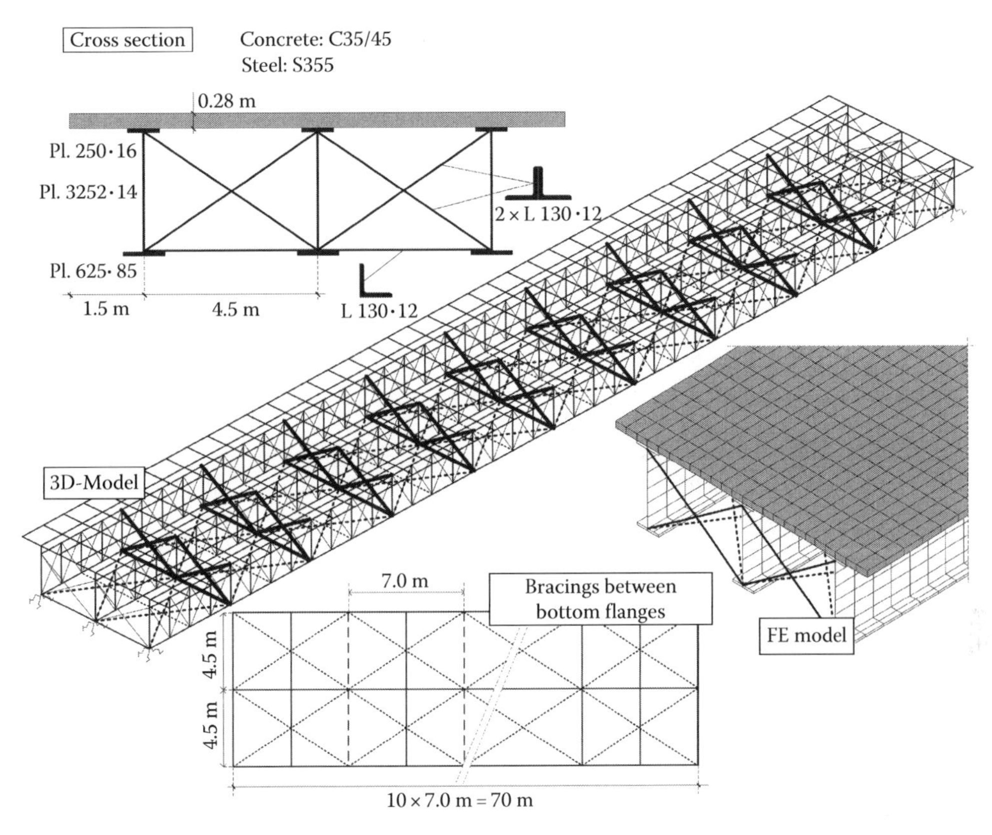

Figure 7.32 Simply supported composite bridge with bracings at bottom flanges; 3D and FE models.

and the rotation of the bridge section is reduced (see the diagrams of w). The load distribution changes and the whole bridge behaves in a way close to a box section bridge. At the same time, the maximum deformation of the extreme girder reduces significantly.

7.1.5 Models for other types of bridges

The previous sections presented models for common composite bridges. Other types are represented by appropriate models. The models for arch bridges include the arches; the stiffening girders; other longitudinal girders, if present; cross girders; and the suspension bars/cables. For global analysis of cable-stayed bridges, pylons, stays, cross girders, and stiffening girders are usually included in the model.

7.2 EFFECTIVE WIDTH OF WIDE FLANGES DUE TO SHEAR LAG

7.2.1 General

When I or T beams are flexed, the compression/tension force in each steel flange is injected into the concrete flange by longitudinal edge shear forces whose magnitude depends on the variation of the bending moment; see Figure 7.33. The longitudinal distribution of the edge shear forces affects the distribution of the normal stress, which decreases toward the outside

Table 7.6 Deflections and stresses for the composite bridge of Figure 7.32

Vertical deflections

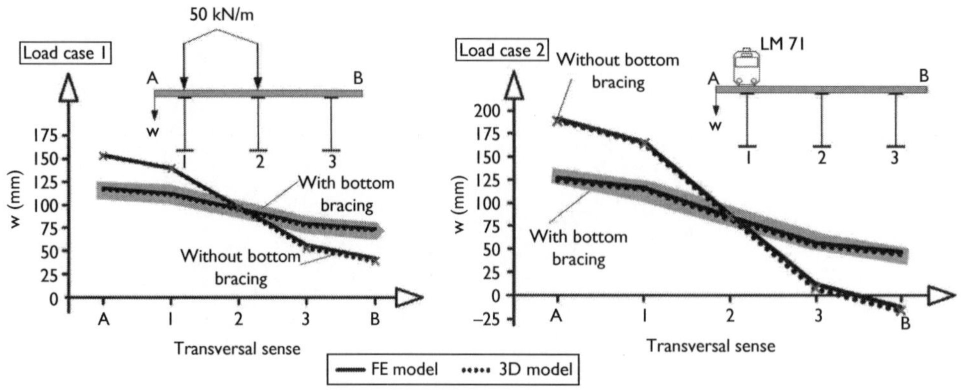

Normal stresses (in N/mm²)

	σ_c	σ_1	σ_2	σ_3
Load Case 1				
Without bottom bracing				
3D model	−12.50	147.4	105.0	68.5
FE model	−11.70	138.0	102.0	71.6
With bottom bracing				
3D model	−8.10	121.9	97.2	78.4
FE model	−8.20	116.0	94.8	79.4
Load Case 2				
Without bottom bracing				
3D model	−18.3	178.0	103.5	29.7
FE model	−16.2	171.0	100.4	28.5
With bottom bracing				
3D model	−10.4	136.5	91.4	74.4
FE model	−10.9	128.5	92.9	82.5

Notes:

σ_c is the minimum concrete stress at the upper face of the concrete.
σ_i is the maximum steel stress at the lower flange of main girders (i = 1, 2, 3).

edge causing distortion of the flange. The decrease of normal stresses away from the loaded edge due to shear distortion is known as *shear lag*.

Simple beam theory cannot capture correctly the true stress distribution along wide flanges. Due to shear lag, these stresses are not any more proportional to the distance from the neutral axis but rather concentrate near the flange-to-web junctions. Accordingly, the theory of elasticity is more appropriate for a correct determination of the stress distribution. However, in order to retain the application of the engineering beam theory, effective^s widths due to shear lag are introduced. The effective^s width is determined by the condition of equal axial forces resulting from (a) the full flange with nonuniform stresses determined by the theory of elasticity and (b) the effective width with uniform maximal stresses. It is noted that this effective width is denoted as effective^s width, that is, with the superscript s, to distinguish it from the effective width due to plate buckling.

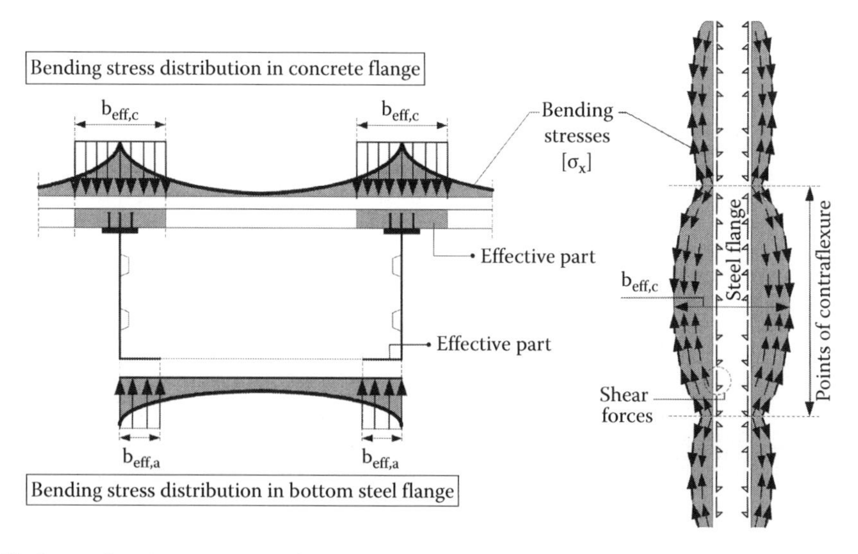

Figure 7.33 Stress distribution in wide flanges and effective^s widths.

In composite plate-girder or truss bridges, the effective^s width refers only to the concrete slab. However, for box girders, effective^s widths must be introduced also for the bottom steel flange. Subsequently, effective^s widths are given for both the concrete slab and the box-girder steel flanges.

7.2.2 Effective^s width of concrete flanges

The effective^s width of the concrete slab may be determined from

$$b_{eff} = b_0 + \sum \beta_i \cdot b_{ei} \tag{7.19a}$$

where (Figure 7.34)

b_0 is the distance between the centers of outstand shear connectors

$$b_{ei} = \frac{L_e}{8} \tag{7.19b}$$

L_e is the distance between points of zero-bending moment given in Figure 7.34, provided that the adjacent internal spans do not differ more than 50% and any cantilever is not larger than ½ the adjacent span

b_i is the distance from the outstand shear connector to a point midway between adjacent webs or distance to the free edge

$$\beta_i = 0.55 + 0.025 \cdot \frac{L_e}{b_{ei}} \leq 1.0 \text{ at end supports for the calculation of } b_{eff,0} \tag{7.19c}$$

$\beta_i = 1.0$ elsewhere (for $b_{eff,1}$ and $b_{eff,2}$)

Generally, the effective^s width and consequently the properties of the cross section vary along the bridge as indicated in Figure 7.34. Effective^s widths at intermediate supports are

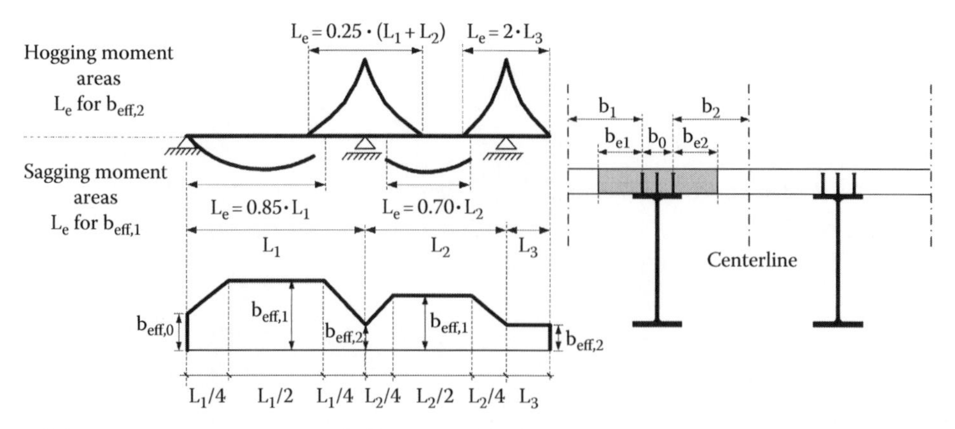

Figure 7.34 Length L_e and distribution of effective^s width of concrete slab along the span. (From EN 1994-2, Design of composite steel and concrete structures, Part 2: Rules for bridges, 2005.)

smaller than those for the span regions. This is because the effect of shear lag is greatest at locations of high shear where the edge forces in the flanges are changing rapidly. However, when elastic global analysis is employed, a uniform effective width may be used, equal to the value of $b_{eff,1}$ at mid-span for span supported to both ends and $b_{eff,2}$ at the support for a cantilever.

During the initial design phase, the distance b_0 between the exterior connectors may be unknown. In such cases, it is allowed to assume $b_0 = 0$.

7.2.3 Effective^s width of steel flanges

In cases of wide steel flanges as in box-girder bridges, effective^s widths must also be determined. The elastic effective^s width for shear lag is determined from

$$b_{eff,i} = \beta \cdot b_{0i}, \quad i = 1,2 \text{ (Figure 7.35)} \tag{7.20}$$

where

b_{0i} is the distance between web and free end for outstand elements (**i=1**) or **half** width of internal elements (**i=2**)

β is a reduction factor given in Table 7.7

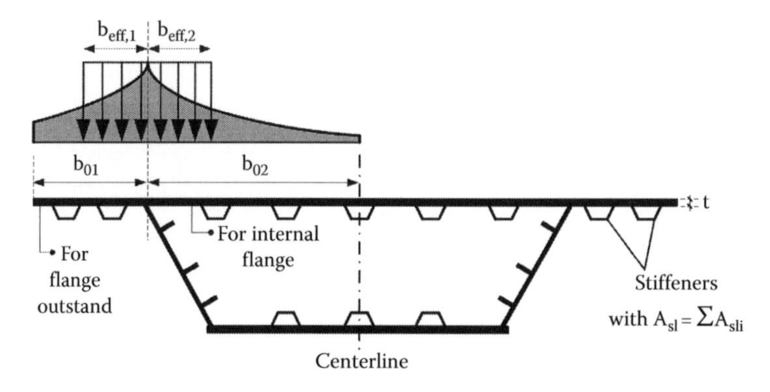

Figure 7.35 Notations for shear lag.

Table 7.7 Effective[s] width factor β for elastic behavior

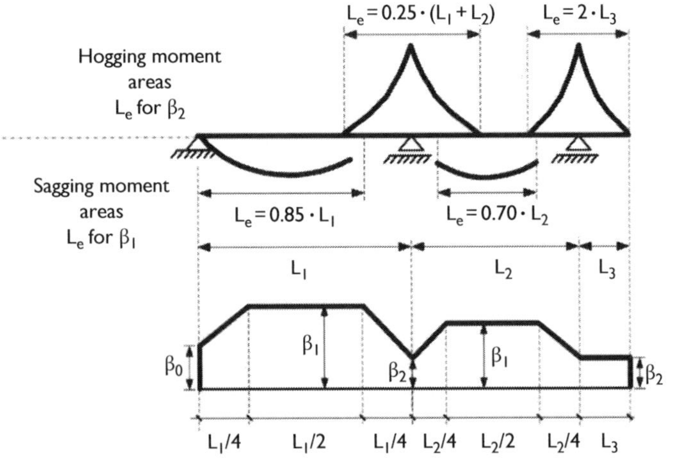

Hogging moment areas L_e for β_2

$L_e = 0.25 \cdot (L_1 + L_2)$ $L_e = 2 \cdot L_3$

Sagging moment areas L_e for β_1

$L_e = 0.85 \cdot L_1$ $L_e = 0.70 \cdot L_2$

L_1 L_2 L_3

β_0 β_1 β_2 β_1 β_2

$L_1/4$ $L_1/2$ $L_1/4$ $L_2/4$ $L_2/2$ $L_2/4$ L_3

Factor κ	*Location for bending*	*β-Value*
$\kappa \leq 0.02$		$\beta = 1.0$
$0.02 < \kappa \leq 0.70$	Sagging bending	$\beta = \beta_1 = \frac{1}{1 + 6.4 \cdot \kappa^2}$
	Hogging bending	$\beta = \beta_2 = \frac{1}{1 + 6.0 \cdot (\kappa - (1/2500 \cdot \kappa)) + 1.6 \cdot \kappa^2}$
$\kappa > 0.70$	Sagging bending	$\beta = \beta_1 = \frac{1}{5.9 \cdot \kappa}$
	Hogging bending	$\beta = \beta_2 = \frac{1}{8.6 \cdot \kappa}$
all κ	End support	$\beta_0 = \left(0.55 + \frac{0.025}{\kappa}\right) \cdot \beta_1 < \beta_1$
all κ	Cantilever	$\beta = \beta_2$ at support and at end

Source: EN 1993-2, Design of steel structures, Steel bridges, 2006.

Notes:

$\kappa_i = \frac{\alpha_{0i} \cdot b_{0i}}{L_e}$, $i = 1,2$ (see Figure 7.35) with $\alpha_{0i} = \sqrt{1 + \frac{A_{sl,i}}{b_{0i} \cdot t}}$.

$A_{sl,i}$ is the area of all longitudinal stiffeners within the width b_{0i}.

If $b_{0i} < L_e/50$, shear lag in steel flanges may be neglected.

The effective length L_e used in Table 7.7 expresses the distance between points of zero-bending moment and is given in Figure 7.34 under the same conditions stated before (adjacent internal spans do not differ more than 50% and any cantilever is not larger than ½ the adjacent span).

For quicker calculations, the reduction factor β can be taken from the diagram in Figure 7.36. As expected, the factor β and subsequently b_{eff} are smaller for hogging moment areas than for the sagging ones. Moreover, it may be seen that the factor κ increases by the addition of longitudinal stiffeners; in this case, the effective[s] width decreases. This is due to the fact that shear deformations occur on the unstiffened steel sheet with little participation of the stiffeners.

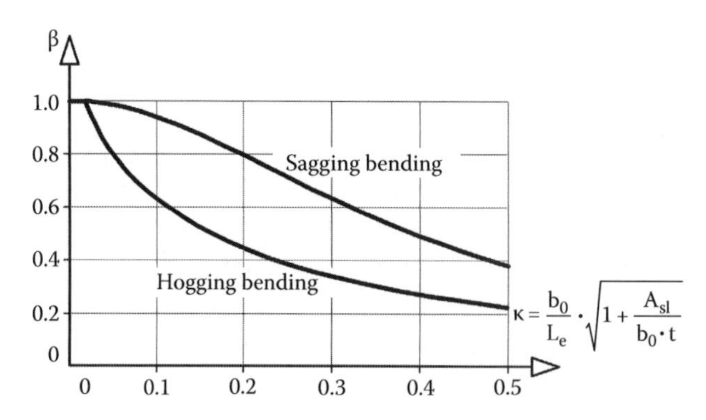

Figure 7.36 Reduction factor β.

Concluding, it may be said that the introduction of the effective[s] width allows for the determination of the flange stresses from the simple beam theory according to

$$\sigma_1 = \frac{M}{W} \tag{7.21}$$

where
M is the acting bending moment
W is the section modulus of the cross section with the effective[s] flange width

However, for stress verifications at serviceability and fatigue limit states, the accurate nonuniform elastic stress distribution across the wide flange as determined by the theory of elasticity and given in Table 7.8 shall be accounted for.

Effective[s] widths due to shear lag as discussed before are based on results of the theory of elasticity and are therefore valid for elastic behavior only. However, when the applied load is increased beyond the value at which the peak stress at the flange-to-web junction reaches the yield stress, the nonlinear stress distribution becomes less pronounced and the effective[s] width increases since the neighboring parts of the flange reach also the yield stress in order to retain equilibrium. In addition, the increased values of the effective[s] widths interact with plate buckling effects. These increased values may optionally be taken into account at ultimate limit states. Accordingly, it may be said that for global analysis and for verifications at serviceability and fatigue limit states where the behavior is elastic, effective[s] widths for steel flanges are to be determined according to Tables 7.7 and 7.8. These values may be conservatively used for verifications at ultimate limit states.

At ultimate limit states, increased values may optimally be used in interaction with plate buckling effects, as presented in Table 7.9. The National Annex may choose the more appropriate method to be used for the calculation of shear lag at ULS. If this is not the case, then according to EN 1993-2, the method 3 is to be recommended.

For small- and medium-span bridges, flanges are not sufficiently wide so that *shear lag* will not usually have a great effect. However, for box-girder bridges with very wide flanges, a considerable reduction of the acting flange width is likely to be imposed. High reductions are expected at SLS and for the hogging moment areas; see Figures 7.37 and 7.38. It is also important to note that during erection (i.e., launching or cantilever method), the effective[s] widths change according to the construction process. This is demonstrated with the following example.

Table 7.8 Elastic stress distribution across a wide flange due to shear lag

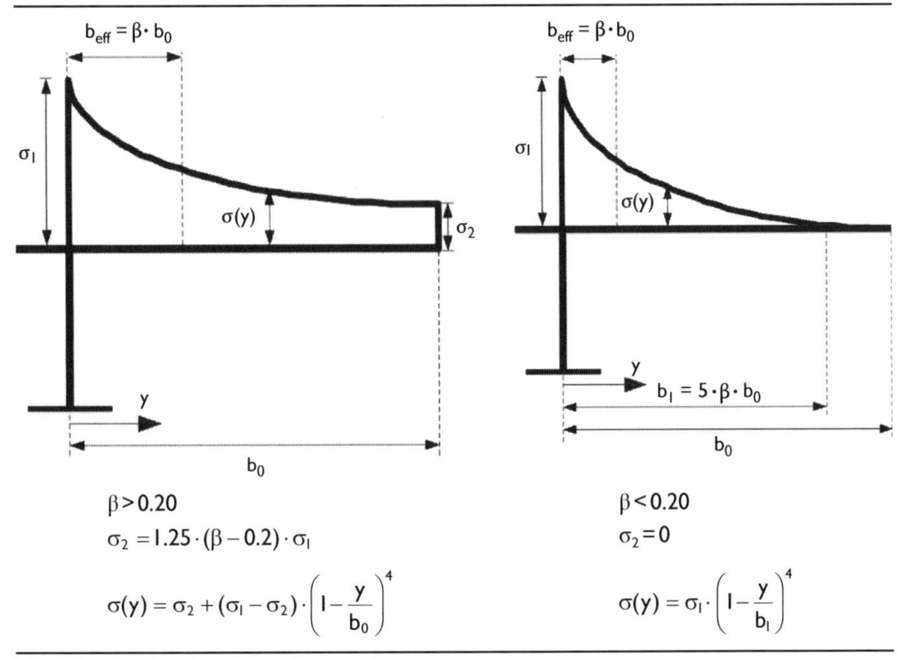

Source: EN 1993-2, Design of steel structures, Steel bridges, 2006.

Note: σ_1 is calculated from Equation 7.21.

Table 7.9 Shear lag calculation methods at ultimate limit states

	Effective[s] width factors
Method 1 (elastic)	β according to Table 7.7
Method 2	β_{ult} calculated as in Table 7.7 but replacing α_0 with $\alpha_0 = \sqrt{\frac{A_{c,eff}}{b_0 \cdot t}}$ where $A_{c,eff}$ is the effective area of compression flanges due to plate buckling
Method 3	Elastic plastic shear lag effects allowing for limited plastic strains may be taken into account using A_{eff} as follows: $A_{eff} = \beta^k \cdot A_{c,eff} \geq \beta \cdot A_{c,eff}$ (7.22) where $A_{c,eff}$ = the effective area of a compressed flange accounting for plate buckling β, κ = factors given in Table 7.7

Source: EN 1993-2, Design of steel structures, Steel bridges, 2006.

Note: The expressions in Methods 2 and 3 may also be applied for flanges in tension in which case $A_{c,eff}$ should be replaced by the gross area of the tension flange.

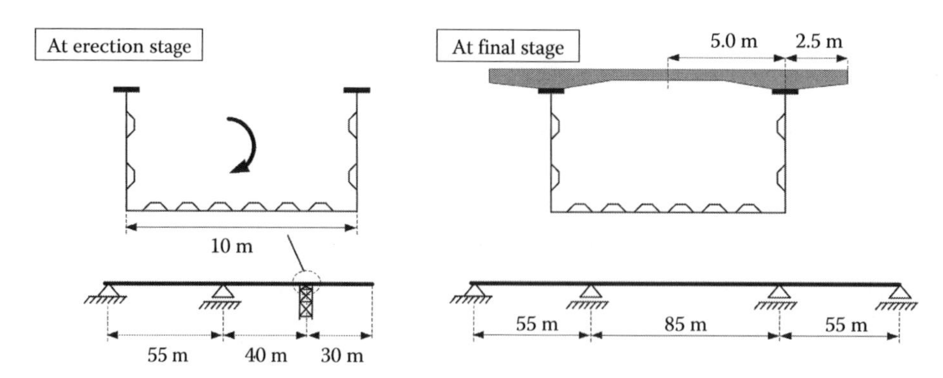

Figure 7.37 Box-girder bridge at erection and final stage.

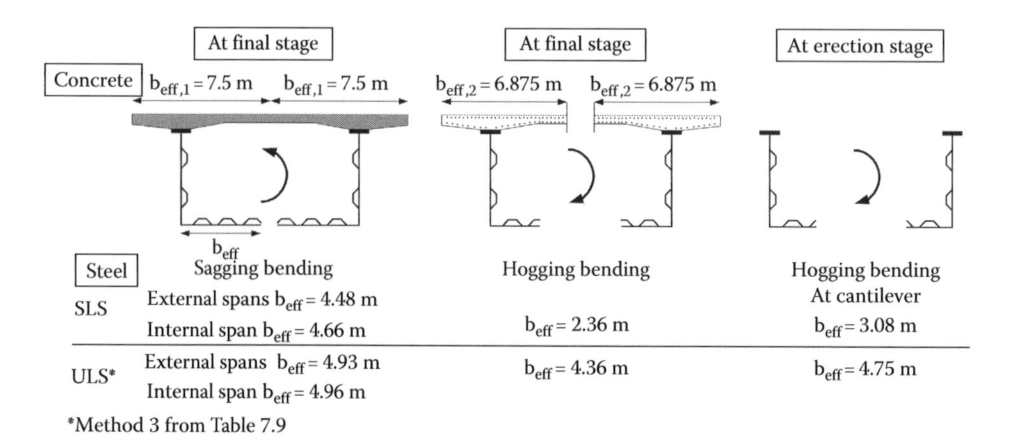

Figure 7.38 Summary of effective^s widths at final and erection stage for the cross sections of Figure 7.37.

EXAMPLE 7.3

The effective^s widths are calculated for both the concrete and the steel flanges of a composite box-girder bridge. The bottom flange has continuous longitudinal stiffeners such that $A_{sl}/(b_0 \cdot t) = 0.6$. The effective^s widths acting with each web will be determined at final stage and indicatively over a temporary support at erection stage.

- **Concrete flange (final stage)**

  For the external spans with L = 55 m

  Since the distance between the shear connectors is not known, it can be assumed that $b_0 = 0$.

  From Figure 7.34 $\Rightarrow L_e = 0.85 \cdot 55 = 46.75$ m

$$b_{e1} = \frac{46.75}{8} = 5.844\ \text{m} > \frac{10}{2} = 5.0\ \text{m} \Rightarrow b_{e1} = 5.0\ \text{m}$$

$$b_{e2} = \frac{46.75}{8} = 5.844\ \text{m} > 2.5\ \text{m} \Rightarrow b_{e2} = 2.5\ \text{m}$$

  Effective width:

$$b_{eff,1} = b_0 + b_{e1} + b_{e2} = 7.5\ \text{m}$$

For the internal span is

$$L_e = 0.7 \cdot 85 = 59.5\,m > 46.75\,m \Rightarrow b_{eff} = 7.5\,m$$

At sagging moment areas, the concrete flange is entirely effective.

At internal support

From Figure 7.34 $\Rightarrow L_e = 0.25 \cdot (55 + 85) = 35\,m$

$$b_{e1} = \frac{35}{8} = 4.375\,m < \frac{10}{2} = 5.0\,m \Rightarrow b_{e1} = 4.375\,m$$

$$b_{e2} = \frac{35}{8} = 4.375\,m > 2.5\,m \Rightarrow b_{e2} = 2.5\,m$$

Effective width:

$$b_{eff,2} = b_0 + b_{e1} + b_{e2} = 6.875\,m$$

At hogging moment areas, the effective part of the concrete flange is reduced.

- **Steel flange (final stage)**

Sagging bending (external spans with L = 55 m)

Table 7.7 $\Rightarrow L_e = 0.85 \cdot 55 = 46.75\,m, \alpha_0 = \sqrt{1 + 0.6} = 1.265, b_0 = 5.0\,m$

$$\kappa = \frac{1.265 \cdot 5.0}{46.75} = 0.135$$

For sagging moment areas,

$$0.02 < \kappa \leq 0.70 \Rightarrow \beta = \beta_1 = \frac{1}{1 + 6.4 \cdot 0.135^2} = 0.896$$

SLS: Equation 7.20 $\Rightarrow b_{eff} = 0.896 \cdot 5.0 = 4.48\,m$

ULS: Table 7.9 $\Rightarrow b_{eff} = 5.0 \cdot 0.896^{0.135} = 4.93\,m > 5.0 \cdot 0.896 = 4.48\,m$

Sagging bending (internal span with L = 85 m)

Table 7.7 $\Rightarrow L_e = 0.7 \cdot 85 = 59.5\,m, \alpha_0 = 1.265, b_0 = 5.0\,m$

$$\kappa = \frac{1.265 \cdot 5.0}{59.5} = 0.106$$

For sagging moment areas,

$$0.02 < \kappa \leq 0.70 \Rightarrow \beta = \beta_1 = \frac{1}{1 + 6.4 \cdot 0.106^2} = 0.932$$

SLS: Equation 7.20 $\Rightarrow b_{eff} = 0.932 \cdot 5.0 = 4.66\,m$

ULS: Table 7.9 $\Rightarrow b_{eff} = 5.0 \cdot 0.932^{0.106} = 4.96\,m \geq 0.932 \cdot 5.0 = 4.66\,m$

Hogging bending

Table 7.7 $\Rightarrow L_e = 0.25 \cdot (55 + 85) = 35\,m, \alpha_0 = 1.265, b_0 = 5.0\,m$

$$\kappa = \frac{1.265 \cdot 5.0}{35} = 0.181$$

For hogging moment areas, $0.02 < \kappa \le 0.70 \Rightarrow$

$$\beta = \beta_2 = \frac{1}{1 + 6.0 \cdot (0.181 - (1/2500 \cdot 0.181)) + 1.6 \cdot 0.181^2} = 0.471$$

SLS: Equation 7.20 $\Rightarrow b_{eff} = 0.471 \cdot 5.0 = 2.36$ m

ULS: Table 7.9 $\Rightarrow b_{eff} = 5.0 \cdot 0.471^{0.181} = 4.36 \text{ m} \ge 0.471 \cdot 5.0 = 2.36$ m

- **Steel flange (erection stage)**

The effective^s width of the bottom flange at the temporary support is investigated. Table 7.7 $\Rightarrow L_3 = 30$ m, $L_e = 2 \cdot 30 = 60$ m, $\alpha_0 = 1.265$, $b_0 = 5.0$ m

$$\kappa = \frac{1.265 \cdot 5.0}{60} = 0.105$$

For hogging moment areas,

$$\beta = \beta_2 = \frac{1}{1 + 6.0 \cdot (0.105 - (1/2500 \cdot 0.105)) + 1.6 \cdot 0.105^2} = 0.615$$

SLS: Equation 7.20 $\Rightarrow b_{eff} = 0.615 \cdot 5.0 = 3.08$ m

ULS: Table 7.9 $\Rightarrow b_{eff} = 5.0 \cdot 0.615^{0.105} = 4.75 \text{m} \ge 0.615 \cdot 5.0 = 3.08$ m

7.3 CROSS-SECTIONAL PROPERTIES

As discussed in the previous sections, the longitudinal girders are composed of the steel beams and the associated slab within the effective^s width. This cross section is transformed in an equivalent cross section of steel material alone, by reducing the concrete area with the modular ratio n_0 for short-term loading or n_L for long-term loading. The cross-sectional properties are summarized in Figure 7.39.

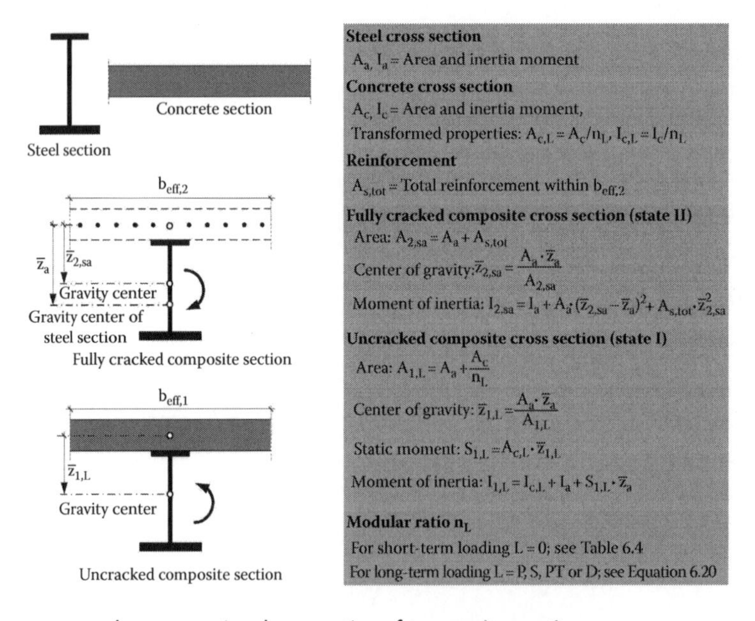

Figure 7.39 Geometry and cross-sectional properties of composite sections.

In the expressions for the cross-sectional properties in Figure 7.39, the slab's reinforcing bars have been considered as concentrated at the slab's center of gravity. This approximation simplifies the calculations without affecting its accuracy.

At hogging moment areas, instead of a fully cracked composite section, a "semi-cracked" section can be used by taking into account the effect of *tension stiffening*; see Section 6.1.6. This can be achieved by the consideration of a slab area that is calculated by Equation 7.14.

For composite box girders, the calculation of the torsional constant has been discussed in Section 7.1.2; see Figure 7.12 and Equation 7.11.

7.4 EFFECTS OF THE RHEOLOGICAL BEHAVIOR OF CONCRETE ON STRUCTURAL SYSTEMS

7.4.1 General

A discussion on the time-dependent behavior of concrete is found in Section 6.1.2. For the concrete members of a composite bridge, creep can be taken into account through the age-adjusted modulus of elasticity given in Equation 7.15. In composite members, the interaction between concrete and steel must be considered through the long-term modular ratio n_L given in Equation 6.20. The modular n_L depends also on the type of loading that may be permanent (P), temporarily permanent (PT), imposed deformations (D), or shrinkage (S). Creep in composite members results in a reduction in bending stiffness with consequences that depend on the static system.

7.4.2 Creep in statically determinate systems

In *statically determinate*, single-span systems, the concrete is under compression and the bending stiffness EI is determined for the uncracked cross section. The stiffness reduction from $E_a \cdot I_{1,0}$ to $E_a \cdot I_{1,P}(t)$ due to creep in the time interval t_0 to t results in

a. Increased deflections
b. No change in the internal acting moments and shear forces
c. Stress redistribution in the cross sections, where the stresses in the steel girders are increased, while the stresses in the concrete slab are decreased; see also Figure 6.6

Figure 7.40 shows how a global positive moment M_0 on the composite section is split into partial moments and a pair of axial forces in the concrete slab and the steel girder that may be determined from the following equations:

$$N_{c,P} = \frac{A_{c,P} \cdot \overline{z}_{1,P}}{I_{1,P}} \cdot M_0 \tag{7.23a}$$

$$M_{c,P} = \frac{I_{c,P}}{I_{1,P}} \cdot M_0 \tag{7.23b}$$

$$N_{a,P} = \frac{A_a \cdot (\overline{z}_a - \overline{z}_{1,P})}{I_{1,P}} \cdot M_0 \tag{7.23c}$$

$$M_{a,P} = \frac{I_a}{I_{1,P}} \cdot M_0 \tag{7.23d}$$

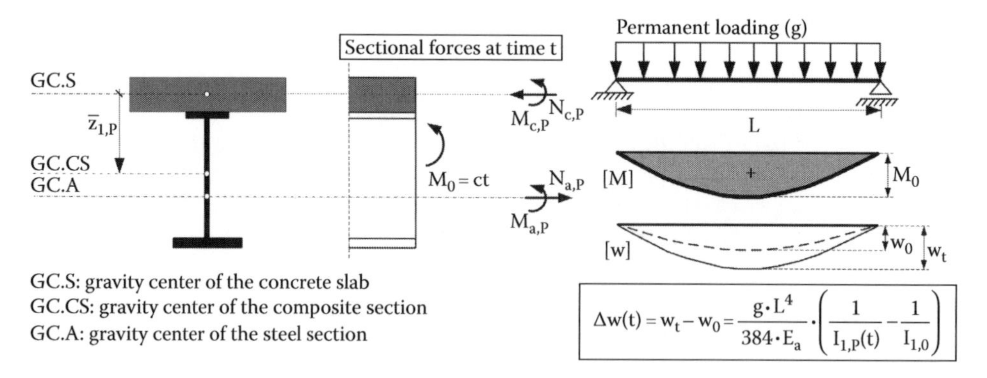

Figure 7.40 Influence of creep in statically determined simple span beams.

Through the time-dependent partial forces and moments given in Equations 7.23, the stresses in concrete and steel can be calculated at any time. In Figure 7.40, one can see the magnification of the deflections that mainly depends on the value of the long-term inertia moment $I_{1,P}$.

7.4.3 Creep and shrinkage in statically indeterminate systems

In *statically indeterminate* continuous beams, the concrete slab is under compression in the span and tension in the internal support regions. The stiffness reduction in the time interval $(t–t_0)$ due to creep has therefore an effect only on the uncracked span regions. The support regions where the concrete is cracked remain unaffected from creep. This means that the stiffness in the support regions increases relatively to the stiffness in the span. Accordingly, the supports attract higher moments while the moments in the spans reduce. Consequently, creep results in continuous beams:

a. Increased deflections
b. Changes in the internal acting moments with an increase of moments at the supports and moment reduction in the spans; see Figure 7.41
c. Corresponding shear forces that induce longitudinal shear forces at the interface between the concrete slab and the steel beam
d. Stress redistribution in the cross sections at sagging moment areas, where the stresses in the steel girders are increased, while the stresses in the concrete slab are decreased

The final stresses at the time t are due to the moments at time t (due to M_P and live loads) and the change of moments $M_{PT}(t, t_0)$ between the times t_0 and t. The M_{PT} moments are called *secondary moments*. A calculation example for a two-span continuous system is demonstrated in Figure 7.41. For the estimation of the secondary moments M_{PT}, the *force method* is used. The initial bending moment diagram $[M_0]$ is calculated by applying the short-term modular ratio n_0 at the noncracked regions. Due to creep, the moment of inertia of the composite parts will decrease from $I_{1,0}$ to $I_{1,P}$. This results in an additional rotation θ_P in the released structure that can be easily calculated by integrating the short-term bending diagram $[M_0]$ with the curvature diagram due to unit moments $\bar{M}$ at the internal support. The rotation θ_P is inconsistent with the actual structure and an additional moment min M_{PT} is developed. The secondary bending moment min M_{PT} produces a rotation at the internal support of the released structure that must eliminate the inconsistency θ_P; this is expressed by the compatibility equation that offers the value for min M_{PT}.

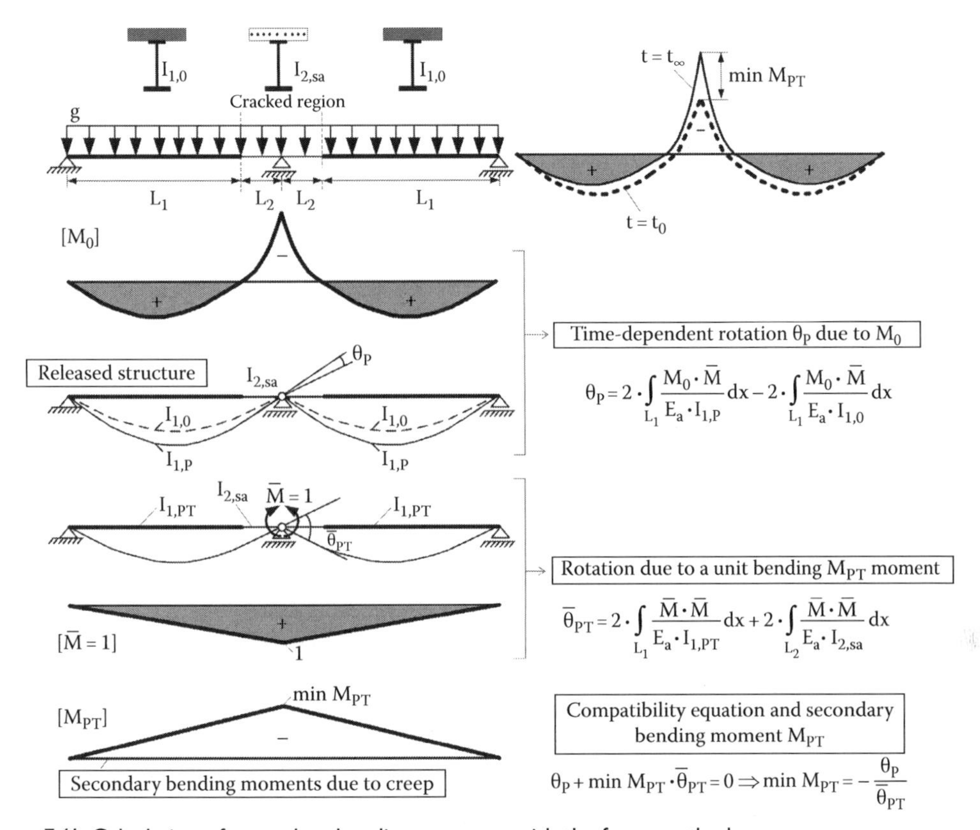

$$\theta_P = 2 \cdot \int_{L_1} \frac{M_0 \cdot \bar{M}}{E_a \cdot I_{1,P}} dx - 2 \cdot \int_{L_1} \frac{M_0 \cdot \bar{M}}{E_a \cdot I_{1,0}} dx$$

$$\bar{\theta}_{PT} = 2 \cdot \int_{L_1} \frac{\bar{M} \cdot \bar{M}}{E_a \cdot I_{1,PT}} dx + 2 \cdot \int_{L_2} \frac{\bar{M} \cdot \bar{M}}{E_a \cdot I_{2,sa}} dx$$

$$\theta_P + \min M_{PT} \cdot \bar{\theta}_{PT} = 0 \Rightarrow \min M_{PT} = -\frac{\theta_P}{\bar{\theta}_{PT}}$$

Figure 7.41 Calculation of secondary bending moments with the force method.

As shown in Figure 7.41, the secondary moments are negative and may be considered as resulting from a concentrated upward support force. Instead of rotations at supports, deflections can be calculated by releasing the internal support in the vertical direction.

It should be noted that the secondary bending moments may even reach half of the value of the initial bending moments at internal supports. Obviously, they should not be neglected. Moreover, the calculation of secondary internal forces with the force method can only be achieved for very simple systems, for example, one-beam models. A more convenient way to calculate secondary effects is through the use of an equivalent temperature. Indeed, when the cross section is subjected to a moment M_0 that causes creep, its curvature changes by a magnitude:

$$\Delta\kappa_t = \frac{M_0}{E_a \cdot I_{1,P}} - \frac{M_0}{E_a \cdot I_{1,0}} \tag{7.24}$$

If the same cross section with a height h were subjected to a linear temperature difference ΔT_{MP} between its upper and lower flanges, a curvature given in the succeeding text would be caused:

$$\Delta\kappa_t = \frac{\Delta T_{MP}}{h} \cdot \alpha_t \tag{7.25}$$

Accordingly, the influence of creep may be accounted for by introducing a linear temperature difference as a "loading," where the upper part is colder and the lower part warmer. This difference may be determined by setting equal the two curvatures and is given by

$$\Delta T_{MP} = \frac{M_0}{E_a} \cdot \frac{h}{\alpha_t} \cdot \left(\frac{1}{I_{1,P}} - \frac{1}{I_{1.0}} \right) \quad (7.26)$$

where

E_a is the modulus of elasticity for steel
α_T is the thermal expansion coefficient for steel
$I_{1,0}$ is the short-term moment of inertia of the composite section; see Figure 7.39
$I_{1,P}$ is the long-term moment of inertia of the composite section (L = P); see Figure 7.39

As shown in Figure 7.42, the temperature difference is introduced only in the uncracked regions with $I = I_{1,PT}$ and **not** in the cracked ones; see creep due to temporarily permanent loads in Section 6.1.2. For simplicity, the temperature difference may be considered with its maximum value, corresponding to the maximum moments. This is obviously a conservative approximation.

The secondary bending moments cause additional normal stresses that have an unfavorable effect at hogging moment areas and a favorable one at the sagging moment areas. The partial internal forces in the uncracked composite section due to secondary bending moments can be calculated by Equations 7.23 by replacing the index P with PT.

REMARK 7.7

- Adding bending moments due to different types of loading leads to false results. For example, adding the short-term moment min M_0 at internal support with the secondary moment min M_{PT} is a mistake. Stresses should be calculated for each bending moment separately and then superpositioned.
- Bridges that are axially restrained develop normal forces N_P and N_{PT}. These forces should be included in the Equations 7.23. A further discussion on the stress calculation of axially restrained systems is found in [7.15].

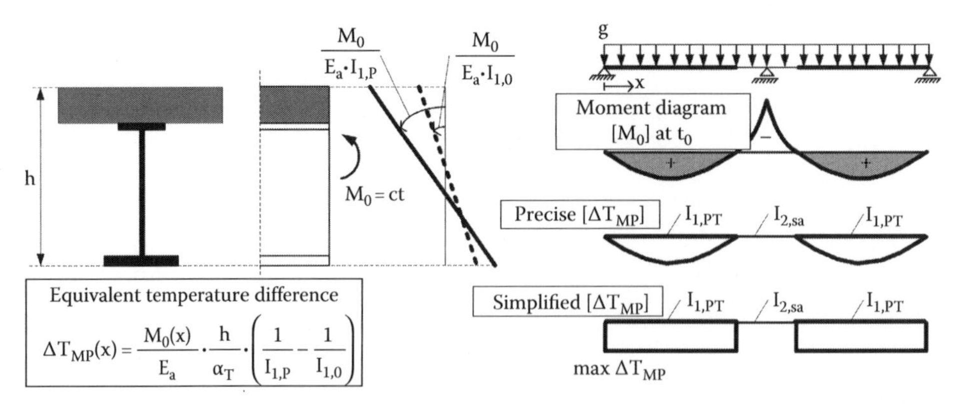

Figure 7.42 Calculation of secondary bending moments with an equivalent temperature difference.

7.4.3.1 Creep due to movements of supports

In continuous bridges, the negative moments at internal supports are in many cases significantly high causing excessive tensile stresses and concrete cracking. In order to reduce the bending moments at hogging moment areas, a support movement δ is imposed as shown in Figure 7.43. The positive moments $[M_D(t_0)]$ at time t_0 can be calculated by applying the short-term cross-sectional properties of the composite girders. Unfortunately, the positive moments decrease due to creep. For systems with **constant** cross-sectional properties along the bridge's axis, the time-dependent bending moments $M_D(t)$ can be estimated with the simple expression of Equation 7.27:

$$M_D(t) = M_D(t_0) \cdot \frac{I_{1,D}}{I_{1,0}} \tag{7.27}$$

where $I_{1,D}$ is the moment of inertia due to creep of type D; for creep due to imposed deformations, see in Section 6.1.2.

When $M_D(t)$ is known, then the partial internal forces can be calculated through Equations 7.23 by replacing the index P with D. Subsequently, concrete stresses can be estimated and cracking can be controlled by imposing an adequate support movement δ.

When the cross-sectional properties of the main girders vary longitudinally, then the simplified method demonstrated in Figure 7.43 **should not be applied**. The secondary hogging moments M_{PT} that cause the reduction of M_D are calculated through the following steps:

1. The bending moments $(M_D(t_0))$ due to the support movement δ are calculated by applying the short-term cross-sectional properties.
2. The equivalent temperature of Equation 7.26 is applied at the noncracked areas that have inertia moment $\mathbf{I_{1,PT}}$.
3. The secondary moments M_{PT} are then calculated.

Attention must be paid to the fact that *in Equation 7.26, the inertia moment $I_{1,P}$ should not be replaced by $I_{1,D}$*. This is because the moments $M_D(t_0)$ are treated as permanent loading and not as imposed deformations. It is also worth mentioning that stresses due to M_D and M_{PT} should be calculated *separately*.

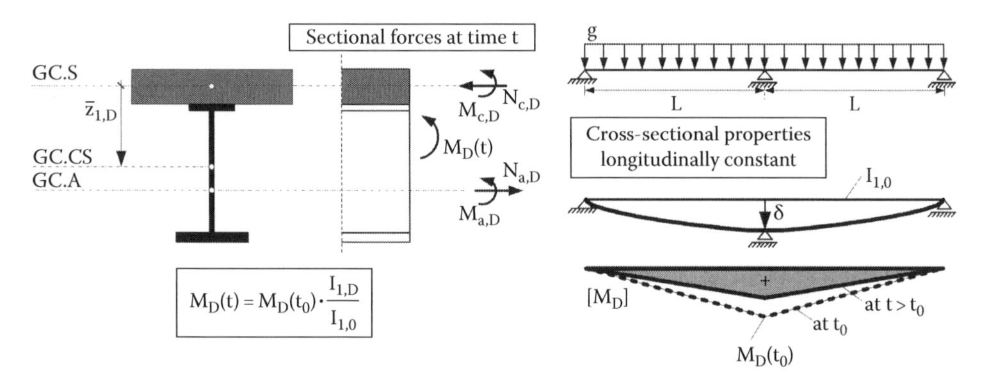

Figure 7.43 Support movement of a continuous two-span composite bridge with nonvariable cross-sectional properties.

7.4.3.2 Shrinkage

The shortening of the concrete slab due to shrinkage is restrained due to its shear connection with the steel girder so that there develops a tension force N_{sh} in it. Since this force is due to restraint, an equal compression force acts in the composite section as well as a moment M_{sh} that equilibrates the pair of forces of the slab and the composite section; see Figure 7.44. Concluding, shrinkage results

- In the concrete slab
  a tension force:

$$N_{sh} = -\varepsilon_{cs} \cdot \frac{\eta_0}{\eta_S} \cdot E_{cm} \cdot A_c \tag{7.28a}$$

- In the composite section
  a compression force:

$$-N_{sh} = \varepsilon_{cs} \cdot \frac{\eta_0}{\eta_S} \cdot E_{cm} \cdot A_c$$

  and a moment:

$$M_{sh} = N_{sh} \cdot \bar{z}_{1,S} \tag{7.28b}$$

  where

  n_0 is the short-term modular ration given in Table 6.4
  n_S is the long-term modular ratio for shrinkage; see Equation 6.20
  $\bar{z}_{1,S}$ is the gravity center of the equivalent section calculated with n_S; see Figure 7.39
  E_{cm} is the concrete's modulus of elasticity given in Table 6.1
  A_c is the sectional area of the concrete slab based on the **geometric width**
  ε_{cs} is the shrinkage strain according to Section 6.1.3

The time-dependent sectional forces and moments are calculated as follows:

$$N_{c,S} = N_{sh} \cdot \left(1 - \frac{A_c}{n_S \cdot A_{1,S}} - \frac{A_c}{n_S \cdot I_{1,S}} \cdot \bar{z}_{1,S}^2\right) \tag{7.29a}$$

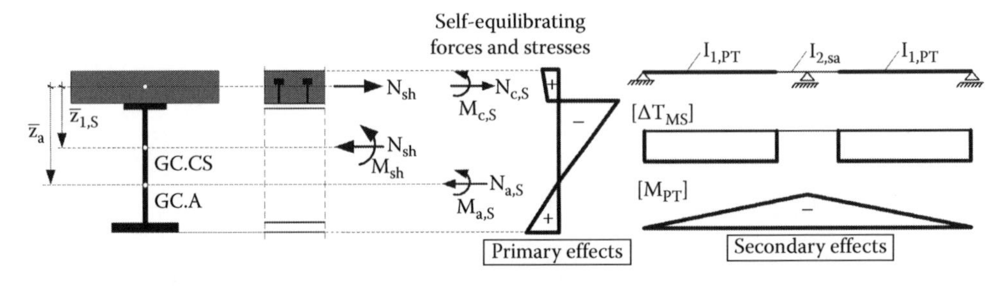

Figure 7.44 Primary and secondary effects due to shrinkage.

$$M_{c,S} = \frac{I_c}{n_S \cdot I_{1,S}} \cdot M_{sh} \tag{7.29b}$$

$$N_{a,S} = N_{sh} \cdot \left[\frac{A_a}{A_{1,S}} - \frac{A_a}{I_{1,S}} \cdot \bar{z}_{1,S} \cdot \left(\bar{z}_a - \bar{z}_{1,S} \right) \right] \tag{7.29c}$$

$$M_{a,S} = M_{sh} \cdot \frac{I_a}{I_{1,S}} \tag{7.29d}$$

The aforementioned self-equilibrating system of forces and moments constitute the *primary effects of shrinkage*; see Figure 7.44. In statically determinate systems, the primary effects of shrinkage result in **non-negligible** values of stresses and deflections.

The secondary internal forces in statically indeterminate systems can be calculated with the force method. But as previously mentioned, a more convenient method is the application of an equivalent temperature difference ΔT_{MS} at the noncracked regions. The temperature difference is calculated with the following equation:

$$\Delta T_{MS} = \frac{M_{sh}}{E_a \cdot I_{1,S}} \cdot \frac{h}{\alpha_t} \tag{7.30}$$

In 3D models, the effects of shrinkage can be calculated by introducing a uniform temperature difference in the longitudinal slab elements equal to

$$\Delta T_{NS} = \frac{\varepsilon_{cs}}{\alpha_T} \tag{7.31}$$

The secondary effects of shrinkage have to be taken into account in both ultimate and serviceability limit states. If **all** cross sections are class 1 or 2, then secondary effects need not to be taken into account at ultimate limit states.

EXAMPLE 7.4

Calculation of the secondary effects for a two-span composite beam due to Figure 7.45.

- Permanent loads (uniformly distributed 40 kN/m, $\varphi(7, \infty) = 2.5$)
- Shrinkage (with $\varepsilon_{cs} = 330 \cdot 10^{-6}$, $\varphi(1, \infty) = 3.6$)
- Support movement ($\delta = 30$ mm)

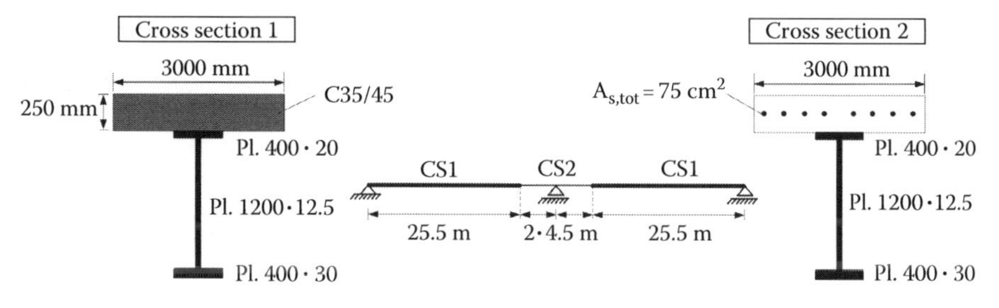

Figure 7.45 Cross sections at cracked and noncracked regions.

Table 7.10 Cross-sectional properties

	n_L	$\bar{z}_L$ (cm)	A_L (cm²)	I_L (cm⁴)
Steel section	—	81.64	350.00	913,809.5
Cross section 1				
Short-term L=0	6.18	18.28	1563.59	2,787,747.54
Long-term L=P	23.18	42.42	673.62	2,051,464.70
Long-term L=PT for φ_t=2.5	14.68	33.19	860.99	2,325,001.35
Long-term L=S	18.42	37.74	757.25	2,189,674.69
Long-term L=PT for φ_t=3.6	18.42	37.74	757.25	2,189,674.69
Long-term D	29.36	47.19	605.49	1,911,522.82
Cross section 2	—	67.24	425	1,325,506.00

The cross-sectional properties can be calculated according to Figure 7.39 and are given in Table 7.10.

- Figure 7.46 shows the "short-term" moment diagram $[M_0]$. According to the simplified method of Figure 7.42, the equivalent temperature difference max ΔT_{MP} is calculated from Equation 7.26 as follows:

$$\max \Delta T_{MP} = \frac{291338}{21000} \cdot \frac{150}{10^{-5}} \cdot \left(\frac{1}{2051464.70} - \frac{1}{2787747.54} \right) = 26.79°C$$

As previously mentioned, max ΔT_{MP} is applied on the uncracked regions with moment of inertia equal to $I_{1,PT}$=2325001.35 cm⁴.

One can see that the secondary bending moment min M_{PT} is significantly high. A less conservative calculation can be achieved by applying a temperature variation that follows the bending moment diagram $[M_0]$.

- The axial force N_{sh} due to shrinkage is according to Equation 7.28a:

$$N_{sh} = -330 \cdot 10^{-6} \cdot \frac{6.18}{18.42} \cdot 3400 \cdot 300 \cdot 25 = -2823.27 \text{ kN}$$

The bending moment M_{sh} is calculated by Equation 7.28b:

$$M_{sh} = 2823.27 \cdot 0.3774 = 1065.50 \text{ kN-m}$$

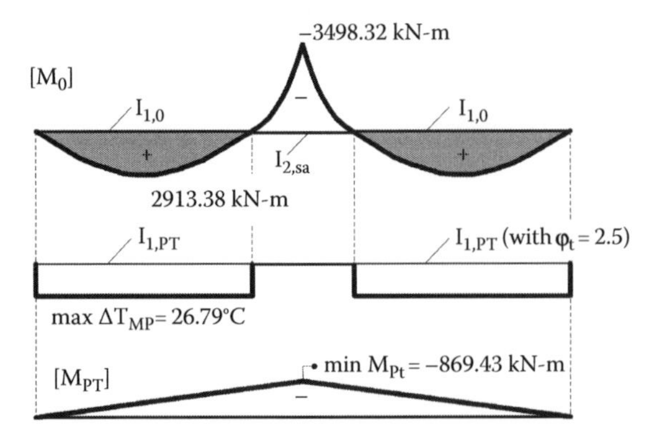

Figure 7.46 Calculation of secondary effects with an equivalent temperature difference for permanent loading (P).

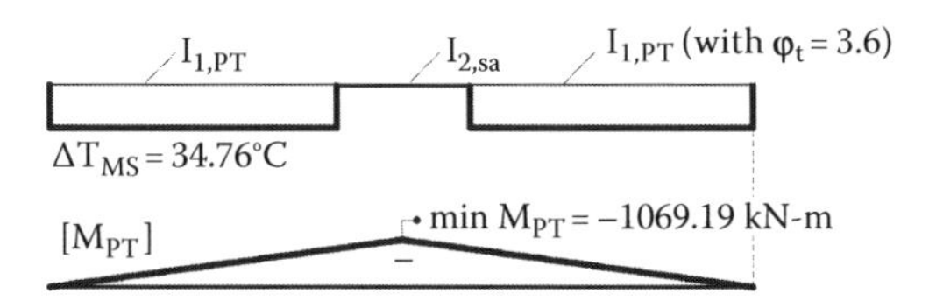

Figure 7.47 Calculation of secondary effects with an equivalent temperature difference for shrinkage (S).

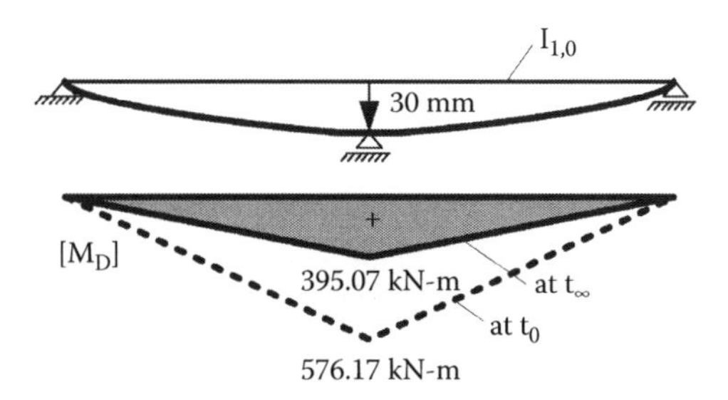

Figure 7.48 Calculation of secondary effects for imposed deformations (D).

The temperature difference that will be applied on the uncracked regions is according to Equation 7.30 (Figure 7.47)

$$\Delta T_{MS} = \frac{106550}{21000 \cdot 2189674.69} \cdot \frac{150}{10^{-5}} = 34.76°C$$

- At internal support, a movement equal to 30 mm is imposed; see Figure 7.48. Since the cross-sectional properties of the beam do not vary longitudinally, the expression in Equation 7.27 can be used. The final value for the positive moment at internal support becomes after the reduction due to creep equal to

$$M_{D,\infty} = 576.17 \cdot \frac{1911522.82}{2787747.54} = 395.07 \text{ kN-m}$$

REMARK 7.8

The tension stiffening effect could be taken into account in the previous calculations by increasing the reinforcement of the fully cracked section (CS2) from 75 cm² to $A_{t,i}$, where $A_{t,i}$ is calculated by Equation 7.14 as follows:

$$A_{t,i} = \frac{75}{1-(0.5 \cdot 0.32/1\% \cdot 50)} = 110.29 \text{ cm}^2; \quad f_{sk} = 500 \text{ N/mm}^2, \quad f_{ctm} = 3.2 \text{ N/mm}^2$$

The inertia moment of the cross section 2 becomes then $I_{2,sa} = 1{,}472{,}822$ cm⁴.

This leads to an increase of the hogging moments ~5%.

It has to be noted that the calculation procedure for the secondary effects with an equivalent temperature can be easily applied on grillage systems. The influence of the secondary effects on elements such as cross girders, diagonals, and bearings can be in this way investigated.

7.5 MODELS FOR SLAB ANALYSIS AND DESIGN IN TRANSVERSE DIRECTION

7.5.1 General

The models presented so far mainly refer to the longitudinal direction of the bridge. However, they do not cover the analysis and design of the concrete slab that spans usually in the transverse direction between the longitudinal steel girders. The analysis of the slab provides the bending moments that allow for the determination of the required reinforcement at ultimate limit states, the control of crack widths, and all other requirements for design. The slab is subjected to distributed loading due to self-weight, traffic, and concentrated wheel loads.

7.5.2 Distributed permanent and variable loads

The analysis for distributed loads may be performed on a strip of unit width. The cross section of the strip is of reinforced concrete with constant or variable thickness in case of haunches. This means that uncracked analysis is performed in transverse direction. The strip is supported by the longitudinal steel girders or the webs of the box for box-girder sections as shown in case b in Figure 7.49. The influence lines for the moments at critical sections, as indicatively shown in Figure 7.50, provide the areas where the traffic loads are to be positioned in order to determine the most unfavorable effects. Accordingly, traffic loads are placed for support moments only over regions of negative values of the influence line, whereas for span moments, only over regions of positive values of the influence line. Design moments are determined by multiplication with the relevant safety factors γ_G and γ_Q. Theoretically, distinction should be made between favorable and unfavorable values of γ_G and γ_Q. This means that γ_G should be taken equal to 1.0 in regions of positive values of the influence line and 1.35 in regions of negative values of the influence line in order to determine the design value of support moments. However, this constitutes an overstatement of the physical probability and many designers do not follow this in practice.

The assumption that the girders provide rigid support represents the conditions near the support region of the girders. However, in the span region, the girders deflect at loading and provide rather a flexible than a rigid support to the slab. Accordingly, the strip should be considered as supported by springs. The spring constant may be determined by consideration of the deflection of the girders due to loading. This deflection is a function of the longitudinal position considered and the overall support conditions of the bridge. For example, for a simply supported girder, the deflection at mid-span due to a uniformly distributed load p is equal to

$$\delta = \frac{5 \cdot p \cdot L^4}{384 \cdot E \cdot I} \tag{7.32}$$

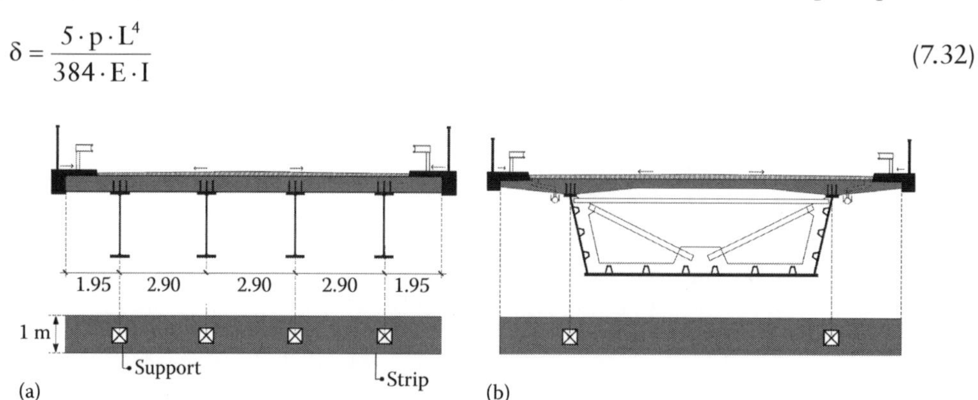

Figure 7.49 Strip model for the analysis of the slab in transverse direction. (a) For multi-girder and (b) box girder bridges.

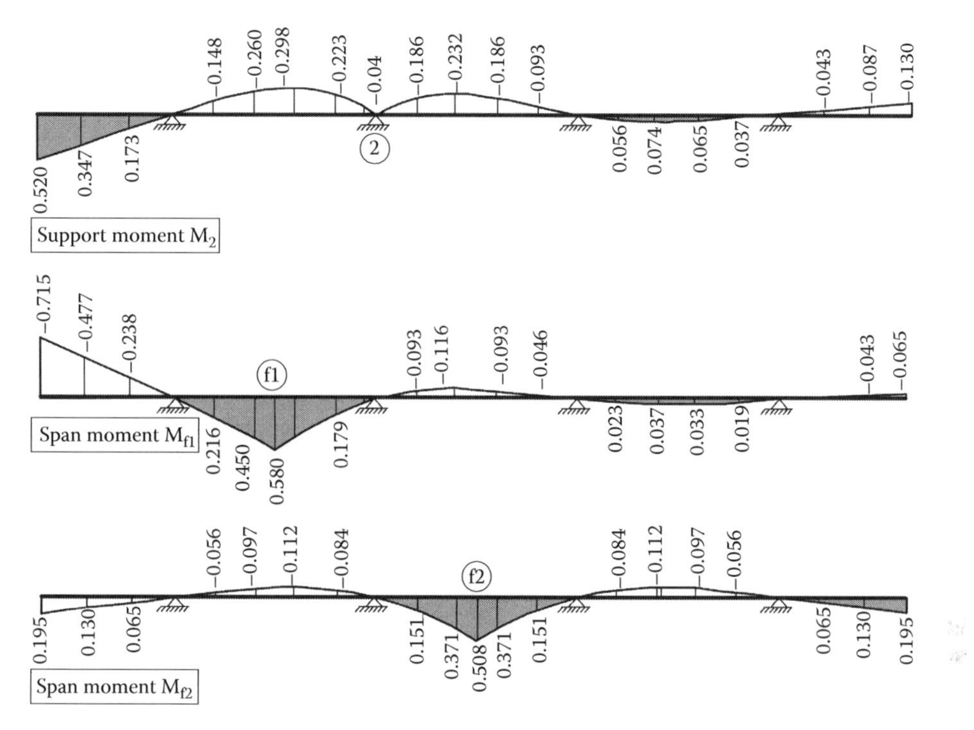

Figure 7.50 Influence lines of moments for the strip model of Figure 7.49a with rigid supports.

The spring constant may be then determined from

$$k = \frac{p \cdot L}{\delta} = \frac{384 \cdot E \cdot I}{5 \cdot L^3} \tag{7.33}$$

where

L is the span length
E·I is the bending stiffness of the girder

Evidently, the flexibility of the spring relative to the flexibility (thickness) of the slab influences considerably the response of the strip to loading. Figure 7.51 shows indicatively the bending moments of a 30 cm thick slab resting on four main girders, for a simply supported 25

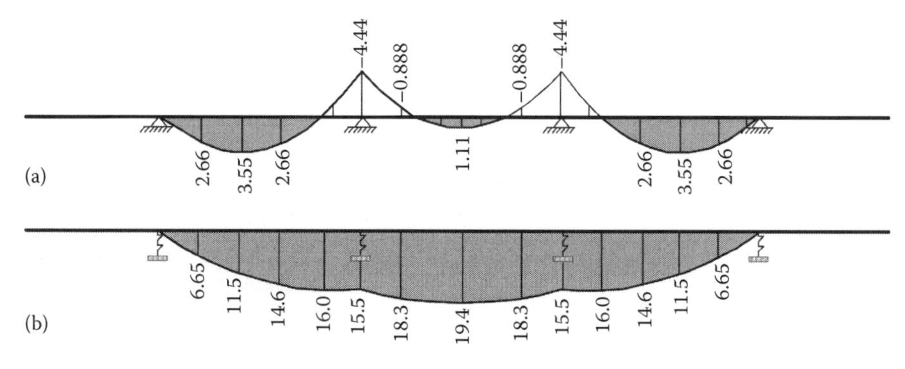

Figure 7.51 Bending moments in kN-m/m of slab (a) on rigid and (b) on flexible supports due to uniformly distributed loading in the central regions.

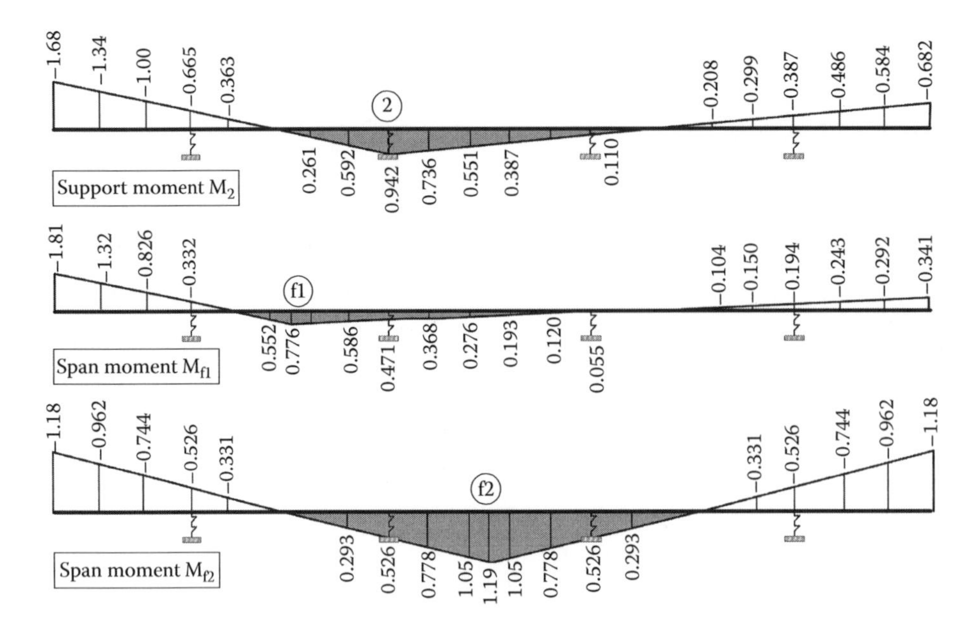

Figure 7.52 Influence lines for the strip model of Figure 7.49a with flexible supports.

m span bridge for girders with $I = 27 \cdot 10^5$ cm^4 under uniformly distributed loading 5.28 kN/m between the girders. It may be seen that the model on rigid supports provides conservative values of the support moments, while the model on flexible supports for span moments.

Besides their values, the influence lines of the bending moments differ for flexible supports. Figure 7.52 shows indicatively the influence lines for the support moment and the two-span moments for a slab with the properties given earlier. It may be seen that the influence lines do not change sign from span to span any more. Consequently, the lanes with traffic loads should be positioned differently compared to the case of rigid supports.

Concluding, it may be said that the bending moments due to distributed permanent and traffic loads may be determined by two strip models: one on rigid supports that expresses the conditions in the support regions of the bridge and one on flexible supports that expresses the conditions in the span regions of the bridge. The first model provides unfavorable bending moments at the supports of the slab, the second at the spans.

7.5.3 Wheel loads from traffic

The strip model is not appropriate for the determination of the slab moments due to wheel loads from traffic. As discussed before, the strip model provides only transverse and not longitudinal moment that exists when a slab is subjected to concentrated forces. In addition, the width of the strip depends on the distance between girders and the contact area of the wheel. Using the theory of elasticity, *Westergaard* [7.35] proposes for a simply supported slab under central load, presented in Figure 7.53, following crude approximation for this width, called *effective width*:

$$b_c = -0.58 \cdot s + 2 \cdot c \tag{7.34}$$

where

s is the distance between girders (=span of the slab)
c is the radius of the circle over which the load is considered to be distributed

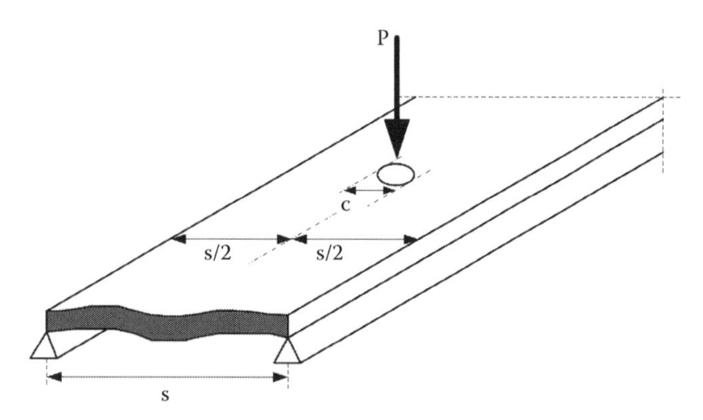

Figure 7.53 Simply supported slab under concentrated central load.

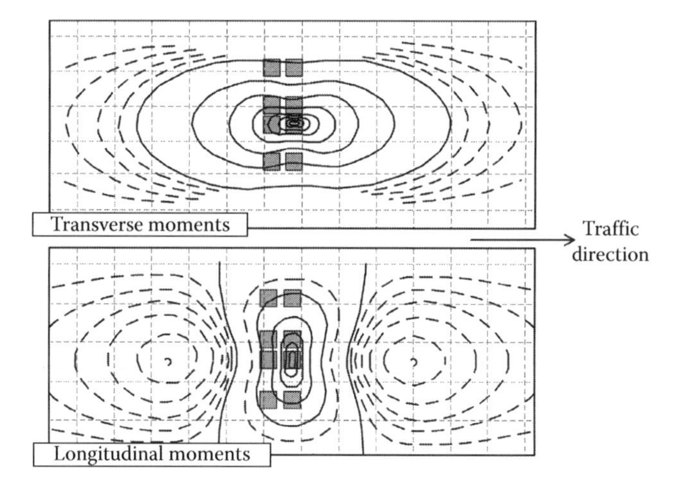

Figure 7.54 Influence areas for bending moments of a simply supported slab and the position of wheel loads to determine maximum values.

However, this simple formula cannot cover the practical situations since the load models considered have more wheels acting on any position and the support conditions of the slab do not always correspond to simple supports. A solution was given by *Pucher* [7.22] who developed charts with influence areas for the moments in transverse and longitudinal directions for elastic slabs under various support conditions, ranging from simple supports to fixed supports. The wheel loads are then placed appropriately within the chart to produce the most unfavorable effects. In fact, the wheel loads are acting in a small contact area, determined from Figure 4.2, and these contact areas are positioned into the charts, as indicatively shown in Figure 7.54. In the United States, similar methods based on *Westergaard* are applied. The final design moments are determined by superposition of the moments due to distributed and wheel loads.

7.5.4 Finite element models

Finite element models are most appropriate for the analysis and design of the deck slab. They are based on the theory of elasticity and provide transverse and longitudinal moments for any type of loading whether distributed or concentrated. The simplest way is to isolate the slab from the overall system and set fixed support conditions at the positions of the girders.

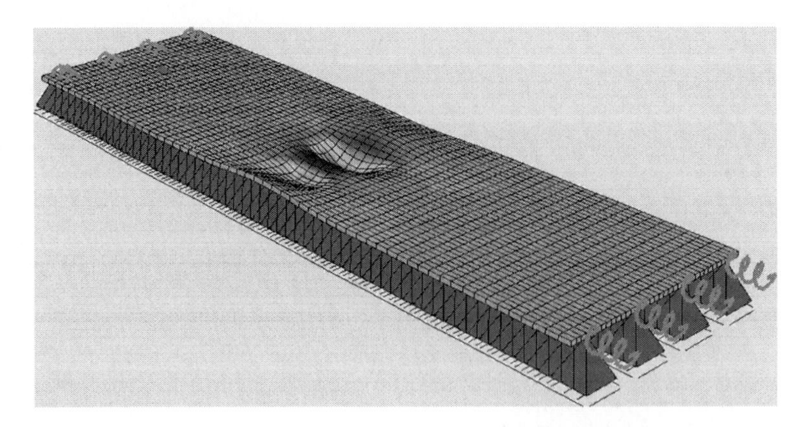

Figure 7.55 Deformed shape of an FE model of a bridge deck on four girders under four wheel loads.

The advantage of this model compared to those previously presented is that it delivers the bending moment distribution with a common and not with separate models for distributed and concentrated loads. Figure 7.55 shows indicatively an FE slab model of a bridge slab supported by four girders. In this example, the length of the bridge is 33 m, the width of the slab 9.44 m, its thickness 22 cm, and the distance between girders is 2.65 m. The figure shows the deformed shape of the deck when subjected to four wheel loads of the tandem system of Load Model 1 presented in Chapter 4. As shown in Table 4.5, each wheel load is 150 kN, the distance between wheel is 2.0 m in transverse and 1.2 m in longitudinal direction. In transverse direction, the two loads are placed at mid-span between the intermediate girders, the other two 1.2 m on the left. The deformed shape shows that after a certain distance the loads do not affect the deck. Consequently, only a small part of the deck in longitudinal direction may be modeled. This length corresponds to the *effective width* proposed by *Westergaard*.

Figure 7.56 shows the bending moments in a part of the deck near the applied loads. It may be seen that the influence length in this example is about 4.0 m, but this clearly depends on

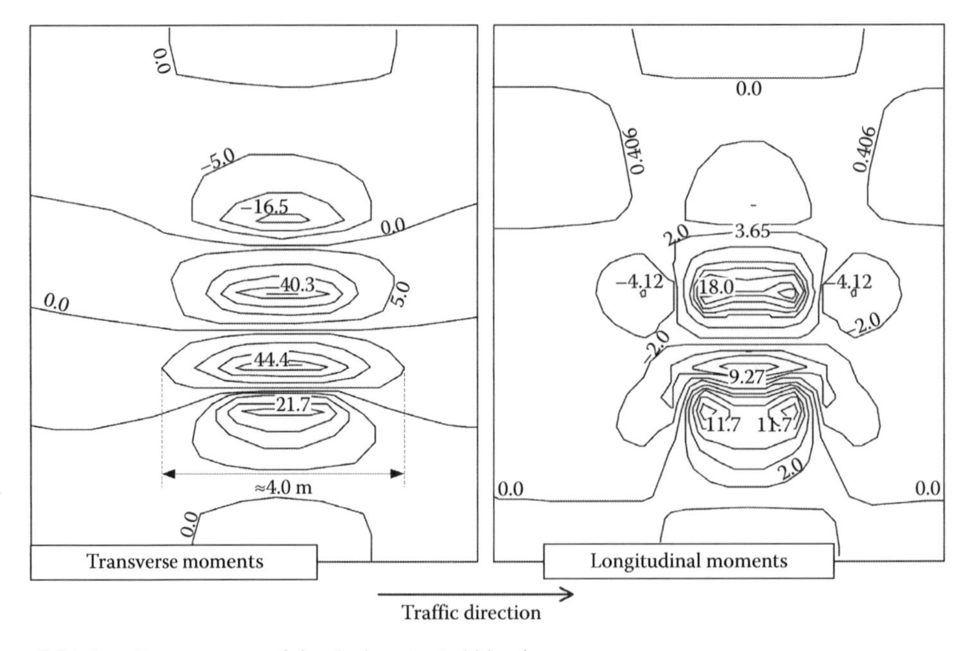

Figure 7.56 Bending moments of the deck; units in kN-m/m.

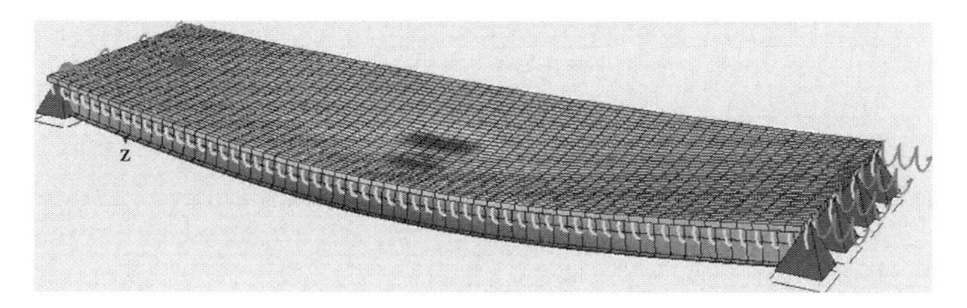

Figure 7.57 Deformed shape of a bridge deck on four girders under four wheel loads in the span.

the distance between the girders. The shape of the transverse moments is similar to the shape for continuous beams, exhibiting positive moments at span and negative over supports. It may be seen that for concentrated forces, the longitudinal moments are quite high and cannot be neglected in design. In this example, the maximal longitudinal span moment is 18 kN-m/m, while the maximal transverse span moment 44.4 kN-m/m, their ratio being about ½.

The FE models with rigid supports are based on similar assumptions as the approximate methods introduced before that were used for a long time in slab analysis. However, they do not consider the flexibility of the longitudinal girders in the span region that corresponds to a rather flexible than rigid support. A further step in FE modeling of the slab is to consider the entire bridge, introducing the girders for simplicity by means of beam elements. As shown in Figure 7.57, the girders deflect under wheel loads in the span region, thus providing flexible supports to the slab.

Figure 7.58 shows the transverse bending moments when the bridge is subjected to uniformly distributed loading 5.28 kN/m^2 between the girders, but not in the cantilevers. It may be seen that the shape of the diagram is similar to those provided by the strip model. The moments in the span region are always positive, either in the span or over the girders, similar to the results of the strip model on flexible supports shown in Figure 7.51. However, near the bridge supports, the moments over the girders become negative since the girders deflect less and get stiffer.

Similar observations may be made for concentrated wheel loads. Figure 7.59 shows that the minimum moment over the girder for wheel loads near supports is –38.9 kN-m/m, more than twice as large compared to the moment for wheel loads in span, –12.4 kN-m/m. Contrary, the maximal moment in span for wheel loads in the span, 58.9 kN-m/m, is approx. 35% larger than the corresponding moment for wheel loads near supports, 43.7 kN-m/m.

Figure 7.60 shows the longitudinal moments for both loading conditions. It may be seen that the moments in span are in both cases as high as 50% of the transverse moments.

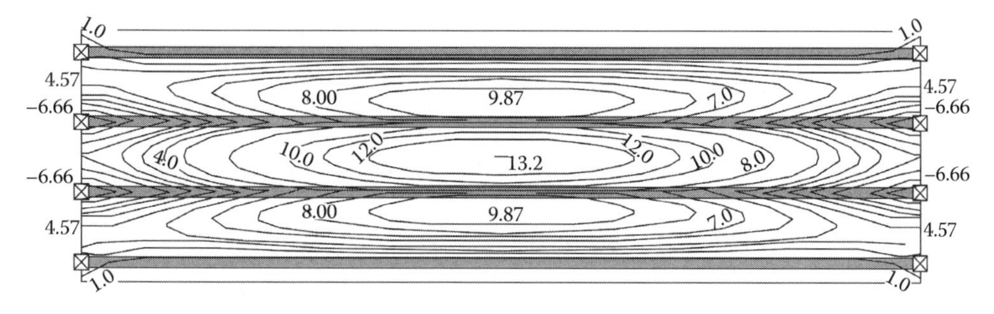

Figure 7.58 Transverse moments of the slab for uniformly distributed loading between the girders; units in kN-m/m.

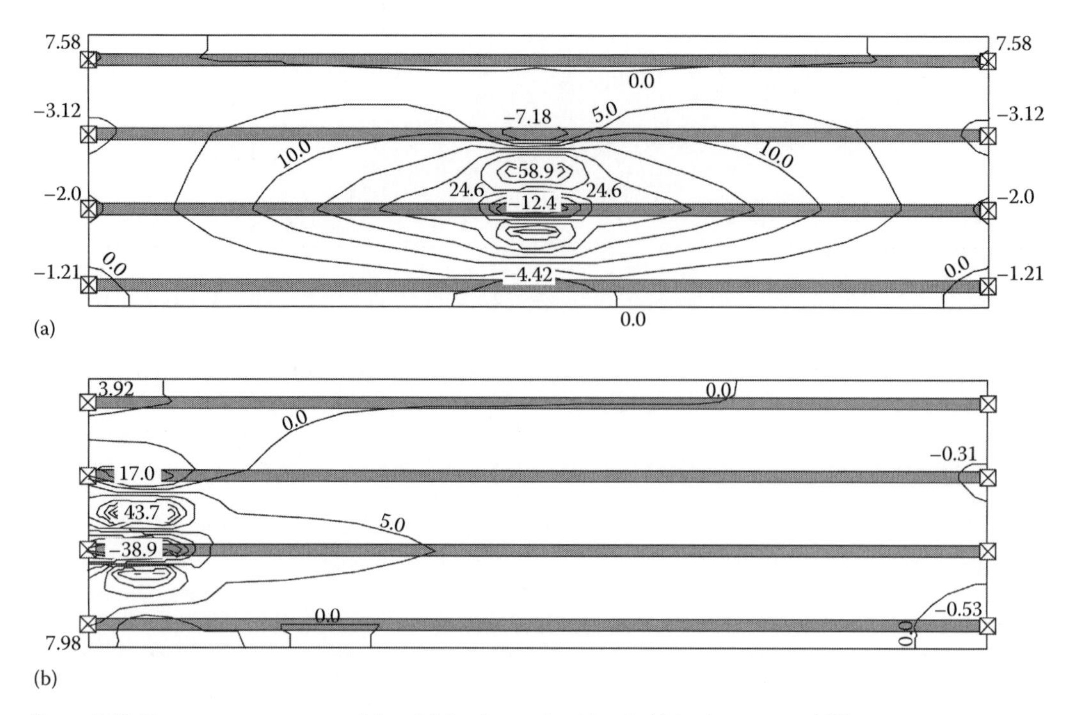

Figure 7.59 Transverse moments of the slab for four wheel loads (a) in the span and (b) near supports; units in kN-m/m.

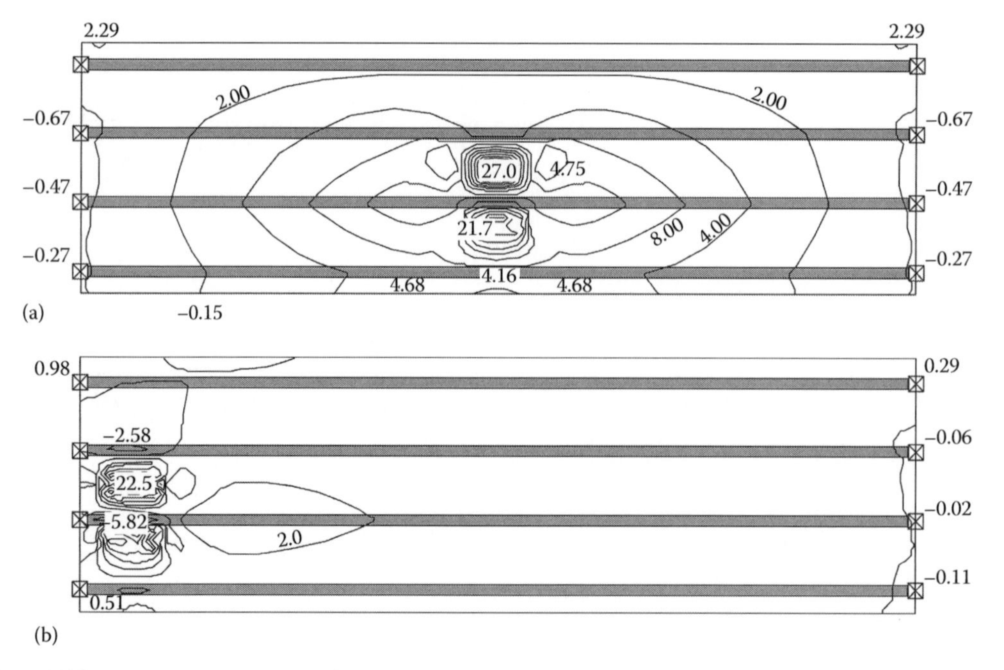

Figure 7.60 Longitudinal moments of the slab for four wheel loads (a) in the span and (b) near supports; units in kN-m/m.

7.6 FINITE ELEMENT MODELS FOR GLOBAL ANALYSIS

Finite element analysis may be employed for global analysis as well. This analysis allows for

a. Tracing of regions of stress concentrations.
   The stresses cannot be determined correctly by application of beam theory in complicated structural details. Therefore, modeling such areas by FEM is useful for stress and fatigue verifications and appropriate strengthening at such details.
b. Code provisions differ in the estimation of the effective widths. Therefore, especially for fatigue verifications, FEM might be used.
c. Calibration of results of other methods.

As seen before, the appropriateness of simplified analysis models may be checked by comparison with FEM results.

Figure 7.61 shows the FE model of a box-girder bridge. The web and the bottom flange of the cross section and the plated diaphragms are represented by shell elements. For the top flange, it is more convenient to use beam elements due to the easy connection with the nodes of the elements of the concrete slab. An additional reason for choosing beam elements for the top flanges is that they are usually classified as compact cross sections; thus, unnecessary plate buckling modes can be avoided.

Stiffeners should be modeled as beam elements. From the internal forces in the stiffeners, the designer can judge about their participation in the structural performance of the cross section. For example, stiffeners that attract low axial forces are noneffective and their position should be changed. Moreover, designers should take a close look at the deformations of the flanges and the webs. Thus, conclusions can be drawn about the stiffness of the stiffeners. This is discussed further in Chapter 8.

Internal braces are modeled as truss or beam elements. It is important to note that truss elements possess only longitudinal degrees of freedoms, and during a buckling analysis, the buckling modes of the bracings will not be analyzed. In such cases, bracings should be modeled as beam elements with rotational releases at the ends.

The concrete slab is usually modeled with shell elements. This is easy to implement but has the disadvantage of neglecting the eccentricity between the gravity centers of the slab and the top flange. In cases of slabs with haunches, volume elements give more accurate results.

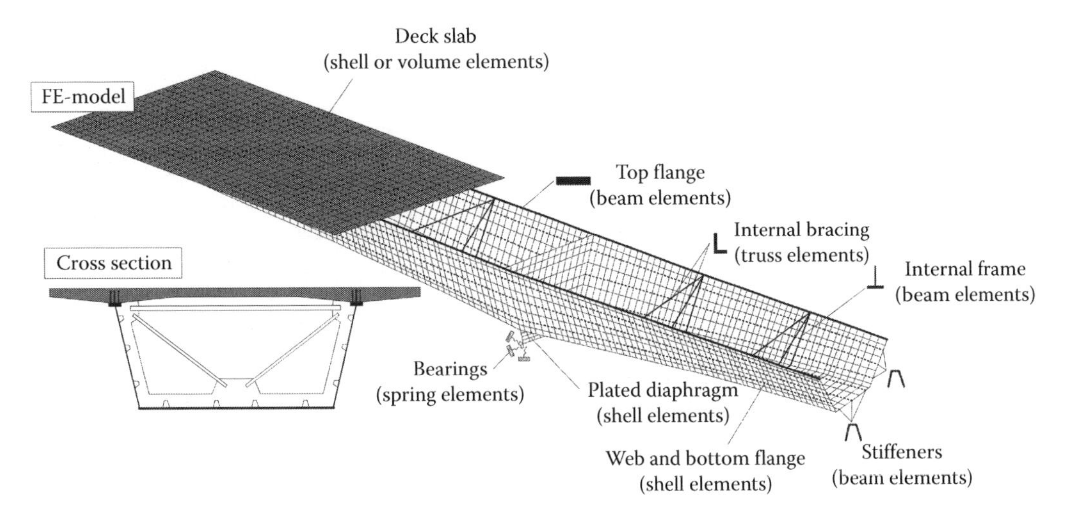

Figure 7.61 FE model for a composite box-girder bridge.

A convenient way to consider concrete cracking at hogging moment areas is to use a reduced modulus of elasticity according to the following equation:

$$E_{c,cr} = \frac{E_s \cdot A_t}{h_c \cdot b_{ref}} \tag{7.35}$$

where

h_c is the depth for the deck slab

E_s is the modulus of elasticity of reinforcing steel

b_{ref} is a reference width (usually equal to 1 m)

A_t is the cross-sectional area of the cracked slab taking into account the tension stiffening effect according to Equation 7.14 for the reference width b_{ref}

As already mentioned in Chapter 6, Poisson's ratio of the cracked concrete is taken as equal to zero.

REMARK 7.9

Many FE programs have shell and volume elements with orthotropic properties. In the longitudinal direction, the elastic modulus of elasticity of the slab elements is determined from Equation 7.35 so that cracking is considered; thus, $E_x = E_{c,cr}$. However, in the transverse direction, concrete cracking may not be of the same magnitude (e.g., in multi-girder bridges of Figure 2.5); therefore, $E_y = E_{cm}$. With orthotropic FEM, both cracking in the longitudinal direction and the increased flexural stiffness in the transverse direction for the distribution of the vertical loads on the main girders can be "realistically" taken into account. For the sagging moment areas, noncracked isotropic elements are used (Figure 7.62).

Interesting information for FE modeling is found also in [7.18] and [7.24].

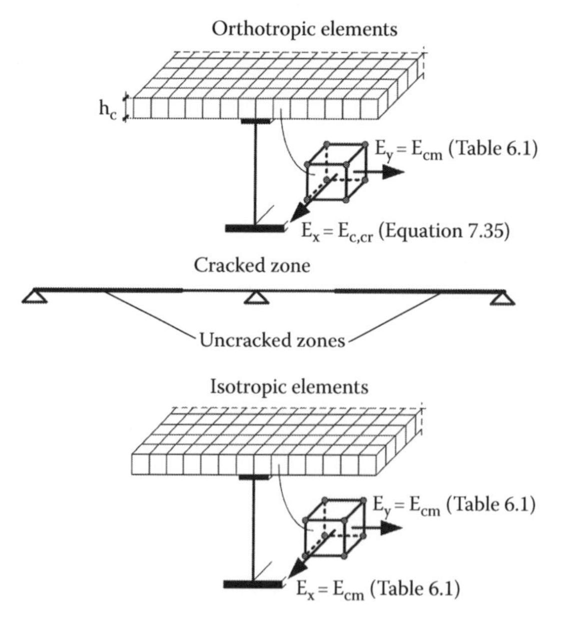

Figure 7.62 Use of orthotropic FEM in cracked regions and isotropic elements in the uncracked ones.

EXAMPLE 7.5

The deck slab of a continuous bridge has a constant depth at hogging moment areas of 27 cm and a reinforcement ratio 2.1%. The slab will be modeled with shell elements whose equivalent modulus of elasticity at hogging moment areas is to be calculated. Materials C35/45, B500B.

C35/45 (Table 6.1): $E_{cm}=3400$ kN/cm², $f_{ctm}=0.32$ kN/cm²

B500B: $E_s=20{,}000$ kN/cm², $f_{sk}=50$ kN/cm²

For a reference width $b_{ref}=1$ m:

Equation 7.14: $A_t = \dfrac{0.021 \cdot 27 \cdot 100}{1-(0.5 \cdot 0.32 / 0.021 \cdot 50)} = 66.89\ \text{cm}^2/\text{m}$

Equation 7.35: $E_{c,eff} = \dfrac{21000 \cdot 66.89}{27 \cdot 100} = 520.3\ \text{kN/cm}^2$

Cracked regions will be modeled with orthotropic C35/45 shell elements with an elastic modulus of elasticity in the longitudinal direction equal to $E_x=520.3$ kN/cm².

REFERENCES

[7.1] Adamakos, Th., Vayas, I., Iliopoulos, A.: Modelling of composite bridges using bar elements–Lateral bracing and cross-frames influence. In: *EUROSTEEL 2011*, Budapest, Hungary, 2011.

[7.2] Adamakos, Th., Vayas, I., Petridis, St., Iliopoulos, A.: Modeling of curved composite I-girder bridges using spatial systems of beam elements. *Journal of Constructional Steel Research* 67(3), 462–470, 2011.

[7.3] Beckman, F., Medlock, R.D.: Skewed bridges and girder movements due to rotations and differential deflections. In: *Proceedings of the World Steel Bridge Symposium*, Orlando, FL, American Institute of Steel Construction, National Steel Bridge Alliance, Chicago, IL, 2005.

[7.4] Beckman, F. et al.: *Shop and Field Issues: A Skewed Perspective. Modern Steel Construction.* American Institute of Steel Construction, Chicago, IL, 2008.

[7.5] Berglund, E. M., Schultz, A. E.: Girder differential deflection and distortion-induced fatigue in skewed steel bridges. *Journal of Bridge Engineering* 11(2), 169–177, 2007.

[7.6] Chang, C. J., White, D. W.: An assessment of modelling strategies for composite curved steel I-girder bridges. *Engineering Structures* 30(11), 2991–3002, 2008.

[7.7] Chavel, B. W.: Construction and detailing methods of horizontally curved steel I-girder bridges. PhD, Department of Civil Engineering, University of Pittsburgh, Pittsburgh, PA, 2008.

[7.8] Coletti, D., Yadlosky, J. M.: Behavior and analysis of curved and skewed steel girder bridge. In: *Proceedings of the World Steel Bridge Symposium 2005*, Orlando, FL, American Institute of Steel Construction, National Steel Bridge Alliance, Chicago, IL, 2005.

[7.9] Courbon, J.: *Application de la resistance de Materiaux au calcul des Ponts*, Dunod, Paris, France, 1950.

[7.10] EN1993-2, CEN (European Committee for Standardization): Design of steel structures. Steel bridges, 2006.

[7.11] EN1994-2, CEN (European Committee for Standardization): Design of composite steel and concrete structures. Part 2: Rules for bridges, 2005.

[7.12] Ghali, A., Favre, R.: *Concrete Structures–Stresses and Deformations.* E&FN Spon, London, U.K., 1994.

[7.13] Gupta, T., Anurag, M.: Effect on support reactions of T-beam skew bridge decks. *ARPN Journal of Engineering and Applied Sciences* 2(1), 1–8, February 2007.

[7.14] Hambly, E. C.: *Bridge Deck Behaviour*, 2nd edn., E&FN. Spon, London, U.K., 1990.

[7.15] Hanswille, G.: Zur Behandlung der Einflüsse aus dem Kriechen und Schwinden des Betons. Forschungsvorhaben: Eurocode 4 Teil 2–Verbundbrücken, Wuppertal, Germany, 1998.
[7.16] Hanswille, G., Stranghöner, N.: *Leitfaden zum DIN Fachbericht 104, Verbundbrücken*, Ernst & Sohn, Berlin, Germany, 2003.
[7.17] Kindmann, R., Frickel, J.: *Elastische und plastische Querschnittstragfähigkeit. Grundlagen, Methoden, Berechnungsverfahren und Beispiele*, Ernst & Sohn, Berlin, Germany, 2002.
[7.18] Kindmann, R., Kraus, M.: *Steel Structures—Design Using FEM*, Ernst & Sohn, Berlin, Germany, 2011.
[7.19] Kolbrunner, C. F., Basler, K.: *Torsion on Structures; An Engineering Approach*, pp. 1–21, 47–50, Springer-Verlag, Berlin, Germany, 1969.
[7.20] Linzell, D. G., Hall, D. H., White, D. W.: Historical perspective on horizontally curved I girder bridge design in the United States. *ASCE Journal of Bridge Engineering* 9(3), 218–229, 2004.
[7.21] O'Brien, E. J., Keogh, D. L.: *Bridge Deck Analysis*, E&FN. Spon, London, U.K., 1999.
[7.22] Pucher, A.: *Influence Surfaces of Elastic Plates*, Springer-Verlag, Vienna, Austria, 1964.
[7.23] Roik, K.: *Vorlesungen über Stahlbau—Grundlagen*, Ernst & Sohn, Berlin, Germany, 1983.
[7.24] Schleicher, W.: *Modellierung und Berechnung von Stahlbrücken*, Ernst & Sohn, Berlin, Germany, 2003.
[7.25] Sherif El-Tawil, Ayman M. Okeil.: Behavior and design of curved composite bridges. Department of Civil and Environmental Engineering, University of Central Florida, Orlando, FL, Final Report, 2002.
[7.26] Sherif El-Tawil, Ayman M. Okeil.: Warping stresses in curved box girder bridges: Case study. *Journal of Bridge Engineering* 9(5), 487–496, September–October, 2004.
[7.27] Unterweger, H.: Global analysis of steel and composite highway bridges—Development of Improved Spatial Beam Models. Technical University Graz, Department of Steel Structures, Graz, Austria.
[7.28] Unterweger, H.: Global analysis of steel and composite structures—Efficiency of simple beam models. Habilitations—thesis, Technical University Graz, Graz, Austria, 2001 (in German).
[7.29] Unterweger, H.: Modeling of bridge decks including cross-section distortion using simplified spatial beam methods. In: *Second European Conference on Steel Structures*, Technical University Prague, Prague, Czech Republic, 1999.
[7.30] Vayas, I., Adamakos, Th., Iliopoulos, A.: Three dimensional modeling for steel-concrete composite bridges using systems of bar elements—Modeling of skewed bridges. *International Journal of Steel Structures* 11(2), 157–169, 2011.
[7.31] Vayas, I., Adamakos, Th., Iliopoulos, A.: Modelling of steel-composite bridges, spatial systems vs. grillages. In: Nin*th International Conference on Steel Concrete Composite and Hybrid Structures*, University of Leeds, Leeds, U.K., 2009.
[7.32] Vayas, I., Iliopoulos, A., Adamakos, Th.: Spatial systems for modelling steel-concrete composite bridges—Comparison of grillage systems and FE models. *Steel Construction Design and Research* 3, 100–111, 2010.
[7.33] Wehnert-Brigdar, A.: Zum Tragverhalten im Grundriss gekrümmter Verbundträger. Dissertation, Fakultät für Bauingenieurwesen, Ruhr—Universität, Bochum, Germany, 2009.
[7.34] West, R.: Recommendations on the use of grillage analysis for slab and pseudo-slab bridge decks. Cement and Concrete Association and Construction Industry Research and Information Association, 1973.
[7.35] Westergaard, H. M.: Computation of stresses in bridge slabs due to wheel loads. *Public Roads* 11, 1–23, March 1930.
[7.36] Whisenhunt, T. W.: Measurement and finite element modeling of the non-composite deflections of steel plate girder bridges. MS thesis, Department of Civil, Construction and Environmental Engineering, North Carolina State University, Raleigh, NC, 2004.
[7.37] Wunderlich, W., Kiener, G.: Statik der Stabtragwerke. Teubner-Verlag, Auflage 1, Germany, 2004.

Chapter 8

Buckling of plated elements

8.1 INTRODUCTION

Box girder bridges were a popular choice during the road building expansion in the 1960s. A serious blow to this use was a sequence of five serious disasters, where new bridges failed or collapsed in 1969 (Danube bridge in Vienna), 1970 (West Gate bridge in Melbourne, Cleddau bridge in Wales), 1971 (Rhine bridge in Koblenz), and 1973 (Zeulenroda); see [8.14]. All accidents happened during construction that was made by incremental launching, where new segments are placed onto the completed portions of the bridge until the bridge superstructure is completed. It was found that failures were primarily due to plate buckling of the bottom flanges that were subjected to compression at the cantilever support (Figure 8.1). These bridges were designed by different codes, so that their revision was set as a priority of the time. Since then, a lot of effort was made to provide safe buckling rules that will be presented in this chapter.

Cross sections of composite bridge girders are composed of plated elements connected at common joints by welding. But also, rolled sections, although homogeneous in production, are composed of plated elements. Figure 8.2 shows the decomposition of an I section and a box section into their plated elements. The common edges provide supports that are considered for simplicity reasons as simple supports. Two types of elements are distinguished: internal elements supported at both edges and outstand elements supported at one edge. For the composite I-girder of Figure 8.2a, element 1 is an internal element and 2 and 3 outstand elements, while for the box girder, all elements are internal. When a cross section is subjected to internal forces and moments or concentrated forces, in-plane direct and shear stresses develop in its plated elements. It could be suggested that the strength of the element is exhausted when the maximum stress reaches its limit value, which is the yield stress f_y for compression and $\tau_y = f_y/\sqrt{3}$ for shear stresses. However, in-plane loaded plates may be subjected to out-of-plane deformations as shown in Figure 8.3 for the bottom flange of a box girder with a concrete top flange. These deformations lead to *plate buckling phenomena* that result in a reduction of the corresponding strength. Like in buckling of struts, the strength reduction depends on the plate slenderness, which is a function of its thickness. Therefore, thinner plates are more prone to plate buckling and exhibit larger reduction in strength. Plate buckling, as Figure 8.3 shows, affects only the compression parts and not the entire cross section and is therefore often referred to as *local buckling*.

A reduction in weight is crucial for an economic design of composite bridges. This leads in practice to thin-walled, and mostly welded, cross sections that are prone to local buckling. As indicated in Table 8.1, such sections are referred to as *class 4 cross sections* and shall be verified for plate buckling.

An increase in strength may be achieved by provision of longitudinal and/or transverse stiffeners as shown in Figure 8.4 (see also Figures 2.21 through 2.25). Stiffening of plates is

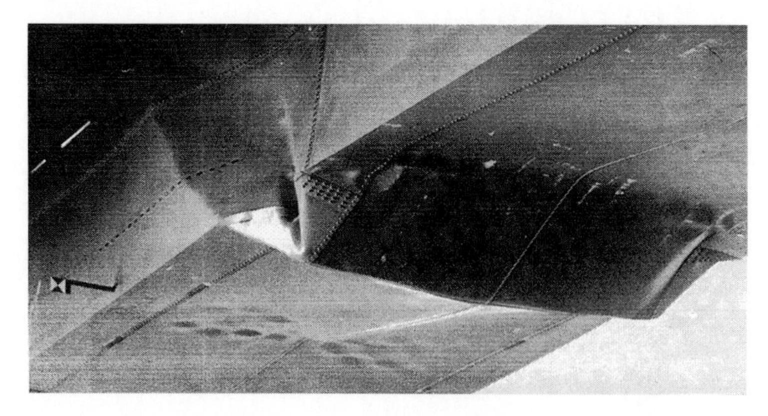

Figure 8.1 Buckling of compressed bottom flange, Danube bridge in Vienna. (From Scheer, J., *Failed bridges. Case Studies, Causes and Consequences*. Ernst und Sohn, Berlin, Germany, 2010.)

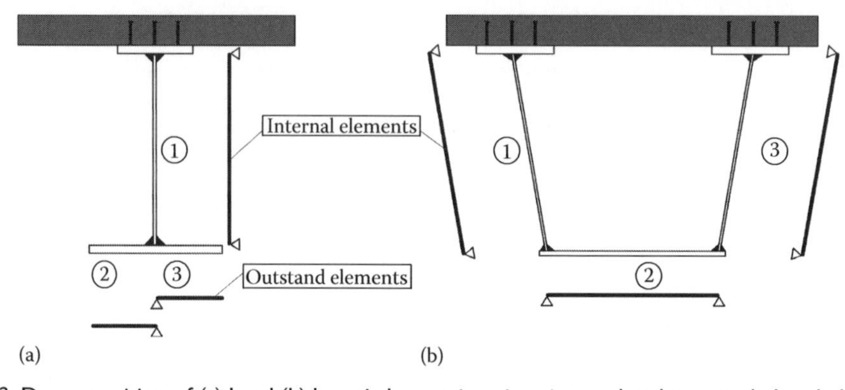

Figure 8.2 Decomposition of (a) I and (b) box girder sections into internal and outstand plated elements.

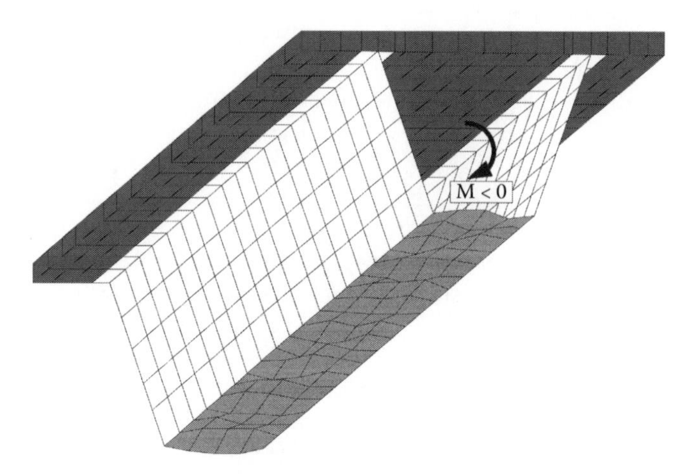

Figure 8.3 Out-of-plane deformations and plate (local) buckling of the bottom flange.

often preferred to a global increase in plate thickness, because the stiffeners may be accommodated in the most stressed regions resulting in less weight and welding of thick sheets is avoided.

Plate buckling refers to flat rectangular panels of thickness t and dimensions a × b, where a is the panel length and b the panel width (see Figure 8.5).

Table 8.1 Design procedures for strut and plate buckling for class 4 cross sections

	Buckling of struts	*Buckling of plates*
Problem	N	σ τ
Step 1: Elastic critical buckling stress	N_{cr}	σ_{cr} τ_{cr}
Step 2: Nondimensional slenderness	$\bar{\lambda} = \sqrt{\dfrac{f_y}{\sigma_{cr}}}$	$\bar{\lambda} = \sqrt{\dfrac{f_y}{\sigma_{cr}}}$ or $\bar{\lambda}_w = \sqrt{\dfrac{f_y/\sqrt{3}}{\tau_{cr}}}$
Step 3: Reduction factor from buckling curves	χ	ρ or χ_w
Step 4: Design strength	$\sigma_{Rd} = \chi \cdot \dfrac{f_y}{\gamma_{M1}}$	$\sigma_{Rd} = \rho \cdot \dfrac{f_y}{\gamma_{M1}}$ or $\left(\tau_{Rd} = \chi_w \cdot \dfrac{f_y/\sqrt{3}}{\gamma_{M1}}\right)$

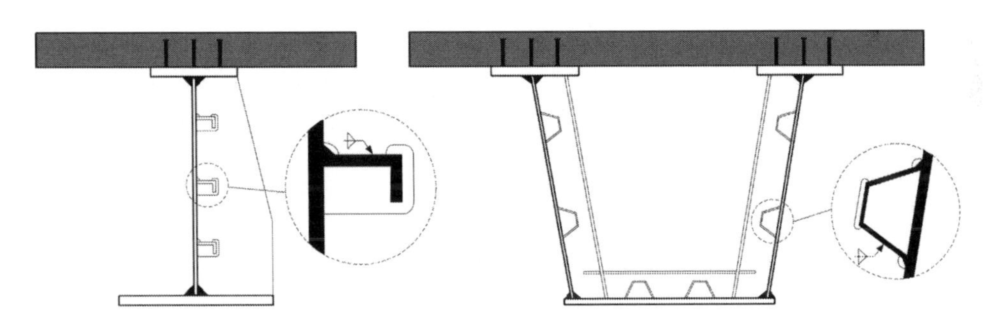

Figure 8.4 Stiffening of plates with transverse and longitudinal stiffeners.

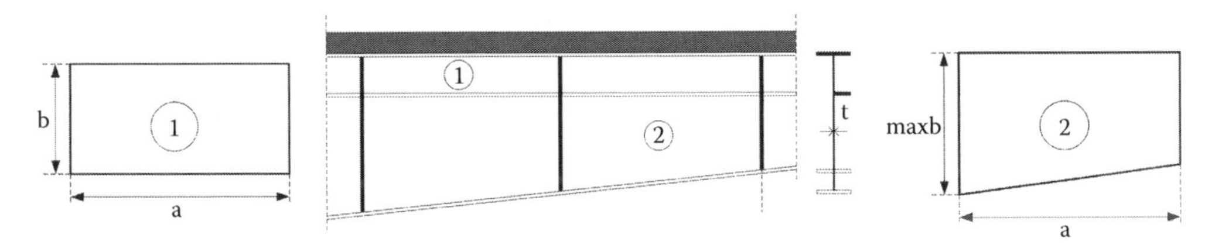

Figure 8.5 Panel dimensions for rectangular panels and at haunches.

Curved panels may be regarded as flat, provided that the curvature radius satisfies

$$r \geq \frac{b^2}{t} \tag{8.1a}$$

Nonrectangular panels, as in case of haunches, may be considered as rectangular, provided a and b are taken as the largest dimensions of the panel (see Figure 8.5). The panels are supported at their edges. The longitudinal edges are at the joints with the neighboring panels of the cross section, as shown in Figure 8.2, and run parallel to the bridge direction. For internal elements, both longitudinal edges are supported, while for outstand elements, one longitudinal edge is supported the other being free. The transverse edges are at the positions of diaphragms or sufficiently stiff transverse stiffeners. When flexible transverse stiffeners are provided, the panel includes these stiffeners as they are not considered as supported edges. Figure 8.6 shows a panel stiffened with two longitudinal and transverse stiffeners. For panels with longitudinal stiffeners, subpanels between stiffeners may be regarded.

The panels are loaded in-plane along their edges. Out-of-plane loading is not considered. Stresses acting at midplane of panels, as shown in Figure 8.7 for the bottom flange of a box girder, are considered. This is due to the fact that *membrane theory* is used for buckling analysis. Compression stresses are positive, tension stresses negative.

This chapter presents the design of members with plated, class 4 elements following the rules of the Eurocodes and especially of EN 1993-1-5 [8.4]. It will be seen that plate buckling

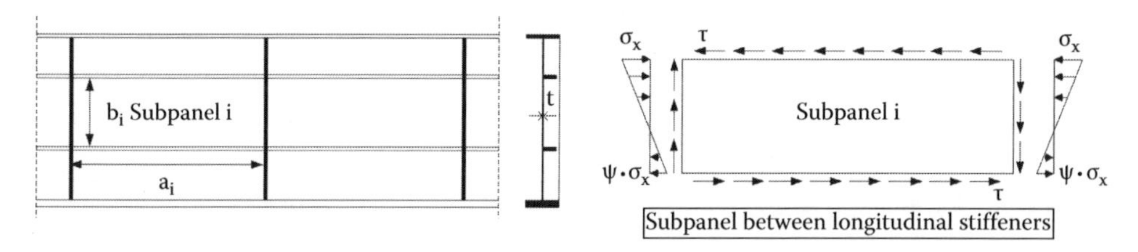

Figure 8.6 Dimensions for rectangular (sub)panels.

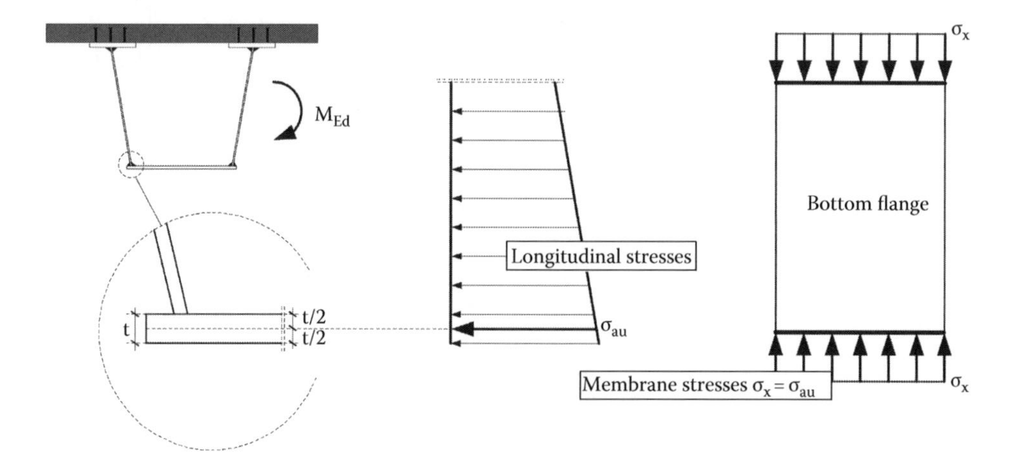

Figure 8.7 Membrane stresses of panels.

behavior is influenced by two effects: *post-critical strength* and *column-like behavior.* For direct stresses, two design methods may be employed [8.5]: the *effective width method* and the *reduced stress method.*

In composite bridges, the slab is connected to the top flange of the steel girder and prevents its out-of-plane deformations. In addition, the bottom flanges of I-girders are usually sufficiently thick not to be classified in class 4. Therefore, the following plated elements are practically considered in design at the final stage:

a. Webs of I- and box girders subjected to direct stresses due to moments and compression
b. Webs of I- and box girders subjected to shear stresses due to vertical shear and torsion
c. Bottom flanges of box girders subjected to direct stresses due to negative moments and compression
d. Bottom flanges of box girders subjected to shear stresses due to torsion
e. Webs of I- and box girders subjected to direct and shear stresses due to concentrated forces, for example, during launching operations

Obviously, at construction stages before placement of the concrete, more elements may be affected by plate buckling since the girders are of pure steel.

As an introduction, it is said that the design procedures for buckling of struts and plates illustrated in Table 8.1 are similar. It may be seen that the first step includes an elastic buckling analysis to determine the elastic critical buckling stress that is subsequently used as reference for the calculation of a nondimensional slenderness. The design strength is finally calculated by determination of reduction factors from relevant buckling curves and application of the factors to the yield stress.

Following these steps, this chapter presents first the determination of critical plate-buckling stresses for single and combined loading conditions, followed by presentation of the plate buckling curves and finally the panel and cross-sectional verifications for internal forces and moments.

8.2 ELASTIC CRITICAL STRESS

8.2.1 Introduction

As outlined before, the first step in the analysis for plate buckling is the determination of the *elastic critical buckling stress.* This is the stress at which an ideal plate without imperfections and elastic behavior becomes unstable. Critical stresses may be determined by the application of linear plate buckling theory, which considers small displacements. In the following, critical buckling stresses of unstiffened and stiffened panels and subpanels under direct and shear loading and various loading combinations will be presented.

8.2.2 Unstiffened panels

As widely known, the equilibrium of a compression strut under *Euler* conditions—elastic behavior, no geometric or structural imperfections, and no load eccentricity—becomes unstable, and the initially straight strut buckles when the load reaches a critical value, which is equal to the Euler load N_{cr}. Similarly, a perfect elastic plate buckles at a critical stress when loaded in-plane. Figure 8.8 shows a plate with *Navier* (simple) support conditions at all edges for uniform compression.

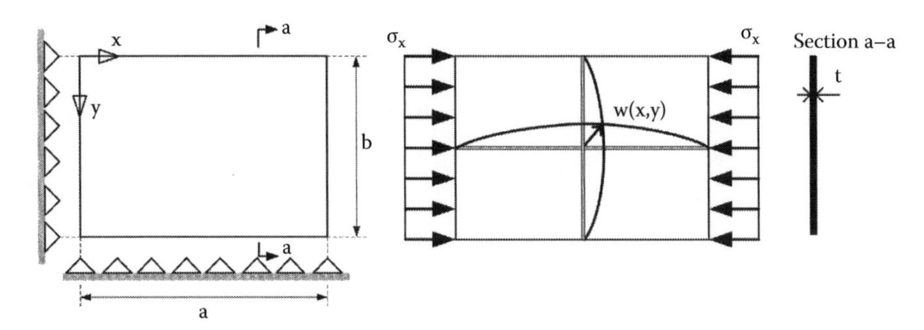

Figure 8.8 Plate panel under uniform compression.

The bending stiffness of this plate is given by

$$D = \frac{E \cdot t^3}{12 \cdot (1 - v^2)} \tag{8.1}$$

where
E is the modulus of elasticity
t is the plate thickness
v is the *Poisson* ratio (=0.3 for steel)

The shape of the plate at the buckling state is expressed by double *Fourier* series as follows:

$$w(x, y) = \sum_{m=1}^{\infty} \sum_{n=1}^{\infty} A_{mn} \cdot \sin \frac{m\pi x}{a} \cdot \sin \frac{n\pi y}{b} \tag{8.2}$$

where
w is the out-of-plane displacement
m and n are the number of sinusoidal half-waves in longitudinal and transverse directions x and correspondingly y.
A_{mn} is the unknown coefficient representing generalized displacements

The critical buckling stress of this plate is determined from

$$\sigma_{cr} = \sigma_x = \frac{D \cdot \pi^2}{t \cdot b^2} \left[m \cdot \left(\frac{b}{a} \right) + \frac{n^2}{m} \cdot \left(\frac{a}{b} \right) \right]^2 \tag{8.3}$$

The minimum value of the critical stress occurs when n = 1 that corresponds to one half-wave in transverse direction. Using Equations 8.1 through 8.3, the critical buckling stress may be written as

$$\sigma_{cr} = k_{\sigma} \cdot \frac{\pi^2 \cdot E}{12 \cdot (1 - v^2)} \cdot \left(\frac{t}{b} \right)^2 \tag{8.4}$$

where k_σ is the *buckling factor* determined from

$$k_\sigma = \left[m \cdot \left(\frac{b}{a} \right) + \frac{1}{m} \left(\frac{a}{b} \right) \right]^2 \tag{8.5}$$

Figure 8.9 shows values of the buckling factor for a simply supported plate under uniform axial compression as a function of the aspect ratio a/b. Evidently, the critical stress is determined from the least value of k_σ. For the loading and support conditions considered, it may be seen that the minimum value of the buckling factor is equal to $k_\sigma = 4$ and is achieved if the aspect ratio is an integer number or if it is larger than 4.

Each value of m corresponds to a buckling mode. Figure 8.10 shows the first two buckling modes of the plate of Figure 8.9 having an aspect ratio $\alpha = a/b = 1.2$. The first mode is the critical one.

Figure 8.9 indicates that the critical number of half-waves increases with increasing value of the aspect ratio. For a plate with an aspect ratio 3, m = 3 provides the least buckling factor, followed by m = 4. Figure 8.11 shows the first buckling mode of such a plate, confirming the earlier observation. It may be seen that the half-wave length for a long plate is approximately equal to its width b. This leads to the conclusion that, contrary

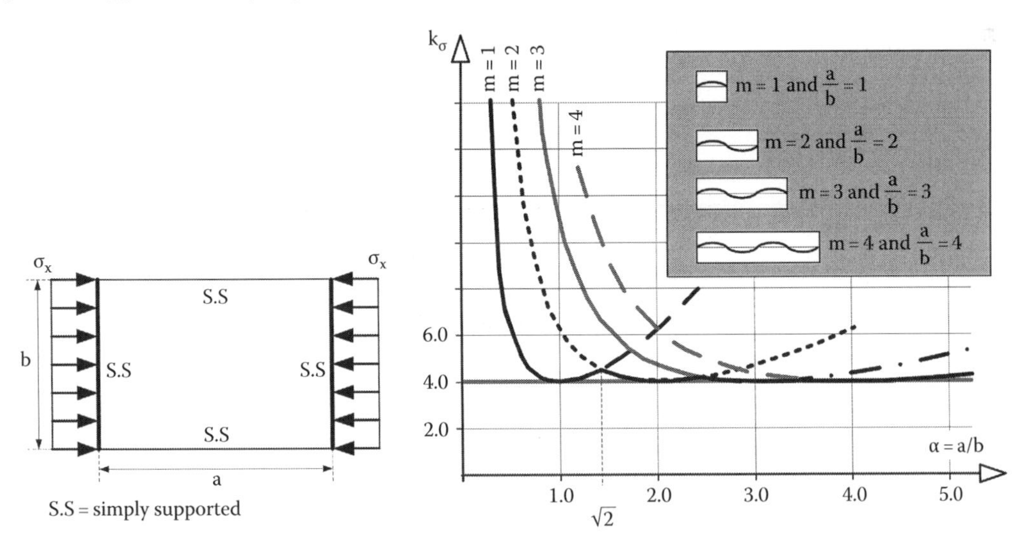

Figure 8.9 Buckling factor k_σ of unstiffened simply supported panels under uniform axial compression.

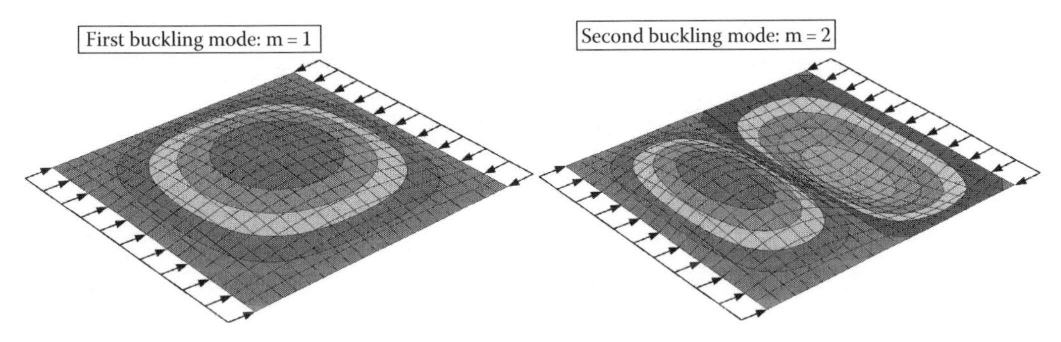

Figure 8.10 First two buckling modes of the panel of Figure 8.9 with aspect ratio 1.2.

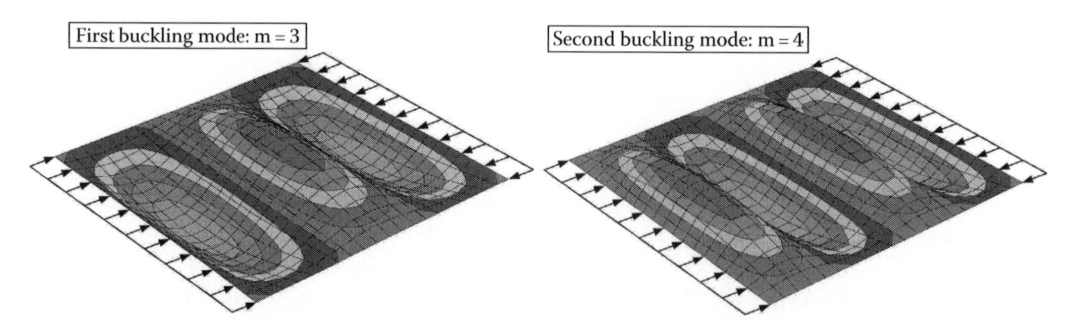

Figure 8.11 First two buckling modes of the panel of Figure 8.9 with aspect ratio 3.

to struts, the buckling length of a plate is determined from its width, not its length. This is of significant practical importance since a long plate corresponds to an unstiffened wall of a cross section.

It may be noted that the critical stress is determined by numerical methods, since it is almost impossible to find analytical solutions. The most usual method employed is the *Rayleigh–Ritz* energy method, where the variations of the strain energy of the plate, ΔU, and the internal work, $\Delta W_{int}(S_{cr})$, for critical stresses S_{cr} are determined (see [8.1]). The critical state corresponds to indifferent equilibrium and therefore to the condition:

$$\Delta U - \Delta W_{int}(S_{cr}) = 0 = \text{minimum} \tag{8.6}$$

The strain energy of an isotropic plate is given by

$$\Delta U_p = \frac{D}{2} \cdot \int_0^a \int_0^b \left\{ \left(\frac{\partial^2 w}{\partial x^2} + \frac{\partial^2 w}{\partial y^2} \right)^2 - 2 \cdot (1 - v) \cdot \left[\frac{\partial^2 w}{\partial x^2} \cdot \frac{\partial^2 w}{\partial y^2} - \left(\frac{\partial^2 w}{\partial x \partial y} \right)^2 \right] \right\} dx \cdot dy \tag{8.7}$$

The internal work of an isotropic plate is given by

$$\Delta W_p = \frac{t}{2} \cdot \int_0^a \int_0^b \left\{ \sigma_x \cdot \left(\frac{\partial w}{\partial x} \right)^2 + \sigma_y \cdot \left(\frac{\partial w}{\partial y} \right)^2 + 2 \cdot \tau \cdot \frac{\partial w}{\partial x} \cdot \frac{\partial w}{\partial y} \right\} dx \cdot dy \tag{8.8}$$

The plate stiffness D and the deflection w from Equations 8.1 to 8.2 are introduced in the earlier expressions, and all necessary derivations and integrations are performed. Then the expressions are substituted in (8.6). This leads to the solution of an eigenvalue problem that provides the buckling modes. For stiffened plates, the same procedure is followed; however, the strain energy and the internal work of the stiffeners must be added in Equations 8.7 and 8.8 as shown later in Section 8.2.3.

The critical buckling stress for steel plates under more general loading and supporting conditions is given by

$$\sigma_{cr,p} = k_\sigma \cdot \sigma_e \tag{8.9}$$

where k_σ is the buckling factor and

$$\sigma_e = \frac{\pi^2 \cdot E}{12 \cdot (1 - v^2) \cdot (b/t)^2} = 189{,}800 \cdot \left(\frac{t}{b}\right)^2 \, [\text{N/mm}^2] = \text{reference stress} \tag{8.10}$$

The critical buckling shear stress is similarly given by

$$\tau_{cr} = k_\tau \cdot \sigma_e \tag{8.11}$$

where k_τ is the buckling factor for shear stresses.

It is reminded that the *Euler stress* of a strut is given by

$$\sigma_{cr} = \frac{\pi^2 \cdot E}{\lambda^2} \tag{8.12}$$

where $\lambda = \frac{L}{i}$ is the slenderness.

Comparing the expressions of the critical plate buckling stress with the *Euler stress* of the strut, it may be observed that the slenderness of the plate is expressed by the ratio b/t, where the width of the plate corresponds to the "length (L)" and the thickness (t) to the "radius of gyration (i)."

The buckling factor depends on

- The support conditions
- The type of stresses (direct or shear stresses)
- The loading conditions—uniform compression, bending, bending and compression—that are expressed by the ratio ψ of the edge stresses

Tables 8.2 and 8.3 give buckling factors for internal and outstand unstiffened panels (see [8.4]).

8.2.3 Stiffened panels

The critical stresses, and ultimately the panel strength, may be increased by provision of stiffeners. In bridges, both longitudinal and transverse stiffeners are provided. The longitudinal stiffeners may have open cross section—flat bars, tees, angles, bulbs—or trapezoidal closed sections as shown in Figure 8.12a. Trapezoidal stiffeners are preferred since they have higher torsional rigidity and are not subjected to lateral torsional buckling when in compression. For fabrication reasons, transverse stiffeners are tees with cutouts in the web to allow passing of the longitudinal stiffeners as shown in Figure 8.12b.

Panels under uniform compression ($\psi = 1$), like bottom flanges of box girders under negative bending, are stiffened by equally distanced longitudinal stiffeners (Figure 8.12b). In webs that are subjected to tension and compression, a small number of individual stiffeners is provided only in the compression zone.

Critical stresses may be determined for stiffened panels by the energy method as introduced before. However, in this case, the strain energy and the internal work of the

Table 8.2 Buckling factors k_σ and k_τ for internal panels

Distribution of direct stresses (compression positive)

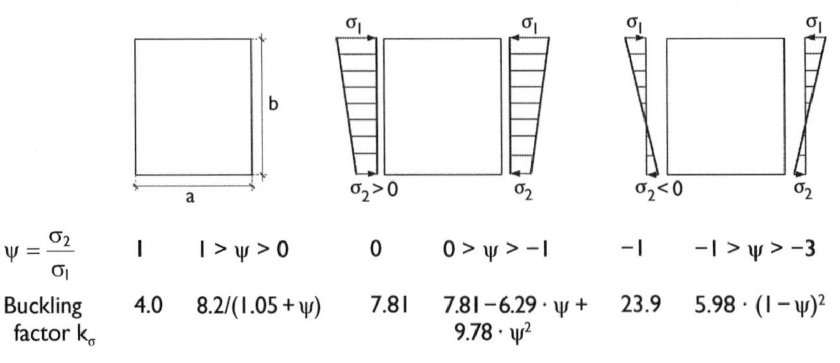

$\psi = \frac{\sigma_2}{\sigma_1}$	1	$1 > \psi > 0$	0	$0 > \psi > -1$	−1	$-1 > \psi > -3$
Buckling factor k_σ	4.0	$8.2/(1.05+\psi)$	7.81	$7.81-6.29\cdot\psi+9.78\cdot\psi^2$	23.9	$5.98\cdot(1-\psi)^2$

Shear stresses

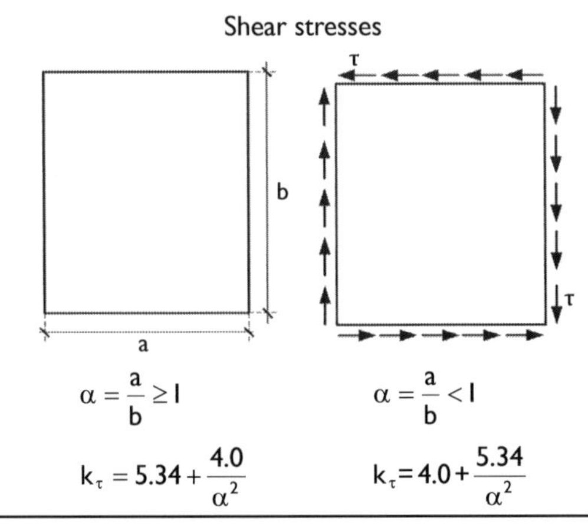

Aspect ratio	$\alpha = \frac{a}{b} \geq 1$	$\alpha = \frac{a}{b} < 1$
Buckling factor k_τ	$k_\tau = 5.34 + \frac{4.0}{\alpha^2}$	$k_\tau = 4.0 + \frac{5.34}{\alpha^2}$

Source: EN 1993-1-5, CEN (European Committee for Standardization): Design of steel structures, Part 1–5: Plated structural elements, 2006.

Table 8.3 Buckling factor k_σ for outstand panels

Maximum compression at free edge

Maximum compression at supported edge

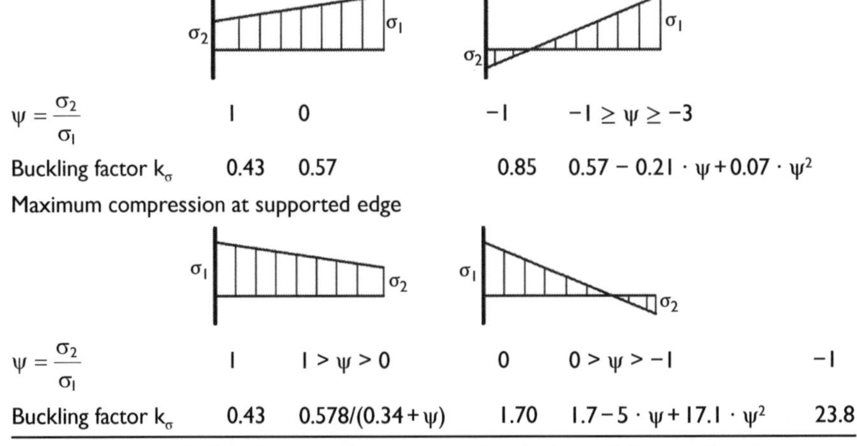

Maximum compression at free edge:

$\psi = \frac{\sigma_2}{\sigma_1}$	1	0	−1	$-1 \geq \psi \geq -3$
Buckling factor k_σ	0.43	0.57	0.85	$0.57-0.21\cdot\psi+0.07\cdot\psi^2$

Maximum compression at supported edge:

$\psi = \frac{\sigma_2}{\sigma_1}$	1	$1 > \psi > 0$	0	$0 > \psi > -1$	−1
Buckling factor k_σ	0.43	$0.578/(0.34+\psi)$	1.70	$1.7-5\cdot\psi+17.1\cdot\psi^2$	23.8

Source: EN 1993-1-5, CEN (European Committee for Standardization): Design of steel structures, Part 1–5: Plated structural elements, 2006.

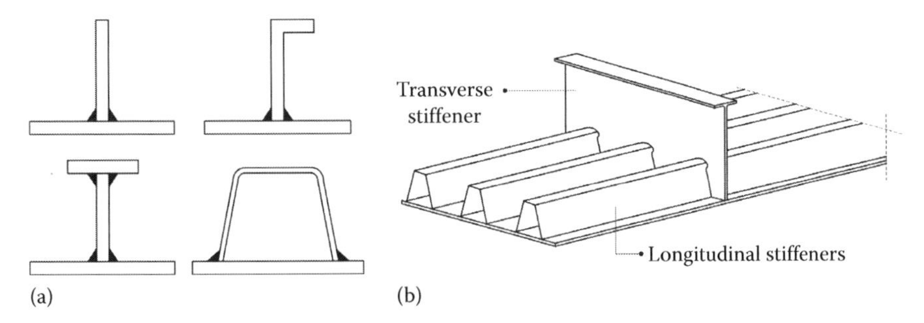

Figure 8.12 (a) Shapes of longitudinal stiffeners and (b) transverse stiffeners with web cutouts.

stiffeners must be added to those of the plate. This may be done either by "smearing" the stiffeners in the plate and writing the relevant expressions for an orthotropic plate or by considering each stiffener individually and adding its strain energy and work to those of the plate. For example, the strain energy and the internal work of n_i longitudinal stiffeners at the positions $y = y_i$, with a stress σ_{xi}, may be written as

$$\Delta U_{sx} = \sum_{i=1}^{n_x} \left\{ \frac{E \cdot I_{sxi}}{2} \cdot \int_0^a \left[\left(\frac{\partial^2 w}{\partial x^2} \right)_{y=y_i} \right]^2 dx \right\} + \sum_{i=1}^{n_x} \left\{ \frac{G \cdot J_{sxi}}{2} \cdot \int_0^a \left[\left(\frac{\partial^2 w}{\partial x \partial y} \right)_{y=y_i} \right]^2 dx \right\} \tag{8.13}$$

$$\Delta W_{sx} = \sum_{i=1}^{n_x} \left\{ \frac{A_{sxi}}{2} \cdot \int_0^a \sigma_{xi} \left(\frac{\partial w}{\partial x} \right)_{y=y_i}^2 dx \right\} \tag{8.14}$$

where

A_{sxi} is the cross-sectional area of the stiffener i
J_{sxi} is the torsional constant of the stiffener i

It is reminded that in practice transverse stiffeners constitute rigid supports for the panel. This is due to the fact that this support is required during the examination of the column-like behavior of the panel is discussed in Section 8.3.3. Accordingly, the panel length is normally equal to the spacing between transverse stiffeners, and the relevant strain energy and internal work are not considered.

The critical and the reference stresses of stiffened plates are determined by Equations 8.9 through 8.11, valid also for unstiffened panels, the only difference being the values of the buckling factors. Buckling factors of stiffened plates may be found in the literature [8.12]. Tables 8.4 and 8.5 are given in EN 1993-1-5 [8.4] and provide values of buckling factors for plates with many equal stiffeners. In Table 8.4, the stiffeners are smeared into an equivalent orthotropic plate. Attention must be paid to the limitations for the stress ratio $\psi \geq 0.5$ (whole panel under compression) and the aspect ratio $\alpha \geq 0.5$.

Tables 8.6 and 8.7 give directly the critical stress for plates with one or two stiffeners in the compression zone. They are derived as a critical stress of the stiffener, considering them as a strut on elastic foundation reflecting the plate effect in transverse direction, known also as the *fictitious column method*.

Table 8.4 Buckling factors for *Navier* panels (simply supported at all edges) with three or more equal distant longitudinal stiffeners considered as orthotropic plates

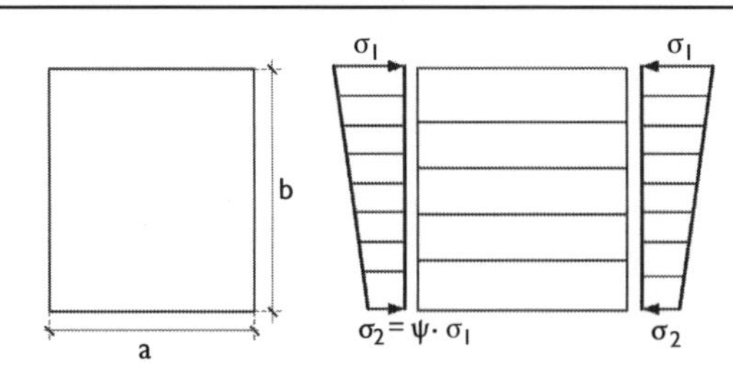

Aspect ratio	*Buckling factor*
$\alpha \leq \sqrt[4]{\gamma}$	$k_{\sigma,p} = \dfrac{2 \cdot \left[(1+\alpha^2)^2 + \gamma - 1\right]}{\alpha^2 \cdot (\psi + 1) \cdot (1+\delta)}$
$\alpha > \sqrt[4]{\gamma}$	$k_{\sigma,p} = \dfrac{4 \cdot \left(1+\sqrt{\gamma}\right)}{(\psi+1) \cdot (1+\delta)}$

Source: EN 1993-1-5, CEN (European Committee for Standardization): Design of steel structures, Part 1–5: Plated structural elements, 2006.

Note: $\psi = \dfrac{\sigma_2}{\sigma_1} \geq 0.5$, ratio of edge stresses; $\alpha = \dfrac{a}{b} \geq 0.5$, aspect ratio; $\sum A_{sl}$, sum of gross area of all the stiffener outstands; I_{sl}, second moment of area of the whole stiffened plate; $A_p = b \cdot t$, gross area of the plate; $\gamma = \dfrac{I_{sl}}{I_p}$, ratio of second moment of areas between stiffened and unstiffened plates; $\delta = \dfrac{\sum A_{sl}}{A_p}$, ratio of gross areas between stiffeners and plate; $I_p = \dfrac{b \cdot t^3}{12 \cdot (1-\nu^2)} = \dfrac{b \cdot t^3}{10.92}$, Second moment of area of the plate.

REMARK 8.1

The buckling factors and stresses given in EN 1993-1-5 (Tables 8.4 through 8.7) assume that

- Transverse stiffeners are rigid. The verification of the rigidity of the transverse stiffeners is discussed in Section 8.11. Flexible transverse stiffeners are not covered by the code, and their use is not recommended.
- Longitudinal stiffeners are not class 4. If not reduced, cross-sectional properties due to local buckling should be used (see Section 8.5).
- The torsional rigidity of the longitudinal stiffeners is zero. Indeed, one can see that in the expressions given in the previous tables, the torsional constant of the longitudinal stiffeners is missing. This is obviously a conservative assumption, which in the case of panels with closed stiffeners can be considerably uneconomical.
- Torsional buckling failure of the longitudinal stiffeners is avoided.
- The panels have no openings.

Table 8.5 Shear buckling factors for *Navier* panels (simply supported at all edges)

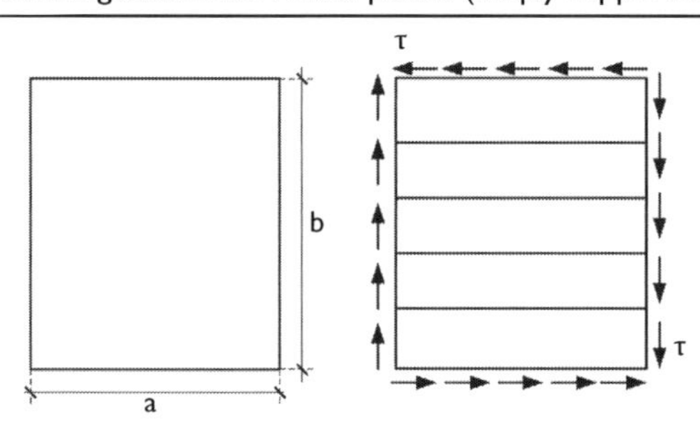

Aspect ratio $\alpha = \frac{a}{b}$	*Buckling factor*
$\alpha \geq 1$	$k_\tau = 5.34 + \frac{4}{\alpha^2} + k_{\tau,st}$
$\alpha < 1$	$k_\tau = 4 + \frac{5.34}{\alpha^2} + k_{\tau,st}$
Factor $k_{\tau,st}$	
Plates with three or more longitudinal stiffeners or with one or two stiffeners if $\alpha \geq 3$	$k_{\tau,st} = \frac{9}{\alpha^2} \cdot \sqrt[4]{\left(\frac{\sum I_{sl}}{t^3 \cdot b}\right)^3} \geq \frac{2.1}{t} \cdot \sqrt[3]{\frac{\sum I_{sl}}{b}}$
Plates with one or two longitudinal stiffeners if $\alpha < 3$	$k_{\tau,st} = 4.1 + \frac{6.3 + 0.18 \cdot \frac{I_{sl}}{t^3 \cdot b}}{\alpha^2} + 2.2 \cdot \sqrt[3]{\left(\frac{I_{sl}}{t^3 \cdot b}\right)}$

Source: EN 1993-1-5, CEN (European Committee for Standardization): Design of steel structures, Part 1–5: Plated structural elements, 2006.

Notes:

I_{sl} = Second moment of area around z–z axis of one stiffener including an associated plate width $15 \cdot \varepsilon \cdot t$ ($\varepsilon = \sqrt{235/f_y}$, f_y in N/mm²).

$\sum I_{sl}$ = Sum of the second moment of area of the stiffeners (if more than two)—not necessarily equally spaced.

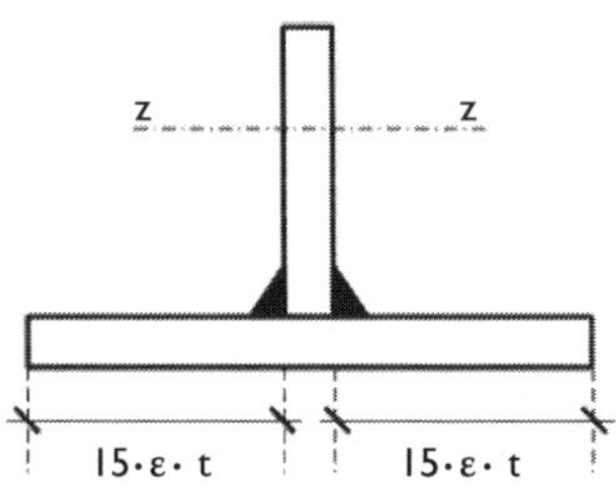

An alternative to closed formulae is the employment of numerical methods. When finite elements are employed, the plate may be modeled by shell elements and the stiffeners by beam or shell elements. Applying linear buckling analysis provides the *critical load factor* α_{cr} by which the applied loads must be multiplied to reach the critical state. The critical stresses are then determined from Equation 8.15, and similarly, for the shear stresses

$$\sigma_{cr} = \alpha_{cr} \cdot \sigma \tag{8.15}$$

where σ is the acting stress.

Table 8.6 Critical stress for *Navier* panels with one longitudinal stiffener in the compression zone under direct stresses

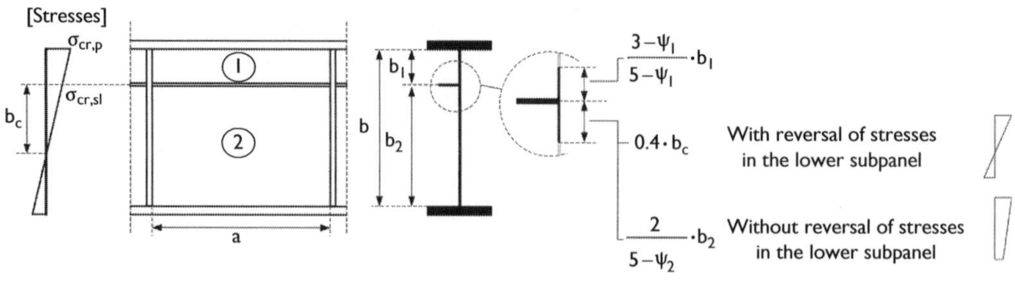

Length a	*Critical stress of stiffener*
$a \geq a_c$	$\sigma_{cr,sl} = \dfrac{1.05 \cdot E_a}{A_{st,1}} \cdot \dfrac{\sqrt{I_{sl,1} \cdot t^3 \cdot b}}{b_1 \cdot b_2}$
$a < a_c$	$\sigma_{cr,sl} = \dfrac{\pi^2 \cdot E_a \cdot I_{sl,1}}{A_{sl,1} \cdot a^2} + \dfrac{E_a \cdot t^3 \cdot b \cdot a^2}{4 \cdot \pi^2 \cdot (1-\nu^2) \cdot A_{sl,1} \cdot b_1^2 \cdot b_2^2}$

Critical plate buckling stress: $\sigma_{cr,p} = \sigma_{cr,sl} \cdot \dfrac{b_c + b_1}{b_c}$

Source: EN 1993-1-5, 2006 Eurocode 3: Design of steel structures, Part 1–5: Plated Structural Elements, 2006.

Note: $A_{sl,1}$, gross area of the stiffener and the adjacent parts of the plate as shown in the figure earlier; ψ, the stress ratio of the subpanel under consideration; $I_{sl,1}$, second moment of area of the stiffener and the adjacent parts of the plate as shown in the figure earlier; b_1, b_2, the distances from the longitudinal edges of the web to the stiffener; $b_1 + b_2 = b$, the width of the whole stiffened plate, $a_c = 4.33 \sqrt[4]{\dfrac{I_{sl,1} \cdot b_1^2 \cdot b_2^2}{t^3 \cdot b}}$, the wavelength of buckling, assuming the rigid transverse stiffeners to be removed.

Table 8.7 Critical stress for *Navier* panels with two longitudinal stiffeners in the compression zone

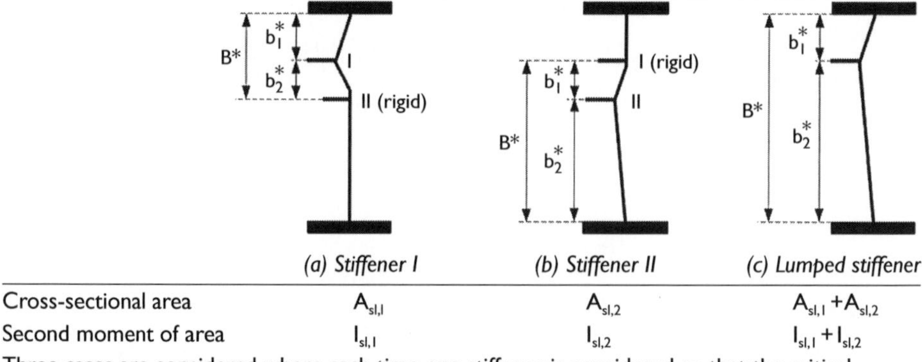

	(a) Stiffener I	*(b) Stiffener II*	*(c) Lumped stiffener*
Cross-sectional area	$A_{sl,1}$	$A_{sl,2}$	$A_{sl,1} + A_{sl,2}$
Second moment of area	$I_{sl,1}$	$I_{sl,2}$	$I_{sl,1} + I_{sl,2}$

Three cases are considered where each time one stiffener is considered so that the critical stress is determined according to Table 8.6. Any stiffeners in the tension zone are ignored.

Case I Stiffener II is considered as rigid. The widths b, b_1, and b_2 of Table 8.6 are set as in Figure (a) earlier.

Case II Stiffener I is considered as rigid. The widths b, b_1, and b_2 of Table 8.6 are set as in Figure (b) earlier.

Case III The two stiffeners are considered as one lumped stiffener that is located at the resultant of the respective forces $\sigma \cdot A_{sl}$ of the individual stiffeners. The widths b, b_1, and b_2 of Table 8.6 are set as in Figure (c) earlier.

The critical stress is the smallest of the three cases.

A powerful tool for the determination of the critical stresses for unstiffened and stiffened rectangular panels is developed in [8.6] and may be downloaded free of charge.

EXAMPLE 8.1

A plate of length a = 3.0 m, width b = 2.0 m, and thickness t = 10 mm shown in Figure 8.13 is subjected to compression with edge stresses $\sigma_{x1} = 10$ N/mm^2 and $\sigma_{x2} = 5$ N/mm^2. The plate is stiffened by three equal distanced longitudinal flat stiffeners of height h = 150 mm and thickness t = 10 mm. The critical stresses are to be determined.

Numerical determination

The load factor for the 1st buckling mode (Figure 8.13) was found by the numerical method as equal to $\alpha_{cr} = 40.23$; the critical stress is then, Equation 8.15:

$$\sigma_{cr} = 40.23 \cdot 10 = 402.3 \text{ N/mm}^2$$

Determination by formulae according to Table 8.4

Equation 8.10, reference stress:

$$\sigma_e = 189,800 \cdot \left(\frac{1}{200}\right)^2 = 4.75 \text{ N/mm}^2$$

Plate

Gross area:

$$A_p = 200 \cdot 1 = 200 \text{ cm}^2$$

Second moment of area:

$$I_p = \frac{200 \cdot 1^3}{10.92} = 18.31 \text{ cm}^4$$

Stiffeners

Area of each stiffener:

$$A_{sl} = 1 \cdot 15 = 15 \text{ cm}^2$$

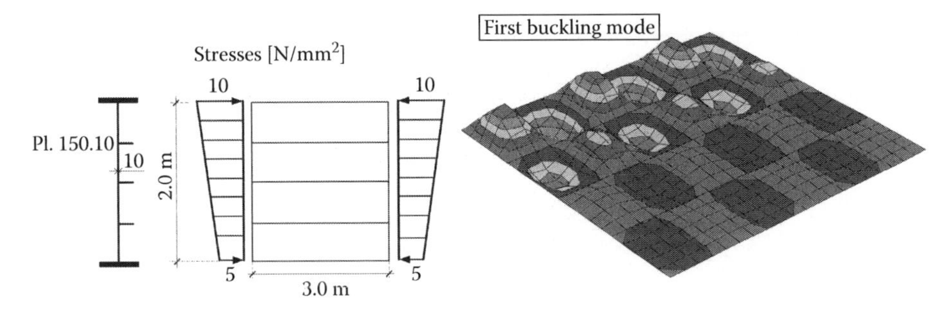

Figure 8.13 Stiffened plate of Example 8.1 and first buckling mode.

Area of three stiffeners:

$$\sum A_{sl} = 3 \cdot 15 = 45\ \text{cm}^2$$

$$\delta = \frac{45}{200} = 0.225$$

Second moment of area of the plate and the three stiffeners: $I_{sl} = 3211.44\ \text{cm}^4$

$$\gamma = \frac{3211.44}{18.31} = 175.39$$

It is

$$\alpha = \frac{300}{200} = 1.5 < \sqrt[4]{\gamma} = \sqrt[4]{175.39} = 3.64$$

Stress ratio:

$$\psi = \frac{5}{10} = 0.5$$

Buckling factor:

$$k_{\sigma,p} = \frac{2 \cdot \left[(1+1.5^2)^2 + 175.39 - 1\right]}{1.5^2 \cdot (0.5+1) \cdot (1+0.225)} = 89.47$$

Equation 8.9, critical stress:

$$\sigma_{cr,p} = 89.47 \cdot 4.75 = 424.98\ \text{N/mm}^2$$

This value is 5.6% higher compared to the numerical one.

EXAMPLE 8.2

The stiffened plate of Example 8.1 (a = 3.0 m, b = 2.0 m, t = 10 mm) is subjected to a shear stress $\tau = 15$ N/mm². The critical stresses are to be determined. Steel S 355.

Numerical determination

The critical load factor for the 1st buckling mode shown in Figure 8.14 was found by the numerical method as equal to $\alpha_{cr} = 21.27$. The critical stress is then, Equation 8.15:

$$\tau_{cr} = 21.27 \cdot 15 = 319.05\ \text{N/mm}^2$$

Determination by formulae according to Table 8.5

Equation 8.10, reference stress:

$$\sigma_e = 189{,}800 \cdot \left(\frac{1}{200}\right)^2 = 4.75\ \text{N/mm}^2$$

$$\text{Steel S 355} \rightarrow \varepsilon = \sqrt{\frac{235}{355}} = 0.81$$

Associated plate width to the stiffener: $2 \cdot 15 \cdot 0.81 \cdot 1 + 1 = 25.3$ cm; see note in Table 8.5.

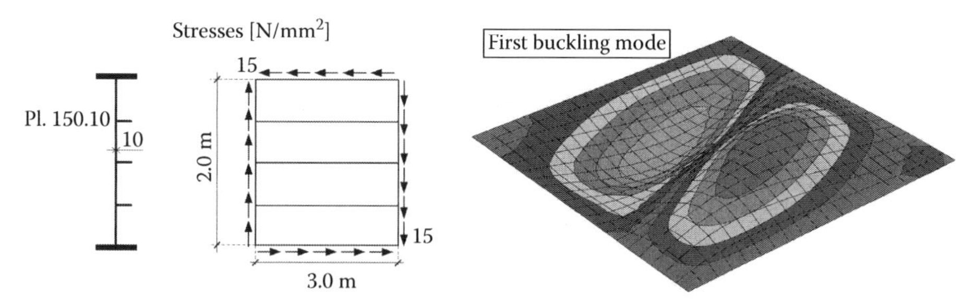

Figure 8.14 Stiffened plate of Example 8.2 and first buckling mode.

Second moment of area of one stiffener including the associated plate width:

$$I_{sl} = 886.04 \text{ cm}^4$$

$$\sum I_{sl} = 3 \cdot 886.04 = 2658.12 \text{ cm}^4$$

Aspect ratio:

$$\alpha = \frac{300}{200} = 1.5$$

$$k_{\tau,st} = \frac{9}{1.5^2} \cdot \sqrt[4]{\left(\frac{2658.12}{1^3 \cdot 200}\right)^3} = 27.84 \geq \frac{2.1}{1} \cdot \sqrt[3]{\frac{2658.12}{200}} = 4.97$$

$$\alpha = 1.5 \geq 1.0 \rightarrow k_\tau = 5.34 + \frac{4.0}{1.5^2} + 27.84 = 34.96$$

$$\tau_{cr} = 34.96 \cdot 4.75 = 166.06 \text{ N/mm}^2$$

Figure 8.14 shows that the first buckling mode includes deformations of the stiffeners. By examination of more buckling modes, it may be observed (Figure 8.15) that in the 5th buckling mode,

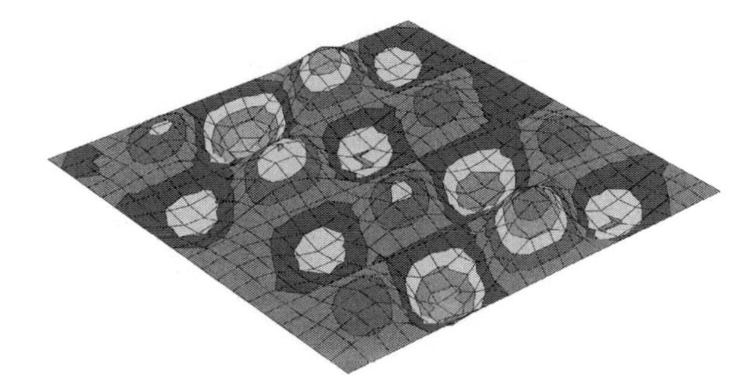

Figure 8.15 Fifth buckling mode of the stiffened plate of Example 8.2.

Table 8.8 Critical shear stresses τ_{cr} [N/mm²]

	Numerical method	*Formulae*
Stiffened panel	319	166
Subpanel between stiffeners	472	414

the stiffeners do not deform so that buckling refers only to the unstiffened subpanels between stiffeners. The critical load factor for this buckling mode is $\alpha_{cr} = 31.45$, and the corresponding critical stress becomes

$$\tau_{cr} = 31.45 \cdot 15 = 471.8 \text{ N/mm}^2$$

The geometric properties of this subpanel are a = 3.0 m, b = 0.5 m, and t = 10 mm.

The critical buckling stress may be also determined analytically.

Reference stress: $\sigma_e = 189{,}800 \cdot \left(\frac{1}{50}\right)^2 = 75.92 \text{ N/mm}^2$

Aspect ratio: $\alpha = \frac{300}{50} = 6 \geq 1$

Buckling factor: $k_\tau = 5.34 + \frac{4}{6^2} + 0 = 5.45$

$$\tau_{cr} = 5.45 \cdot 75.92 = 413.76 \text{ N/mm}^2$$

The critical shear stresses are summarized in Table 8.8. It may be seen that the minimum values refer to the stiffened panel and not the unstiffened subpanels.

EXAMPLE 8.3

The stiffened plate of Example 8.2 (b = 2.0 m, t = 10 mm) is subjected to a shear stress $\tau = 15$ N/mm². The critical stresses are to be determined when the length of the plate varies from 2.0 to 9.0 m.

The critical shear stresses determined numerically and from Table 8.5 are shown in Table 8.9. It may be seen that the formulae provide too conservative values. The critical stress for the unstiffened subpanels for all considered lengths is the same as determined before in Example 8.2. It may be observed that the critical stress for the complete panel is always smaller than the one for the subpanels.

Table 8.9 Critical shear stresses τ_{cr} [N/mm²] for different plate lengths

Plate length	*a = 2 m*	*a = 3 m*	*a = 4 m*	*a = 9 m*
τ_{cr} from Table 8.5	342	166	104	50
τ_{cr} numerical	381	319	238	86

8.2.4 Stiffened and unstiffened panels: Combined loading conditions

The critical state for combined loading $\sigma_{x,Ed}$, $\sigma_{z,Ed}$, and τ_{Ed}, is reached when the stresses reach following values:

$$\sigma_{cr,x} = \alpha_{cr,x} \cdot \sigma_{x,Ed}, \quad \sigma_{cr,z} = \alpha_{cr,z} \cdot \sigma_{z,Ed}, \quad \tau_{cr} = \alpha_{cr,\tau} \cdot \tau_{Ed} \tag{8.16}$$

where α_{cr} is the critical load factor and the index Ed defines the design values of the applied stresses.

α_{cr} may be determined by application of the *Rayleigh–Ritz energy method* as outlined before. However, if values of this factor, $\alpha_{cr,x}$, $\alpha_{cr,z}$, and $\alpha_{cr,\tau}$, are known for individual stress conditions, the multiplication factor α_{cr} for combined loading may be determined approximately from the following interaction formula:

$$\frac{1}{\alpha_{cr}} = \frac{1+\psi_x}{4\cdot\alpha_{cr,x}} + \frac{1+\psi_z}{4\cdot\alpha_{cr,z}} + \sqrt{\left(\frac{1+\psi_x}{4\cdot\alpha_{cr,x}} + \frac{1+\psi_z}{4\cdot\alpha_{cr,z}}\right)^2 + \frac{1-\psi_x}{2\cdot\alpha_{cr,x}^2} + \frac{1-\psi_z}{2\cdot\alpha_{cr,z}^2} + \frac{1}{\alpha_{cr,\tau}^2}} \tag{8.17}$$

where

$$\alpha_{cr,x} = \frac{\sigma_{cr,x}}{\sigma_{x,Ed}}, \quad \alpha_{cr,z} = \frac{\sigma_{cr,z}}{\sigma_{z,Ed}} \text{ and } \alpha_{cr,\tau} = \frac{\sigma_{cr,\tau}}{\tau_{Ed}}$$

ψ_x is the longitudinal direct stress ratio
ψ_z is the transverse direct stress ratio

The load factors $\alpha_{cr,x}$ and $\alpha_{cr,z}$ should be calculated taking $\sigma_{x,Ed}$ and accordingly $\sigma_{z,Ed}$ as the greatest compressive stress.

It has to be noted that in cases of tensile direct stresses σ_x throughout the panel, $a_{cr,x}$ is taken equal to infinity (∞). Load factors with negative values should not be used in Equation 8.17. Panels that are wholly in tension still need to be checked since shear buckling may still be significant, for example, bottom flanges of box girder bridges at spans with shear due to torsion.

EXAMPLE 8.4

The critical stresses for an unstiffened plate with the geometric properties a = 3.0 m, b = 2.0 m, and t = 40 mm subjected to combined loading as shown in Figure 8.16 are to be determined.

Numerical determination

The numerical calculation provides a critical load factor for the first buckling mode (Figure 8.16) α_{cr} = 17.18.

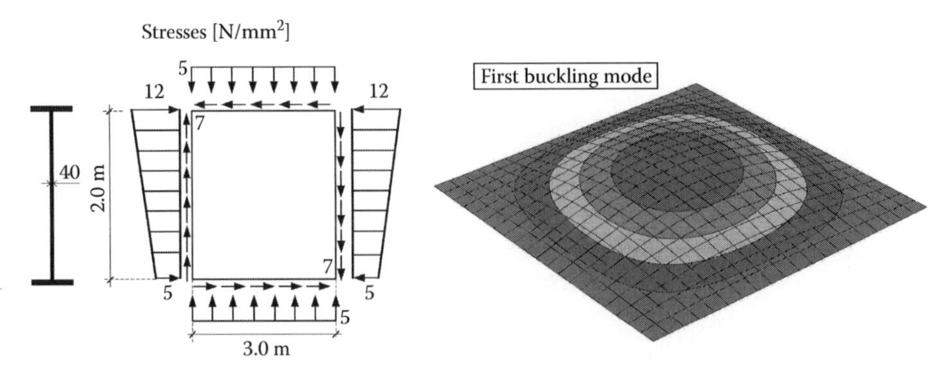

Figure 8.16 Unstiffened plate of Example 8.4 under combined load.

Accordingly, Equation 8.16:

$$\sigma_{cr,x} = 17.18 \cdot 12 = 206.2\,\text{N/mm}^2, \quad \sigma_{cr,z} = 17.18 \cdot 5 = 85.9\,\text{N/mm}^2$$

$$\tau_{cr} = 17.18 \cdot 7 = 120.3\,\text{N/mm}^2$$

Determination by formulae according to Tables 8.2 and 8.5

From Equation 8.10, reference stress:

$$\sigma_e = 189{,}800 \cdot \left(\frac{4}{200}\right)^2 = 75.92\,\text{N/mm}^2$$

Stresses σ_x

It is

$$\psi_x = \frac{5}{12} = 0.42 < 1 \rightarrow k_{\sigma x} = \frac{8.2}{1.05 + 0.42} = 5.58; \text{ (Table 8.2)}$$

$$\sigma_{cr,x} = 5.58 \cdot 75.92 = 423.6\,\text{N/mm}^2$$

$$\alpha_{cr,x} = \frac{423.6}{12} = 35.3$$

Stresses σ_z

The critical stress is determined by the same procedure as for σ_x, but taking as width of the panel its length b = 3.0 m.

Reference stress: $\sigma_e = 189{,}800 \cdot \left(\frac{4}{300}\right)^2 = 33.74\,\text{N/mm}^2$

$$\psi_z = 1 \rightarrow k_{\sigma z} = 4 \rightarrow \sigma_{cr,z} = 4 \cdot 33.74 = 135\,\text{N/mm}^2; \text{ (Table 8.2)}$$

and $\alpha_{cr,z} = \frac{135}{5} = 27$

Stresses τ

$$\alpha = \frac{300}{200} = 1.5 \geq 1 \rightarrow k_\tau = 5.34 + \frac{4.0}{1.5^2} = 7.12 \rightarrow \tau_{cr} = 7.12 \cdot 75.92 = 540.6\,\text{N/mm}^2$$

$$\alpha_{cr,\tau} = \frac{540.6}{7} = 77.2$$

Application of Equation 8.17:

$$\frac{1}{\alpha_{cr}} = \frac{1+0.42}{4 \cdot 35.3} + \frac{1+1}{4 \cdot 27} + \sqrt{\left(\frac{1+0.42}{4 \cdot 35.3} + \frac{1+1}{4 \cdot 27}\right)^2 + \frac{1-0.42}{2 \cdot 35.3^2} + \frac{1-1}{2 \cdot 27^2} + \frac{1}{77.2^2}}$$

From which, $\alpha_{cr} = 15.76$ (deviation from numerical method is -8.2%).

REMARK 8.2

In cases of excessively stiffened plates, FE analysis may give buckling modes followed by load factors with negative values. This does not mean that the software calculates wrong. A negative value for the load factor signifies that all the loads of that loading case need to act in the direction opposite to that in which they have been applied to cause buckling. The designer should neglect such buckling modes.

8.3 STRENGTH OF PLATES

8.3.1 General

Like in struts, the ultimate strength of plates differs from the critical buckling strength. Table 8.1 shows that the ultimate strength is determined by buckling curves, the critical buckling strength being only a reference value for the calculation of the nondimensional slenderness that is the abscissa of the buckling curve. The difference between critical stresses and ultimate strength is due to the fact that

a. The assumption made to determine the critical stresses is not valid in real plates, as such plates
   - Are not plane before loading but have geometric imperfections.
   - They have also structural imperfections (initial welding stresses) due to the fabrication processes.
   - They do not behave indefinitely elastically but are entering into the elastic–plastic state at higher levels of loading.
b. Plates possess considerable post-buckling strength.
c. Plates under compression with many stiffeners or short plates exhibit column-like behavior and may lose the post-buckling strength.

These aspects will be discussed in the following sections.

8.3.2 Postbuckling plate behavior: Plate buckling curves

As outlined before, an absolutely plane elastic plate remains plane for applied loads smaller than the critical buckling load. If the load is increased beyond the critical value, the plate starts to buckle. As shown in Figure 8.17, buckling is accompanied with a reduction in in-plane rigidity and increase of out-of-plane deformations. However, unlike struts, plates are supported at the longitudinal edges, so that after buckling a catenary action (Z forces in Figure 8.17) develops in transverse direction, which "supports" transversely the plate and allows a further increase of the longitudinal load and a stable loading path. A plate has therefore *post-buckling strength*, which is accounted for in design. Figure 8.17 shows that a plate with geometric imperfections behaves at large displacements similar to a perfect plate, implying its low sensitivity to geometric imperfections.

The description of the post-buckling behavior may be done by application of nonlinear plate buckling theory that involves large displacement analysis. For elastic plates, the *Karman–Marguerre* [8.17] solutions of the relevant equations exist. However, the behavior of real plates differs from the one for elastic plates due to inelastic material behavior. Figure 8.18 shows such load–deflection curves for plates with yielding material, with and without geometric imperfections indicating that the stiffness reduction is not only due to out-of-plane displacements but also due to material yielding.

Figure 8.18 indicates that the ultimate strength of the plate is different from the critical buckling stress. The ultimate strength may be determined by appropriate buckling curves that provide, like in buckling of struts, a reduction factor to be applied to the yield stress. In analogy to the nondimensional slenderness of struts $\bar{\lambda}=\sqrt{f_y/\sigma_{cr}}$, the nondimensional slenderness of plates may be determined by the following relations:

For direct stresses:

$$\bar{\lambda}_p = \sqrt{\frac{f_y}{\sigma_{cr,p}}} \tag{8.18}$$

For shear stresses:

$$\bar{\lambda}_w = \sqrt{\frac{f_y/\sqrt{3}}{\tau_{cr}}} = 0.76 \cdot \sqrt{\frac{f_y}{\tau_{cr}}} \tag{8.19}$$

where $\sigma_{cr,p}$ and τ_{cr} are the critical buckling stresses.

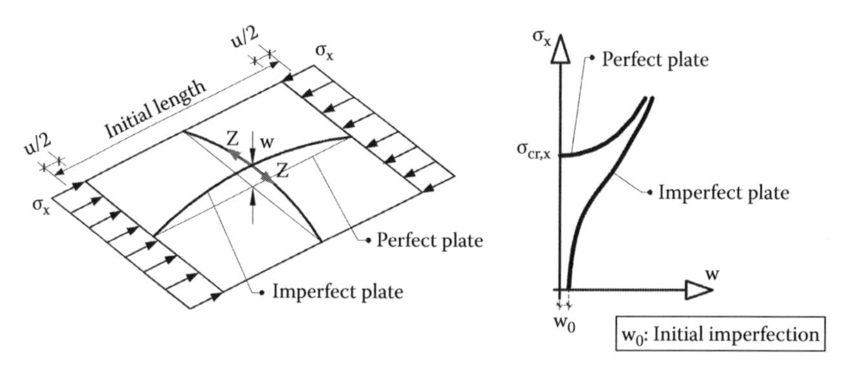

Figure 8.17 Axially compressed elastic plate and load–deflection curve.

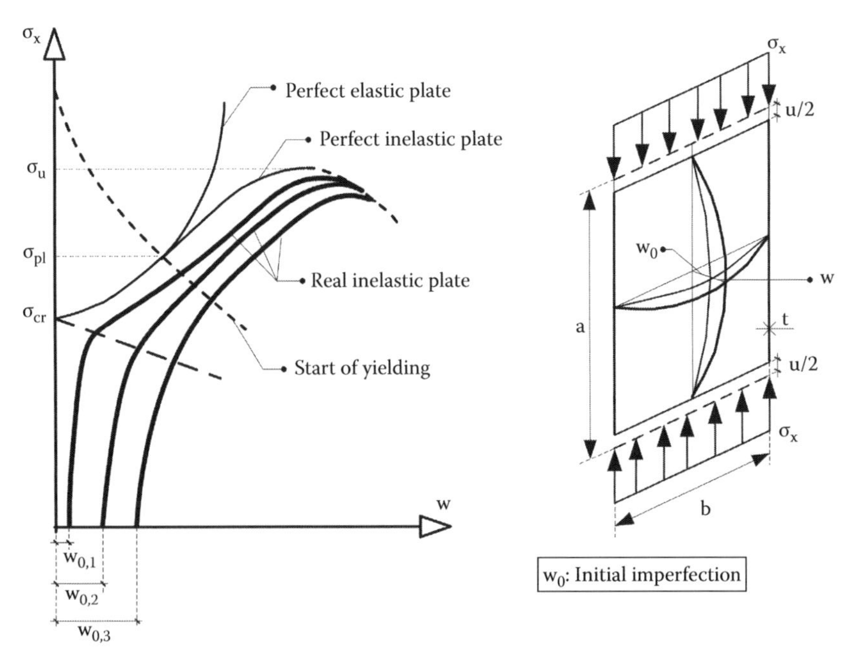

Figure 8.18 Load–deflection curves of inelastic plates.

The nondimensional slenderness may be determined by substitution of $\sigma_{cr,p}$ of Equations 8.9 and 8.10 in Equation 8.18:

$$\bar{\lambda}_p = \sqrt{\frac{235}{k_\sigma \cdot 189{,}800 \cdot \left(\frac{t}{b}\right)^2}} \cdot \sqrt{\frac{f_y}{235}} = \frac{b}{28.4 \cdot t \cdot \varepsilon \cdot \sqrt{k_\sigma}} \tag{8.20}$$

where $\varepsilon = \sqrt{235/f_y}$ (= 0.81 for the usual steel grade in bridges S 355).

Similarly, the nondimensional slenderness for shear is given by

$$\bar{\lambda}_w = 0.76 \cdot \sqrt{\frac{235}{k_\tau \cdot 189{,}800 \cdot \left(\frac{t}{b}\right)^2}} \cdot \sqrt{\frac{f_y}{235}} = \frac{b}{37.4 \cdot t \cdot \varepsilon \cdot \sqrt{k_\tau}} \tag{8.21}$$

Further on the buckling factor for a plate with a large aspect ratio, α, is equal to $k_\tau = 5.34$ (see Table 8.5) so that the earlier expression becomes

$$\bar{\lambda}_w = \frac{b}{86.4 \cdot t \cdot \varepsilon} \tag{8.22}$$

Buckling curves provide reduction factors as a function of $\bar{\lambda}$. EN1993-1-5 [8.4] uses for internal and outstand panels modified *Winter curves* [8.18] as buckling curves that are given in the following:

Internal panels:

$$\rho = 1 \quad \text{for } \bar{\lambda}_p \le 0.673$$

$$\rho = \frac{\bar{\lambda}_p - 0.055 \cdot (3 + \psi)}{\bar{\lambda}_p^2} \le 1 \quad \text{for } \bar{\lambda}_p > 0.673 \text{ and } (3 + \psi) \ge 0 \tag{8.23}$$

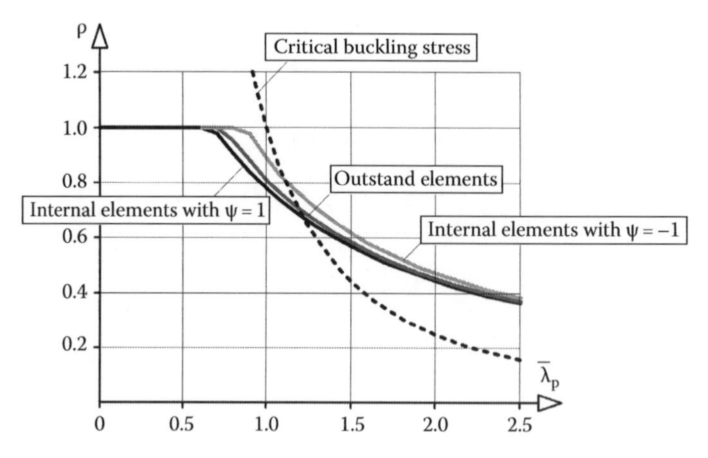

Figure 8.19 Plate buckling curves for direct stresses.

Outstand panels:

$$\rho = 1 \quad \text{for } \bar{\lambda}_p \leq 0.748$$

$$\rho = \frac{\bar{\lambda}_p - 0.188}{\bar{\lambda}_p^2} \leq 1 \quad \text{for } \bar{\lambda}_p > 0.748 \tag{8.24}$$

Figure 8.19 illustrates plate buckling curves for direct stresses for various ratios of applied stresses as well as the *Euler hyperbola* that expresses the critical buckling stress. The following may be observed:

- At low slenderness, the strength is influenced by yielding.
- The influence of post-buckling strength is pronounced at high slenderness, and the buckling curves exhibit higher values than the Euler hyperbola.
- Imperfections affect the strength at intermediate slenderness.

Reduction factors for *shear buckling* are given in Table 8.10. The corresponding shear buckling curves are shown in Figure 8.20, together with the critical curve. Similar observations regarding yielding and post-buckling strength, like for direct stresses, may be made. Table 8.10 implies that a reduction in shear strength due to shear buckling shall be accounted for plates

Table 8.10 Reduction factors χ_w for shear buckling resistance

	Rigid end posts, Figure 8.30	*Nonrigid end posts, Figure 8.30*
$0.83/\eta > \bar{\lambda}_w$	η	η
$0.83/\eta \leq \bar{\lambda}_w < 1.08$	$0.83/\bar{\lambda}_w \leq 1$	$0.83/\bar{\lambda}_w \leq 1$
$\bar{\lambda}_w \geq 1.08$	$1.37/(0.7 + \bar{\lambda}_w)$	$0.83/\bar{\lambda}_w \leq 1$

Notes:

EN1993-1-5 recommends $\eta = 1.2$. For steel grades higher than S460 $\eta = 1.0$. Different values may be found in the National Annex. η expresses the strength increase due to strain hardening.

$\bar{\lambda}_w$ is determined from Equation 8.22 for plates without or Equation 8.21 with longitudinal stiffeners. In the latter case, k_τ is the smallest value between the stiffened panel and all subpanels between longitudinal stiffeners.

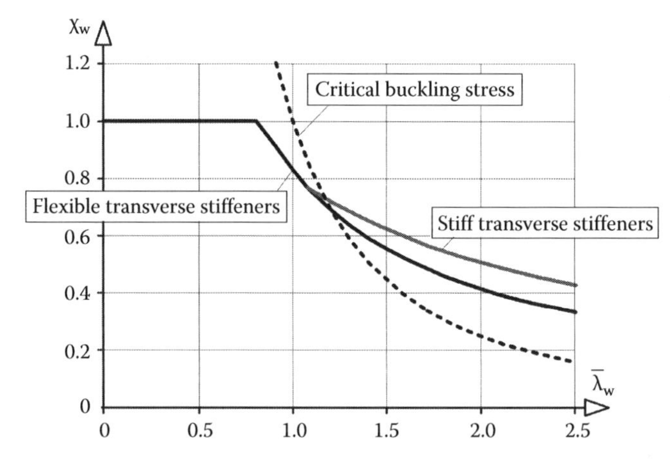

Figure 8.20 Shear buckling curves (n = 1.0, ε = 1).

of slenderness $\bar{\lambda}_w > 0.83/\eta$. This suggests, considering Equations 8.21 and 8.22, that shear buckling shall be accounted for when the width-(h_w, height of web)-to-thickness (t) ratios exceed the following values (in parenthesis values for steel grades equal or higher than S 460):

Unstiffened panels:

$$\frac{h_w}{t} > 60 \cdot \varepsilon \left(\frac{h_w}{t} > 72 \cdot \varepsilon \right) \tag{8.25}$$

Longitudinally stiffened panels:

$$\frac{h_w}{t} > 25.8 \cdot \varepsilon \cdot \sqrt{k_\tau} \left(\frac{h_w}{t} > 31 \cdot \varepsilon \cdot \sqrt{k_\tau} \right) \tag{8.26}$$

REMARK 8.3

In order to provide a better insight into the buckling curves of Figure 8.20, the following approach is presented.

In Figure 8.39, one can see the stress field of a girder web after shear buckling. For pure shear, the principal stresses are calculated as follows:

$$\sigma_1 = \frac{\tau}{\tan\varphi} \quad \text{(rotated principal tensile stress)} \tag{R8.1}$$

$$\sigma_2 = -\tau \cdot \tan\varphi \quad \text{(rotated principal compressive stress)} \tag{R8.2}$$

where

τ is the shear stress

φ is the angle between horizontal and principal stresses

Experimental tests indicated that the principal stress σ_2 does not exceed the elastic shear stress for shear buckling τ_{cr}. This leads to the following conservative assumption:

$$\sigma_2 = -\tau_{cr} = -k_\tau \cdot \sigma_e = -k_\tau \cdot \frac{\pi^2 \cdot E_a}{12 \cdot (1 - v^2)} \cdot \left(\frac{t}{b}\right)^2 \text{; see Equation 8.10} \quad \text{(R8.3)}$$

The principal tensile stress becomes then equal to

$$\sigma_1 = \frac{\tau^2}{\tau_{cr}} \quad \text{(R8.4)}$$

The ultimate strength of the web is based on the *von Mises* criterion and is reached when

$$\sigma_1^2 + \sigma_2^2 - \sigma_1 \cdot \sigma_2 = f_y^2 \quad \text{(R8.5)}$$

Substituting Equations R8.3 and R8.4 with Equation R8.5 gives the following shear strength:

$$\frac{\tau_u}{\tau_y} = \chi_w = \frac{\sqrt[3]{4}}{\bar{\lambda}_w} \cdot \sqrt{\sqrt{1 - \frac{1}{4 \cdot \bar{\lambda}_w^4}} - \frac{1}{2 \cdot \sqrt{3} \cdot \bar{\lambda}_w^2}} \quad \text{(R8.6)}$$

where

$$\tau_y = \frac{f_y}{\sqrt{3}}$$

$$\bar{\lambda}_w = \sqrt{\frac{\tau_y}{\tau_{cr}}}$$

The expression in (R8.6) generally overestimates the shear resistance of girder webs with no rigid end posts (see Figure 8.30). However, it matches reasonably with the experimental results for cases with rigid end posts (see [8.10]). Equation R8.6 is the "background equation" for the reduction factors given in Table 8.10. Modifications have been applied so that the scattering of the test results has been considered.

8.3.3 Column-like behavior

As mentioned before, the post-buckling plate strength is based on the catenary action in the transverse direction (see Figure 8.17). However, under certain conditions, this catenary action cannot be activated. This is the case of stiffened plates or unstiffened plates with low aspect ratio ($\alpha < 1$) subjected to compression. Indeed, under these conditions, the plate curvature in transverse direction is low, as indicatively shown in Figure 8.21a, so that the catenary action is very weak. This suggests that the plate behaves like a column, that is, a plated element without longitudinal supports (Figure 8.21b).

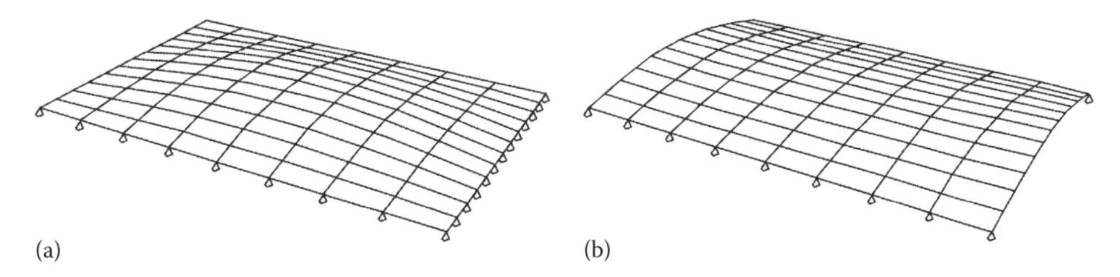

Figure 8.21 First buckling mode of a short plate (aspect ratio $\alpha=0.5$) (a) with and (b) without longitudinal supports (column buckling behavior).

Again, linear buckling theory is used to check if *column-like behavior* is present or not. The relevant criterion is the ratio:

$$\frac{\sigma_{cr,p}}{\sigma_{cr,c}}(\geq 1) \tag{8.27}$$

where

$\sigma_{cr,p}$ is the critical plate buckling stress

$\sigma_{cr,c}$ is the critical column buckling stress, that is, the Euler stress of the same plate in which the supports along the longitudinal edges are removed.

For large values of $\sigma_{cr,p}/\sigma_{cr,c}$, the element behaves more than a plate and exhibits post-buckling strength. However, as this ratio approaches unity, the plate behaves more like a column and loses its post-buckling strength.

For plates with distributed loading along their edges, the critical column buckling stress $\sigma_{cr,c}$ may be determined from Expressions 8.28 and 8.29.

Unstiffened plates

The critical column buckling stress for unstiffened plates under uniform compression may be obtained from

$$\sigma_{cr,c,\psi=1} = \frac{\pi^2 \cdot E \cdot t^2}{12 \cdot (1-\nu^2) \cdot a^2} \tag{8.28}$$

For nonuniform compression ($\psi \neq 1$), $\sigma_{cr,c}$ is the largest compression stress across the panel and according to own calculations may be obtained from

$$\sigma_{cr,c,\psi\neq 1} = \sigma_{cr,c,\psi=1} \cdot (1.5 - 0.5 \cdot \psi) \tag{8.29}$$

Stiffened plates

The *Euler stress* of the most compressed longitudinal stiffener is given by

$$\sigma_{cr,sl} = \frac{\pi^2 \cdot E \cdot I_{sl,1}}{A_{sl,1} \cdot a^2} \tag{8.30}$$

where $I_{sl,1}$ and $A_{sl,1}$ are the second moment of area and the cross-sectional area of the stiffener, respectively, including adjacent parts of the plate as shown in Table 8.11, column 2.

However, Figure 8.22 indicates that the aforementioned stress is not the highest compression stress of the plate and cannot be directly compared to the corresponding one for plate

Table 8.11 Widths b_1, b_2, and b_3 of Figure 8.22

	Width for gross area	*Width for effective area*	*Condition for* ψ_i
$b_{1,inf}$	$\frac{3-\psi_1}{5-\psi_1}\cdot b_1$	$\frac{3-\psi_1}{5-\psi_1}\cdot b_{1,eff}$	$\psi_1 = \frac{\sigma_{cr,sl,1}}{\sigma_{cr,p}} > 0$
$b_{2,sup}$	$\frac{2}{5-\psi_2}\cdot b_2$	$\frac{2}{5-\psi_2}\cdot b_{2,eff}$	$\psi_2 = \frac{\sigma_2}{\sigma_{cr,sl,1}} > 0$
$b_{2,inf}$	$\frac{3-\psi_2}{5-\psi_2}\cdot b_2$	$\frac{3-\psi_2}{5-\psi_2}\cdot b_{2,eff}$	$\psi_2 > 0$
$b_{3,sup}$	$0.4\cdot b_{3c}$	$0.4\cdot b_{3c,eff}$	$\psi_3 = \frac{\sigma_3}{\sigma_2} < 0$

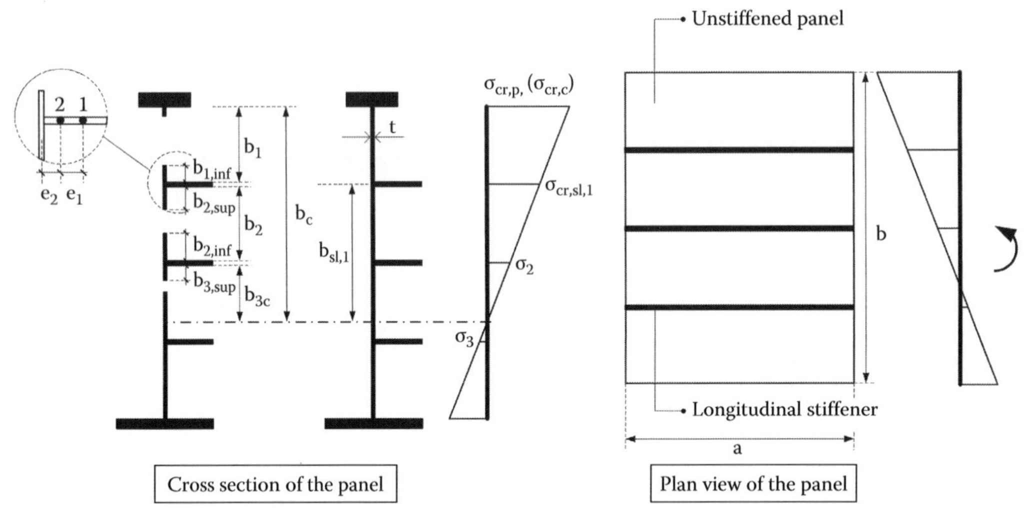

Notation

Point 1: Gravity center of the stiffener

Point 2: Gravity center of the column (stiffener + associated plate width)

$e = \max(e_1, e_2)$

Figure 8.22 Notation for Equations 8.28 through 8.31.

buckling. Therefore, using simple geometric relations, the critical column buckling stress to be introduced in Equation 8.27 is obtained from

$$\sigma_{cr,c} = \frac{b_c}{b_{sl,1}}\cdot\sigma_{cr,sl} \tag{8.31}$$

where

- b_c is the distance from the position of zero direct stress to the most compressive panel fiber; it is different from that in Table 8.6 (see Remark 8.5)
- $b_{sl,1}$ is the distance from the position of zero direct stress to the stiffener

REMARK 8.4

The column buckling load $\sigma_{cr,sl}$ of Equation 8.30 gives conservative results since it is based on the following assumptions:

- The compression force acting on the stiffener is considered constant. In reality, axial forces in stiffeners follow the bending diagram and are longitudinally variable.
- The longitudinal stiffener is considered simply supported ignoring the rotational restraint at its ends.

REMARK 8.5

EN 1993-1-5 uses the same definition of b_c in Equation 8.31, which corresponds to Figure 8.22, and in Table 8.6 for the determination of the critical stress for panels with one longitudinal stiffener in the compression zone. The designer should be very careful during the calculation of b_c in each case.

Column-like behavior is accounted for by modifying the reduction factor ρ for plate buckling as follows:

$$\rho_c = (\rho - \chi_c) \cdot \xi \cdot (2 - \xi) + \chi_c \tag{8.32}$$

where

$$\xi = \frac{\sigma_{cr,p}}{\sigma_{cr,c}} - 1 \quad \text{and} \quad 0 \le \xi \le 1 \tag{8.33}$$

Equation 8.32 indicates that for $\xi = 1$, that is, in cases where $\sigma_{cr,p} \ge 2 \cdot \sigma_{cr,c}$, a column-like behavior does not exist so that it is $\rho_c = \rho$, while for $\xi = 0$, the plate behaves rather like a column and does not exhibit post-buckling strength. The plate buckling stress $\sigma_{cr,p}$ cannot be smaller than $\sigma_{cr,c}$, and consequently a lower limit of zero is placed on ξ.

The notation for Equation 8.32 is the following:

χ_c is the reduction factor for column buckling as a function of the nondimensional slenderness $\bar{\lambda}_c$ determined in the following:

$$\bar{\lambda}_c = \sqrt{\frac{f_y}{\sigma_{cr,c}}} \quad \text{for unstiffened plates} \tag{8.34}$$

$$\bar{\lambda}_c = \sqrt{\frac{\beta_{A,c} \cdot f_y}{\sigma_{cr,c}}} \quad \text{for stiffened plates} \tag{8.35}$$

$\beta_{A,c} = 1.0$ for classes 1, 2, and 3 stiffeners

$\beta_{A,c} = \dfrac{A_{sl,1,eff}}{A_{sl,1}}$ is the ratio of effective to gross area of the stiffener, for class 4 stiffeners

It is noted that it is not recommended to use class 4 stiffeners that are themselves prone to local buckling. However, this is not always feasible.

The reduction factor χ_c for column buckling is determined from the European column-buckling curves according to the following relations:

$$\chi_c = \frac{1}{\Phi + \sqrt{\Phi^2 - \bar{\lambda}_c^2}} \leq 1 \tag{8.36}$$

$$\Phi = 0.5 \cdot \left[1 + \alpha \cdot (\bar{\lambda}_c - 0.2) + \bar{\lambda}_c^2\right] \tag{8.37}$$

The imperfection factor α shall be determined as follows:

Unstiffened panels	$\alpha = 0.21$ that corresponds to the European buckling curve a
Stiffened panels	$\alpha_e = \alpha + \frac{0.09}{i/e}$ $i = \sqrt{\frac{I_{sl,I}}{A_{sl,I}}}$ radius of gyration of the most compressed stiffener $e = \max(e_1, e_2)$ as indicated in Figure 8.22 e_1 = distance between gravity centers of the stiffener + associated plate and the stiffener alone e_2 = distance of the gravity center of the stiffener + associated panel from midsurface of the panel $\alpha = 0.34$ (buckling curve b) for closed stiffeners $\alpha = 0.49$ (buckling curve c) for open stiffeners

EXAMPLE 8.5

The reduction factor of an internal unstiffened panel with a = 1.0 m, b = 2.1 m, and t = 16 mm when subjected to uniform longitudinal compression is to be determined. Steel grade S 355.
Reference stress, Equation 8.10:

$$\sigma_e = 189{,}800 \cdot \left(\frac{16}{2100}\right)^2 = 11.02 \text{ N/mm}^2$$

Aspect ratio:

$$\alpha = \frac{100}{210} = 0.476$$

Buckling factor (m = 1), Equation 8.5:

$$k_\sigma = \left(\frac{1}{0.476} + 0.476\right)^2 = 6.64$$

Critical plate buckling stress, Equation 8.9:

$$\sigma_{cr,p} = 6.64 \cdot 11.02 = 73.17 \text{ N/mm}^2$$

Nondimensional plate slenderness, Equation 8.18:

$$\bar{\lambda}_p = \sqrt{\frac{355}{73.17}} = 2.20 > 0.673$$

Uniform compression: $\psi = 1$
Reduction factor for plate buckling, Equation 8.23:

$$\rho = \frac{2.2 - 0.055 \cdot (3+1)}{2.2^2} = 0.41$$

It is $\alpha = 0.476 < 1.0$ so that column-like behavior shall be considered.
Euler column stress, Equation 8.28:

$$\sigma_{cr,c} = \frac{\pi^2 \cdot 21{,}000 \cdot 1.6^2}{12 \cdot (1-0.3^2) \cdot 100^2} \cdot 10 = 48.5 \text{ N/mm}^2$$

Nondimensional column slenderness, Equation 8.34:

$$\bar{\lambda}_c = \sqrt{\frac{355}{48.5}} = 2.7$$

Unstiffened panel: European buckling curve a with $\alpha = 0.21$, Equation 8.37:

$$\Phi = 0.5 \cdot \left[1 + 0.21 \cdot (2.7 - 0.2) + 2.7^2\right] = 4.41$$

Reduction factor for column buckling, Equation 8.36:

$$\chi_c = \frac{1}{4.41 + \sqrt{4.41^2 - 2.7^2}} = 0.127$$

Equation 8.33:

$$\xi = \frac{73.17}{48.5} - 1 = 0.51$$

Final reduction factor, Equation 8.32:

$$\rho_c = (0.41 - 0.127) \cdot 0.51 \cdot (2 - 0.51) + 0.127 = 0.342$$

It may be seen that for this small aspect ratio, the reduction factor is reduced due to column-like behavior from 0.41 to 0.34.

EXAMPLE 8.6

The reduction factors of an internal panel with a = 3.0 m, b = 2.1 m, and t = 16 mm (Figure 8.23), subjected to nonuniform longitudinal compression with $\psi = 0$, shall be determined. The panel is stiffened by two flats of height h = 150 mm and thickness t = 20 mm that are positioned at 1/3 of the panel width. Steel grade S 355.

Subpanel 1

Width:

$$b = \frac{210}{3} = 70 \text{ cm}$$

Reference stress, Equation 8.10:

$$\sigma_e = 189,800 \cdot \left(\frac{16}{700}\right)^2 = 99.2 \text{ N/mm}^2$$

Stress ratio: $\psi = 140/210 = 0.67$

Buckling factor (Table 8.2):

$$k_\sigma = \frac{8.2}{1.05 + 0.67} = 4.77$$

Critical plate buckling stress, Equation 8.9:

$$\sigma_{cr,p} = 4.77 \cdot 99.2 = 473.2 \text{ N/mm}^2$$

Nondimensional plate slenderness, Equation 8.18:

$$\bar{\lambda}_p = \sqrt{\frac{355}{473.2}} = 0.87 > 0.673$$

Reduction factor for plate buckling, Equation 8.23:

$$\rho = \frac{0.87 - 0.055 \cdot (3 + 0.67)}{0.87^2} = 0.88$$

Aspect ratio $\alpha = \dfrac{300}{70} = 4.29 > 1$ so that column-like behavior is not relevant.

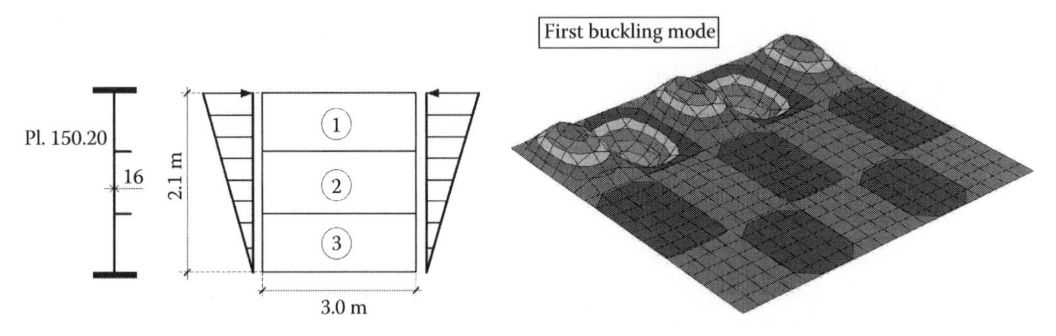

Figure 8.23 Stiffened panel of Example 8.6 and first buckling mode.

Subpanel 2

This subpanel has identical geometric properties with subpanel 1, the difference being in the stress ratio, which is $\psi = 70/140 = 0.5$. Repeating the calculations, the reduction factor is derived as $\rho = 0.93$.

Complete stiffened panel

Plate buckling

Reference stress: $\sigma_e = 11.02$ N/mm²

The critical buckling factor is determined numerically as equal to $k_\sigma = 52.6$.

Critical plate buckling stress: $\sigma_{cr,p} = 52.6 \cdot 11.02 = 579.65$ N/mm²

Nondimensional plate slenderness: $\bar{\lambda}_p = \sqrt{\dfrac{355}{579.65}} = 0.78 > 0.673$

Reduction factor for plate buckling: $\rho = \dfrac{0.78 - 0.055 \cdot (3 + 0)}{0.78^2} = 1.28 > 1.0 \rightarrow \rho = 1.0$

Column-like behavior

Gross area of the most compressed stiffener:

$$A_{st} = 15 \cdot 2 = 30\,\text{cm}^2$$

Second moment of area of the stiffener:

$$I_{st} = \frac{15^3 \cdot 2}{12} = 562.5\ \text{cm}^4$$

Calculation of the associated plate width of the most compressed stiffener

With the notation of Figure 8.22, it is for the upper subpanel 1 $\psi_1 = 0.67$ and for the lower subpanel $\psi_2 = 0.50$.

The participating plate widths are accordingly (Table 8.11),

$$b_{1,inf} = \frac{3 - 0.67}{5 - 0.67} \cdot 70 = 37.7\ \text{cm}$$

$$b_{2,sup} = \frac{2}{5 - 0.5} \cdot 70 = 31.1\,\text{cm}$$

The stiffener under consideration has a T section, composed of the flat and an associated plate width of $37.7 + 31.1 + 2 = 70.8$ cm. This T section has the following properties:

$$A_{sl,1} = 144.1\ \text{cm}^2, \quad I_{sl,1} = 2382\ \text{cm}^4, \quad i = 4.1\ \text{cm}$$

Euler stress of the stiffener, Equation 8.30:

$$\sigma_{cr,sl} = \frac{\pi^2 \cdot 21{,}000 \cdot 2382}{144.1 \cdot 300^2} = 38.03\ \text{kN/cm}^2 = 380.3\ \text{N/mm}^2$$

Critical column buckling stress of the panel, Equation 8.31:

$$\sigma_{cr,c} = \frac{210}{140} \cdot 380.3 = 570.4\ \text{N/mm}^2$$

The stiffener is not classified as class 4:

$$(b/t = 150/20 = 7.5 < 14 \cdot \varepsilon = 11.34) \rightarrow \beta_{A,c} = 1$$

Column slenderness, Equation 8.34:

$$\bar{\lambda}_c = \sqrt{\frac{355}{570.4}} = 0.79$$

Figure 8.22: $e_1 = 6.48$ cm, $e_2 = 1.82$ cm

$$e = \max(6.48, 1.82) = 6.48 \text{ cm}$$

Imperfection factor for open stiffener section:

$$\alpha_e = 0.49 + \frac{0.09}{4.1/6.48} = 0.63$$

Equation 8.37:

$$\Phi = 0.5 \cdot \left[1 + 0.63 \cdot (0.79 - 0.2) + 0.79^2\right] = 1.0$$

Reduction factor, Equation 8.36:

$$\chi_c = \frac{1}{1.0 + \sqrt{(1.0^2 - 0.79^2)}} = 0.62$$

Equation 8.33:

$$\xi = \frac{579.65}{570.4} - 1 = 0.016$$

Final reduction factor, Equation 8.32:

$$\rho_c = (1.0 - 0.62) \cdot 0.016 \cdot (2 - 0.016) + 0.62 = 0.63$$

Note: The critical plate buckling stress may be determined by the formulae given in Table 8.6 and the procedure described in Table 8.7 (two longitudinal stiffeners in the compression zone).

Case I
The following figures are obtained with reference to Table 8.7:

$$b_1 = 700 \text{ mm}, \quad b_2 = 700 \text{ mm} \rightarrow b = b_1 + b_2 = 1400 \text{ mm}$$

Stress ratios for the upper/lower subpanel:

$$\psi_1/\psi_2 = 0.67/0.5$$

Adjacent widths of the plate for the two subpanels according to Table 8.6 (without reversal of stresses): 37.7/31.1 cm

The stiffener has a T section consisting of the stiffener and 70.8 cm associated plate width. The properties of this T section are

$$A_{sl,1} = 144.1\ cm^2, \quad I_{sl,1} = 2382\ cm^4$$

The reference length according to Table 8.6 is

$$a_c = 4.33 \cdot \sqrt[4]{\frac{2382 \cdot 70^2 \cdot 70^2}{1.6^3 \cdot 140}} = 432.7\ cm > a = 300\ cm$$

The critical stiffener stress is

$$\sigma_{cr,sl} = \frac{\pi^2 \cdot 21{,}000 \cdot 2382}{144.1 \cdot 300^2} + \frac{21{,}000 \cdot 1.6^3 \cdot 300^2 \cdot 140}{4 \cdot \pi^2 \cdot (1 - 0.3^2) \cdot 144.1 \cdot 70^2 \cdot 70^2}$$

$$= 38.03 + 8.69 = 46.72\ kN/cm^2 = 467.2\ N/mm^2$$

where the first term is the *Euler stress* of the stiffener alone and the second the transverse contribution of the plate.

Width of compression zone $b_c = 140$ mm:

$$\sigma_{cr,p} = \frac{140 + 70}{140} \cdot 467.2 = 700.8\ N/mm^2;\ \text{see Table 8.6}$$

Case 2

Repeating this methodology for the second stiffener following the procedure of Table 8.7 with $b_c = 70$ mm gives

$$\sigma_{cr,p} = \frac{70 + 70}{70} \cdot 467.2 = 934.4\ N/mm^2;\ \text{see Table 8.6}$$

Case 3

The consideration of an equivalent lumped stiffener with $b_1 = 1011$ mm, $b_2 = 1089$ mm, $b = 2100$ mm (centre of force for the two stiffeners) gives the following cross-sectional properties:

$$A_{sl,lumped} = 2 \cdot 144.1 = 288.2\ cm^2,$$

$$I_{sl,lumped} = 2 \cdot 2382 = 4764\ cm^4$$

The reference length according to Table 8.6 is

$$a_c = 4.33 \cdot \sqrt[4]{\frac{4764 \cdot 101.1^2 \cdot 108.9^2}{1.6^3 \cdot 210}} = 697\ cm > a = 300\ cm$$

The critical stiffener stress is

$$\sigma_{cr,sl} = \frac{\pi^2 \cdot 21{,}000 \cdot 4764}{288.2 \cdot 300^2} + \frac{21{,}000 \cdot 1.6^3 \cdot 300^2 \cdot 210}{4 \cdot \pi^2 \cdot (1-0.3^2) \cdot 288.2 \cdot 101.1^2 \cdot 108.9^2}$$

$$= 38.03 + 1.30 = 39.33\ \text{kN/cm}^2 = 393.3\ \text{N/mm}^2$$

The critical panel stress is with $b_c = 1089$ mm:

$$\sigma_{cr,p} = \frac{108.9 + 101.1}{108.9} \cdot 393.3 = 758.3\ \text{N/mm}^2\text{; see Table 8.6}$$

The critical plate buckling stress is finally the lowest value, 700.8 N/mm². This is higher than the numerically determined.

8.4 DESIGN BY THE REDUCED STRESS METHOD

The *reduced stress method* is a simplified design method. It is based on the assumption that the resistance of the cross section is exhausted when its most unfavorable wall reaches its design strength as given in the following. The beneficial load shedding from overstressed panels is not taken into account, and this makes the method in most cases conservative. In reduced stress method, the cross sections may be classified as class 3 cross sections. The relevant design procedure is illustrated in Figure 8.24 and may be summarized as follows:

Step 1: Performance of static analysis, possibly including 2nd-order effects (e.g., in cable-stayed bridges) in order to determine internal forces and moments.

Step 2: Isolation of panels for each wall, considering simple support conditions at their joint edges.

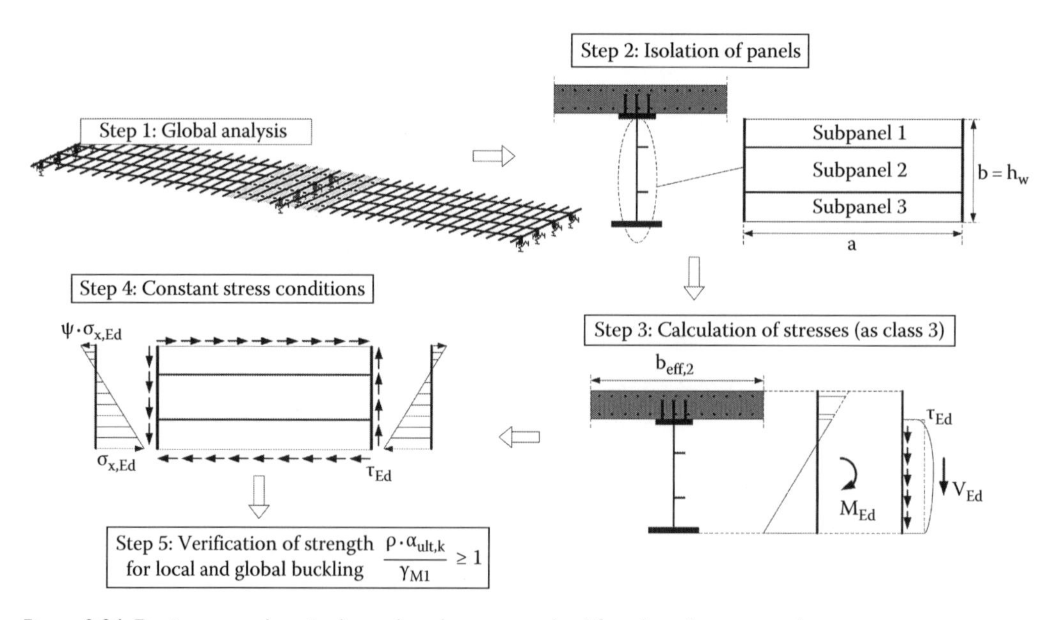

Figure 8.24 Design procedure in the reduced stress method for class 4 cross sections.

Step 3: Definition of design stresses $\sigma_{x,Ed}$, $\sigma_{z,Ed}$, and τ_{Ed} from internal forces and moments, based on gross cross-sectional properties.

Step 4: Consideration of constant stress conditions at panel edges. For stress gradients along the panel, use the stresses at a distance s = min {0.4·a or 0.5·b} of the most stressed panel end. The distance s is chosen by the code because failure is mainly dependent on the stresses within the middle portion of the buckling waveform and not at its boundaries.

Step 5: Verification of each panel and subpanel separately, according to the design conditions given in the following.

The panel may be verified by means of the following condition:

$$\frac{\rho \cdot \alpha_{ult,k}}{\gamma_{M1}} \geq 1 \tag{8.38}$$

where $\alpha_{ult,k}$ is the minimum multiplier of the design stresses that leads to yielding.

Using the *von Mises* criterion, this multiplier may be obtained from

$$\frac{1}{\alpha_{ult,k}^2} = \left(\frac{\sigma_{x,Ed}}{f_y}\right)^2 + \left(\frac{\sigma_{z,Ed}}{f_y}\right)^2 - \frac{\sigma_{x,Ed} \cdot \sigma_{z,Ed}}{f_y^2} + 3 \cdot \left(\frac{\tau_{Ed}}{f_y}\right)^2 \tag{8.39}$$

The reduction factor is equal to $\rho = \min(\rho_x, \rho_z, \chi_w)$ and is determined as a function of $\bar{\lambda}_p$ from Equation 8.40:

$$\bar{\lambda}_p = \sqrt{\frac{\alpha_{ult,k}}{\alpha_{cr}}} \tag{8.40}$$

where α_{cr} is the minimum multiplier of design stresses to reach the critical first buckling mode of the plate under the combined stresses. It may be approximately determined from Equation 8.17 if only the multipliers for the individual components are known.

Substituting Equation 8.39 with Equation 8.38 offers the following verification:

$$\left(\frac{\sigma_{x,Ed}}{f_y/\gamma_{M1}}\right)^2 + \left(\frac{\sigma_{z,Ed}}{f_y/\gamma_{M1}}\right)^2 - \frac{\sigma_{x,Ed} \cdot \sigma_{z,Ed}}{f_y^2/\gamma_{M1}^2} + 3 \cdot \left(\frac{\tau_{Ed}}{f_y/\gamma_{M1}}\right)^2 \leq \rho^2 \tag{8.41}$$

A less conservative approach is given by the code in Equation 8.42:

$$\left(\frac{\sigma_{x,Ed}}{\rho_x \cdot f_y/\gamma_{M1}}\right)^2 + \left(\frac{\sigma_{z,Ed}}{\rho_z \cdot f_y/\gamma_{M1}}\right)^2 - \frac{\sigma_{x,Ed} \cdot \sigma_{z,Ed}}{\rho_x \cdot \rho_z \cdot f_y^2/\gamma_{M1}^2} + 3 \cdot \left(\frac{\tau_{Ed}}{\chi_w \cdot f_y/\gamma_{M1}}\right)^2 \leq 1.0 \tag{8.42}$$

It is important to note that in the case of stiffened panels, the final reduction factors ρ_x and ρ_z should be obtained as an interpolation between plate and buckling column behaviors (see Equations 8.32 and 8.33).

EXAMPLE 8.7

The panel of Example 8.5 is to be verified if the acting stress is $\sigma_{x,Ed} = 110$ N/mm^2.
Equation 8.39:

$$\frac{1}{\alpha_{ult,k}^2} = \left(\frac{110}{355}\right)^2 \rightarrow \alpha_{ult,k} = 3.23$$

Plate buckling behavior
Load factor:

$$\alpha_{cr,p} = \frac{73.17}{110} = 0.67$$

Equation 8.40:

$$\bar{\lambda}_p = \sqrt{\frac{3.23}{0.67}} = 2.2 > 0.673$$

Reduction factor for plate buckling, Equation 8.23:

$$\rho = \frac{2.2 - 0.055 \cdot (3+1)}{2.2^2} = 0.41$$

Column buckling behavior
Load factor:

$$\alpha_{cr,c} = \frac{48.5}{110} = 0.44$$

Equation 8.40:

$$\bar{\lambda}_c = \sqrt{\frac{3.23}{0.44}} = 2.7$$

Unstiffened Panel: European buckling curve a with $\alpha = 0.21$
Equation 8.37:

$$\Phi = 0.5 \cdot \left[1 + 0.21 \cdot (2.7 - 0.2) + 2.7^2\right] = 4.41$$

Reduction factor for column buckling, Equation 8.36:

$$\chi_c = \frac{1}{4.41 + \sqrt{4.41^2 - 2.7^2}} = 0.127$$

Equation 8.33:

$$\xi = \frac{73.17}{48.5} - 1 = 0.51$$

Final reduction factor, Equation 8.32:

$$\rho_c = (0.41 - 0.127) \cdot 0.51 \cdot (2 - 0.51) + 0.127 = 0.342$$

Equation 8.42:

$$\frac{110}{0.342 \cdot 355 / 1.1} = 1 \text{ (sufficient)}$$

EXAMPLE 8.8

The panel of Example 8.6 (Figure 8.23) is to be verified if the acting stress at the edge is $\sigma_{x,Ed} = 200$ N/mm^2.

Upper subpanel

Equation 8.39:

$$\frac{1}{\alpha_{ult,k}^2} = \left(\frac{200}{355}\right)^2 \rightarrow \alpha_{ult,k} = 1.78$$

Plate buckling behavior

Load factor:

$$\alpha_{cr,p} = \frac{473.2}{200} = 2.36$$

Equation 8.40:

$$\bar{\lambda}_p = \sqrt{\frac{1.78}{2.36}} = 0.87 > 0.673$$

Reduction factor for plate buckling, Equation 8.23:

$$\rho = \frac{0.87 - 0.055 \cdot (3 + 0.67)}{0.87^2} = 0.88$$

Column buckling behavior

Aspect ratio: $\alpha = \frac{300}{70} = 4.29 \rightarrow$ Column buckling behavior is not relevant.

Equation 8.42:

$$\frac{200}{0.88 \cdot 355/1.1} = 0.7 < 1 \text{ (sufficient)}$$

Complete stiffened panel

Plate buckling behavior

Load factor:

$$\alpha_{cr,p} = \frac{579.65}{200} = 2.9$$

Equation 8.40:

$$\bar{\lambda}_p = \sqrt{\frac{1.78}{2.9}} = 0.78 > 0.673$$

Reduction factor for plate buckling:

$$\rho = \frac{0.78 - 0.055 \cdot (3+0)}{0.78^2} = 1.28 > 1.0 \rightarrow \rho = 1.0$$

Column buckling behavior

$$\alpha_{cr,c} = \frac{570.4}{200} = 2.85$$

Equation 8.40:

$$\bar{\lambda}_c = \sqrt{\frac{1.78}{2.85}} = 0.79$$

Figure 8.22: $e_1 = 6.48$ cm, $e_2 = 1.82$ cm

$$e = \max(6.48, 1.82) = 6.48 \text{ cm}$$

Imperfection factor for open stiffener section:

$$\alpha_e = 0.49 + \frac{0.09}{4.1/6.48} = 0.63$$

Equation 8.37:

$$\Phi = 0.5 \cdot \left[1 + 0.63 \cdot (0.79 - 0.2) + 0.79^2\right] = 1.0$$

Reduction factor, Equation 8.36:

$$\chi_c = \frac{1}{1.0 + \sqrt{(1.0^2 - 0.79^2)}} = 0.62$$

Equation 8.33:

$$\xi = \frac{579.65}{570.4} - 1 = 0.016$$

Final reduction factor, Equation 8.32:

$$\rho_c = (1.0 - 0.62) \cdot 0.016 \cdot (2 - 0.016) + 0.62 = 0.63$$

Equation 8.42:

$$\frac{200}{0.63 \cdot 355 / 1.1} = 0.98 \leq 1.0 \text{ (sufficient)}$$

In Example 8.6, it was seen that the critical buckling stress of the upper subpanel is lower than the corresponding stress of the full stiffened panel (473.2 vs. 579.65 N/mm²). Figure 8.23 confirms it since the first buckling mode involves buckling of the upper subpanel only. However, this example shows that the resistance of the complete panel is lower than the resistance of the subpanel. This is due to the fact that the stiffened panel exhibits column-like behavior ($\xi \approx 0$), while the subpanel does not. It may be concluded that the results of linear plate buckling theory cannot be directly transferred to strength. They merely serve as reference values for the resistance determination.

It should also be noted that the reduction factors ρ and ρ_c have the same values with those in Example 8.6. This is due to the fact that no interaction with shear or transverse stresses exists.

EXAMPLE 8.9

A cross section of a box girder bridge shown in Figure 8.25 and cross frames at a distance of 4.5 m shall be verified in the support region. The relevant design internal forces and moments at a distance $0.4 \cdot a = 0.4 \cdot 4.5 = 1.8$ m < $0.5 \cdot b = 0.5 \cdot 4.15 = 2.07$ m from the support are as follows: bending moment $M_{Ed} = -39{,}000$ k-Nm, shear force $V_{Ed} = 1600$ kN, and torque $M_{T,Ed} = 7550$ kN-m. The bottom flange is stiffened by four longitudinal trapezoidal stiffeners, the web by two stiffeners L100 × 50 × 8. The stiffeners are equally distanced. The flange stiffeners participate in the compression resistance, the web stiffeners do not. The latter are placed merely to increase the buckling strength of the web. Steel grade S 355.

Design stresses

The bending moment produces tension in the concrete slab so that the stresses are determined for the structural steel cross section. For simplicity reasons in this example, the contribution of the slab reinforcement was neglected. The stiffeners of the bottom flange participate in the

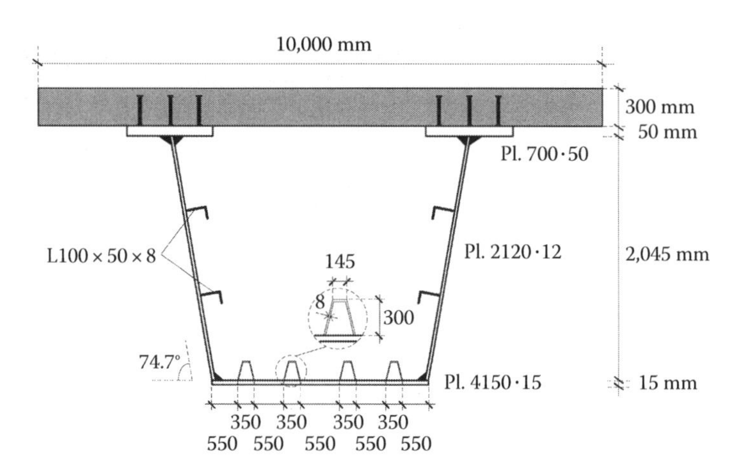

Figure 8.25 Cross section of box girder of Example 8.9 and dimensions of the trapezoidal stiffeners of the bottom flange.

resistance to direct stresses. In order to take into account this contribution, their area is smeared over the entire bottom flange width. This is done by enhancing the thickness t of the bottom flange according to the relation $t_{smeared} = t + \sum_1^4 A_{st}/b$, where A_{st} is the area of one stiffener and b is the width of the bottom flange. The equivalent smeared thickness is 20.1 mm. The stiffeners of the web are placed only to enhance its buckling resistance so that they do not enhance its thickness (see also Section 8.11.5).

The properties of the box girder cross section are as follows:

Second moment of area: $I_y = 18.35 \cdot 10^6\ cm^4$

Distance of neutral axis from bottom flange: z = 96.3 cm

Area of the cross-section trapezoid between mid-lines of plates: $A_0 = 105{,}757\ cm^2$

Stresses in the flange

Direct stresses:

$$\sigma_{Ed} = \frac{3.9 \cdot 10^6}{18.35 \cdot 10^6} \cdot 96.3 = 20.47\ kN/cm^2\ \text{(compression positive)}$$

Shear stresses:

$$\tau_{Ed} = \frac{M_{T,Ed}}{2 \cdot A_0 \cdot t} = \frac{7.55 \cdot 10^5}{2 \cdot 105757 \cdot 1.5} = 2.38\ kN/cm^2$$

where t is 1.5 cm, that is, the enhancement in the flange thickness does not refer to the shear stresses due to torsion.

Stresses in the web

Maximal compression stress:

$$\sigma_{Ed} = 20.47\ kN/cm^2\ (\text{as in the flange})$$

Shear stresses from torsion:

$$\tau_{Ed} = \frac{M_{T,Ed}}{2 \cdot A_0 \cdot t} = \frac{7.55 \cdot 10^5}{2 \cdot 105757 \cdot 1.2} = 2.97\ kN/cm^2$$

Shear stresses from shear force:

$$\tau_{Ed} = \frac{V_{Ed}}{2 \cdot A_w \cdot \sin\alpha} = \frac{1600}{2 \cdot 212 \cdot 1.2 \cdot \sin 74.7^\circ} = 3.26\ kN/cm^2$$

where 74.7° is the angle of inclination of the web.

The resultant shear stresses in the web are $\tau_{Ed} = 2.97 + 3.26 = 6.23\ kN/cm^2$.

The previously mentioned shear stresses refer obviously to one web, since the shear flow due to torsion adds in one web and subtracts in the other web to the shear flow due to shear force.

Verification of the bottom flange

Subpanel 1 between stiffeners

Direct stresses

Width b = 55 cm, t = 1.5 cm

Reference stress, Equation 8.10:

$$\sigma_e = 14.12 \text{ kN/cm}^2$$

Stress ratio: $\psi = 1$

Buckling factor (Table 8.2): $k_\sigma = 4$

Critical plate buckling stress, Equation 8.9:

$$\sigma_{cr,p} = 4 \cdot 14.12 = 56.48 \text{ kN/cm}^2$$

Load factor:

$$\alpha_{cr,x} = \frac{56.48}{20.47} = 2.76$$

The subpanel is long ($\alpha > 1$), so that column-like behavior is not considered.

Shear stresses

$$\alpha = \frac{450}{55} = 8.18 > 1 \rightarrow k_\tau = 5.34 + \frac{4}{8.18^2} = 5.4 \text{ (see Table 8.5)}$$

Equation 8.11:

$$\tau_{cr} = 5.4 \cdot 14.12 = 76.25 \text{ kN/cm}^2$$

$$\alpha_{cr,\tau} = \frac{76.25}{2.38} = 32.04$$

Reduction factors

Equation 8.17:

$$\frac{1}{\alpha_{cr}} = \frac{1+1}{4 \cdot 2.76} + \sqrt{\left(\frac{1+1}{4 \cdot 2.76}\right)^2 + \frac{1}{32.04^2}} \rightarrow \alpha_{cr} = 2.02$$

Equation 8.39:

$$\frac{1}{\alpha_{ult,k}^2} = \left(\frac{20.47}{35.5}\right)^2 + 3 \cdot \left(\frac{2.38}{35.5}\right)^2 \rightarrow \alpha_{ult,k} = 1.7$$

Equation 8.40:

$$\bar{\lambda}_p = \sqrt{\frac{1.7}{2.02}} = 0.92 > 0.673$$

Equation 8.23:

$$\rho = \frac{0.92 - 0.055 \cdot (3+1)}{0.92^2} = 0.83$$

$$\text{Table 8.10}: \bar{\lambda}_w = \bar{\lambda}_p = 0.92 \rightarrow \chi_w = \frac{0.83}{0.92} = 0.9$$

Verification

Equation 8.42:

$$\left(\frac{20.47}{0.83 \cdot 35.5/1.1}\right)^2 + 3 \cdot \left(\frac{2.38}{0.9 \cdot 35.5/1.1}\right)^2 = 0.58 \leq 1.0 \text{ (sufficient)}$$

Subpanel 2 between webs of stiffener

Width b = 35 cm, t = 1.5 cm

It is $b/t = 35/1.5 = 23.3 < 42 \cdot \varepsilon = 42 \cdot 0.81 = 34 \rightarrow$ no class 4 wall with reduction for plate buckling.

Subpanel 3 inclined web of stiffener

Width b = 31.9 cm, t = 0.8 cm

The repetition of the calculations as for subpanel 1 gives $\rho = 0.87$, $\chi_w = 0.97$.

Equation 8.42:

$$\left(\frac{20.47}{0.87 \cdot 35.5/1.1}\right)^2 + 3 \cdot \left(\frac{2.38}{0.97 \cdot 35.5/1.1}\right)^2 = 0.55 \leq 1.0 \text{ (sufficient)}$$

Subpanel 4 top part of stiffener

Width b = 14.5 cm, t = 0.8 cm

It is $b/t = 14.5/0.8 = 18.1 < 42 \cdot \varepsilon \cdot = 42 \cdot 0.81 = 34 \rightarrow$ no class 4 wall with reduction for plate buckling.

Complete stiffened panel

Plate buckling behavior

Geometric properties:

a = 450 cm, b = 415 cm, t = 1.5 cm

Reference stress, Equation 8.10:

$$\sigma_e = 0.248 \text{ kN/cm}^2$$

Stress ratio: $\psi = 1$

Table 8.4:

$$\text{Gross area: } A_p = 415 \cdot 1.5 = 622.5 \text{ cm}^2$$

Second moment of area:

$$I_p = \frac{415 \cdot 1.5^3}{10.92} = 128.3 \text{ cm}^4$$

Properties of the longitudinal stiffeners:

$A_{sl} = 61.08$ cm², $I_{sl} = 5629.3$ cm⁴, and $z_{sl} = 12.14$ cm (distance of center of gravity from top)

Area of four stiffeners:

$$\sum A_{sl} = 4 \cdot 61.08 = 244.32 \text{ cm}^2$$

$$\delta = \frac{244.32}{622.5} = 0.392$$

Second moment of area of the plate and the four stiffeners:

$$\sum I_{sl} = 83{,}400 \text{ cm}^4$$

$$\gamma = \frac{83{,}400}{128.3} = 650.23$$

It is

$$\alpha = \frac{450}{415} = 1.08 < \sqrt[4]{\gamma} = \sqrt[4]{650.23} = 5.05$$

Buckling factor:

$$k_{\sigma,p} = \frac{2 \cdot \left[\left(1 + 1.08^2\right)^2 + 650.23 - 1\right]}{1.08^2 \cdot (1+1) \cdot (1 + 0.392)} = 402.75$$

Critical stress, Equation 8.9:

$$\sigma_{cr,p} = 402.75 \cdot 0.248 = 99.88 \text{ kN/cm}^2$$

Load factor:

$$\alpha_{cr,x} = \frac{99.88}{20.47} = 4.88$$

For the shear stresses:

Associated plate width at each side of the stiffeners webs (Table 8.5):

$15 \cdot \varepsilon \cdot t = 15 \cdot 0.81 \cdot 1.5 = 18.2 \text{ cm} > 35/2 = 17.5 \text{ cm}$

This means that the total width of the associated plate is $35 + 2 \cdot 18.2 = 71.4$ cm.

The second moment of area of the stiffener and the associated plate is

$$I_{sl,1} = 18{,}640.3 \text{ cm}^4$$

And for the four stiffeners:

$$\sum I_{sl,1} = 4 \cdot 18640.3 = 74561.2 \text{ cm}^4$$

$$k_{\tau,st} = \frac{9}{1.08^2} \cdot \sqrt[4]{\left(\frac{74561.2}{1.5^3 \cdot 415}\right)^3} = 152.1 \geq \frac{2.1}{1.5} \cdot \sqrt[3]{\frac{74561.2}{415}} = 7.9$$

The shear buckling factor is determined from Table 8.5, for $\alpha = 1.08 > 1$:

$$k_\tau = 5.34 + \frac{4.0}{1.08^2} + 152.1 = 160.87$$

Critical shear buckling stress, Equation 8.11:

$$\tau_{cr} = 160.87 \cdot 0.248 = 39.9 \text{ kN/cm}^2$$

$$\alpha_{cr,\tau} = \frac{39.9}{2.38} = 16.4$$

Equation 8.17:

$$\frac{1}{\alpha_{cr}} = \frac{1+1}{4 \cdot 4.88} + \sqrt{\left(\frac{1+1}{4 \cdot 4.88}\right)^2 + \frac{1}{16.4^2}} \rightarrow \alpha_{cr} = 4.51$$

Equations 8.23 and 8.40:

$$\bar{\lambda}_p = \sqrt{\frac{1.7}{4.51}} = 0.61 \leq 0.673 \rightarrow \rho = 1$$

Column buckling behavior

Associated plate width of the stiffeners

For all subpanels, it is $\psi_1 = 1.0$ and $\psi_2 = 1.0$.

The participating plate widths adjacent to each stiffener web are accordingly (Table 8.11),

$$b_{1,inf} = \frac{3-1}{5-1} \cdot 55 = 0.5 \cdot 55 = 27.5 \text{ cm}$$

$$b_{2,sup} = \frac{2}{5-1} \cdot 35 = 0.5 \cdot 35 = 17.5 \text{ cm}$$

Each trapezoidal stiffener has therefore an associated plate width of $2 \cdot (27.5 + 17.5) = 90$ cm.

This section has the following properties:

$$A_{sl,1} = 196 \text{ cm}^2, \quad I_{sl,1} = 20{,}215 \text{ cm}^4, \quad i = 10.15 \text{ cm}$$

Euler stress of the stiffener, Equation 8.30:

$$\sigma_{cr,sl} = \frac{\pi^2 \cdot 21{,}000 \cdot 20{,}215}{196 \cdot 450^2} = 105.5 \text{ kN/cm}^2$$

Critical column buckling stress of the panel, Equation 8.31:

$$\sigma_{cr,c} = 105.5 \text{ kN/cm}^2$$

Load factor:

$$\alpha_{cr,c} = \frac{105.5}{20.47} = 5.15$$

Equation 8.17:

$$\frac{1}{\alpha_{cr}} = \frac{1+1}{4 \cdot 5.15} + \sqrt{\left(\frac{1+1}{4 \cdot 5.15}\right)^2 + \frac{1}{16.4^2}} \rightarrow \alpha_{cr} = 5.2$$

It may be observed that $\sigma_{cr,c} > \sigma_{cr,p}$. This result is physical meaningless since the critical stress of the plate without longitudinal supports may not be larger than the one with such supports is corrected later in the derivation of the factor ξ.

Note: In order to determine the effective cross section of the stiffener, only the direct stresses should be taken into account for the calculation of the reduction factors ρ [EN 1993-1-5/4.3.(3)]. Therefore, the previous reduction factors calculated for subpanels 1–4 should not be used for the determinations of $A_{sl,1,eff}$ and $I_{sl,1,eff}$.

Subpanel 1 between stiffeners

Width b = 55 cm, t = 1.5 cm

Reference stress, Equation 8.10:

$$\sigma_e = 14.12 \text{ kN/cm}^2$$

Stress ratio: $\psi = 1$

Buckling factor (Table 8.2): $k_\sigma = 4$

Critical plate buckling stress, Equation 8.9:

$$\sigma_{cr,p} = 4 \cdot 14.12 = 56.48 \text{ kN/cm}^2$$

Nondimensional plate slenderness, Equation 8.18:

$$\bar{\lambda}_p = \sqrt{\frac{35.5}{56.48}} = 0.793 > 0.673$$

Reduction factor for plate buckling, Equation 8.23:

$$\rho = \frac{0.793 - 0.055 \cdot (3+1)}{0.793^2} = 0.911$$

Subpanel 3 inclined web of stiffener

Width b = 31.9 cm, t = 0.8 cm

The repetition of the calculations as for subpanel 1 gives $\rho = 0.864$.

The reduction factors for the inclined wall of the stiffener and the plate between stiffeners are equal to $\rho = 0.864$ and $\rho = 0.911$, respectively. Therefore, the relevant widths must be reduced due to plate buckling to 0.864·31.9=27.6 cm and 0.911·55=50.1 cm, respectively. The effective area of the stiffener and the associated plate is accordingly

$$A_{sl,I,eff} = (35 + 50.1) \cdot 1.5 + 2 \cdot 27.6 \cdot 0.8 + 14.5 \cdot 0.8 = 183.41 \text{ cm}^2$$

$$I_{sl,I,eff} = 19450.3 \text{ cm}^2 \rightarrow i = \sqrt{\frac{19450.3}{183.41}} = 10.3 \text{ cm}$$

Figure 8.22: $e_1 = 12.8$ cm, $e_2 = 5.80$ cm
$e = \max(12.8, 5.80) = 12.8$ cm

Imperfection factor for closed stiffener section:

$$\alpha_e = 0.34 + \frac{0.09}{10.3/12.8} = 0.45$$

$$\beta_{A,c} = \frac{183.41}{196} = 0.94 \text{ (class 4 stiffeners)}$$

Equation 8.35:

$$\bar{\lambda}_c = \sqrt{\frac{0.94 \cdot 35.5}{105.5}} = 0.56$$

Equation 8.37:

$$\Phi = 0.5 \cdot \left[1 + 0.45 \cdot (0.56 - 0.2) + 0.56^2\right] = 0.74$$

Reduction factor, Equation 8.36:

$$\chi_c = \frac{1}{0.74 + \sqrt{0.74^2 - 0.56^2}} = 0.82$$

Equation 8.33:

$$\xi = \frac{99.88}{105.5} - 1 = -0.05 < 0 \rightarrow \xi = 0$$

Final reduction factor, Equation 8.32:

$$\rho_c = \chi_c = 0.82$$

The nondimensionless slenderness for shear is taken as $\bar{\lambda}_w = \bar{\lambda}_p$ determined previously for the plate buckling interaction of direct and shear stresses without consideration of the column-like buckling behavior due to the fact that the latter affects only direct but not shear stresses. This means that the slenderness used for shear buckling corresponds to the plate supported along its longitudinal edges. This issue is not clarified in the Code, that is, whether this assumption should be made or conservatively $\bar{\lambda}_w = \bar{\lambda}_c$ is to be adopted.

Accordingly it is, $\bar{\lambda}_w = 0.61 < \dfrac{0.83}{1.2} = 0.69$

From Table 8.10: $\chi_w = 1.2$

$$\text{Equation 8.42:}\quad \left(\frac{20.47}{0.82 \cdot 35.5/1.1}\right)^2 + 3 \cdot \left(\frac{2.38}{1.2 \cdot 35.5/1.1}\right)^2 = 0.61 \le 1.0 \text{ (sufficient)}$$

It is noted that a further strength reduction due to shear lag as presented in Section 7.3 might be necessary. In this example, no such reduction of the effective[s] width of the bottom flange was considered to be present.

Verification of the web

The clear width of the web between flanges is $h_w = 212.0$ cm. The width of the tension zone is 112.2 cm and of the compression zone 99.8 cm. Therefore, the lower stiffener is in compression and the upper stiffener in tension. The distances of the two stiffeners from the neutral axis are 41.53 cm and 29.13 cm for the upper and the lower stiffeners, respectively.

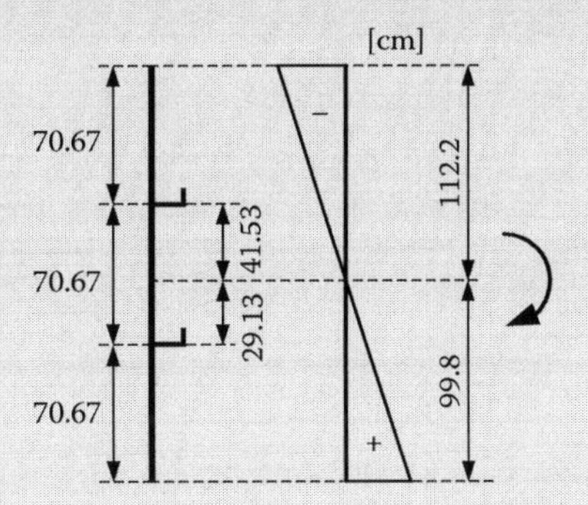

Lower subpanel between lower stiffener and bottom flange (most compressed)

Plate buckling behavior

Width b = 212/3 = 70.67 cm, t = 1.2 cm

Stress ratio:

$$\psi = \frac{29.13}{99.8} = 0.29$$

Reference stress, Equation 8.10:

$$\sigma_e = 5.47 \text{ kN/cm}^2$$

Buckling factor (Table 8.2):

$$k_\sigma = \frac{8.2}{1.05 + 0.29} = 6.11$$

Critical plate buckling stress, Equation 8.9:

$$\sigma_{cr,p} = 6.11 \cdot 5.47 = 33.42 \text{ kN/cm}^2$$

Load factor:

$$\alpha_{cr,x} = \frac{33.42}{20.47} = 1.63$$

The subpanel is long ($\alpha > 1$) so that column-like behavior is not considered.

For the shear stresses:

Aspect ratio:

$$\alpha = \frac{450}{70.67} = 6.36 > 1$$

Buckling factor (Table 8.5):

$$k_\tau = 5.34 + \frac{4.0}{6.36^2} = 5.44$$

Critical shear buckling stress, Equation 8.11:

$$\tau_{cr} = 5.44 \cdot 5.47 = 29.8 \text{ kN/cm}^2$$

Load factor:

$$\alpha_{cr,\tau} = \frac{29.8}{6.23} = 4.78$$

Reduction factors:

Equation 8.17:

$$\frac{1}{\alpha_{cr}} = \frac{1+0.29}{4 \cdot 1.63} + \sqrt{\left(\frac{1+0.29}{4 \cdot 1.63}\right)^2 + \frac{1}{4.78^2}} \rightarrow \alpha_{cr} = 2.05$$

Equation 8.39:

$$\frac{1}{\alpha_{ult,k}^2} = \left(\frac{20.47}{35.5}\right)^2 + 3 \cdot \left(\frac{6.23}{35.5}\right)^2 \rightarrow \alpha_{ult,k} = 1.53$$

Equation 8.40:

$$\bar{\lambda}_p = \sqrt{\frac{2.05}{1.53}} = 1.16 > 0.673$$

Equation 8.23:

$$\rho = \frac{1.16 - 0.055 \cdot (3 + 0.29)}{1.16^2} = 0.73$$

Table 8.10: $\bar{\lambda}_w = \bar{\lambda}_p = 1.16 > 1.08 \rightarrow \chi_w = \dfrac{0.83}{1.16} = 0.72$ (nonrigid end posts)

Verification

Equation 8.42:

$$\left(\frac{20.47}{0.73\cdot 35.5/1.1}\right)^2+3\cdot\left(\frac{6.23}{0.72\cdot 35.5/1.1}\right)^2=0.97\leq 1.0\ \text{(sufficient)}$$

Complete stiffened panel

Plate buckling behavior

Geometric properties:

$a=450$ cm, $b=h_w=212.0$ cm, $t=1.2$ cm

The critical plate buckling stress will be determined according to Table 8.6 since only one stiffener is in the compression zone:

Distances $b_1=70.67$ cm, $b_2=212.0-70.67=141.33$ cm, $b_c=29.13$ cm

Stress ratio for the lower subpanel: $\psi_1=0.29$, determined before.

Associated plate widths for the stiffener (Table 8.6):

Lower panel:

$$b_{1,\text{inf}}=\frac{3-0.29}{5-0.29}\cdot 70.67=40.66\ \text{cm}$$

Intermediate panel:

$$b_{2,\text{inf}}=0.4\cdot 29.13=11.65\,\text{cm}\ \text{(due to reversal of stresses)}$$

The associated plate width is therefore $40.66+11.65=52.31$ cm.

Properties of the longitudinal stiffeners $L100\times 50\times 8$:

$A_{st}=11.5\ \text{cm}^2$, $I_{st}=116\ \text{cm}^4$, and $z_s=3.59$ cm (distance of the center of gravity from the top)

Properties of the longitudinal stiffener + associated plate width:

$A_{sl,1}=74\ \text{cm}^2$, $I_{sl,1}=590\ \text{cm}^4$, $i=2.8$ cm

$$\alpha_c=4.33\cdot\sqrt[4]{\frac{590\cdot 70.67^2\cdot 141.33^2}{1.2^3\cdot 212}}=487.5\ \text{cm}>a=450\ \text{cm}$$

Critical stress of the stiffener (Table 8.6):

$$\sigma_{cr,sl}=\frac{\pi^2\cdot 21{,}000\cdot 590}{74\cdot 450^2}+\frac{21{,}000\cdot 1.2^3\cdot 212\cdot 450^2}{4\cdot\pi^2\cdot(1-0.3^2)\cdot 74\cdot 70.67^2\cdot 141.33^2}=8.2+5.9=14.1\,\text{kN/cm}^2$$

The first term of the earlier relation is the *Euler* stress of the stiffener, while the second is the contribution of the plate in transverse direction.

Critical plate buckling stress (Table 8.6):

$$\sigma_{cr,p}=14.1\cdot\frac{29.13+70.67}{29.13}=48.3\ \text{kN/cm}^2$$

Load factor:

$$\alpha_{cr,x}=\frac{48.3}{20.47}=2.36$$

For the shear stresses:

Aspect ratio:

$$\alpha = \frac{450}{212} = 2.12$$

Reference stress, Equation 8.10:

$$\sigma_e = 0.608\ \text{kN/cm}^2$$

Associated plate width at each side of the stiffeners webs (Table 8.5):

$$15 \cdot \varepsilon \cdot t = 15 \cdot 0.81 \cdot 1.2 = 14.6\ \text{cm}$$

This means that the total width of the associated plate is $2 \cdot 14.6 + 0.8 = 30$ cm.

The second moment of area of the stiffener and the associated plate is $I_{sl,1} = 540\ \text{cm}^4$

Table 8.5: $\alpha < 3 \rightarrow k_{\tau,st} = 4.1 + \dfrac{6.3 + 0.18 \cdot \left(\dfrac{540}{1.2^3 \cdot 212}\right)}{2.12^2} + 2.2 \cdot \sqrt[3]{\dfrac{540}{1.2^3 \cdot 212}} = 8.1$

The shear buckling factor is determined from Table 8.5:

$$k_\tau = 5.34 + \frac{4.0}{2.12^2} + 8.10 = 14.3$$

Critical shear buckling stress, Equation 8.11:

$$\tau_{cr} = 14.3 \cdot 0.608 = 8.69\ \text{kN/cm}^2$$

Load factor:

$$\alpha_{cr,\tau} = \frac{8.69}{6.23} = 1.39$$

Stress ratio for the web panel:

$$\psi = -\frac{112.2}{99.8} = -1.12$$

Equation 8.17:

$$\frac{1}{\alpha_{cr}} = \frac{1-1.12}{4 \cdot 2.36} + \sqrt{\left(\frac{1-1.12}{4 \cdot 2.36}\right)^2 + \frac{1}{1.39^2}} \rightarrow \alpha_{cr} = 1.41$$

Equation 8.39:

$$\frac{1}{\alpha_{ult,k}^2} = \left(\frac{20.47}{35.5}\right)^2 + 3 \cdot \left(\frac{6.23}{35.5}\right)^2 \rightarrow \alpha_{ult,k} = 1.53$$

Equation 8.40:

$$\bar{\lambda}_p = \sqrt{\frac{1.53}{1.41}} = 1.04 > 0.673$$

Equation 8.23:

$$\rho = \frac{1.04 - 0.055 \cdot (3 - 1.12)}{1.04^2} = 0.87$$

Column-like behavior

The *Euler stress* of the stiffener and associated plate widths, Equation 8.30 (Table 8.11 and Figure 8.22):

$$\sigma_{cr,sl} = \frac{\pi^2 \cdot 21{,}000 \cdot 590}{74 \cdot 450^2} = 8.2\ \text{kN/cm}^2$$

Critical column buckling stress of the panel, Equation 8.31; $b_c = 99.8$ cm, $b_{sl,1} = 29.13$ cm:

$$\sigma_{cr,c} = 8.2 \cdot \frac{99.8}{29.13} = 28.1\ \text{kN/cm}^2$$

Equation 8.35:

$$\bar{\lambda}_c = \sqrt{\frac{1 \cdot 35.5}{28.1}} = 1.12\ (\beta_{A,c} = 1,\ \text{stiffeners are not class 4})$$

Figure 8.22: $e_1 = 5.94$ cm, $e_2 = 1.07$ cm

$e = \max(5.94;\ 1.07) = 5.94$ cm

Imperfection factor for open stiffener section:

$$\alpha_e = 0.49 + \frac{0.09}{2.8/5.94} = 0.68$$

Equation 8.37:

$$\Phi = 0.5 \cdot \left[1 + 0.68 \cdot (1.12 - 0.2) + 1.12^2\right] = 1.44$$

Reduction factor, Equation 8.36:

$$\chi_c = \frac{1}{1.44 + \sqrt{1.44^2 - 1.12^2}} = 0.43$$

Equation 8.33:

$$\xi = \frac{48.3}{28.1} - 1 = 0.72$$

Final reduction factor, Equation 8.32:

$$\rho_c = (0.87 - 0.43) \cdot 0.72 \cdot (2 - 0.72) + 0.43 = 0.84$$

The slenderness to be used for shear is $\bar{\lambda}_w = \bar{\lambda}_p = 1.04$ that was determined previously for the plate buckling interaction of direct and shear stresses.

Table 8.10: $\chi_w = \frac{0.83}{1.04} = 0.80$

Verification

Equation 8.42:

$$\left(\frac{20.47}{0.84 \cdot 35.5/1.1}\right)^2 + 3 \cdot \left(\frac{6.23}{0.80 \cdot 35.5/1.1}\right)^2 = 0.74 \leq 1 \text{ (sufficient)}$$

8.5 EFFECTIVE WIDTH METHOD

8.5.1 General

The reduced stress method is conservative since it limits the stresses in all panels separately. Therefore, the effective width method allows shedding of direct stresses between panels and subpanels, which results in an enhanced resistance of the complete cross section. The effective width method is used for panels subjected to direct stresses that result in from global bending moments and axial forces. It is combined with the relevant verifications for shear forces as presented in the following paragraphs to check combined effects.

The effective width method is used under the following restrictions:

- The panels should be rectangular.
- The flanges should be parallel (to within 10^0).
- Skewed stiffeners should not be allowed.
- The diameter of any unstiffened open hole or cutout should not exceed 5% of the panel's width.
- No flange-induced buckling should occur (see Section 8.10).
- Members should be of uniform cross sections.

8.5.2 Unstiffened panels

The effective width method starts from the observation that the stress distribution across a panel is nonuniform in the post-buckling range. Figure 8.26 shows that there takes place a redistribution of stresses from the middle buckled part to the stiffer edges. In the effective^p width model [8.17], the nonuniform stress distribution over the entire width b is substituted by a uniform stress distribution over a reduced effective width b_{eff}, a procedure similar to the effective^s width model. The effective width is determined from the condition that the acting axial force is equal in both cases, according to the relation

$$\sigma_m \cdot b = \int_0^b \sigma_x(y) \cdot dy = \sigma_{max} \cdot b_{eff} \tag{8.43}$$

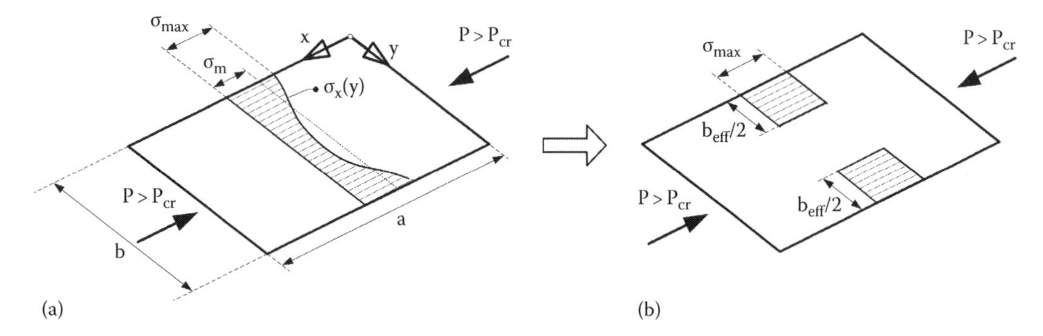

Figure 8.26 Panel under compression: (a) real stress distribution and (b) stress distribution.

At the ultimate load, it is $\sigma_m = \rho \cdot f_y$ and $\sigma_{max} = f_y$. The effective width is determined by substituting in Equation 8.43 and is given by

$$b_{eff} = \rho \cdot b \leq b \tag{8.44}$$

where ρ is the reduction factor due to plate buckling from Equations 8.23 and 8.24 as a function of the nondimensional slenderness $\bar{\lambda}_p$.

However, if the maximum compression stress $\sigma_{com,Ed}$ is smaller than f_y, the nondimensional slenderness may be reduced according to

$$\bar{\lambda}_{p,red} = \bar{\lambda}_p \cdot \sqrt{\frac{\sigma_{com,Ed}}{f_y / \gamma_{M0}}} \tag{8.45}$$

The effective width of unstiffened internal and outstand panels is given in Tables 8.12 and 8.13, respectively. It is noted that the effective width is introduced only in class 4 plated elements for which it is $\rho < 1$.

8.5.3 Longitudinally stiffened panels

For stiffened panels, an effective area rather than an effective width is determined. The effective area is composed of full areas at the stiff edges and effective areas of the central buckled parts. The effective^p area of the compression zone of a stiffened panel may be taken as:

$$A_{c,eff} = \rho_c \cdot A_{c,eff,loc} + \sum b_{edge,eff} \cdot t \tag{8.46}$$

where ρ_c is the reduction factor for global buckling of the stiffened panel, ignoring local buckling of subpanels.

The area $A_{c,eff,loc}$ is composed of the effective^p area of all the stiffeners and subpanels that are full or partly in compression, except the effective width adjacent to the two longitudinally supported edges, as indicatively shown in Figure 8.27:

$$A_{c,eff,loc} = A_{s\ell,eff} + \sum_c \rho_{loc} \cdot b_{c,loc} \cdot t \tag{8.47}$$

Table 8.12 Effective[p] widths for internal elements

Stress distribution (compression positive)	*Effective[p] width b_{eff}*
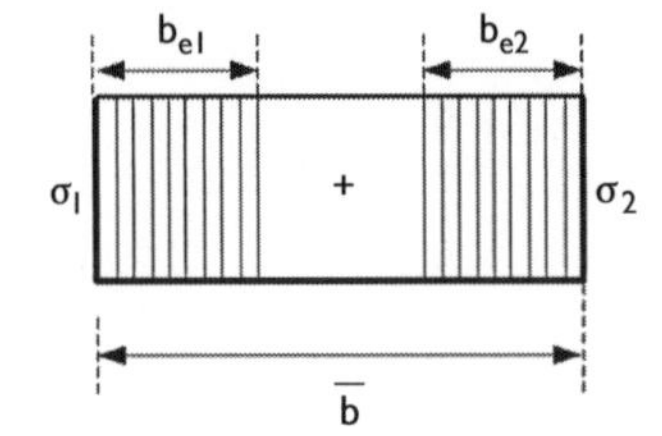 b_{e1}, b_{e2}, σ_1, +, σ_2, $\bar{b}$	$\psi = 1$ $b_{eff} = \rho \cdot \bar{b}$ $b_{e1} = 0.5 \cdot b_{eff}$ $b_{e2} = 0.5 \cdot b_{eff}$
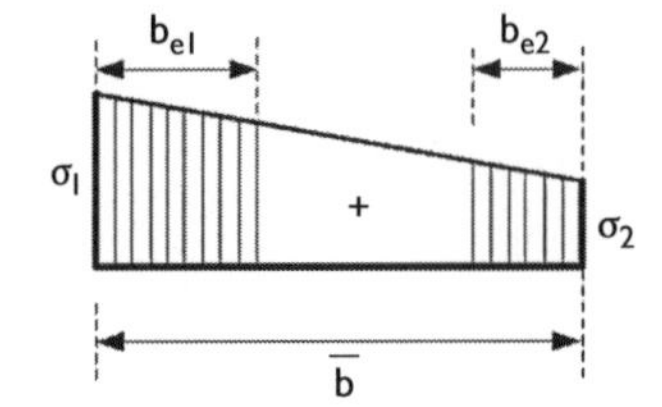 b_{e1}, b_{e2}, σ_1, +, σ_2, $\bar{b}$	$1 > \psi \geq 0$ $b_{eff} = \rho \cdot \bar{b}$ $b_{e1} = \frac{2}{5-\psi} \cdot b_{eff}$ $b_{e2} = b_{eff} - b_{e1}$
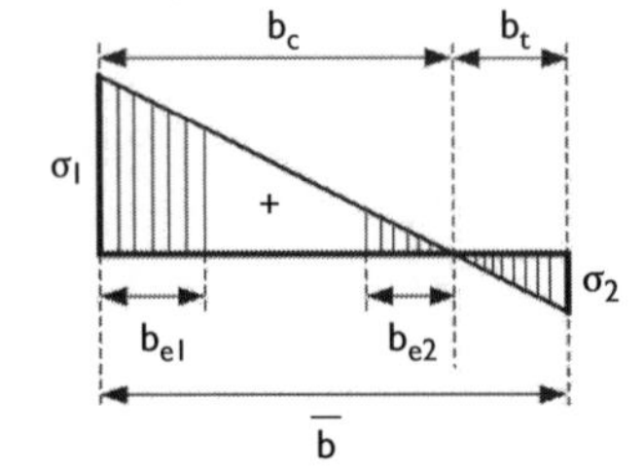 b_c, b_t, σ_1, +, σ_2, b_{e1}, b_{e2}, $\bar{b}$	$\psi < 0$ $b_{eff} = \rho \cdot b_c = \frac{\rho \cdot \bar{b}}{1-\psi}$ $b_{e1} = 0.4 \cdot b_{eff}$ $b_{e2} = 0.6 \cdot b_{eff}$

Source: EN 1993-1-5 Eurocode 3: Design of steel structures, Part 1–5: Plated Structural Elements, 2006.

Note: $\psi = \sigma_2/\sigma_1$.

For $\xi < 1$, Equation 8.33, ρ is substituted by ρ_c of Equation 8.32.

where

$\sum_c$ applies to the compression part of the stiffened panel, *except the edge parts of width* $b_{edge,eff}$

$A_{s\ell,eff}$ is the sum of the effective[p] areas of the longitudinal stiffeners *in the compression zone*

$b_{c,loc}$ is the width of the compression part of each subpanel

ρ_{loc} is the reduction factor for each subpanel

REMARK 8.6

For plates where the stress reverses and becomes tensile, the tensile area should be included in the calculation of $A_{c,eff,loc}$.

For wide flanges, the interaction between shear lag and plate buckling shall be accounted for by further reduction of the effective[p] area of Equation 8.46 in accordance to Table 7.9

$$A_{eff} = A_{c,eff} \cdot (\beta \text{ or } \beta_{ult} \text{ or } \beta^{\kappa}) \tag{8.48}$$

where β, β_{ult}, and β^k are the effective[s] width factors at ultimate limit states presented in Table 7.7.

Table 8.13 Effective^{p} widths for outstand elements

Stress distribution (compression positive)	*Effective^{p} width b_{eff}*
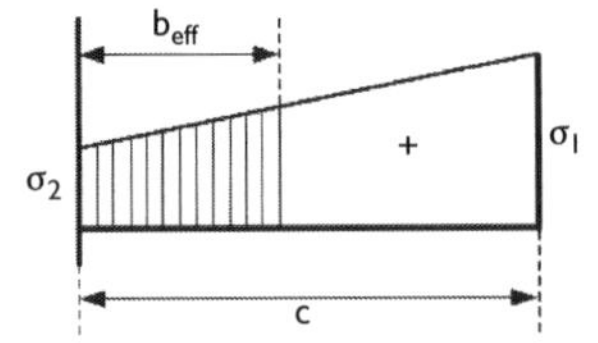	$1 > \psi \geq 0$ $b_{eff} = \rho \cdot c$
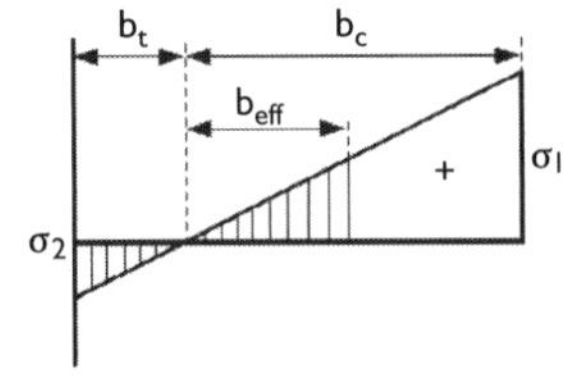	$\psi < 0$ $b_{eff} = \rho \cdot b_c = \rho \cdot c/(1 - \psi)$
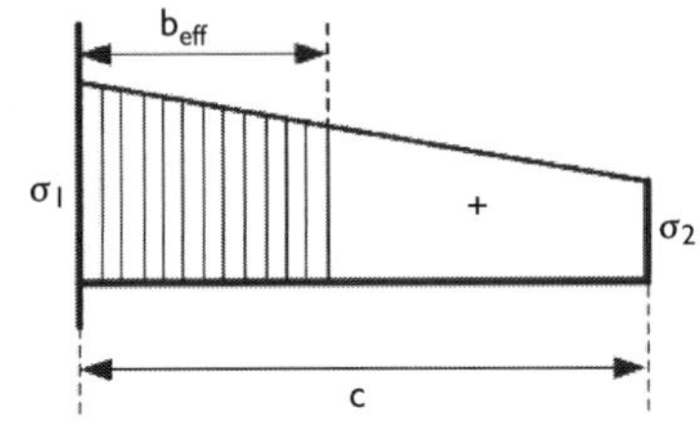	$1 > \psi \geq 0$ $b_{eff} = \rho \cdot c$
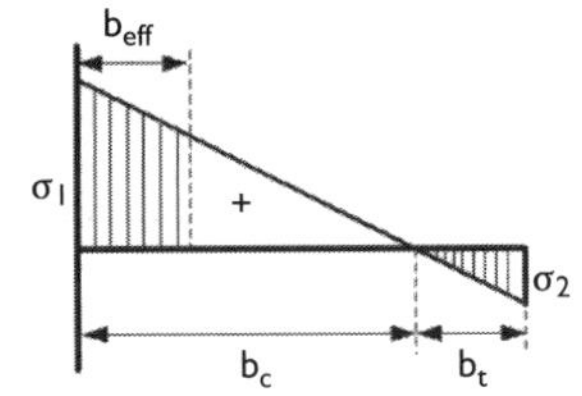	$\psi < 0$ $b_{eff} = \rho \cdot b_c = \rho \cdot c/(1 - \psi)$

Source: EN 1993-1-5 Eurocode 3: Design of steel structures, Part 1–5: Plated Structural Elements, 2006.

Note: $\psi = \sigma_2/\sigma_1$.

For $\xi < 1$, Equation 8.33, ρ is substituted by ρ_c of Equation 8.32.

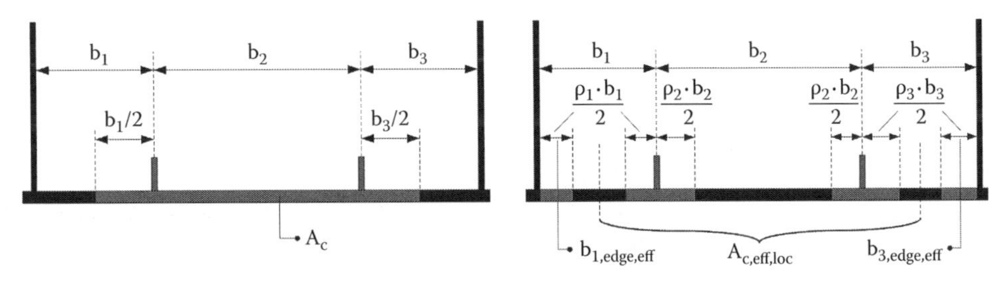

Figure 8.27 Example of a stiffened panel under uniform compression. Gross area A_c and effective^{p} area $A_{c,eff,loc}$.

8.6 MEMBER VERIFICATION FOR AXIAL COMPRESSION AND BENDING

Cross sections with unstiffened class 4 walls or stiffened walls subjected to axial compression and uniaxial bending may be verified by introduction of the properties of the effective rather than the gross cross section. The effective cross section is determined separately for axial compression and for bending. As outlined before, an important parameter for the determination of effective widths for unstiffened walls or subpanels and effective areas for stiffened walls is the ratio ψ between the edge stresses (see Tables 8.12 and 8.13).

For axial compression, the stress ratio is obviously $\psi=1$ for all parts of the cross section. Therefore, the determination of the effective cross section is straightforward. For nonsymmetric sections, seldom in composite bridge sections, the centroid of the effective cross section may be shifted in relation to the centroid of the gross cross section. This fact should be taken into account in design as indicated in Equation 8.49.

For uniaxial bending, the stress distribution and accordingly the stress ratio are determined separately for the flanges and the webs. The stress distribution for the flanges is based on the properties of the gross cross section, possibly accounting for the effects of shear lag for both compression and tension flanges. Following this distribution, a new cross section is found consisting of the effective area of the compression flange and the gross area of the tension flange and the webs (Figure 8.28a). For this new cross section, a new stress distribution is found providing the stress ratios and the effective area of the webs (Figure 8.28b).

The following is noted:

a. The stress distribution follows elastic analysis.
b. Effective cross sections are generally different for axial compression and for bending.
c. For I-girder composite bridges, the effective areas refer only to the webs, while for box girder composite bridges to the bottom flange and the webs.
d. The properties of the effective cross section may be used for flexural or lateral torsional buckling member verifications.
e. For construction stages, the stresses from the various stages may be calculated for the cross section with effective flanges only (Figure 8.28a). The resulting stresses from all stages may then be used to determine the effective area of the webs. The final stress distribution is then used for the effective section (Figure 8.28b).

Member verifications are performed according to the following relation:

$$\eta_1 = \frac{N_{Ed}}{\left(\dfrac{f_y \cdot A_{eff}}{\gamma_{M0}}\right)} + \frac{M_{Ed} + N_{Ed} \cdot e_N}{\left(\dfrac{f_y \cdot W_{eff}}{\gamma_{M0}}\right)} \leq 1.0 \tag{8.49}$$

where

A_{eff} is the effective cross-sectional area for axial compression
W_{eff} is the elastic section modulus of the effective cross section for pure bending
e_N is the shift of the neutral axis between gross and effective sections for axial compression
N_{Ed} is the design axial force
M_{Ed} is the design bending moment

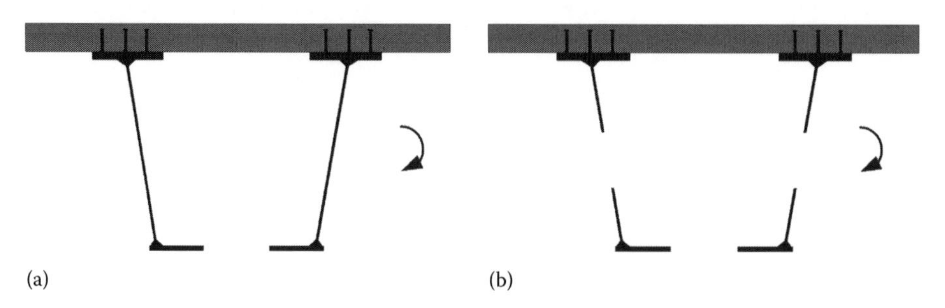

Figure 8.28 (a) Cross section with effective flange and full webs and (b) effective cross section.

For panels with variable stresses along their length, the verification of Equation 8.49 should be done at a distance s = min{0.4·a or 0.5·b} of the most stressed panel end. This check needs to be repeated at the end of the panel using gross-sectional properties.

It is noted that N_{Ed} and M_{Ed} may possibly be accounted for by second-order analysis, where relevant. This refers mostly to arch or stay cable bridges that are subjected to significant compression forces.

REMARK 8.7

The verification of members under compression and bending with the approach of Section 8.6 may be in some cases laborious and consequently impractical. Indeed, the shift e_N in the neutral axis leads to a change in the applied moment, which will in turn cause a change in stress distribution. In such cases, designers will find the *reduced stress method* of Section 8.4 much more convenient.

8.7 RESISTANCE TO SHEAR

In webs subjected to shear forces, a tension field is developed in the post-buckling state. This tension field is anchored in the flanges as shown in Figure 8.29. The girder behaves then quasi like a truss with the tension field acting as tension diagonals. Accordingly, the flanges anchoring the tension field contribute also to the shear resistance and not only the web as considered in the reduced stress method.

Flange contribution:

$$V_{bf,Rd} = \frac{2 \cdot M_{f,Rd}}{c}$$

The design shear resistance is then obtained from

$$V_{b,Rd} = V_{bw,Rd} + V_{bf,Rd} \leq \frac{\eta \cdot f_{yw} \cdot h_w \cdot t}{\sqrt{3} \cdot \gamma_{M1}} \tag{8.50}$$

For η, see Table 8.10.

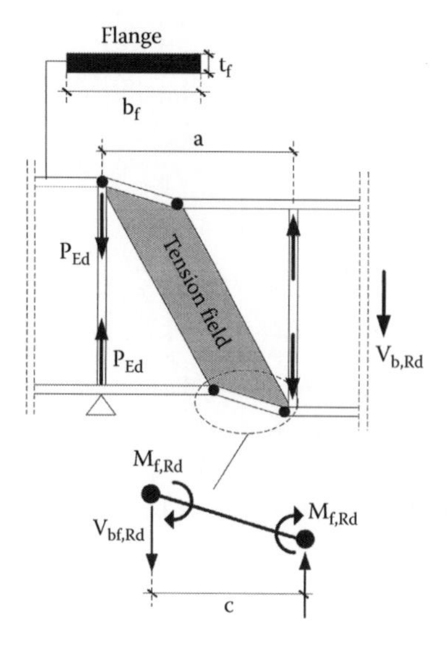

Figure 8.29 Tension field in web under shear force.

In the earlier relation, the first term expresses the contribution of the web and the second of the flanges. The contribution of the web is given by

$$V_{bw,Rd} = \frac{\chi_w \cdot f_{yw} \cdot h_w \cdot t}{\sqrt{3} \cdot \gamma_{M1}} \tag{8.51}$$

where

χ_w is the reduction factor for shear as determined from Table 8.10

f_{yw} is the yield strength of the web and the other notation as shown in Figure 8.30

The contribution of the flange is given by

$$V_{bf,Rd} = \frac{b_f \cdot t_f^2 \cdot f_{yf}}{c \cdot \gamma_{M1}} \cdot \left[1 - \left(\frac{M_{Ed}}{M_{f,Rd}}\right)^2\right] \tag{8.52}$$

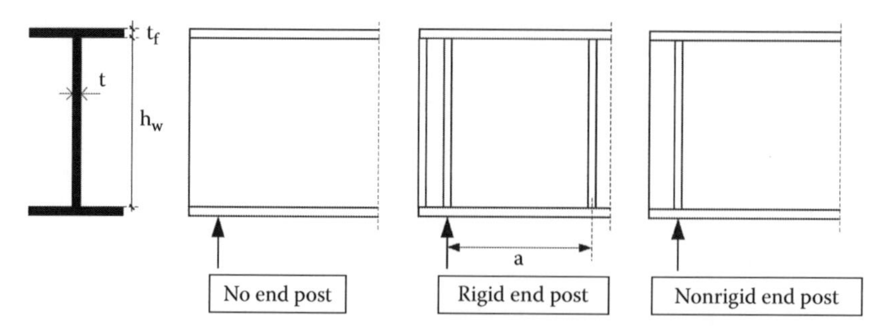

Figure 8.30 Notations and types of support conditions.

where

b_f and t_f are taken for the flange leading to the lowest resistance (that has the lowest product $b_f \cdot t_f \cdot f_{yf}$). However, if one flange is connected to a concrete slab and acts compositely, the width and thickness of the noncomposite flange should be taken.

b_f being taken as not larger than $15 \cdot \varepsilon \cdot t_f$ on each side of the web

$M_{f,Rd} = \dfrac{M_{f,k}}{\gamma_{M0}}$ is the design-bending resistance of the cross section consisting of the flanges only. If a β-factor smaller than 1 applies for the determination of $M_{pl,Rd}$ of the composite section (Figure 9.8), then $M_{f,Rd}$ should be reduced by this factor.

$M_{f,k} = \min\{N_{Rf1}, N_{Rf2}\} \cdot e$ N_{Rf1}, N_{Rf2} axial resistance of the effective^p area of the flanges, taking into account the contribution of concrete in compression for flanges acting compositely with concrete

e distance between centroids of flanges

$$c = a \cdot \left(0.25 + \frac{1.6 \cdot b_f \cdot t_f^2 \cdot f_{yf}}{t \cdot h_w^2 \cdot f_{yw}}\right) \tag{8.53}$$

a distance between rigid transverse stiffeners.

When an axial force N_{Ed} is present, the value of $M_{f,Rd}$ should be reduced by the factor

$$\left(1 - \frac{N_{Ed}}{((A_{f1} + A_{f2}) \cdot f_{yf})/\gamma_{M2}}\right) \tag{8.54}$$

For a composite section, N_{Ed} is the axial force acting on the composite section.

Obviously, the contribution of the flanges is high at end supports and small at intermediate supports where high-bending moments act simultaneously with shear.

The verification for shear is performed according to the following relation:

$$\eta_3 = \frac{V_{Ed}}{V_{b,Rd}} \le 1.0 \tag{8.55}$$

REMARK 8.8

The tension field shown in Figure 8.29 may transfer nonnegligible axial forces to the shear connectors and consequently lead to a local damage of the concrete plate (e.g., pullout failure). This may occur in cases of cross sections with "very strong" upper flanges (i.e., hybrid girders where $f_{yf} > f_{yw}$ or high values of t_f). Therefore, it is advisable to ignore the flange contribution during the calculation of $V_{b,Rd}$.

8.8 RESISTANCE TO CONCENTRATED TRANSVERSE FORCES

In bridges, web stiffeners are provided at supports. However, there are cases where the transfer of concentrated forces through an unstiffened web may not be avoided. This occurs typically in bridges that are constructed by launching, where all sections, whether stiffened or not, are subjected to concentrated support forces (Figure 8.31).

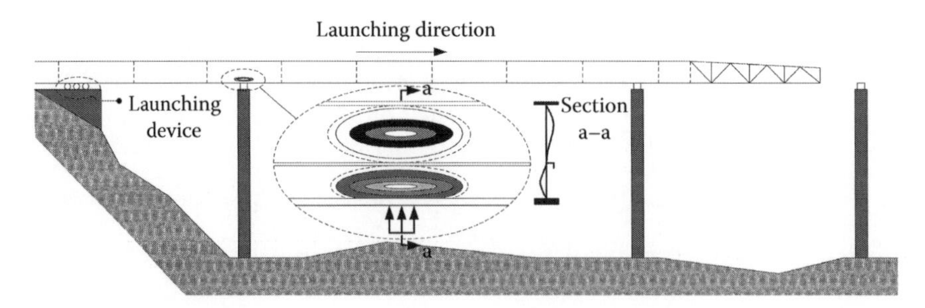

Figure 8.31 Bridge during launching.

The design buckling resistance of girders to concentrated transverse forces may be obtained from

$$F_{Rd} = \frac{f_{yw} \cdot L_{eff} \cdot t_w}{\gamma_{M1}} \tag{8.56}$$

where

t_w is the web thickness
f_{yw} is the yield stress of the web
L_{eff} is the effective length for resistance to concentrated forces determined from

$$L_{eff} = \chi_F \cdot l_y \tag{8.57}$$

where

l_y is the effective loaded length obtained from Table 8.14

$$\chi_F = \frac{0.5}{\bar{\lambda}_F} \le 1.0 \tag{8.58}$$

$$\bar{\lambda}_F = \sqrt{\frac{l_y \cdot t_w \cdot f_{yw}}{F_{cr}}} \tag{8.59}$$

$$F_{cr} = 0.9 \cdot k_F \cdot E_a \cdot \frac{t_w^3}{h_w} \tag{8.60}$$

For webs without longitudinal stiffeners, the coefficient k_F is taken from Figure 8.32 as a function of the loading conditions and the length of the panel between transverse stiffeners.

The length of stiff bearing s_s should not be taken as larger than h_w.

For webs with longitudinal stiffeners, the buckling coefficient k_F for the case of Figure 8.32a may be obtained from Equation 8.61, provided that the following geometric conditions apply: $0.05 \le \frac{b_1}{h_w} \le 0.3$ and $\frac{b_1}{a} \le 0.3$:

$$k_F = 6 + 2 \cdot \left[\frac{h_w}{a}\right]^2 + \left[5.44 \cdot \frac{b_1}{a} - 0.21\right] \cdot \sqrt{\gamma_s} \tag{8.61}$$

Table 8.14 Effective loaded length l_y

End posts as in Figure 8.32a and b	*End posts as in Figure 8.32c*
$l_y = s_s + 2 \cdot t_f \left(1 + \sqrt{m_1 + m_2}\right)$ where $l_y \leq a$	$l_y = \min(l_{y1}, l_{y2})$ $l_{y1} = l_e + t_f \cdot \sqrt{\frac{m_1}{2} + \left(\frac{l_e}{t_f}\right)^2 + m_2}$ $l_{y2} = l_e + t_f \cdot \sqrt{m_1 + m_2}$ $l_e = \frac{k_F \cdot E_a \cdot t_w^2}{2 \cdot f_{yw} \cdot h_w} \leq s_s + c$
$m_1 = \frac{f_{yf} \cdot b_f}{f_{yw} \cdot t_w}$ For box girders $b_f \leq 30 \cdot \varepsilon \cdot t_f$	
$m_2 = 0.02 \cdot \left(\frac{h_w}{t_f}\right)^2$ if $\bar{\lambda}_F > 0.5$	
$m_2 = 0$ if $\bar{\lambda}_F \leq 0.5$	

Notes:
– Length of stiff bearing s_s from Figure 8.32
– The estimation of the bearing length l_y is based on the collapse mechanism of Figure 8.34. It is assumed that four plastic hinges are formed in the flange. The internal work due to the plastification of the flange is $4 \cdot M_{pl,f,Rd} \cdot \theta$ where $M_{pl,f,Rd} = f_{yf} \cdot b_f \cdot t_f^2/4$ and $\theta = 2 \cdot \Delta/s_y$.
Therefore, the internal work is written as $W_{int} = f_{yf} \cdot b_f \cdot t_f^2 \cdot 2 \cdot \Delta/s_y$.
The external works is equal to $W_{ext} = [l_y - (s_s + 2 \cdot t_f) - 0.5 \cdot s_y] \cdot f_{yw} \cdot t_w$.
From Figure 8.34, one can see that $l_y = s_s + 2 \cdot t_f + s_y$.
Thus, $W_{ext} = 0.5 \cdot s_y \cdot f_{yw} \cdot t_w$.
Equating the work done externally to work done internally gives

$$s_y = 2 \cdot t_f \cdot \sqrt{m_1} = 2 \cdot t_f \cdot \sqrt{\frac{b_f}{t_w} \cdot \frac{f_{yf}}{f_{yw}}}$$

The bearing length becomes then $l_y = s_s + 2 \cdot t_f \cdot (1 + \sqrt{m_1})$.
The earlier expression is offered from EN 1993-2 for $\bar{\lambda}_F \leq 0.5$.

For slender webs ($\bar{\lambda}_F > 0.5$), an additional factor $m_2 = 0.02 \cdot \left(\frac{h_w}{t_w}\right)^2$ is introduced, which is based on experimental observations.

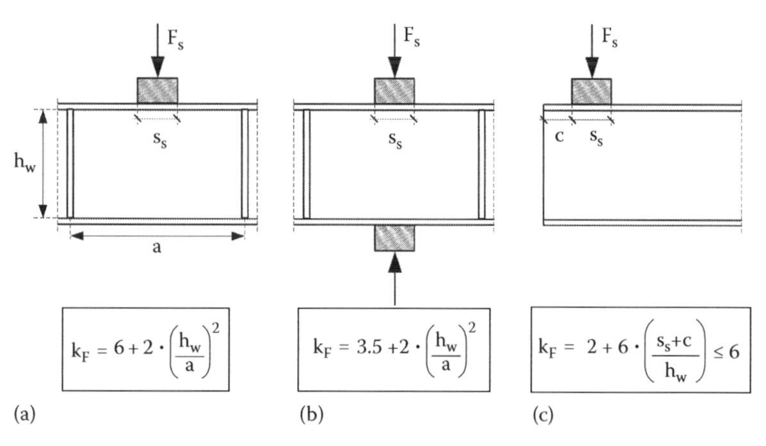

Figure 8.32 Buckling coefficients k_F for concentrated forces in unstiffened panels. (a) Load applied through one flange, (b) load applied through one flange and directly transferred to the other flange, and (c) load applied through one flange close to an unstiffened end.

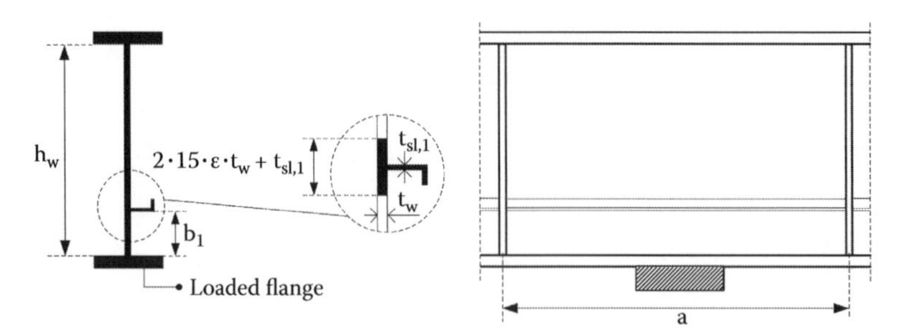

Figure 8.33 Notation for webs with longitudinal stiffeners.

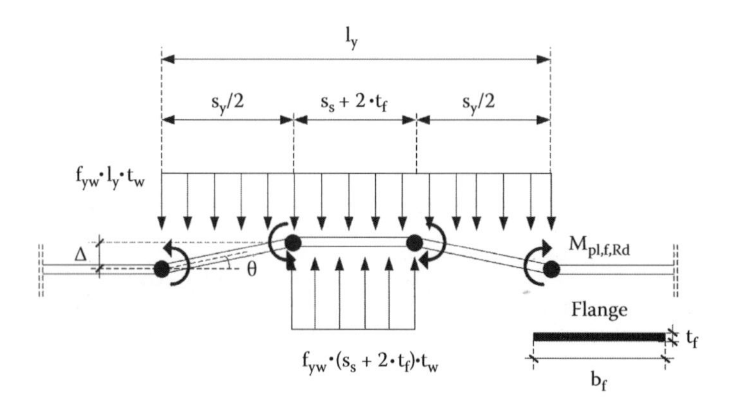

Figure 8.34 Flange collapse mechanism.

where

b_1 is the height of the loaded subpanel, equal as the clear distance between the loaded flange and the closest stiffener:

$$\gamma_s = 10.9 \cdot \frac{I_{sl,1}}{h_w \cdot t_w^3} \leq 13 \cdot \left[\frac{a}{h_w}\right]^3 + 210 \cdot \left[0.3 - \frac{b_1}{a}\right] \tag{8.62}$$

$I_{sl,1}$ is the second moment of area of the closest stiffener, including an effective width of the plate as indicated in Figure 8.33. Different values of k_F may be found in the National Annex.

The length of stiff bearing s_s is the distance over which the applied force is effectively distributed from the flange to the web. The dispersion through solid steel material is at a slope 1:1, as shown in Figure 8.35. For several closely spaced concentrated forces at

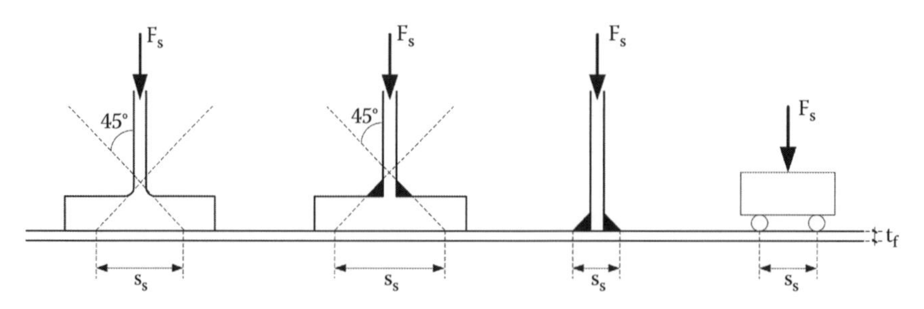

Figure 8.35 Length of stiff bearing s_s.

center-to-center distance s_s, verifications for each individual force and for the sum of forces at distance s_s are required.
The verification is performed as follows:

$$\eta_2 = \frac{F_{s,Ed}}{F_{Rd}} \leq 1.0 \tag{8.63}$$

REMARK 8.9

By comparing Equation 8.61 with the k_F values in Figure 8.32, one can observe that for small values of b_l, the resistance F_{Rd} of an unstiffened panel may be greater than the resistance of a stiffened one. Indeed, this happens for small values of b_l:

$$k_F \text{ from Equation (8.61)} < k_F \text{ from Figure 8.32 case a} \rightarrow \frac{b_l}{a} < 0.039$$

This is physically not meaningless. A stiffener that is close to the loaded flange may not be capable of stabilizing the web. However, designers should in such cases conservatively choose for k_F, the value for the unstiffened panel. Other parameters such as the type of the stiffeners (i.e., closed or opened) and their distribution along the web are unfortunately not considered in the procedure of determining k_F according to EN 1993-1-5 (see [8.13], [8.9]). In all cases, it is therefore recommended to calculate k_F from an FE analysis.

8.9 INTERACTION

8.9.1 Interaction N, M, V

Members under the simultaneous action of axial compression, bending moments, and shear forces should be verified by Equation 8.49 for compression and bending as described in Section 8.6 and by Equation 8.55 for shear as described in Section 8.7. Additionally, they shall be verified for the interaction of all action effects as given in the following. The verifications for this interaction include all cross sections, except those at a distance smaller than the web height, $h_w/2$, from internal supports with vertical stiffeners. This is because in the code, it is assumed that there is a bearing stiffener present. However, an additional verification at support is recommended.

For sections subjected to small axial compression forces, the interaction relation is given by Equation 8.64. The axial compression is considered small if it does not completely suppress the tension-bending stresses so that the entire web is in compression. The interaction shall be considered only in cases when $\bar{n}_3 > 0.5$ and $\bar{\eta}_1 \geq \frac{M_{f,Rd}}{M_{N,pl,Rd}}$, where $\bar{\eta}_1$ and $\bar{\eta}_3$ are given in the following. Otherwise, no interaction should be examined. The first condition indicates high shear in the web, the second high bending that may not be resisted by the flanges alone:

$$\bar{\eta}_1 + \left(1 - \frac{M_{f,Rd}}{M_{N,pl,Rd}}\right) \cdot (2 \cdot \bar{\eta}_3 - 1)^2 \leq 1.0 \tag{8.64}$$

where

$$\bar{\eta}_1 = \frac{M_{Ed}}{M_{N,pl,Rd}} \tag{8.65}$$

$$\bar{\eta}_3 = \frac{V_{Ed}}{V_{bw,Rd}} \tag{8.66}$$

The plastic bending moment accounting for the presence of axial force may be taken as

$$M_{N,pl,Rd} = M_{pl,Rd} \cdot \left(1 - \frac{N_{Ed}}{A_g \cdot f_y / \gamma_{M1}}\right) \tag{8.67}$$

It is noted that in Equation 8.67, the plastic bending moment is taken into account irrespectively of the section class.

$M_{f,Rd}$ is the design-bending resistance of the cross section consisting of the flanges only as described in Section 8.7. In presence of axial forces, the value of $M_{f,Rd}$ should be reduced by the factor of Equation 8.54.

For sections subjected to high axial compression so that the whole web is in compression, $M_{f,Rd}$ is set to zero.

Flanges of box girders may also be verified by Equation 8.64 taking $M_{f,Rd}=0$. The shear stresses in the panel result in global torque plus ½ of the maximum shear stress across the flange from shear forces. In addition, the edge subpanels between the web and the extreme longitudinal stiffeners should be verified adding the shear stresses from global torque with the average shear stresses in the subpanel from shear.

8.9.2 Interaction N, M, F_s

For concentrated force F_s acting in the compression flange, the following interaction should be verified in addition to Equation 8.64:

$$\eta_2 + 0.8 \cdot \eta_1 \leq 1.4 \tag{8.68}$$

REMARK 8.10

In Equation 8.68, one can observe that the influence of shear is not considered in the interaction with the patch load. This is due to experimental observations that led to this exclusion [8.9].

8.10 FLANGE-INDUCED BUCKLING

In modern I-girder bridges, the flanges may be formed from very thick plates. When these thick flanges are in compression, significant deviation compression forces develop in the web due to the curvature (Figure 8.36) that may lead to buckling of the flange in the plane of the web. To prevent this possibility, the following geometric limitation is set:

$$\frac{h_w}{t_w} \leq k \cdot \frac{E_a}{f_{yf}} \cdot \sqrt{\frac{A_w}{A_{f,eff}}} \tag{8.69}$$

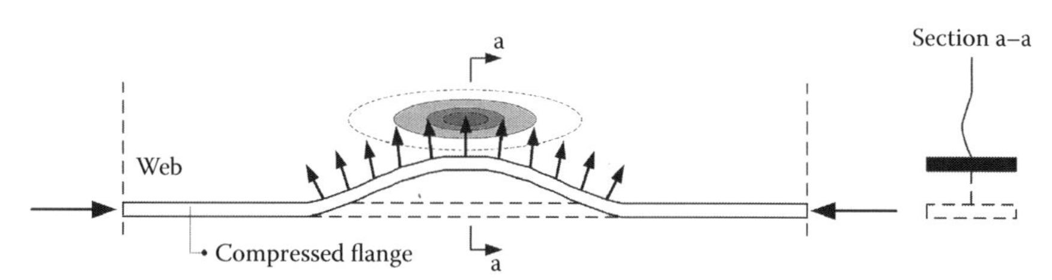

Figure 8.36 Flange-induced buckling.

where

A_w is the web area
$A_{f,eff}$ is the effective area of the compression flange
k = 0.4 if the plastic moment resistance is utilized
k = 0.55 if the elastic moment resistance is utilized

For girders curved in elevation, the right side of Equation 8.69 should be divided by the factor

$$1+\frac{E_a \cdot h_w}{3 \cdot r \cdot f_{yf}}$$

where r is the radius of curvature of the compression flange.

The flange-induced buckling may occur in cases of highly curved beams with large compressed flanges and slender webs. In such cases, Equation 8.69 may be decisive when determining the web dimensions.

8.11 DESIGN OF STIFFENERS AND DETAILING

8.11.1 Introduction

As noted in the introduction, most of the failures occurred in box girder bridges in the 1970s were due to insufficient construction detailing. Most failures took place in splice regions of flange plates or were due to lateral torsional buckling of longitudinal or open-section stiffeners. These issues were studied extensively experimentally and numerically and led to the formulation of code provisions as presented in the following. Transverse stiffeners limit the buckling length of the longitudinal ones. In addition, they reduce the distortional frame deformations of box girder walls if placed at small distances.

8.11.2 Intermediate transverse stiffeners in compression panels

As noted before, in the current construction practice, transverse stiffeners are selected sufficient rigid to provide rigid supports of the adjacent panels. The sufficiency in rigidity of transverse stiffeners may be checked by application of linear plate buckling theory. Figure 8.37 shows the first buckling mode of a compression panel with two longitudinal stiffeners and two types of transverse stiffeners with different rigidities. It may be seen that the first stiffener (Figure 8.37a) deforms, while the second remains straight. Accordingly, its rigidity should be sufficient. However, the information provided by linear plate buckling theory is only limited as it does not refer to the post-buckling state and does not provide the stress state. Therefore, second-order analysis for the stiffener with imperfections is employed instead.

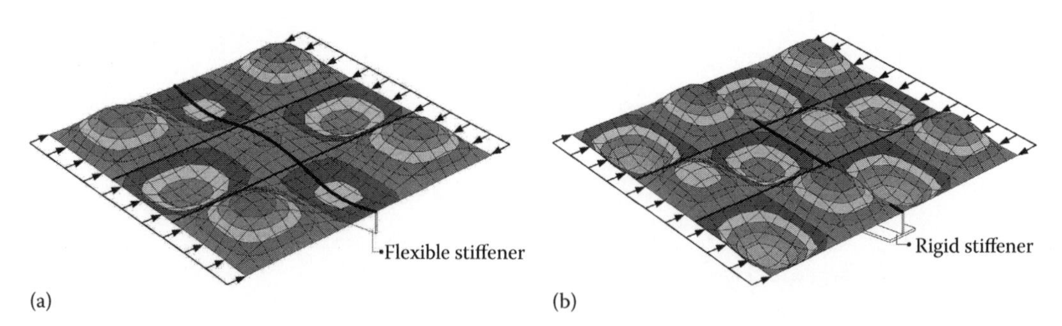

Figure 8.37 First buckling mode of a panel under uniform compression (a) with flexible and (b) with rigid transverse stiffeners.

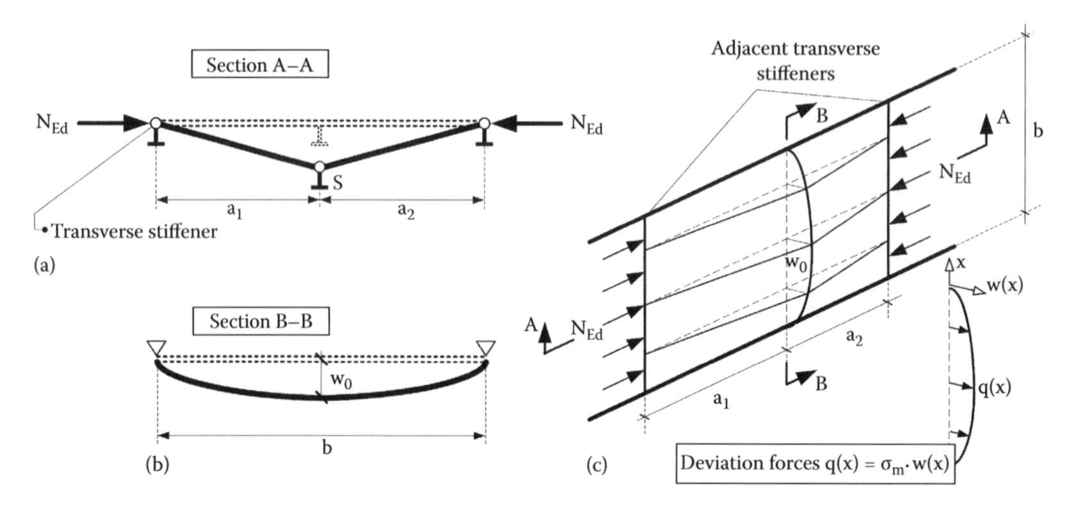

Figure 8.38 Imperfection of transverse stiffener. (a) Longitudinal section, (b) transverse section, and (c) panel with geometrical imperfections.

Transverse stiffeners are then treated as simply supported beams with sinusoidal imperfection w_0 given in the following (Figure 8.38b). The span of the beam is b, equal to the width of the panel under consideration. Obviously, b is equal to h_w for web stiffeners or to the width of the bottom flange for flange stiffeners in box girders.
The deviation forces due to imperfection result in a uniform transverse loading that is equal to

$$q = \frac{\pi}{4} \cdot \sigma_m \cdot (w_0 + w_{el}) \tag{8.70}$$

where

σ_m is defined in Equation 8.73
$w_0 = s/300$ is the value of the imperfection (see Figure 8.38b)
w_{el} is the additional deflection to be calculated iteratively or set equal to b/300
s is the min (a_1, a_2, b)
a_1, a_2 are the lengths of the adjacent panels
b is the panel width (=span of transverse stiffener)

The verification for stiffeners subjected to compression forces N_{Ed} (e.g., due to tension field action; see forces P_{Ed} in Figure 8.29 or external load) is performed by limiting the resulting maximal stress to the design yield stress according to Relation 8.71, which implies also that the additional deflection is smaller than b/300:

$$\sigma_{max} = \frac{N_{Ed}}{A} + \frac{q \cdot b^2/8}{W} \frac{1}{1 - N_{Ed}/N_{cr,st}} \leq \frac{f_y}{\gamma_{M1}} \quad (8.71)$$

N_{Ed} is the maximal design compression force of the adjacent panels but not smaller than the maximal compression stress times half the effective[p] compression area of the panel including the longitudinal stiffeners. When axial forces in adjacent panels differ, the larger of the two is taken into consideration (see Example 8.12).

$N_{cr,st}$ is the elastic critical buckling (Euler) load of the stiffener.

It is noted that any transverse loading present should be added to the deviation forces q in Equation 8.71.

Transverse stiffeners not subjected to compression forces may be verified alternatively as having a minimum second moment of area given by

$$I_{st,min} = \frac{\sigma_m}{E_a} \cdot \left(\frac{b}{\pi}\right)^4 \left(1 + w_0 \cdot \frac{300}{b} \cdot u\right) \quad (8.72)$$

where

$$\sigma_m = \frac{\sigma_{cr,c}}{\sigma_{cr,p}} \cdot \frac{N_{Ed}}{b} \cdot \left(\frac{1}{a_1} + \frac{1}{a_2}\right) \quad (8.73)$$

$$u = \frac{\pi^2 \cdot E_a \cdot e_{max}}{\left(f_y \cdot 300 \cdot b/\gamma_{M1}\right)} \geq 1.0 \quad (8.74)$$

e_{max} is the distance of the extreme fiber of the stiffener to the centroid of the stiffener

$\sigma_{cr,c}/\sigma_{cr,p}$ is the ratio of column-like critical buckling stress to the plate critical buckling stress $(0.5 \leq \sigma_{cr,c}/\sigma_{cr,p} \leq 1)$

REMARK 8.11

In EN 1993-1-5, it is not clear for which panel length should the buckling stresses $\sigma_{cr,c}$ and $\sigma_{cr,p}$ be calculated. A conservative simplification would be to take the ratio $\sigma_{cr,c}/\sigma_{cr,p}$ equal to 1 [8.2]. In some cases, this can be significantly uneconomical. Therefore, it is recommended to calculate the buckling stresses for the shorter panel. This maximizes the ratio $\sigma_{cr,c}/\sigma_{cr,p}$ and keeps the calculation at a safe level [8.9].

REMARK 8.12

The stiffener to be checked is assumed to have straight and rigid adjacent stiffeners. The deviation forces q(x) can be then approximated by the following equation:

$$q(x) = w(x) \cdot \frac{\sigma_{cr,c}}{\sigma_{cr,p}} \cdot \left[\frac{N_{Ed}}{b} \cdot \left(\frac{1}{a_1} + \frac{1}{a_2}\right)\right] \rightarrow q(x) = w(x) \cdot \sigma_m \quad (R8.7)$$

where

w(x) is the initial sinusoidal bow with max $w = w_0$

N_{Ed} the compressive force due to the membrane action in the adjacent panels

The maximum value of the deviation forces is obviously equal to

$$\max q = \sigma_m \cdot \max w(x) \rightarrow \max q = \sigma_m \cdot (w_0 + \Delta) \tag{R8.8}$$

where Δ is the deflection at mid-height.

By taking the previous in consideration, the maximum stress and the maximum deflection are written as follows:

$$\sigma_{max} = \frac{M_{max} \cdot e_{max}}{I_{st}} = \frac{\max q \cdot b^2 \cdot e_{max}}{\pi^2 \cdot I_{st}} = \frac{\sigma_m \cdot (w_0 + \Delta) \cdot b^2 \cdot e_{max}}{\pi^2 \cdot I_{st}} \tag{R8.9}$$

$$\Delta = \frac{\max q \cdot b^4}{\pi^2 \cdot E_a \cdot I_{st}} = \frac{\sigma_m \cdot (w_0 + \Delta) \cdot b^4}{\pi^2 \cdot E_a \cdot I_{st}} = \frac{\sigma_{max} \cdot b^2}{\pi^2 \cdot E_a \cdot e_{max}} \tag{R8.10}$$

where

M_{max} is the maximum value of the bending moment due to deviation forces
e_{max} is the distance of the extreme fiber of the stiffener to the centroid of the stiffener

By introducing (R8.10) in (R8.9), the stiffener's second moment of inertia becomes

$$I_{st} = \frac{\sigma_m}{E_a} \cdot \left(\frac{b}{\pi}\right)^4 \cdot \left(1 + w_0 \cdot \frac{\pi^2 \cdot E_a \cdot e_{max}}{b^2 \cdot \sigma_{max}}\right) \tag{R8.11}$$

or

$$I_{st} = \frac{\sigma_m}{E_a} \cdot \left(\frac{b}{\pi}\right)^4 \cdot \left(1 + \frac{w_0}{w}\right) \tag{R8.12}$$

The minimum allowable values for I_{st} can be determined by introducing $\sigma_{max} = f_y/\gamma_{M1}$ and $w = b/300$ in Equations R8.11 and R8.12, respectively. This leads to Equation 8.72 given in EN 1993-1-5.

8.11.3 Shear in transverse stiffeners

Rigid end posts

The cross-sectional shear resistance at supports is usually calculated assuming a rigid end post (see Figure 8.30). In such cases, the bearing stiffeners should be capable of resisting the membrane forces in the web (see N_H in Figure 8.39) by acting as beams spanning between the flanges. Membrane forces are developed due to the tension field action in the adjacent panels and act only when the shear stress τ reaches the critical value τ_{cr} [8.7], [8.16].

Rigid end posts may comprise two transverse stiffeners on both sides of the web. This results in a short beam with an I cross-section the flanges of which are the stiffeners and the web the strip of web panel between them (Figure 8.39a). Alternatively, a rolled section may be used instead.

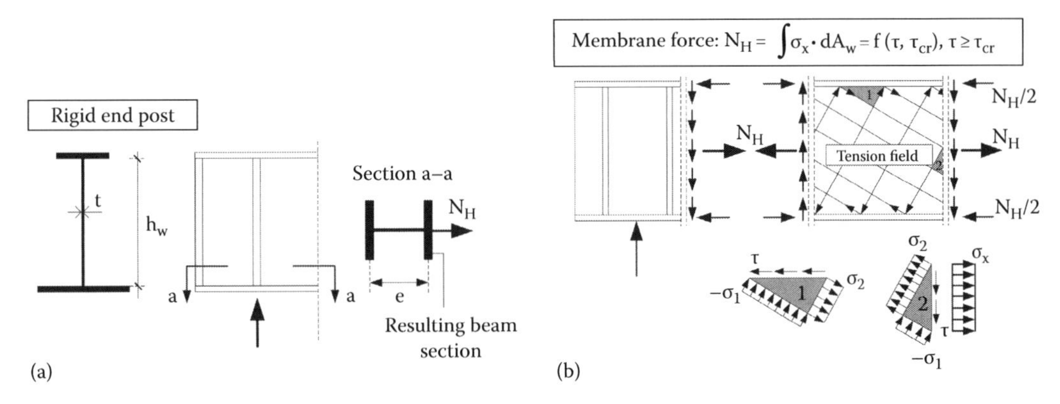

Figure 8.39 (a) Rigid end post and (b) tension field action and membrane force.

Each double-sided stiffener consisting of flat plates should have a minimum cross-sectional area:

$$\min A_{st} = \frac{4 \cdot h_w \cdot t^2}{e} \tag{8.75}$$

where
- t is the panel thickness
- e is the center distance between the stiffeners

In addition, it must be $e > 0.1 \cdot h_w$.

If the end posts are not made of flat stiffeners, its section modulus for bending around a horizontal axis perpendicular to the web must be at least equal to

$$W_{min} = 4 \cdot h_w \cdot t^2 \tag{8.76}$$

Rigid end posts are checked for buckling due to compression and bending, using buckling curve c and for buckling length $0.75 \cdot h_w$. Compression results from the reaction force while bending from the longitudinal membrane stresses in the plane of the web.

REMARK 8.13

Designing a bearing stiffener as a rigid end post is not always feasible. Moreover, rigid end post conditions do not offer economical advantages for web slenderness $\bar{\lambda}_w < 1.08$ (see Table 8.10). For this reason, EN 1993-1-5 offers a second alternative by providing a single double-sided bearing stiffener and a vertical stiffener adjacent to the support so that the subpanel resists the maximum shear when designed with a nonrigid end post. Panels beyond the adjacent stiffener are then designed with rigid end-post conditions (Figure 8.40).

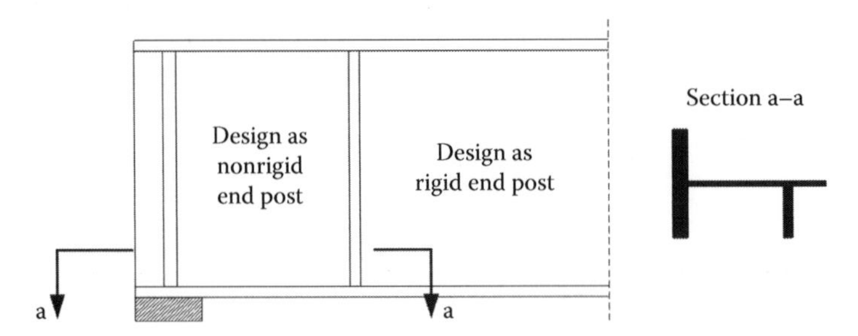

Figure 8.40 Alternative solution to avoid rigid end post.

Intermediate transverse stiffeners

The effective section of intermediate transverse stiffeners acting as rigid supports for web panels must have a second moment of area I_{st} that satisfies the conditions:

$$I_{st} \geq 1.5 \cdot h_w^3 \cdot \frac{t^3}{a^2} \quad \text{if } \frac{a}{h_w} < \sqrt{2} \tag{8.76a}$$

$$I_{st} \geq 0.75 \cdot h_w \cdot t^3 \quad \text{if } \frac{a}{h_w} \geq \sqrt{2} \tag{8.76b}$$

Intermediate transverse stiffeners are checked for buckling to an axial compression force, using buckling curve c and for a minimum buckling length equal to $0.75 \cdot h_w$. The compression force is given by

$$P_{Ed} = V_{Ed} - \frac{\chi_w \cdot f_{yw} \cdot h_w \cdot t}{\sqrt{3} \cdot \gamma_{M1}} \tag{8.77}$$

where

χ_w is calculated for the web panel between adjacent transverse stiffeners assuming the stiffener under consideration is removed (i.e., taking twice the panel length)

V_{Ed} is the shear force at distance $0.5 \cdot h_w$ from the edge of the panel with the largest shear force

It is worth mentioning that the force P_{Ed} acts on the plane of the web and is activated due to the shear tension field action. Designers should be careful in cases of asymmetric stiffeners and take into account the resulting eccentricity by verifying the stiffener against axial load and bending. Despite the fact that EN 1993-1-5 allows a verification based on the plastic resistance of the stiffener, a check based on the elastic properties is recommended Moreover, lateral–torsional buckling of the stiffener should be excluded.

The notation P in Equation 8.77 has been chosen so that it can be distinguished from the force N_{Ed} presented in 8.11.2.

8.11.4 Torsional requirements for open section stiffeners

To avoid torsional buckling around the plane of the panel they are stiffening, longitudinal and transverse stiffeners must comply with the following requirement:

$$\frac{I_T}{I_p} \geq 5.3 \cdot \frac{f_y}{E_a} \tag{8.78}$$

where

I_p is the polar second moment of area of the stiffener around the edge fixed to the plate
I_T is the *St. Venant* constant of the stiffener alone

The earlier relation gives for a flat stiffener of height h and thickness t:

$$I_T = \frac{h \cdot t^3}{3} \quad \text{and} \quad I_P = \frac{h^3 \cdot t}{12} + h \cdot t \cdot \left(\frac{h}{2}\right)^2 = \frac{h^3 \cdot t}{3}$$

Substituting in Equation 8.78, the condition for the flat may be written as

$$\left(\frac{t}{h}\right)^2 \geq 5.3 \cdot \frac{f_y}{E}$$

or

$$\frac{h}{t} \leq 13 \cdot \varepsilon \tag{8.79}$$

where

$$\varepsilon = \sqrt{\frac{235}{f_y\,[\mathrm{N/mm^2}]}}$$

Stiffeners with warping rigidity may alternatively fulfill the condition:

$$\sigma_{cr} \geq \theta \cdot f_y \tag{8.80}$$

where

θ is a parameter with a recommended value 6
$\sigma_{cr} = \frac{1}{I_p}\left(G_a \cdot I_T + \frac{\pi^2 \cdot E_a \cdot I_w}{a^2}\right)$ is the critical stress for torsional buckling around the edge fixed to the plate
a is the length between transverse stiffeners

For angles or tees, σ_{cr} is equal to the critical stress of the double section (i.e., a U or an I section of double height) without restraint from the plate.

8.11.5 Discontinuous longitudinal stiffeners

Longitudinal stiffeners in compression flanges pass usually through openings made in the transverse stiffeners and are welded to them in order to provide continuity (see Figure 8.4). Continuity may be also provided by fully splicing them. Accordingly, longitudinal stiffeners in flanges are considered in the global analysis and stress calculations. Example 8.9 presents a practical method on how to include them in analysis and design; by "smearing" them into the flange, they stiffen and accordingly notionally increase the thickness of this panel.

Oppositely, longitudinal stiffeners in webs may be discontinuous. In such a case, they are not considered in global analysis and stress calculations but are merely used for buckling verification to increase the panel buckling strength as indicatively shown in Example 8.9.

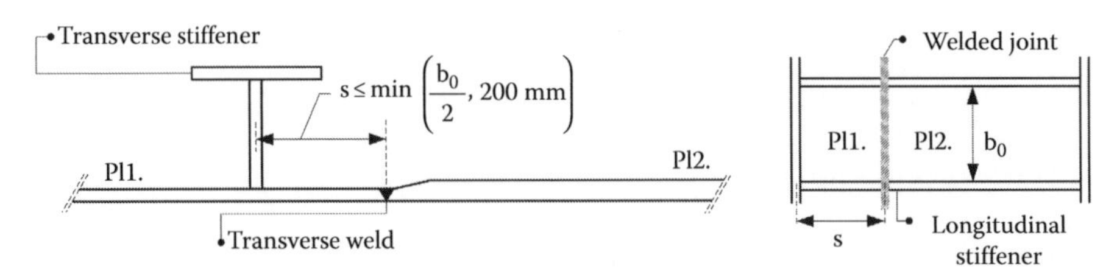

Figure 8.41 Splice of plate sheets with change in thickness and transverse stiffener.

8.11.6 Splices of plate sheets

The plate sheets for flanges and webs are delivered in certain lengths and must be welded on site to form the entire bridge. Their thickness may vary along the bridge to accommodate the gradient of action effects. The thickness of the thicker sheet is gradually reduced to the thinner to allow an execution of a butt weld and alleviate fatigue problems (Figure 8.41). One side of the sheet, for bottom flanges, the lower side, does not change in level that results in an eccentricity between the middle planes of adjacent sheets. At splices of sheets with changes in thickness, transverse stiffeners should be provided. The effects of eccentricity need not be taken into account when these transverse stiffeners are stiffening the thinner sheet and their distance to the splice is lower than min ($b_0/2$, 200 m), where b_0 is the width of subpanels between stiffeners.

8.11.7 Cutouts in stiffeners

Cutouts in longitudinal stiffeners are provided at positions of splices of plate sheets to avoid weld concentrations (Figure 8.42). Large cutouts in stiffeners may result in buckling of the compressed sheet in the unsupported region. A typical example was the *Rhine bridge* in Koblenz in which the cutout in the compression bottom flange had a length $l = 460$ mm or $l = 42 \cdot t$ for a plate thickness of 11 mm. Buckling of this sheet resulted in the collapse of this bridge during the cantilevering erection. As a consequence, maximum dimensions for cutouts both in height and length are specified. The height shall be limited to ¼ of the height of the stiffener but no more than 40 mm. The maximum length is limited to

- $l \leq 6 \cdot t_{min}$ for flat stiffeners in compression
- $l \leq 8 \cdot t_{min}$ for other stiffeners in compression
- $l \leq 15 \cdot t_{min}$ for stiffeners without compression

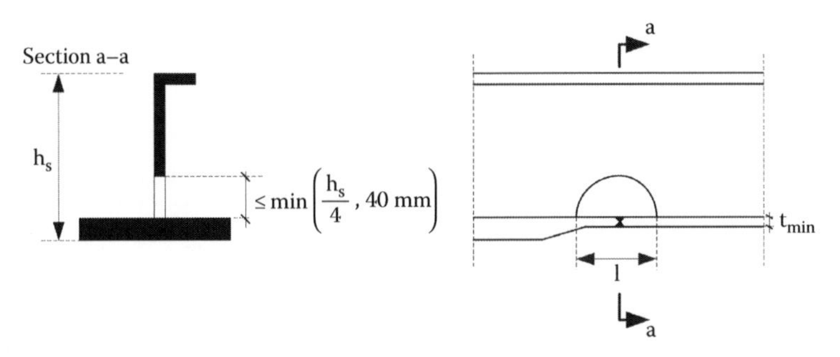

Figure 8.42 Cutouts in longitudinal stiffeners.

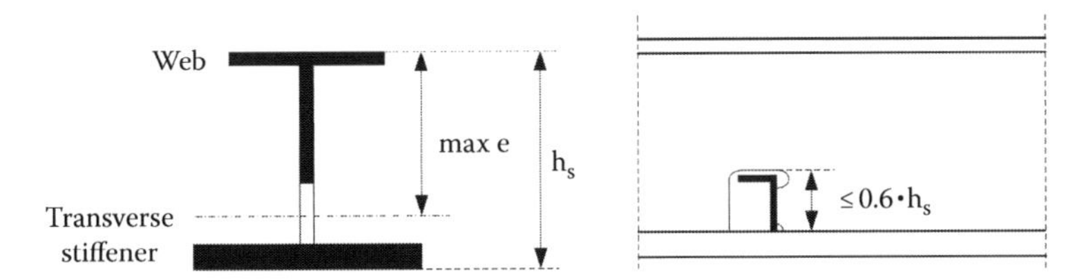

Figure 8.43 Cutouts in transverse stiffeners.

If the acting compression stress of the panel $\sigma_{x,Ed}$ is lower than the buckling resistance $\sigma_{x,Rd}$, then the earlier figures may be enhanced by the factor $\sqrt{\sigma_{x,Rd}/\sigma_{x,Ed}}$, but the resulting length should not exceed $15 \cdot t_{min}$.

8.11.8 Transverse stiffeners

Cutouts in transverse stiffeners are provided to allow continuity of the longitudinal stiffeners that pass through them. Their maximum dimensions are shown in Figure 8.43. In addition, the web of the stiffener in the cutout region should resist the following shear force:

$$V = \frac{I_{net}}{\max e} \cdot \frac{f_{yk}}{\gamma_{M0}} \cdot \frac{\pi}{b_G} \tag{8.81}$$

where

I_{net} is the second moment of area of the net section of the transverse stiffener
max e is the maximum distance from the neutral axis of the net section
b_G is the span of the stiffener

8.11.9 Web to flange welds

When the web is able to resist the acting shear without contribution of the flanges, that is, when $V_{Ed} \leq V_{bw,Rd}$ from Equation 8.51, the web to flange welds may be designed to resist the nominal shear force. Otherwise, they shall be designed for the full shear resistance of the web $n \cdot f_{yw} \cdot t/(\sqrt{3} \cdot \gamma_{M1})$.

EXAMPLE 8.10

The cross section of the box girder bridge in Example 8.9 shall be verified by the effective width method.

Effective area of the bottom flange

a. Effective width of the edge subpanels

The edge subpanels between the web and the edge stiffeners have width b = 55 cm and thickness t = 1.5 cm.

Reference stress, Equation 8.10:

$\sigma_e = 14.12\ kN/cm^2$

Stress ratio: $\psi = 1$
Buckling factor (Table 8.2): $k_\sigma = 4$
Critical plate buckling stress, Equation 8.9:

$$\sigma_{cr,p} = 4 \cdot 14.12 = 56.48 \text{ kN/cm}^2$$

Nondimensional plate slenderness, Equation 8.18:

$$\bar{\lambda}_p = \sqrt{\frac{35.5}{56.48}} = 0.793 > 0.673$$

Reduction factor for plate buckling, Equation 8.23:

$$\rho = \frac{0.793 - 0.055 \cdot (3 + 1)}{0.793^2} = 0.911$$

Effective width (Table 8.12):

$$b_{eff} = 0.911 \cdot 55 = 50.11 \text{cm}$$

Figure 8.27:

$$b_{l,edge,eff} = 0.5 \cdot 50.11 = 25.06 \text{cm}$$

b. Reduction factors for other subpanels of plate
For the subpanels between the webs of the stiffeners ($b = 35$ cm, $t = 1.5$ cm), it was found from Example 8.9 that there is no reduction due to plate buckling so that $\rho_{loc} = 1$.
For the subpanels between the stiffeners ($b = 55$ cm, $t = 1.5$ cm), it was found from Example 8.9 that $\rho_{loc} = 0.911$. Therefore, from Table 8.12, $b_{eff} = 0.911 \cdot 55 = 50.11$ cm
Accordingly, the effective area of the bottom plate is, Equation 8.47,

$$\sum_c \rho_{loc} \cdot b_{c,loc} \cdot t = (4 \cdot 50.11 + 4 \cdot 35) \cdot 1.5 = 510.7 \text{ cm}^2$$

The effective parts of the subpanels that are supported by the webs were excluded.

c. Effective area of the stiffeners
The reduction factor for the inclined web of stiffener ($b = 31.9$ cm, $t = 0.8$ cm) was found in Example 8.9 $\rho = 0.864$.
For the top part of stiffener ($b = 14.5$ cm, $t = 0.8$ cm), it was found from Example 8.9 that $\rho = 1.0$.
The effective area of the stiffeners is therefore

$$A_{sl,eff,1} = (2 \cdot 0.864 \cdot 31.9 + 14.5) \cdot 0.8 = 55.7 \text{ cm}^2$$

Effective area of all stiffeners:

$$A_{sl,eff} = 4 \cdot 55.7 = 222.8 \text{ cm}^2$$

d. Effective area of the stiffened panel

The effective area of the stiffeners and subpanels is equal to, Equation 8.47,

$$A_{c,eff,loc} = 222.8 + 510.7 = 733.5\ \text{cm}^2$$

The reduction factor of the stiffened panel is calculated as follows.

Plate buckling

From Example 8.9, it was found that

$$\sigma_{cr,p} = 99.88\,\text{kN/ cm}^2$$

Nondimensional plate slenderness, Equation 8.18:

$$\bar{\lambda}_p = \sqrt{\frac{35.5}{99.88}} = 0.596 \leq 0.673$$

Reduction factor for plate buckling, Equation 8.23: $\rho = 1$

Column-like behavior

From Example 8.9, it was found that

$$\sigma_{cr,c} = 105.5\ \text{kN/cm}^2$$

During the subpanel verifications, it was found that the reduction factors for the inclined wall of the stiffener and the plate between stiffeners are equal to $\rho = 0.864$ and $\rho = 0.911$, respectively. Therefore, the relevant widths must be reduced due to plate buckling to $0.864 \cdot 31.9 = 27.6$ cm and $0.911 \cdot 55 = 50.1$ cm, respectively. The effective area of the stiffener and the associated plate is accordingly

$$A_{sl,1,eff} = (35 + 50.1) \cdot 1.5 + 2 \cdot 27.6 \cdot 0.8 + 14.5 \cdot 0.8 = 183.41\,\text{cm}^2$$

$$\rightarrow \beta_{A,c} = \frac{183.41}{196} = 0.936$$

Column slenderness, Equation 8.35:

$$\bar{\lambda}_c = \sqrt{\frac{0.936 \cdot 35.5}{105.5}} = 0.55$$

$$I_{sl,1,eff} = 19{,}450.3\ \text{cm}^4 \rightarrow i = \sqrt{\frac{19450.3}{183.41}} = 10.3\ \text{cm}$$

Figure 8.22: $e_1 = 12.8$ cm, $e_2 = 5.80$ cm

$e = \max(12.8, 5.80) = 12.8$ cm

Imperfection factor for closed stiffener section:

$$\alpha_e = 0.34 + \frac{0.09}{10.3/12.8} = 0.45$$

Equation 8.37:

$$\Phi = 0.5 \cdot \left[1 + 0.45 \cdot (0.55 - 0.2) + 0.55^2\right] = 0.73$$

Reduction factor, Equation 8.36:

$$\chi_c = \frac{1}{0.73 + \sqrt{0.73^2 - 0.55^2}} = 0.826$$

Equation 8.33:

$$\xi = \frac{99.88}{105.5} - 1 = -0.05 < 0 \rightarrow \xi = 0$$

Final reduction factor, Equation 8.32:

$$\rho_c = \chi_c = 0.826$$

The effective area of the stiffened panel is then, Equation 8.46,

$$A_{c,eff} = 0.826 \cdot 733.5 + 2 \cdot 25.06 \cdot 1.5 = 681.1\,\text{cm}^2$$

Effective area of the web

The stress ratio of the web will be determined for a new cross section consisting of the effective flange area and the gross web area. In order to determine the properties of this cross section, the equivalent thickness of the bottom flange is found from

$$t_{eq} = \frac{A_{c,eff}}{b} = \frac{681.1}{415} = 1.64\,\text{cm}$$

The properties of this cross section are as follows:

Second moment of area:

$$I_y = 17.371 \cdot 10^6\,\text{cm}^4$$

The clear width of the web between flanges is $h_w = 212.0$ cm. The width of the tension zone is 100.4 cm and of the compression zone 111.6 cm. Therefore, the lower stiffener is in compression of the upper stiffener in tension.

The stiffeners are equally distanced. All subpanels between the stiffeners have therefore a width of

$$b = \frac{212}{3} = 70.67\,\text{cm}.$$

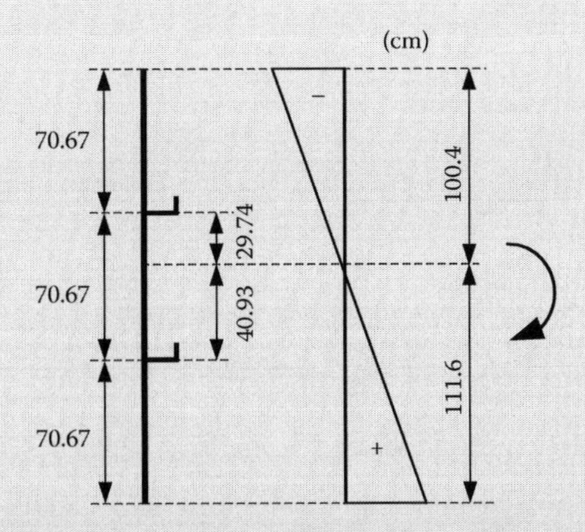

a. Effective width for the intermediate panel
   b = 70.67 cm, t = 1.2 cm
   Stress ratio: $\psi = -0.73$
   Reference stress, Equation 8.10:

   $$\sigma_e = 5.47 \text{ kN/cm}^2$$

   Buckling factor (Table 8.2):

   $$k_\sigma = 7.81 - 6.29 \cdot (-0.73) + 9.78 \cdot (-0.73)^2 = 17.61$$

   Critical plate buckling stress, Equation 8.9:

   $$\sigma_{cr,p} = 17.61 \cdot 5.47 = 96.32 \text{ kN/cm}^2$$

   Nondimensional plate slenderness, Equation 8.18:

   $$\bar{\lambda}_p = \sqrt{\frac{35.5}{96.32}} = 0.61 \leq 0.673 \rightarrow \rho = 1$$

   Effective width (Table 8.12):

   $$b_{eff} = 1.0 \cdot 40.93 = 40.93 \text{ cm}$$

b. Effective widths for the compression edge subpanel
   The lower edge subpanels between the web and the compression stiffener have a width b = 70.67 cm and thickness t = 1.2 cm.
   Stress ratio: $\psi = 0.37$
   Reference stress, Equation 8.10:

   $$\sigma_e = 5.47 \text{ kN/cm}^2$$

Buckling factor (Table 8.2):

$$k_\sigma = \frac{8.2}{1.05 + 0.37} = 5.77$$

Critical plate buckling stress, Equation 8.9:

$$\sigma_{cr,p} = 5.77 \cdot 5.47 = 31.6 \text{ kN/cm}^2$$

Nondimensional plate slenderness, Equation 8.18:

$$\bar{\lambda}_p = \sqrt{\frac{35.5}{31.6}} = 1.06 > 0.673$$

Reduction factor for plate buckling, Equation 8.23:

$$\rho = \frac{1.06 - 0.055 \cdot (3 + 0.37)}{1.06^2} = 0.78$$

Effective width (Table 8.12):

$$b_{eff} = 0.78 \cdot 70.67 = 55.1 \text{ cm}$$

Effective width of intermediate subpanel, Figure 8.27:

$$b_{I,edge,eff} = \frac{0.78 \cdot 70.67}{2} = 27.55 \text{ cm}$$

c. Reduction factor for the stiffened panel

Plate buckling

Geometric properties:

$$a = 450 \text{ cm}, \; b = h_w = 212.0 \text{ cm}, \; t = 1.2 \text{ cm}$$

The critical plate buckling stress will be determined according to Table 8.6 since only one stiffener is in the compression zone. With the notation of this table, it is

$$b_I = 70.67 \text{ cm}, \quad b_2 = 2 \cdot 70.67 = 141.34 \text{ cm}, \quad b_c = 111.6 - 70.67 = 40.93 \text{ cm}$$

Stress ratio for the lower subpanel:

$$\psi_I = \frac{40.93}{111.6} = 0.37$$

Stress ratio for the intermediate subpanel:

$$\psi_2 = \frac{-29.74}{40.93} = -0.73$$

Associated plate widths for the stiffener (Table 8.6):
Lower subpanel:

$$\frac{3-\psi_I}{5-\psi_I} \cdot b_I = \frac{3-0.37}{5-0.37} \cdot 70.67 = 42.5 \text{ cm}$$

Intermediate subpanel: $0.4 \cdot b_c = 0.4 \cdot 40.93 = 16.4$ cm (due to reversal of stresses)
The associated plate width is therefore 42.5 + 16.4 = 58.9 cm.

Properties of the longitudinal stiffeners L100 × 50 × 8:

$A_{st} = 11.5 \text{ cm}^2$, $I_{st} = 116 \text{ cm}^4$, $z_s = 3.59$ cm (distance of center of gravity from top)

Properties of the longitudinal stiffener + associated plate width:

$$A_{sl,I} = 82.2 \text{ cm}^2, \quad I_{sl,I} = 599.6 \text{ cm}^4, \quad i = 2.7 \text{ cm}$$

$$\text{Table 8.6: } a_c = 4.33 \cdot \sqrt[4]{\frac{599.6 \cdot 70.67^2 \cdot 141.34^2}{212 \cdot 1.2^3}} = 489.5 \text{ cm} > a = 450 \text{ cm}$$

Critical stress of the stiffener (Table 8.6):

$$\sigma_{cr,sl} = \frac{\pi^2 \cdot 21000 \cdot 599.6}{82.2 \cdot 450^2} + \frac{21000 \cdot 1.2^3 \cdot 212 \cdot 450^2}{4 \cdot \pi^2 \cdot (1-0.3^2) \cdot 82.2 \cdot 70.67^2 \cdot 141.34^2}$$

$$\rightarrow \sigma_{cr,sl} = 7.5 + 5.3 = 12.8 \text{ kN/cm}^2$$

Critical plate buckling stress (Table 8.6):

$$\sigma_{cr,p} = 12.8 \cdot \frac{40.93 + 70.67}{40.93} = 34.9 \text{ kN/cm}^2$$

Nondimensional plate slenderness, Equation 8.18:

$$\bar{\lambda}_p = \sqrt{\frac{35.5}{39.4}} = 0.95 > 0.673$$

Stress ratio for the web panel:

$$\psi = -\frac{100.4}{111.6} = -0.9$$

Reduction factor for plate buckling, Equation 8.23:

$$\rho = \frac{0.95 - 0.055 \cdot (3 - 0.9)}{0.95^2} = 0.92$$

Column-like behavior

Critical stress of the stiffener according to Equation 8.30 and cross sectional properties according to Figure 8.22 and Table 8.11:

$$\sigma_{cr,sl} = \frac{\pi^2 \cdot 21000 \cdot 672}{92 \cdot 450^2} = 7.5 \text{ kN/cm}^2$$

Critical column buckling stress of the panel, Equation 8.31:

$$\sigma_{cr,c} = 7.5 \cdot \frac{111.6}{40.93} = 20.4 \text{ kN/cm}^2$$

The stiffener is not class 4 → $\beta_{A,C} = 1$

Column slenderness, Equation 8.35:

$$\bar{\lambda}_c = \sqrt{\frac{1 \cdot 35.5}{20.4}} = 1.32$$

Figure 8.22: $e_1 = 6.05$ cm, $e_2 = 0.96$ cm

e = max (6.05, 0.96) = 6.05 cm

Imperfection factor for open stiffener section:

$$\alpha_e = 0.49 + \frac{0.09}{2.7/6.05} = 0.69$$

Equation 8.37:

$$\Phi = 0.5 \cdot [1 + 0.69 \cdot (1.32 - 0.2) + 1.32^2] = 1.76$$

Reduction factor, Equation 8.36:

$$\chi_c = \frac{1}{1.76 + \sqrt{1.76^2 - 1.32^2}} = 0.34$$

Equation 8.33:

$$\xi = \frac{34.9}{20.4} - 1 = 0.71$$

Final reduction factor, Equation 8.32:

$$\rho_c = (0.92 - 0.34) \cdot 0.71 \cdot (2 - 0.71) + 0.34 = 0.87$$

d. Effective area of the compression zone
The effective area of the web is, Equation 8.47,

$$A_{c,eff,loc} = \sum_c \rho_{loc} \cdot b_{c,loc} \cdot t = \left(40.93 + \frac{55.1}{2}\right) \cdot 1.2 = 82.2 \text{ cm}^2$$

This area does not include the area of the stiffeners, since it does not participate in the compression resistance, but is used only for increasing the buckling resistance of the panel; $A_{sl,eff} = 0$ in Equation 8.47.

e. Effective area of the web
The effective area of the compression zone includes the area near the compression stiffener and the area of the edge zone, Equation 8.46:

$$A_{c,eff} = 0.87 \cdot 82.2 + 27.55 \cdot 1.2 = 104.6 \text{ cm}^2$$

Area of the tension zone:

$$A_{ten} = 100.4 \cdot 1.2 = 120.5 \text{ cm}^2$$

Effective area of the web:

$$A_{eff} = 104.6 + 120.5 = 225.1\,\text{cm}^2$$

In order to determine the properties of the effective cross section, a reduced thickness of the web is considered:

$$t_{eq} = \frac{225.1}{212} = 1.06\,\text{cm}$$

Effective cross section

The properties of the effective cross section with equivalent thickness for the flange 1.64 cm and for the web 1.06 cm are as follows:

Second moment of the area:

$$I_y = 16.736 \cdot 10^6\,\text{cm}^4$$

Distance of neutral axis from the top fiber: z = 102 cm

Distance of neutral axis from the bottom fiber: z = 110 cm

Section modulus of the effective cross section:

$$W_{eff} = \frac{16.736 \cdot 10^6}{110} = 152.1 \cdot 10^3\,\text{cm}^3$$

Bending verification

Equation 8.49:

$$\eta_1 = \frac{3.9 \cdot 10^6}{152.1 \cdot 10^3 \cdot 35.5/1.0} = 0.72 \le 1 \text{ (sufficient)}$$

Resistance to shear of the flange

The shear of the flange is due to torsion that may be determined from the second formula of *Bredt*.

Shear flow in the flange due to torsion:

$$v_{Ed} = \frac{M_T}{2 \cdot A_0} = \frac{7.55 \cdot 10^5}{2 \cdot 105757} = 3.57\,\text{kN/cm}$$

Design shear force of the flange:

$$V_{Ed} = v_{Ed} \cdot b = 3.57 \cdot 415 = 1481.55\,\text{kN}$$

The reduction factor for shear was determined from Example 8.9 as $\chi_w = 1.0$.

Shear resistance of the flange, Equation 8.51:

$$V_{w,Rd} = \frac{1 \cdot 415 \cdot 1.5 \cdot 35.5}{\sqrt{3} \cdot 1.1} = 11598.8\,\text{kN} > V_{Ed}$$

$$= 1481.55\,\text{kN (sufficient, Equation (8.55))}$$

It is (Equation 8.66) $\bar{\eta}_3 = \frac{1481.55}{11,598.8} < 0.5$ so that no interaction with the bending moment needs to be considered.

Resistance to shear of the web

The webs resist shear due to shear force and torsion.

Shear due to shear force:

$$V_{Ed,V} = \frac{V_{Ed}}{2 \cdot \sin\alpha} = \frac{1600}{2 \cdot \sin 74.7} = 829.5 \text{ kN}$$

Shear due to torsion:

$$V_{Ed,T} = v_{Ed} \cdot b = 3.57 \cdot 212 = 756.8 \text{ kN}$$

The two shear forces are added in one web and subtracted in the other.
Shear force in the most stressed web:

$$V_{Ed} = V_{Ed,V} + V_{Ed,T} = 1586.3 \text{ kN}$$

The buckling stress for shear was determined from Example 8.9 as $\tau_{cr} = 8.69$ kN/cm^2.

Slenderness, Equation 8.19:

$$\bar{\lambda}_w = 0.76 \cdot \sqrt{\frac{35.5}{8.69}} = 1.53$$

Reduction factor (Table 8.10):

$$\chi_w = \frac{0.83}{1.53} = 0.54 \text{ (for nonrigid end post)}$$

Shear resistance due to contribution of the web, Equation 8.51:

$$V_{bw,Rd} = \frac{0.54 \cdot 212 \cdot 1.2 \cdot 35.5}{\sqrt{3} \cdot 1.1} = 2559.7 \text{ kN} > V_{Ed} = 1586.3 \text{ kN (sufficient)}$$

It is (Equation 8.66) $\bar{\eta}_3 = \frac{1586.3}{2559.7} = 0.62 > 0.5$ so that the interaction with the bending moment needs to be considered.

Interaction bending moment and shear

The plastic moment of the box section is found as $M_{pl,Rd} = 5.754 \cdot 10^6$ kN-cm.

There is no axial force, so that $M_{N,pl,Rd} = M_{pl,Rd}$.

Equation 8.65:

$$\bar{\eta}_1 = \frac{3.9 \cdot 10^6}{5.754 \cdot 10^6} = 0.68$$

For a box girder, it is $M_{f,Rd} = 0$.
Interaction, Equation 8.64:

$$0.68 + (1 - 0) \cdot (2 \cdot 0.62 - 1)^2 = 0.74 \leq 1 \text{ (sufficient)}$$

REMARK 8.14

The effective width method has attracted criticism [8.3], [8.8], [8.9] due to the fact that it does not cover typical loading situations, which can occur during launching. This can be the case of very long launching bearings. Moreover, in certain cases, there may not be any considerable economic advantages compared to the reduced stress method (compare Examples 8.9 and 8.10).

EXAMPLE 8.11

The longitudinal web stiffeners of Examples 8.9 and 8.10 should be verified for lateral torsional buckling. Cross section L100 × 50 × 8, steel grade S 355.

Properties of the cross section:

$A_{st} = 11.5\ cm^2$, $I_{st} = 116\ cm^4$, $z_s = 3.59\ cm$, $I_z = 19.5\ cm^4$, $I_T = 2.5\ cm^4$

Distance of the stiffener's centroid from the web plate $z = 10 - 3.59 = 6.41$ cm.

Distance between stiffener's centroid and midline of its web $y = 0.73$ cm.

Second moments of area of the stiffener around the edge fixed to the plate:

$$I_y = 116 + 11.5 \cdot 6.41^2 = 587\ cm^4$$

$$I_z = 19.5 + 11.5 \cdot 0.73^2 = 25.6\ cm^4$$

Polar second moment of area of the stiffener around the edge fixed to the plate:

$$I_p = 587 + 25.6 = 612.6\ cm^4$$

Torsional requirement, Equation 8.78:

$$\frac{2.5}{612.6} = 0.0041 < 5.3 \cdot \frac{355}{210{,}000} = 0.009$$

The criterion is not satisfied so that the criterion of Equation 8.80 taking into account the warping rigidity will be examined:

$$I_w = \frac{1}{3} \cdot (h_{st} - 0.5 \cdot t_{st})^2 \cdot b_{st}^3 \cdot t_{st} = \frac{1}{3} \cdot (10 - 0.5 \cdot 0.8)^2 \cdot 5^3 \cdot 0.8 = 3072\ cm^6$$

The length between transverse stiffeners is a = 450 cm. However, since the transverse stiffeners provide fixity to the longitudinal one, the buckling length will be taken as half this distance:

$$\sigma_{cr} = \frac{1}{612.6} \cdot \left(\frac{21{,}000 \cdot 2.5}{2 \cdot (1 + 0.3)} + \frac{\pi^2 \cdot 21{,}000 \cdot 3072}{225^2} \right)$$

$$= 53.5\ kN/cm^2 < 6 \cdot 35.5 = 213\ kN/cm^2$$

Accordingly, criterion (8.80) is also not satisfied.

In a further step, σ_{cr} is determined taking into account restraining from the web plate that provides a continuous elastic torsional support. The relevant rigidity is given by

$$c_\theta = \frac{4 \cdot E_a \cdot I_{plate}}{b} = \frac{E_a \cdot t^3}{3 \cdot b \cdot (1 - \nu^2)}$$

where b is the distance between stiffeners.

Considering the stiffener as a strut on elastic support, its critical stress is obtained from

$$\sigma_{cr} = \frac{1}{I_p} \cdot (2 \cdot \sqrt{c_\theta \cdot E_a \cdot I_w} + G_a \cdot I_T)$$

that is valid if the length of the stiffener is larger than $l_{cr} = \pi \cdot \sqrt[4]{\frac{E_a \cdot I_w}{c_\theta}}$.

For the stiffener under consideration, it is

$$c_\theta = \frac{21{,}000 \cdot 1.2^3}{3 \cdot 70.67 \cdot (1 - 0.3^2)} = 188 \text{ kN-m/m}$$

The critical length is

$$l_{cr} = \pi \cdot \sqrt[4]{\frac{21{,}000 \cdot 3072}{188}} = 76.0 \text{ cm} < 450 \text{ cm}$$

The critical stress is then

$$\sigma_{cr} = \frac{1}{612.6} \cdot (2 \cdot \sqrt{188 \cdot 21{,}000 \cdot 3072} + (21{,}000/2.6) \cdot 2.5) = 392 \text{ kN/cm}^2$$

and

$$\sigma_{cr} > 6 \cdot 35.5 = 213 \text{ kN/cm}^2 \text{ (sufficient)}$$

EXAMPLE 8.12

In Examples 8.9 and 8.10, the transverse stiffeners of the bottom flange have a T shape with flange plate 200 × 20 mm and a web plate 300 × 15 mm. The distance of the stiffeners is 450 mm. The stiffeners shall be verified.

The stiffeners are not subjected to axial forces so that they will be verified by Equation 8.72.

Maximal compression stress of the flange, Example 8.9:

$$\sigma_{max} = 20.47 \text{ kN/cm}^2$$

Effective area of the flange, from Example 8.10:

$$A_{c,eff} = 681.1 \text{ cm}^2$$

Compression force of the adjacent panels:

$$N_{Ed} = 20.47 \cdot \frac{681.1}{2} = 6971.1\ \text{kN}$$

Critical plate buckling stress, Example 8.9:

$$\sigma_{cr,p} = 99.88\ \text{kN/cm}^2$$

Critical column buckling stress, Example 8.9:

$$\sigma_{cr,c} = 105.5\ \text{kN/cm}^2$$

Length of stiffener = width of compression flange: $b = 415$ cm
Length of adjacent panels: $a_1 = a_2 = 450$ cm

Equation 8.73:

$$\sigma_m = \frac{105.5 \cdot 6971.1}{99.88 \cdot 415} \cdot \frac{2}{450} = 0.0788\ \text{kN/cm}^2$$

The stiffener includes adjacent parts of the flange plate with width $15 \cdot \varepsilon \cdot t (= 15 \cdot 0.81 \cdot 1.5 = 18.23\ \text{cm})$ on each side of the web. For this section, it is $I_{st} = 26{,}899\ \text{cm}^4$, and the distance of its centroid from the extreme fiber is $e_{max} = 18.8$ cm.

Equation 8.74:

$$u = \frac{\pi^2 \cdot 21{,}000 \cdot 18.8}{35.5 \cdot 300 \cdot 415/1.1} = 0.97 < 1 \rightarrow u = 1$$

Imperfection: $w_0 = \min(415, 450)/300 = 1.38$ cm

Equation 8.72:

$$I_{st,min} = \frac{0.0788}{21{,}000} \cdot \left(\frac{415}{\pi}\right)^4 \cdot \left(1 + 1.38 \cdot \frac{300}{415} \cdot 1\right) = 2287.1\ \text{cm}^4$$

and $I_{st} = 26899\ \text{cm}^4 > I_{st,min}$ (sufficient)

REFERENCES

[8.1] Bazant, Z. P., Cedolin, L.: *Stability of Structures. Elastic, Inelastic, Fracture and Damage Theories*, Dover Publications, Mineola, NY, 2003.

[8.2] Beg, D.: Plated and box girder stiffener design in view of Eurocode 3—Part 1.5. In: *Sixth National Conference on Metal Structures*, Ioannina, Greece, 2008.

[8.3] Braun, B., Kuhlmann, U.: *Reduced Stress Design of Plates under biaxial Loading*. Steel Construction 5, No. 1, pp. 33–40, Ernst & Sohn, Berlin, Germany, 2012.

[8.4] EN 1993-1-5, CEN (European Committee for Standardization): Design of steel structures, Part 1–5: Plated structural elements, 2006.
[8.5] EN 1993-2, CEN (European Committee for Standardization): Design of steel structures—Part 2: Steel Bridges, 2006.
[8.6] EBPlate Version 2.01, *Centre Technique Industriel de la Construction Metallique (CTICM)*. CTICM: www.cticm.com
[8.7] Grote, H.: *Zum Einfluss des Beulens auf die Tragfähigkeit von Walzprofilen aus hochfestem Stahl*. Reihe 4, Nr. 195, Fortschritt-Berichte VDI—Düsseldorf, Germany, 2003.
[8.8] Hanswille, G.: *The New German Code for Composite Bridges. Institute for Steel and Composite Structures*, University of Wuppertal, Wuppertal, Germany, 2004.
[8.9] Hendy, C. R., Murphy, C. J.: *Designers' Guide to EN 1993-2 Eurocode 3: Design of steel structures. Part 2: Steel bridges*, Thomas Telford, London, U.K., 2007.
[8.10] Johansson, B., Maquoi, R., Sedlacek, G., Müller, C., Schneider, R.: *New Design Rules for Plated Structures in Eurocode 3*. Stahlbau 68, Heft 11, pp. 857–879, Ernst und Sohn, Berlin, Germany, 1999.
[8.11] Kindmann, R.: *Brücken und Kranbahnträger. Brücken, Plattenbeulen, Betriebsfestigkeit. Ruhr*—Universität Bochum, Lehrstuhl für Stahl- und Verbundbau, 2004.
[8.12] Klöppel, K., Sheer, J.: *Buckling Coefficients of Stiffened Rectangular Plates*, Ernst und Sohn, Berlin, Germany, 1960.
[8.13] Kuhlmann, U., Seitz, M.: Longitudinally stiffened girder webs subjected to patch loading. In: *Proceedings of the Steel-Bridge 2004 Conference*, Millau, France, 2004.
[8.14] Scheer, J.: *Failed bridges. Case Studies, Causes and Consequences*, Ernst & Sohn, Berlin, Germany, 2010.
[8.15] Scheer, J., Vayas, I.: *Traglastversuche an längsgestauchten und schubbeanspruchten versteiften Platten*. Der Stahlbau 52, pp. 207–213, Ernst und Sohn, Berlin, Germany, 1983.
[8.16] Sedlacek, G., Eisel, H., Hensen, W., Kühn, B., Paschen, M.: *Leitfaden zum DIN Fachbericht 103*—Stahlbrücken, Ernst und Sohn, Berlin, Germany, 2003.
[8.17] Von Karman, T.: *Festigkeitsproblem im Maschinenbau*. Encyclopaedie der Mathematischen Wissencshaften, 1910.
[8.18] Winter, G.: Stress distribution in and equivalent width of flanges of wide thin-wall steel beams, NACA Technical, Note 784, 1940.

Chapter 9

Ultimate limit states

9.1 CLASSIFICATION OF CROSS SECTIONS

The purpose of cross-sectional classification is to examine whether their bending resistance may be determined by elastic or plastic analysis and whether their walls are subjected to local buckling so that additional verifications to plate buckling according to Chapter 8 will be required. Four classes of cross sections are distinguished as indicated in Table 9.1:

Class 1: Sections develop their plastic bending resistance and have sufficient rotation capacity.
Class 2: Sections develop their plastic bending resistance but have limited rotation capacity.
Class 3: Sections develop their elastic bending resistance.
Class 4: Sections are subjected to local buckling and have a resistance lower than the elastic resistance.

Since local buckling is crucial for cross sections in developing their strength and ductility, the width to thickness ratios (c/t) of the compressed walls serves as the criterion for the classification. To classify a cross section under a combination of an axial force and a bending moment (N, M), a plastic stress distribution is considered first. Subsequently, the c/t ratios are examined for each wall separately, to classify them in class 1 or 2. If the walls fail to be classified in class 1 or 2, an elastic stress distribution is considered to examine if the walls satisfy the limits for class 3. If they do not, these walls belong to class 4. The entire cross section is then classified in accordance with the largest class of its walls. Class 4 cross sections can be classified as class 3 if plate buckling verifications are made by the *reduced stress method* in accordance with Section 8.4. Alternatively, the *effective width method* in accordance with Section 8.5 is employed. The reduced stress method refers mainly to cross sections with longitudinally stiffened walls. Cross sections with unstiffened walls are almost always verified for plate buckling with the effective width method.

Tables 9.2 and 9.3 give the limiting c/t ratios for internal elements supported at two edges and external elements supported at one edge only. It may be seen that for rolled sections, the width c is composed of the straight part of the element and for welded elements of the clear part between weld toes. For external elements, c/t limits are given only for compression, since they mirror stress conditions due to uniaxial bending of the cross section that is usually relevant for bridge sections. The distinction between classes 1 and 2 is in practice of limited importance for bridges, since analysis by plastic hinge theory that applies only for class 1 and not for class 2 may be employed only for accidental loadings.

Tables 9.2 and 9.3 may be used for the classification of steel girders at construction stages before concrete casting. After concrete casting, the top flanges are rigidly connected to the

Table 9.1 Classes of cross sections

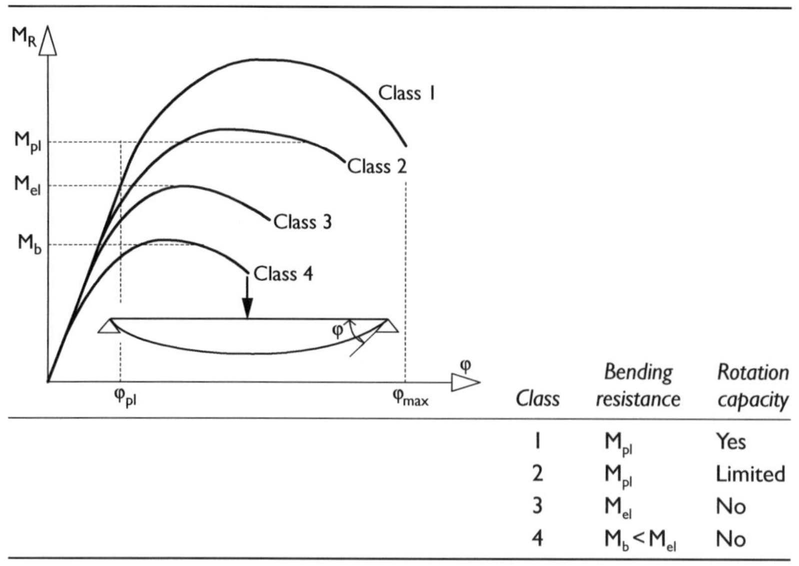

Class	Bending resistance	Rotation capacity
1	M_{pl}	Yes
2	M_{pl}	Limited
3	M_{el}	No
4	$M_b < M_{el}$	No

Notes:

Cross sections where plate buckling verifications are required:

Class 4 treated as class 3: Class 4 cross sections where plate buckling verifications are made by the reduced stress method; see Section 8.4.

Class 4: Plate buckling verifications are made by the effective width method; see Section 8.5.

concrete slab through the shear connectors. In such cases, the steel flange attached to the slab may be classified as class 1 or 2, although it could be class 3 or 4, provided the spacing of connectors is appropriately selected (see Table 9.3), since concrete prevents its local buckling. Accordingly, the classification of composite bridge sections at the service stage is relevant mainly for hogging bending where the bottom flange is in compression and the web partly in compression. In sagging bending, the cross section is usually class 1 or 2, since the compression flange is connected to the slab, and therefore, class 1 and the compression part of the web are usually small due to the position of the neutral axis near the top flange.

It is important to note that the class of a composite cross section depends also on the sequence of construction and the effects due to creep and shrinkage. Therefore, the classification should be conducted for short- and long-term design; see Tables 9.12 and 9.13. If, for example, the web is classified as class 3 for the short-term effects, then it may be changed to class 4 for the long-term ones. Indeed, the factor ψ in Table 9.2 is determined from the direct stresses that are time dependent due to creep and shrinkage; see Example 9.4. Moreover, the classification of a box-girder cross section is also influenced from the shear lag effect on the wide flanges.

Table 9.4 gives the classification limits for the bottom flanges of steel cross sections encased in concrete (*filler-beam decks*). Obviously, classification takes place only for the hogging moment areas. It is noted that for the hogging moments of continuous filler-beam decks with class 1 cross sections, a redistribution at ultimate limit state (ULS) other than fatigue up to 15% is allowed.

Cross sections with class 1 or 2 flanges and class 3 web may be classified as class 2 provided that only an effective part of the web in accordance with Figure 9.1 is considered. The effective part of the web in compression extends $20 \cdot \varepsilon \cdot t_w$ from the plastic neutral axis of the *effective section* and $20 \cdot \varepsilon \cdot t_w$ from the compression flange, the remaining part being not effective.

Table 9.2 Classification of internal elements

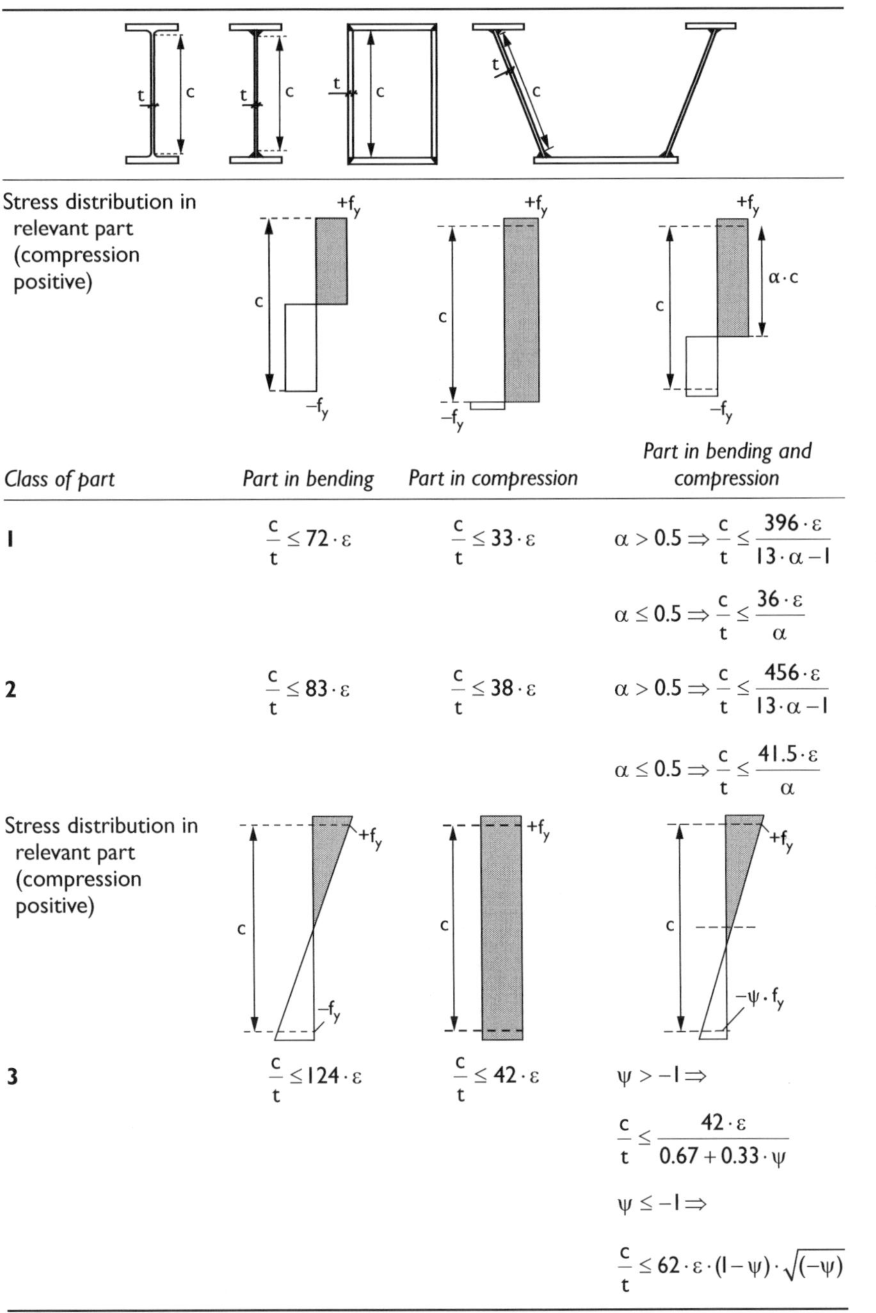

Class of part	Part in bending	Part in compression	Part in bending and compression
Stress distribution in relevant part (compression positive)	(see figure)	(see figure)	(see figure)
1	$\frac{c}{t} \leq 72 \cdot \varepsilon$	$\frac{c}{t} \leq 33 \cdot \varepsilon$	$\alpha > 0.5 \Rightarrow \frac{c}{t} \leq \frac{396 \cdot \varepsilon}{13 \cdot \alpha - 1}$ $\alpha \leq 0.5 \Rightarrow \frac{c}{t} \leq \frac{36 \cdot \varepsilon}{\alpha}$
2	$\frac{c}{t} \leq 83 \cdot \varepsilon$	$\frac{c}{t} \leq 38 \cdot \varepsilon$	$\alpha > 0.5 \Rightarrow \frac{c}{t} \leq \frac{456 \cdot \varepsilon}{13 \cdot \alpha - 1}$ $\alpha \leq 0.5 \Rightarrow \frac{c}{t} \leq \frac{41.5 \cdot \varepsilon}{\alpha}$
Stress distribution in relevant part (compression positive)	(see figure)	(see figure)	(see figure)
3	$\frac{c}{t} \leq 124 \cdot \varepsilon$	$\frac{c}{t} \leq 42 \cdot \varepsilon$	$\psi > -1 \Rightarrow$ $\frac{c}{t} \leq \frac{42 \cdot \varepsilon}{0.67 + 0.33 \cdot \psi}$ $\psi \leq -1 \Rightarrow$ $\frac{c}{t} \leq 62 \cdot \varepsilon \cdot (1 - \psi) \cdot \sqrt{(-\psi)}$

Source: EN 1993-1-1, Design of steel structures, Part 1-1: General rules and rules for buildings, 2005.

Note: $\varepsilon = \sqrt{\frac{235}{f_y}}$ f_y in [MPa] for buckling analysis or $\varepsilon = \sqrt{\frac{235}{\sigma_{com}}}$ for section design, where σ_{com} is the maximum compression stress in the part in [MPa].

Table 9.3 Classification of external elements

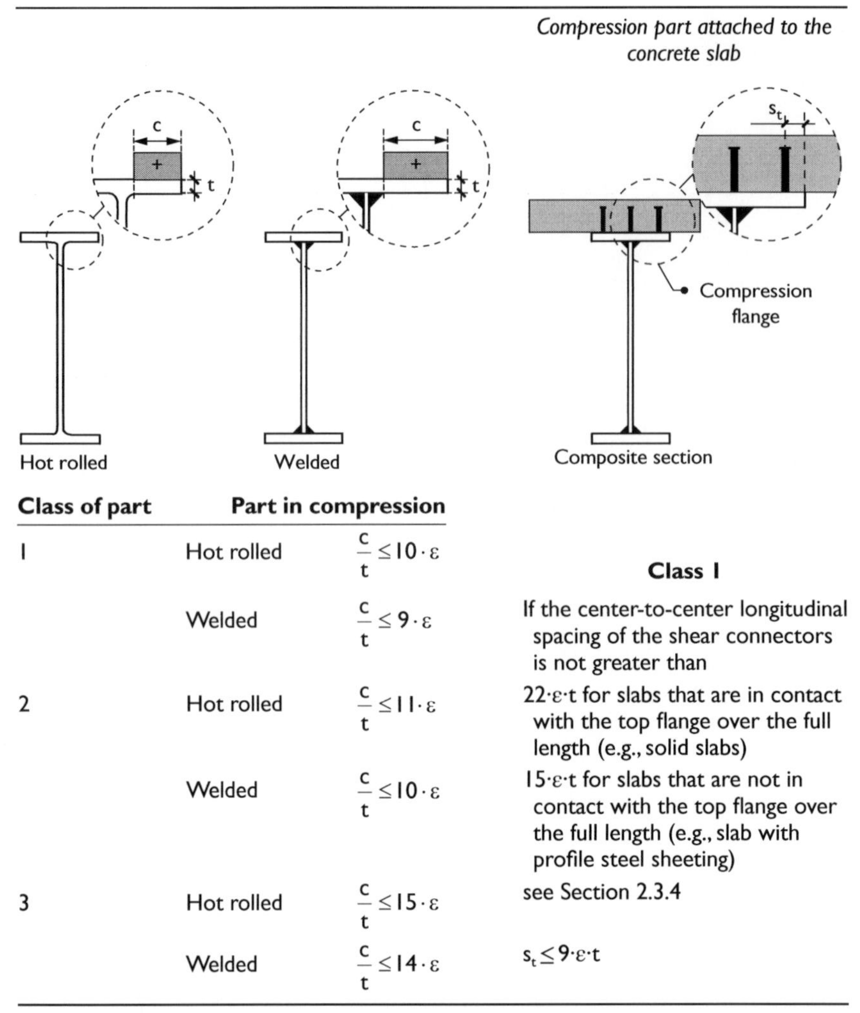

Class of part	Part in compression		Compression part attached to the concrete slab
1	Hot rolled	$\frac{c}{t} \leq 10 \cdot \varepsilon$	**Class 1**
	Welded	$\frac{c}{t} \leq 9 \cdot \varepsilon$	If the center-to-center longitudinal spacing of the shear connectors is not greater than
2	Hot rolled	$\frac{c}{t} \leq 11 \cdot \varepsilon$	$22 \cdot \varepsilon \cdot t$ for slabs that are in contact with the top flange over the full length (e.g., solid slabs)
	Welded	$\frac{c}{t} \leq 10 \cdot \varepsilon$	$15 \cdot \varepsilon \cdot t$ for slabs that are not in contact with the top flange over the full length (e.g., slab with profile steel sheeting)
3	Hot rolled	$\frac{c}{t} \leq 15 \cdot \varepsilon$	see Section 2.3.4
	Welded	$\frac{c}{t} \leq 14 \cdot \varepsilon$	$s_t \leq 9 \cdot \varepsilon \cdot t$

Sources: EN 1993-1-1, Design of steel structures, Part 1-1: General rules and rules for buildings, 2005; EN1994-2, Design of composite steel and concrete structures. Part 2: Rules for bridges, 2005.

Note: ε as in Table 9.2.

Table 9.4 Classification of filler-beam decks

Class of cross section	*Type of steel girder*	*Limit*
1	Hot rolled or welded	$\frac{c}{t} \leq 9 \cdot \varepsilon$
2		$\frac{c}{t} \leq 14 \cdot \varepsilon$
3		$\frac{c}{t} \leq 20 \cdot \varepsilon$

c

Hot rolled

t c

Welded

Source: EN1994-2, Design of composite steel and concrete structures. Part 2: Rules for bridges, 2005.

Note: ε as in Table 9.2.

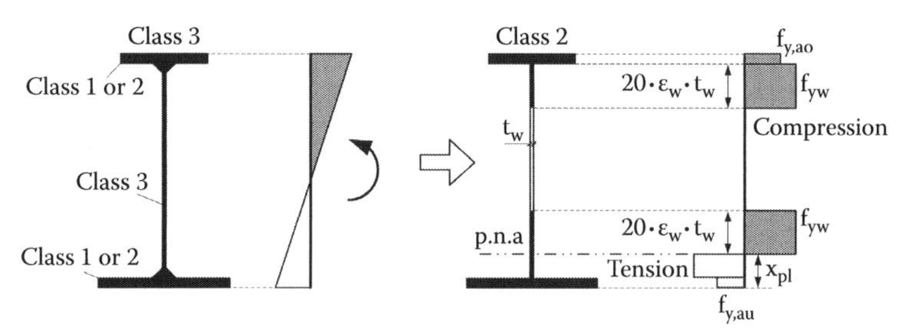

Figure 9.1 Effective class 2 web that was initially class 3. (From EN 1993-1-1, Design of steel structures, Part 1-1: General rules and rules for buildings, 2005.)

REMARK 9.1

Treating class 3 cross sections as class 2 according to Figure 9.1 is a design facilitation for avoiding complicated stress calculations. For composite cross sections, secondary internal forces due to the effects of creep and shrinkage for the ULS verifications *should not be neglected; use Tables 5.6 or 5.8*. The reason for this is that no plastifications will occur to absorb the secondary stresses. Unfortunately, this is not adequately explained in EN 1994-2.

The purpose of classification is to determine the bending resistance of cross sections, whether of pure steel or composite. In composite bridges, the cross sections turn during construction stages from pure steel to composite. However, the classification for *persistent design* situations refers to the final composite section. This is due to the fact that the level of loading during construction stages is generally low so that the bending resistance of the pure steel section is not exhausted. At such low levels of moments, the moment–rotation response is linear (see Table 9.1) and no local buckling occurs. On the contrary, during construction stages that constitute *transient design* situations, for example, during incremental launching of the steel girder or during concreting, the steel girder might be loaded near its capacity. Therefore, only for such design situations, a classification of the steel girder might be required. *Concluding, it may be said that for a bridge design, the classification and the corresponding verifications depending on the class refer to the composite section.* With these remarks, Table 9.5 gives a short presentation of the cross-sectional classes most usually found in composite bridges as well as the relevant section in which

Table 9.5 Most usual bridge section classes

Bridge section	*Top flange*	*Bottom flange*	*Web(s)*	*Plate buckling Verifications*	*Cross section*	*Verifications described in*
Plate girder	1 or 2	1 or 2	1 or 2	—	1 or 2	9.7
	1 or 2	1 or 2	3	—	2 with effective web	9.8
	1 or 2	1 or 2	3	—	3	9.9
	1 or 2	1 or 2	4	Reduced stress method	Treated as class 3	9.10
	1 or 2	1 or 2	4	Effective width method	4	9.11
Box girder	1 or 2	3	3	—	3	9.9
	1 or 2	4	3 or 4	Reduced stress method	Treated as class 3	9.10 or 9.12
	1 or 2	4	3 or 4	Effective width method	4	9.11 or 9.12

the required verifications are described. It may be seen that the flanges in plate girders are usually class 1 or 2 so that they are fully effective in bending. Oppositely, geometric considerations and detailing practice reveal that bottom flanges of box girders are wide and thin. They are therefore mostly class 4 and are stiffened by longitudinal stiffeners in hogging bending regions.

EXAMPLE 9.1

A bridge section is composed of two main girders. The steel girder is a welded plate girder (Figure 9.2). On top of the girder is a 30 cm thick concrete slab with an effective width of 6.0 m. The area of the reinforcement is at the top 145 cm² and at the bottom 93 cm² (Figure 9.3). The cross section shall be classified (a) for transient design situations during launching and (b) for persistent design situations at service stages. Steel grade for the steel girder S355, strength class of concrete C35/45, reinforcement B500B.

Yield stress of structural steel (Table 6.14):

Web t = 18 mm: $f_y = 345\ \text{N/mm}^2$

Flange t = 80 mm: $f_y = 325\ \text{N/mm}^2$

Accordingly, it is for the web $\varepsilon = \sqrt{\dfrac{235}{345}} = 0.825$ and for the flange $\varepsilon = \sqrt{\dfrac{235}{325}} = 0.85$.

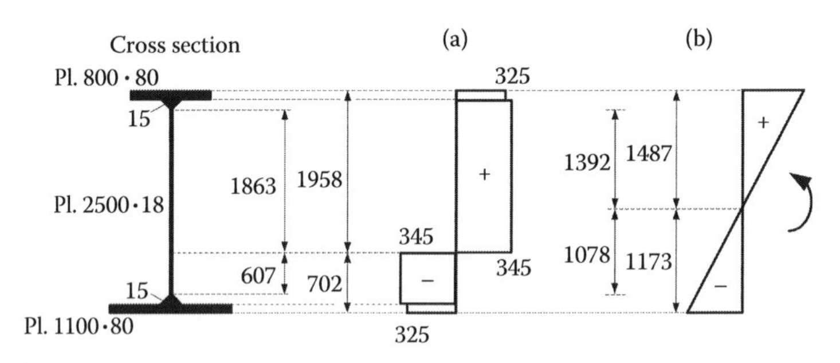

Figure 9.2 Girder section at construction stages and (a) plastic and (b) elastic stress distribution for the steel girder for sagging bending. Sign of stresses (+, compression; −, tension). Stresses in N/mm².

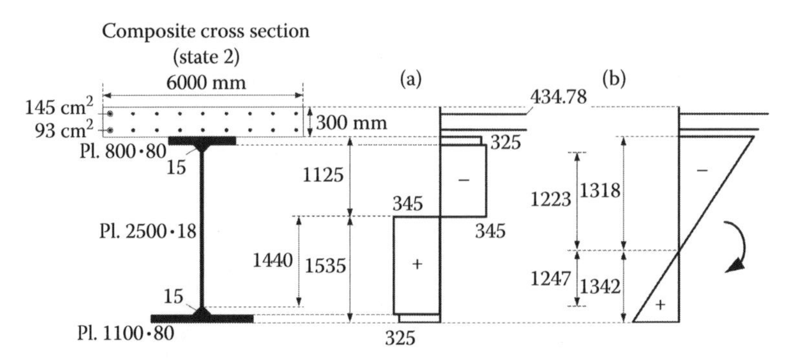

Figure 9.3 Girder section at service stages and (a) plastic and (b) elastic stress distribution for hogging bending, stresses in N/mm².

TRANSIENT DESIGN SITUATIONS AT CONSTRUCTION STAGES DURING LAUNCHING

For such situations, the concrete slab is not present, and the cross section is composed only of the steel girder. Accordingly, the classification refers to the pure steel girder that might be stressed up to its capacity and shall be made for positive and negative bending.

Sagging bending

For sagging bending, the classification refers to the top flange that is under compression and the web that is partly under compression. A plastic stress distribution as shown in Figure 9.2 is considered first.

Top flange

The top flange is an outstand element subjected to pure compression:

$$c = \frac{800 - 18}{2} - 15 = 376 \text{ mm}$$

Table 9.3:

$$\frac{c}{t} = \frac{376}{80} = 4.7 \leq 9 \cdot 0.85 = 7.65$$

The top flange is class 1.

Web

The web is an internal element subjected to partial compression:

$$c = 2500 - 2 \cdot 15 = 2470 \text{ mm}$$

The distance of the plastic neutral axis from the top fiber is 1958 mm. The compression part of the web is $1958 - 80 - 15 = 1863 \text{ mm} \rightarrow \alpha = \frac{1863}{2470} = 0.75 > 0.5$.
Table 9.2:

$$\frac{c}{t} = \frac{2470}{18} = 137.22 > \frac{456 \cdot 0.825}{13 \cdot 0.75 - 1} = 42.99$$

$\rightarrow$ the web is not class 2 and therefore also not class 1.

Accordingly, an elastic stress distribution is considered to determine if the web is class 3 or 4. For the elastic distribution, the distance of the center of gravity (neutral axis) from the top fiber is 1487 mm and from the bottom fiber 1173 mm. The part of the web under compression is $1487 - 80 - 15 = 1392$ mm and under tension $1173 - 80 - 15 = 1078$ mm. The stress ratio is accordingly $\psi = \frac{-1078}{1392} = -0.77 > -1$.

Table 9.2:

$$\frac{c}{t} = \frac{2470}{18} = 137.22 > \frac{42 \cdot 0.825}{0.67 + 0.33 \cdot (-0.77)} = 83.31$$

$\rightarrow$ the web is not class 3, and therefore, it is class 4.

Finally, the section is class 4 for sagging bending.

Hogging bending

For hogging bending, the classification refers to the bottom flange that is under compression and the web that is partly under compression. The stress distribution is the same as for sagging bending, as shown in Figure 9.2, the difference being the sign of the stresses that is reversed now. A plastic stress distribution is considered first.

Bottom flange

The bottom flange is an outstand element subjected to pure compression:

$$c = \frac{1100 - 18}{2} - 15 = 526 \text{ mm}$$

Table 9.3:

$$\frac{c}{t} = \frac{526}{80} = 6.58 < 9 \cdot 0.85 = 7.65$$

The bottom flange is class 1.

Web

For plastic stress distribution, the compression part of the web is 607 mm:

$$\alpha = \frac{607}{2470} = 0.245 < 0.5$$

Table 9.2:

$$\frac{c}{t} = \frac{2470}{18} = 137.22 > \frac{36 \cdot 0.825}{0.245} = 121.22 \rightarrow \text{not class 1}$$

$$\frac{c}{t} = \frac{2470}{18} = 137.22 \leq \frac{41.5 \cdot 0.825}{0.245} = 139.74 \rightarrow \text{the web is class 2}$$

Accordingly, the section is class 2 for hogging bending.

PERSISTENT DESIGN SITUATIONS AT SERVICE STAGES

At service stages, the concrete is casted and the cross section is composite. The cases of sagging and hogging bending will be considered.

Sagging bending

For sagging bending, the classification refers to the web that is potentially under compression. The compression top flange is protected from local buckling due to its attachment to the concrete slab and is classified as class 1; see also requirements for the shear connection in Table 9.3. The distance of the neutral axis from the top fiber of the slab considering a plastic stress distribution is 337 mm, which is smaller than 300 + 80 = 380 mm. Accordingly, the neutral axis is within the top flange. Consequently, the entire web is in tension and needs no classification. The cross section is therefore class 1.

Hogging bending

Design yield stress of reinforcing steel:

$$f_{sd} = \frac{500}{1.15} = 434.78\ \text{N/mm}^2$$

For hogging bending, the bottom flange and part of the web are in compression. The bottom flange was already found to be class 1. Therefore, only the web will be classified.

Plastic stress distribution

The width of the web is $c = 2500 - 2 \cdot 15 = 2470$ mm.

The plastic neutral axis is found to have a distance 1425 mm from the extreme fiber of the slab. The height of compression stresses is therefore 2960 − 1425 = 1535 mm.

The compression part of the web is 1535 − 80 − 15 = 1440 mm:

$$\alpha = \frac{1440}{2470} = 0.58 > 0.5$$

Table 9.2:

$$\frac{c}{t} = \frac{2470}{18} = 137.22 > \frac{456 \cdot 0.825}{13 \cdot 0.58 - 1} = 57.52 \rightarrow \text{web is not class 2.}$$

Accordingly, an elastic stress distribution is considered to determine if the web is class 3 or 4.

Elastic stress distribution

For the elastic distribution, the distance of the center of gravity (neutral axis) from the top fiber is 1618 mm and from the bottom fiber 1342 mm. The part of the web under compression is 1342 − 80 − 15 = 1247 mm and under tension 1618 − 300 − 80 − 15 = 1223 mm. The stress ratio is accordingly $\psi = \frac{-1223}{1247} = -0.98 > -1$.

Table 9.2:

$$\frac{c}{t} = \frac{2470}{18} = 137.22 > \frac{42 \cdot 0.825}{0.67 + 0.33 \cdot (-0.98)} = 99.97$$

→ the web is not class 3 and therefore it is class 4.

Finally, the section is class 4 for hogging bending.

9.2 RESISTANCE TO TENSION: ALLOWANCE FOR FASTENER HOLES IN BENDING CAPACITY

The resistance to tension is smaller between the plastic resistance of the gross section and the ultimate resistance of the net section, the latter considering deductions for fastener holes and other openings. Accordingly, it is

$$N_{t,Rd} = \min\{N_{pl,Rd}, N_{u,Rd}\} \tag{9.1}$$

where

$$N_{pl,Rd} = \frac{A \cdot f_y}{\gamma_{M0}} \tag{9.2}$$

$$N_{u,Rd} = \frac{0.9 \cdot A_{net} \cdot f_u}{\gamma_{M2}} \tag{9.3}$$

f_y is the yield stress
f_u is the ultimate strength
A is the gross section area
A_{net} is the net section area
γ_{M0}, γ_{M2} are the partial factors of safety with recommended values 1.0 and correspondingly 1.25 (see Table 5.20)

For non-staggered fastener holes, the failure plane is perpendicular to the member axis crossing all holes so that the net area is the gross area less the sum of the sectional areas of the holes. For staggered holes, the failure plane may have a zigzag shape. To account for its larger length, the net area is determined deducing all holes crossing the failure plane but adding a component $t \cdot s^2/(4 \cdot p)$ for consecutive staggered holes, where p is the distance between centers of holes perpendicular to the member axis and s parallel to it. Table 9.6 presents an example calculation.

Table 9.6 Example calculation of net section area

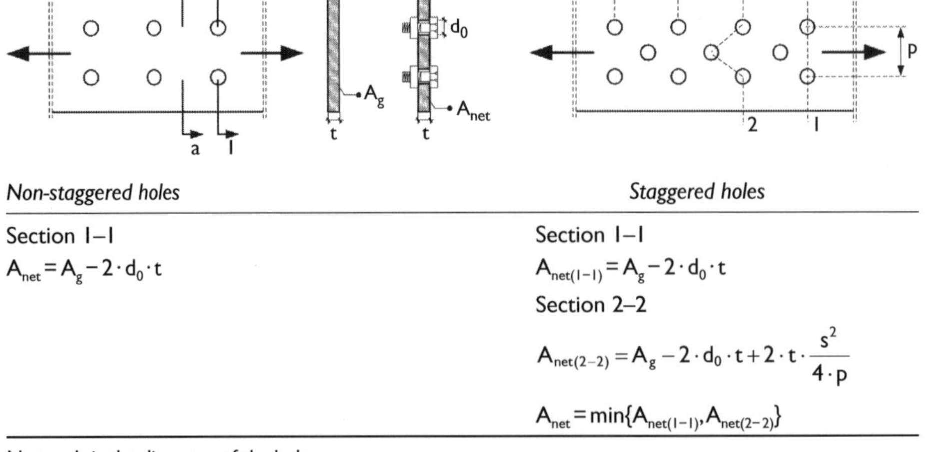

Non-staggered holes	*Staggered holes*
Section I–I	Section I–I
$A_{net} = A_g - 2 \cdot d_0 \cdot t$	$A_{net(I-I)} = A_g - 2 \cdot d_0 \cdot t$
	Section 2–2
	$A_{net(2-2)} = A_g - 2 \cdot d_0 \cdot t + 2 \cdot t \cdot \frac{s^2}{4 \cdot p}$
	$A_{net} = \min\{A_{net(I-I)}, A_{net(2-2)}\}$

Note: d_0 is the diameter of the hole.

If connections are of category C, that is, slip resistant at ULS using preloaded bolts, the member resistance to tension at fasteners holes is obtained from

$$N_{net,Rd} = \frac{A_{net} \cdot f_y}{\gamma_{M0}} \tag{9.4}$$

To ensure ductile member behavior, the plastic resistance should be smaller than the ultimate resistance, that is, $N_{pl,Rd} \leq N_{u,Rd}$. Considering Equations 9.2 and 9.3, this may be written as

$$\frac{A \cdot f_y}{\gamma_{M0}} \leq \frac{0.9 \cdot A_{net} \cdot f_u}{\gamma_{M2}} \Rightarrow \frac{A_{net}}{A} \geq \frac{1}{0.9} \cdot \frac{f_y}{f_u} \cdot \frac{\gamma_{M2}}{\gamma_{M0}} \tag{9.5}$$

Consideration of ductile behavior is important for determining the bending capacity of plate girders at bolted splice regions. Ductile behavior for the tension flange is ensured if

$$\frac{A_{f,net}}{A_f} \geq \frac{1}{0.9} \cdot \frac{f_y}{f_u} \cdot \frac{\gamma_{M2}}{\gamma_{M0}} \tag{9.6}$$

where A_f and $A_{f,net}$ are the gross and net section areas of the tension flange.

If condition (9.6) is fulfilled, the section modulus of the girders may be determined on basis of the gross cross section ignoring fastener holes. However, if this condition is not fulfilled, the girder cross section is composed of the gross section of the compression flange and the web and the net section of the tension flange. A similar rule applies for the tension zone of the web, where fastener holes should be allowed for if condition (9.6) is not satisfied, where A_f and $A_{f,net}$ are the gross and net section areas of the tension flange and the tension zone of the web. In compression zones, fastener holes need not be allowed for, that is, there is no need to determine net sections.

9.3 RESISTANCE OF STEEL MEMBERS AND CROSS SECTIONS TO COMPRESSION

Steel members in composite bridges to which this section refers may be chords and diagonals of truss girders, cross bracings, or plan bracings in the top or bottom flange used for lateral stability of main girders at construction stages or to provide a quasi-closed cross section; see Figure 7.15.

The resistance of steel cross sections to compression may be obtained from

$$N_{c,Rd} = \frac{A \cdot f_y}{\gamma_{M0}} \quad \text{for classes 1, 2, and 3 cross sections} \tag{9.7}$$

$$N_{c,Rd} = \frac{A_{eff} \cdot f_y}{\gamma_{M1}} \quad \text{for class 4 cross sections} \tag{9.8}$$

where

- A is the area of the gross cross section not allowing for fastener holes
- A_{eff} is the area of the effective cross section allowing for local buckling as described in Chapter 8

The buckling resistance of steel members to compression may be obtained from

$$N_{b,Rd} = \frac{\chi \cdot A \cdot f_y}{\gamma_{M1}} \text{ for classes 1, 2, and 3 cross sections} \tag{9.9}$$

$$N_{b,Rd} = \frac{\chi \cdot A_{eff} \cdot f_y}{\gamma_{M1}} \text{ for class 4 cross sections} \tag{9.10}$$

where

χ is the reduction factor for flexural buckling

$\chi = 1$ for $\bar{\lambda} \leq 0.2$

$$\chi = \frac{1}{\Phi + \sqrt{\Phi^2 - \bar{\lambda}^2}} \text{ for } \bar{\lambda} > 0.2 \tag{9.11}$$

$$\Phi = 0.5 \cdot [1 + \alpha \cdot (\bar{\lambda} - 0.2) + \bar{\lambda}^2] \tag{9.12}$$

$\bar{\lambda}$ is the nondimensional slenderness

$$\bar{\lambda} = \sqrt{\frac{A \cdot f_y}{N_{cr}}} \text{ for classes 1, 2, and 3 cross sections} \tag{9.13}$$

$A = A_{eff}$ for class 4 cross sections

α is an *imperfection factor* in dependence of the relevant *buckling curve* (Table 9.7)

$$N_{cr} = \frac{\pi^2 \cdot E \cdot I}{l_k^2} \text{ is the Euler buckling load} \tag{9.14}$$

l_k is the buckling length for the axis considered

$\gamma_{M1} = 1.1$ (recommended value) for bridges

Five European buckling curves with different imperfection factors are distinguished. The imperfections are equivalent geometric imperfections that unite geometric and structural imperfections (*residual stresses*). Both imperfections are different for various section shapes and affect differently the buckling response about the strong and weak axis of the cross section. Accordingly, the *buckling curves* are associated to shapes and dimensions of cross sections and the axis about which buckling is considered as shown in Table 9.8.

Table 9.7 Imperfection factors (α) for European buckling curves

Buckling curve	a_0	a	b	c	d
Imperfection factor α	0.13	0.21	0.34	0.49	0.76

Source: EN 1993-1-1, Design of steel structures, Part 1-1: General rules and rules for buildings, 2005.

Table 9.8 Selection of buckling curves

Cross sections	*Limits*		*Buckling about axis*	*Buckling curve*	
				S235–S420	*S460*
Rolled sections 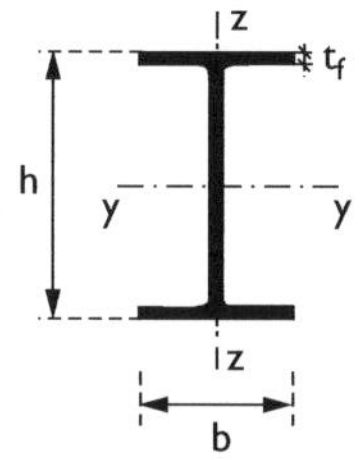	$h/b > 1.2$	$t_f \leq 40$ mm	y–y	a	a_0
			z–z	b	a_0
		$40 \text{ mm} < t_f \leq 100$ mm	y–y	b	a
			z–z	c	a
	$h/b \leq 1.2$	$t_f \leq 100$ mm	y–y	b	a
			z–z	c	a
		$t_f > 100$ mm	y–y	d	c
			z–z	d	c
Welded I sections 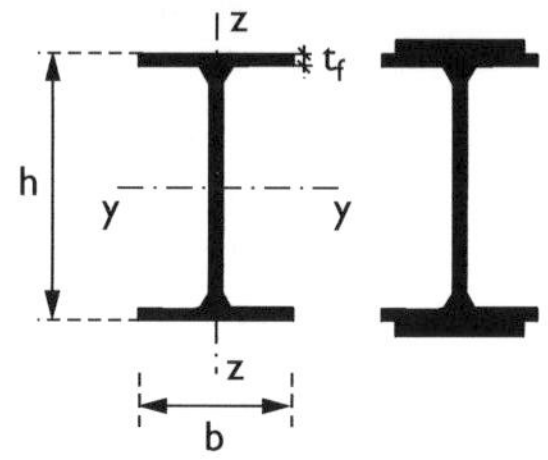		$t_f \leq 40$ mm	y–y	b	b
			z–z	c	c
		$t_f > 40$ mm	y–y	c	c
			z–z	d	d
Hollow sections 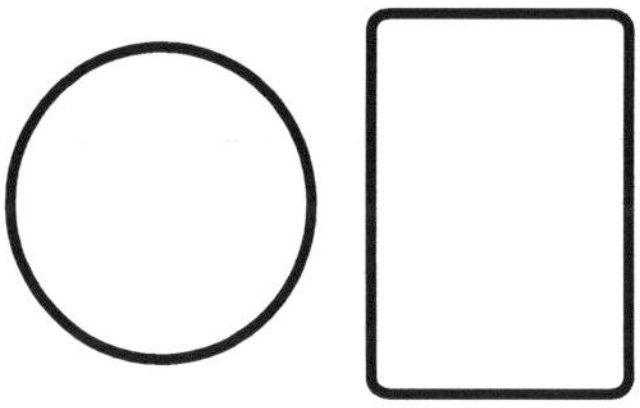	Hot finished		Any	a	a_0
	Cold formed		Any	c	c
Welded box sections 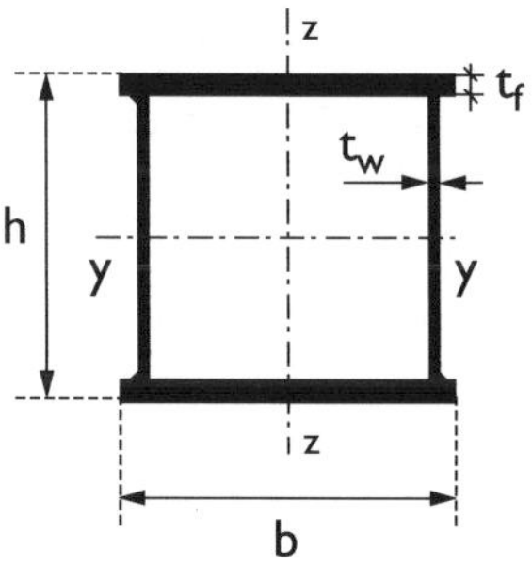	Generally (except as follows)		Any	b	b
	Thick welds: $a > 0.5 \cdot t_f$; $\frac{b}{t_f} < 30$, $\frac{h}{t_w} < 30$		Any	c	c
L sections 			Any	b	b

Source: EN 1993-1-1, Design of steel structures, Part 1-1: General rules and rules for buildings, 2005.

9.4 RESISTANCE TO SHEAR DUE TO VERTICAL SHEAR AND TORSION

Vertical shear is resisted by the steel sections alone and due to equilibrium conditions only by the webs, as indicatively shown in Figure 9.4a and b. Shear stresses develop also in the flanges, but are usually not considered for ULS design since they are not relevant for the provision of vertical equilibrium. The shear stresses in box-girder flanges due to vertical shear are considered also in serviceability limit state (SLS) stress verifications.

Torsion due eccentric loading is resisted by differential bending of the steel girders in multiple I-girder bridges, as outlined in Chapter 7. Accordingly, torsion does not directly affect the main girders for such types of bridge decks. However, in box-girder bridges, torsion is resisted primarily by means of *St. Venant* torsion due to the high torsion rigidity of the box section. Under this condition, box girders resist torsional moments $M_{T,Ed}$ by development of a constant shear flow in their walls that according to the first formula of *Bredt* is equal to

$$v_{MT,Ed} = \frac{M_{T,Ed}}{2 \cdot A_0} [\text{in kN/m}] \tag{9.15}$$

where A_0 is the area enclosed between centerlines of the walls.

This shear flow results in shear forces in the walls determined from

$$V_{MT,Ed} = v_{MT,Ed} \cdot b [\text{in kN}] \tag{9.16}$$

where b is the length of the relevant wall.

In case of combined vertical shear and torsion, the shear forces are added in one web and subtracted in the other; see Figure 9.4c. In the flanges, they are only due to torsion. Concluding, it may be said that vertical shear results in shear forces in the webs of plate and box girders and torsion shear forces in all walls of box girders.

The resistance to shear at ULS must be checked for webs and for bottom steel flanges of box girders. Shear forces in the concrete flange due to torsion are resisted by the diaphragm action of the slab that is much thicker than the walls of the steel girder and is usually not verified. However, the horizontal shear in the concrete flange due to torsion is transferred by the shear connectors as Figure 9.4d shows. Accordingly, shear connectors are designed for the simultaneous action of both longitudinal and transverse shear forces, the former due to vertical shear and the latter due to torsion.

The shear resistance of steel walls, whether webs of I and box girders or bottom flanges of box girders, may be determined at ULSs as the plastic shear resistance in accordance

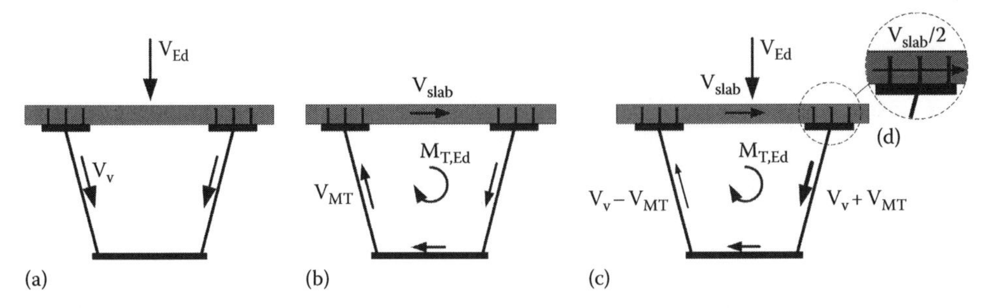

Figure 9.4 Shear forces in box girders due to (a) vertical shear, (b) torsion, (c) resultant forces due to vertical shear and torsion in box girders, and (d) shear forces in the connectors due to torsion in box girders.

with Equation 9.19 or as the shear buckling resistance in accordance with Section 8.7. Shear buckling is not relevant if the width to thickness ratio of the walls is limited to:

For walls without longitudinal stiffeners according to Equation 8.25:
Steel grades up to S420:

$$\frac{h_w}{t_w} \leq 60 \cdot \varepsilon \tag{9.17a}$$

Steel grades S460 and higher:

$$\frac{h_w}{t_w} \leq 72 \cdot \varepsilon \tag{9.17b}$$

For walls with longitudinal stiffeners according to Equation 8.26:
Steel grades up to S420:

$$\frac{h_w}{t_w} \leq 25.8 \cdot \varepsilon \cdot \sqrt{\kappa_\tau} \tag{9.18a}$$

Steel grades S460 and higher:

$$\frac{h_w}{t_w} \leq 31 \cdot \varepsilon \cdot \sqrt{\kappa_\tau} \tag{9.18b}$$

where according to Figure 9.5,
h_w and t_w are the width and correspondingly the thickness of the walls under consideration
κ_τ is the relevant shear buckling factor determined from Table 8.5

The plastic resistance may be obtained from

$$V_{pl,Rd} = \frac{f_y}{\sqrt{3} \cdot \gamma_{M0}} \cdot A_v \tag{9.19}$$

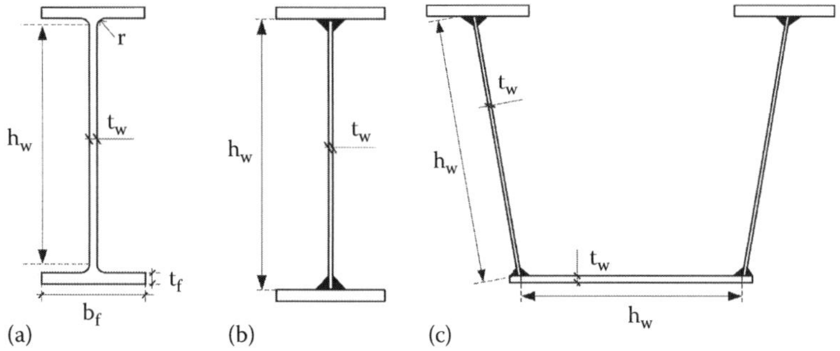

Figure 9.5 Notation for the width and thickness of walls for the shear area. (a) Rolled I sections, (b) welded I sections, and (c) box sections.

A_v is the *shear area* to be determined from:
Webs of rolled I and H sections (see Figure 9.5a):

$$A_v = A_a - 2 \cdot b_f \cdot t_f + (t_w + 2 \cdot r) \cdot t_f \geq \eta \cdot h_w \cdot t_w \tag{9.20}$$

where according to notes of Table 8.10,
$\eta = 1.2$ for steel grades up to S420
$\eta = 1.0$ for steel grades S460 and higher

Webs of welded I and H sections (see Figure 9.5b):

$$A_v = h_w \cdot t_w \tag{9.21}$$

Webs and bottom flanges of box sections (see Figure 9.5c):

$$A_v = h_w \cdot t_w \tag{9.22}$$

Walls not complying with the limits given by conditions (9.17) and (9.18) must be verified for shear buckling in accordance with Section 8.7. If the verification is performed by the effective width method, the shear buckling resistance may be determined in accordance with Equation 8.50; see Remark 8.8. If the verification is performed by the reduced stress method, the shear buckling resistance is obtained from

$$V_{bw,Rd} = \tau_{Rd} \cdot A_v \tag{9.23a}$$

where

$$\tau_{Rd} = \chi_w \cdot \frac{f_{yw}}{\sqrt{3} \cdot \gamma_{M1}} \tag{9.23b}$$

For χ_w, see Table 8.10.
Equation 9.23 is equivalent to Equation 8.51.

9.5 RESISTANCE TO BENDING OF STEEL CROSS SECTIONS

During construction stages and before casting of concrete main girders are composed of their steel section. The bending resistance of such sections or other pure steel sections, like transverse beams not connected to a concrete slab, may be determined as follows:

$$M_{c,Rd} = M_{pl,Rd} = \frac{f_y \cdot W_{pl}}{\gamma_{M0}} \text{ for cross sections of classes 1 and 2} \tag{9.24}$$

$$M_{c,Rd} = M_{el,Rd} = \frac{f_y \cdot W_{el,min}}{\gamma_{M0}} \text{ for cross sections of class 3} \tag{9.25}$$

$$M_{c,Rd} = \frac{f_y \cdot W_{eff,min}}{\gamma_{M0}} \text{ for cross sections of class 4} \tag{9.26}$$

In the aforementioned relations, the section moduli, possibly accounting for fastener holes in tension areas if required (see Section 9.2), are as follows:

W_{pl} is the plastic section modulus of the cross section
$W_{el,min}$ is the elastic section modulus of the cross section for the fiber with the maximum stress
$W_{eff,min}$ is the section modulus of the effective cross section for the fiber with the maximum stress

It is noted that a cross section may have a different class for sagging and hogging bending so that different formulae apply to the bending resistance and that the properties of the effective cross section may be determined in accordance with Section 8.5.

Cross sections that are designed for plate buckling in accordance with the reduced stress method are considered as class 3 sections. Such sections are verified in accordance with Section 8.4.

9.6 INTERACTION OF BENDING WITH SHEAR FOR STEEL CROSS SECTIONS

In the presence of shear forces V_{Ed}, resulting from either vertical shear or torsion, part of the material strength of the walls is "exploited" to resist the shear. The relevant wall is then not able to develop the full yield strength to resist bending moments (or axial forces). Accordingly, the bending resistance may be determined for a cross section with the same geometry but with reduced yield strength of the walls that resist shear forces as shown in Figure 9.6. The reduced yield strength may be determined from

$$f_{y,red} = (1-\rho)\cdot f_{yd} \tag{9.27a}$$

where f_{yd} design strength of steel (see Table 9.9).

$$\rho = \left(\frac{2\cdot V_{Ed}}{V_{Rd}} - 1\right)^2 \tag{9.27b}$$

where
V_{Ed} is the design shear force of the relevant wall resulting from vertical shear and torsion
V_{Rd} is the design shear resistance of the wall, either the plastic resistance $V_{pl,Rd}$ (Equation 9.19) or the shear buckling resistance $V_{bw,Rd}$ determined from Equation 9.23

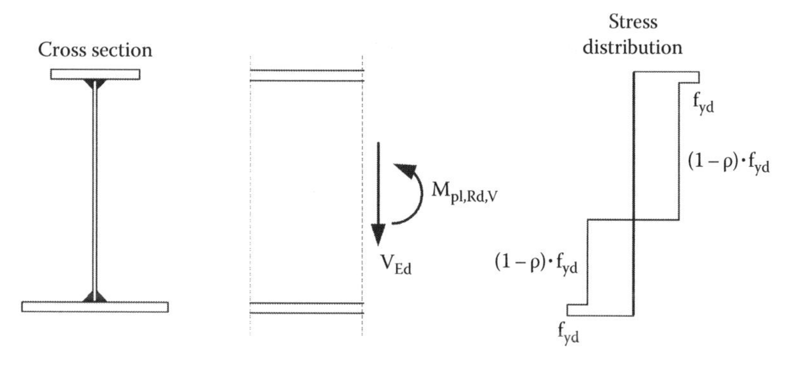

Figure 9.6 Plastic stress distribution allowing for the interaction with shear.

Table 9.9 Design strengths of materials at ULS for class 1 or 2 cross sections

Concrete	In compression	$0.85 \cdot f_{cd} = 0.85 \cdot \frac{f_{ck}}{\gamma_c} = 0.85 \cdot \frac{f_{ck}}{1.5}$
	In tension	0
Structural steel	In tension and compression	$f_{yd} = \frac{f_{yk}}{\gamma_{M0}} = \frac{f_{yk}}{1.0}$
		[or $f_{y,red} = (1-\rho) \cdot f_{yd}$ for high shear]
Reinforcement	Compression or tension	$f_{sd} = \frac{f_{sk}}{\gamma_s} = \frac{f_{sk}}{1.15}$

It is noted that *due to high shear, the position of the plastic neutral axis may change.* This should be accounted for in the classification of the web.

Where conditions (9.28a or 9.28b) apply, interaction is not accounted for and the yield strength need not to be reduced:

$$\frac{V_{Ed}}{V_{pl,Rd}} \leq 0.5 \tag{9.28a}$$

or

$$\frac{V_{Ed}}{V_{bw,Rd}} \leq 0.5 \tag{9.28b}$$

where

$V_{pl,Rd}$ is the plastic shear resistance of the wall according to Equation 9.19
$V_{bw,Rd}$ is the buckling resistance of the wall according to Equation 9.23

Cross sections that are designed for plate buckling by the reduced stress method are verified following the procedure described in Section 8.4, which takes into account the interaction of bending with shear through the von Mises criterion.

9.7 CLASS 1 AND 2 CROSS SECTIONS

9.7.1 General

Class 1 or 2 cross sections may be checked in the level of internal forces and moments, by comparison of the acting internal forces and moments with the corresponding plastic design resistances. Accordingly, this section refers to the determination of the plastic bending resistance for sagging and hogging bending allowing for shear or axial forces.

9.7.1.1 Sagging bending

For sagging bending I, girder composite sections are usually class 1 or 2, as explained in Section 9.1. The cross section is composed of the steel girder, the concrete flange, and

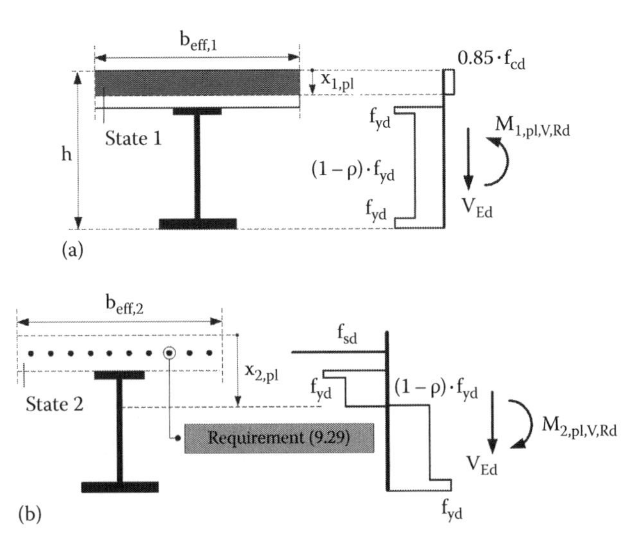

Figure 9.7 Plastic stress distribution to determine $M_{pl,V,Rd}$: (a) for sagging bending and shear and (b) for hogging bending and shear.

the relevant reinforcement within the effective width of the slab; see Section 7.2.2. For full shear connection between the steel beam and the concrete slab, the bending resistance may be determined considering a plastic stress distribution, where all materials are stressed up to their design strength given in Table 9.9. This stress distribution leads to the determination of the plastic moment of the section $M_{1,pl,Rd}$ (index 1 means that the contribution of concrete in bending resistance is taken into account; state 1). For sagging bending, the contribution of the reinforcement is usually small and may be neglected. In areas of design shear larger than 50% of the corresponding shear resistance (Equations 9.19 or 9.23), the design strength of structural steel is reduced by the factor $(1-\rho)$; see Equations 9.27 and Table 9.9. The relevant design moment resistance for sagging bending may be assigned as $M_{1,pl,V,Rd}$; $M_{2,pl,V,Rd}$ is for hogging bending (state 2 for the deck slab). Figure 9.7 shows indicatively plastic stress distributions for the determination of the plastic bending resistance of cross sections subjected to high shear, where the design strength in the web is reduced. These stresses, when integrated over the relevant areas, give no axial force in the section. For a simultaneous presence of a design axial force N_{Ed}, the stress distributions have to be modified to result in, by appropriate integration, the axial force N_{Ed}. In bolted splices, allowance for fastener holes must be taken into account in tension areas as for steel beams.

9.7.1.2 Hogging bending

Concrete under tension does not participate in the bending resistance. Accordingly, for hogging bending, the cross section is composed of the steel section and the reinforcement only. In regions of hogging bending, a minimum ratio of tension reinforcement shall be provided as follows:

$$\rho_s = \frac{A_s}{A_c} \geq \min \rho_s = \delta \cdot \frac{f_y}{235} \cdot \frac{f_{ctm}}{f_{sk}} \cdot \sqrt{k_c} \tag{9.29}$$

where

δ is 1.0 or 1.1 for cross-sectional class 1 or correspondingly 2
A_s is the area of tension reinforcement of ductility class B or C; see Section 6.3
A_c is the area of the concrete flange within the effective width $b_{eff,2}$
f_y is the yield strength of structural steel in [N/mm²]
f_{ctm} is the mean tensile strength of concrete in [N/mm²]; see Table 6.1
k_c is a coefficient; see Section 10.3.2. Conservatively, $k_c = 1$

REMARK 9.2

minρ_s is the minimum reinforcement that is required for a ductile flexural failure without fracture of the reinforcement. Reinforcing rebars should be adequately anchored according to the provisions of EN 1992-1-1 [9.2]. Welded meshes should be avoided due to brittle behavior.

A plastic stress distribution may be realized if the strain does not exceed the ultimate material strain. This strain limitation does not affect structural steel or reinforcing steel of ductility classes B and C since the ultimate strain is sufficiently high. However, the limit concrete strain is smaller and may be reached if the steel beam is of high-strength steel S420 or S460 and the distance of the plastic neutral axis from the extreme concrete fiber is large. Therefore, the design resistance to sagging moment has to be reduced by the factor-β. Figure 9.8 shows that the reduction starts when $x_{1,pl}/h > 0.15$, that is, when the depth of the plastic neutral axis is larger than 15% of the overall depth of the cross section. For $x_{1,pl}/h > 0.40$, the bending resistance should be determined either by nonlinear methods considering the stress–strain curves of the materials or taken equal to the elastic resistance.

REMARK 9.3

In case of hybrid girders with flanges of S420 or S460 and web of S355, the reduction factor-β of Figure 9.8 should also be applied.

As discussed in Chapter 7, elastic global analysis is employed in composite bridges, although some inelastic rotation, that is, some moment redistribution, should be developed at the internal supports of continuous beams to allow for the development of the plastic moment resistance in the span regions. However, this inelastic rotation capacity is limited if the cross section at these supports is of class 3 or 4. In addition, larger rotations and higher redistributions are needed if adjacent spans have quite unequal lengths. Therefore, for continuous beams, *the bending resistance of class 1 or 2 cross section in the span regions with*

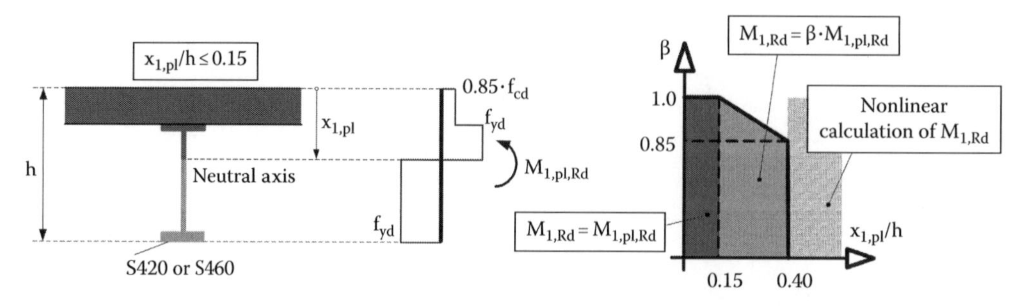

Figure 9.8 Reduction factor-β for the plastic sagging moment for steel qualities S420 and S460.

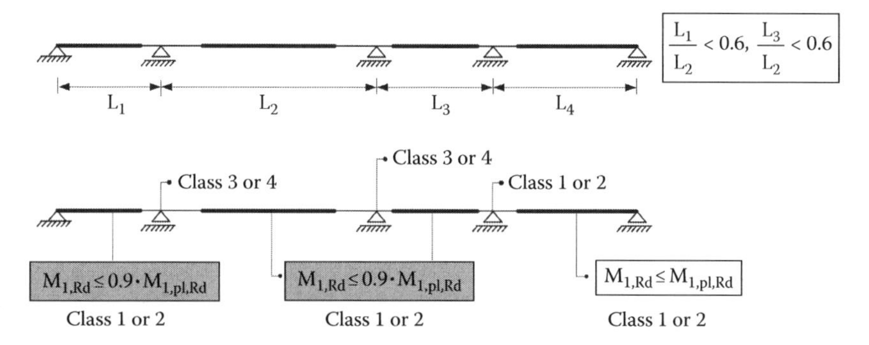

Figure 9.9 Moment resistance limitation for spans of continuous beams.

Table 9.10 Verifications of class 1 or 2 cross sections for bending and shear

Field of application	*Shear*	*Bending*
Spans of simple or continuous beams except last case below	$V_{Ed} \leq 0.5 \cdot V_{Rd}$	$M_{Ed} \leq \beta \cdot M_{1,pl,Rd}$
	$V_{Ed} > 0.5 \cdot V_{Rd}$	$M_{Ed} \leq \beta \cdot M_{1,pl,V,Rd}$
Internal supports of continuous beams	$V_{Ed} \leq 0.5 \cdot V_{Rd}$	$M_{Ed} \leq M_{2,pl,Rd}$
	$V_{Ed} > 0.5 \cdot V_{Rd}$	$M_{Ed} \leq M_{2,pl,V,Rd}$
Spans of continuous beams with cross sections at adjacent supports class 3 or 4 and ratios of adjacent spans ≤ 0.6	$V_{Ed} \leq 0.5 \cdot V_{Rd}$	$M_{Ed} \leq 0.9 \cdot \beta \cdot M_{1,pl,Rd}$
	$V_{Ed} > 0.5 \cdot V_{Rd}$	$M_{Ed} \leq 0.9 \cdot \beta \cdot M_{1,pl,V,Rd}$

Notes:

V_{Rd} is the plastic resistance (Equation 9.19) or the shear buckling resistance (Equation 9.23).

$M_{1,pl,Rd}$ and $M_{2,pl,Rd}$ are the plastic moments for sagging and hogging bending.

$M_{1,pl,V,Rd}$ and $M_{2,pl,V,Rd}$ are the reduced plastic moments for sagging and hogging bending due to shear.

For β, see Figure 9.8.

sagging bending should not exceed 90% of the design resistance moment if the cross sections at adjacent internal supports are class 3 or class 4 and the ratio of the shorter to the longer spans adjacent to that support is less than 0.6; see Figure 9.9. Table 9.10 summarizes the resistance of class 1 or 2 cross sections.

It is reminded that if *all cross sections of a bridge are class 1 or 2 and not susceptible to lateral torsional buckling* (see Section 9.13), secondary internal forces due to creep, shrinkage, and temperature are ignored in the combination of actions for verifications at ULSs; see Table 5.7.

EXAMPLE 9.2

Verify the cross section of Figure 9.10 for a design hogging moment M_{Ed} = 50,000 kN-m and a design shear force V_{Ed} = 3,500 kN. The reinforcement area is A_{so} = 58 cm², A_{su} = 38 cm². Grade of steel girder S355, strength class of concrete C35/45, grade of reinforcement B500B.

Design yield stress of structural steel (Table 6.14):

Web t = 18 mm:

$$f_{yd} = \frac{345}{1.0} = 345 \text{ N/mm}^2$$

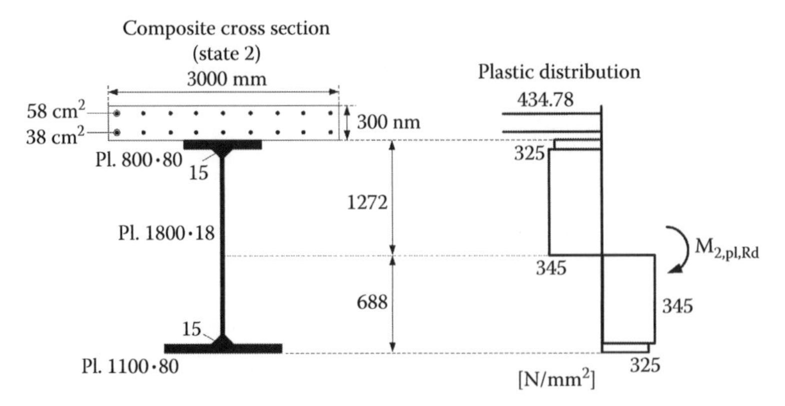

Figure 9.10 Cross section of Example 9.2 and plastic stress distribution for classification.

Flange t = 80 mm:

$$f_{yd} = \frac{325}{1.0} = 325 \text{ N/mm}^2$$

Figure 9.10 shows the plastic stress distribution. The distance of the neutral axis from the top concrete fiber is found to be $x_{2,pl}$ = 1572 mm. The height of the compression zone is then 2260 − 1572 = 688 mm.

Design yield stress of reinforcing steel:

$$f_{sd} = \frac{500}{1.15} = 434.78 \text{ N/mm}^2$$

CROSS-SECTIONAL CLASSIFICATION

Bottom flange

$$f_y = 325 \text{ N/mm}^2 \rightarrow \varepsilon = \sqrt{\frac{235}{325}} = 0.85$$

The bottom flange is an outstand element subjected to pure compression:

$$c = \frac{1100 - 18}{2} - 15 = 526 \text{ mm}$$

Table 9.3:

$$\frac{c}{t} = \frac{526}{80} = 6.58 \le 9 \cdot 0.85 = 7.65$$

The bottom flange is class 1.

Web

$$f_y = 345\ \text{N/mm}^2 \rightarrow \varepsilon = \sqrt{\frac{235}{345}} = 0.825$$

The width of the web is c = 1800 − 2 · 15 = 1770 mm.

The compression part of the web is $688-80-15=593\ \text{mm} \rightarrow \alpha = \frac{593}{1770} = 0.34 < 0.5$.

Table 9.2:

$$\frac{c}{t} = \frac{1770}{18} = 98.33 > \frac{36 \cdot 0.825}{0.34} = 87.35 \rightarrow \text{the web is not class 1}$$

$$\frac{c}{t} = 98.33 \leq \frac{41.5 \cdot 0.825}{0.34} = 100.70 \rightarrow \text{the web is class 2}$$

Accordingly, the section is class 2 for hogging bending.

Ductility reinforcement in the slab

Required minimum reinforcement ratio according to Equation 9.29:

$$\min \rho_s = 1.1 \cdot \frac{345}{235} \cdot \frac{3.2}{500} \cdot \sqrt{1.0} = 0.0103$$

where it is set on the safe side $f_{yk} = 345\ \text{N/mm}^2$ of the web, which is larger than that of the flange and for $k_c = 1$.

Required reinforcement area:

$$A_s = 0.0103 \cdot (300 \cdot 30) = 92.7\ \text{cm}^2$$

The actual reinforcement is 58 + 38 = 96 cm², larger than the minimum required.

SHEAR RESISTANCE

The web does not have longitudinal stiffeners.

$$\text{Equation 9.17a: } \frac{h_w}{t_w} = \frac{1800}{18} = 100 > 60 \cdot 0.825 = 49.5$$

Accordingly, the shear buckling resistance is relevant and will be calculated in accordance with Chapter 8. Transverse vertical stiffeners are supposed to be provided at a distance a = 3.0 m.

Aspect ratio of the panel (Table 8.5):

$$\alpha = \frac{3.0}{1.8} = 1.67$$

Reference stress, Equation 8.10:

$$\sigma_e = 18980 \cdot \left(\frac{1.8}{180}\right)^2 = 1.898\ \text{kN/cm}^2$$

Buckling factor for shear (Table 8.5):

$$k_\tau = 5.34 + \frac{4}{1.67^2} = 6.77$$

Critical buckling shear stress, Equation 8.11:

$$\tau_{cr} = 6.77 \cdot 1.898 = 12.85 \text{ kN/cm}^2$$

Nondimensional slenderness, Equation 8.19:

$$\bar{\lambda}_w = 0.76 \cdot \sqrt{\frac{34.5}{12.85}} = 1.25 > 1.08$$

Reduction factor for nonrigid end posts (Table 8.10):

$$\chi_w = \frac{0.83}{1.25} = 0.66$$

Shear resistance of web, Equation 8.51:

$$V_{bw,Rd} = \frac{0.66 \cdot 34.5 \cdot 180 \cdot 1.8}{\sqrt{3} \cdot 1.1} = 3872 \text{ kN}$$

Shear verification: $V_{Ed} = 3500$ kN $< V_{bw,Rd}$ (sufficient)
Note: Remark 8.8

BENDING RESISTANCE

It is $\frac{V_{Ed}}{V_{Rd}} = \frac{3500}{3872} = 0.90 > 0.5$.

Accordingly, there is a moment–shear interaction.

Reduction factor for shear stress in the web, Equation 9.27b:

$$\rho = (2 \cdot 0.90 - 1)^2 = 0.64$$

The design stress in the web is then

$$345 \cdot (1 - 0.64) = 124.2 \text{ N/mm}^2$$

The bending resistance is then found as $M_{2,pl,V,Rd} = 51840.1$ kN-m.

Note that the full plastic resistance is $M_{2,pl,Rd} = 55624.18$ kN-m, that is, the reduction due to interaction with shear is only 6.6%.

Bending check, Table 9.10 row 4: $M_{Ed} = 50000$ kN-m $< M_{2,pl,V,Rd} = 51840.1$ kN-m

It is noted that for the new stress distribution, the distance of the plastic neutral axis from the top slab fiber is 2010 mm. It may be proved that the web and accordingly the cross section can be classified as class 1.

9.8 CROSS SECTIONS WITH CLASS 3 WEBS THAT MAY BE TREATED AS CLASS 2 SECTIONS (HOLE-IN-WEB METHOD)

Cross sections with class 1 or 2 flanges and class 3 web may be classified and designed as class 2 provided that only an effective part of the web is considered, *hole-in-web method*. The plastic bending resistance of such sections, possibly reduced by the presence of shear or axial force, is determined by consideration of part of the web as not effective. Figure 9.11 shows plastic stress distributions for the determination of their plastic bending resistance. The verifications of such sections are similar to those for sections 1 and 2 as presented in the previous section.

Plastic neutral axis for *sagging* bending:

$$x_{1,pl} = h - t_{fu} + \frac{N_{pl,fu} - N_{pl,c} - N_{pl,fo} - N_{pl,20\varepsilon,d}}{(1-\rho)\cdot f_{yw,d}\cdot t_w} \tag{9.30a}$$

Plastic neutral axis for *hogging* bending:

$$x_{2,pl} = h_c + t_{fo} + \frac{N_{pl,fu} + N_{pl,20\varepsilon,d} - N_{pl,fo} - N_{pl,s}}{(1-\rho)\cdot f_{yw,d}\cdot t_w} \tag{9.30b}$$

where

$N_{pl,fo} = f_{y,fo,d}\cdot A_{fo}$

$N_{pl,fu} = f_{y,fu,d}\cdot A_{fu}$

$N_{pl,c} = 0.85\cdot f_{cd}\cdot b_{eff,1}\cdot h_c$

$N_{pl,s} = f_{sd}\cdot A_s$

$\varepsilon_w = \sqrt{235/f_{yw}[\text{MPa}]}$

$N_{pl,20\varepsilon,d} = (1-\rho)\cdot f_{yw}\cdot 40\cdot \varepsilon_w \cdot t_w^2$

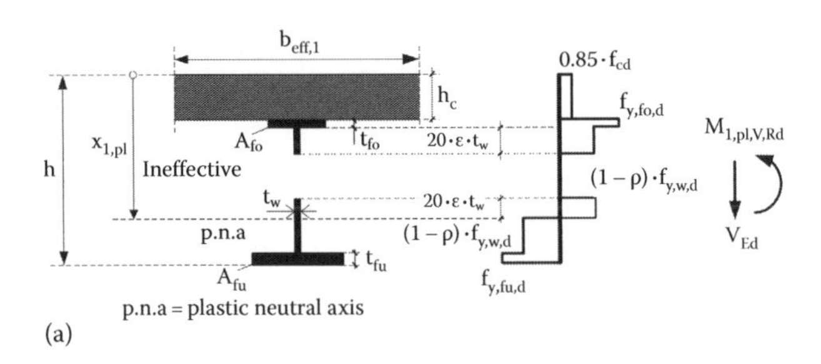

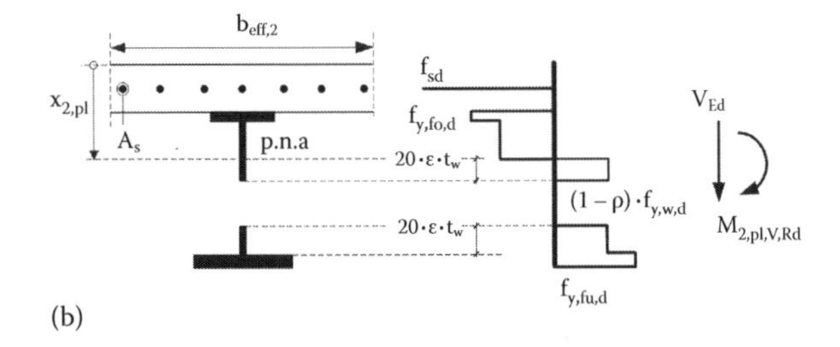

Figure 9.11 Cross sections with class 3 webs treated as class 2 and plastic stress distribution to determine $M_{pl,V,Rd}$: (a) for sagging bending and shear and (b) for hogging bending and shear.

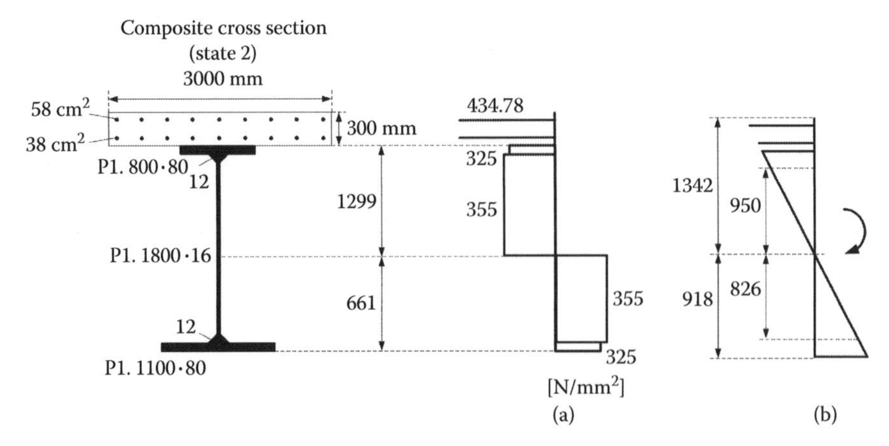

Figure 9.12 Cross section of Example 9.3 with (a) plastic and (b) elastic stress distribution for classification.

EXAMPLE 9.3

Verify the cross section of Figure 9.12 for a design hogging moment $M_{Ed} = 50{,}000$ kN-m and a design shear force $V_{Ed} = 1450$ kN. The reinforcement area is $A_{so} = 58$ cm², $A_{su} = 38$ cm². Grade of steel girder S355, strength class of concrete C35/45, grade of reinforcement B500B.

Design yield stress of structural steel (Table 6.14):

Web t = 16 mm:

$$f_{yd} = \frac{355}{1.0} = 355 \text{ N/mm}^2$$

Flange t = 80 mm:

$$f_{yd} = \frac{325}{1.0} = 325 \text{ N/mm}^2$$

Design yield stress of reinforcing steel:

$$f_{sd} = \frac{500}{1.15} = 434.78 \text{ N/mm}^2$$

Figure 9.12 shows the plastic stress distribution. The distance of the neutral axis from the top concrete fiber is found to be $x_{2,pl} = 1599$ mm. The height of the compression zone is then $2260 - 1599 = 661$ mm.

CROSS-SECTIONAL CLASSIFICATION

Bottom flange

$$f_y = 325 \text{ N/mm}^2 \rightarrow \varepsilon = \sqrt{\frac{235}{325}} = 0.85$$

The bottom flange is an outstand element subjected to pure compression:

$$c = \frac{1100 - 16}{2} - 12 = 530 \text{ mm}$$

Table 9.3:

$$\frac{c}{t} = \frac{530}{80} = 6.63 \leq 9 \cdot 0.85 = 7.65$$

The bottom flange is class 1.

Web

$$f_y = 355\ \text{N/mm}^2 \rightarrow \varepsilon = \sqrt{\frac{235}{355}} = 0.81$$

The width of the web is c = 1800 − 2 · 12 = 1776 mm.

The compressed part of the web is 661 − 80 − 12 = 569 mm $\rightarrow \alpha = \frac{569}{1776} = 0.32 \leq 0.5$

Table 9.2:

$$\frac{c}{t} = \frac{1776}{16} = 111 > \frac{41.5 \cdot 0.81}{0.32} = 105 \rightarrow \text{web is not class 2}$$

Accordingly, an elastic stress distribution is considered to determine if the web is class 3 or 4.

Elastic stress distribution; see Figure 9.12.

For the elastic distribution, the distance of the center of gravity (neutral axis) from the top fiber is 1342 mm and from the bottom fiber 918 mm. The part of the web under compression is 918 − 80 − 12 = 826 mm and under tension 1342 − 300 − 80 − 12 = 950 mm. The stress ratio is accordingly $\psi = -\frac{950}{826} = -1.15 \leq -1$.

Table 9.2: $\frac{c}{t} = \frac{1776}{16} = 111 \leq 62 \cdot 0.81 \cdot (1 + 1.15) \cdot \sqrt{1.15} = 115.8 \rightarrow$ The web is class 3, and therefore, the cross section is class 3.

Note: The stress ratio should be calculated by taking into account the concreting sequence; $\psi = \frac{\sigma_{a,w,Ed,o} + \sigma_{2,w,Ed,o}}{\sigma_{a,w,Ed,u} + \sigma_{2,w,Ed,u}}$ where $\sigma_{a,w,Ed}$ are the stresses acting on the pure steel cross section and $\sigma_{2,w,Ed}$ are the stresses acting on the fully cracked section. A classification taking into account the construction sequence is demonstrated in Example 9.4.

SLAB REINFORCEMENT

See Example 9.2.

SHEAR RESISTANCE

The web does not have longitudinal stiffeners.

Equation 9.17a:

$$\frac{h_w}{t_w} = \frac{1800}{16} = 112.5 > 60 \cdot 0.81 = 48.6$$

Accordingly, the shear buckling resistance is relevant and will be calculated in accordance with Chapter 8. Transverse vertical stiffeners are supposed to be provided at a distance a = 3.0 m.

Aspect ratio of the panel (Table 8.5):

$$\alpha = \frac{3.0}{1.8} = 1.67$$

Reference stress, Equation 8.10:

$$\sigma_e = 18980 \cdot \left(\frac{1.6}{180}\right)^2 = 1.5 \text{ kN/cm}^2$$

Buckling factor for shear (Table 8.5):

$$k_\tau = 5.34 + \frac{4}{1.67^2} = 6.77$$

Critical buckling shear stress, Equation 8.11:

$$\tau_{cr} = 6.77 \cdot 1.5 = 10.16 \text{ kN/cm}^2$$

Nondimensional slenderness, Equation 8.19:

$$\bar{\lambda}_w = 0.76 \cdot \sqrt{\frac{35.5}{10.16}} = 1.42 > 1.08$$

Reduction factor for nonrigid end posts (Table 8.10):

$$\chi_w = \frac{0.83}{1.42} = 0.58$$

Shear resistance of web, Equation 8.51:

$$V_{bw,Rd} = \frac{0.58 \cdot 35.5 \cdot 180 \cdot 1.6}{\sqrt{3} \cdot 1.1} = 3112.4 \text{ kN}$$

Shear verification:

$$V_{Ed} = 1450 \text{ kN} \leq V_{bw,Rd} \text{ (sufficient)}$$

It is $\frac{V_{Ed}}{V_{Rd}} = \frac{1450}{3112.4} = 0.47 < 0.5$; therefore, there is no moment–shear interaction.

BENDING RESISTANCE

The flanges are class 1 and the web class 3. Accordingly, the section will be treated as class 2 by determination of the plastic bending resistance considering part of the web as noneffective (Figure 9.13).

The position of the plastic neutral axis is calculated from Figure 9.11 as follows:

$$N_{pl,s} = \frac{50}{1.15} \cdot 96 = 4173.9 \text{ kN}, \quad N_{pl,fo} = 32.5 \cdot 640 = 20800 \text{ kN}$$

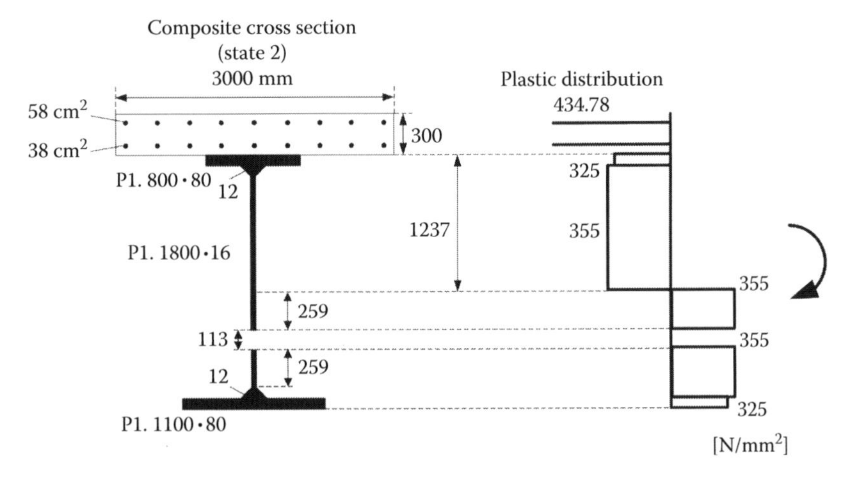

Figure 9.13 Plastic stress distribution with effective part of the web.

$N_{pl,fu} = 32.5 \cdot 880 = 28600$ kN, $N_{pl,20\varepsilon,d} = 35.5 \cdot 40 \cdot 0.81 \cdot 1.6^2 = 2944.5$ kN

$$x_{2,pl} = 30 + 8 + \frac{28600 + 2944.5 - 20800 - 4173.9}{35.5 \cdot 1.6} = 153.7\,\text{cm}$$

Width of effective parts: $20 \cdot 0.81 \cdot 16 = 259$ mm.

This gives a bending resistance $M_{2,pl,Rd} = 55093.56$ kN-m.

Bending check, Table 9.10 row 3: $M_{Ed} = 50000$ kN-m $\leq M_{2,pl,Rd}$ (sufficient)

9.9 CLASS 3 CROSS SECTIONS

Opposite to class 1 or 2 sections, verifications for class 3 cross sections at ULSs are performed in the level of stresses rather in the level of internal forces and moments. Stresses under factored loads are limited to the design material strengths, the latter being determined with larger resistance safety factors compared to SLSs. Design stresses are determined by elastic stress analysis of the gross cross section, possibly allowing for shear lag as presented in Section 7.2. The relevant combinations of actions are presented in Tables 5.6 and 5.8. It may be seen that the *secondary effects of creep and shrinkage have to be taken into account. Stresses at construction stages must also be accounted for.* Direct, shear, and *von Mises* design stresses are limited to the relevant design resistances as presented in Table 9.11.

The stress points to be verified are usually (Table 9.12)

- For concrete, the top fiber of the deck slab "co" (row 1 in Table 9.11)
- For reinforcement, the upper layer "so"
- For structural steel,
  - The extreme fiber of the top flange "ao"
  - The extreme fiber of the bottom flange "au"
  - The point in the web where shear stresses have their maximum value "wm"
  - Extreme points "wo" and "wu" in the web where *von Mises* stresses must be limited as direct and shear stresses coexist

Table 9.11 Stress design for class 3 cross sections (shear buckling not relevant)

	Material	*Stresses*	*Verification*
1	Concrete	Compression	$\sigma_{c,Ed,0} \leq \frac{f_{ck}}{\gamma_c} = \frac{f_{ck}}{1.5}$
2	Structural steel	Direct stresses	$\sigma_{a,Ed} \leq f_{yd} = \frac{f_{yk}}{\gamma_{M0}} = \frac{f_{yk}}{1.0}$
3		Shear stresses	$\tau_{a,Ed} \leq \frac{f_{yk}}{\sqrt{3}\cdot\gamma_{M0}} = \frac{f_{yk}}{\sqrt{3}\cdot 1.0}$
4		*von Mises* stresses	$\left(\frac{\sigma_{a,Ed}}{f_{yk}/\gamma_{M0}}\right)^2 + 3\cdot\left(\frac{\tau_{a,Ed}}{f_{yk}/\gamma_{M0}}\right)^2 \leq 1.0$
5	Reinforcement	Tension (or compression)	$\sigma_{s,Ed} \leq f_{sd} = \frac{f_{sk}}{\gamma_s} = \frac{f_{sk}}{1.15}$

Note: The von Mises stress refers to points where direct and shear stresses coexist, for example, in webs.

As already mentioned, the loading history of a bridge due to concreting sequence is for the stress verifications of primary importance; see Figure 2.50. Stresses should be calculated for combinations of *short-term* and *long-term* actions. The short-term design refers to the beginning of the bridge's life where creep and shrinkage are not well developed and can for simplification be neglected. Therefore, global analysis and stresses are based on the short-term modular ratio n_0 given in Table 6.4. On the contrary, the long-term design refers to a time in which creep and shrinkage have been fully developed and must be taken into account, t = 30,000 days.

The short-term design is usually critical for the concrete stresses (Equation 9.31). The short-term design considering the loading history is demonstrated in Table 9.12. Bending moments acting on the same composite cross section can be combined, and the final stress value is calculated. Moreover, preloading of the pure steel girders (e.g., due to wet concrete) is considered at both sagging and hogging bending areas (Equation 9.32).

Long-term design usually leads to critical verifications for the reinforcing and the structural steel. Tensile stresses in concrete due to shrinkage may be developed at the bottom fiber of the deck slab; this may need to be investigated especially in the case of small-span bridges where compressive stresses are not high. Considering loading history is more laborious than in the case of short-term design due to the concrete's rheological behavior. This is shown in Table 9.13 and is explained as follows:

For sagging bending areas:

- Stresses due to preloading of the pure steel girders are not shown in the figures of Table 9.13 but are obviously to be taken into account.
- The design bending moment $M_{1,Ed,i,0}^{Permanent}$ of the loading case (i) is calculated without the effects of creep for a system with short-term cross-sectional properties. $M_{1,Ed,i,0}^{Permanent}$ is introduced when concrete has an age t_{0i} and causes creep of type P (Section 6.1.2). Therefore, stresses at time $t = \infty$ due to $M_{1,Ed,i,0}^{Permanent}$ are based on the long-term inertia moment $I_{1,P}(t_{0i}, \infty)$ calculated with a creep factor $\phi(t_{0i}, \infty)$; see Figure 7.39. It is noted that $M_{1,Ed,i,0}^{Permanent}$ refers to the design moments due to permanent loadings G and imposed deformations D, if any.
- In continuous systems, additional deformations and rotations due to creep are restrained due to the system's static indeterminacy. At the position of $M_{1,Ed,i,0}^{Permanent}$, a secondary bending moment $M_{1,Ed,i,PT}$ is developed calculated according to Section 7.4.3.

Table 9.12 Short-term stress calculation for class 3 cross sections considering loading history

Notation for stresses (i = number of loading case)

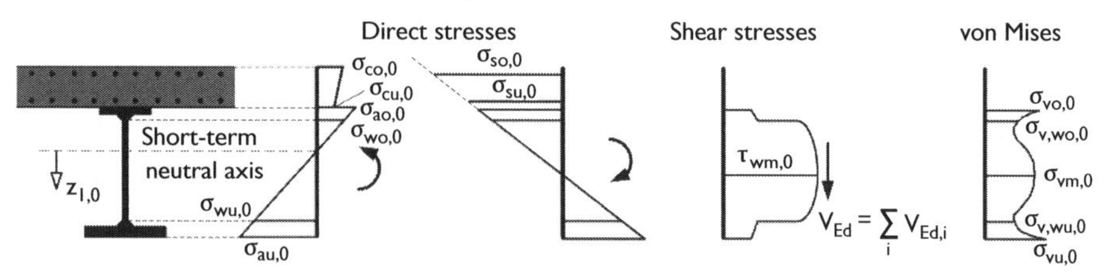

Loading history for calculating short-team direct stresses

Sagging bending areas	Hogging bending areas
Bending moments acting on pure steel (I_a)	Bending moments acting on pure steel (I_a)

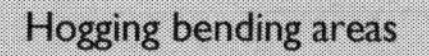

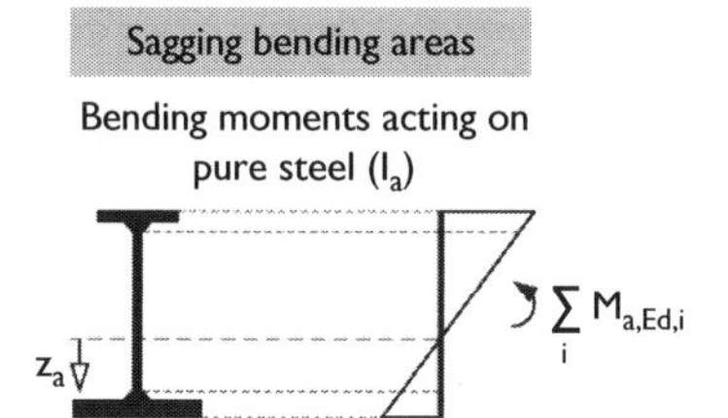

Bending moments acting on uncracked composite section ($I_{1,0}$)

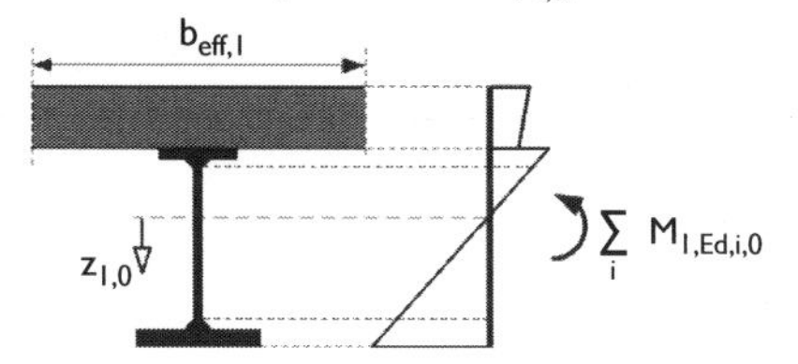

Bending moments acting on fully cracked composite section ($I_{2,sa}$)

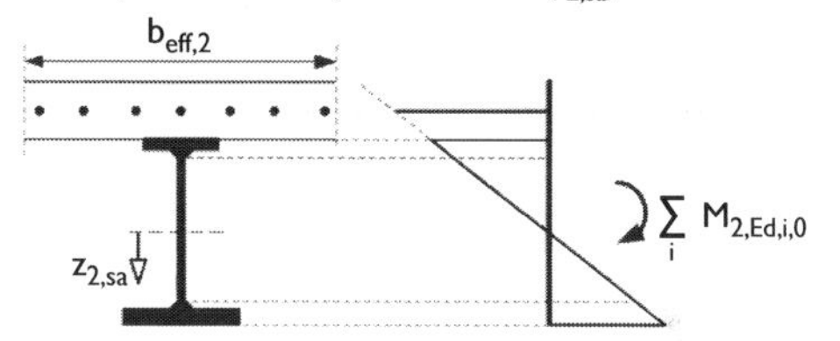

Direct stresses for sagging bending

Concrete: $\sigma_{c,Ed,0} = \sum_i M_{I,Ed,i,0} \cdot \frac{z_{c,0}}{n_0 \cdot I_{I,0}}$ for fibers "co" and "cu" (9.31)

Structural steel:

$$\sigma_{a,Ed,0} = \sum_i M_{a,Ed,i} \cdot \frac{z_a}{I_a} + \sum_i M_{I,Ed,i,0} \cdot \frac{z_{a,0}}{I_{I,0}} \text{ for fibers "ao, au, wo, wu"} \quad (9.32)$$

Direct stresses for hogging bending

Reinforcement:

$$\sigma_{s,Ed,0} = \sum_i M_{2,Ed,i,0} \cdot \frac{z_{s,sa}}{I_{2,sa}} \text{ for fibers "so" and "su"} \quad (9.33)$$

Structural steel:

$$\sigma_{a,Ed,0} = \sum_i M_{a,Ed,i} \cdot \frac{z_a}{I_a} + \sum_i M_{2,Ed,i,0} \cdot \frac{z_{a,sa}}{I_{2,sa}} \text{ for fibers "ao, au, wo, wu"} \quad (9.34)$$

Shear stresses:

$$\tau_{a,Ed,0} = \sum_i V_{Ed,i} / A_w \quad (9.35)$$

Table 9.13 Long-term stress calculation for class 3 cross sections considering loading history

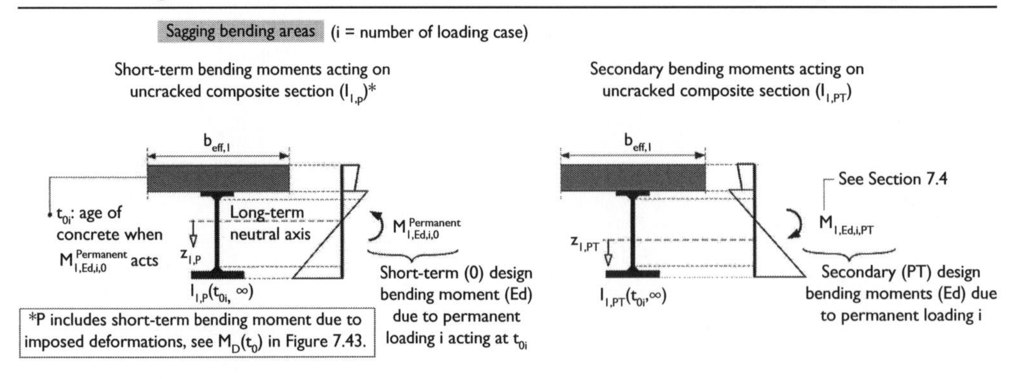

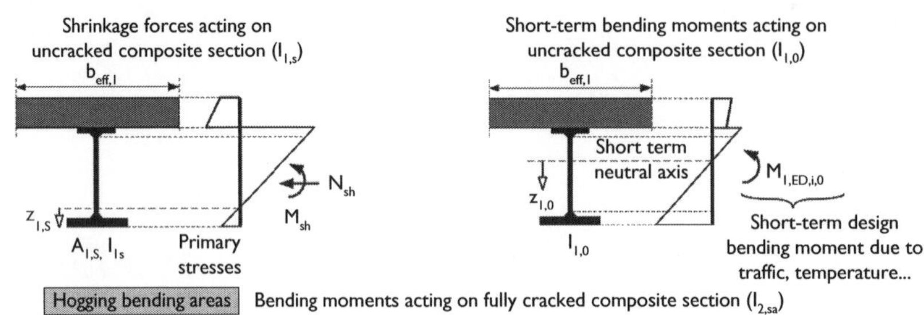

Hogging bending areas Bending moments acting on fully cracked composite section ($I_{2,sa}$)

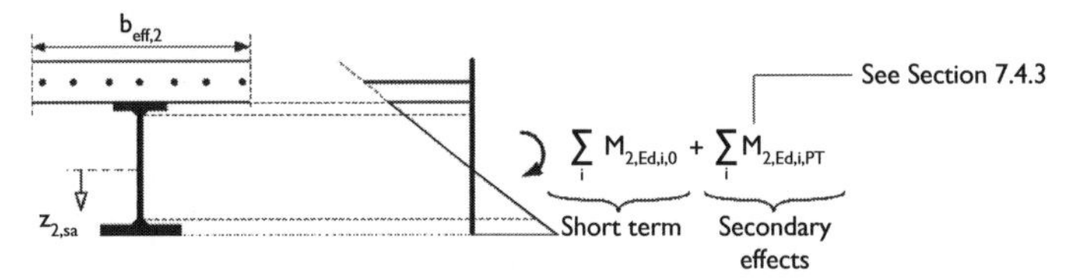

Direct stresses for sagging bending

Concrete: according to EN. 1994-2, the verification can be omitted.

Structural steel: for fibers "ao, au, wo, wu,"

$$\sigma_{a,Ed,\infty} = \sum_i M_{a,Ed,i} \cdot \frac{z_a}{I_a} + \sum_i M^{Permanent}_{I,Ed,i,0} \cdot \frac{z_{a,P}(t_{0i},\infty)}{I_{I,P}(t_{0i},\infty)}$$

$$+ \sum_i \left(M_{I,Ed,i,PT} \cdot \frac{z_{a,PT}(t_{0i},\infty)}{I_{I,PT}(t_{0i},\infty)} \right) + \left(\frac{N_{sh}}{A_{I,S}} + M_{sh} \cdot \frac{z_{a,S}}{I_{I,S}} \right) + \sum_i M_{I,Ed,i,0} \cdot \frac{z_{a,0}}{I_{I,0}} \quad (9.36)$$

Direct stresses for hogging bending

Reinforcement: for fibers "so" and "su,"

$$\sigma_{s,Ed,\infty} = \sum_i M_{2,Ed,i,0} \cdot \frac{z_{s,sa}}{I_{2,sa}} + \sum_i M_{2,Ed,i,PT} \cdot \frac{z_{s,sa}}{I_{2,sa}} \quad (9.37)$$

Structural steel: for fibers "ao, au, wo, wu,"

$$\sigma_{a,Ed,\infty} = \sum_i M_{a,Ed,i} \cdot \frac{z_a}{I_a} + \sum_i M_{2,Ed,i,PT} \cdot \frac{z_{a,sa}}{I_{2,sa}} + \sum_i M_{2,Ed,i,0} \cdot \frac{z_{a,sa}}{I_{2,sa}} \quad (9.38)$$

Shear stresses:

$$\tau_{a,Ed,\infty} = \frac{\left(\sum_i V_{Ed,i} + \sum_i V_{Ed,i,PT} \right)}{A_w} \quad (9.39)$$

This bending moment is permanent but with a magnitude that continuously changes, creep of type PT (Section 6.1.2). Stresses due to $M_{1,Ed,i,PT}$ are calculated based on the inertia moment $I_{1,PT}(t_{0i}, \infty)$.

- Shrinkage forces are calculated from Equations 7.28a and 7.28b. They cause self-equilibrating stresses (primary effects). Secondary bending moments due to shrinkage (see Section 7.4.3) are treated as $M_{1,Ed,i,PT}$ moments acting on cross sections with $I_{1,PT}(t_{0S}=1 \text{ day}, \infty)$.
- Bending moments due to nonpermanent actions (e.g., traffic, temperature, wind) are calculated from a system with short-term cross-sectional properties. They act on composite sections with moment of inertia equal to $I_{1,0}$.
- Long-term stresses are then combined and $\sigma_{c,\infty}$ and $\sigma_{a,\infty}$ are calculated; Equations 9.36 through 9.39.

For hogging bending areas:

- Stresses due to preloading of the steel girder are not shown in the figures of Table 9.13, but they are obviously taken into account.
- All the bending moments acting on the fully cracked cross section are added, and stresses in structural steel and reinforcement are calculated.

REMARK 9.4

- Secondary bending moments due to creep are hogging moments that lead to a considerable increase of stresses at internal supports and decrease them at spans; see Figure 7.41. However, since they result from the same source, a unique safety factor, 1.35 as for permanent loads, is applied although the effects are favorable at internal supports and unfavorable at spans.
- Primary effects of shrinkage act on the uncracked areas and have an unfavorable influence on the stresses of structural steel; see Equation 9.38. Secondary bending moments due to shrinkage are in most cases considerably high (Figure 7.44). They tend to increase the stresses at internal supports and reduce them at spans. In all cases, shrinkage effects are multiplied with a safety factor equal to 1.0. Shrinkage stresses are included in the fourth term of Equation 9.36.
- It was mentioned that imposed deformations can be treated as permanent loadings causing creep of type P *and not D*. Indeed, the short-term sagging moment due to a support settlement ($M_D(t_0)$ in Figure 7.43) is considered as a permanent action causing creep of type P. The stresses due to this action are integrated in the second term of Equation 9.36. The secondary hogging moment due to the support settlement can be calculated with the equivalent temperature of Equation 7.26 by substituting M_0 with $M_D(t_0)$. The corresponding secondary stresses are included in the third term of Equation 9.36.
- Cross sections in bridges are not loaded instantaneously. Therefore, there is no objective rule for defining the age of concrete at loading time t_0. For example, the age of concrete of a composite section may be 14 days at the beginning of casting and 28 days at the end of it. Many designers consider t_0 as the average value, that is, (14+28)/2=21 days. However, creep will start developing from the very beginning. Conservatively, the minimum value of concrete's age can be applied.

- It was noted that for the short-term design (Table 9.12), creep and shrinkage are not taken into account. In part of the literature [9.8], short-term design is represented by the age of concrete at the time of traffic opening, and therefore, creep and shrinkage are not neglected. This approach is not followed in this book because it makes calculations considerably laborious and no significant differences arise.
- For bridges casted in several stages, EN 1994-2 permits the use of one mean value t_0 for the determination of the creep coefficient. This simplification is allowed due to the lack of adequate information on the timetable of construction at the design stage. However, it is recommended to avoid the code's suggestion since it has no theoretical background and can lead to unsafe results for stresses, secondary moments, and deformations. Indeed, the mean value t_0 for a continuous bridge is unlikely to be less than a month. For the majority of the segments, the first loading may occur at an age as low as a week; thus, adopting a mean value will underestimate creep factor $\phi(t, t_0)$ considerably. Designers should be able to make safe assumptions for the age of concrete at loading; see Example 9.4.
- For the majority of bridges, verifications for the extreme cases of short- and long-term design are considered as adequate. However, verifications for intermediate periods (between t_0 and t_∞) may be considered from the designer as necessary. Guidance may also be given in the National Annex.
- It is reminded that the web should be classified both for short- and long-term effects by taking into account the construction sequence and shear lag effect; see Example 9.4.

Table 9.14 Stress design for class 3 cross sections (shear buckling is relevant)

	Material	*Stresses*	*Verification*
1	Concrete	Compression	$\sigma_{c,Ed,0} \leq \frac{f_{ck}}{\gamma_c} = \frac{f_{ck}}{1.5}$
2	Reinforcement	Tension (or compression)	$\sigma_{s,Ed} \leq f_{sd} = \frac{f_{sk}}{\gamma_s} = \frac{f_{sk}}{1.15}$
3	Structural steel	Direct stresses	$\sigma_{a,Ed} \leq f_{yd} = \frac{f_{yk}}{\gamma_{M0}} = \frac{f_{yk}}{1.0}$
4		Shear stresses	$\tau_{a,Ed} \leq \frac{\chi_w \cdot f_{yk}}{\sqrt{3} \cdot \gamma_{M1}} = \frac{\chi_w \cdot f_{yk}}{\sqrt{3} \cdot 1.1}$
5		Direct and shear stresses (von Mises)	$\left(\frac{\sigma_{a,Ed}}{f_{yk}/\gamma_{M0}}\right)^2 + 3 \cdot \left(\frac{\tau_{a,Ed}}{\chi_w \cdot f_{yk}/\gamma_{M1}}\right)^2 = \left(\frac{\sigma_{a,Ed}}{f_{yk}/1.0}\right)^2 + 3 \cdot \left(\frac{\tau_{a,Ed}}{\chi_w \cdot f_{yk}/1.1}\right)^2 \leq 1.0$

Notes:

The von Mises stress refers to points where direct and shear stresses coexist, for example, in webs.

Additional stresses, if any, in concrete and in reinforcement due to local bending of the deck slab should be taken into account.

Warping stresses should be neglected if the distortional effects do not exceed 10% of the bending ones.

Table 9.11 applies to class 3 cross sections, where the web is not prone to shear buckling. If shear buckling of the web is relevant, the shear stresses in the web must be limited to the shear buckling strength. Additionally, the interaction between direct and shear buckling strength, usually by application of the reduced stress method, must be examined. In this case, the design stresses that are introduced in the interaction do not refer to specific points of the web but to the average design shear stress and the *maximal* direct stress in the web panel. Table 9.14 presents the required verifications.

9.10 CLASS 4 CROSS SECTIONS THAT ARE TREATED AS CLASS 3 CROSS SECTIONS

Class 4 cross sections in which plate buckling verifications are performed by the reduced stress method (see Section 8.4) may be treated as class 3 cross sections. This category applies usually to cross sections with longitudinally stiffened walls, since in the absence of longitudinal stiffeners, the effective width method (see Section 8.5) may be more easily applied. In addition, in the presence of torsion, the effective width method does not provide interaction relations including torsion so that the reduced stress method is applied. Consequently, this section refers more often to box-girder bridges with longitudinal stiffeners in the bottom flange and the webs. The verification procedure for such types of cross sections with reference to the section of Figure 9.14 is as follows:

Step 1: Performance of static analysis, possibly including second-order effects (e.g., in cable-stayed bridges), to determine internal forces and moments.

Step 2: Definition of design direct and shear stresses from internal forces and moments, from elastic section analysis based on gross section properties with shear lag effects; see Section 7.2.

Step 3: Limitation of stresses at extreme points for parts of the section where plate buckling is not relevant. In Figure 9.14, these are the tensile stress of the top girder flange and the tensile stress of the upper reinforcement layer.

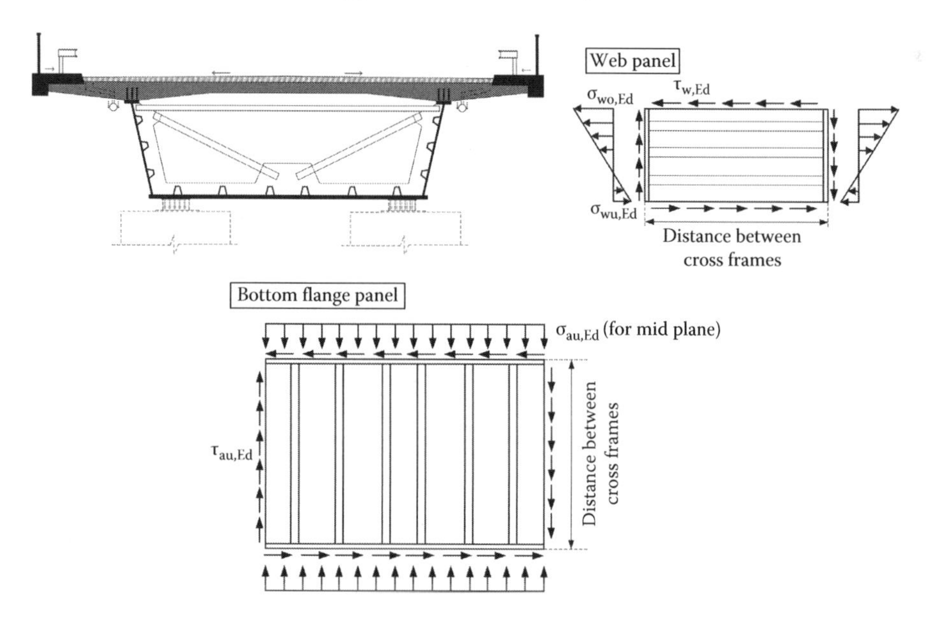

Figure 9.14 Stresses in box girders under hogging bending, vertical shear and torsion, and isolation of panels.

Step 4: Isolation of panels for each wall, considering simple support conditions at their joint edges. The transverse edges coincide with the position of transverse stiffeners or cross frames.

Step 5: Limitation of stresses of panels or subpanels by the reduced stress method. In Figure 9.14, these are the webs and the bottom flange.

REMARK 9.5

At sagging moment areas, bottom flanges are under tension. However, in some cases, buckling verifications due to interaction of tension with high shear forces may be necessary (i.e., in curved bridges).

Table 9.15 presents the required design relations. It may be seen that stress limitations in structural steel refer to individual *points* with extreme stresses for section parts that are not

Table 9.15 Design for class 4 cross sections treated as class 3 sections (reduced stress method)

	Material	*Stresses*	*Verification*
1	Concrete	Compression	$\sigma_{c,Ed,0} \le \frac{f_{ck}}{\gamma_c} = \frac{f_{ck}}{1.5}$
2	Reinforcement	Tension (or compression)	$\sigma_{s,Ed} \le f_{sd} = \frac{f_{sk}}{\gamma_s} = \frac{f_{sk}}{1.15}$
3	Structural steel stress points in panels *with no plate buckling* verifications	Direct stresses	$\sigma_{a,Ed} \le f_{yd} = \frac{f_{yk}}{\gamma_{M0}} = \frac{f_{yk}}{1.0}$
4		Shear stresses	$\tau_{Ed} \le \frac{\chi_w \cdot f_{yk}}{\sqrt{3} \cdot \gamma_{M1}} = \frac{f_{yk}}{\sqrt{3} \cdot 1.1}$
5		*von Mises stresses*	$\left(\frac{\max \sigma_{a,Ed}}{f_{yk}/\gamma_{M0}}\right)^2 + 3 \cdot \left(\frac{\tau_{Ed}}{\chi_w \cdot f_{yk}/\gamma_{M1}}\right)^2 \le 1.0$
6	Structural steel stresses in panels *with plate buckling* verifications	Direct stresses	$\max \sigma_{a,Ed} \le f_{yd} = \frac{\rho_x \cdot f_{yk}}{\gamma_{M1}} = \frac{\rho_x \cdot f_{yk}}{1.1}$
7		Shear stresses	$\tau_{a,Ed} \le \frac{\chi_w \cdot f_{yk}}{\sqrt{3} \cdot \gamma_{M1}} = \frac{\chi_w \cdot f_{yk}}{\sqrt{3} \cdot 1.1}$
8		*von Mises stresses*	$\left(\frac{\max \sigma_{a,Ed}}{\rho_x \cdot f_{yk}/\gamma_{M1}}\right)^2 + 3 \cdot \left(\frac{\tau_{a,Ed}}{\chi_w \cdot f_{yk}/\gamma_{M1}}\right)^2 = \left(\frac{\max \sigma_{a,Ed}}{\rho_x \cdot f_{yk}/1.1}\right)^2 + 3 \cdot \left(\frac{\tau_{a,Ed}}{\chi_w \cdot f_{yk}/1.1}\right)^2 \le 1.0$ (see also Equation 8.42 or 8.41)

Notes:

$\max\sigma_{a,Ed}$ is the maximum compression stress in the panel under consideration. Direct stresses refer to midplane of panels.

Additional stresses, if any, in concrete and in reinforcement due to local bending of the deck slab should be taken into account.

Warping stresses should be neglected if the distortional effects do not exceed 10% of the bending ones.

prone to plate buckling. In the example of Figure 9.14, this may be the stress in the extreme fiber of the tension flange. Oppositely, stress limitations for panels subjected to plate buckling verifications refer not to individual points but to the *maximum* compression stress in the panel and the average shear stress. Direct stresses in panels for plate buckling verifications refer to midplanes of the panels and not at extreme fibers. Accordingly, the direct stress in the bottom flange in the example of Figure 9.14 is determined at its midplane and not at the bottom fiber.

In composite bridges, *hybrid steel girders* are used quite often; see Example 9.4. These are steel girders with flanges and web(s) made of different steel grades. According to EN 1993-1-5, flanges may have a yield strength f_{yf} up to $2 \cdot f_{yw}$, where f_{yw} is the yield strength of the web. This limitation is based on available tests in this field.

EXAMPLE 9.4

A two-span road bridge has a composite box cross section (Figure 9.16). The bottom flange is stiffened by four longitudinal stiffeners. The upper reinforcement area of the slab is $A_{so} = 300\ cm^2$ and the lower $A_{su} = 200\ cm^2$. After placing the steel girder, the concrete is casted in segments as shown in Figure 9.15.

The self-weight of the steel girder is $g_l = 30.4$ kN/m, the dry self-weight of the slab is $g_c = 11 \cdot 0.3 \cdot 25 = 82.5$ kN/m, and the self-weight of the superstructure is $g_2 = 3.4\ kN/m^2$ (waterproofing layer 3 cm, asphalt 8 cm, two concrete pavement $0.2 \times 0.5\ m^2$ all with specific weight 25 kN/m²), safety barriers 1.3 kN/m, and cornices 0.5 kN/m.

The cross sections of the bridge at the internal support and the midspan shall be verified at ULSs. Steel grade of the flanges is S420 at span and S460 at internal support. Steel grades for the webs are S355 and S460 correspondingly. Strength class of concrete C35/45, grade of reinforcement B500B.

The thickness of the concrete slab and the depth of the steel girder actually vary along the bridge width. However, in this example, they are both taken as constant for simplification of the calculations of the composite girder.

The resistance of the cross section will be verified for the internal forces arising at internal support and at span (x = 27.5 m).

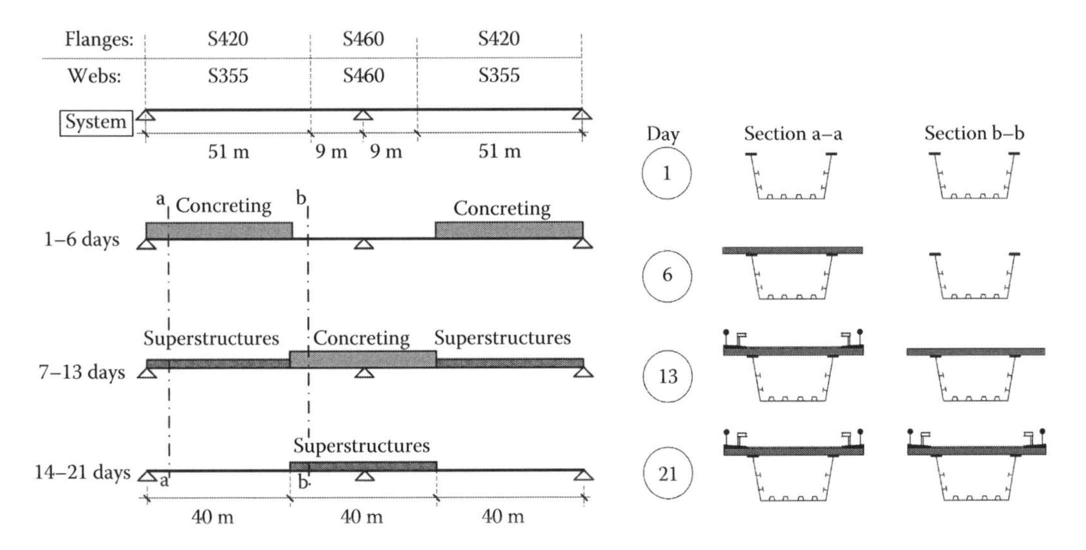

Figure 9.15 Structural system and concreting sequence.

Assumption: The reduction factors ρ for plate buckling are calculated according to Chapter 8. They are assumed to be equal to 0.90 both at erection and at final stage. The shear buckling reduction factor χ_w for the webs is assumed to be equal to 0.80.

CROSS-SECTIONAL CLASSIFICATION

Design yield stress of structural steel (Table 6.14):

Web at span t=25 mm:

$$f_{yd} = \frac{345}{1.0} = 345\ \text{N/mm}^2$$

Top flange at span t=50 mm:

$$f_{yd} = \frac{390}{1.0} = 390\ \text{N/mm}^2$$

Bottom flange at span t=15 mm:

$$f_{yd} = \frac{420}{1.0} = 420\ \text{N/mm}^2$$

Web at internal support t=25 mm:

$$f_{yd} = \frac{440}{1.0} = 440\ \text{N/mm}^2$$

Top flange at internal support t=50 mm:

$$f_{yd} = \frac{430}{1.0} = 430\ \text{N/mm}^2$$

Bottom flange at internal support t=15 mm:

$$f_{yd} = \frac{460}{1.0} = 460\ \text{N/mm}^2$$

In order to define the elastic cross-sectional properties, the cross-sectional area of the longitudinal stiffeners in the bottom flange is "smeared" over the flange width. The resulting additional thickness is $t = \frac{A_{stiffeners}}{b_{flange}} = 20$ mm. Accordingly, the elastic properties are determined for a total thickness of the bottom flange 15+20=35 mm. The web stiffeners do not have any significant influence on the total thickness of the web that remains equal to 25 mm.

Note: Longitudinal stiffeners may possibly have much lower thickness than the web or the flange that they support and therefore higher yielding stress. However, keeping the yielding stress of the supported plate is a conservative calculation approach. Longitudinal stiffeners in webs are allowed to be considered only to enhance the resistance to plate buckling and therefore not to be taken into account for the determination of the cross-sectional properties.

Classification at internal support

Web

The web should be classified for both short- and long-term design. However, the internal forces for the long-term design are much more critical, and only the "long-term classification" is shown.

The long-term stresses in the web are calculated in page 389 by taking into account the concreting sequence, creep, shrinkage, and shear lag. The maximum tensile stress is $\sigma_{wo,Ed,\infty}=37.53$ kN/cm^2 and the compressive one $\sigma_{wu,Ed,\infty}=34.37$ kN/cm^2.

The long-term stress ratio is accordingly $\psi_\infty=-\dfrac{37.53}{34.37}=-1.09<-1$:

$$f_y=440\ \text{N/mm}^2 \rightarrow \varepsilon=\sqrt{\frac{235}{440}}=0.73$$

From geometric considerations, the width of the web, neglecting the weld thicknesses, is determined as c=2915 mm.

Table 9.2:

$$\frac{c}{t}=\frac{2915}{25}=116.6>62\cdot 0.73\cdot(1+1.09)\cdot\sqrt{1.09}=98.75.\ \rightarrow \text{the web is class 4.}$$

Note: Classification of a stiffened plate is conducted with the nominal thickness value and not the smeared one.

Bottom flange

The bottom flange is subjected to compression:

$$f_y=460\ \text{MPa}\rightarrow\varepsilon=\sqrt{\frac{235}{460}}=0.71$$

Table 9.2: $\dfrac{c}{t}=\dfrac{4000}{15}=266.67>42\cdot 0.71=29.82\rightarrow$ The flange is class 4, and then the cross section is class 4.

Short-term classification at midspan

Web

The maximum tensile stress is $\sigma_{wu,Ed,0}=19.94$ kN/cm^2 and the compressive one $\sigma_{wo,Ed,0}=12.96$ kN/cm^2; see page 391:

$$f_y=345\ \text{MPa}\rightarrow\varepsilon=\sqrt{\frac{235}{345}}=0.83$$

The short-term stress ratio is accordingly

$$\psi_0=-\frac{19.94}{12.96}=-1.54<-1$$

Table 9.2:

$$\frac{c}{t}=\frac{2915}{25}=116.6<62\cdot 0.83\cdot(1+1.54)\cdot\sqrt{1.54}=162.2\rightarrow \text{the web is class 3}$$

The composite section for the short-term design is class 3.

Long-term classification at midspan

Web

The maximum tensile stress is $\sigma_{wu,Ed,\infty}=19.1$ kN/cm^2 and the compressive one $\sigma_{wo,Ed,\infty}=16.67$ kN/cm^2; see page 395.

The short-term stress ratio is accordingly

$$\psi_{\infty} = -\frac{19.1}{16.67} = -1.15 < -1.$$

Table 9.2:

$$\frac{c}{t} = \frac{2915}{25} = 116.6 < 62 \cdot 0.83 \cdot (1+1.15) \cdot \sqrt{1.15} = 118.65 \rightarrow \text{the web is class 3}$$

The composite section for the long-term design is class 3.

For the cross section at internal support, the bottom flange and the web are class 4. Therefore, the cross section is class 4. At midspan, the composite section is class 3 for both the short- and long-term design. However, since plate buckling of the bottom flange and the web is performed by the *reduced stress method* (see Section 8.4), the cross sections are treated as class 3. Accordingly, sections will be designed on the basis of stresses taking into account construction stages for short- and long-term design as demonstrated in Tables 9.12 and 9.13. Construction stages shall be generally considered for the entire bridge, independent on whether other cross sections, for example, at spans, are class 1 or 2. This is due to the fact that in order to neglect construction stages, creep and shrinkage *all* cross sections must be class 1 or 2.

MODULAR RATIOS

The short-term modular ratio is given in Table 6.4: $n_0 = 6.18$.

The creep factors are calculated according to Equation 6.9 and the long-term modular ratios from Equation 6.20. The results are given in Table 9.16.

The torsional modular ratios for the sagging bending areas are $n_{0G,1} = 5.7$ (Table 6.8) and for the hogging ones $n_{0G,2} = 4.75$. It is assumed that permanent loadings cause no torsion so that long-term modular ratios $n_{L,G}$ are not taken into account.

SHEAR LAG IN CONCRETE FLANGES

For sagging bending

According to Figure 7.34, $L_e = 0.85 \cdot 60 = 51$ m.

Considering for simplicity $b_0 = 0$, the full widths adjacent to the webs are

$$b_1 = 7/2 = 3.5 \text{ m} \quad \text{and} \quad b_2 = 2.0 \text{ m}$$

From Equation 7.19b,

$$b_{e1} = 51/8 = 6.375 \text{ m} > b_1 \quad \text{and} \quad b_{e2} = 51/8 = 6.375 \text{ m} > b_2$$

The concrete flanges are not reduced due to shear lag; therefore, for the whole section, $b_{eff,1} = 11.0$ m.

Table 9.16 Long-term modular ratios and creep factors

t_0 in days	$\phi(t_0, \infty)$	P	S	PT
1	2.66	—	15.22	15.22
7	1.86	18.82	—	12.50
14	1.63	17.26	—	11.72

For hogging bending

According to Figure 7.34,

$$L_e = 0.25 \cdot (60 + 60) = 30 \text{ m}$$

From Equation 7.19b,

$$b_{e1} = 30/8 = 3.75 \text{ m} > b_1 \quad \text{and} \quad b_{e2} = 30/8 = 3.75 \text{ m} > b_2$$

The concrete flanges are not reduced due to shear lag; therefore, $b_{eff,2} = 11.0$ m

SHEAR LAG IN BOTTOM FLANGE

For sagging bending

According to Table 7.7,

$$L_e = 0.85 \cdot 60 = 51 \text{ m}$$

It is $b_0 = 4/2 = 2.0 \text{ m} > L_e/50 = 1.02 \text{ m} \rightarrow$ Shear lag must not be neglected.

It is $A_{sl}/(b_0 \cdot t) = 20/15 = 1.33 \rightarrow \alpha_0 = \sqrt{1 + 1.33} = 1.53$:

$$\kappa = \frac{1.53 \cdot 2.0}{51} = 0.06 \Rightarrow 0.02 < \kappa \leq 0.7 \rightarrow \beta = \beta_1 = \frac{1}{1 + 6.4 \cdot 0.06^2} = 0.98$$

Effective[s] width, Equation 7.20:

$$b_{eff} = 0.98 \cdot 2 = 1.96 \text{ m}$$

To take into account the effective[s] width, the smeared thickness of the bottom flange is taken as $t_{eq} = 3.5 \cdot 1.96/2 = 3.4$ cm.

For hogging bending

According to Table 7.7:

$$L_e = 0.25 \cdot (60 + 60) = 30 \text{ m}$$

It is $b_0 = 4/2 = 2.0 \text{ m} > L_e/50 = 0.6 \text{ m} \rightarrow$ Shear lag must not be neglected.

It is $A_{sl}/(b_0 \cdot t) = 20/15 = 1.33 \rightarrow \alpha_0 = \sqrt{1 + 1.33} = 1.53$.

$$\kappa = \frac{1.53 \cdot 2.0}{30} = 0.102 \Rightarrow 0.02 < \kappa \leq 0.7 \rightarrow$$

$$\beta = \beta_2 = \frac{1}{1 + 6.0 \cdot \left(0.102 - \frac{1}{2500 \cdot 0.102}\right) + 1.6 \cdot 0.102^2} = 0.62$$

Effective[s] width, Equation 7.20: $b_{eff} = 0.62 \cdot 2 = 1.24$ m

To take into account the effective[s] width, the smeared thickness of the bottom flange is taken as $t_{eq} = 3.5 \cdot 1.24/2 = 2.17$ cm.

The aforementioned values reflect elastic effective[s] widths. For cross-sectional verifications at ULSs, plastic effective[s] widths will be used. Following method 3, (see Table 7.9) the following effective widths will be used for verifications.

For sagging bending

Effective[s] width:

$$b_{eff} = 0.98^{0.06} \cdot 2 = 2.0\,\text{m} > 1.96\,\text{m}$$

The smeared thickness of the bottom flange is taken as $t_{eq} = 3.5$ cm.

For hogging bending

Effective[s] width:

$$b_{eff} = 0.62^{0.102} \cdot 2 = 1.90\,\text{m} > 1.24\,\text{m}$$

The smeared thickness of the bottom flange is taken as $t_{eq} = 3.5 \cdot 1.9/2 = 3.4$ cm.

In sagging and bending moment areas at ULS, t_{eq} is approximately equal to 35 mm. There will be no reduction of the cross-sectional properties due to shear lag in the bottom flange.

CROSS-SECTIONAL PROPERTIES (TABLES 9.17 AND 9.18)

LOADINGS

Self-weight steel	$1.05^{a} \cdot 30.4 = 31.92$ kN/m
Wet concrete	$26 \cdot 11 \cdot 0.3 = 85.8$ kN/m
Hardened concrete	$25 \cdot 11 \cdot 0.3 = 82.5$ kN/m
Superstructure	$3.4 \cdot 11 = 37.4$ kN/m
Safety barriers	1.3 kN/m
Cornices	0.5 kN/m
Thermal $\Delta T_{M,heat}$	Table 4.14 for waterproofed surface: $1.1 \cdot 15 = 16.5$°C
Thermal ΔT_{cool}	Table 4.14 for waterproofed surface: $0.9 \cdot 18 = 16.2$°C
Shrinkage	$\varepsilon_{cs}(1, \infty) = 300 \cdot 10^{-6}$
Secondary effects	As follows
Traffic loads (group gr1a)	From Figure 9.17, the loads in longitudinal direction are calculated as For half girder A: $q_{UDL,A,k} = 31.92$ kN/m and axle loads $P_{TS,A,k} = 336.5$ kN For half girder B: $q_{UDL,B,k} = 4.58$ kN/m and axle loads $P_{TS,B,k} = 163.5$ kN

[a] A 5% increase for the self-weight of the steel cross section was considered so that the weights of bracings, studs, and stiffeners are taken into account.

Table 9.17 Cross-sectional properties for ULS with $t_{au} = 35$ mm

Sagging and hogging bending					
Steel girder					
A_a	3872.77	cm^2			
z_{ao}	142.34	cm			
z_{au}	114.16	cm			
I_a	$0.45454 \cdot 10^8$	cm^4			
Sagging bending			*Hogging bending*		
Short term with $n_0 = 6.18$			*Fully cracked section*		
$A_{I,0}$	9712.58	cm^2	$A_{2,sa}$	4372.77	cm^2
$z_{co,0}$	77.64	cm	$z_{ao,sa}$	124.14	cm
$z_{ao,0}$	47.64	cm	$z_{au,sa}$	132.36	cm
$z_{au,0}$	208.86	cm	$z_{so,sa}$	148.14	cm
$I_{I,0}$	$1.037 \cdot 10^8$	cm^4	$z_{su,sa}$	130.14	cm
			$I_{2,sa}$	$0.567 \cdot 10^8$	cm^4
Sagging bending for $t_0 = 1$ day (shrinkage)					
Long term with $n_S = 15.22$			*Long term with $n_{PT} = 15.22$*		
$A_{I,S}$	6540.97	cm^2	$A_{I,PT}$	6540.97	cm^2
$z_{co,S}$	108.02	cm	$z_{co,PT}$	108.02	cm
$z_{ao,S}$	78.02	cm	$z_{ao,PT}$	78.02	cm
$z_{au,S}$	178.48	cm	$z_{au,PT}$	178.48	cm
$I_{I,S}$	$0.8493 \cdot 10^8$	cm^4	$I_{I,PT}$	$0.8493 \cdot 10^8$	cm^4
Sagging bending for $t_0 = 7$ days					
Long term with $n_P = 18.82$			*Long term with $n_{PT} = 12.50$*		
$A_{I,P}$	6126.22	cm^2	$A_{I,PT}$	7012.77	cm^2
$z_{co,P}$	114.32	cm	$z_{co,PT}$	101.76	cm
$z_{ao,P}$	84.32	cm	$z_{ao,PT}$	71.76	cm
$z_{au,P}$	172.18	cm	$z_{au,PT}$	184.74	cm
$I_{I,P}$	$0.8107 \cdot 10^8$	cm^4	$I_{I,PT}$	$0.8878 \cdot 10^8$	cm^4
Sagging bending for $t_0 = 14$ days					
Long term with $n_P = 17.26$			*Long term with $n_{PT} = 11.72$*		
$A_{I,P}$	6284.71	cm^2	$A_{I,PT}$	7188.47	cm^2
$z_{co,P}$	111.81	cm	$z_{co,PT}$	99.64	cm
$z_{ao,P}$	81.81	cm	$z_{ao,PT}$	69.64	cm
$z_{au,P}$	174.69	cm	$z_{au,PT}$	186.86	cm
$I_{I,P}$	$0.8261 \cdot 10^8$	cm^4	$I_{I,PT}$	$0.9008 \cdot 10^8$	cm^4

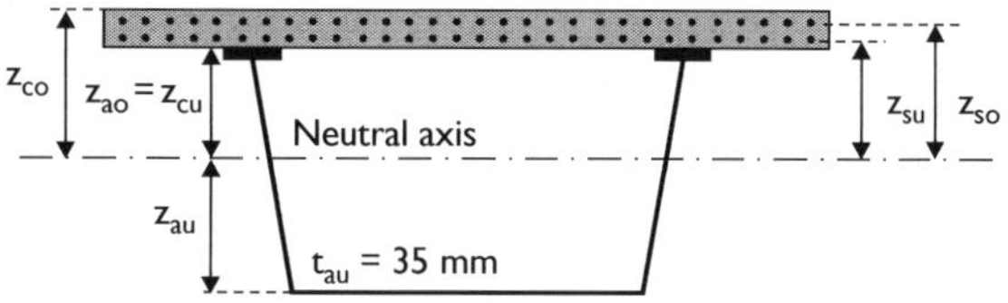

GRILLAGE MODEL: INTERNAL FORCES

A grillage as described in Figure 7.23 is used for the calculation of the internal forces.

For the bracings, the equivalent shear area is calculated from Table 7.3 as follows:

$$f_Q = \frac{7+4}{7} = 1.57$$

From Figure 9.16 for the diagonals, $h \approx 2.2$ m, $b = 7$ m, $d = 4.13$ m, and $A_d = 68.5$ cm^2.

Table 9.18 Torsional constants for ULS

Sagging bending			*Hogging bending*		
Steel girder			Steel girder		
$I_{T,a}$	16,855.8	cm^4	$I_{T,a}$	16,855.8	cm^4
Composite section ($n_{0G,1}=5.7$)			Fully cracked section ($n_{0G,2}=4.75$)		
$I_{T,1}$	$1.628\cdot10^8$	cm^4	$I_{T,2}$	$1.382\cdot10^8$	cm^4

Note: Torsional constants for the composite section were calculated according to Figure 7.12. For the sagging bending areas, $\nu_c=0.2$ and $h_{c,G}=30/5.7=5.26$ cm. For the hogging bending areas, $\nu_c=0$ and $h_{c,G}=0.5\cdot30/4.75=3.16$ cm. One can see that the torsional rigidity of the cross section at internal support is 15% lower than the torsional rigidity at spans.

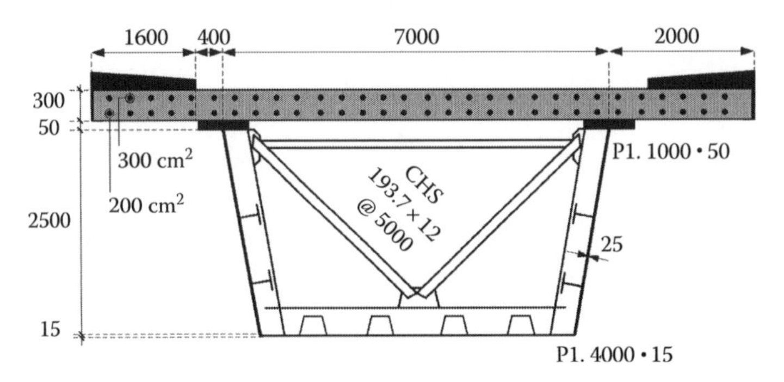

Figure 9.16 Cross section of the bridge at the intermediate support.

Shear area:

$$A_s = 1.57\cdot\frac{21000}{8100}\cdot\frac{220^2\cdot700\cdot68.5}{2\cdot413^3} = 67\,cm^2$$

The aforementioned value is introduced as A_z in the fictitious transverse beams of Figure 7.23. The torsional constants of Table 9.18 are introduced as I_T in the fictitious central girder.

The loading history and the distribution of the cross-sectional properties along the bridge are explained as follows:

- The self-weights of steel and wet concrete of the first concreting phase (1–6 days) act on pure steel cross section. The edge girders have a bending stiffness equal to $E_a\cdot I_a/2$ (Figure 9.18).
- The next concreting phase (7–13 days) takes place at internal region with the wet concrete acting on the pure steel cross section. At the same time, superstructures are placed on the composite parts of the system. The age of concrete is conservatively taken equal to 7 days. For the calculation of stresses at sagging moment areas, the composite girders are accounted for a bending stiffness equal to $E_a\cdot I_{1,0}/2$ with $t_0=7$ days; see Remark 9.4. For the stress calculation, the secondary effects due to creep must be taken into account. The maximum short-term sagging moment is equal to 7050.2 kN-m; see Figure 9.19. The equivalent temperature that acts on the composite parts with a bending stiffness $E_a\cdot I_{1,PT}/2$

with $t_0 = 7$ days is calculated from Equation 7.26; this leads to the bending diagram due to secondary effects in Figure 9.19:

$$\Delta T_{MP} = \frac{2 \cdot 705020}{21000} \cdot \frac{286.5}{10^{-5}} \cdot \left(\frac{1}{0.8107 \cdot 10^8} - \frac{1}{1.037 \cdot 10^8} \right) = 5.18°C$$

- The entire system now consists of composite cross sections. The superstructures for the internal region are placed between 14 and 21 days. The short-term bending moment diagram is shown in Figure 9.20. The maximum short-term sagging moment is equal to 1098.45 kN-m. The age of concrete of the edge composite parts is taken equal to $t_0 =$ 14 days, and the equivalent temperature is equal to

$$\Delta T_{MP} = \frac{2 \cdot 109845}{21000} \cdot \frac{286.5}{10^{-5}} \cdot \left(\frac{1}{0.8261 \cdot 10^8} - \frac{1}{1.037 \cdot 10^8} \right) = 0.74°C$$

For the internal *uncracked* parts, the age of concrete is taken equal to $t_0 = 7$ days. The maximum bending moment for this segment is equal to 1098 kN-m. Thus,

$$\Delta T_{MP} = \frac{2.109800}{21000} \cdot \frac{286.5}{10^{-5}} \cdot \left(\frac{1}{0.8107 \cdot 10^8} - \frac{1}{1.037 \cdot 10^8} \right) = 0.81°C$$

The aforementioned temperature differences are acting on the composite parts with bending stiffnesses $E_a \cdot I_{I,PT}(14, \infty)/2$ and $E_a \cdot I_{I,PT}(7, \infty)/2$ accordingly. For the support region, the state II bending stiffness $E_a \cdot I_{2,sa}$ is applied.

Note 2: Some designers take advantage from the weight reduction of concrete due to hardening from 26 to 25 kN/m³ by considering an opposite load equal to $(25–26) \cdot h_c = -h_c$ (kN/m²) acting on the final system. This is not followed in this example.

- Shrinkage is calculated according to Figure 7.44.

$$\text{Equation 7.28a: } N_{sh} = -300 \cdot 10^{-6} \cdot \frac{6.18}{15.22} \cdot 3400 \cdot 33000 = 13667.46 \text{ kN}$$

From Table 9.17 for $t_0 = 1$ day:

$$\bar{z}_{i,S} = 108.02 - 15 = 93.02 \text{ cm}$$

$$\text{Equation 7.28b: } M_{sh} = 13667.46 \cdot 93.02/100 = 12713.47 \text{ kN-m}$$

$$\text{Equation 7.30: } \Delta T_{MS} = \frac{1271347}{21000 \cdot 0.8493 \cdot 10^8} \cdot \frac{286.5}{10^{-5}} = 20.42°C$$

For the calculation of the secondary effects, the aforementioned temperature is applied on the *uncracked* composite parts with $I_{I,PT} = I_{I,S}$.

- The application of thermal actions is demonstrated in Figure 9.22. Thermal actions are considered as short-term, and therefore, creep is neglected.
- For the traffic actions, the group load gr1a of Table 4.7 is considered; see also Figure 9.17. It is noted that a torsional constant $I_{T,1} = 1.628 \cdot 10^8$ cm⁴ is accounted for the sagging moment areas. For the cracked region, $I_{T,2} = 1.382 \cdot 10^8$ cm⁴. The torsional bending diagram taken from the central girder of the grillage model is shown in Figure 9.23.

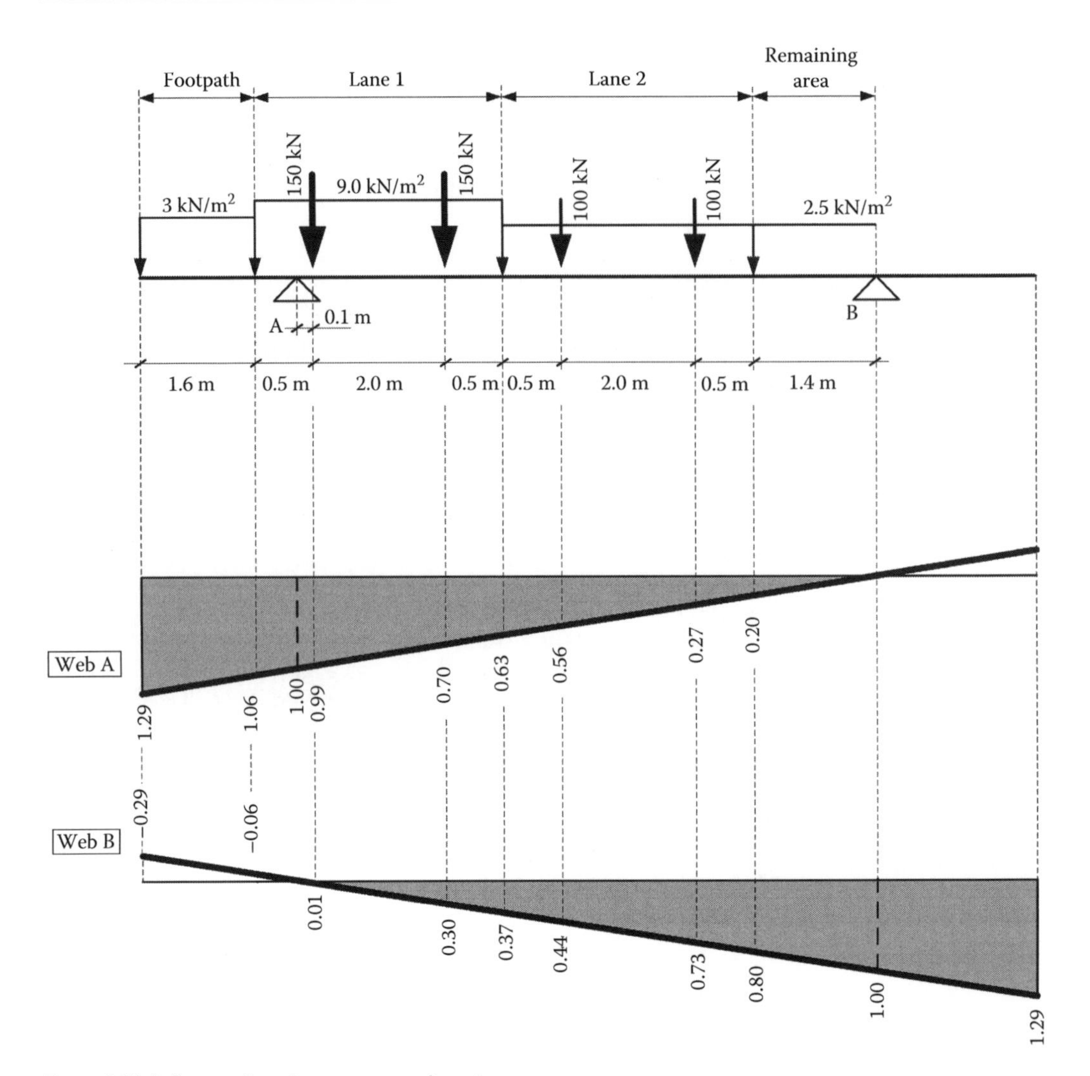

Figure 9.17 Influence lines in transverse direction.

DESIGN ACTIONS AND STRESS: CALCULATIONS

The basic combination at ULS with traffic loads as leading action is exemplarily investigated. Combinations with other leading variables should be verified as well.

Internal support

Short-term design

Obviously, the hogging bending moments are less critical than those for the long-term design due to the development of the secondary moments. Therefore, short-term design is not shown.

Long-term design

The bending moment that acts on the pure steel cross section is taken from Figures 9.18 and 9.19 and is equal to

$$\min M_{a,k,\infty} = -7104.5 - 13202.6 - 9121.4 - 803.6 = -30232.1 \text{ kN-m}$$

$$\min M_{a,Ed,\infty} = 1.35 \cdot (-30232.1) = -40813.3 \text{ kN-m}$$

The basic combination at ULS according to Equation 5.11a for the actions acting on the fully cracked cross section (structural steel + reinforcement) with the traffic loads gr1a as leading variable (Table 5.6, line 1) offers the following results:

$$\min M_{2,Ed,\infty} = 1.35 \cdot (-2479.0 - 250.79) + 1.0 \cdot (-6934.3) + 1.5 \cdot 0.6 \cdot (-8397.2) + 1.35 \cdot (-9227.8) = -30634.5 \text{ kN-m}$$

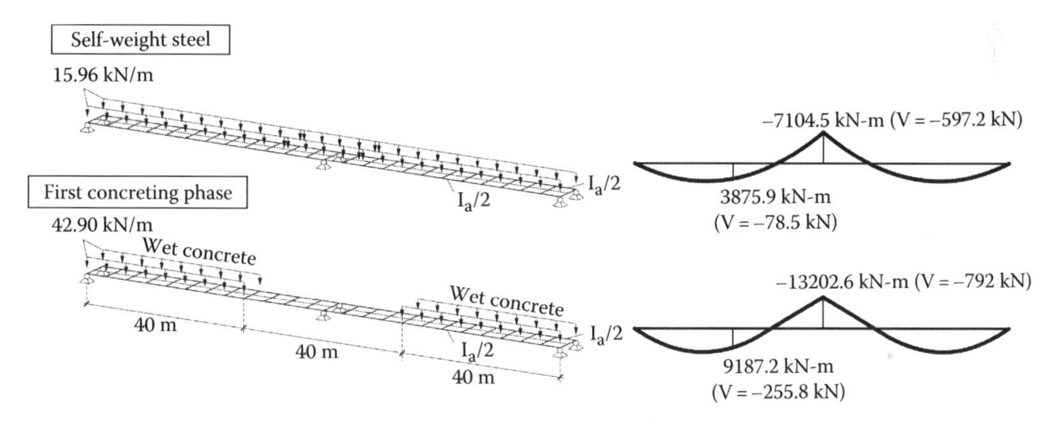

Figure 9.18 Loads acting on the pure steel system and results for the half girders.

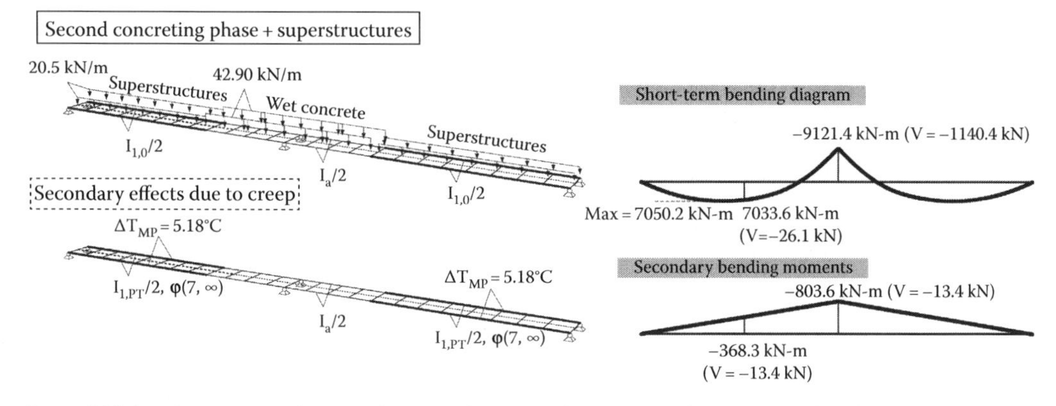

Figure 9.19 Loads acting on the mixed system (pure steel + composite) and results for the half girders.

The coexisting shear and torsion are calculated as follows:

$$\min V_{Ed,\infty} = 1.35 \cdot (-597.2 - 792 - 1140.4 - 13.4 - 383 - 4.2) + 1.0 \cdot (-115.6)$$
$$+1.5 \cdot 0.6 \cdot (-139.9) + 1.35 \cdot (-1168) = -5774.08 \text{ kN}$$

$$\min M_{T,Ed} = 1.35 \cdot (-2167.2) = 2925.72 \text{ kN-m}$$

Stresses on structural steel, Equation 9.38:

For the top flange, the tensile stress is

$$\sigma_{ao,Ed,\infty} = 4081330 \cdot \frac{142.34}{0.45454 \cdot 10^8/2} + 3063450 \cdot \frac{124.14}{0.567 \cdot 10^8/2} = 25.56 + 13.41$$

$$= 38.97 \text{ kN/cm}^2 \leq \frac{43}{1.0} = 43 \text{ kN/cm}^2 \text{ (sufficient), Table 9.11, row 2}$$

For the bottom flange, the absolute value of the compressive stress is

$$\sigma_{au,Ed,\infty} = 4081330 \cdot \frac{114.16}{0.45454 \cdot 10^8/2} + 3063450 \cdot \frac{132.36}{0.567 \cdot 10^8/2} = 20.50 + 14.3$$

$$= 34.8 \text{ kN/cm}^2 \leq \frac{0.90 \cdot 46}{1.1} = 37.64 \text{ kN/cm}^2 \text{ (sufficient), Table 9.15, row 6}$$

Actually, the design stress should be determined at the midplane of the bottom flange due to the fact that this flange is designed for plate buckling; see note of Table 9.15. In this example, this stress was determined at the extreme fiber considering that the difference between the two stresses is negligible due to the small thickness of the bottom flange (15 mm).

The shear stress in the bottom flange due to torsion is calculated from $\tau_{T,au} = \frac{M_{T,Ed}}{2 \cdot A_o \cdot t_{au}}$ Equation 9.15, where $A_o \approx \frac{7+4}{2} \cdot 2.5 = 13.75 \text{m}^2$.

Therefore,

$$\tau_{T,au} = \frac{292572}{2 \cdot 13.75 \cdot 10^4 \cdot 1.5} = 0.7 \text{ kN/cm}^2 \approx 0$$

No verification for interaction between direct and shear stresses is necessary.

For the web, shear stresses are considered constant along its length. The maximum shear stress due to vertical shear is equal to

$$\tau_{V,w,\infty} = \frac{V_{Ed,\infty}}{A_w \cdot \sin\alpha} = \frac{5774.08}{291.5 \cdot 2.5 \cdot \sin(59.08^\circ)} = 9.23 \text{ kN/cm}^2$$

The shear stress due to torsion is calculated as follows:

$$\tau_{T,w} = \frac{292572}{2 \cdot 13.75 \cdot 10^4 \cdot 2.5} = 0.43 \text{ kN/cm}^2$$

The shear stress in the web due to torsion is added to the shear stress due to vertical shear. The maximum total shear is calculated equal to

$$\tau_{w,Ed,\infty} = \tau_{V,w,\infty} + \tau_{T,w} = 9.23 + 0.43 = 9.66\ \text{kN/cm}^2 \leq 0.8 \cdot \frac{44}{\sqrt{3} \cdot 1.1}$$

$$= 18.48\ \text{kN/cm}^2 \text{(sufficient), Table 9.15, row 7}$$

The maximum tensile stress in the web is equal to

$$\sigma_{wo,Ed,\infty} = 4081330 \cdot \frac{142.34 - 5}{0.45454 \cdot 10^8/2} + 3063450 \cdot \frac{124.14 - 5}{0.567 \cdot 10^8/2} = 24.66 + 12.87$$

$$= 37.53\ \text{kN/cm}^2 < \frac{44}{1.0} = 44\ \text{kN/cm}^2\ \text{(sufficient)}$$

von Mises stress verification:

$$\left(\frac{37.53}{44/1.0}\right)^2 + 3 \cdot \left(\frac{9.66}{0.8 \cdot 44/1.1}\right)^2 = 0.73 + 0.27 = 1.00\ \text{(sufficient), Table 9.14, row 5}$$

For the compressed part of the web, the maximum value for the direct stresses is

$$\sigma_{wu,Ed,\infty} = 4081330 \cdot \frac{114.16 - 1.5}{0.45454 \cdot 10^8/2} + 3063450 \cdot \frac{132.36 - 1.5}{0.567 \cdot 10^8/2} = 20.23 + 14.14$$

$$= 34.37\ \text{kN/cm}^2 < \frac{0.90 \cdot 44}{1.1} = 36\ \text{kN/cm}^2\ \text{(sufficient); see Table 9.15, row 6}$$

The interaction between shear and direct stresses is verified through Table 9.15, row 8:

$$\left(\frac{34.37}{0.9 \cdot 44/1.1}\right)^2 + 3 \cdot \left(\frac{9.66}{0.8 \cdot 44/1.1}\right)^2 = 0.91 + 0.27 = 1.18 > 1\ \text{(not sufficient)}$$

Note: If the number of longitudinal stiffeners in the web is increased, then the buckling factor ρ may become equal to one. In this case, the previous check becomes sufficient. The local increase of longitudinal stiffeners at internal support will not have any significant effect on the cost of the structure.

Stresses in reinforcing steel:
For the top reinforcement layer, Equation 9.37 (Table 9.13):

$$\sigma_{so,Ed,\infty} = 3063450 \cdot \frac{148.14}{0.567 \cdot 10^8/2} = 16.00\ \text{kN/cm}^2 < \frac{50}{1.15}$$

$$= 43.48\ \text{kN/cm}^2\ \text{(sufficient), Table 9.15, row 2}$$

There are no additional stresses due to local bending of the deck slab since the TS is located away from the internal support.

Span (x = 27.5 m)

The bending that acts on the steel cross section is taken from Figure 9.18 and is equal to

$$M_{a,k} = 3875.9 + 9187.2 = 13063.1 \text{ kN-m}$$

$$M_{a,Ed} = 1.35 \cdot 13063.1 = 17635.19 \text{ kN-m}$$

Short-term design

The basic combination at ULS according to (5.11a) for the actions acting on uncracked cross section with leading variable the traffic loads gr1a (Table 5.6, line 1) offers the following results:

$$M_{I,Ed,0} = 1.35 \cdot (7033.6 + 743) + 1.5 \cdot 0.6 \cdot 3920 + 1.35 \cdot 10293.7 = 27922.91 \text{ kN-m}$$

The coexisting shear force is calculated as follows:

$$V_{I,Ed,0} = 1.35 \cdot (-78.5 - 255.8 - 26.1 + 27) + 1.5 \cdot 0.6 \cdot 142.6 + 1.35 \cdot (-203.4)$$
$$= -596.34 \text{ kN}$$

The coexisting torsional moment is equal to

$$M_{T,I,Ed} = 1.35 \cdot 354.8 = 478.98 \text{ kN-m}$$

For the concrete slab, the maximum value of compressive stress is calculated from Equation 9.31 and is equal to

$$\sigma_{co,Ed,glob,0} = 2792291 \cdot \frac{77.64}{6.18 \cdot 1.037 \cdot 10^8/2} = 0.68 \text{ kN/cm}^2$$

The TS is located where the maximum sagging moment arises. The additional concrete stress due to local bending of the deck slab is 0.24 kN/cm^2 (FE calculation):

$$\sigma_{co,Ed,0} = \sigma_{co,Ed,glob,0} + \sigma_{co,Ed,loc,0} = 0.68 + 1.35 \cdot 0.24 = 0.98 \text{ kN/cm}^2 < f_{cd} = \frac{3.5}{1.5}$$
$$= 2.33 \text{ kN/cm}^2 \text{ (sufficient), Table 9.15, row 1}$$

For the top flange, the maximum value of the compression stress is equal to (Equation 9.32)

$$\sigma_{ao,Ed,0} = 1763519 \cdot \frac{142.34}{0.45454 \cdot 10^8/2} + 2792291 \cdot \frac{47.64}{1.037 \cdot 10^8/2} = 11.04 + 2.57$$
$$= 13.61 \text{ kN/cm}^2 < \frac{39}{1.0} = 39 \text{ kN/cm}^2 \text{ (sufficient), Table 9.15, row 3}$$

For the bottom flange, the tensile stress due to permanent actions is equal to

$$\sigma_{au,Ed,0} = 1763519 \cdot \frac{114.16}{0.45454 \cdot 10^8/2} + 2792291 \cdot \frac{208.86}{1.037 \cdot 10^8/2} = 8.86 + 11.24$$

$$= 20.1 \text{ kN/cm}^2 < \frac{42}{1.0} = 42 \text{ kN/cm}^2 \text{ (sufficient)}$$

Torsion at span causes at bottom flange negligible shear stresses.

The maximum shear stress due to shear in the web is equal to

$$\tau_{V,w} = \frac{V_{l,Ed,0}}{A_w \cdot \sin\alpha} = \frac{596.34}{291.5 \cdot 2.5 \cdot \sin(59.08°)} = 0.95 \text{ kN/cm}^2$$

The torsional bending moment at x=27.5 m is small, and the corresponding shear stresses are negligible ($\tau_{T,w} \approx 0$). Therefore,

$$\tau_{w,Ed,0} = 0.95 \text{ kN/cm}^2 \leq 0.8 \cdot \frac{34.5}{\sqrt{3} \cdot 1.1} = 14.48 \text{ kN/cm}^2$$

The maximum value of tensile stress in the web is equal to

$$\sigma_{wu,Ed,0} = 1763519 \cdot \frac{114.16 - 1.5}{0.45454 \cdot 10^8/2} + 2792291 \cdot \frac{208.86 - 1.5}{1.037 \cdot 10^8/2} = 8.74 + 11.2$$

$$= 19.94 \text{ kN/cm}^2 \leq \frac{34.5}{1.0} = 34.5 \text{ kN/cm}^2 \text{ (sufficient)}$$

von Mises stress verification (Table 9.14, row 5):

$$\left(\frac{19.94}{34.5/1.0}\right)^2 + 3 \cdot \left(\frac{0.95}{0.8 \cdot 34.5/1.1}\right)^2 = 0.33 < 1 \text{ (sufficient)}$$

The maximum value of the compressive stress in the web is equal to

$$\sigma_{wo,Ed,0} = 1763519 \cdot \frac{142.34 - 5}{0.45454 \cdot 10^8/2} + 2792291 \cdot \frac{47.64 - 5}{1.037 \cdot 10^8/2} = 10.66 + 2.30$$

$$= 12.96 \text{ kN/cm}^2 < 0.9 \cdot \frac{34.5}{1.1} = 28.23 \text{ kN/cm}^2 \text{ (sufficient)}$$

von Mises stress verification (Table 9.15, row 8):

$$\left(\frac{12.96}{0.9 \cdot 34.5/1.1}\right)^2 + 3 \cdot \left(\frac{0.95}{0.8 \cdot 34.5/1.1}\right)^2 = 0.21 < 1 \text{ (sufficient)}$$

Long-term design

The basic combination at ULS according to (5.11a) for the actions acting on the uncracked cross section with leading variable the traffic loads gr1a offers the following results:

Permanent bending moment due to permanent loads acting on the composite cross section of type P with age of concrete 7 days; see short-term bending diagram in Figure 9.19:

$$M_{I,Ed,7} = 1.35 \cdot 7033.6 = 9495.36 \text{ kN-m}$$

Temporarily permanent bending moment acting on the composite cross section of type PT with age of concrete 7 days; see diagram of secondary bending moments in Figure 9.19:

$$M_{I,Ed,PT,7} = 1.35 \cdot (-368.3) = -497.21 \text{ kN-m}$$

Permanent bending moment acting on the composite cross section of type P with age of concrete 14 days; see short-term bending diagram in Figure 9.20:

$$M_{I,Ed,14} = 1.35 \cdot 743 = 1003 \text{ kN-m}$$

Temporarily permanent bending moment acting on the composite cross section of type PT with age of concrete 14 days; see diagram of secondary bending moments in Figure 9.20:

$$M_{I,Ed,PT,14} = 1.35 \cdot (-114.9) = -155.11 \text{ kN-m}$$

Secondary bending moment due to shrinkage; see Figure 9.21:

$$M_{I,Ed,PT,I} = 1.0 \cdot (-3178.2) = -3178.2 \text{ kN-m}$$

Short-term bending moments due to traffic and thermal actions taken from Figures 9.22 and 9.23:

$$M_{I,Ed,0} = 1.35 \cdot 10293.7 + 0.6 \cdot 1.5 \cdot 3920 = 17424.5 \text{ kN-m}$$

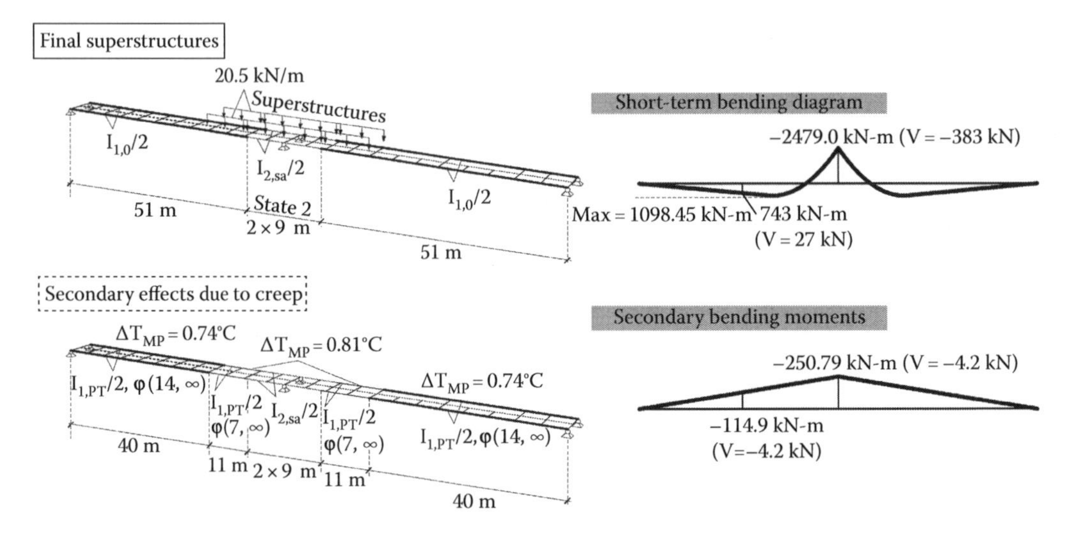

Figure 9.20 Loads acting on composite system and results for the half girders.

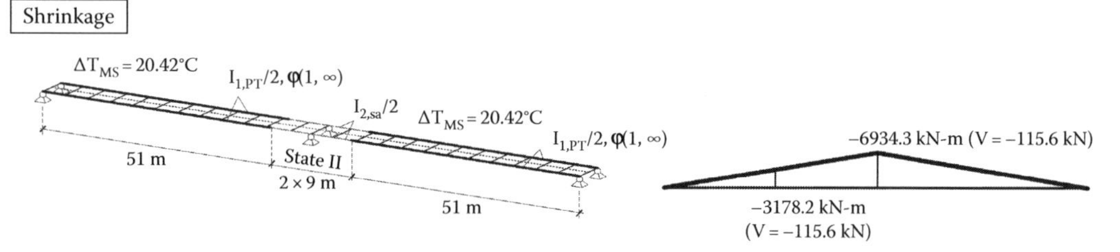

Figure 9.21 Shrinkage and results for the half girders.

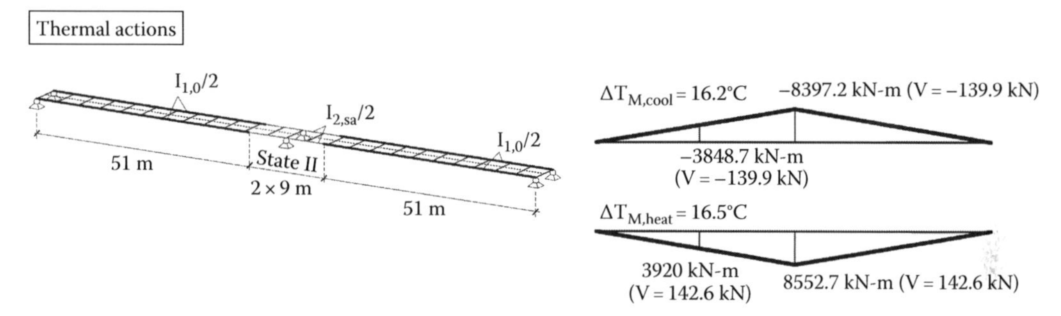

Figure 9.22 Thermal actions and results for the half girders.

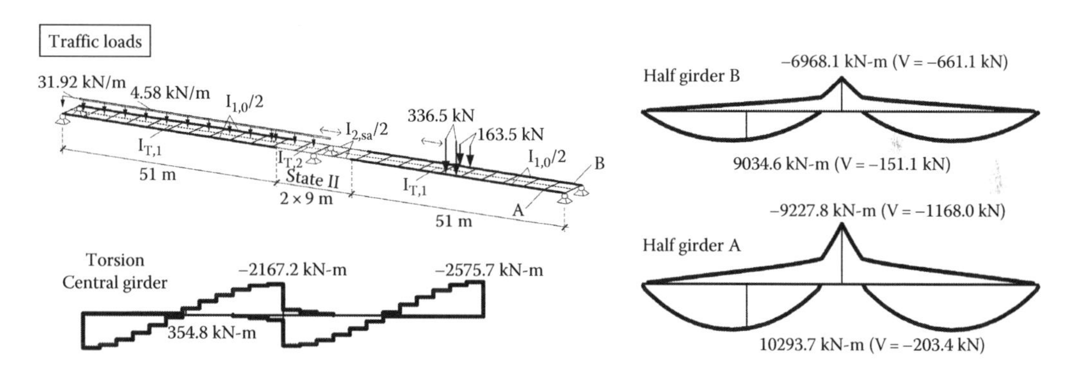

Figure 9.23 Traffic actions and results for the half girders.

The coexisting shear force is calculated as follows:

$$V_{I,Ed,\infty} = 1.35 \cdot (-78.5 - 255.8 - 26.1 - 13.4 - 4.2 + 27) + 1.0 \cdot (-115.6) + 1.5 \cdot 0.6 \cdot 142.6$$
$$+ 1.35 \cdot (-203.4) = -620.1 \text{ kN}$$

The coexisting torsional moment is equal to

$$M_{T,I,Ed} = 1.35 \cdot 354.8 = 478.98 \text{ kN-m}$$

For the top flange, the maximum value of the compression stress is equal to (Equation 9.36)

$$\sigma_{ao,Ed,\infty} = 17635.19 \cdot \frac{142.34}{0.45454 \cdot 10^8/2} + 949536 \cdot \frac{84.32}{0.8107 \cdot 10^8/2}$$

$$+(-49721) \cdot \frac{71.76}{0.8878 \cdot 10^8/2} + 100300 \cdot \frac{81.81}{0.8261 \cdot 10^8/2}$$

$$+(-15511) \cdot \frac{69.64}{0.9008 \cdot 10^8/2} + (-317820) \cdot \frac{78.02}{0.8493 \cdot 10^8/2}$$

$$+1742450 \cdot \frac{47.64}{1.037 \cdot 10^8/2} + \left(\frac{13667.46}{6540.97} + \frac{1271347 \cdot 78.02}{0.8493 \cdot 10^8} \right)$$

$$= 17.39 \text{ kN/cm}^2 \leq \frac{39}{1.0} = 39 \text{ kN/cm}^2 \text{ (sufficient)}$$

One can see that the top flange stress due to long-term effects has been increased from 13.61 kN/cm^2 to 17.39 kN/cm^2 (28%). The compressive stress due to preloading of the pure steel cross section is 11.04 kN/cm^2 (63.5% of the total stress). The top flange stress due to the primary effects of shrinkage (last term in bracket) is 3.26 kN/cm^2 (18.8% of the total stress).

For the bottom flange, the maximum tensile stress is equal to (Equation 9.36)

$$\sigma_{au,Ed,\infty} = 17635.19 \cdot \frac{114.16}{0.45454 \cdot 10^8/2} + 949536 \cdot \frac{172.18}{0.8107 \cdot 10^8/2}$$

$$+(-49721) \cdot \frac{184.84}{0.8878 \cdot 10^8/2} + 100300 \cdot \frac{174.69}{0.8261 \cdot 10^8/2}$$

$$+(-15511) \cdot \frac{186.86}{0.9008 \cdot 10^8/2} + (-317820) \cdot \frac{178.48}{0.8493 \cdot 10^8/2}$$

$$+1742450 \cdot \frac{208.86}{1.037 \cdot 10^8/2} + \left(-\frac{13667.46}{6540.97} + \frac{1271347 \cdot 178.48}{0.8493 \cdot 10^8} \right)$$

$$= 19.31 \text{ kN/cm}^2 \leq \frac{42}{1.0} = 42 \text{ kN/cm}^2 \text{ (sufficient)}$$

One can see that the bottom flange stress due to long-term effects has been reduced from 20.1 kN/cm^2 to 19.31 kN/cm^2 (−3.9%). The stress due to preloading of the pure steel cross section is 8.86 kN/cm^2 (45.8% of the total stress). The bottom flange stress due to the primary effects of shrinkage (last term in bracket) is 0.58 kN/cm^2 (only 3% of the total stress). Obviously, shrinkage is not that critical for the bottom flange as for the top one.

Torsion at span causes at the bottom flange negligible shear stresses.

The coexisting shear stress due to shear in the web is equal to

$$\tau_{V,w} = \frac{V_{l,Ed,\infty}}{A_w \cdot \sin\alpha} = \frac{620.1}{291.5 \cdot 2.5 \cdot \sin(59.08)} = 0.99 \text{ kN/cm}^2$$

The torsional bending moment at x = 27.5 m is small, and the corresponding shear stresses are negligible ($\tau_{T,w} \approx 0$). Therefore,

$$\tau_{w,Ed,\infty} = 0.99 \text{ kN/cm}^2 \leq 0.8 \cdot \frac{34.5}{\sqrt{3} \cdot 1.1} = 14.48 \text{ kN/cm}^2 \text{ (sufficient)}$$

The maximum value of tensile stress in the web is equal to

$$\begin{aligned}
\sigma_{wu,Ed,\infty} = {} & 17635.19 \cdot \frac{114.16 - 1.5}{0.45454 \cdot 10^8/2} + 949536 \cdot \frac{172.18 - 1.5}{0.8107 \cdot 10^8/2} \\
& + (-49721) \cdot \frac{184.84 - 1.5}{0.8878 \cdot 10^8/2} + 100300 \cdot \frac{174.69 - 1.5}{0.8261 \cdot 10^8/2} \\
& + (-15511) \cdot \frac{186.86 - 1.5}{0.9008 \cdot 10^8/2} + (-317820) \cdot \frac{178.48 - 1.5}{0.8493 \cdot 10^8/2} \\
& + 1742450 \cdot \frac{208.86 - 1.5}{1.037 \cdot 10^8/2} + \left(-\frac{13667.46}{6540.97} + \frac{1271347 \cdot (178.48 - 1.5)}{0.8493 \cdot 10^8} \right) \\
= {} & 19.1 \text{ kN/cm}^2 < \frac{34.5}{1.0} = 34.5 \text{ kN/cm}^2 \text{ (sufficient)}
\end{aligned}$$

von Mises stress verification:

$$\left(\frac{19.1}{34.5/1.0} \right)^2 + 3 \cdot \left(\frac{0.99}{0.8 \cdot 34.5/1.1} \right)^2 = 0.31 < 1 \text{ (sufficient)}$$

The maximum compressive stress in the web is

$$\begin{aligned}
\sigma_{wo,Ed,\infty} = {} & 17635.19 \cdot \frac{142.34 - 5}{0.45454 \cdot 10^8/2} + 949536 \cdot \frac{84.32 - 5}{0.8107 \cdot 10^8/2} \\
& + (-49721) \cdot \frac{71.76 - 5}{0.8878 \cdot 10^8/2} + 100300 \cdot \frac{81.81 - 5}{0.8261 \cdot 10^8/2} \\
& + (-15511) \cdot \frac{69.64 - 5}{0.9008 \cdot 10^8/2} + (-317820) \cdot \frac{78.02 - 5}{0.8493 \cdot 10^8/2} \\
& + 1742450 \cdot \frac{47.64 - 5}{1.037 \cdot 10^8/2} + \left(\frac{13667.46}{6540.97} + \frac{1271347 \cdot (78.02 - 5)}{0.8493 \cdot 10^8} \right) \\
= {} & 16.67 \text{ kN/cm}^2 < 0.9 \cdot \frac{34.5}{1.0} = 28.23 \text{ kN/cm}^2 \text{ (sufficient)}
\end{aligned}$$

von Mises stress verification:

$$\left(\frac{16.67}{0.9 \cdot 34.5/1.1} \right)^2 + 3 \cdot \left(\frac{0.99}{0.8 \cdot 34.5/1.1} \right)^2 = 0.35 < 1 \text{ (sufficient)}$$

Obviously, the total depth of the cross section at spans can be significantly reduced. In real cases, the total depth varies parabolically along the bridge as in Figure 2.21. Therefore, stresses should be verified for many different positions and not only at midspan and the supports.

Note: The use of different modular ratios n_L depending on the age of concrete is from the theoretical point of view the correct approach. However, it makes calculations laborious especially in bridges with many concrete segments. In [9.11], it is shown that stresses σ_{co}, σ_{cu}, and σ_{au} are considerably robust against the deviations of modular ratios; for deviations $|\Delta n_L|/n_L \leq 20\%$, the maximum change for the concrete stresses is less than 10% and for the tension stress in the bottom flange less than 3%. The stresses in the upper part of the steel cross section σ_{ao} and σ_{wo} are very sensible against the n_L deviations and in some cases may govern the design. The position of the elastic neutral axis is also quite sensible against the deviations of n_L, and this may be critical for the classification of the cross section. It is suggested that for concrete segments with modular ratios with a relative difference less than 20%, the modular ratio with the highest value can be chosen as the representative one. For greater differences in modular ratios, generalizations should be avoided. In Table 9.16, the deviations for both n_P and n_{PT} for 7 and 14 days are smaller than 20%, so calculations could be done for all of the segments with $n_P = 18.82$ and $n_{PT} = 12.5$; this would reduce the computation load significantly.

9.11 CLASS 4 CROSS SECTIONS

This section refers to class 4 cross sections where plate buckling verifications are made by the *effective width method*. The procedure for this section class was presented in detail in Section 8.6; see Figure 8.28. The necessary steps are shown once more as follows:

Step 1: Determination of direct stresses on the basis of the gross section
Step 2: Determination of the effective area of the compression flange
Step 3: Determination of the direct stresses based on the gross section of the web(s) and effective section of the compression flange
Step 4: Determination of the effective area of the web(s) and the final effective cross section
Step 5: Section verification for bending moments and axial forces
Step 6: Resistance to shear
Step 7: Verification for the interaction bending moments, axial forces, and shear forces

Table 9.19 illustrates the procedure for the same cross section as in Figure 8.28. The design relations of Table 9.19 indicate that the procedure leads to cross-sectional verifications. However, plate buckling verifications to determined effective areas are performed for panels that have a specific length. Accordingly, member verifications are included in design. In the presence of torsion, the webs resist different shear forces. Verifications for shear and its interactions should therefore be based on the larger shear force. Example 8.10 demonstrates the application of the effective width method on the cross section of a box-girder bridge.

9.12 CLASS 4 CROSS SECTIONS COMPOSED OF THE FLANGES

In the previous sections, it was seen that cross sections with class 1 or 2 flanges and class 3 webs may be treated as class 2 sections by introduction of an effective area for the web. A practical, although conservative, alternative is to assign the entire bending resistance to the flanges and all shear resistance to the web. The bending resistance of the cross section composed of the two flanges is only determined as $M_{f,Rd}$. Since the web is resisting shear force only and the flanges bending moment only, there is no need to examine interaction.

Table 9.19 Design procedure for class 4 sections

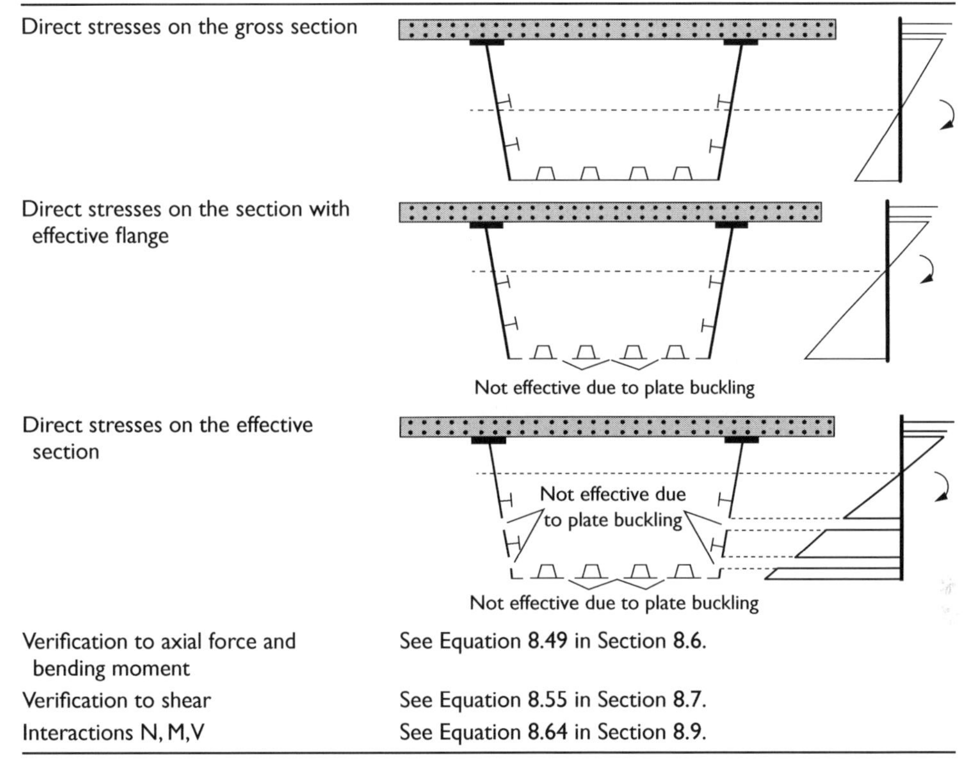

Direct stresses on the gross section	
Direct stresses on the section with effective flange	
Direct stresses on the effective section	
Verification to axial force and bending moment	See Equation 8.49 in Section 8.6.
Verification to shear	See Equation 8.55 in Section 8.7.
Interactions N, M, V	See Equation 8.64 in Section 8.9.

Notes:

A_{eff} is calculated in presence of compression force.

A_{eff} and W_{eff} are calculated separately.

Concentrated forces and their interaction are verified separately; see Sections 8.8 and 8.9.2.

This methodology is especially advantageous for cross sections with class 4 webs; plate buckling verifications for direct stresses for the web are avoided. Webs are then verified only for shear buckling.

For *sagging* bending, both flanges are class 1 and $M_{1,f,Rd}$ is determined with a plastic stress distribution.

The resistance to *hogging bending* $M_{2,f,Rd}$ is determined as follows:

- For class 1 or 2 compression flanges with a plastic stress distribution, so that $M_{2,f,Rd} = M_{2,pl,f,Rd}$.
- For class 3 compression flanges with an elastic stress distribution, with limiting stresses f_{yd} for structural steel in compression or tension and f_{sd} for reinforcement. It is then $M_{2,f,Rd} = M_{2,el,f,Rd}$. In such case, the loading history should be taken into account.
- For class 4 compression flanges, as usually in box girders, with an elastic stress distribution on the section with effective compression flange area, with limiting stresses as for class 3 sections and effective width method in Section 8.5.
- For class 4 compression flanges with an elastic stress distribution on the section with full compression flange area, with limiting stresses f_{yd} for structural steel in tension, the reduced stress $\sigma_{Rd} = \rho \cdot f_{yd}$ in accordance with Section 8.4 (reduced stress method) for structural steel in compression and f_{sd} for reinforcement. Again the loading history should not be omitted.

9.13 LATERAL TORSIONAL BUCKLING

9.13.1 Introduction

Compression flanges of in-plane loaded girders are susceptible to instability in the form of lateral out-of-plane deformations. Lateral out-of-plane deformations due to instability may also occur in truss, arched, arch-and-tie bridges, etc., where the entire compression chord deflects laterally. The former type of instability, where the compression part of the cross section is susceptible to lateral deformations, is named *lateral torsional buckling*; the latter, where the entire section deforms laterally, is called *lateral or flexural buckling.* Resistance to lateral or lateral torsional buckling may be enhanced by addition of bracing that provides lateral support. Depending on the conditions, the lateral support may be rigid or flexible. Rigid support is generally provided by the connection of bracing elements directly to the compression flange, whereas elastic support provides cross frames or bracing elements in the tension flange or the web. Flanges connected to the concrete slab are not susceptible to lateral torsional buckling due to the lateral restraint provided by the slab. For lateral buckling, bracing elements are usually connected directly to the member, for example, the arch, that may deform laterally.

Geometric and loading conditions change during construction stages in composite bridges. Top flanges are not supported before concrete casting by the concrete deck and may deform laterally, whereas at service conditions, only bottom flanges at internal supports are susceptible to lateral deformations. Lateral torsional buckling may be verified by different methods that are presented in the following.

9.13.2 General method

The general method is applicable to both lateral and lateral torsional buckling. It requires the performance of a linear buckling analysis for the entire 3D system in order to obtain the critical load factor α_{crit}. This factor is the load amplifier of the in-plane design loads at which the fundamental buckling mode for lateral or lateral torsional buckling occurs; see Figure 7.30. Since the overall system is considered, the general method applies not only to the lateral stability problems of compression flanges for cross sections but also to the lateral stability of entire systems as arches, trusses, etc.

The nondimensional out-of-plane slenderness of the system is determined from

$$\bar{\lambda}_{op} = \sqrt{\frac{\alpha_{ult,k}}{\alpha_{crit}}} \tag{9.40}$$

where

$\alpha_{ult,k}$ is the load amplifier of the design loads to reach the *characteristic* resistance of the most critical section neglecting any out-of-plane effects. If necessary, second-order bending moments should be included

α_{crit} is the load amplifier of the *in-plane* design loads to reach the fundamental buckling mode for lateral or lateral torsional buckling. For the calculation of α_{crit}, the 3D model described in Section 7.1.4 should be used

The reduction factor χ_{op} for lateral or lateral torsional buckling may be determined as a function of $\bar{\lambda}_{op}$ by the following condition:

$$\chi_{op} = \min(\chi, \chi_{LT}) \tag{9.41}$$

where

χ is the reduction factor for lateral buckling to be determined by Equation 9.11, as described in Section 9.3

χ_{LT} is the reduction factor for lateral torsional buckling that is determined by a similar equation as follows:

$$\chi_{LT} = \frac{1}{\Phi_{LT} + \sqrt{\Phi_{LT}^2 - \bar{\lambda}_{LT}^2}} \quad \text{for } \bar{\lambda}_{LT} > 0.2 \text{ else } \chi_{LT} = 1 \tag{9.42}$$

$$\Phi_{LT} = 0.5 \cdot \left[1 + \alpha \cdot (\bar{\lambda}_{LT} - 0.2) + \bar{\lambda}_{LT}^2\right] \tag{9.43}$$

with $\bar{\lambda}_{LT} = \bar{\lambda}_{op}$

The imperfection factor α is determined from Table 9.7 *using the buckling curves of Table 9.20*. It is mentioned that EN 1993-1-1 also gives alternative expressions for Equations 9.42 and 9.43 that are accompanied by the application of different buckling curves.

The buckling verification may be written as

$$\frac{\chi_{op} \cdot \alpha_{ult,k}}{\gamma_{M1}} \geq 1.0 \tag{9.44}$$

REMARK 9.6

- When first-order global analysis is conducted, $\alpha_{ult,k}$ can be calculated from the capacity expressions of EN 1993-1-1 corresponding to the most critical cross section. For a cross section under a bending moment M_{Ed} and an axial force N_{Ed}, the simplest way for calculating the amplifier $a_{ult,k}$ is achieved with the following interaction:

$$\frac{N_{Ed}}{N_{Rk}} + \frac{M_{Ed}}{M_{Rk}} = \frac{1}{\alpha_{ult,k}} \rightarrow \alpha_{ult,k} = \left(\frac{N_{Ed}}{N_{Rk}} + \frac{M_{Ed}}{M_{Rk}}\right)^{-1} \tag{R9.1}$$

- Expression R9.1 should not be used when second-order global analysis is conducted. This is because internal forces do not have a linear relation with the imposed loads. In such a case, the imposed loads should be progressively increased until the cross section reaches its characteristic strength.

Table 9.20 Selection of buckling curves for lateral torsional buckling of I girders

Type of cross section	*Height to flange width ratios*	*Buckling curve*
Rolled I sections	$h/b \leq 2$	a
	$h/b > 2$	b
Welded I sections	$h/b \leq 2$	c
	$h/b > 2$	d
Other cross sections	All	d

EXAMPLE 9.5

The lateral stability of the girders of Figure 7.30 at construction stages during concrete casting shall be determined. Steel grade S355.

The design load due to self-weight of the girder and the concrete weight is a uniformly distributed load of 24.19 kN/m.

The design bending moment is $\max M_{Ed} = 24.19 \cdot \frac{25^2}{8} = 1889.84$ kN-m.

It may be easily proven that the cross section is class 1 with characteristic moment of resistance $M_{Rk}=5881.12$ kN-m.

Equation R9.1: $\alpha_{ult,k} = \frac{5881.12}{1889.84} = 3.11$

For lateral torsional buckling from Table 9.20, rolled section with $h/b>2 \rightarrow$ buckling curve b $\rightarrow$ Table 9.7, $\alpha=0.34$

For lateral buckling from Table 9.8 for $h/b>1.2 \rightarrow$ buckling curve b

Accordingly, χ_{LT} is equal to $\chi_{op}=\chi_{LT}$.

From the analysis of Figure 7.30, the nondimensional out-of-plane slenderness, the reduction factor, and the resistance against lateral torsional buckling are calculated as follows:

- First analysis ($\alpha_{crit}=0.58$):

$$\bar{\lambda}_{LT} = \sqrt{\frac{3.11}{0.58}} = 2.32 \rightarrow \Phi_{LT} = 0.5 \cdot \left[1 + 0.34 \cdot (2.32 - 0.2) + 2.32^2\right] = 3.55$$

$$\rightarrow \chi_{LT} = \frac{1}{3.55 + \sqrt{3.55^2 - 2.32^2}} = 0.16 \rightarrow \frac{0.16 \cdot 3.11}{1.1} = 0.45 < 1.0 \text{ (not sufficient)}$$

However, it was obvious that the verification is not sufficient because $\alpha_{crit} < 1$.

- Second analysis ($\alpha_{crit}=2.95$):

$$\bar{\lambda}_{LT} = 1.03 \rightarrow \Phi_{LT} = 1.17 \rightarrow \chi_{LT} = 0.58 \rightarrow \frac{0.58 \cdot 3.11}{1.1} = 1.64 > 1.0 \text{ (sufficient)}$$

- Third analysis ($\alpha_{crit}=8.41$):

$$\bar{\lambda}_{LT} = 0.61 \rightarrow \Phi_{LT} = 0.76 \rightarrow \chi_{LT} = 0.83 \rightarrow \frac{0.83 \cdot 3.11}{1.1} = 2.35 > 1.0 \text{ (sufficient)}$$

- Final analysis ($\alpha_{crit}= 10.48$):
  The buckling mode refers to the lateral buckling of the angle cleat L150 · 10 with $\alpha_{crit}> 10$. For verification, see Section 9.13.5.

9.13.3 Simplified method: Rigid lateral supports

9.13.3.1 Verification during concreting stages

The simplified method may be used to verify the resistance to lateral torsional buckling of a compression flange only and *not for lateral buckling of full systems*. This method isolates from the cross section the compression flange including 1/3 of the compressed part of the web (Figure 9.24) and treats it as a compression member subjected to out-of-plane flexural buckling. The supports are considered as rigid when the compression flange is directly supported by stiff bracing elements.

The nondimensional slenderness is found by the following expressions:

$$\bar{\lambda}_{LT} = \sqrt{\frac{f_y \cdot A_{eff}}{N_{cr}}} \tag{9.45}$$

$$A_{eff} = A_f + \frac{A_{wc}}{3} \tag{9.46}$$

$$N_{cr} = \frac{\pi^2 \cdot E \cdot I_{eff,z}}{L^2} \tag{9.47}$$

where

A_f is the area of the flange or its effective area for class 4 flanges
A_{wc} is the compression zone of the web but not more than the effective width of the web adjacent to the compression flange
N_{cr} is the critical buckling, Euler, load of the column for out-of-plane buckling
$I_{eff,z}$ is the transverse second moment of area of the flange and the effective web
L is the length between the "rigid" supports

The reduction factor χ_{LT} is determined by Equation 9.42 as a function of $\bar{\lambda}_{LT}$. The lateral torsional buckling verification is then written as

$$\max N_{f,Ed} \leq \chi_{LT} \cdot \frac{A_{eff} \cdot f_y}{\gamma_{M1}} \tag{9.48}$$

where $\max N_{f,Ed}$ is the *maximum* design force developed inside L. It can be taken from a first-order analysis of the 3D model; see Section 7.1.4.

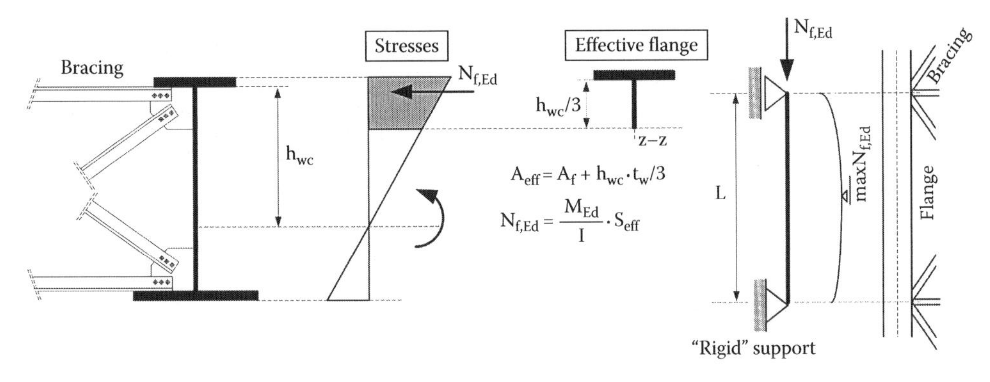

Figure 9.24 Modeling of the compression flange as a T-section column on rigid supports.

Expression 9.47 is used when braces are considered to act as rigid supports for the compression flange. This means that the effective length is restricted to the distance between the braces. However, Figure 7.30 shows that buckling modes at erection stages involve rotations and transverse displacements of the compressed flanges for much longer spans. Therefore, taking L equal to the distance of the braces may be in some cases unsafe; see Example 9.6. A better alternative is to calculate the Euler load N_{cr} from the following expression:

$$N_{cr} = a_{crit} \cdot N_{f,Ed} \tag{9.49}$$

where a_{crit} is the load factor calculated from a buckling analysis of the 3D model.

EXAMPLE 9.6

The lateral stability of the girders in Figure 7.30 will be verified for the third analysis with the simplified method. Steel grade S355.

The maximum compression force is taken from the 3D model at midspan:

$$\max N_{f,Ed} = 2113.23 \text{ kN}$$

Note: The total depth of the equivalent T section is equal to

$$h = 4 + \left(\frac{92.8}{2}\right) \cdot \frac{1}{3} = 4 + 15.47 = 19.47 \text{ cm}, \quad \frac{h_{wc}}{3} = 15.47 \text{ cm}$$

The cross-sectional area of the T section is

$$A_{eff} = 30.2 \cdot 4 + 15.47 \cdot 2.1 = 153.29 \text{ cm}^2$$

The bending moment at midspan is

$$\max M_{Ed} = 24.19 \cdot 25^2/8 = 1889.84 \text{ kN-m}$$

The maximum compression force is calculated as follows:

$$\max N_{f,Ed} = \frac{188984}{722300} \cdot \left(\frac{100.8}{2} - 4.06\right) \cdot 153.29 = 1858.6 \text{ kN}$$

The 3D model offers a 13.7% more conservative value than the exact one.

The Euler load is calculated from Equation 9.49:

$$N_{cr} = 8.41 \cdot 2113.23 = 17772.3 \text{ kN}$$

$$I_{eff,z} = \frac{30.2^3 \cdot 4}{12} = 9181.2 \text{ cm}^4$$

Equation 9.45: $\bar{\lambda}_{LT} = \sqrt{\dfrac{35.5 \cdot 153.29}{17772.3}} = 0.55$

From Table 9.20 for h/b > 2 → buckling curve b → Table 9.7, $\alpha = 0.34$

Equation 9.43: $\Phi_{LT} = 0.5 \cdot [1 + 0.34 \cdot (0.55 - 0.2) + 0.55^2] = 0.71$

Equation 9.42: $\chi_{LT} = \dfrac{1}{0.71 + \sqrt{0.71^2 - 0.55^2}} = 0.86$

Equation 9.48: $\max N_{f,Ed} = 2113.23 \text{ kN} \leq 0.86 \cdot \dfrac{153.29 \cdot 35.5}{1.1} = 4254.49 \text{ kN}$

If plane bracings are considered as rigid supports, then L = 2.5 m, and this leads to less conservative results. Indeed,

$$N_{cr} = \frac{\pi^2 \cdot 21000 \cdot 9181.2}{250^2} = 30415.71 \text{ kN} \rightarrow \bar{\lambda}_{LT} = 0.42 \rightarrow \chi_{LT} = 0.92$$

$$\rightarrow \max N_{f,Ed} = 2113.23 \text{ kN} \leq 0.92 \cdot \frac{153.29 \cdot 35.5}{1.1} = 4551.32 \text{ kN}$$

REMARK 9.7

- Neglecting the web during the calculation of A_{eff} in Equation 9.46 will reduce the slenderness value from Equation 9.45; therefore, it may be an *unsafe* approach.
- As explained in Section 7.1.4, the effective flanges of the 3D model are composed of the corresponding flanges and one-third of the web. This is not totally compatible with the simplified method in which one-third of the compressed part of the web is required. However, this difference causes negligible deviations, and the approximation for the web in the 3D model is considered as adequately precise; see also comparative analysis for a_{crit} in Tables 7.4 and 7.5.

9.13.3.2 Verification at hogging moment areas of continuous plate-girder bridges

At the final stage of continuous I plate-girder bridges, lateral torsional buckling may occur in the hogging moment areas. The bottom flange needs to be supported by braces acting as "rigid supports." In such a case, the Euler load for the effective flange should not be calculated from Equation 9.47 but from the following procedures:

$$N_{cr} = m \cdot \frac{\pi^2 \cdot E_a \cdot I_{eff,z}}{L^2} \tag{9.50a}$$

$$m = \min(m_1, m_2) \tag{9.50b}$$

$$m_1 = 1 + 0.44 \cdot (1 + \mu) \cdot \Phi^{1.5} + \frac{(3 + 2 \cdot \Phi) \cdot \gamma}{350 - 50 \cdot \mu} \tag{9.50c}$$

$$m_2 = 1 + 0.44 \cdot (1 + \mu) \cdot \Phi^{1.5} + \left[0.195 + \Phi \cdot (0.05 + 0.01 \cdot \mu)\right] \cdot \sqrt{\gamma} \tag{9.50d}$$

$$\mu = \frac{V_2}{V_1} \quad \text{for } V_1 > V_2 \tag{9.50e}$$

$$\Phi = \frac{2 \cdot \left(1 - \frac{M_2}{M_1}\right)}{1 + \mu} \quad \text{for } M_1 > M_2 > 0 \tag{9.50f}$$

M_1 is the maximum absolute value of the hogging moment. M_2 is the absolute value of the hogging moment at the location of the bracing (Figure 9.25a). V_2 and V_1 are the coexisting shear forces. The factors m_1, m_2 take into account the variation of the bending moment that is assumed to be parabolically distributed.

A U-frame action may be activated by transverse bending of the web as in Figure 9.25b. This is expressed by the γ-factor that is calculated as follows:

$$\gamma = \frac{c \cdot L^4}{E_a \cdot I_{eff,z}} \tag{9.50g}$$

where

$$c = \frac{3 \cdot E_a \cdot I}{h_a^3} = \frac{3 \cdot E_a \cdot t_w^3}{12 \cdot (1 - \nu_a^2) \cdot h_a^3} = \frac{E_a \cdot t_w^3}{3.64 \cdot h_a^3} \tag{9.50h}$$

However, this stiffness is generally very low, and the γ-factor may be taken equal to zero.

When web stiffeners are present and rigidly connected to the top steel flanges, the spring constants C and c may be determined by Equations 9.55 and 9.56, where the second moment of area I_v includes the transverse stiffeners and the adjacent parts of the stiffened plate at width $15 \cdot \varepsilon \cdot t_w$ as indicated in Section 9.13.4 (see also last point of Remark 9.8).

It is important to note that the Euler load calculated with Equation 9.50a is valid *only when the bending moment does not reverse within length L*. This is valid for the stability checks from internal support till the position of the adjacent bracing (L in Figure 9.25a). For the length L_{rem}, there is a moment reversal, and a conservative assumption is unavoidable; this is $M_2 = 0$, $V_2 = V_1$, and $c = 0$ that leads to $m = 1.88$.

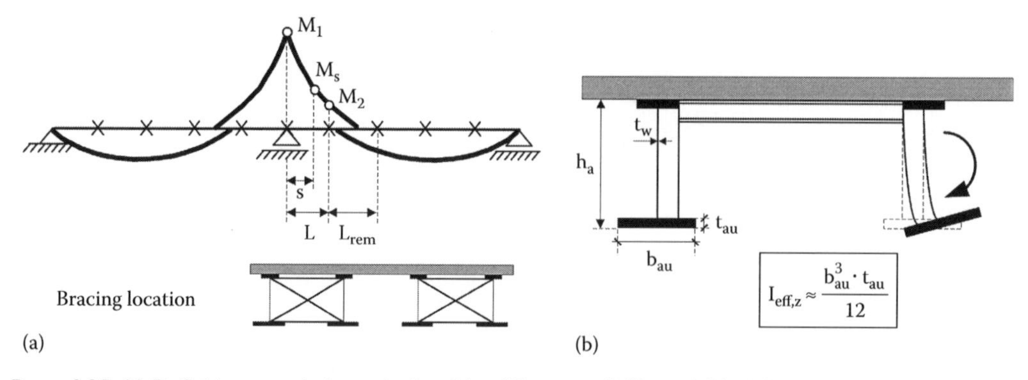

Figure 9.25 (a) Definitions needed to calculate N_{cr} of Equation 9.50a and (b) U-frame action.

Once N_{cr} is known the slenderness value can be calculated from Equation 9.45. Finally, the reduction factor for lateral torsional buckling can be calculated from Equations 9.42 and 9.43. However, according to EN 1993-2, the resistance verification for the beam is required to be conducted at a distance $s = 0.20 \cdot L/\sqrt{m}$ from the support. At this section, the bending moment can be approximately estimated after a linear interpolation:

$$M_s \approx M_1 - \frac{s}{L} \cdot (M_1 - M_2) \tag{9.51}$$

The slenderness at this section is different than $\bar{\lambda}_{LT}$ and becomes

$$\bar{\lambda}_s = \bar{\lambda}_{LT} \cdot \sqrt{\frac{M_1}{M_s}} \tag{9.52}$$

A new reduction factor $\chi_{LT,s}$ referring to $\bar{\lambda}_s$ should then be calculated. The new verification is expressed as follows:

$$M_s \le \chi_{LT,s} \cdot M_{Rd} \tag{9.53}$$

REMARK 9.8

- It was mentioned that for the calculation of the Euler load, the U-frame action can be taken into account. This leads to less conservative results, but it has the disadvantage that tension forces may be transferred to the shear studs. A careful design for avoiding local failure in the deck plate will be necessary. Conservatively, the γ-factor can be set equal to zero.
- EN 1993-2 requires that the design of the main girders should be conducted at a distance s from the adjacent support. This is because the highest stresses from the transverse buckling of the bottom flange arise at some distance away from the rigid support. Comparing the moment M_1 at support with the resistance with respect to lateral torsional buckling is too conservative; however, in part of the literature, it is recommended, that is, [9.10].
- The procedure previously described leads to the following expression for the slenderness of the compressed flange (see Figure 9.25):

$$\bar{\lambda}_{LT} = 1.1 \cdot \frac{L}{b_{au}} \cdot \sqrt{\frac{f_{yk}}{E_a \cdot m}} \cdot \sqrt{1 + \frac{A_{wc}}{3 \cdot b_{au} \cdot t_{au}}} \tag{R9.2}$$

- In the case of integral bridges, coexisting axial compression forces may arise. A linear interaction should be used:

$$\frac{N_{Ed}}{N_{b,Rd}} + \frac{M_{Ed}}{M_{b,Rd}} \le 1 \tag{R9.3}$$

where $N_{b,Rd}$ is the flexural resistance according to Equation 9.9 or 9.10 for buckling around the minor axis.

- It was noted that factors m_1 and m_2 in Equations 9.50c and 9.50d are based on a parabolic bending distribution with negligible axial loading. In a different case, the aforementioned expression should be avoided. An improved method is found in [9.1].
- The spring constant in Equation 9.50h does not include the rigidity of the transverse stiffeners. A less conservative approach is to estimate c by the use of Equations 9.54 and 9.56.

EXAMPLE 9.7

A two-span composite bridge has a rigid cross bracing at a distance 5 m from the internal support ($L = 5$ m in Figure 9.25a). Bracings every 12 m follow ($L_{rem} = 12$ m in Figure 9.25a). The cross section of the main girders is shown in Figure 9.10; it is the same along the bridge. The hogging moment at the internal support is $M_1 = 50{,}000$ kN-m and at the bracing $M_2 = 30{,}250$ kN-m. The shear at bracing is 75% of the value at internal support. Steel S355.

Check at internal support ($L = 5$ m, Figure 9.25); recommended:

$$A_f = 110 \cdot 8 = 880 \text{ cm}^2$$

$$I_{eff,z} = 110^3 \cdot 8/12 = 887333.3 \text{ cm}^4$$

Equation 9.50e: $\mu = \dfrac{V_2}{V_1} = 0.75$

Equation 9.50f: $\Phi = \dfrac{2 \cdot \left(1 - \dfrac{30250}{50000}\right)}{1 + 0.75} = 0.45$

Web stiffeners are not attached to the top flange, and U-frame action cannot be activated. Therefore, $\gamma = 0$.

Equation 9.50c: $m_1 = 1 + 0.44 \cdot (1 + 0.75) \cdot 0.45^{1.5} + 0 = 1.23$

Equation 9.50d: $m_2 = 1.23$

Equation 9.50b: $m = 1.23$

Equation 9.50a: $N_{cr} = 1.23 \cdot \dfrac{\pi^2 \cdot 21000 \cdot 887333.3}{500^2} = 903921 \text{ kN}$

The compression zone of the web is calculated from the plastic distribution of stresses since the fully cracked cross section is class 2. Therefore, $h_{wc} = 68.8 - 8 = 60.8$ cm.

Equation 9.46: $A_{eff} = 880 + \dfrac{60.8 \cdot 1.8}{3} = 916.48 \text{ cm}^2$

At internal support:

Equation 9.45: $\bar{\lambda}_{LT} = \sqrt{\dfrac{32.5 \cdot 916.48}{903921}} = 0.18 < 0.2$

Equation 9.42: $\chi_{LT} = 1$

The resistance is not reduced due to lateral torsional buckling:

$$M_{b,Rd} = M_{2,pl,V,Rd} = 47817.7 \text{ kN-m} < M_1 = 50000 \text{ kN-m (not sufficient)}$$

Note: The hogging bending resistance $M_{2,pl,V,Rd}$ (Example 9.2) was calculated with γ_{M1} = 1.1.

At a distance $s = 0.20 \cdot 500/\sqrt{1.23} = 90.2$ cm from the internal support, the design bending moment is approximately calculated from Equation 9.51:

$$M_s \approx 50000 - \frac{90.2}{500} \cdot (50000 - 30250) = 46437.1 \text{ kN-m}$$

The slenderness is estimated from Equation 9.52:

$$\bar{\lambda}_s = 0.18 \cdot \sqrt{\frac{50000}{46437.1}} = 0.19 < 0.2 \rightarrow \chi_{LT} = 1$$

Again, the resistance is not reduced due to lateral torsional buckling. The shear force is approximately equal to V_1, and therefore, the same value of $M_{2,pl,V,R}d$ is used:

$$1.0 \cdot 47{,}817.7 \text{ kN-m} > 46{,}437.1 \text{ kN-m (sufficient)}$$

Check in the area where bending reverses (L_{rem} = 12 m, Figure 9.25):
The bending moment reverses and Equation 9.50a is not valid. Conservatively, it is assumed that $M_2 = 0$, $V_2 = V_1$, $c = 0 \rightarrow m = 1.88$.

$$\text{Equation 9.50a: } N_{cr} = 1.88 \cdot \frac{\pi^2 \cdot 21000 \cdot 887333.3}{1200^2} = 239861.6 \text{ kN}$$

$$\text{Equation 9.45: } \bar{\lambda}_{LT} = \sqrt{\frac{32.5 \cdot 916.48}{239861.6}} = 0.35$$

Using curve d (Table 9.7): $\alpha = 0.76$

$$\text{Equation 9.43: } \Phi_{LT} = 0.5 \cdot [1 + 0.76 \cdot (0.35 - 0.2) + 0.35^2] = 0.62$$

$$\text{Equation 9.42: } \chi_{LT} = \frac{1}{0.62 + \sqrt{0.62^2 - 0.35^2}} = 0.88$$

The resistance is reduced due to lateral torsional buckling:

$$M_{b,Rd} = 0.88 \cdot 47817.7 = 42079.58 \text{ kN-m} > 30250 \text{ kN-m (sufficient)}$$

It may be seen that cross bracings at the *service stage* are required only at the intermediate support and at two positions adjacent to it to ensure lateral torsional buckling stability of the *bottom flanges*. However, more cross bracings, as indicated in Figure 9.25a, are required *at construction stages* to ensure lateral torsional buckling stability of the *top flanges*.

9.13.4 Simplified method: Flexible lateral supports

Lateral supports do not always exist at the level of the compression flange. The compression flange is then laterally supported indirectly through the stiffness of adjacent elements, mostly in the form of U frames, and then considered as flexibly supported. Figure 9.26 illustrates typical examples. Figures 9.26a and b show I-girder bridges at service state, where the lateral stability of the bottom flange at internal supports is provided by activating a U-frame action. Figure 9.26c shows a half-through bridge where the lateral stability of the top flange is provided by a flexible U steel frame, same as in Figure 9.26d.

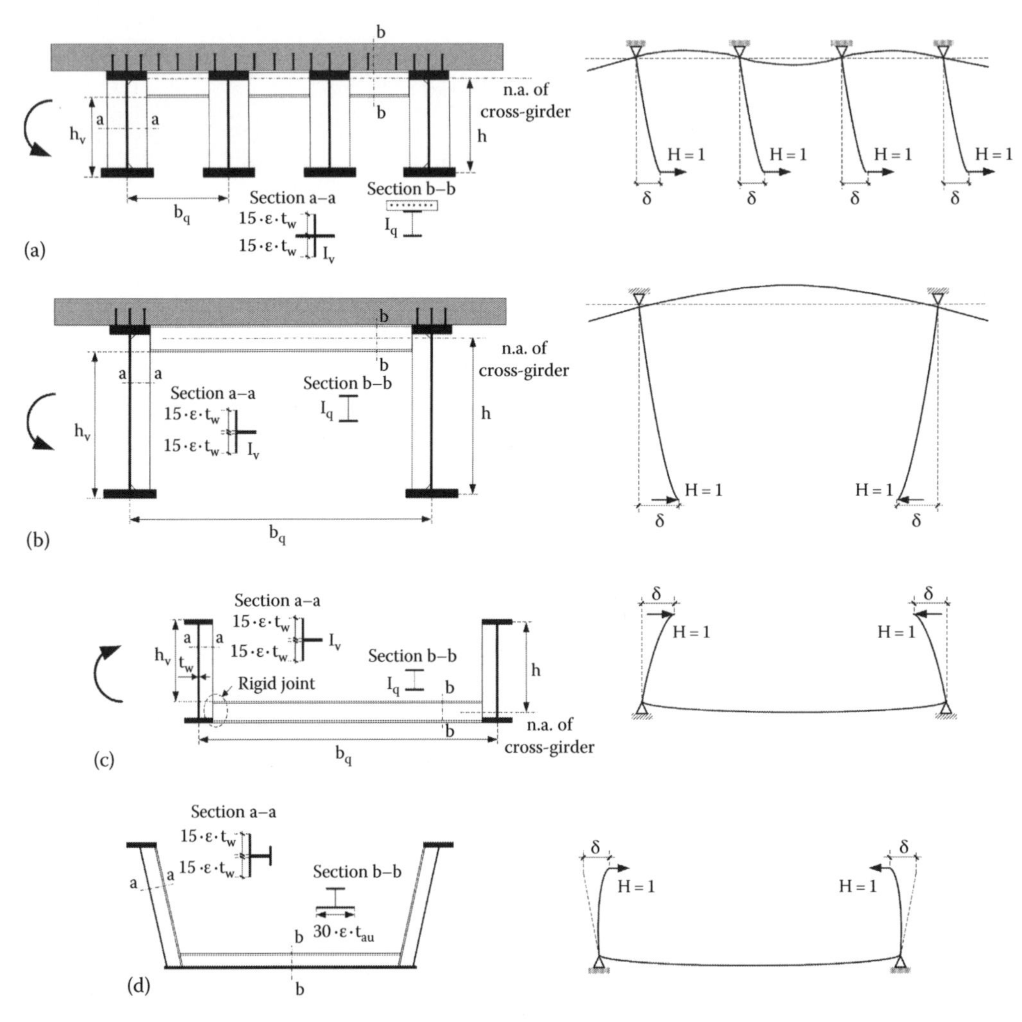

Figure 9.26 Examples of flexible supports and relevant models. (a) Multiple I-girders, (b) twin I-girders, (c) half-trough section, and (d) opened box girder.

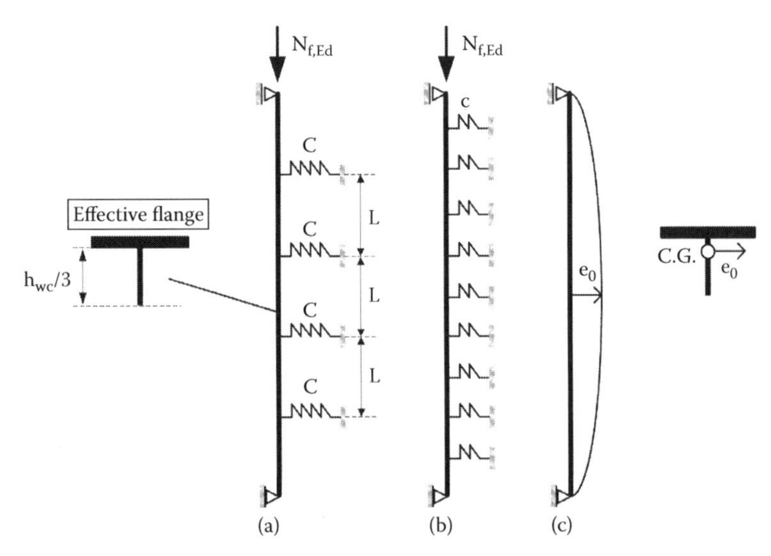

Figure 9.27 Modeling of the compression flange as a T-section column on flexible supports. (a) With discrete springs, (b) with a continuous spring, and (c) with imperfections.

The compression flange is modeled as a column on spring supports as illustrated in Figure 9.27. The spring constant C may be determined by application of transverse forces H = 1 at the ends of the cross frames as shown in Figure 9.26. It is equal to

$$C = H/\delta \ [\text{kN/m}] \tag{9.54}$$

where

H = 1 kN are the lateral forces

δ is the corresponding lateral displacements in m. Different directions for the forces H which may lead to the maximum displacement δ (minimum C value) should be investigated

As shown in Figure 9.26, the cross sections of the web or flange beam elements of this model are composed of the transverse stiffeners and the adjacent parts of the stiffened plate at a distance $15 \cdot \varepsilon \cdot t_w$ for the web sheet and $15 \cdot \varepsilon \cdot t_{au}$ of the bottom flange sheet (Figure 9.26d). The transverse girders are composed of their own cross section. Transverse girders may be composite (Figure 9.26a) or pure steel (Figure 9.26b through d). The spring stiffness C for the configurations of Figure 9.26a through c is estimated from the following expression:

$$C = \frac{E_a \cdot I_v}{\dfrac{h_v^3}{3} + \dfrac{h^2 \cdot b_q \cdot I_v}{n \cdot I_q}} \ [\text{kN/m}] \tag{9.55}$$

where

n = 3 only for the internal girders in Figure (9.26a)

n = 2 in all other cases

For configurations different than those in Figure 9.26a through c, Equation 9.54 should be employed, for example, in Figure 9.26d.

The springs may be "smeared" over the length L, so that the column is supported continuously by springs with a constant

$$c = C/L \ [kN/m^2] \tag{9.56}$$

The distance of the lateral supports L is the distance of the cross frames.

The cross section of the column is a T section composed as before by the compression flange and 1/3 of the compressed part of the web. The design procedure is the same as before, the only difference being the magnitude of the critical buckling load N_{cr} that must be determined by linear buckling analysis. The critical buckling load of an axially compressed column on continuous elastic supports (Figure 9.27b) may be determined from

$$N_{crit} = 2 \cdot \sqrt{c \cdot E_a \cdot I_{eff,z}} \tag{9.57}$$

$I_{eff,z}$ is the transverse second moment of area of the effective flange.

It is important to note that *Equation 9.57 is based on the assumption of rigid end supports* (Figure 9.27). This can be the case of stiff X bracings, heavily stiffened plated diaphragms, or concrete cross girders. However, in some bridges, flexible end supports are present. The buckling load then should be calculated through linear buckling analysis of 3D or FE model by taking into account the reduced rigidity of the end supports. Moreover, rigid connections between the structural elements are assumed (Figure 9.26c). Flexible bolted connections may enhance the buckling lengths.

EXAMPLE 9.8

The lateral stability of the top flange of the box-girder bridge of Example 9.4 shall be verified at construction stages. Cross frames are placed every 5 m. X cross bracings are provided only at supports.

Before casting of the concrete slab, the top flanges when in compression are susceptible to lateral torsional buckling. Accordingly, lateral torsional buckling may occur at spans where the top flange is in compression. The lateral support of the top flange is flexible and is provided by the stiffness of the cross frames; see Figure 9.26d. The cross frames of the girder are shown in Figure 9.28. Their cross sections consist of the T stiffeners and an adjacent part of the flange and web sheet. The participating effective widths are calculated as follows and are shown in Figure 9.28 (sections a–a and b–b):

Flange sheet:

$$30 \cdot \varepsilon_{S420} \cdot t_{au} = 30 \cdot 0.748 \cdot 1.5 = 33.7 \, cm$$

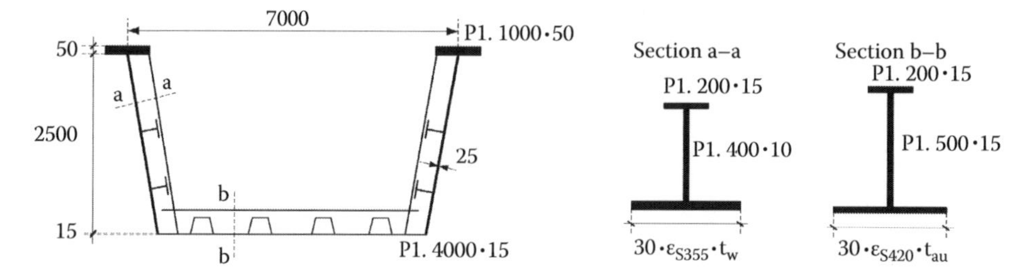

Figure 9.28 Cross section and cross frames of the hybrid steel girder.

Web sheet:

$$30 \cdot \varepsilon_{S355} \cdot t_w = 30 \cdot 0.825 \cdot 2.5 = 61.8 \text{ cm}$$

ε_{S420} and ε_{S355} were calculated with yielding strengths of Table 6.14.

The cross-sectional properties for the cross girder in section a–a of Figure 9.28 are

$$A = 224.5 \text{ cm}^2, \quad I_y = 56570.70 \text{ cm}^4, \quad z_{ao} = 33.35 \text{ cm}, \quad z_{au} = 10.65 \text{ cm}$$

The cross-sectional properties for the cross girder in section b–b of Figure 9.28 are

$$A = 155.55 \text{ cm}^2, \quad I_y = 67249.60 \text{ cm}^4, \quad z_{ao} = 29.9 \text{ cm}, \quad z_{au} = 23.1 \text{ cm}$$

Figure 9.26d illustrates the analysis model for the cross frames, where two horizontal forces $H = 1$ kN act at the level of the top flange. The lateral displacement due to these forces is determined as $\delta = 0.16$ mm.

Spring stiffness of the lateral supports, Equation 9.54:

$$C = 1/(0.16 \cdot 10^{-3}) = 6250 \text{ kN/m}$$

The largest positive moments before hardening of concrete are due to self-weight of the steel girder and self-weight of concrete at edge segments. These are the load cases shown in Figure 9.18. The highest design moment at span is equal to

$$\max M_{a,Ed} = 1.35 \cdot (4{,}068.95 + 9{,}949.8) = 18{,}925.31 \text{ kN-m}$$

The reduction of the cross-sectional properties of the steel girder due to plate buckling of the web is negligible, and therefore, values are taken from Table 9.17. Otherwise for the calculation of the cross-sectional properties, the effective width method should be used; see Section 8.5. The second moment of area is $I_a = 0.45454 \cdot 10^8$ cm^4, the distance of the neutral axis from the top fiber of the top flange $z_{ao} = 142.34$ cm.

$$\text{Compression part of the web: } h_{wc} = \frac{(142.34 - 5)}{\sin(59.08°)} = 160.14 \text{ cm}$$

The verification is performed for a beam under compression on elastic spring supports (Figure 9.27b). The cross section of the beam has a T cross section and is composed of the flange and 1/3 of the compressed part of the web. The participating web height is then 160.14/3 = 53.38 cm. The cross-sectional properties of the effective flange are

$$A_{eff} = 100 \cdot 5 + \frac{160.14}{3} \cdot 2.5 = 633.45 \text{ cm}^2, \quad I_{eff,z} = 416726 \text{ cm}^4, \quad z_{eff,ao} = 7.23 \text{ cm}$$

$z_{eff,ao}$ is the distance of the neutral axis from the top fiber of the steel flange.

Distance of this axis from the neutral axis of the steel girder:

$$142.34 - 7.23 = 135.11 \text{ cm}$$

Maximal direct stress at this fiber due to design moment:

$$\sigma = \frac{1892531}{0.45454 \cdot 10^8} \cdot 135.11 = 5.63 \text{ kN/cm}^2$$

Axial force on the T section, Figure 9.24:

$$\max N_{f,Ed} = 5.63 \cdot 633.45 = 3566.32 \text{ kN}$$

Equation 9.56: $c = 6250/5 = 1250 \text{ kN/m}^2$

The critical buckling load is calculated from Equation 9.57:

$$N_{crit} = 2 \cdot \sqrt{1250 \cdot 10^{-4} \cdot 21000 \cdot 416726} = 66148.49 \text{ kN}$$

Equation 9.45: $\bar{\lambda}_{LT} = \sqrt{\dfrac{34.5 \cdot 133.45 + 39 \cdot 500}{66148.49}} = 0.60$

Using curve d (Table 9.7): $\alpha = 0.76$

Equation 9.43: $\Phi_{LT} = 0.5 \cdot [1 + 0.76 \cdot (0.6 - 0.2) + 0.6^2] = 0.83$

Equation 9.42: $\chi_{LT} = \dfrac{1}{0.83 + \sqrt{0.83^2 - 0.6^2}} = 0.71$

Lateral torsional buckling verification, Equation 9.48:

$$3566.32 \text{ kN} \le 0.71 \cdot \frac{34.5 \cdot 133.45 + 39 \cdot 500}{1.1} = 15558.05 \text{ kN (sufficient)}$$

It is important to note that the support shown in Figure 9.16 has been considered as a rigid one; if this assumption is correct, it is checked in Example 9.9. In the presence of flexible end supports, the buckling load would be significantly lower.

It may be seen that cross frames or cross bracings of box girders serve several functions:

- They provide transverse supports on bottom flange and web panels and their longitudinal stiffeners and enhance their plate buckling resistance, as shown in Chapter 8.
- They provide elastic support on top flanges at construction stages and enhance their resistance to lateral torsional buckling.
- They limit the cross-sectional distortional deformations and allow modeling of the sections as beam elements.

9.13.5 Resistance and rigidity of supporting members

Supporting members, whether cross frames, cross bracings, or transverse bracings, have to resist the deviation forces of the primary system resulting from initial bow imperfections (Figure 9.27c). Rewriting the Euler formula, the critical buckling length of the system on elastic supports is given by

$$l_k = \pi \cdot \sqrt{\frac{E_a \cdot I_{eff,z}}{N_{cr}}} \tag{9.58}$$

Such a system on elastic supports, subjected to a compression force $N_{f,Ed}$, is shown in Figure 9.27. The imperfection is taken as $e_0 = l_k/640$. Instead of the system with imperfections, an equivalent system without imperfections but with a lateral uniform force q_{Ed} is considered. Equating the moments at midspan of the two systems gives the equivalent uniformly distributed load

$$q_{Ed} \cdot \frac{l_k^2}{8} = \frac{N_{f,Ed} \cdot e_0}{1 - \dfrac{N_{f,Ed}}{N_{cr}}} = \frac{N_{f,Ed} \cdot l_k/640}{1 - \dfrac{N_{f,Ed}}{N_{cr}}} \rightarrow q_{Ed} = \frac{N_{f,Ed}}{l_k \cdot 80} \cdot \frac{1}{1 - \dfrac{N_{f,Ed}}{N_{cr}}} \tag{9.59}$$

The distance of the springs is L; the lateral spring forces are then $F_{Ed} = q_{Ed} \cdot L$. Introducing in the previous equation, the lateral forces of the supporting elements may be determined as equal to

$$F_{Ed} = \frac{N_{f,Ed}}{l_k \cdot 80} \cdot \frac{L}{1 - \dfrac{N_{f,Ed}}{N_{cr}}} \tag{9.60a}$$

According to EN 1993-2, the aforementioned relation holds if $l_k > 1.2 \cdot L$. Otherwise the lateral forces are given by

$$F_{Ed} = \frac{N_{f,Ed}}{100} \quad \text{if } l_k < 1.2 \cdot L \tag{9.60b}$$

F_{Ed} is applied to the bracing member for buckling verification.

It has been said that the knowledge of the rigidity of the bracing members is of primary importance for a "realistic" calculation of the buckling length for the lateral torsion verification of the main girder. The buckling load for rigid restraints is estimated from Equation 9.47 while for flexible supports from Equation 9.57. The minimum rigidity for classifying a supporting member as a rigid one can be calculated by equating the two expressions as follows:

$$2 \cdot \sqrt{\min c \cdot E_a \cdot I_{eff,z}} = \frac{\pi^2 \cdot E \cdot I_{eff,z}}{L^2} \rightarrow \min c = \frac{\pi^4 \cdot E_a \cdot I_{eff,z}}{4 \cdot L^4} \tag{9.61}$$

Normally, the rigidity verification of the supporting member should be the first one. Then resistance verifications follow.

EXAMPLE 9.9

The rigidity of the supporting members for the bridge in Example 9.4 should be verified:

- For the spans (Figure 9.28)
- For the end supports (Figure 9.16)

For the spans, the stiffness c has been calculated in Example 9.8: c = 1250 kN/m^2.

From Equation 9.61:

$$\min c = \frac{\pi^4 \cdot 21000 \cdot 416726}{4 \cdot 500^4} = 3.4\ \text{kN/cm}^2 > c = 0.125\ \text{kN/cm}^2$$

The U frames should be considered as flexible supports.

At end supports, the cross frames are triangulated by the CHS sections (see Figure 9.16) so that they may be considered as rigid. Indeed the application of a unit load on the end cross frame including the CHS 193.7 × 12 gives a negligible deformation of $\delta = 0.00231$ mm.

Equation 9.54: $C = 1/(0.00231 \cdot 10^{-3}) = 432900.43$ kN/m

Equation 9.56: $c = (432900.43/5) \cdot 10^{-4} = 8.65\ \text{kN/cm}^2 > \min c = 3.4\ \text{kN/cm}^2$

The bracing configuration of Figure 9.16 can be considered as a rigid support.

9.14 DESIGN OF THE CONCRETE DECK SLAB

The concrete deck is usually a reinforced concrete slab subjected to bending moments and shear forces from self-weight and traffic loads and must be designed as a reinforced concrete section at ultimate and serviceability limits in accordance with the provisions of Eurocode 2 [9.2], [9.3]. The necessary verifications are the following:

- Bending and vertical shear resistance for the ULS combinations of actions
- Punching shear for the concentrated wheel loads
- Limitations of the crack widths for frequent SLS combination of actions
- Stress limitations for the characteristic SLS combination of actions
- Minimum reinforcement
- Shear resistance of the joints between adjacent concreting segments
- Shear transmission between casted in situ concrete and precast elements.

Calculations of the internal forces can be conducted either with the strip method described in Section 7.5.2 or the FE method of Section 7.5.4. For the reinforced concrete design, reference is made to the relevant literature. However, the following comments are noticeable:

- For the majority of the deck slabs, there will be no need for adding shear reinforcement, except those required from construction detailing.
- Punching shear due to the concentrated wheel loads is rarely critical for the design of the deck slab. However, local bending moments due to the wheel loads should be taken into account by adding them to the global ones. Local bending moments are not of negligible magnitude.
- The deck slab should be verified both against LM1 and LM2.

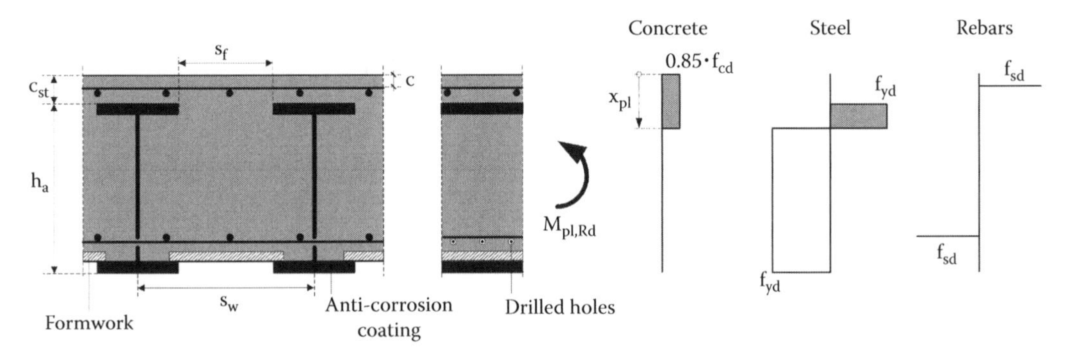

Figure 9.29 Block stress diagrams for the calculation of the plastic bending resistance of filler-beam deck slabs.

- EN 1992-1-1 (clause 5.8.3.1) imposes a demanding slenderness criterion that should be taken into account during the design of the deck slab. If the slab's slenderness λ is higher than the minimum slenderness λ_{lim} required from the code, second-order bending moments in the slab should be calculated. For multiple plate-girder bridges with a slab of 25 cm and main girders at a distance of 3.5–4 m, the code's slenderness criterion is usually not fulfilled.
- The bending resistance of filler-beam deck slabs can be calculated assuming a plastic distribution of stresses (for class 1 or 2) as shown in Figure 9.29. Due to the encasement in concrete, there is no danger of lateral torsional buckling failure at the final stage.

For the estimation of the shear resistance, the contribution of concrete can be taken into account. The partial shear forces $V_{c,Ed}$ and $V_{a,Ed}$ (Figure 9.30) are calculated based on the distribution of the partial bending moments $M_{pl,s,Rd}$, $M_{pl,fw,Rd}$, and $M_{pl,V,Rd}$. The shear force acting on concrete depends on the plastic resistance of the reinforced concrete cross section and is estimated from the following equation:

$$V_{c,Ed} = V_{Ed} \cdot \frac{M_{pl,s,Rd}}{M_{pl,Rd}} \tag{9.62}$$

The shear force of Equation 9.62 should be lower or equal to the shear resistance of the concrete section according to EN 1992-1-1. The shear force that acts on the steel cross section is obviously equal to

$$V_{a,Ed} = V_{Ed} \cdot \frac{M_{pl,fw,Rd} + M_{pl,V,Rd}}{M_{pl,Rd}} = V_{Ed} - V_{c,Ed} \tag{9.63}$$

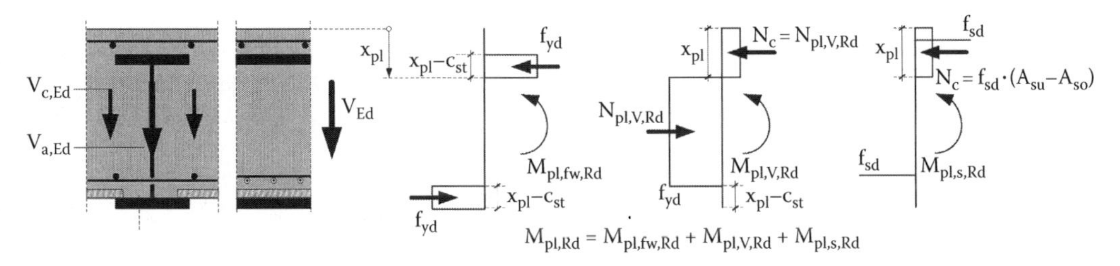

Figure 9.30 Block stress diagrams for the calculation of the shear resistance of filler-beam deck slabs.

The shear resistance of the steel section is calculated according to Equation 9.19. Due to encasement in concrete shear buckling can be considered as irrelevant.

The main restrictions that are imposed by the code for filler-beam deck bridges are the following:

- The steel beams should not be curved in plan.
- The deck skew should not be higher than 30°.
- The steel girder's depth h_a (Figure 9.29) should lie between 210 and 1100 mm.
- The beams' spacing s_w (Figure 9.29) should fulfill

$$s_w \leq \min\left[h_a/3 + 600\text{ mm}, 750\text{ mm}\right]$$

- The concrete cover c_{st} (Figure 9.29) should fulfill

$$c_{st} \geq 70\text{ mm and } c_{st} \leq \min\left[150\text{ mm}, h_a/3, x_{pl} - t_f\right]$$

- The distance s_f (Figure 9.29) should not be smaller than 150 mm, so that concrete can be adequately compacted.
- The concrete should be of normal density and the steel's surface descaled.
- The bottom layer of transverse reinforcement passes through drilled holes (not punched) and has a diameter not less than 16 mm and a spacing not more than 300 mm.

It is pointed out that cross sections of steel girders that are fully encased in concrete are not covered from the regulations of EN 1994-2.

REFERENCES

[9.1] Collin, P., Möller, M., Johansson, B.: Lateral torsional buckling of continuous bridge girders. *Journal of Constructional Steel Research* 45(2), 217–235, 1998.

[9.2] EN 1992-1-1, CEN (European Committee for Standardization): Design of concrete structures, Part 1–1: General rules and rules for buildings, 2004.

[9.3] EN 1992-2, CEN (European Committee for Standardization): Design of concrete structures—Part 2: Concrete bridges—Design and detailing rules, 2005.

[9.4] EN 1993-1-1, CEN (European Committee for Standardization): Design of steel structures, Part 1–1: General rules and rules for buildings, 2005.

[9.5] EN 1993-1-5, CEN (European Committee for Standardization): Design of steel structures, Part 1–5: Plated structural elements, 2006.

[9.6] EN 1993-2, CEN (European Committee for Standardization): Design of steel structures—Part 2: Steel bridges, 2006.

[9.7] EN1994-2, CEN (European Committee for Standardization): Design of composite steel and concrete structures. Part 2: Rules for bridges, 2005.

[9.8] Eurocode 3 and 4. *Guidance Book*. Application to steel-concrete composite road bridges. Sètra, 2007.

[9.9] Hendy, C. R., Jones, R. P.: Lateral buckling of steel plate girders for bridges with flexible lateral restraints or torsional restraints. *Workshop on Eurocode 4–2, Composite Bridges*. Stockholm, Sweden, 2011.

[9.10] Hendy, C. R., Murphy, C. J.: Designers' guide to EN 1993-2 Eurocode 3: Design of steel structures. Part 2: Steel bridges. Thomas Telford, London, U.K., 2007.

[9.11] Iliopoulos, A.: Creep factors for the computation of composite girders. *Stahlbau* 75(5), 375–379, 2006 (in German).

Chapter 10

Serviceability limit states

10.1 INTRODUCTION

Serviceability limit states (SLSs) concern the functioning of the structure and its structural members under normal use, the comfort of people, and the structural appearance, associated with inappropriately high deformations or cracking rather than aesthetics, and refer to

- Stress limitations for structural steel, reinforcement, and concrete
- Web breathing
- Design of shear connectors (see Chapter 12)
- Control of cracking for concrete
- Deflections and vibrations control

10.2 STRESS ANALYSIS AND LIMITATIONS

Stress analysis refers to the determination of internal forces and moments and subsequently of stresses for all cross sections and their fibers along the bridge. Stress analysis is similar to this at ultimate limit state (ULS) for cross sections of classes 3 and 4 (see Sections 9.9 and 9.10 and Example 9.4). It is based on the following considerations:

- Internal forces and moments are determined for SLSs by means of elastic global analysis (Chapter 7). Second-order global analysis as indicated by Equation 7.18 may be required to be performed.
- The combinations of actions to be examined are described in Table 5.12.
- The effects of creep and shrinkage in concrete are taken into account by the use of appropriate modular ratios of concrete in accordance with Equation 6.20.
- The primary effects of creep and shrinkage are taken into account in the uncracked regions in accordance with Section 7.4.
- The secondary effects of creep and shrinkage are taken into account by introduction of equivalent temperature changes in the uncracked regions as additional "loads" in accordance with Equations 7.26 and 7.30 correspondingly.
- Shear lag is taken into account by introduction of elastic effective[s] widths for the concrete flange and the steel bottom flange of box girders in accordance with Section 7.2. It is reminded that the effective widths for the steel plates are greater at ULS.
- The effects of plate buckling for class 3 or 4 cross section are normally ignored, unless the effective area of a compression wall of a cross section is smaller than 50% of its gross section or the reduction factors ρ for plate buckling are smaller than 0.5.
- The sequence of construction is taken into account (see Tables 9.12 and 9.13).

- Concrete in tension is neglected so that in regions of hogging moment, the properties of the fully cracked section in accordance with Section 7.3 are considered.
- The effect of tension stiffening during the stress calculation in reinforcing steel is taken into account.
- Tensile stresses due to shrinkage and hydration of cement (see Section 6.1.5) are investigated so that cracking at unexpected areas, that is, at spans, is avoided.
- Stresses in concrete and in reinforcing steel due to local bending of the deck slab due to concentrated wheel loads should be added to those due to global effects (see Section 7.5.4).

For girder-type bridges, the response is linear not only in respect to material behavior but also in respect to geometry. Accordingly, stress analysis that includes as discussed before global analysis and stress determination along the bridge may be performed for each individual action or "load" separately, presumed that the extent of cracked regions is correctly estimated (see Example 10.1). **The combinations, presented in Table 5.12, refer to the stresses rather than the actions.** The resulting stresses are the design stresses for the relevant SLS combination and are assigned as $\sigma_{Ed,ser}$ and $\tau_{Ed,ser}$ for structural steel, $\sigma_{s,Ed,ser}$ for reinforcement, and $\sigma_{c,Ed,ser}$ for concrete.

For some types of bridges, for example, cable stayed, the effects of deformations may not be ignored and second-order global analysis has to be employed. **The combination rules for such analysis apply directly to the actions.**

REMARK 10.1

It is reminded that

- The shear lag effects at SLS for wide steel flanges are more pronounced than for ULS (see Section 7.2.3). Effective^s widths for box-girder bridges at hogging bending areas at SLS may be up to 50% of the effective^s widths at ULS (see Figure 7.38). In such cases, omitting the SLS stress verifications may lead to irreversible deformations due to yielding of the structural steel.
- According to EN 1993-2, distortional effects may be neglected when the effects from distortion do not exceed 10% of the bending effects.

The design serviceability stresses must be appropriately limited. Stress limitations refer to structural steel, reinforcement, and concrete as described in the following.

10.2.1 Structural steel

To ensure elastic behavior under service loads, the design stresses of structural steel for the **characteristic** SLS combination of actions (Table 5.12, Equation 5.23) must be limited to the yield strength of steel as follows:

Direct stresses:

$$\sigma_{Ed,ser} \leq \frac{f_y}{\gamma_{M,ser}} \tag{10.1}$$

Shear stresses:

$$\tau_{Ed,ser} \leq \frac{f_y}{\gamma_{M,ser} \cdot \sqrt{3}} \tag{10.2}$$

von Mises stresses:

$$\sqrt{\sigma_{Ed,ser}^2 + 3 \cdot \tau_{Ed,ser}^2} \leq \frac{f_y}{\gamma_{M,ser}} \tag{10.3}$$

The partial resistance factor for serviceability is, according to Table 5.20, equal to $\gamma_{M,ser} = 1.0$ (recommended value).

It is noted that highest direct stresses appear in the extreme flange fibers and highest von Mises stresses usually at the flange–web intersection. Therefore, the stress points to be verified are the extreme fibers of the flanges and the flange–web intersections.

10.2.2 Reinforcement

To ensure elastic behavior and avoid excessive cracking or deformation, tensile stress in the reinforcement for the **characteristic** SLS combination must be limited to the following:

$$\sigma_{s,Ed,ser} + \Delta\sigma_s \leq k_3 \cdot f_{sk} \tag{10.4a}$$

When the stress is caused by an imposed deformation, for example, prestress by imposed deformations or for settlements, it should be verified instead:

$$\sigma_{s,Ed,ser} + \Delta\sigma_s \leq k_4 \cdot f_{sk} \tag{10.4b}$$

where

$\sigma_{s,Ed,ser}$ is the tension stress in the reinforcement under the characteristic SLS combination combining global **and local effects** (i.e., due to wheel loads; see Sections 7.5.3 and 7.5.4)
$\Delta\sigma_s$ is the additional stress due to tension stiffening (Equation 10.4c)
f_{sk} is the characteristic strength of the reinforcement

The additional stress due to tension stiffening is estimated as follows:

$$\Delta\sigma_s = \frac{0.4 \cdot f_{ctm}}{\alpha_{st} \cdot \rho_s} \tag{10.4c}$$

where

ρ_s is the reinforcement ratio ($=A_{s,tot}/A_c$)

$$\alpha_{st} = \frac{A_{2,sa} \cdot I_{2,sa}}{A_a \cdot I_a} \tag{10.4d}$$

f_{ctm} is the mean tensile strength of concrete (Table 6.1)
A_a, I_a are the area and the second moment of area of the steel section
$A_{2,sa}$, $I_{2,sa}$ are the area and the second moment of area of the fully cracked composite section (steel section + reinforcement) (Figure 7.39)

The recommended values of the strength parameters are $k_3 = 0.8$ and $k_4 = 1.0$. Obviously, the aforementioned verifications are made in the cracked regions where the reinforcement is in tension.

10.2.3 Concrete

The compressive stress of concrete must be limited in order to avoid micro-cracking. Accordingly, the following stress limitations exist for the **characteristic** SLS combination for exposure classes XD, XF, and XS (see Section 5.6):

$$\sigma_{c,Ed,ser} \leq k_1 \cdot f_{ck} \tag{10.5a}$$

For composite cross sections at sagging moment areas, the following stress limitation for the **quasi-permanent** SLS combination applies so that creep remains linear:

$$\sigma_{c,Ed,ser} \leq k_2 \cdot f_{ck} \tag{10.5b}$$

where

$\sigma_{c,Ed,ser}$ is the compression stress in the concrete under the characteristic or quasi-permanent SLS combination combining global **and local effects** (i.e., due to wheel loads; see Sections 7.5.3 and 7.5.4)

f_{ck} is the characteristic strength of the concrete

The recommended values of the strength parameters are $k_1 = 0.6$ and $k_2 = 0.45$.

REMARK 10.2

- For composite cross sections of class 3 or 4 at sagging moment areas, Equations 10.5a and b are usually covered from the ULS verification for concrete in Tables 9.13 through 9.15.
- Concrete stresses are calculated for short-term design so that the favorable effects from creep and shrinkage are ignored.

10.3 CRACKING OF CONCRETE

10.3.1 General

A concrete slab supported by longitudinal steel girders has two main functions in composite bridges. In transverse direction, it acts as a concrete plate for distributing the loads to the girders, while in the longitudinal direction it constitutes, through its effectives width, the top flange of the girder. Since concrete under tension inevitably cracks, cracking of concrete affects both transverse and longitudinal function. Crack control in transverse direction refers to reinforced concrete cross sections, possibly prestressed by transverse tendons. In the longitudinal direction, it refers to the concrete flange of the steel girder.

Cracking of concrete under tension is inevitable. However, when applying certain rules, as presented in the following, cracks widths will be limited and durability will not be substantially affected. At the final stage at hogging moment areas, a minimum reinforcement amount is placed so that yielding of reinforcing steel immediately after cracking is avoided. Cracking is controlled by limiting the spacing or the diameters of the rebars. The necessary theoretical background on cracking of concrete is found in Section 6.1.6.

10.3.2 Minimum reinforcement

For concrete members (e.g., deck slab in transverse direction or concrete cross frames), Equation 6.29c applies. For composite members under hogging bending, a similar expression is used:

$$A_{s,min} = \frac{f_{ct,eff} \cdot k \cdot k_s \cdot k_c \cdot A_{ct}}{\sigma_s} \tag{10.6}$$

where

k_s is the 0.9 reduction factor accounting for the reduction of tensile force in the deck slab due to local slip of the shear connection

k is the 0.8 reduction factor allowing for the effect of nonuniform self-equilibrating stresses (see residual stresses in Figure 6.13)

$f_{ct,eff}$ is the mean value of the tensile strength of concrete that may be taken as f_{ctm} (see Table 6.1) or in case of uncertainty for the age of concrete at cracking as 3 MPa

$k_c = \frac{1}{1 + h_c/(2 \cdot \bar{z}_{1,0})} + 0.3 \leq 1.0$ is a coefficient that takes into account the nonlinear distribution of stresses immediately prior to cracking. For tension members, for example, tension chords of trusses and ties in arch-and-tie bridges, $k_c = 1$

$\bar{z}_{1,0}$ is the distance between centroids of the concrete flange and the composite section calculated with the short-term modular ratio n_0 (see Figure 7.39)

σ_s is the maximum permissible stress in the reinforcement after cracking that may be set to f_{sk} or to a lower value depending on the bar diameter to satisfy the crack width limits in accordance with Table 6.10

A_{ct} is the area of the tension zone prior to cracking that may be set equal to A_c, the area of the concrete flange within the effective width $b_{eff,2}$ (see Section 7.2.2)

At least half of the minimum reinforcement should be placed adjacent to the top fiber of the concrete flange.

10.3.3 Limitation of crack width

As outlined before, cracking of concrete in tension is unavoidable. However, crack widths may be limited if certain rules concerning size and spacing of reinforcing bars are respected. These rules concern both transverse and longitudinal directions of the slab. The limiting crack widths depend generally on the exposure class of the bridge. However, they are more or less fixed to 0.3 mm for reinforced concrete slabs and 0.2 mm for slabs prestressed by tendons. If the slab is prestressed in transverse direction, the 0.2 mm limit is valid also for the longitudinal direction.

REMARK 10.3

According to EN 1992-2, the recommended exposure class for a concrete surface protected by waterproofing is XC3 (see Section 5.6.1). For this exposure class, the recommended crack width is 0.3 mm.

In case that minimum reinforcement is provided in accordance with Section 10.3.2, the control of crack width may be achieved by observing rules concerning bar sizes or bar spacing in accordance with Tables 6.10 and 6.11. **High-bond bars** are used for reinforcement.

When the maximum bar diameter is limited for controlling the crack width of concrete members, Equation 6.44 applies. For composite girders, the bar diameter should be modified as follows:

$$\Phi_s = \Phi_s^* \cdot \frac{f_{ct,eff}}{f_{ct,0}} \tag{10.7}$$

where

Φ_s^* is the maximum bar diameter from Table 6.10

$f_{ct,0}$ is the reference strength of concrete (=2.9 N/mm^2)

The steel stress σ_s in reinforcement to be used in Tables 6.10 and 6.11 is determined for the fully cracked section (state 2) for the **quasi-permanent combination of actions for long-term design** by taking into account the stress enhancement due to tension stiffening according to Equation 10.4c.

Filler-beam decks are verified against cracking similar to reinforced concrete sections according to the provisions of EN 1992-1-1 (see Section 6.1.6).

10.3.4 Thermal cracking during concreting (determination of cracked regions)

It was mentioned in Section 7.1.3 that defining the cracked regions along the bridge is of primary importance for the quality of the results taken from the global analysis. With the *simplified method*, the casting sequence is not taken into account and actually cracked parts of the bridge may be treated as uncracked. On the other hand, the *general method* is more laborious, but it is in any case recommended.

Strictly, when the general method is applied, the casting sequence and the rheological behavior of concrete should be carefully taken into account. However, this needs a detailed knowledge of the construction's time plan that during design is rarely known. But even if the time plan is known during the design phase, a time-consuming step-by-step analysis with the incremental method of Section 6.1.2 would be necessary. The cracked regions can be practically located through a short-term design by neglecting creep and shrinkage according to the recommendations of Remark 7.3. Indeed, a negligible part of drying shrinkage is developed during the time period of construction (see Figure 6.10). Autogenous shrinkage develops more rapidly, but due to coexisting creep, its effect is reduced. Moreover, a great part of creep and shrinkage is developed before composite action is activated, and this justifies the use of short-term cross-sectional properties in finding the cracked regions throughout the bridge.

The hydration of cement has been presented in Section 6.1.5. This can be one of the main causes of cracking evolution during the erection stage, known also as *early thermal cracking*. The phenomenon develops during the concrete's curing period (≈5–7 days), but if cracks arise, these are irreversible. Therefore, temperature effects due to cement's hydration should be taken into account. According to EN 1994-2, a temperature difference of 20°C between steel and concrete should be assumed. This is equivalent to a free strain of $\varepsilon_H = -200 \cdot 10^{-6}$ for the concrete slab. The shortening of the concrete slab due to ε_H is restrained due to its shear connection with the steel girder so that there develops a tension force N_H in it. Since this force is due to restraint, an equal compression force acts in the composite section as well as a moment M_H that equilibrates the pair of forces of the slab and the composite section, similar to Figure 7.44 for shrinkage. This results in the following:

- In the concrete slab
  - A tension force:

$$N_H = -\varepsilon_H \cdot E_{cm} \cdot A_c \tag{10.8a}$$

- In the composite section
  - A compression force:

$$-N_H = \varepsilon_H \cdot E_{cm} \cdot A_c$$

  - A moment:

$$M_H = N_H \cdot \overline{z}_{1,0} \tag{10.8b}$$

where
n_0 is the short-term modular ratio given in Table 6.4
$\overline{z}_{1,0}$ is the gravity center of the equivalent section calculated with n_0 (see Figure 7.39)
E_{cm} is the concrete's modulus of elasticity given in Table 6.1
A_c is the sectional area of the concrete plate based on the **geometric width**

The primary effects in the deck slab (residual stresses) due to the hydration of cement are calculated as follows:

$$\sigma_{co,H} = \frac{N_H}{A_c} - \frac{N_H}{n_0 \cdot A_{1,0}} - M_H \cdot \frac{z_{co,0}}{n_0 \cdot I_{1,0}} \tag{10.9a}$$

$$\sigma_{cu,H} = \frac{N_H}{A_c} - \frac{N_H}{n_0 \cdot A_{1,0}} - M_H \cdot \frac{z_{cu,0}}{n_0 \cdot I_{1,0}} \tag{10.9b}$$

where
$z_{co,0}$ is the vertical distance between the gravity center of the composite section for short-term loading and the top fiber of the concrete slab
$z_{cu,0}$ is the vertical distance between the gravity center of the composite section for short-term loading and the bottom fiber of the concrete slab

The secondary effects due to the hydration of cement can be estimated through an equivalent temperature as that of Equation 7.30. Thus,

$$\Delta T_H = \frac{M_H \cdot h}{E_a \cdot I_{1,0} \cdot \alpha_t} \tag{10.10}$$

where
h is the total depth of the composite section
$I_{1,0}$ is the second moment of area of the composite section for short-term loading
α_t is the coefficient for thermal expansion taken equal to 10^{-5}/°C

The application of Equation 10.10 leads to additional tensile stresses in concrete.

It is important to note that hydration of cement has a non-negligible effect **only on the length of the cracked regions.** In these regions, cracking should be controlled by placing the minimum reinforcement and limiting the diameter or the spacing of the bars according to Sections 10.3.2 and 10.3.3. Stresses and secondary internal forces due to the hydration of cement are not considered for limiting stresses at SLS or ULS.

EXAMPLE 10.1

The cracked regions of the box-girder bridge in Example 9.4 will be defined according to the general method of EN 1994-2. The concreting sequence is shown in Figure 9.15.

Concrete C35/45: from Table 6.1, $f_{ctm} = 3.2$ MPa

From Table 9.17, the cross-sectional properties of the composite section for short-term loading are

$A_{1,0}$	9712.58 cm^2
$z_{co,0}$	77.64 cm
$z_{cu,0}$	47.64 cm
$I_{1,0}$	$1.037 \cdot 10^8$ cm^4

As a first step, an uncracked analysis considering all loads as short term is conducted. The loads and the corresponding bending moments acting on the composite parts of the bridge are shown as follows. Bending moments acting on pure steel cross sections do not have an influence on the length of the cracked regions and thus are not taken into account (see the jump of the bending moment in the first diagram at the end of the edge segments in Figure 10.1).

Note: The determination of the cracked regions is based on the maximum bending moments acting on the half girder A (Figure 9.17). A less conservative approach would be to determine the cracked regions by adding the bending moments and conduct the stress calculations for the whole cross section.

The hydration of cement is taken into account as an imposed strain in the deck slab equal to $\varepsilon_H = -200 \cdot 10^{-6}$. Therefore, the corresponding tensile force in the slab is estimated from Equation 10.8a:

$$N_H = 200 \cdot 10^{-6} \cdot 3,400 \cdot (1,100 \cdot 300) = 22,440 \text{ kN}$$

Equation 10.8b (Table 9.17):

$$M_H = 22,440 \cdot \frac{(77.64 - 15)}{100} = 14,056.4 \text{ kN-m}$$

The primary stresses are considered constant along the bridge and equal to

Equation 10.9a:

$$\sigma_{co,H} = \frac{22,440}{33,000} - \frac{22,440}{6.18 \cdot 9,712.58} - 1,405,640 \cdot \frac{77.64}{6.18 \cdot 1.037 \cdot 10^8} = 0.14 \text{ kN/cm}^2$$

Equation 10.9b:

$$\sigma_{cu,H} = \frac{22,440}{33,000} - \frac{22,440}{6.18 \cdot 9,712.58} - 1,405,640 \cdot \frac{47.64}{6.18 \cdot 1.037 \cdot 10^8} = 0.20 \text{ kN/cm}^2$$

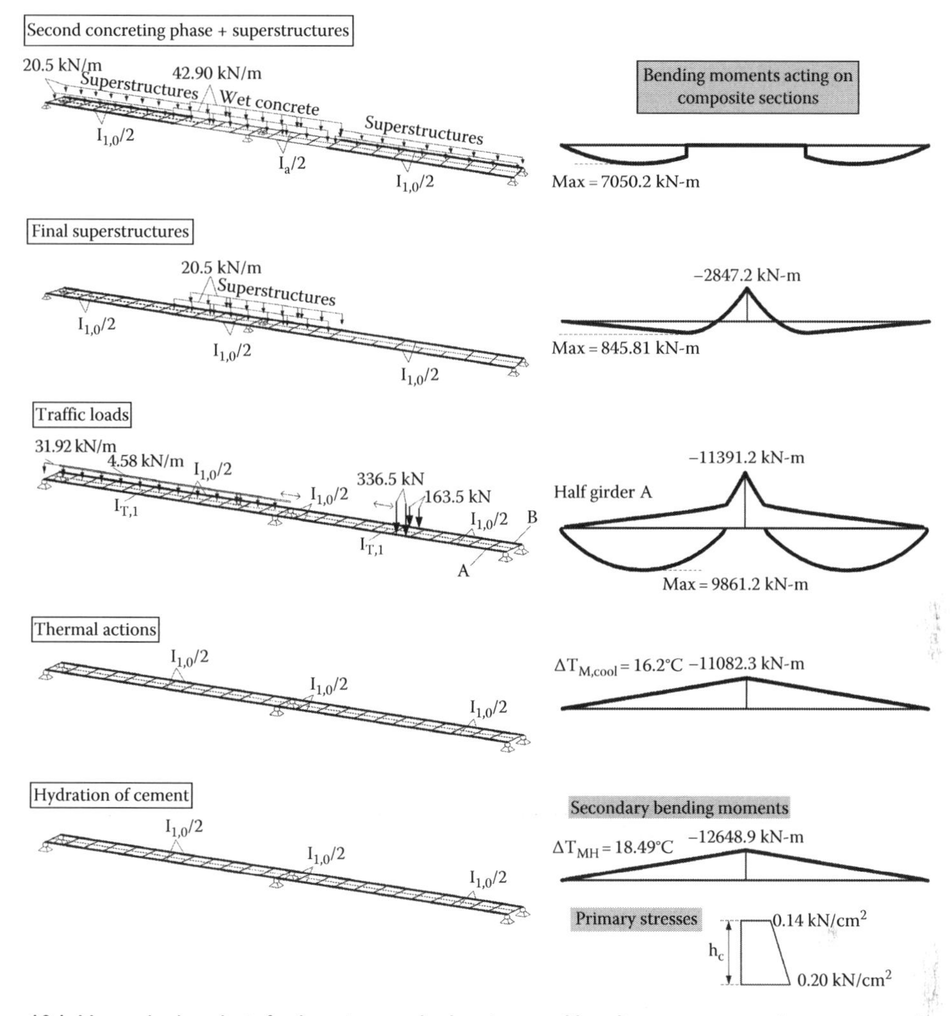

Figure 10.1 Uncracked analysis for locating cracked regions and bending moments acting on composite sections (bending moments refer to half girder A).

Secondary effects due to hydration of the cement should be calculated for each segment separately. A tolerable simplification is to apply the equivalent temperature of Equation 10.10 along the uncracked system as shown in Figure 10.1:

Equation 10.10:

$$\Delta T_H = \frac{1,405,640 \cdot 286.5}{21,000 \cdot 1.037 \cdot 10^8 \cdot 10^{-5}} = 18.49°C$$

Figure 10.2 shows the longitudinal distribution of the maximum tensile stress in concrete after one uncracked and two cracked analyses. The fist cracked length was determined equal to 2·36 m. After the application of a 72 m cracked length, a second global analysis was conducted and concrete stresses were recalculated. The cracked length was reduced to 2·31 m. The third cracked analysis offered almost identical concrete stresses, and thus, convergence was achieved.

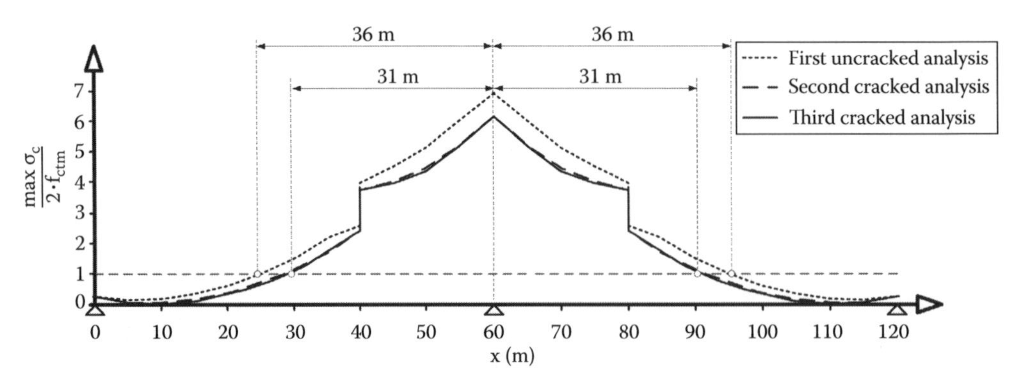

Figure 10.2 Longitudinal tensile concrete stress variation.

The cracked lengths calculated with the general method are quite robust against deviations of N_H and M_H. Therefore, high accuracy calculations are from practical point of view unnecessary. One can also observe that the cracked length is almost equal to 52% of the total length. This is a great difference compared to the simplified method of EN 1994-2 with which a 15% cracked length can be assumed. Obviously with a different construction sequence, the cracked lengths can be further reduced.

It is pointed out that the procedure presented in Example 10.1 is strictly applied for the determination of the cracked length and **not** for stress limitation.

Note: In case the designer wishes to take into account autogenous shrinkage, this can be calculated with Equation 6.23. The shrinkage strain can be added to ε_H and Equation 10.10 can be applied.

EXAMPLE 10.2

The stresses at SLS for the box-girder bridge in Example 9.4 are verified. In addition, the cracking of concrete at intermediate support is controlled. The cracked lengths of the simplified method are applied.

The effective width[s] of the concrete flanges is independent of the type of the limit state (ULS or SLS). Therefore, according to Example 9.4, there is no reduction due to shear lag. In contrast, the effective width[s] of the bottom flange is highly dependent on the type of the limit state especially, for the case of hogging bending. Indeed, taking a closer look in Example 9.4, one can see that the effective[s] width of the bottom flange for hogging bending at ULS is $2 \cdot 1.90 = 3.80$ m and at SLS $2 \cdot 1.24 = 2.48$ m (35% smaller!!!). The latter leads to an equivalent thickness of the bottom flange equal to 21.7 mm for hogging bending; the corresponding cross-sectional properties are summarized in Table 10.1. For sagging bending, the thickness remains practically unchanged and approximately equal to 35 mm so that the values in Table 9.17 are adopted.

Internal support

Stresses in reinforcing steel

The design bending moment is calculated according to the characteristic combination of actions (Equation 5.23) with leading variable of the traffic loads for long-term design.

Table 10.1 Cross-sectional properties for SLS with $t_{au,eq}=21.7$ mm for *hogging bending* for the cross section of Figure 9.16

Steel girder		*Fully cracked section*	
A_a	3323.6 cm²	$A_{2,sa}$	3823.6 cm²
z_{ao}	124.74 cm	$z_{ao,sa}$	106.47 cm
z_{au}	131.76 cm	$z_{au,sa}$	150.03 cm
I_a	0.37562·10⁸ cm⁴	$z_{so,sa}$	130.47 cm
		$z_{su,sa}$	112.47 cm
		$I_{2,sa}$	0.46049·10⁸ cm⁴

Bending moment acting on fully cracked section:

$$\min M_{2,Ed,ser,\infty} = -2,479.0 - 250.79 - 6,934.3 - 9,227.8 + 0.6 \cdot (-8,397.2)$$

$$= -23,930.21\,\text{kN-m}$$

The stress in the upper layer of reinforcement is equal to

$$\sigma_{so,\infty} = 2,393,021 \cdot \frac{130.47}{0.46049 \cdot 10^8/2} = 13.56\,\text{kN/cm}^2$$

The enhancement due to tension stiffening is calculated as follows:

Equation 10.4d:

$$\alpha_{st} = \frac{3823.6 \cdot 0.46049}{3323.6 \cdot 0.37562} = 1.41$$

Equation 10.4c:

$$\Delta\sigma_s = \frac{0.4 \cdot 0.32}{1.41 \cdot 0.015} = 6.05\,\text{kN/cm}^2$$

The total stress becomes (Equation 10.4a)

$$\sigma_{so,Ed,ser,\infty} = 13.56 + 6.05 = 19.61\,\text{kN/cm}^2 \le 0.8 \cdot \frac{50}{1.0} = 40\,\text{kN/cm}^2 \text{ (sufficient)}$$

Stresses in structural steel

The design bending moment is calculated according to the characteristic combination of actions (Equation 5.23) with leading variable of the traffic loads for long-term design.

Bending moment acting on pure steel cross section:

$$\min M_{a,Ed,ser,\infty} = -7,104.5 - 13,202.6 - 9,121.4 - 803.6 = -30,232.1\,\text{kN-m}$$

Bending moment acting on fully cracked section:

$$\min M_{2,Ed,ser,\infty} = -23,930.21\,\text{kN-m}$$

The coexisting shear and torsion are equal to

$$V_{Ed,ser,\infty} = -4297.74\,\text{kN}$$

$$M_{T,Ed,ser} = -2167.2\,\text{kN-m}$$

The tensile stress in the upper flange is equal to (Equation 10.1 and Table 10.1)

$$\sigma_{ao,Ed,ser,\infty} = 3,023,210 \cdot \frac{124.74}{0.37562 \cdot 10^8/2} + 2,393,021 \cdot \frac{106.47}{0.46049 \cdot 10^8/2} = 20.1 + 11.07$$

$$= 31.17\,\text{kN/cm}^2 < \frac{43}{1.0} = 43\ \text{kN/cm}^2\ \text{(sufficient)}$$

The compressive stress in the bottom flange is equal to (Equation 10.1 and Table 10.1)

$$\sigma_{au,Ed,ser,\infty} = 3,023,210 \cdot \frac{131.76}{0.37562 \cdot 10^8/2} + 2,393,021 \cdot \frac{150.03}{0.46049 \cdot 10^8/2} = 21.2 + 15.6$$

$$= 36.8\ \text{kN/cm}^2 < \frac{46}{1.0} = 46\ \text{kN/cm}^2\ \text{(sufficient)}$$

Strength reduction due to plate buckling was omitted because all $\rho = 0.9 > 0.5$ (see Example 9.4). Shear stresses in the bottom flange are negligible, so no verification for σ–τ interaction is necessary.

The maximum shear stress in the webs due to shear is equal to

$$\tau_{V,Ed,ser,\infty} = \frac{4297.74}{291.5 \cdot 2.5 \cdot \sin(59.08°)} = 6.88\ \text{kN/cm}^2$$

The shear stress due to torsion is equal to

$$\tau_{T,Ed,ser,\infty} = \frac{216,720}{2 \cdot 13.75 \cdot 10^4 \cdot 2.5} = 0.32\ \text{kN/cm}^2$$

The total shear stress in the web becomes (Equation 10.2)

$$\tau_{Ed,ser,\infty} = 6.88 + 0.32 = 7.2\ \text{kN/cm}^2 < \frac{44}{\sqrt{3} \cdot 1.0} = 25.4\ \text{kN/cm}^2\ \text{(sufficient)}$$

The maximum tensile stress in the web is equal to (Equation 10.1)

$$\sigma_{wo,Ed,ser,\infty} = 3,023,210 \cdot \frac{124.74 - 5}{0.37562 \cdot 10^8/2} + 2,393,021 \cdot \frac{106.47 - 5}{0.46049 \cdot 10^8/2} = 19.27 + 10.55$$

$$= 29.82\ \text{kN/cm}^2 < \frac{44}{1.0} = 44\ \text{kN/cm}^2\ \text{(sufficient)}$$

The von Mises stress is calculated from Equation 10.3:

$$\sigma_{v,wo,ser,\infty} = \sqrt{29.82^2 + 3 \cdot 7.2^2} = 32.32\ \text{kN/cm}^2 < \frac{44}{1.0} = 44\ \text{kN/cm}^2\ \text{(sufficient)}$$

The maximum compressive stress in the web is equal to (Equation 10.1)

$$\sigma_{wu,Ed,ser,\infty} = 3,023,210 \cdot \frac{131.76 - 1.5}{0.37562 \cdot 10^8/2} + 2,393,021 \cdot \frac{150.03 - 1.5}{0.46049 \cdot 10^8/2} = 20.97 + 15.43$$

$$= 36.4\ \text{kN/cm}^2 < \frac{44}{1.0} = 44\ \text{kN/cm}^2\ \text{(sufficient)}$$

The von Mises stress is calculated from Equation 10.3:

$$\sigma_{v,wu,ser,\infty} = \sqrt{36.4^2 + 3 \cdot 7.2^2} = 38.47\ \text{kN/cm}^2 < \frac{44}{1.0} = 44\ \text{kN/cm}^2\ \text{(sufficient)}$$

Stresses in concrete

The maximum bending moment acting on the composite section at span (x=27.5 m) for short-term loading is calculated for the quasi-permanent combination of actions (Equation 5.25):

$$\max M_{I,ser,0} = 7033.6 + 743 + 0.5 \cdot 3920 = 9736.6\ \text{kN-m}$$

The maximum compressive stress in concrete is calculated as follows (Equation 10.5b and Table 9.17):

$$\sigma_{co,Ed,ser,0} = 973,660 \cdot \frac{77.64}{6.18 \cdot 1.037 \cdot 10^8/2} = 0.24\ \text{kN/cm}^2 < 0.45 \cdot 3.5 = 1.58\ \text{kN/cm}^2$$

The aforementioned verification is sufficient and guarantees a linear creep evolution.

The maximum bending moment acting on the composite section at span for short-term loading is calculated for the characteristic combination of actions (Equation 5.23):

$$\max M_{I,ser,0} = 7,033.6 + 743 + 0.6 \cdot 3,920 + 10,293.7 = 20,422.3\ \text{kN-m}$$

The maximum compressive stress in concrete is calculated as follows:

$$\sigma_{co,Ed,glob,0} = 2,042,230 \cdot \frac{77.64}{6.18 \cdot 1.037 \cdot 10^8/2} = 0.49\ \text{kN/cm}^2$$

The previous stress refers to the global effects. An additional stress due to local bending of the deck slab from the wheel loads is calculated according to Section 7.5.4 and is found 0.24 kN/cm^2. The total compression stress becomes (Equation 10.5a)

$$\sigma_{co,Ed,ser,0} = 0.49 + 0.24 = 0.73\ \text{kN/cm}^2 < 0.6 \cdot 3.5 = 2.1\ \text{kN/cm}^2$$

The direct stresses in structural steel are calculated both for short- and long-term designs based on the loading sequence shown in Figures 9.18 through 9.23. The design stresses are based on the characteristic combination with leading variable of the traffic load group gr1a (see Table 4.7). Plate buckling is not taken into account since all plate buckling factors are higher than 0.5. The results are summarized in Tables 10.2 and 10.3. The third column refers to the second moment of area (I) on which the stress calculation is based. Shear stresses due to vertical shear and torsion are negligible and therefore are not presented. Finally, multiplying the stresses in the

Table 10.2 Direct stresses in structural steel at midspan for short-term design in kN/cm² (Half girder A, characteristic combination)

Loading from	*Figure*	*I ($10^8 \cdot cm^4$)*	*M (kN-m)*	σ_{ao}	σ_{au}	σ_{wo}	σ_{wu}
Pure steel	9.18	$I_a = 0.45454$	3,875.9	2.43	1.95	2.34	1.92
First concreting	9.18	$I_a = 0.45454$	9,187.2	5.75	4.61	5.55	4.55
Second concreting + superstructures	9.19	$I_{1,0} = 1.037$	7,033.6	0.65	2.83	0.58	2.81
Superstructures	9.20	$I_{1,0} = 1.037$	743.0	0.07	0.30	0.06	0.30
0.6·Thermal	9.22	$I_{1,0} = 1.037$	2,352.0	0.22	0.95	0.19	0.94
Traffic (gr1a)	9.23	$I_{1,0} = 1.037$	10,293.7	0.95	4.15	0.85	4.12
			$\Sigma\sigma$	10.06	14.79	9.57	14.64
			$f_{yk}/\gamma_{M,ser}$	39.0	42.0	34.5	34.5

Table 10.3 Direct stresses in structural steel at midspan for long-term design in kN/cm² (Half girder A, characteristic combination)

Loading from	*Figure*	*I ($10^8 \cdot cm^4$)*	*M (kN-m)*	σ_{ao}	σ_{au}	σ_{wo}	σ_{wu}
Pure steel	9.18	$I_a = 0.45454$	3,875.9	2.43	1.95	2.34	1.92
First concreting	9.18	$I_a = 0.45454$	9,187.2	5.75	4.61	5.55	4.55
Second concreting + superstructures	9.19	$I_{1,P} = 0.8107$	7,033.6	1.46	2.99	1.38	2.96
Superstructures	9.20	$I_{1,P} = 0.8261$	743.0	0.15	0.31	0.14	0.31
0.6·Thermal	9.22	$I_{1,0} = 1.037$	2,352.0	0.22	0.95	0.19	0.94
Traffic (gr1a)	9.23	$I_{1,0} = 1.037$	10,293.7	0.95	4.15	0.85	4.12
Secondary from second concreting	9.19	$I_{1,PT} = 0.8878$	−368.3	−0.06	−0.15	−0.06	−0.15
Secondary from superstructures	9.20	$I_{1,PT} = 0.9008$	−114.9	−0.02	−0.05	−0.02	−0.05
Shrinkage − primary effects	7.44	$I_{1,S} = 0.8493$	—	3.26	0.58	3.18	0.56
Shrinkage + secondary effects	9.21	$I_{1,PT} = 0.8493$	−3,178.2	−0.58	−1.34	−0.55	−1.32
			$\Sigma\sigma$	13.55	14.00	13.01	13.84
			$f_{yk}/\gamma_{M,ser}$	39.0	42.0	34.5	34.5

Note: For the short-term design, creep and shrinkage have been neglected (see also Example 9.4). In part of the literature, shrinkage is included in the calculation of the short-term steel stresses at spans at the time of traffic opening. For example, assuming that when the bridge is delivered to traffic, the shrinkage strain is 30% of its long-term value; thus, primary stresses due to shrinkage become.

$\sigma_{ao,S} = 0.3 \cdot 3.26 = 0.98$ kN/cm².

$\sigma_{au,S} = 0.3 \cdot 0.58 = 0.17$ kN/cm².

$\sigma_{wo,S} = 0.3 \cdot 3.18 = 0.95$ kN/cm².

$\sigma_{wu,S} = 0.3 \cdot 0.56 = 0.17$ kN/cm².

The short-term design stresses are then equal to

$\sigma_{ao} = 0.98 + 10.06 = 11.04$ kN/cm².

$\sigma_{au} = 0.17 + 14.79 = 14.96$ kN/cm².

$\sigma_{wo} = 0.95 + 9.57 = 10.52$ kN/cm².

$\sigma_{wu} = 0.17 + 14.64 = 14.81$ kN/cm².

following text with the partial factors for ULS leads to the design stresses of Example 9.4. Obviously, the stress calculation at ULS is more critical than that for the SLS.

The time when the bridge is delivered to traffic is rarely known during design. Therefore, a conservative assumption for the magnitude of the "short-term shrinkage" may be necessary; however, this should not be greater than 50% of its long-term value.

Minimum reinforcement at intermediate support

The age of concrete during cracking is unknown and $f_{ct,eff}$ was taken as equal to 3 MPa. From Table 6.10, the stress in reinforcing steel after cracking for a crack width equal to 0.3 mm for a reinforcing bar diameter of 20 mm is $\sigma_s = 220$ MPa.

The sectional area of the fully cracked section according to Table 10.1 is 3823.6 cm². Assuming that the reinforcement is located in the gravity center of the slab, the elastic neutral axis for the cracked section is equal to 106.47 + 30/2 = 121.47 cm. The cross-sectional area of the uncracked section **at internal support** for a short-term modular ratio $n_0 = 6.18$ is calculated as follows:

$$A_{I,0} = \frac{30 \cdot 1100}{6.18} + 3823.6 = 9163.41 \text{ cm}^2$$

The vertical distance between the gravity center of the uncracked deck slab and the gravity center of the composite section is

$$\bar{z}_{I,0} = \frac{3823.6 \cdot 121.47}{9163.41} = 50.69 \text{ cm}$$

The k_c factor then becomes

$$k_c = \frac{1}{1 + 30/(2 \cdot 50.69)} + 0.3 = 1.07 > 1.0 \rightarrow k_c = 1.0$$

From Equation 10.6:

$$A_{s,min} = \frac{0.3 \cdot 0.8 \cdot 0.9 \cdot 1 \cdot (30 \cdot 1100)}{22} = 324 \text{ cm}^2 < 500 \text{ cm}^2 \text{ (sufficient)}$$

According to the code, at least half of $A_{s,min}$ should be placed in the upper layer:

$$A_{so,min} = 0.5 \cdot 324 \text{ cm}^2 = 162 \text{ cm}^2 < 300 \text{ cm}^2 \text{ (sufficient)}$$

Limitation of crack width

The design bending moment acting on fully cracked section is calculated for the quasi-permanent combination of actions (see Equation 5.25 in Table 5.12). For example, the traffic group gr1a is selected as leading variable. Obviously, the long-term design offers the most unfavorable results:

$$\min M_{2,Ed,ser,\infty} = -2,479.0 - 250.79 - 6,934.3 + 0 \cdot (-9,227.8) + 0.5 \cdot (-8,397.2)$$

$$= -13,862.7 \text{ kN-m}$$

The stress in the upper layer of reinforcement is estimated by taking into account the tension stiffening effect as previously found:

$$\sigma_{so,Ed,ser,\infty} = 1,386,270 \cdot \frac{130.47}{0.46049 \cdot 10^8/2} + 6.05 = 13.91 \text{ kN/cm}^2$$

Note: Local bending moments at internal support are zero because min $M_{2,Ed,ser,\infty}$ arises when the wheel loads are located at the spans.

For this stress, Table 6.10 gives a maximum bar diameter of 32 mm. The maximum modified bar diameter is calculated as follows:

$$\max \Phi_s = \frac{3}{2.9} \cdot 32 = 33.1\,\text{mm}$$

The selected bar diameter was 20 mm < 33.1 mm, and therefore, crack control is satisfactory. Alternatively, Table 6.11 gives a maximum bar spacing above 300 mm.

10.4 WEB BREATHING

Slender webs may slightly buckle each time traffic passes over the bridge. These cyclic out-of-plane deformations, which are similar to the chests' movements during breathing, result in secondary bending stresses that may lead to fatigue cracks in the flange–web or in the web-to-transverse stiffener junction. Instead of calculating these deformations and the corresponding stresses and performing a fatigue analysis, detailing rules limiting the web slenderness are given, except for road bridges where plate buckling verifications are performed by the reduced stress method (see Section 8.4). For railway bridges and road bridges where plate buckling verifications are performed by the effective width method (Section 8.5), the following limitations must be examined:

$$\frac{b}{t_w} \leq 30 + 4.0 \cdot L \quad \text{but} \quad \frac{b}{t} \leq 300 \text{ for road bridges} \tag{10.11a}$$

$$\frac{b}{t_w} \leq 55 + 3.3 \cdot L \quad \text{but} \quad \frac{b}{t} \leq 250 \text{ for railway bridges} \tag{10.11b}$$

where
- b is the height of the web or the height of the largest subpanel for longitudinally stiffened webs
- t_w is the thickness of the web
- L is the span of the bridge in meters, but not less than 20 m

Otherwise, the following stress limitation should be applied:

$$\sqrt{\left(\frac{\sigma_{x,Ed,ser}}{k_\sigma \cdot \sigma_E}\right)^2 + \left(\frac{1.1 \cdot \tau_{Ed,ser}}{k_\tau \cdot \sigma_E}\right)^2} \leq 1.1 \tag{10.12}$$

where $\sigma_{x,Ed,ser}$ and $\tau_{Ed,ser}$ are the design stresses at the flange–web junction under the **frequent** SLS combination.

It is mentioned that the limitation of Equation 10.12 should be fulfilled both for short- and long-term design.

EXAMPLE 10.3

The cross sections of the bridge in Example 9.4 shall be checked in respect to web breathing.

The depth of the largest subpanel (Figure 9.16) is 835 mm.

Equation 10.11a:

$$\frac{835}{25} = 33.4 \leq 30 + 4.0 \cdot 60 = 270 \quad \text{and} \quad \frac{b}{t} \leq 300 \text{ (sufficient)}$$

The aforementioned verification is valid if longitudinal stiffeners are considered as adequately rigid. Otherwise, b = 2917 mm and a more conservative check follows:

Equation 10.11a:

$$\frac{2915}{25} = 116.6 \leq 30 + 4.0 \cdot 60 = 270 \quad \text{and} \quad \frac{b}{t} \leq 300 \text{ (sufficient)}$$

In case of rigid longitudinal stiffeners, excessive breathing would arise if

$$\frac{835}{t_w} > 270 \rightarrow t_w < 3.1 \text{ mm}$$

In the case of flexible longitudinal stiffeners for a web thickness, $t_w < 10.8$ mm.

It is obvious that for the majority of bridges, web breathing will not govern the design.

10.5 DEFLECTIONS

10.5.1 General

Deflections are to be determined with due consideration of construction stages and time-dependent effects. There exists no code-prescribed deflection limit for road and pedestrian bridges so that such limits must be agreed with the owner of the bridge. Problems arise mainly not due to deflections, δ, but due to slopes, φ, that is, variations of deflections. Figure 10.3 illustrates that the most adverse effects to the bridge and the vehicle are for vehicle A that enters the bridge, where deflections are small and slopes large, and not for vehicle B at the middle of the bridge, where deflections are large but slopes small. Figure 10.3 shows that the largest slopes occur at common supports of consecutive single-span bridges. From this point, continuous systems are preferable. Large slopes give rise to impacts and introduce dynamic effects in the bridge and the vehicles affecting the dynamic behavior of the system, especially at high speeds. These effects may be critical for high-speed train lines. Large differential deformations both in longitudinal and transverse direction may adversely affect the drainage of water from the bridge. Finally, large slope variations lead to high strains and may result in cracks in asphaltic surfaces.

Excessive deformations are controlled by enhancing the stiffness and by precambering. The value of precamber **for highway bridges** may be selected such that the deflections of the deck are within ±1/2 of the deflections due to frequent traffic loads.

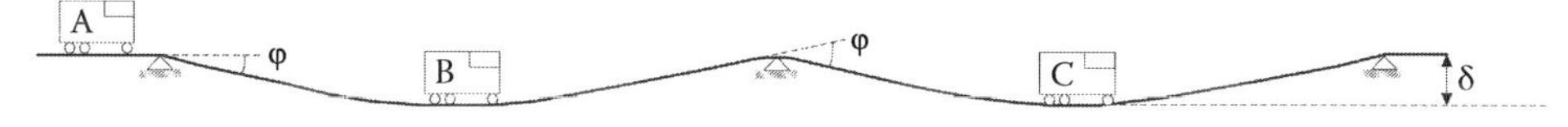

Figure 10.3 Deformations and slopes of consecutive single-span bridges.

Table 10.4 Maximum vertical deflections to avoid excessive track maintenance

Speed range	*Permissible deflection*
V < 80 km/h	$\delta_{stat} \leq L/600$
80 ≤ V ≤ 200 km/h	$\delta_{stat} \leq L/(15 \cdot V - 400)$
V > 200 km/h	$\delta_{stat} \leq L/2600$ and $\delta_{dyn} \leq$ value from authorities

Notes:

δ_{stat} is the maximum vertical deflection measured along any track due to LM 71 (SW/0) (see Section 4.6.2).

δ_{dyn} is the maximum vertical deflection measured along any track due to high-speed trains (i.e., real trains defined by the authorities and/or the universal dynamic trains HSLM designed for international lines found in EN 1991-2).

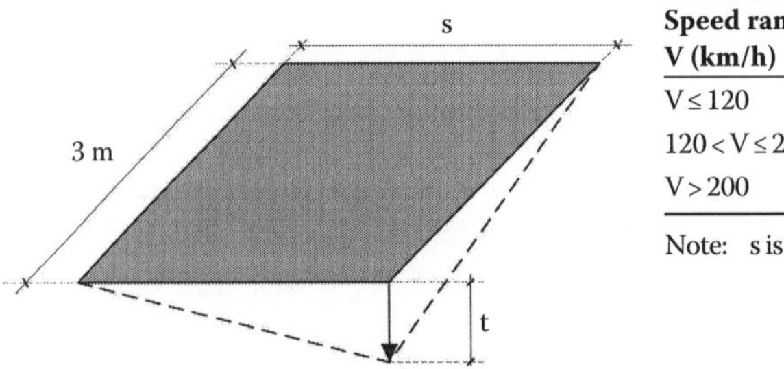

Speed range V (km/h)	Maximum twist (mm/3 m)
V ≤ 120	4.5
120 < V ≤ 200	3.0
V > 200	1.5

Note: s is the track gauge (=1.35 m).

Figure 10.4 Definition of deck twist and recommended limiting values.

In the case of **railway bridges,** the simplified limits for vertical deflections in Table 10.4 are recommended [10.1]. If these limits are respected, then the angular rotations φ of Figure 10.3 need not be investigated.

The twist of the bridge deck is very important for rail traffic safety. Therefore, it should be checked considering the characteristic values of LM 71, SW/0, or SW/2 multiplied by the weight factor α in Section 4.6.2 and dynamic factor Φ (Equations 4.6 and 4.7). Centrifugal forces according to Section 4.6.4 must be considered as well. For speeds higher than 200 km/h, the high speed load models HSLM should be also included. The maximum recommended values for the deck twist t are given in Figure 10.4. One can see that the twist deck refers to a length of 3m.

It is pointed out that the limiting values in Figure 10.4 refer to the maximum deck twist that is caused by the corresponding load models. In case of additional deck twists, for example, due to temperature effects, the total deck twist should be limited to 7.5 mm/3 m.

EXAMPLE 10.4

The precamber of the bridge of Example 9.4 is to be determined. The cracked lengths of the simplified method will be used.

The concreting sequence, the long-term effects, and the shear lag effect (for SLS) are taken into account. In Table 10.5, the deflections for short-term design are summarized for three different positions along the bridge; x = 15, 30, and 45 m. One can see that the first two loading cases act on the pure steel system ($I = I_a$ for all cross sections) (see Figure 9.18). In the sagging

Table 10.5 Short-term deflections, and total deflections for the frequent combination

Loading from	Second moment area ($\cdot 10^8$ cm^4) x=0–40 m 80–120 m	40–51 m 69–80 m	2·15%L 51–69 m	Deflection in mm x=15	30	45
Pure steel	$I_a = 0.45454$	$I_a = 0.45454$	$I_a = 0.3756$	21.2	25.5	12.3
First concreting	$I_a = 0.45454$	$I_a = 0.45454$	$I_a = 0.3756$	50.0	57.8	25.5
Second concreting + superstructures	$I_{1,0} = 1.037$	$I_a = 0.45454$	$I_a = 0.3756$	12.2	15.3	8.2
Superstructures	$I_{1,0} = 1.037$	$I_{1,0} = 1.037$	$I_{2,sa} = 0.46049$	1.7	2.7	1.9
0.5·Thermal	$I_{1,0} = 1.037$	$I_{1,0} = 1.037$	$I_{2,sa} = 0.46049$	6.3	6.8	3.5
0.75·TS	$I_{1,0} = 1.037$	$I_{1,0} = 1.037$	$I_{2,sa} = 0.46049$	4.6	6.3	4.1
0.4·(uniformly distributed load (UDL) + q_{fk})	$I_{1,0} = 1.037$	$I_{1,0} = 1.037$	$I_{2,sa} = 0.46049$	6.9	9.3	6.2
			$\Sigma\delta_0$	102.9	123.7	61.7

moment areas, the second moment of area is taken from Table 9.17. In the hogging moment area, the second moment of area is reduced due to the shear lag effect on the bottom flange and is taken from Table 10.1. In the second concreting phase (Figure 9.19), the composite region of the bridge has $I = I_{1,0}$, while the remaining one $I = I_a$. The final part of the superstructures acts on a composite system (see Figure 9.20). Concrete in the hogging moment areas (2·15%·60 = 18 m) is considered as fully cracked and the second moment of area is equal to $I_{2,sa}$ of Table 10.1. Thermal effects and traffic loads act on the same composite system but multiplied with the frequent combination factors.

In Table 10.6, the first two rows refer to the pure steel system and remain unchanged. In the mixed system of Figure 9.19 (third row), the composite regions have a reduced second moment of area due to creep equal to $I = I_{1,P}$ calculated with a creep factor $\varphi(7, \infty)$. The fourth row refers to the system of Figure 9.20. The age of concrete of the edge composite regions has an age of 14 days at the time of loading, and the long-term inertia moment $I_{1,P}$ is calculated with a creep factor $\varphi(14, \infty)$. The remaining uncracked regions (40–51 m and 69–80 m) are assumed to have an age for concrete equal to 7 days. The cracked region 51–69 m is accounted for an inertia

Table 10.6 Long-term deflections and total deflections for the frequent combination

Loading from	Second moment area ($\cdot 10^8$ cm^4) x = 0–40 m 80–120 m	40–51 m 69–80 m	2·15%L 51–69 m	Deflection in mm x=15	30	45
Pure steel	$I_a = 0.45454$	$I_a = 0.45454$	$I_a = 0.3756$	21.2	25.5	12.3
First concreting	$I_a = 0.45454$	$I_a = 0.45454$	$I_a = 0.3756$	50.0	57.8	25.5
Second concreting + superstructures	$I_{1,P} = 0.8107$	$I_a = 0.45454$	$I_a = 0.3756$	19.9	26.3	15.7
Superstructures	$I_{1,P} = 0.8261$	$I_{1,P} = 0.8107$	$I_{2,sa} = 0.46049$	2.0	3.0	2.1
0.5·Thermal	$I_{1,0} = 1.037$	$I_{1,0} = 1.037$	$I_{2,sa} = 0.46049$	6.3	6.8	3.5
0.75·TS	$I_{1,0} = 1.037$	$I_{1,0} = 1.037$	$I_{2,sa} = 0.46049$	4.6	6.3	4.1
0.4·(UDL + q_{fk})	$I_{1,0} = 1.037$	$I_{1,0} = 1.037$	$I_{2,sa} = 0.46049$	6.9	9.3	6.2
Shrinkage—primary effects	$I_{1,S} = 0.8493$	$I_{1,S} = 0.8493$	$I_{2,sa} = 0.46049$	25.8	32.4	19.7
			$\Sigma\delta_\infty$	115.5	141.9	76.8

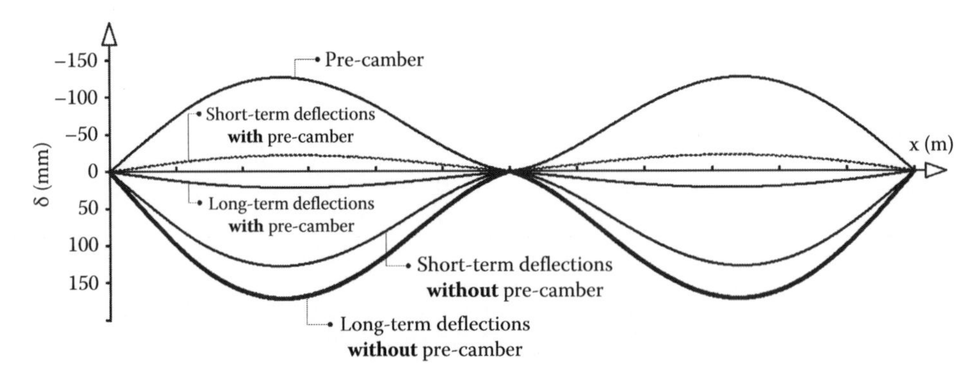

Figure 10.5 Short- and long-term deflections (δ) due to frequent traffic loads and selected precamber.

moment equal to $I_{2,sa}$ from Table 10.1. The short-term actions temperature and traffic are calculated as in Table 10.5 with their frequent values. Shrinkage is calculated by applying the $I_{I,S}$ second moment of area on the uncracked regions and the equivalent temperature of Figure 9.21. One can observe that the maximum deflection due to shrinkage is 22.8% (!!!) of the total long-term deflection.

Precamber is selected to be the average of deflections due to permanent loading for short- and long-term design and is shown in Figure 10.5. This leads to negative values of short-term deflections (min $\delta_0 = -21.9$ mm), which means that before creep and shrinkage are fully developed, the deck will be lightly curved toward the upside. The deck will deflect due to long-term effects and frequent traffic about 22 mm at the midspan.

Note: For simplification, the cracked lengths of the simplified method were used. Normally the cracked lengths of the general method in Example 10.1 should be adopted. These would lead to higher values of deformations and precamber.

10.5.2 Filler-beam decks

For the calculation of deflections of filler-beam deck bridges, cracking of concrete has to be taken into account carefully along the length of the bridge. EN 1994-2 states that the average second moment of area I_{eff} can be applied:

$$I_{eff} = \frac{I_1 + I_2}{2} \tag{10.13}$$

where

I_1 is the second moment of area of the uncracked composite section
I_2 is the second moment of area of the cracked composite section

For the estimation of I_2, the code allows an additional simplification by taking as neutral axis the plastic neutral axis of the composite cross section. The neutral axis for the fully cracked section (state 2) is however an elastic one and is calculated from the equilibrium of the internal forces [10.9]. This leads to the recommended procedure presented in Figure 10.6. The second moment of area of the fully cracked cross section then becomes

$$I_2 = \frac{b_c \cdot c^3}{3 \cdot n_0} + I_{sa} + A_{sa} \cdot \left(\overline{z}_2 - \overline{z}_{sa}\right)^2 \tag{10.14}$$

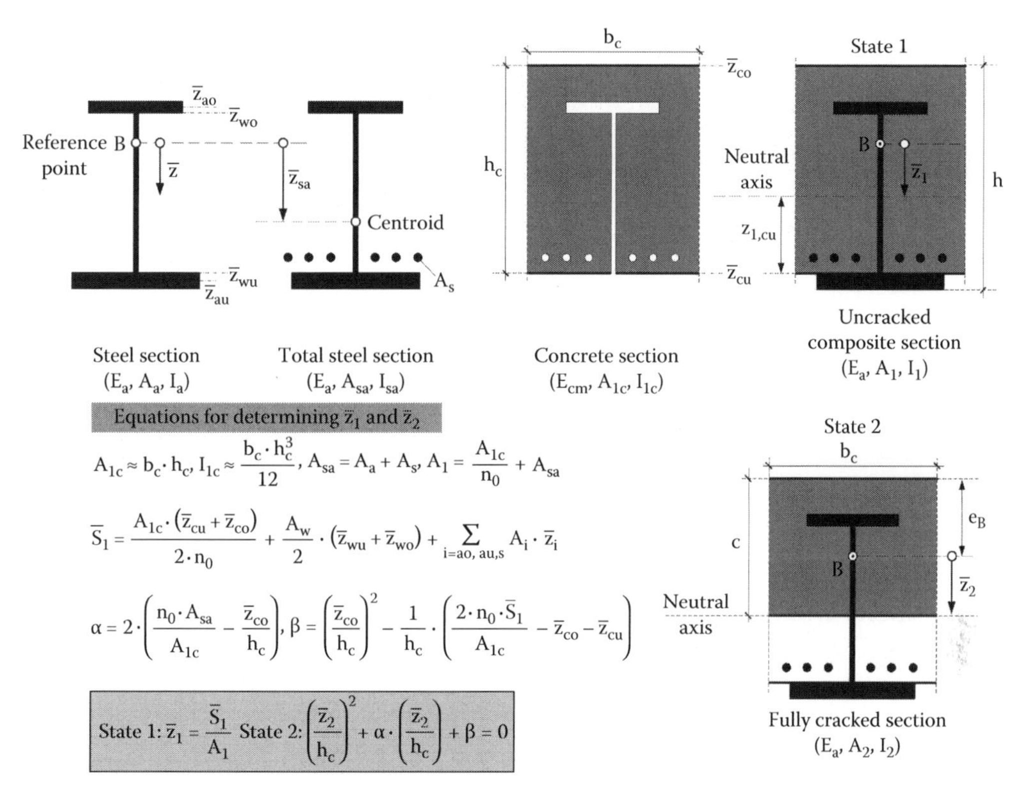

Figure 10.6 Determination of the elastic neutral axis for the cracked cross section (state 2).

where

$\bar{z}_2$ is the coordinate of the neutral axis of the fully cracked cross section calculated from the second-order equation in Figure 10.6

A_{sa} is the cross-sectional area of the total steel cross section (structural + reinforcing steel)

I_{sa} is the inertia moment of the total steel cross section (structural + reinforcing steel)

$\bar{z}_{sa}$ is the vertical distance between the centroid of the total steel cross section and the reference point B

n_0 is the short-term modular ratio from Table 6.4

c is the depth of the compression zone ($=e_B + \bar{z}_2$)

Notation

The reference point B can be arbitrary but inside the web's depth.

In Figure 10.7, one can see that the stiffness of a filler beam is not constant but dependent on the acting bending moment. Instead of using the average inertia moment of Equation 10.13, deflections can be calculated by integrating the curvatures of Equation 6.34a. For the uncracked regions, the curvatures are

$$\kappa_1 = \frac{M_{Ed,ser}}{E_a \cdot I_1} \tag{10.15}$$

The uncracked regions are those with $M_{Ed,ser} \leq M_{cr} = f_{ctm} \cdot n_0 \cdot I_1 / z_{1,cu}$ where f_{ctm} is the mean tensile strength of concrete in Table 6.1.

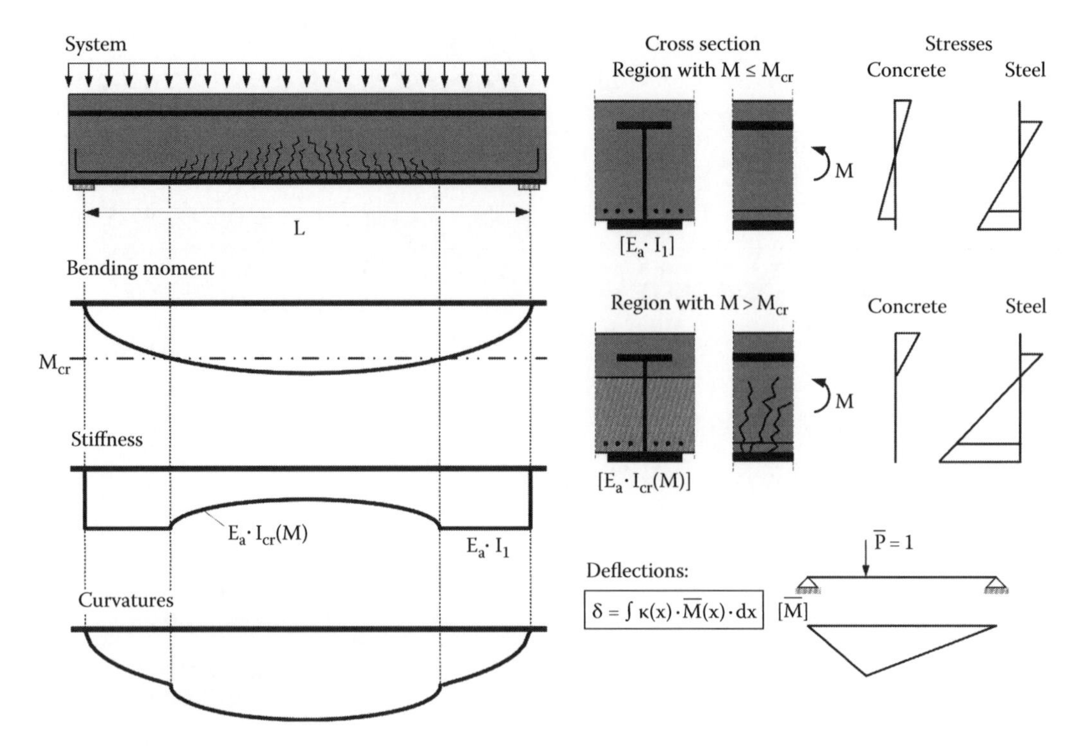

Figure 10.7 Flexural rigidity distribution of a simply supported filler-beam deck bridge. (From Iliopoulos, A., *Stahlbau, Ernst und Sohn*, 78(8), 555, 2009.)

For the fully cracked sections in regions with $M_{Ed,ser} > M_{cr}$, the curvatures along the length of the bridge are

$$\kappa_2 = \frac{M_{Ed,ser}}{E_a \cdot I_2} \tag{10.16}$$

The final values of the curvatures are calculated with Equations 6.34a and b. Deflections can be calculated by integrating the curvatures along the length of the bridge. The integration of the curvatures can be avoided by applying the following equivalent temperature on the uncracked system:

$$\kappa - \kappa_1 = \frac{\alpha_t \cdot \Delta T_{cr}}{h} \rightarrow \zeta \cdot (\kappa_2 - \kappa_1) = \frac{\alpha_t \cdot \Delta T_{cr}}{h}$$

$$\rightarrow \Delta T_{cr}(x) = \frac{\zeta(x) \cdot (\kappa_2 - \kappa_1) \cdot h}{\alpha_t} \tag{10.17}$$

where

κ_1, κ_2 are the curvatures for the uncracked and the fully cracked composite section according to Equations 10.15 and 10.16

h is the total depth of the composite cross section

α_t is the coefficient of thermal expansion for steel

$\zeta(x)$ is the factor calculated from Equation 6.34b following the longitudinal variation of the bending moment $M_{Ed,ser}$

The application of equivalent temperature difference ΔT_{cr} on the uncracked system gives the enhancement of the deflections due to cracking.

10.6 VIBRATIONS

Vibrations on bridges are mainly caused by traffic or wind. For large-span bridges, usually over 200 m, the aerodynamic stability in respect to vortex shedding, buffeting, fluttering, and galloping must be examined by calculation, wind tunnel tests, or combinations of them. For cable-stayed bridges, vibrations may also concern the cables, especially under wind and rain conditions.

Concerning traffic loading, vibrations of the structure including its interaction with traffic are usually examined in railway bridges during passing of high-speed trains. EN 1990-A2 [10.2] asks for a minimum value of the fundamental frequency f_{b0} of the deck in transverse direction with a recommended value 1.2 Hz. Furthermore, it requires a limitation of the vertical accelerations b_v in the wagon for comfort purposes with recommended values between 1.0 and 2.0 m/s^2 depending on the level of comfort. These accelerations are determined by dynamic analysis including the train–bridge interaction. Alternatively, the vertical bending displacements due to LM71 including the weight factor α and the dynamic factor Φ must be limited to certain values depending on the span length, the train velocity, the number of spans, and the type of static system (single span, continuous). For bridges with spans up to 120 m, a chart is given in [10.2] providing the maximal allowed displacements for bridges with 3 or more consecutive spans. These values may be appropriately adapted to other systems.

In Figure 10.8, two vibration modes due to vertical flexure and torsion of simply supported plate-girder bridge are demonstrated. Torsional vibration modes have a negative influence (especially due to fatigue) and their contribution to the dynamic response of the bridge should be limited.

Most susceptible to vertical and horizontal vibrations are very light pedestrian bridges. Vibrations caused by individuals or groups of people crossing the bridge may lead to discomfort up to panic of persons. For detailed information, reference is made to the relevant specifications and the literature.

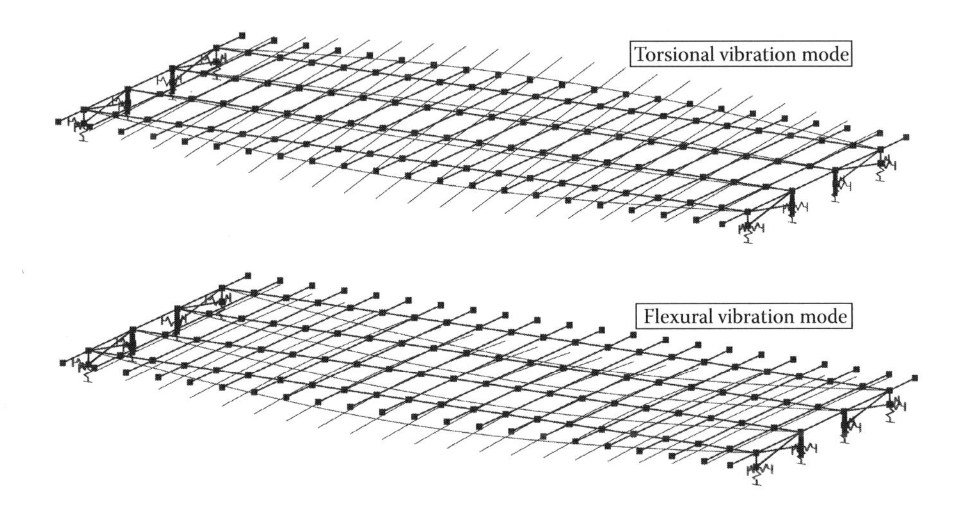

Figure 10.8 Vibration modes of a simply supported plate-girder bridge.

REFERENCES

[10.1] Calgaro, J.-A., Tschumi, M., Gulvanessian, H.: *Designer's Guide to Eurocode 1: Actions on Bridges*. Thomas Telford, London, U.K., 2010.

[10.2] EN 1990, CEN (European Committee for Standardization): Basis of structural design Annexure 2: Application on bridges (normative), Brussels, Belgium, 2004.

[10.3] EN 1993-1-1, CEN (European Committee for Standardization): Design of steel structures, Part 1–1: General rules and rules for buildings, Brussels, Belgium, 2005.

[10.4] EN 1993-1-5, CEN (European Committee for Standardization): Design of steel structures, Part 1–5: Plated structural elements, Brussels, Belgium, 2006.

[10.5] EN 1993-2, CEN (European Committee for Standardization): Design of steel structures–Part 2: Steel bridges, Brussels, Belgium, 2006.

[10.6] EN 1992-1-1, CEN (European Committee for Standardization): Design of concrete structures, Part 1-1: General rules and rules for buildings, Brussels, Belgium, 2004.

[10.7] EN 1992-2, CEN (European Committee for Standardization): Design of concrete structures–Part 2: Concrete bridges–Design and detailing rules, Brussels, Belgium, 2005.

[10.8] EN1994-2, CEN (European Committee for Standardization): Design of composite steel and concrete structures. Part 2: Rules for bridges, Brussels, Belgium, 2005.

[10.9] Iliopoulos, A.: A proposal for the calculation of deformations in concrete–encased steel beams. *Stahlbau*, 78(8), 555–561, 2009 (in German).

[10.10] Svensson, H.: *Schrägkabelbrücken*, Ernst und Sohn, Berlin, Germany, 2011.

[10.11] Vayas, I.: *Verbundkonstruktionen auf der Grundlage von Eurocode 4*, Ernst & Sohn, Berlin, Germany, 1999.

Chapter 11

Fatigue

11.1 GENERAL

Fatigue is a process in which damage is accumulated in the material undergoing fluctuating loading. Damage takes the form of cracks in the material that develop slowly at early stages of loading and accelerate very quickly toward the end (Figure 11.1). Microcracks start to develop at points of stress concentration at nominal stresses that may be well below the elastic limit. These cracks grow slowly under continuing fatigue loading but start to accelerate when the local stresses near the crack front increase due to cross-sectional reduction. Fracture occurs when the remaining section area is not able to support the applied load. Evidently, tension stresses are more significant than compression stresses. Fatigue is a local phenomenon that takes place at regions of stress concentration such as rapid changes of cross sections, at section reductions due to bolted connections or in welding regions, where the material undergoes metallurgic changes (see also Section 11.11).

Road and railway bridges are subjected to fatigue traffic loading, while long-span or very flexible bridges may be subjected to fatigue wind loading. Short to medium span bridges are more susceptible to fatigue, since the ratio between traffic to permanent loading is large compared to long-span bridges.

11.2 FATIGUE RESISTANCE TO CONSTANT AMPLITUDE LOADING

The fatigue resistance does not depend on the yield or tensile strength of the material. For constant amplitude cycles (Figure 11.2), it depends on the number of loading cycles, N, and the applied stress range that is given by

$$\Delta\sigma = \sigma_{max} - \sigma_{min} \tag{11.1}$$

In nonwelded details or stress-relieved details, the influence of compression stresses is not as large as for tension stresses. The stress range for such details is calculated from

$$\Delta\sigma = \sigma_{max} - 0.6 \cdot \sigma_{min} = \sigma_{max} + 0.6 \cdot |\sigma_{min}| \tag{11.2}$$

The fatigue resistance is accordingly expressed by the so-called S-N (or *Wöhler*) curves that relate the applied stress range $\Delta\sigma$ or $\Delta\tau$ with the number of cycles N (Figure 11.3). The nominal fatigue resistance is the stress range $\Delta\sigma_c$ or $\Delta\tau_c$ for two million cycles ($N_C = 2 \cdot 10^6$).

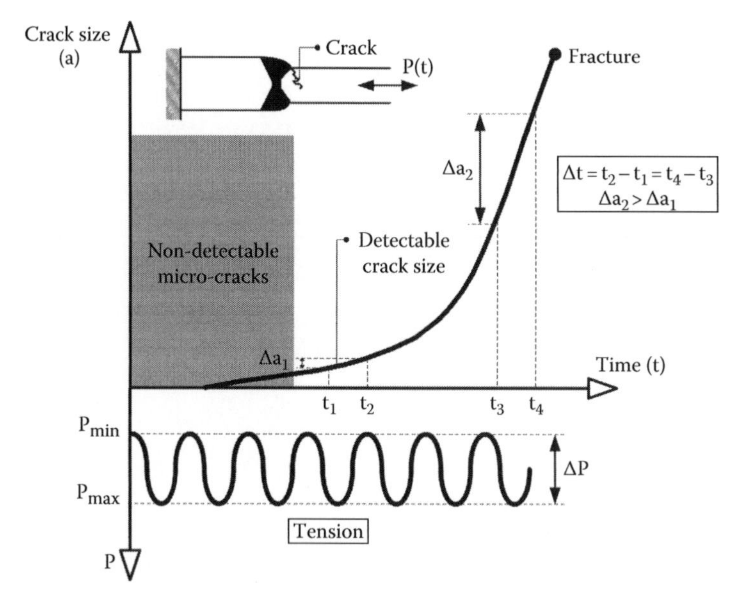

Figure 11.1 Development of fatigue crack.

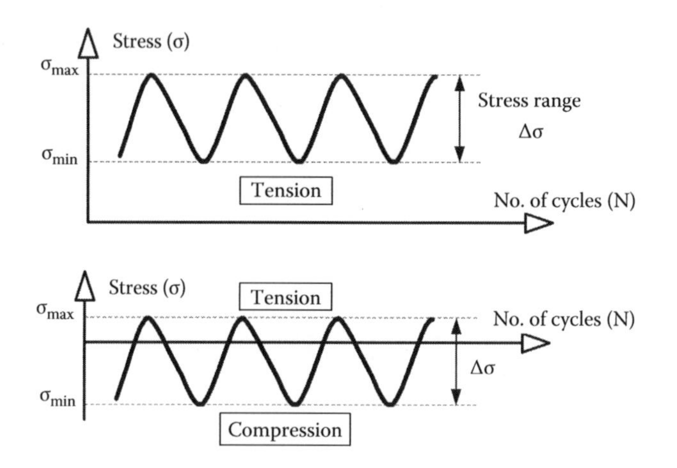

Figure 11.2 Constant amplitude stress history.

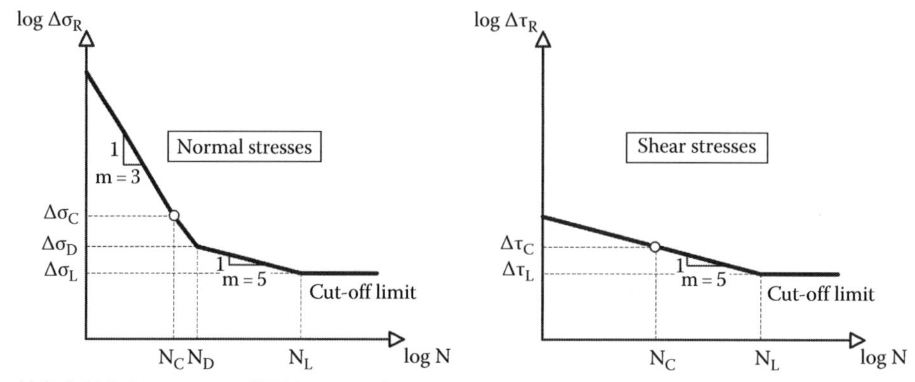

Figure 11.3 S-N fatigue curves (*Wöhler* curves).

Figure 11.3 indicates that in logarithmic scale, the number of cycles is linearly related to the applied stress range and is expressed by the following relations:

$$\log N = \log \alpha - m \cdot \log \Delta\sigma_R \tag{11.3}$$

$$\log N = \log \alpha - m \cdot \log \Delta\tau_R \tag{11.4}$$

where α and m are constants.

Numerous experimental investigations [11.14] showed that the inclination of the straight line, m, for construction details has certain fixed values. For direct stress ranges, the inclination is $m = 3$ up to 5 million cycles, $N_D = 5 \cdot 10^6$. It then changes to $m = 5$ for higher number of cycles and up to 100 million cycles, $N_L = 10^8$. For even larger number of cycles, there exists a *cut-off limit* for the fatigue resistance. That means that if the applied stress range is lower than $\Delta\sigma_L$, no fatigue damage occurs, independent of the number of cycles. Similar conditions exist for shear stresses. The inclination here is $m = 5$ up to the cutoff limit of 100 million cycles.

In regions of constant inclination, Equation 11.3, and similarly 11.4, may be rewritten as follows:

$$\left(\frac{\Delta\sigma_1}{\Delta\sigma_2}\right)^m = \frac{N_2}{N_1} \tag{11.5}$$

In the region $m = 3$, it is

$$\left(\frac{\Delta\sigma_D}{\Delta\sigma_C}\right)^3 = \frac{N_C}{N_D} = \frac{2 \cdot 10^6}{5 \cdot 10^6} \rightarrow \Delta\sigma_D = 0.737 \cdot \Delta\sigma_C \tag{11.6}$$

while in the region $m = 5$, it is

$$\left(\frac{\Delta\sigma_L}{\Delta\sigma_D}\right)^5 = \frac{N_D}{N_L} = \frac{5 \cdot 10^6}{1 \cdot 10^8} \rightarrow \Delta\sigma_L = 0.549 \cdot \Delta\sigma_D = 0.405 \cdot \Delta\sigma_C \tag{11.7}$$

The corresponding conditions for **shear stresses** may be written as

$$\left(\frac{\Delta\tau_L}{\Delta\tau_C}\right)^5 = \frac{N_C}{N_L} = \frac{2 \cdot 10^6}{1 \cdot 10^8} \rightarrow \Delta\tau_L = 0.457 \cdot \Delta\tau_C \tag{11.8}$$

The S-N curve gives the fatigue resistance as the number of cycles to failure N_i for a certain applied stress range $\Delta\sigma_i$. If the stress range $\Delta\sigma_i$ is applied at a lower number of cycles, $n_i < N_i$, no fatigue failure occurs but some damage is done. This damage may be calculated from

$$D = \frac{n_i}{N_i} \tag{11.9}$$

D has the obvious limits

$$0 \leq D \leq 1 \tag{11.10}$$

where D = 0 indicates no damage, while D = 1 indicates failure.

11.3 FATIGUE RESISTANCE TO VARIABLE AMPLITUDE LOADING

The applied loading is generally of variable amplitude. Such loading histories as indicatively shown in Figure 11.4 pose a problem in defining the number and amplitude of the cycles and are converted for fatigue analysis purposes to constant amplitude loading. This may be done by a cycle counting method, the *reservoir method* being the most used one.

The basis of the reservoir method is shown in Figure 11.5 using the stress history of Figure 11.4. The line of the stress history is extended so that the peak stress levels repeat themselves and the regions between peaks are filled with water to form a reservoir. Subsequently a tap is opened at the lowest trough (T_1) to drain the reservoir. This corresponds to one cycle with stress range $\Delta\sigma_1$. The remaining level of water is now lowered

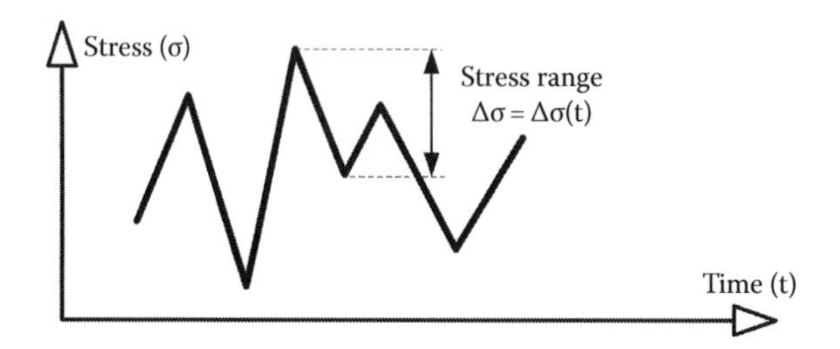

Figure 11.4 Variable amplitude stress history.

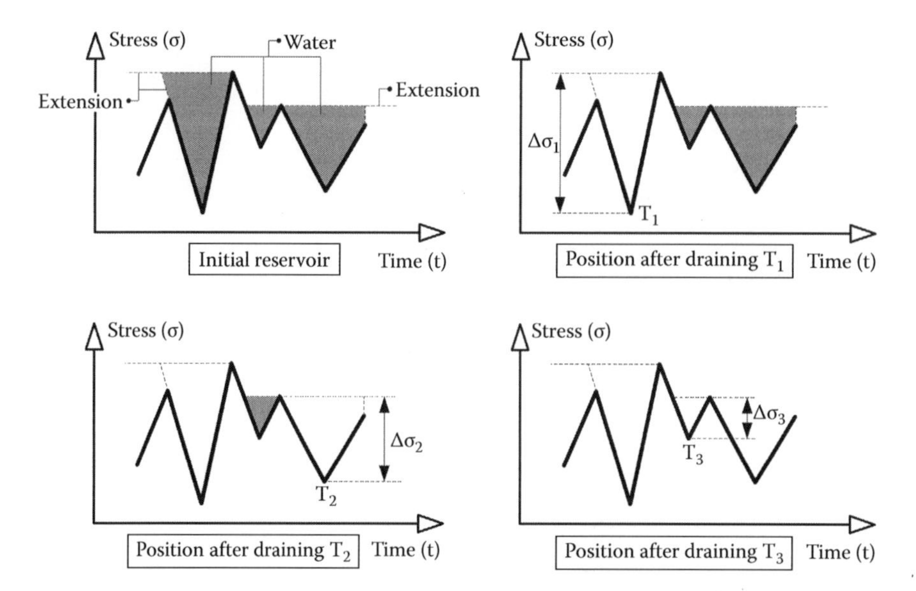

Figure 11.5 Application of the *reservoir method*.

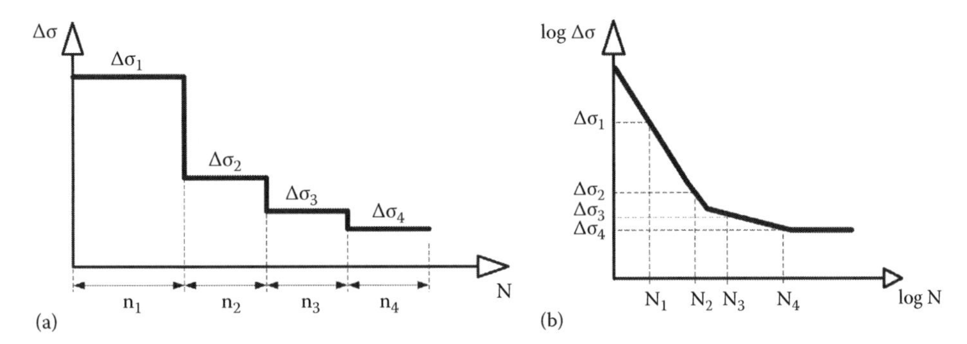

Figure 11.6 (a) Stress histogram for variable amplitude loading and (b) number of cycles to failure versus stress ranges.

to the next highest peak. A tap is now opened to the next lowest trough (T_2) providing one cycle with stress range $\Delta\sigma_2$, and the water is allowed to drain. The procedure is repeated, that is, opening the tap from the lowest remaining trough and counting the stress ranges, until the reservoir is empty. Every stress range has n = 1 cycles unless during opening a tap two or more equal stress ranges appear. If this is not the case, stress ranges may be grouped together, for example, every 10 MPa, so that more cycles correspond to one group.

The next step is to define the stress histogram that gives the stress ranges $\Delta\sigma_i$ with the corresponding number of cycles n_i (Figure 11.6a).

The fatigue damage done by each stress range is expressed by Equation 11.9. The total damage produced by the entire stress histogram is determined in accordance with the *Palmgren–Miner* law of linear damage [11.10], [11.11], as the sum of the damage done by each individual stress ranges independently. The law of linear damage accumulation is written as

$$D = \frac{n_1}{N_1} + \frac{n_2}{N_2} + \cdots + \frac{n_k}{N_k} = \sum_{i=1}^{k} \frac{n_i}{N_i} \tag{11.11}$$

where (Figure 11.6)
n_i is the number of cycles for stress ranges $\Delta\sigma_i$
N_i is the number of cycles to failure for stress ranges $\Delta\sigma_i$

Stress ranges smaller than $\Delta\sigma_L$ are not included in the summation.

11.4 DETAIL CATEGORIES

As outlined in Section 11.1, fatigue is a local phenomenon that depends on stress concentrations and therefore on the shape of the construction detail. Accordingly, the fatigue resistance depends on the *detail category*. Each detail category is associated with a figure that gives the fatigue resistance at $N_C = 2 \cdot 10^6$ (2 million cycles). For example, detail category 71 means that the fatigue resistance at 2 million cycles is 71 MPa. The relevant S-N curves for direct and shear stress ranges proposed in EN 1993-1-9 [11.4] are illustrated in Figures 11.7 and 11.8. The fatigue resistance of shear connectors is 90 MPa with a slope m = 8. The fatigue resistance for reinforcement (straight or bent bars) in tension is 162.5 MPa at

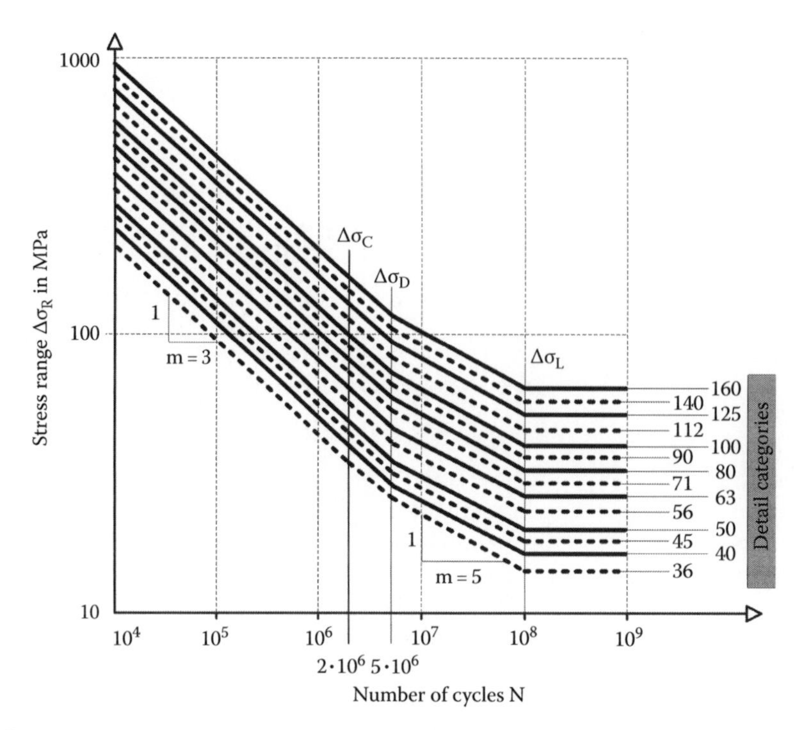

Figure 11.7 Fatigue resistance curves of steel for direct stress ranges.

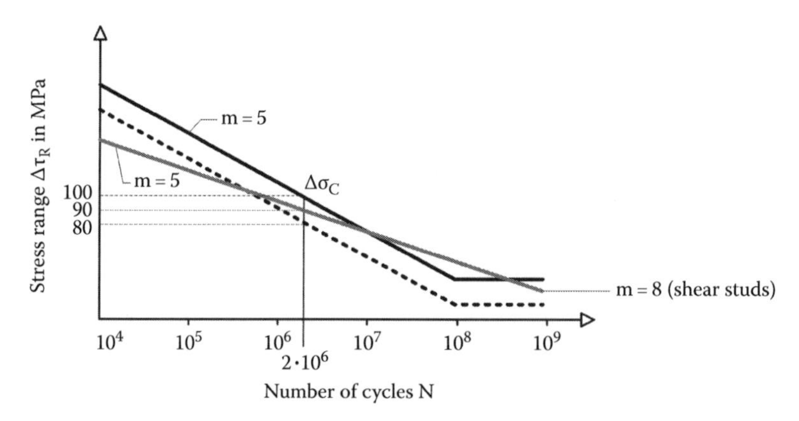

Figure 11.8 Fatigue resistance curves for steel and shear studs for shear stress ranges.

1 million cycles with a slope m = 9. The relevant S-N curve is also shown in Figure 11.8. It may be seen that there exist only two detail categories for shear stresses, namely, 100 and 80, while many more for direct stresses with the lowest category being 36. This, together with the steeper inclination of the fatigue curves, indicates that direct stresses are more detrimental to fatigue damage compared with shear stresses.

The main detail categories for common bridge applications according to EN 1993-1-9 are given in Figures 11.9 through 11.12. More details are found in the code. For transverse butt welds, there is an important size effect for plate thicknesses t > 25 mm, expressed by the reduction factor $k_s = (25/t)^{0.2}$. In this way, gross stress concentrations due to abrupt changes and hot spots that are not included in the basic detail categories are taken into account.

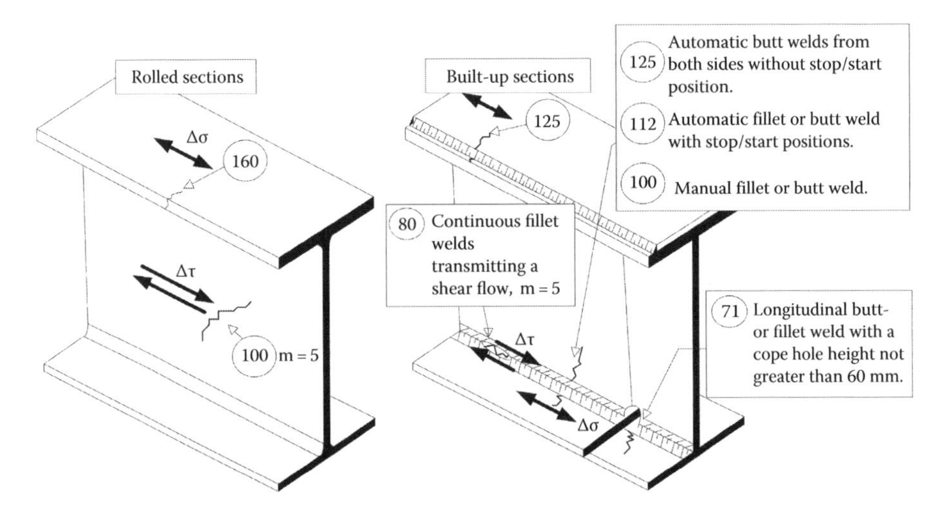

Figure 11.9 Detail categories for rolled and built-up welded sections.

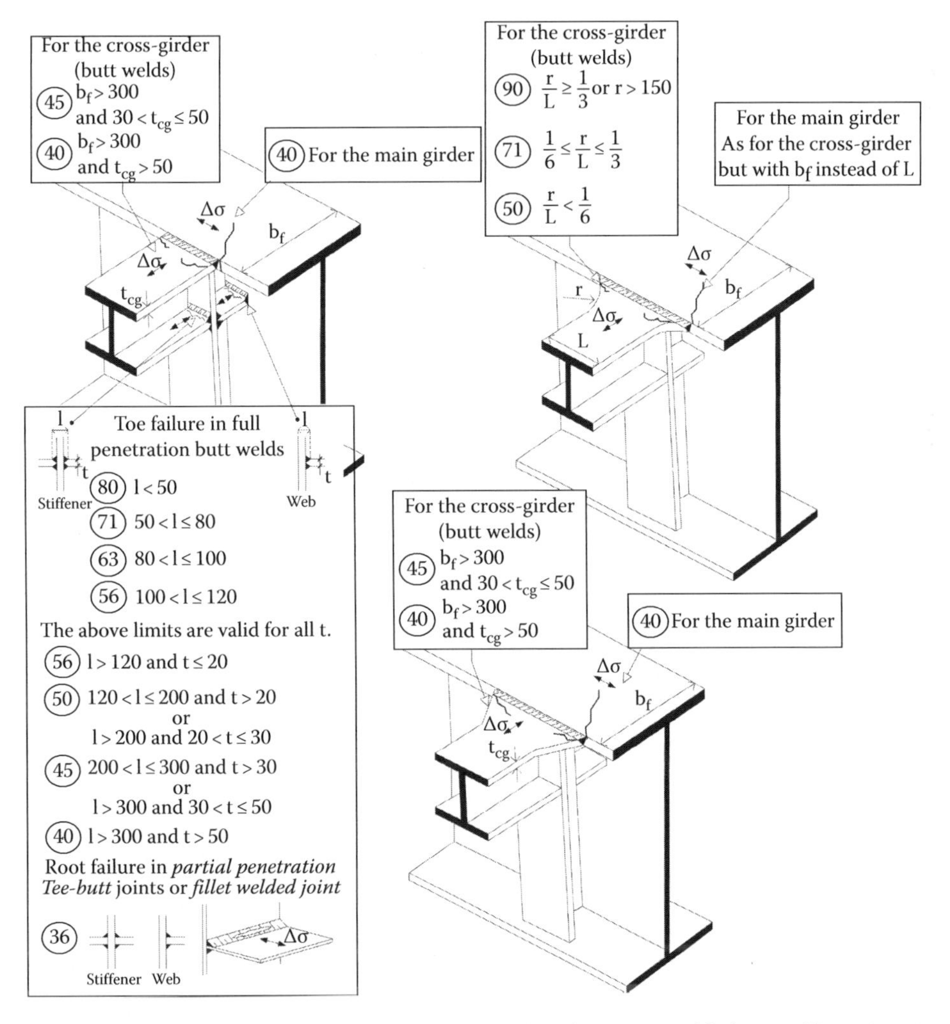

Figure 11.10 Detail categories for transverse butt welds and load-carrying welded joints (dimensions in mm).

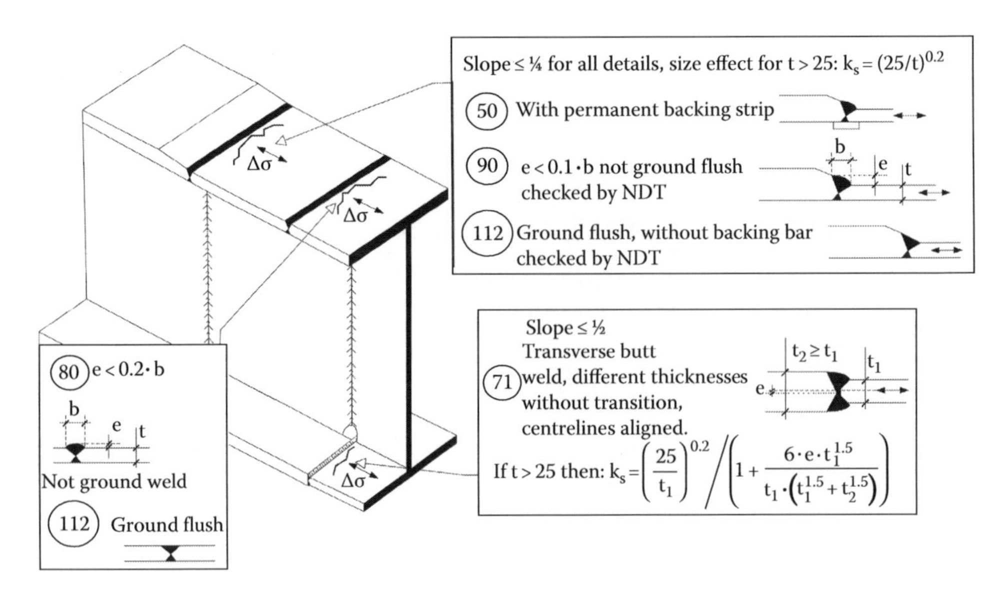

Figure 11.11 Detail categories for transverse butt welds.

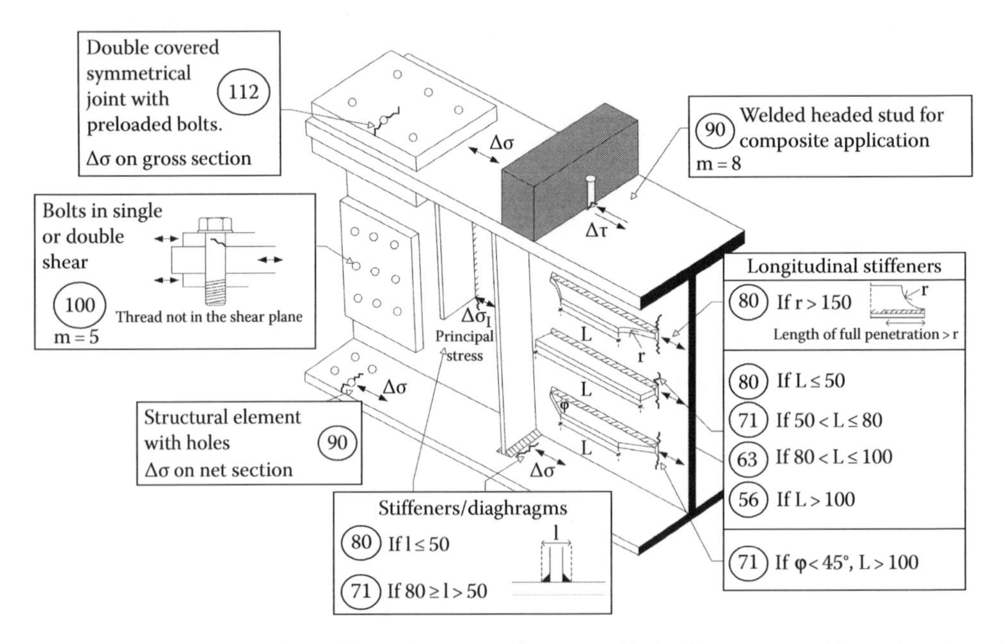

Figure 11.12 Detail categories for weld attachments, stiffeners, and bolted connections (dimensions in mm).

REMARK 11.1

In Example 7.2, it has been shown that in cases of significant torsional loadings, warping stresses should not be neglected. Theoretically, warping stresses should be taken into account during the calculation of the direct stress ranges $\Delta\sigma$. However, after concreting, composite members possess high torsional rigidity and warping stresses become small; they usually do not have any considerable influence on the fatigue verifications and can be omitted.

11.5 FATIGUE LOAD MODELS AND SIMPLIFIED FATIGUE ANALYSIS

Vehicles running on bridges have various shapes with different numbers of axles, axle loads, and axle spacing. Consequently, various load models are proposed by EN 1991-2 that may serve several purposes, for example, the assessment of remaining fatigue life of existing bridges. However, for road bridges, fatigue assessment may be done by a simplified procedure that is based on a single vehicle model. This is called *fatigue load model* 3 (FLM3) and is illustrated in Figure 11.13. The axle weight is 120 kN, the total weight is 480 kN, and the contact surface of the wheels is 0.40×0.40 m^2. For bridges longer than 40 m, a second vehicle running on the same lane may be considered, with a center-to-center distance from the first larger than 40 m. The axle loading of the second vehicle is 36 kN, that is, 30% of that of FLM3. The vehicle moves along the bridge to produce maximum and minimum effects and is placed centrally on the appropriate notional lanes that are identified in the design. Horizontal forces acting simultaneously with the vertical ones, such as centrifugal forces, should be taken into account.

REMARK 11.2

In EN 1991-2, the reader will find five load models for fatigue verifications:

1. FLM1: It is derived from LM1 with 70% of the characteristic values of axle loads and 30% of the characteristic values of uniformly distributed loads. This load model is in general very conservative [11.1].
2. FLM2: It consists of a set of lorries, called *frequent lorries*, and is applied instead of FLM1 in the case of short influence lines. It is also a conservative model.
3. FLM3: As shown in Figure 11.13, it is assumed that after a conventional number of crossings, the same fatigue damage is reached as in the case of real traffic during the design lifetime of the bridge.
4. FLM4: It consists of five *equivalent lorries* that reproduce more accurately the traffic effects on European roads than FLM3. The contribution of each lorry in the final fatigue model is based on probabilistic methods.
5. FLM5: It is based on the direct use of recorded traffic data.

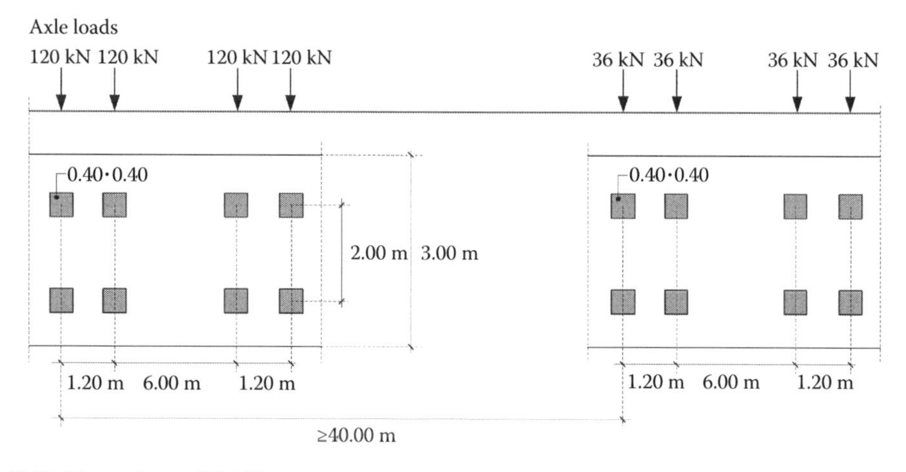

Figure 11.13 Dimensions of FLM3.

FLM1 and FLM2 are used in order to check whether the fatigue life of steel bridges may be considered as unlimited when constant stress amplitude is given, for example, S-N curves. However, they are considered as inappropriate for the most common verifications.

The fatigue load model for railway bridges is load model 71 (LM71) and where required SW/0 (see Section 4.6.2), taking the factor $\alpha=1$ and including the dynamic factors Φ_2 or Φ_3 from Equations 4.6 through 4.7. By comparing the fatigue damage due to single vehicle for road bridges and LM71 for railway bridges with the corresponding damage due to real traffic, appropriate calibration factors were determined and were introduced in design in the form of damage equivalent factors λ. These factors were calibrated for spans up to 80 m for road and 100 m for railway bridges. For larger spans, an additional calibration is required. However, large-span bridges are less susceptible to fatigue due to the fact that permanent loads prevail so that stress ranges due to traffic are not so significant.

For structural steel, the characteristic serviceability limit state (SLS) combination, excluding traffic loads, is relevant for fatigue assessment of the effects of fatigue traffic loading being added. Secondary effects of creep and shrinkage together with the effects of staged construction are taken into account if unfavorable. Temperature is taken by the temperature difference component ΔT_M that has in accordance with Section 4.7.3 positive or negative values. One of the two values is considered that produces the most unfavorable effects. If this is not obvious, both temperature values should be considered. Table 11.1 illustrates the combinations of actions for road and railway bridges.

Refined analysis models should be used for fatigue assessment. "Secondary" moments, for example, in truss girders, should be accounted for since they may produce significant stress ranges that are detrimental for fatigue. In case of box girders, effects of warping and cross-sectional distortion should be allowed. Internal forces and moments are determined by cracked elastic global analysis for composite bridges non-prestressed by tendons. Their extreme values are determined by adding the values due to all actions except fatigue loads to the minimum and maximum values due to fatigue loads. The extreme design moments, and similarly the extreme shear forces, are determined from

$$M_{min,f,Ed} = M_{perm} + M_{min,f} \tag{11.12}$$

$$M_{max,f,Ed} = M_{perm} + M_{max,f} \tag{11.13}$$

where

M_{perm} are the moments due to all actions in the combination except fatigue traffic loads
$M_{min,f}$ are the minimum moments due to fatigue loading
$M_{max,f}$ are the maximum moments due to fatigue loading

Table 11.1 Characteristic SLS combination of actions for fatigue assessment of structural steel

Actions →	*Noncyclic*				*Cyclic*
Type of bridge ↓	*Permanent loads G*	*Creep C_{sec}*	*Shrinkage S_{sec}*	*Temperature $T(\Delta T_M)$*	*Fatigue*
Road	1.0	1.0	1.0	0.6	FLM3
Railway	1.0	1.0	1.0	0.6	LM71

REMARK 11.3

It has to be noted that in the noncyclic part of Equations 11.12 and 11.13, there should be a differentiation between the bending moments $M_{a,Ed}$ acting on pure steel sections before concrete hardening and those acting on the composite sections $M_{com,Ed}$ after concrete hardening since they act on different cross sections and produce different stresses. The noncyclic part becomes then

$$M_{perm} = M_{a,Ed} + M_{com,Ed} \tag{R11.1}$$

where
com = 1 for sagging bending (state 1)
com = 2 for hogging bending (state 2)

If positive and negative temperature values are taken into account, M_{perm} has also maximum and minimum values. In such a case, the earlier equations are written as

$$M_{min,f,Ed} = M_{perm,max} + M_{min,f} \tag{11.14}$$

$$M_{max,f,Ed} = M_{perm,max} + M_{max,f} \tag{11.15}$$

or

$$M_{min,f,Ed} = M_{perm,min} + M_{min,f} \tag{11.16}$$

$$M_{max,f,Ed} = M_{perm,min} + M_{max,f} \tag{11.17}$$

where
$M_{perm,max}$ are the moments due to all actions in the combination, except fatigue traffic loads, with temperature value leading to maximum moments
$M_{perm,min}$ are the moments due to all actions in the combination, except fatigue traffic loads, with temperature value leading to minimum moments

Maximum and minimum stresses are determined in the sections from the earlier internal forces and moments by carefully considering the construction sequence (see Remark 11.3). Stresses are determined on the basis of the uncracked section if concrete is in compression. For concrete in tension, stresses are determined for the cracked section, with due consideration of the effect of tension stiffening for the stresses in reinforcement (see Section 6.1.6). Since fatigue is influenced mainly by the stress range and therefore the difference between minimum and maximum stresses developed during the movement of the fatigue vehicles along the bridge, it might be sufficient to consider only internal forces and moments from these vehicles. However, the influence of other actions in the fatigue combination is important to detect regions where concrete is in tension or compression in order to correctly determine the stiffness properties for analysis and the stresses in the cross section.

For the verifications of the headed studs and concrete in road bridges, the passage of the special vehicle FLM3 produces the required maximum and minimum internal forces. For railway bridges, similar procedure is followed with LM71. Noncyclic internal forces can be excluded.

Table 11.2 Frequent SLS combination of actions for fatigue assessment of reinforcement with annex NN, EN 1992-2

Actions →	*Noncyclic*				*Cyclic*
Type of bridge ↓	*Permanent and secondary effects* G, C_{sec}, S_{sec}	*Traffic*		*Temperature* $T\ (\Delta T_M)$	*Fatigue*
Road	1.0	TS	0.75	0.5	$\beta \cdot$FLM3
		UDL	0.4		
		q_{fk}	0.4		

Notes:

$\beta = 1.75$ for verification at internal supports in continuous bridges.
$\beta = 1.40$ for verification in other areas.
For railway bridges, see Annex NN in 1992-2.

For the reinforcement verifications, Equations 11.12 through 11.17 apply but with the noncyclic part determined from the *frequent combination* of Table 11.2 with the traffic as the leading noncyclic action.

11.6 FATIGUE VERIFICATION FOR STRUCTURAL STEEL

11.6.1 Simplified fatigue assessment

In the *simplified fatigue assessment*, important parameters influencing fatigue resistance are taken into account by a *damage equivalent factor* λ, the values of which are calibrated for road bridges with span up to 80 m and for railway bridges with span up to 100 m. λ is obtained from

$$\lambda = \lambda_1 \cdot \lambda_2 \cdot \lambda_3 \cdot \lambda_4 \leq \lambda_{max} \tag{11.18}$$

where

λ_1 is a factor accounting for the length of the critical influence line
λ_2 is a factor accounting for traffic volume
λ_3 is a factor accounting for the design life of the bridge
λ_4 is a factor accounting for traffic in other lanes
λ_{max} is the maximum value of λ depending on the fatigue limit N_D

Values of the earlier factors for road and railway bridges are given in the following.

11.6.1.1 Road bridges

- Factor λ_1 (Figure 11.14)
  At midspan:

$$\lambda_1 = 2.55 - 0.7 \cdot \frac{L - 10}{70} \tag{11.19}$$

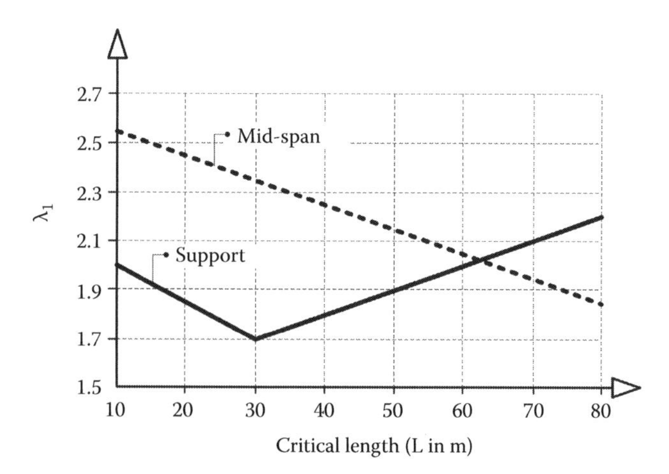

Figure 11.14 Factors λ_1 for road bridges.

At support:

$$\lambda_1 = 2.0 - 0.3 \cdot \frac{L-10}{20} \quad \text{for } 10\ \text{m} \le L \le 30\ \text{m} \tag{11.20a}$$

$$\lambda_1 = 1.7 + 0.5 \cdot \frac{L-30}{50} \quad \text{for } 30\ \text{m} \le L \le 80\ \text{m} \tag{11.20b}$$

where L is the critical length of the relevant influence line in [m] that is given in Table 11.3.

Table 11.3 Critical length L

Effect	*System/member*	*Position*	*Critical length L*
Moment	Simply supported		Span length L_i
	Continuous	Span	Span length L_i of the considered span
		Internal support	Mean of the two spans L_i and L_j adjacent to the support
	Cross girders or frames supporting longitudinal stiffeners		Sum of the two adjacent spans of the stiffeners
Shear force	Simply supported or continuous	Span	$0.4 \cdot$ Span under consideration L_i
		Support	Span under consideration L_i
Support reactions		End support	Span under consideration L_i
		Internal support	Sum of the two adjacent spans $L_i + L_j$
	Arch bridges	Hangers	Twice the distance of hangers
		Arch	Half the span of the arch

Internal support
Span
Span
End support
$0.15 \cdot L_1$ $0.15 \cdot L_2$ $0.15 \cdot L_2$
L_1 L_2

- Factor λ_2
  This factor may be obtained from

$$\lambda_2 = \frac{Q_{m1}}{Q_0} \cdot \left(\frac{N_{obs}}{N_0} \right)^{1/m} \tag{11.21}$$

where Q_{m1} is the average gross weight in kN of the lorries in the slow lane, with

$$Q_{m1} = \left(\frac{\sum n_i \cdot Q_i^m}{\sum n_i} \right)^{1/m}$$

$Q_0 = 480$ kN (weight of FLM3 vehicle)
$N_0 = 0.5 \cdot 10^6$
N_{obs} is the total number of lorries per year in the slow lane (Table 11.4)
Q_i the gross weight in kN of lorry i in the slow lane
n_i is the number of lorries of weight Q_i in the slow lane
$m = 5$

- Factor λ_3
  This factor may be obtained from

$$\lambda_3 = \left(\frac{t_{Ld}}{100} \right)^{1/m} \tag{11.22}$$

where
t_{Ld} is the design life of the bridge in years (usually 100 years)
$m = 5$

- Factor λ_4
  This factor may be obtained from

$$\lambda_4 = \left[1 + \frac{N_2}{N_1} \cdot \left(\frac{n_2 \cdot Q_{m2}}{n_1 \cdot Q_{m1}} \right)^m + \frac{N_3}{N_1} \cdot \left(\frac{n_3 \cdot Q_{m3}}{n_1 \cdot Q_{m1}} \right)^m + \cdots + \frac{N_k}{N_1} \cdot \left(\frac{n_k \cdot Q_{mk}}{n_1 \cdot Q_{m1}} \right)^m \right]^{1/m} \tag{11.23}$$

Table 11.4 Values of N_{obs}

	Traffic categories	*N_{obs} per year and slow lane*
1	Roads and motorways with two or more lanes per direction with high flow rates of lorries	$2 \cdot 10^6$
2	Roads and motorways with medium flow rates of lorries	$0.5 \cdot 10^6$
3	Main roads with low flow rates of lorries	$0.125 \cdot 10^6$
4	Local roads with low flow rates of lorries	$0.05 \cdot 10^6$

where

k is the number of lanes with heavy traffic

N_j is the number of lorries per year in lane j

Q_{mj} is the average gross weight of lorries in lane j

n_j is the value of the relevant influence line in the middle of the lane (Figure 11.15. For multi-girder bridges, the transverse influence line can be calculated with the *Courbon method* described in Section 7.1.2; see also Figure 7.4)

$m = 5$

- Factor λ_{max} (Figure 11.16)
  At midspan:

$$\lambda_{max} = 2.5 - 0.5 \cdot \frac{L-10}{15} \quad \text{for } L \leq 25 \text{ m else } \lambda_{max} = 2 \tag{11.24}$$

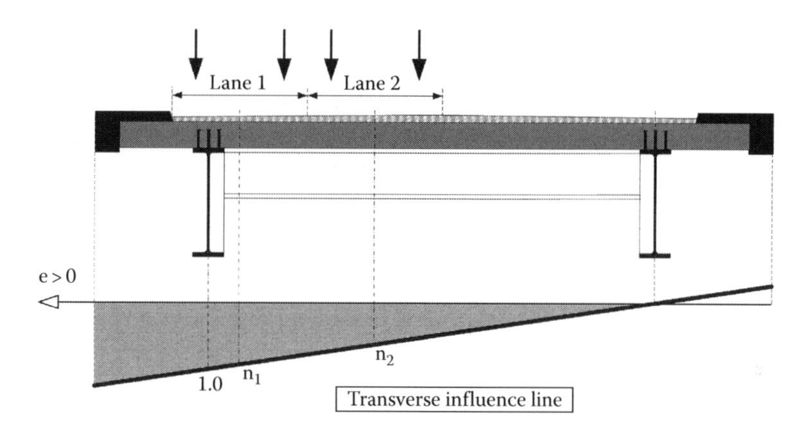

Figure 11.15 Definition of n_j.

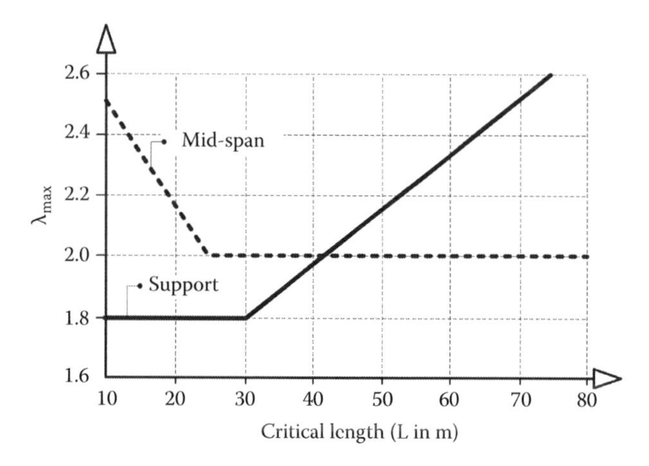

Figure 11.16 Factor λ_{max} for road bridges.

At support:

$$\lambda_{max} = 1.8 + 0.9 \cdot \frac{L - 30}{50} \quad \text{for } L \geq 30 \text{ m else } \lambda_{max} = 1.8 \tag{11.25}$$

11.6.1.2 Railway bridges

- Factor λ_1 (Figure 11.17)
  This factor depends on the type of traffic such as freight trains, passenger trains, or high-speed trains. EN 1991-2 gives eight train types as well as a mixed traffic that corresponds to a combination of train types. Figure 11.17 presents the envelope of all standard train types and values for mixed traffic. The critical length L is determined similarly as for road bridges from Table 11.3.
  Factors λ_1 for multiple unit and underground and rail traffic with 25t axles are found in EN 1993-2 [11.5].
- Factor λ_2
  This is obtained from Table 11.5.
- Factor λ_3
  This is obtained from Table 11.6.
- Factor λ_4
  This is obtained from Table 11.7.
- Factor λ_{max}
  For railway bridges, $\lambda_{max} = 1.4$.

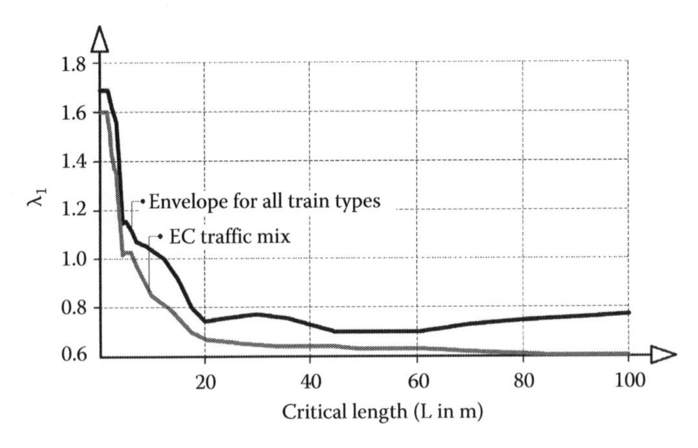

Figure 11.17 Factors λ_1.

Table 11.5 Factor λ_2

Annual traffic weight per track in million tons	*5*	*10*	*15*	*20*	*25*	*30*	*35*	*40*	*50*
λ_2	0.72	0.83	0.90	0.96	1.00	1.04	1.07	1.10	1.15

Table 11.6 Factor λ_3

Design life in years	*50*	*60*	*70*	*80*	*90*	*100*	*120*
λ_3	0.87	0.90	0.93	0.96	0.98	1.00	1.04

Table 11.7 Factor λ_4

$\Delta\sigma_I/\Delta\sigma_{I+2}$	*1.00*	*0.90*	*0.80*	*0.70*	*0.60*	*0.50*
λ_4	1.00	0.91	0.84	0.77	0.72	0.71
$\Delta\sigma_I$	Is the stress range at the considered point due to LM71 on one track					
$\Delta\sigma_{I+2}$	Is the stress range at the same point due to LM71 on any two tracks					

Note: Table 11.7 applies if $\Delta\sigma_I$ and $\Delta\sigma_{I+2}$ have the same sign.

11.6.2 Stress range and fatigue assessment

The stress ranges at a point are determined as the difference between maximum and minimum stresses and are multiplied for road bridges with the damage equivalent factor λ and for railway bridges additionally with the dynamic factor Φ_2 or Φ_3 (see Equations 4.6 or 4.7). This converts the reference stress range $|\sigma_{max,f,Ed} - \sigma_{min,f,Ed}|$, which is based on the extreme bending moments of Equations 11.12 and 11.13 into an equivalent one $\Delta\sigma_E$, which can be compared against the fatigue strength $\Delta\sigma_R$ related to $2 \cdot 10^6$ cycles.

11.6.2.1 Road bridges

$$\Delta\sigma_{E2} = \lambda \cdot |\sigma_{max,f,Ed} - \sigma_{min,f,Ed}| \tag{11.26}$$

$$\Delta\tau_{E2} = \lambda \cdot |\tau_{max,f,Ed} - \tau_{min,f,Ed}| \tag{11.27}$$

REMARK 11.4

In the vicinity of the expansion joints ($D \leq 6$ m in Figure 4.3), the equivalent stress ranges have to be multiplied with the factor $(1 + \Delta\Phi_{fat})$; $\Delta\Phi_{fat}$ is given by Equation 4.1. Fatigue verifications in expansion joint areas may become critical since the stress ranges are increased up to 30%.

11.6.2.2 Railway bridges

$$\Delta\sigma_{E2} = \lambda \cdot \Phi_i \cdot |\sigma_{max,f,Ed} - \sigma_{min,f,Ed}| \tag{11.28}$$

$$\Delta\tau_{E2} = \lambda \cdot \Phi_i \cdot |\tau_{max,f,Ed} - \tau_{min,f,Ed}| \tag{11.29}$$

where

- $i = 2$ for a carefully maintained track (see Equation 4.6)
- $i = 3$ for a track with regular maintenance (see Equation 4.7)

Shear stresses result in shear forces and torsion and are resisted by the steel girder web or the flange for torsion. Shear stress ranges may be then directly determined from the differences between maximum and minimum stresses due to fatigue loading. In absence of torsion, the stress range may be obtained from

$$\left|\tau_{max,f,Ed} - \tau_{min,f,Ed}\right| = \left|V_{max,f,Ed} - V_{min,f,Ed}\right| \cdot \frac{S}{I_a \cdot t_w} \tag{11.30}$$

REMARK 11.5

The calculation of the shear stress range should be based on the elastic stress distribution and not the average one. Therefore, Equation 11.30 is recommended instead of the following:

$$\left|\tau_{max,f,Ed} - \tau_{min,f,Ed}\right| = \left|V_{max,f,Ed} - V_{min,f,Ed}\right| \cdot \frac{1}{h_w \cdot t_w}$$

Similar conditions apply for direct stresses, if the concrete is in tension or in compression for both the minimum and the maximum moments, $M_{min,f,Ed}$ and $M_{max,f,Ed}$, from Equations 11.12 and 11.13, so that the cross section is in both cases either cracked or uncracked. The stress range at the point under consideration in the absence of axial force may then be obtained from Uncracked sections:

$$\left|\sigma_{max,f,Ed} - \sigma_{min,f,Ed}\right| = \left|M_{max,f,Ed} - M_{min,f,Ed}\right| \cdot \frac{z_{1,0}}{I_{1,0}} \tag{11.31}$$

Cracked sections:

$$\left|\sigma_{max,f,Ed} - \sigma_{min,f,Ed}\right| = \left|M_{max,f,Ed} - M_{min,f,Ed}\right| \cdot \frac{z_{2,sa}}{I_{2,sa}} \tag{11.32}$$

where

- $M_{max,f,Ed}$, $M_{min,f,Ed}$ are the maximum and minimum bending moments calculated with the characteristic combination
- $I_{1,0}$ is the second moment of area of the uncracked section for short-term loading (see Figure 7.39)
- $I_{2,sa}$ is the second moment of area of the fully cracked section (see Figure 7.39)
- $z_{1,0}$ and $z_{2,sa}$ are the distances of the considered point from the centroid of the uncracked or cracked section

REMARK 11.6

For the fatigue assessment of steel in hogging moment areas, it is advisable to consider concrete as fully cracked by neglecting the effect of tension stiffening on the cross-sectional properties. This offers a conservative steel stress calculation, which is expressed by application of Equation 11.32.

However, if the moments $M_{min,f,Ed}$ and $M_{max,f,Ed}$ are of different signs and produce, in absence of axial force, tension or compression in the concrete slab, minimum and maximum stresses for the total moment must be determined separately, one on the basis of the cracked section and the other on the uncracked section. At construction stages, some load cases may apply only at the steel section. The relevant stresses on the steel girder are determined on the basis of the steel section, and they are equal for maximum and minimum noncyclic moments, so that they do not contribute in the stress range (see Remark 11.7). Accordingly, stress ranges in this case may be determined from

$$\left|\sigma_{max,f,Ed} - \sigma_{min,f,Ed}\right| = M_{max,f,Ed} \cdot \frac{z_{1,0}}{I_{1,0}} - M_{min,f,Ed} \cdot \frac{z_{2,sa}}{I_{2,sa}} \quad (11.33)$$

REMARK 11.7

In Remark 11.3, it has been noted that the noncyclic bending moment M_{perm} consists of the moment part acting on steel $M_{a,Ed}$ and the part acting on the composite section $M_{1,Ed}$ (uncracked) or $M_{2,Ed}$ (cracked). The final expression for the stress range $|\sigma_{max,f,Ed} - \sigma_{min,f,Ed}|$ depends on the signs of $\sigma_{min,f,Ed}$ and $\sigma_{max,f,Ed}$ as follows [11.12]:

$$\text{Case 1: } M_{min,f,Ed} > 0 \quad \text{and} \quad M_{max,f,Ed} > 0 \rightarrow \left|\sigma_{max,f,Ed} - \sigma_{min,f,Ed}\right| = \Delta M_f \cdot \frac{z_{1,0}}{I_{1,0}} \quad (R11.2)$$

$$\text{Case 2: } M_{min,f,Ed} < 0 \quad \text{and} \quad M_{max,f,Ed} < 0 \rightarrow \left|\sigma_{max,f,Ed} - \sigma_{min,f,Ed}\right| = \Delta M_f \cdot \frac{z_{2,sa}}{I_{2,sa}} \quad (R11.3)$$

$$\text{Case 3: } M_{min,f,Ed} < 0 \quad \text{and} \quad M_{max,f,Ed} > 0 \rightarrow \left|\sigma_{max,f,Ed} - \sigma_{min,f,Ed}\right|$$

$$= M_{1,Ed} \cdot \frac{z_{1,0}}{I_{1,0}} - M_{2,Ed} \cdot \frac{z_{2,sa}}{I_{2,sa}} + M_{max,f} \cdot \frac{z_{1,0}}{I_{1,0}} - M_{min,f} \cdot \frac{z_{2,sa}}{I_{2,sa}} \quad (R11.4)$$

One can see that in all three cases, the stress range does not depend on the loads acting on steel sections $M_{a,Ed}$.

The fatigue verification for structural steel includes checks for direct stresses, shear stresses, and their combination at points where they coexist. They may be written as

$$\frac{\gamma_{Ff} \cdot \Delta\sigma_{E2}}{\Delta\sigma_C / \gamma_{Mf,a}} \leq 1 \quad (11.34)$$

$$\frac{\gamma_{Ff} \cdot \Delta\tau_{E2}}{\Delta\tau_C / \gamma_{Mf,a}} \leq 1 \quad (11.35)$$

$$\left(\frac{\gamma_{Ff} \cdot \Delta\sigma_{E2}}{\Delta\sigma_C / \gamma_{Mf,a}}\right)^3 + \left(\frac{\gamma_{Ff} \cdot \Delta\tau_{E2}}{\Delta\tau_C / \gamma_{Mf,a}}\right)^5 \leq 1 \quad (11.36)$$

where

$\gamma_{Ff} = 1.0$
$\Delta\sigma_{E2}$, $\Delta\tau_{E2}$ are the equivalent stress ranges from Equations 11.26 to 11.29 for $2 \cdot 10^6$ cycles
$\Delta\sigma_C$, $\Delta\tau_C$ are the fatigue strengths for $N_c = 2 \cdot 10^6$ cycles
$\gamma_{Mf,a}$ is the partial safety factor for fatigue strength according to Table 11.8.

Table 11.8 Partial safety resistance factors $\gamma_{Mf,a}$ for steel

		Consequences of failure	
	Assessment method	*Low*	*High*
$\gamma_{Mf,a}$ for steel	Damage tolerant	1.00	1.15
	Safe life	1.15	1.35

$\gamma_{Mf,a}$ covers uncertainties associated with discontinuities, the size of the detail, the welding processes, and the residual stresses due to nonuniform temperature variations during welding (see Section 11.11).

$\gamma_{Mf,a}$ depends on the *consequences of failure* and the required reliability. Larger safety factors are foreseen for main than for secondary members, recognizing the fact that failure in main members may lead to total collapse, while if secondary members fail, such a collapse may be avoided by appropriate redistributions. The required reliability is influenced by appropriate detailing (e.g., provision of crack arresting holes) that limits the possibility of crack propagation after crack formation and the possibility for inspection of a detail and by regular inspection. This method is called *damage-tolerant method* and is associated with lower safety factors. If this is not applied, the so-called *safe life method* is employed for which higher safety factors are foreseen. Table 11.8 gives the recommended values of γ_{Mf} factors for steel, designated as $\gamma_{Mf,a}$.

It should be noted that in addition to the verifications of Equations 11.34 through 11.36, the stress ranges of the direct stress $\Delta\sigma$ and the shear stress ranges $\Delta\tau$ due to frequent loads $(\psi_1 \cdot Q_k)$ must be limited to $1.5 \cdot f_y$ and $1.5 \cdot f_y/\sqrt{3}$ correspondingly. For hybrid girders, $\Delta\sigma_{frequent} \leq 1.5 \cdot f_{yf}$ and $\Delta\tau_{frequent} \leq 1.5 \cdot f_{yw}/\sqrt{3}$.

REMARK 11.7

- In EN 1993-1-9 [11.4], it is stated that fatigue verifications for some details should be based on the principal stresses instead of using the combined check in Equation 11.36. A typical case is when vertical stiffeners do not terminate in the flange (see Figure 11.12). Recommendations for such cases are also found in EN 1993-1-1. It has to be noted that the calculation of principal stresses is more trustworthy when structural elements are modeled with finite elements.
- The reader will find in EN 1993-2 the following expression for combining local and global stress ranges:

$$\Delta\sigma_{E2} = \lambda_{glob} \cdot \Delta\sigma_{f,E,glob} + \lambda_{loc} \cdot \Delta\sigma_{f,E,loc} \text{ (road bridges)} \quad \text{(R11.5)}$$

$$\Delta\sigma_{E2} = \lambda_{glob} \cdot \Phi_{i,glob} \cdot \Delta\sigma_{f,E,glob} + \lambda_{loc} \cdot \Phi_{i,loc} \cdot \Delta\sigma_{f,E,loc} \text{ (rail bridges)} \quad \text{(R11.6)}$$

$i = 2$ or 3 according to Equations 4.6 and 4.7.

Local effects may arise from concentrated wheel loads causing secondary bending moments. This is quite common in steel bridges with orthotropic deck.

Stresses in fillet welds are calculated according to Figure 11.18. Stress ranges for the components σ_{wf} (transverse to the weld's axis) and τ_{wf} (longitudinal to the weld's axis) should be also verified according to Equations 11.34 through 11.36.

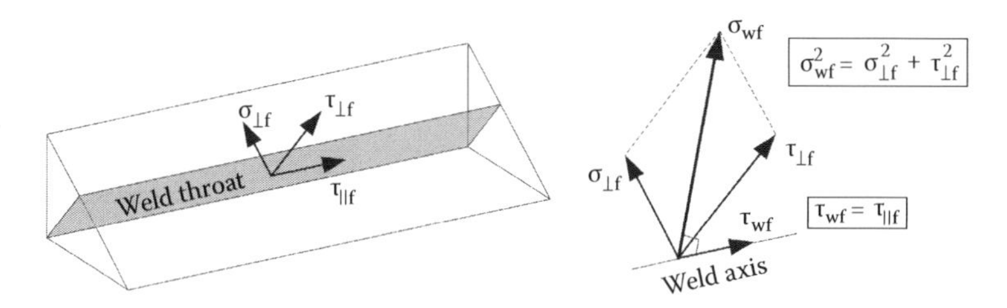

Figure 11.18 Stress components in fillet welds.

11.7 FATIGUE VERIFICATION FOR HEADED STUDS

11.7.1 General

There are different types of fatigue failure for the case of headed studs (Figure 11.19), [11.8], [11.13]. For hogging bending, the top flange is in tension and the crack starts at the weld but proceeds into the base material away from the weld itself. In sagging moment areas (compressed flange), two different types of failures may occur: failure type A or B. In type A, the fatigue crack is developed in the shear stud, directly at the transition to the weld. In type B, the crack occurs in the base steel, directly at the transition to the weld. It is worth mentioning that the encasement of the shear studs in concrete makes inspection impossible. Moreover, it has been shown that damage accumulation affects the static strength of the headed studs at ultimate limit states (ULSs) due to cyclic preloading [11.7], [11.8]. Therefore, a conservative design is recommended especially in hogging moment areas (see Remark 11.8).

11.7.2 Stress range and fatigue assessment

The fatigue verifications for shear connectors are similar to those for structural steel and are based on the experimental observations of Figure 11.19. The fatigue resistance of shear stud connectors is 90 MPa at 2 million cycles, and the slope of the fatigue curve is

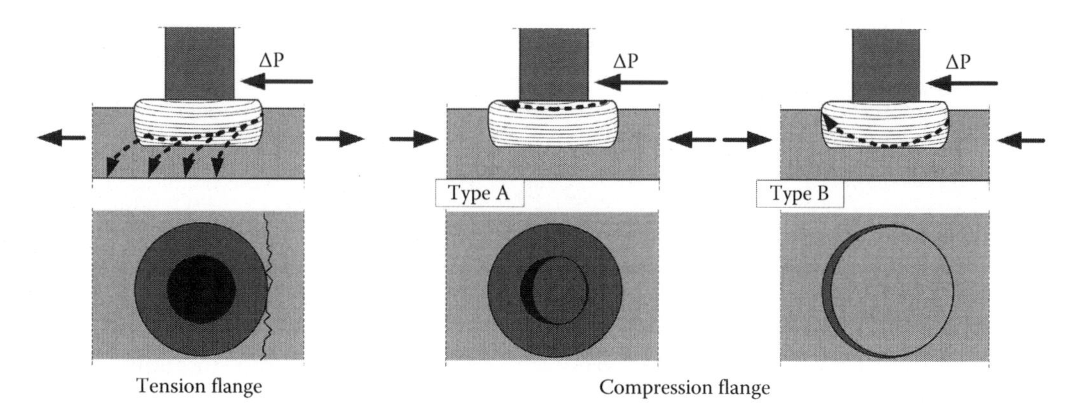

Figure 11.19 Fatigue failure of headed studs and crack propagation.

m = 8 (Figure 11.8). The design of shear studs is based on shear stress ranges. The damage equivalent factor is obtained from Equation 11.37:

$$\lambda_v = \lambda_{v,1} \cdot \lambda_{v,2} \cdot \lambda_{v,3} \cdot \lambda_{v,4} \tag{11.37}$$

where

$\lambda_{v,1} = 1.55$ for road bridges with spans up to 100 m
$\lambda_{v,1} = 0.9 - \frac{L}{133}$ for $L \le 20$ m else $\lambda_{v1} = 0.75$ for railway bridges
$\lambda_{v,2}$ from Equation 11.21 with m = 8
$\lambda_{v,3}$ from Equation 11.22 with m = 8
$\lambda_{v,4}$ from Equation 11.23 with m = 8

Since stresses in the connectors are determined always on the basis of the **uncracked** section (Remark 11.8), the stress ranges may be obtained from the shear due to fatigue traffic loads **only** in accordance with

$$\left|\tau_{max,f,Ed} - \tau_{min,f,Ed}\right| = \left|V_{max,f} - V_{min,f}\right| \cdot \frac{S_{1,0} \cdot e_L}{I_{1,0} \cdot n \cdot A_d} \tag{11.38}$$

where

$V_{max,f}$, $V_{min,f}$ are the maximum and minimum shear forces due to fatigue loading only
$S_{1,0}$ is the first moment of area (static moment) of the concrete slab and the reinforcement in respect to the centroid of the uncracked composite section
$I_{1,0}$ is the second moment of area of the uncracked section for short-term loading
e_L is the longitudinal spacing of shear connectors
n is the number of shear connectors in one section
A_d is the section area of the connector shank ($=\pi \cdot d^2/4$)

Stress ranges considering the damage equivalent factor and the dynamic factor are then obtained from

Road bridges:

$$\Delta\tau_{E2} = \lambda_v \cdot \left|\tau_{max,f,Ed} - \tau_{min,f,Ed}\right| \tag{11.39}$$

Railway bridges:

$$\Delta\tau_{E2} = \lambda_v \cdot \Phi_2(\text{or} \cdot \Phi_3) \cdot \left|\tau_{max,f,Ed} - \tau_{min,f,Ed}\right| \tag{11.40}$$

For shear connectors on tension flanges, the stress range of the top fiber of the tension flange is to be additionally verified treating it in the detail category 80. In addition, the interaction between direct and shear stress ranges in the top fiber of the steel section and the connectors correspondingly shall be verified. The design procedure is summarized in Table 11.9.

Table 11.9 Fatigue verification for shear connectors

	Compression flange (sagging bending)	*Tension flange (hogging bending)*
Shear stress ranges for connectors	$\Delta\tau_{E2}$ from Equations 11.39 or 11.40	$\Delta\tau_{E2}$ from Equations 11.39 or 11.40
Direct stress ranges on top fiber of steel flange		Equation 11.32
Damage equivalent factor	Equation 11.37	Equation 11.37
Resistance at 2 million cycles	$\Delta\tau_C = 90$ MPa	$\Delta\tau_C = 90$ MPa $\Delta\sigma_C = 80$ MPa
Verifications	$\frac{\gamma_{Ff} \cdot \Delta\tau_{E2}}{\Delta\tau_C / \gamma_{Mf,v}} \leq 1$	$\frac{\gamma_{Ff} \cdot \Delta\tau_{E2}}{\Delta\tau_C / \gamma_{Mf,v}} \leq 1$ and $\frac{\gamma_{Ff} \cdot \Delta\sigma_{E2}}{\Delta\sigma_C / \gamma_{Mf,a}} \leq 1$ with $\Delta\sigma_{E2} = \lambda_a \cdot \Delta\sigma_C$ For λ_a see Equation 11.18 $\frac{\gamma_{Ff} \cdot \Delta\sigma_{E2}}{\Delta\sigma_C / \gamma_{Mf,a}} + \frac{\gamma_{Ff} \cdot \Delta\tau_{E2}}{\Delta\tau_C / \gamma_{Mf,v}} \leq \mathbf{1.3}$

Note: The safety factor for shear studs, designated as $\gamma_{Mf,v}$, has a recommended value 1.0 recognizing that it is a secondary element and has ductile behavior that allows redistributions. For $\gamma_{Mf,a}$, see Table 11.8. The earlier verifications are valid only if the maximum longitudinal shear force per headed stud does not exceed 75% of its design shear resistance P_{Rd} under the characteristic combination of actions. The interaction between normal and shear stresses in hogging moment areas should be verified for $\max\Delta\sigma_{E2}$ and the corresponding $\Delta\tau_{E2}$, $\max\Delta\tau_{E2}$ and the corresponding $\Delta\sigma_{E2}$, $\min\Delta\sigma_{E2}$ and the corresponding $\Delta\tau_{E2}$, and $\min\Delta\tau_{E2}$ and the corresponding $\Delta\sigma_{E2}$, leading to four verifications.

REMARK 11.8

- In Figure 11.19, it was shown that in tension flanges, the fatigue crack starts at the weld and ends in the base steel. Therefore, in the verifications of Table 11.9, the shear stress $\Delta\tau$ at the steel–concrete interface and the normal stress range $\Delta\sigma$ in the tension flange are included. In compression flanges, cracks are "captivated" in the weld material leaving base steel intact; therefore, $\Delta\sigma$ is missing.
- In hogging moment areas, it is recommended to calculate the shear stress $\Delta\tau$ by considering an uncracked deck plate. This is a conservative simplification for taking into account the shear flow increase due to the tension stiffening effect. In contrast, $\Delta\sigma$ in the top flange of the steel girder should be calculated with the cracked cross-sectional properties so that steel stresses are not underestimated.

11.8 FATIGUE VERIFICATION FOR REINFORCING STEEL

11.8.1 Fatigue assessment

Fatigue verification for reinforcing steel is in accordance with Annex NN of EN 1992-2 and is made similarly as for structural steel on the basis of the following relation:

$$\gamma_{Fs,f} \cdot \Delta\sigma_{s,equ} \leq \frac{\Delta\sigma_{R,s}(N^*)}{\gamma_{Ms,f}} \tag{11.41}$$

where

$$\Delta\sigma_{s,equ} = \lambda_s \cdot |\sigma_{s,max,f,Ed} - \sigma_{s,min,f,Ed}| \text{ damage equivalent stress for road bridges} \quad (11.42)$$

$$\Delta\sigma_{s,equ} = \lambda_s \cdot \Phi_i \cdot |\sigma_{s,max,f,Ed} - \sigma_{s,min,f,Ed}| \text{ damage equivalent stress for railway bridges} \quad (11.43)$$

$\gamma_{Fs,F} = 1.0$ is the partial safety factor for action effects
$\gamma_{Ms,F} = 1.15$ is the partial safety factor for fatigue strength
$\Delta\sigma_{R,s} = 162.5$ MPa is the fatigue strength for straight and bent reinforcing bars at $N^* = 1$ million cycles
$\sigma_{s,max,f,Ed}$, $\sigma_{s,min,f,Ed}$ are maximum and minimum stresses calculated with the frequent combination of Table 11.2
λ_s are the damage equivalent factors for reinforcement
$i = 2$ for a carefully maintained track according to Equation 4.6
$i = 3$ for a track with regular maintenance according to Equation 4.7

Unlike structural steel, fatigue resistance for reinforcement is defined at one instead of two million cycles so that modifications are made to fatigue loading and the damage equivalent factors as a result of a calibration procedure that are discussed in the following.

11.8.1.1 Road bridges

The damage equivalent factor is determined by the multiplication of four partial factors $\lambda_{s,i}$ that express the same influencing factors as for structural steel (Equation 11.18) and an additional factor Φ_{fat} that expresses the influence of the pavement roughness:

$$\lambda_s = \varphi_{fat} \cdot \lambda_{s,1} \cdot \lambda_{s,2} \cdot \lambda_{s,3} \cdot \lambda_{s,4} \quad (11.44)$$

with $\Phi_{fat} = 1.2$ for regularly maintained surfaces, else = 1.4.

For straight or bent bars, the factor $\lambda_{s,1}$ accounting for the length of the critical influence line may be determined from the following graph (Figure 11.20):

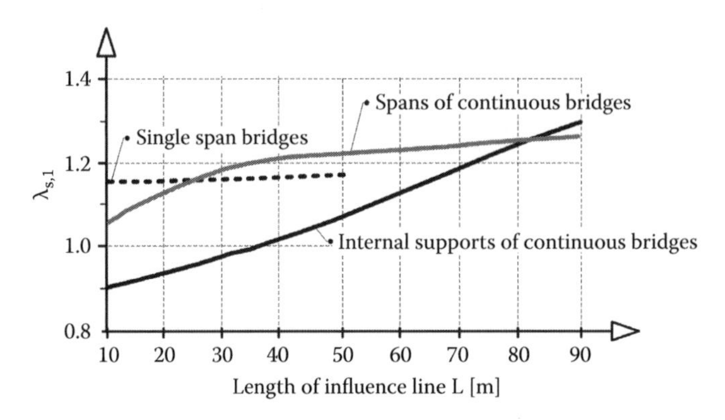

Figure 11.20 $\lambda_{s,1}$ values.

The factor $\lambda_{s,2}$ accounting for traffic volume is determined from

$$\lambda_{s,2} = \bar{Q} \cdot \sqrt[k_2]{\frac{N_{obs}}{2 \cdot 10^6}} \tag{11.45}$$

where

N_{obs} is the total number of lorries per year in the slow lane (Table 11.4)
$k_2 = 9$ is the slope of the fatigue curve for straight or bent bars
$\bar{Q}$ is a coefficient for the type of traffic with values 1.0, 0.94, or 0.82 for long distance, medium distance, or local traffic. Long distance means hundreds of kilometers, medium distance means 50–100 km, and local traffic means distances less than 50 km

The factor $\lambda_{s,3}$ accounting for the design life of the bridge is determined from

$$\lambda_{s,3} = \sqrt[k_2]{\frac{N_{years}}{100}} \tag{11.46}$$

where N_{years} is the design life of the bridge in years.

The factor $\lambda_{s,4}$ accounting for traffic in other lanes may be determined from

$$\lambda_{s,4} = \sqrt[k_2]{\frac{\sum N_{obs,i}}{N_{obs,1}}} \tag{11.47}$$

where

$N_{obs,i}$ is the number of lorries crossing lane i per year
$N_{obs,1}$ is the number of lorries crossing the slow lane per year

11.8.1.2 Railway bridges

The traffic model for railway bridges is LM71 (or SW/0 when required) without accounting for factor α. For traffic on more than one track, LM71 is applied maximal on two tracks. The damage equivalent factor is determined in accordance with

$$\lambda_s = \lambda_{s,1} \cdot \lambda_{s,2} \cdot \lambda_{s,3} \cdot \lambda_{s,4} \tag{11.48}$$

The factor $\lambda_{s,1}$ may be determined for loaded lengths of the influence line between 2 and 20 m by following the expression

$$\lambda_{s,1}(L) = \lambda_{s,1}(2\text{ m}) + \left[\lambda_{s,1}(20\text{ m}) - \lambda_{s,1}(2\text{ m})\right] \cdot (\log L - 0.3) \tag{11.49}$$

where

$\lambda_{s,1}(2\text{ m})$ is the value of $\lambda_{s,1}$ for $L = 2$ m (Table 11.10)
$\lambda_{s,1}(20\text{ m})$ is the value of $\lambda_{s,1}$ for $L = 20$ m (Table 11.10)
$\lambda_{s,1}(L)$ is the value of $\lambda_{s,1}$ for $2\text{ m} < L < 20\text{ m}$

Table 11.10 Factors $\lambda_{s,1}$ for straight or bent bars

	Normal traffic		*Heavy traffic*	
	$L \leq 2$ m	$L \geq 20$ m	$L \leq 2$ m	$L \geq 20$ m
Simple-span bridges	0.90	0.65	0.95	0.70
Internal span of continuous bridges	0.95	0.50	1.05	0.55
End span of continuous bridges	0.90	0.65	1.00	0.65
Internal supports of continuous bridges	0.85	0.70	0.85	0.75

Note: For light traffic, the values for normal traffic may be used as an approximation.

The factor $\lambda_{s,2}$ accounting for traffic volume is determined from

$$\lambda_{s,2} = \sqrt[k_2]{\frac{V}{25 \cdot 10^6}} \tag{11.50}$$

where

V is the traffic volume in million tons/year/track

$k_2 = 9$ is the slope of the fatigue curve for straight or bent bars

The factor $\lambda_{s,3}$ accounting for the design life of the bridge may be determined from Equation 11.22 with $m = k_2$.

The factor $\lambda_{s,4}$ accounts for loading on more than one track. Only two, the most unfavorable, tracks are loaded. The factor $\lambda_{s,4}$ may then be obtained from

$$\lambda_{s,4} = \sqrt[k_2]{n + (1-n) \cdot s_1^{k_2} + (1-n) \cdot s_2^{k_2}} \tag{11.51}$$

where

$$s_1 = \frac{\Delta\sigma_1}{\Delta\sigma_{1+2}}, \quad s_2 = \frac{\Delta\sigma_2}{\Delta\sigma_{1+2}}$$

n is the traffic proportion simultaneously crossing the bridge with recommended value 0.12

$\Delta\sigma_1$, $\Delta\sigma_2$ is the stress range in the cross section under consideration due to LM71 on one track

$\Delta\sigma_{1+2}$ is the stress range in the same cross section under consideration due to LM71 on two tracks simultaneously

$k_2 = 9$ is the slope of the fatigue curve for straight or bent bars

11.8.2 Stress ranges

In cracked regions, the stress in reinforcement is determined accounting for the influence of tension stiffening. Stresses and stress ranges are determined accounting for the loading type in the concrete slab as it is demonstrated in Figure 11.21. When the slab is constantly under

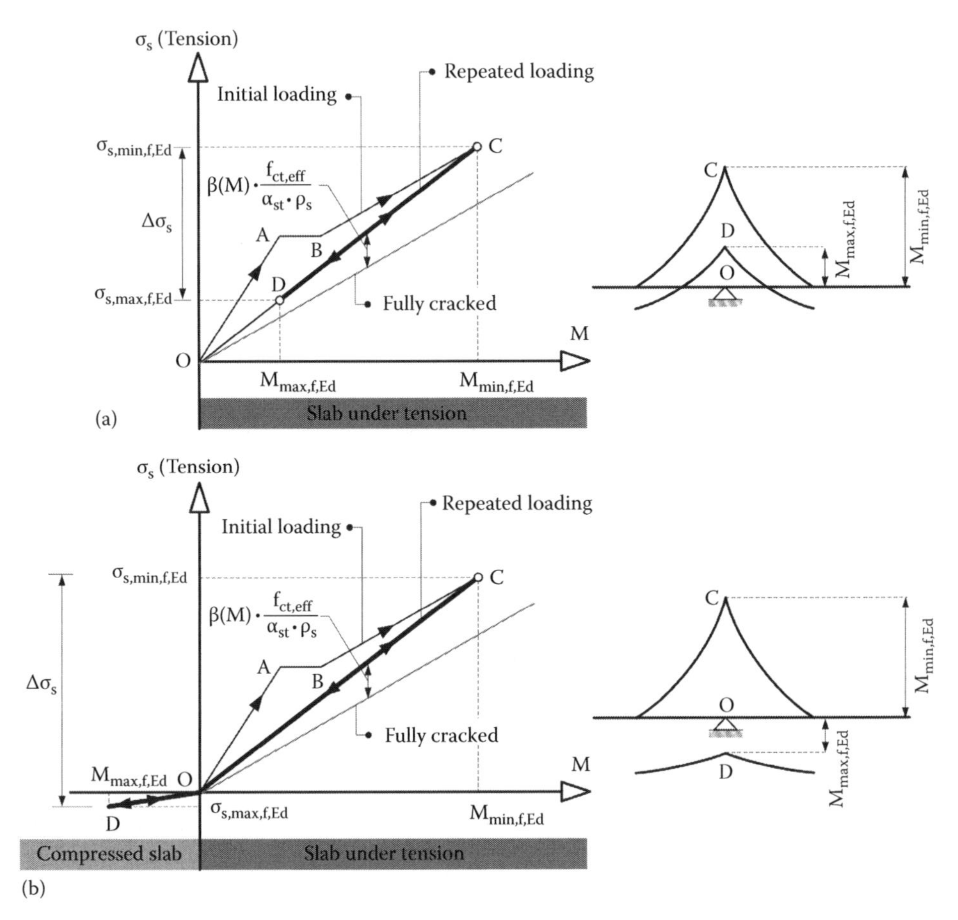

Figure 11.21 Stress ranges in reinforcement. (a) Slab constantly under tension and (b) slab under tension–compression.

tension (Figure 11.21a), the route CD expresses the fatigue behavior of the reinforcement. Starting from O route, OABC defines the stress σ_s by taking into account the tension stiffening effect. In point C, the hogging moment reaches its minimum value $M_{min,f,Ed}$ and then it is decreased till $M_{max,f,Ed}$. Thereafter, stress σ_s is increased again and "oscillates" between C and D. In some cases, sagging bending may occur (Figure 11.21b). The deck slab changes from cracked to uncracked and the opposite. This fatigue behavior is expressed by the route DOC. It is worth mentioning that the linear approximation for the fatigue behavior does not realistically express the real behavior of the concrete slab under repeated loading but it offers a conservative approach that simplifies the calculations. This is shown in Equation 11.53 subsequently. One can also observe that the route OC does not coincide with the extension of OD in Figure 11.21b. This means that during repeated loading, the tensile strength of concrete is neglected.

The stress ranges for

- Case 1 (slab constantly under tension; see Figure 11.21a)
  The minimum and maximum total moments acting on the **composite section,** $M_{min,f,Ed}$ and $M_{max,f,Ed}$, produce tension in the slab and the reinforcement. $M_{min,f,Ed}$ and $M_{max,f,Ed}$ are the minimum and maximum moments calculated for the frequent combination of

Table 11.2 in case of road bridges. In railway bridges, stresses are only due to LM71 (or SW/0 when required). The maximum tensile stress in the reinforcement due to maximum moment, accounting for tension stiffening, is determined from

$$\sigma_{s,max,f,Ed} = \left|M_{max,f,Ed}\right| \cdot \frac{z_{2,s}}{I_{2,sa}} + \beta \cdot \frac{f_{ctm}}{\alpha_{st} \cdot \rho_s} \tag{11.52a}$$

where

$\beta = 0.2$ is the β-factor of Figure 11.21
f_{ctm} is the mean tensile strength of concrete (Table 6.1)
$I_{2,sa}$ is the second moment of area of the fully cracked section
$z_{2,s}$ is the distance of reinforcement from the centroid of the fully cracked section

$$\alpha_{st} = \frac{A_{2,sa} \cdot I_{2,sa}}{A_a \cdot I_a} \tag{11.52b}$$

A_a, I_a are the area and the second moment of area of the steel section
$A_{2,sa}$, $I_{2,sa}$ are the area and the second moment of area of the fully cracked section (steel section + reinforcement) (see Figure 7.39)

REMARK 11.9

In Figure 11.21, one can see that after the initial loading, the contribution of tension stiffening becomes nonconstant and dependent on the magnitude of the hogging moment M; $\beta = \beta(M)$. For the fatigue verifications, **a fixed value equal to 0.2** is used so that the decreasing of tension stiffening due to fatigue loading is considered [11.7], [11.12].

The minimum tensile stress in the reinforcement according to the configuration of Figure 11.21a is

$$\sigma_{s,min,f,Ed} = \sigma_{s,max,f,Ed} \cdot \frac{M_{min,f,Ed}}{M_{max,f,Ed}} \tag{11.53}$$

The stress range may be obtained from the combination of Equations 11.52a and 11.53 as follows:

$$\sigma_{s,max,f,Ed} - \sigma_{s,min,f,Ed} = \left(\left|M_{max,f,Ed}\right| \cdot \frac{z_{2,s}}{I_{2,sa}} + 0.2 \cdot \frac{f_{ctm}}{\alpha_{st} \cdot \rho_s}\right) \cdot \left|\frac{M_{max,f,Ed} - M_{min,f,Ed}}{M_{max,f,Ed}}\right| \tag{11.54}$$

- Case 2 (slab under tension–compression; see Figure 11.21b)
  The design moment on the *composite* section $M_{min,f,Ed}$ (Equation 11.12) produces tension, while $M_{max,f,Ed}$ (Equation 11.13) produces compression in the reinforcement.
  The stress range may be obtained from

$$\sigma_{s,max,f,Ed} - \sigma_{s,min,f,Ed} = M_{max,f,Ed} \cdot \frac{z_{1,s}}{I_{1,0}} + \left|M_{min,f,Ed}\right| \cdot \frac{z_{2,s}}{I_{2,sa}} + 0.2 \cdot \frac{f_{ctm}}{\alpha_{st} \cdot \rho_s} \tag{11.55}$$

where $z_{1,s}$ is the distance of reinforcement from the centroid of the short-term uncracked section.

- Case 3 (slab constantly under compression)
  The minimum and maximum design moments acting on the *composite* section, $M_{min,f,Ed}$ and $M_{max,f,Ed}$, produce compression in the slab and the reinforcement. Stresses on reinforcement are determined on the basis of the uncracked section for short-term loading. The stress range may be obtained from

$$\sigma_{s,max,f,Ed} - \sigma_{s,min,f,Ed} = \left(M_{max,f,Ed} - M_{min,f,Ed}\right) \cdot \frac{z_{1,s}}{I_{1,0}} \tag{11.56}$$

REMARK 11.10

1. By taking into account, the construction sequence Equations 11.54, 11.55, and 11.56 are rewritten as follows [11.12]:
   a. Case 1—$M_{max,f,Ed} < 0$ and $M_{min,f,Ed} < 0$

$$\left|\sigma_{max,f,Ed} - \sigma_{min,f,Ed}\right| = \left(\left|M_{2,Ed} + M_{min,f}\right| \cdot \frac{z_{2,s}}{I_{2,sa}} + 0.2 \cdot \frac{f_{ct,eff}}{\alpha_{st} \cdot \rho_s}\right) \cdot \left(1 - \frac{M_{2,Ed} + M_{max,f}}{M_{2,Ed} + M_{min,f}}\right) \tag{R11.7}$$

   b. Case 2—$M_{max,f,E} > 0$ and $M_{min,f,E} < 0$

$$\left|\sigma_{max,f,Ed} - \sigma_{min,f,Ed}\right| = \left(\left|M_{1,Ed} + M_{max,f}\right| \cdot \frac{z_{1,s}}{I_{1,0}}\right) - \left[\left(M_{2,Ed} + M_{min,f}\right) \cdot \frac{z_{2,s}}{I_{2,sa}} + 0.2 \cdot \frac{f_{ct,eff}}{\alpha_{st} \cdot \rho_s}\right] \tag{R11.8}$$

   c. Case 3—$M_{max,f,Ed} > 0$ and $M_{min,f,Ed} > 0$

$$\left|\sigma_{max,f,Ed} - \sigma_{min,f,Ed}\right| = \Delta M_f \cdot \frac{z_{1,s}}{I_{1,0}} \tag{R11.9}$$

   where
   $M_{1,Ed}$ is the design bending moment due to all actions except fatique traffic loads acting on the uncracked composite section (state 1)
   $M_{2,Ed}$ is the design bending moment due to all actions except fatique traffic loads acting on the fully cracked composite section (state 2)
   $M_{min,f}$ is the minimum moment due to fatigue loading
   $M_{max,f}$ is the maximum moment due to fatigue loading

2. Deck slabs are also subjected to **local fatigue loadings** due to the presence of the concentrated wheel loads. Therefore, the fatigue verification in Equation 11.41 should be fulfilled in combination with Equations 11.42 for road bridges and 11.43 for railway bridges. Fatigue due to local effects may be surprisingly high for the reinforcement located in slabs near composite crossbeams (see Figure 2.17). Additional bending moments due to wheel loads can be calculated using *Pucher's* method. However, FE methods as described in Section 7.5 are recommended.

3. It is important to note that the damage equivalent factor $\lambda_{s,1}$ for local effects is calculated for a different influence length than for global effects. If the length of the influence line is difficult to be found, then a conservative assumption is unavoidable. $\lambda_{s,2}$, $\lambda_{s,3}$, and $\lambda_{s,4}$ are the same for both local and global effects.

11.9 FATIGUE VERIFICATION FOR CONCRETE

Fatigue verification should be made for concrete under compression. Stresses refer, therefore, only to the uncracked section. The fatigue strength of concrete is given by

$$f_{cd,fat} = 0.85 \cdot e^{s \cdot \left(1-\sqrt{28/t_0}\right)} \cdot f_{cd} \cdot \left(1 - \frac{f_{ck}}{250}\right) \tag{11.57}$$

where

t_0 is the concrete age at first loading in days
f_{ck} is the concrete strength in MPa
s is a factor in dependence of the hardening rate of concrete. It is equal to 0.2, 0.25, or 0.38 for cements of rapid, normal, or slow hardening rate

The verification procedure is summarized in Table 11.11.

REMARK 11.11

- For the great majority of composite bridges, the fatigue verification for concrete will not be critical. This may not be the case in small span composite bridges where traffic loads govern the design.
- The verification procedure in Table 11.11 is according to EN 1992-1-1 and reference for this is given in EN 1994-2. The fatigue assessment of compressed concrete found in Annex NN of EN 1992-2 is **not** valid for composite bridges.

Table 11.11 Fatigue verification procedure for concrete

	Nonfatigue loading	*Fatigue loading*
Moments	All loads except fatigue loading	—
	M_{perm} (frequent combination in Table 11.2)	
Stresses (compression)	$\sigma_{c,max} = M_{perm} \cdot \frac{z_c}{I_{I,0} \cdot n_0}$, $\sigma_{c,min} = M_{perm} \cdot \frac{z_c}{I_{I,0} \cdot n_0}$	
	Tension stresses are set to zero (0).	
Verifications	$\frac{\sigma_{c,max}}{f_{cd,fat}} \le 0.5 + 0.45 \cdot \frac{\sigma_{c,min}}{f_{cd,fat}}$ and	(11.58)
	$\frac{\sigma_{c,max}}{f_{cd,fat}} \le 0.9$ if $f_{ck} \le 50$ MPa	(11.59)

Note: $I_{I,0}$, second moment of area of the uncracked section; z_c, distance of the extreme concrete fiber from the centroid of the uncracked section; n_0, short-term modular ratio of concrete; $\sigma_{c,min}$ should be taken as zero if negative (in tension).

11.10 POSSIBILITIES OF OMITTING FATIGUE ASSESSMENT

Fatigue assessment may be omitted when the following conditions are met:

- Pedestrian bridges, canal bridges, etc., that are predominantly statically loaded and are unlikely to be excited by wind loading
- Secondary parts of road and railway bridges that are not stressed by traffic loads nor are excited by wind
- Main girders and their attachments of road bridges when the detail category is at least 71 and the span larger than 45 m

Accordingly, the following parts of road bridges with spans larger than 45 m are not required to be assessed for fatigue:

- Main girders from rolled sections in unspliced regions (detail categories 100–160, Table 8.1 in EN 1993-2, details 1–7)
- Plate and box main girders from built-up welded sections in unspliced regions (detail categories 71–125, Table 8.2 in EN 1993-2, details 1–11)
- Main girders in bolted splice regions with preloaded, fitted, or injection bolts, including bolts in shear (detail categories 80–112, Table 8.1 in EN 1993-2, details 8–12 and 15)
- For nominal stress ranges $\gamma_{Ff} \cdot \Delta\sigma \leq 26/\gamma_{Mf}$ [MPa]
- For numbers of loading cycles $N \leq 2 \cdot 10^6 \cdot \left[\dfrac{36/\gamma_{Mf}}{\gamma_{Ff} \cdot \Delta\sigma_{E2}}\right]$
- For detail categories in which ranges $\gamma_{Ff} \cdot \Delta\sigma \leq \Delta\sigma_D/\gamma_{Mf}$

EXAMPLE 11.1

A road bridge carrying three lanes, out of which two are slow, has a concrete deck supported by two plate girders with dimensions shown in Figure 11.22. The bridge is continuous with two spans of 25 m each. It is part of a long-distance motorway with high flow rates of lorries. The girders are spliced by welding at a distance 16.25 m from the end support. The concrete is cast in one phase. The self-weight of superstructures is 37 kN/m length, equally divided in the two main girders.

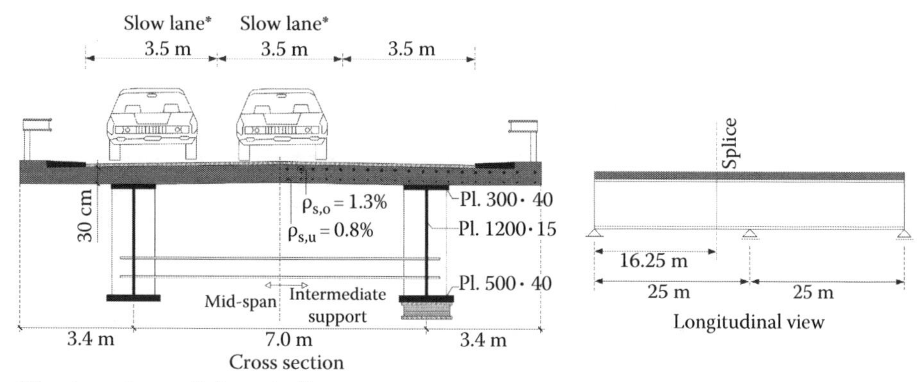

Figure 11.22 Cross section of the road bridge of Example 11.1—longitudinal view and splice position.

The fatigue resistance of the bridge at the splice cross section is to be verified for a design life of 100 years.

Concrete C35/45. Steel grades for structural steel, reinforcing steel, and shear connectors are not relevant for fatigue assessment.

Global analysis

A grillage model is used for global analysis. The longitudinal beam elements represent the composite girders that are composed of the steel girders and the associated effective width of the slab. The transverse beams represent the concrete slab (see Section 7.1.3). Cracked analysis is employed, with cracked sections in the region $0.15 \cdot 25 = 3.75$ m on the left and right side of the internal support. The properties of the uncracked section for short-time loading and the cracked section for hogging bending are given in Table 11.12. The longitudinal reinforcement ratio at hogging moment areas is 2.1%. For the top layer, it is 1.3% and for the bottom layer 0.8%.

The section properties in Table 11.12 were calculated with an effective width of 312.5 cm, which corresponds to hogging bending (see Figure 7.34). However, the cross section is not always under negative moments. Adopting the smallest value of b_{eff} for the cross-sectional properties is a conservative approach and in similar cases is recommended. Alternatively, different effective width values depending on the sign of M_{perm} should be used; this makes calculation quite laborious, but it may lead to a more economical design due to lower stresses in sagging bending areas.

The self-weight of the steel girder and the concrete slab act at the steel section of the girders and are not considered for fatigue. Global analysis is performed for the load cases that are relevant for fatigue, which are the self-weight of superstructures, G_2; temperature difference $\Delta T_{M,cool} = 18°C$ ($\Delta T_{M,heat}$ is less critical); and the secondary effects of shrinkage, S_{sec}, and of creep, C_{sec}, for 30,000 days. For traffic loads, load model 3 (FLM3) is used. This is positioned in longitudinal and transverse direction in such a way to deliver the most adverse maximum and minimum moments of the composite girders at the cross section under consideration. For the correct position in longitudinal direction, the influence line of the moment is determined as shown in Figure 11.23. The four axle loads of FLM3 are then placed separately for maximum and minimum moments. In transverse direction, two slow lanes of 3.0 m width each are considered and the wheel loads applied (Table 11.13).

Table 11.12 Properties of uncracked and cracked section of the composite girders

	A [m²]	I [m⁴]	z_{ao} [m]	z_{au} [m]	z_s [m]	z_c [m]
Steel section	$500 \cdot 10^{-4}$	$1.40 \cdot 10^{-2}$	−0.739	0.541	–	–
Uncracked (short term)	$1995.54 \cdot 10^{-4}$	$4.47 \cdot 10^{-2}$	−0.073	1.207	−0.313	−0.373
Cracked (steel + reinf.)	$696.9 \cdot 10^{-4}$	$2.58 \cdot 10^{-2}$	−0.482	0.798	−0.722	−0.782

Note: I, second moment of area; z, distances of selected fibers from centroid of the section (+, downward; −, upward); ao, top fiber of top flange; au, bottom fiber of bottom flange; s, top reinforcement layer; c, top fiber of concrete slab.

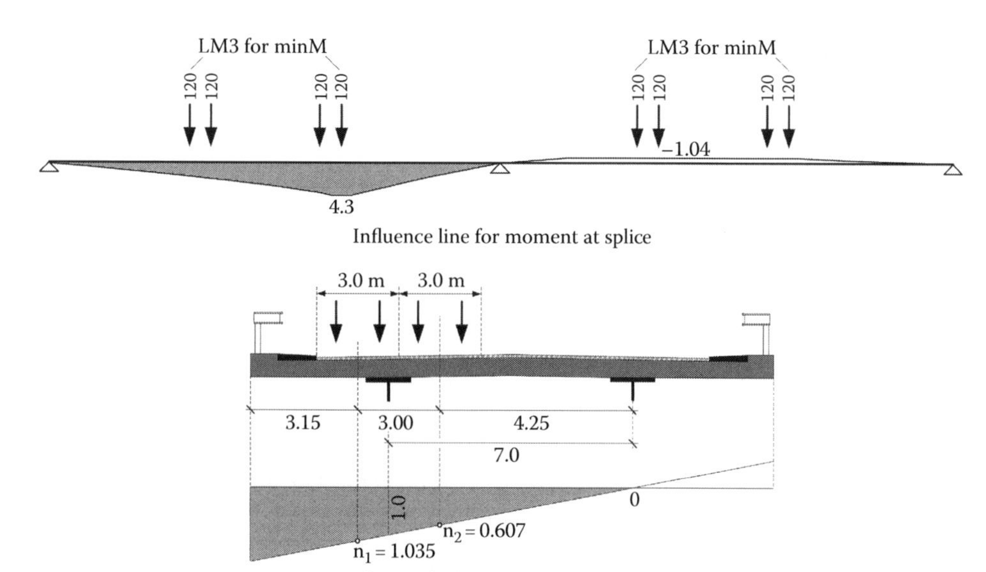

Figure 11.23 Position of LM3 in longitudinal and transverse direction.

Table 11.13 Moments and shear forces of girder at splice section

Action *Notation*	*Superstructures* G_2	*Temperature difference* $\Delta T_{M,cool}$	*Shrinkage secondary* S_{sec}	*Creep secondary* C_{sec}	*FLM3* *Max*	 *Min*
Moment [kN-m]	600	−1336.67	−961	−671	1434	−449
Shear force [kN]	−113	−68.33	−52	−37	−248	−33

Fatigue assessment of steel girder

Beforehand, direct stress and the shear stress ranges due to frequent loads ($\psi_1 \cdot Q_k$) are calculated and limited to $1.5 \cdot f_y$ and $1.5 \cdot f_y/\sqrt{3}$ correspondingly. For the fatigue assessment, stresses and stress ranges will be determined at the top and bottom fibers of the steel girder, points "ao" and "au."

The resulting moments for the combinations of actions relevant for fatigue (Table 11.1) are as follows:

- Combination $G + C_{sec} + S_{sec} + 0.6 \cdot \Delta TM_{cool} + FLM3$
  This combination gives minimum values of moments due to permanent loads:

  $M_{perm,min} = 600 - 802 - 961 - 671 = -1834$ kN-m

  $M_{min,f} = -449$ kN-m

  $M_{max,f} = 1434$ kN-m

  Equation 11.16: $M_{min,f,Ed} = M_{perm,min} + M_{min,f} = -1834 - 449 = -2283$ kN-m

  Equation 11.17: $M_{max,f,Ed} = M_{perm,min} + M_{max,f} = -1834 + 1434 = -400$ kN-m

Both moments are negative and cause tension in the slab so that stresses are determined for the cracked section. Stress ranges may then be determined in accordance with Equation 11.32.

Top flange, point "ao"

$$\left|\sigma_{max,f,Ed} - \sigma_{min,f,Ed}\right| = \left|-2283 - (-400)\right| \cdot \frac{0.482}{2.58 \cdot 10^{-2}} \cdot \frac{1}{10^4} = 3.52 \text{ kN/cm}^2$$

Bottom flange, point "au"

$$\left|\sigma_{max,f,Ed} - \sigma_{min,f,Ed}\right| = \left|-2283 - (-400)\right| \cdot \frac{0.798}{2.58 \cdot 10^{-2}} \cdot \frac{1}{10^4} = 5.82 \text{ kN/cm}^2$$

Alternatively, Equation R11.3 can be used.

- Combination $G + C_{sec} + S_{sec} + FLM3$
This combination gives maximum values of moments due to permanent loads:

$$M_{perm,max} = 600 - 961 - 671 = -1032 \text{ kN-m}$$

$$M_{min,f} = -449 \text{ kN-m}$$

$$M_{max,f} = 1434 \text{ kN-m}$$

Equation 11.14: $M_{min,f,Ed} = M_{perm,max} + M_{min,f} = -1032 - 449 = -1481$ kN-m

Equation 11.15: $M_{max,f,Ed} = M_{perm,max} + M_{max,f} = -1032 + 1434 = 402$ kN-m

Minimum moment causes tension, and maximum moment compression in the slab so that stresses are determined for the cracked and correspondingly uncracked section. Stress ranges are then determined in accordance with Equation 11.33.

Top flange, point "ao"

$$\left|\sigma_{max,f,Ed} - \sigma_{min,f,Ed}\right| = \left[402 \cdot \frac{0.073}{4.47 \cdot 10^{-2}} - \left(-1481 \cdot \frac{0.482}{2.58 \cdot 10^{-2}}\right)\right] \cdot \frac{1}{10^4} = 2.83 \text{ kN/cm}^2$$

Bottom flange, point "au"

$$\left|\sigma_{max,f,Ed} - \sigma_{min,f,Ed}\right| = \left[402 \cdot \frac{1.207}{4.47 \cdot 10^{-2}} - \left(-1481 \cdot \frac{0.798}{2.58 \cdot 10^{-2}}\right)\right] \cdot \frac{1}{10^4} = 5.67 \text{ kN/cm}^2$$

Alternatively, Equation R11.4 could be used.

Subsequently, the damage equivalent factors λ_i will be determined.

- Factor λ_1
The system is a continuous beam. The critical length for the moments is equal to the span length L_i of the considered span (Table 11.3). → L = 25 m:

Equation 11.19: $\lambda_1 = 2.55 - 0.7 \cdot \dfrac{25 - 10}{70} = 2.4$

- Factor λ_2

  The average gross weight of lorries in the slow lane may be determined from fatigue load model 4 (FLM4) that is represented from the distribution in Table 11.14:

$$Q_{ml} = \left(\frac{20 \cdot 200^5 + 5 \cdot 310^5 + 50 \cdot 490^5 + 15 \cdot 390^5 + 10 \cdot 450^5}{100} \right)^{1/5} = 445.4 \text{ kN}$$

  Total number of lorries per year in the slow lane (Table 11.4): $N_{obs} = 2.0 \cdot 10^6$:

  Equation 11.21: $\lambda_2 = \frac{445.4}{480} \cdot \left(\frac{2 \cdot 10^6}{0.5 \cdot 10^6} \right)^{1/5} = 1.224$

- Factor λ_3

  Design life of the bridge $t_{Ld} = 100$ years:

  Equation 11.22: $\lambda_3 = \left(\frac{100}{100} \right)^{1/5} = 1$

- Factor λ_4

  Number of lanes with heavy traffic k=2

  The values of the influence line for slow lanes (Figure 11.23) are assumed to be determined from a linear distribution among the main girders. The average gross weight of lorries and the number of lorries per year are assumed to be equal for both slow lanes:

  Equation 11.23: $\lambda_4 = \left[1 + \frac{N_{slow}}{N_{slow}} \cdot \left(\frac{0.607 \cdot 445.4}{1.035 \cdot 445.4} \right)^5 \right]^{1/5} = 1.01$

- Factor λ_{max}, Figure 11.16

  Equation 11.24: $\lambda_{max} = 2.5 - 0.5 \cdot \frac{25 - 10}{15} = 2$

  Damage equivalent factor, Equation 11.18: $\lambda = 2.4 \cdot 1.224 \cdot 1 \cdot 1.01 = 2.97$

  It is $\lambda > \lambda_{max}$ so that finally $\lambda = 2$.

- Damage equivalent stress ranges, Equation 11.26

  Top flange "ao": $\Delta\sigma_{E2} = 2 \cdot 3.52 = 7.04$ kN/cm^2

  Bottom flange "au": $\Delta\sigma_{E2} = 2 \cdot 5.82 = 11.64$ kN/cm^2

- Fatigue resistance

  The flange plates at the splice position are welded from both sides by transverse butt welds. The welds are ground flush to plate surface and tested by nondestructive methods. The detail may be then classified in detail category 112. The thickness of the flange plates is 40 mm>25 mm (see Figure 11.11).

  Reduction factor for size effect:

$$k_s = \left(\frac{25}{40} \right)^{0.2} = 0.91$$

  Fatigue resistance:

$$\Delta\sigma_C = 0.91 \cdot 11.2 = 10.19 \text{ kN/cm}^2$$

Table 11.14 Weights of equivalent lorries from FLM4

Vehicle type		*Traffic type*		
		Lorry percentage (n_i)		
Lorry	*Total weight* Q_i *[kN]*	*Long distance (hundreds of km)*	*Medium distance (50–100 km)*	*Local traffic (<50 km)*
No. 1	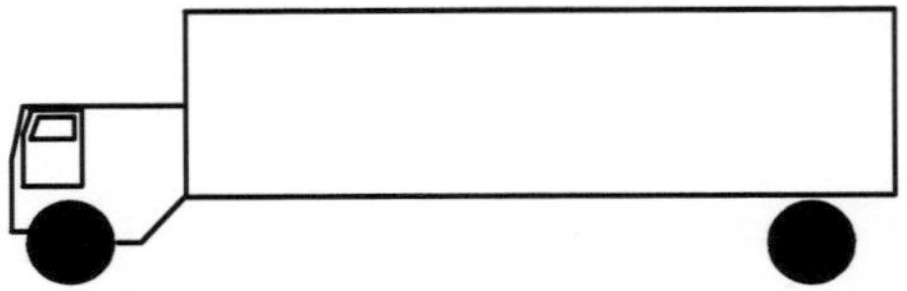 200	20.0	40.0	80.0
No. 2	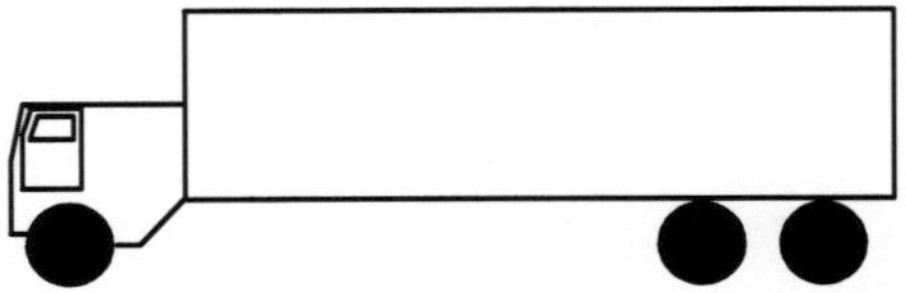 310	5.0	10.0	5.0
No. 3	 490	50.0	30.0	5.0
No. 4	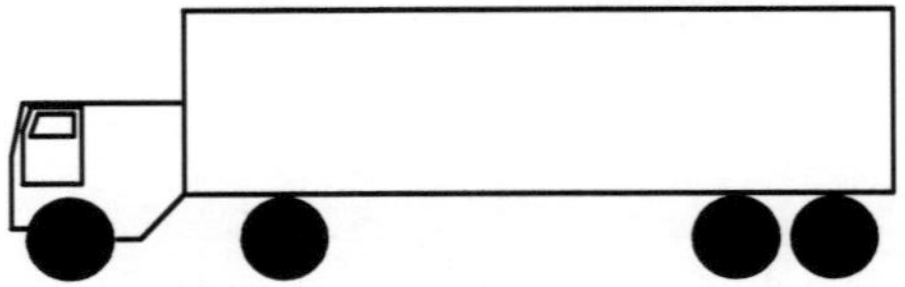 390	15.0	15.0	5.0
No. 5	 450	10.0	5.0	5.0

- Fatigue assessment

  Partial safety factor for action effects: $\gamma_{Ff} = 1.0$

  Main girder with high failure consequences. Safe life assessment method

  Partial safety factor for fatigue resistance (Table 11.8): $\gamma_{Mf,a} = 1.35$

  Equation 11.34: Top flange: $1.0 \cdot 7.04 \leq \frac{10.19}{1.35} = 7.55$ kN/cm² (sufficient)

  Equation 11.34: Bottom flange: $1.0 \cdot 11.64 > \frac{10.19}{1.35} = 7.55$ kN/cm² (not sufficient)

Fatigue assessment of shear connectors

- Stress ranges

  The maximum longitudinal shear force per headed stud under the characteristic combination of actions should be calculated and compared to the 75% of the studs' shear resistance (see Table 12.1 and notes in Table 11.9). It is assumed that this shear force is below $75\% \cdot P_{Rd}$.

  Diameter of shear connectors 22 mm: $A_d = \pi \cdot \frac{0.022^2}{4} = 3.8 \cdot 10^{-4}\ \text{m}^2$

  In the splice region, n=4 connectors are placed in the cross section at a longitudinal spacing $e_L = 350$ mm.

  Modular ratio of concrete for short-term loading (Table 6.4): $n_0 = 6.18$

  Static moment of the slab ($b_{eff} = 312.5$ cm) + reinforcement:

$$S_{c,0} = \left[\frac{3.125 \cdot 0.3}{6.18} + (2.1\% \cdot 3.125 \cdot 0.3)\right] \cdot \left(0.373 - \frac{0.3}{2}\right) = 3.82 \cdot 10^{-2}\ \text{m}^3$$

Note: In the static moment $S_{c,0}$, the reinforcement was considered as concentrated in the centroid of the slab; this is a common simplification. If the reinforcement is neglected, a 12% smaller static moment is calculated. This leads to underestimation of shear stresses.

Stress range, Equation 11.38:

$$\left|\tau_{max,f,Ed} - \tau_{min,f,Ed}\right| = \left|-248 - (-33)\right| \cdot \frac{3.82 \cdot 10^{-2}}{4.47 \cdot 10^{-2}} \cdot \frac{0.35}{4 \cdot 3.8 \cdot 10^{-4}} \cdot \frac{1}{10^4} = 4.23\ \text{kN/cm}^2$$

- Damage equivalent factors

  Factor $\lambda_{v,1} = 1.55$

$$Q_{ml} = \left(\frac{20 \cdot 200^8 + 5 \cdot 310^8 + 50 \cdot 490^8 + 15 \cdot 390^8 + 10 \cdot 450^8}{100}\right)^{1/8} = 457.4\ \text{kN}$$

Factor $\lambda_{v,2}$, Equation 11.21: $\lambda_{v2} = \frac{457.4}{480} \cdot \left(\frac{2 \cdot 10^6}{0.5 \cdot 10^6}\right)^{1/8} = 1.13$

Factor $\lambda_{v,3}$, Equation 11.22: $\lambda_{v3} = \left(\frac{100}{100}\right)^{1/8} = 1$

Factor $\lambda_{v,4}$, Equation 11.23: $\lambda_{v4} = \left[1 + \frac{N_{slow}}{N_{slow}} \cdot \left(\frac{0.607 \cdot 445.4}{1.035 \cdot 445.4}\right)^8\right]^{1/8} = 1$

Damage equivalent factor, Equation 11.37: $\lambda_v = 1.55 \cdot 1.13 \cdot 1 \cdot 1 = 1.75$

- Damage equivalent stress range
  Equation 11.39: $\Delta\tau_{E2} = 1.75 \cdot 4.23 = 7.4\ \text{kN/cm}^2$
- Fatigue assessment
  Fatigue strength:

  $\Delta\tau_c = 90\ \text{MPa}$

  Table 11.9: $1.0 \cdot 7.4 < \frac{9}{1.0} = 9\ \text{kN/cm}^2$ (sufficient)

  For the direct stresses, the cross section is considered as fully cracked:
- Combination $G + C_{sec} + S_{sec} + 0.6 \cdot \Delta TM_{cool} + FLM3$
  Equation 11.32:

$$\left|\sigma_{max,f,Ed} - \sigma_{min,f,Ed}\right| = \left|-400 - (-2283)\right| \cdot \frac{0.482}{2.58 \cdot 10^{-2}} \cdot \frac{1}{10^4} = 3.52\ \text{kN/cm}^2$$

Table 11.9: $\Delta\sigma_C = 80\ \text{MPa}$

From previous calculations: $\lambda_a = 2.0$

$$\Delta\sigma_{E2} = 2.0 \cdot 3.52 = 7.04\ \text{kN/cm}^2 \rightarrow 1.0 \cdot 7.04 > \frac{8.0}{1.35} = 5.93\ \text{kN/cm}^2 \text{ (not sufficient)}$$

Interaction is not checked. In case of sufficient verifications, the characteristic combination $G + C_{sec} + S_{sec} + FLM3$ should be investigated as well.

Note: It is reminded that shear studs should be verified for all four possible combinations for different values of $\Delta\sigma$ and $\Delta\tau$ (see notes in Table 11.9).

Fatigue assessment of reinforcing steel

Global effects

For the **noncyclic** part, the frequent combination in Table 11.2 is applied. Indicatively, the following four combinations will be verified:

Combination Ia:

$$G + C_{sec} + S_{sec} + 0.5 \cdot \Delta TM_{cool} + (0.75 \cdot TS + 0.40 \cdot UDL + 0.4 \cdot q_{fk}) + 1.40 \cdot FLM3$$

$$M_{perm,min} = 600 - 668.3 - 961 - 671 - 3200.4 = -4900.7\ \text{kN-m}$$

$$M_{min,f} = 1.4 \cdot (-449) = -628.6\ \text{kN-m}$$

$$M_{max,f} = 1.4 \cdot 1434 = 2007.6\ \text{kN-m}$$

Equation 11.16: $M_{min,f,Ed} = -4900.7 - 628.6 = -5529.3\ \text{kN-m}$

Equation 11.17: $M_{max,f,Ed} = -4900.7 + 2007.6 = -2893.1\ \text{kN-m}$

Combination 1b:

$$G + C_{sec} + S_{sec} + \left(0.75 \cdot TS + 0.40 \cdot UDL + 0.4 \cdot q_{fk}\right) + 1.40 \cdot FLM3$$

$$M_{perm,max} = 600 - 961 - 671 - 3200.4 = -4232.4 \text{ kN-m}$$

$$M_{min,f} = 1.4 \cdot \left(-449\right) = -628.6 \text{ kN-m}$$

$$M_{max,f} = 1.4 \cdot 1434 = 2007.6 \text{ kN-m}$$

Equation 11.14: $M_{min,f,Ed} = -4232.4 - 628.6 = -4861$ kN-m

Equation 11.15: $M_{max,f,Ed} = -4232.4 + 2007.6 = -2224.8$ kN-m

Combination 2a:

$$G + C_{sec} + S_{sec} + 0.5 \cdot \Delta TM_{cool} + 1.40 \cdot FLM3$$

$$M_{perm,min} = 600 - 668.3 - 961 - 671 = -1700.3 \text{ kN-m}$$

$$M_{min,f} = 1.4 \cdot (-449) = -628.6 \text{ kN-m}$$

$$M_{max,f} = 1.4 \cdot 1434 = 2007.6 \text{ kN-m}$$

Equation 11.16: $M_{min,f,Ed} = -1700.3 - 628.6 = -2328.9$ kN-m

Equation 11.17: $M_{max,f,Ed} = -1700.3 + 2007.6 = 307.3$ kN-m

Combination 2b:

$$G + C_{sec} + S_{sec} + 1.40 \cdot FLM3$$

$$M_{perm,max} = 600 - 961 - 671 = -1032 \text{ kN-m}$$

$$M_{min,f} = 1.4 \cdot \left(-449\right) = -628.6 \text{ kN-m}$$

$$M_{max,f} = 1.4 \cdot 1434 = 2007.6 \text{ kN-m}$$

Equation 11.14: $M_{min,f,Ed} = -1032 - 628.6 = -1660.6$ kN-m

Equation 11.15: $M_{max,f,Ed} = -1032 + 2007.6 - 975.6$ kN-m

Notes:

- In the previous combinations, the noncyclic traffic load is treated as the leading variable action for the frequent combination given in Table 11.2. However, fatigue may be developed without the influence of the traffic load gr1a. For this reason, gr1a in combinations 2a and 2b has been omitted.
- In combinations 1a and 1b, FLM3 has been multiplied with 1.40 because the position for which the verification is conducted is outside the cracked length (=15%·L) (see notes in Table 11.2).
- It may be seen that for both combinations 1a and 1b, tension in the reinforcement is produced. The stress ranges will be therefore determined by the procedure of case 1 in accordance with Section 11.8.2. For combinations 2a and 2b, one moment produces tension, and the other compression in the reinforcement. The stress ranges will be therefore determined by the procedure of case 2.

Equation 11.52b: $\alpha_{st} = \dfrac{696.9 \cdot 2.58}{500 \cdot 1.4} = 2.57$

Case 1 for combination 1a (slab constantly under tension), Equation 11.54:

$$\sigma_{s,max,f,Ed} - \sigma_{s,min,f,Ed} = \left(|-2893.1| \cdot \frac{0.722}{2.58 \cdot 10^{-2}} \cdot \frac{1}{10^3} + 0.2 \cdot \frac{3.2}{2.57 \cdot 0.021} \right) \cdot \left| \frac{-2893.1 - (-5529.3)}{-2893.1} \right| = (80.96 + 11.86) \cdot 0.91$$

$$= 84.47 \text{ MPa} = 8.45 \text{ kN/cm}^2$$

Case 1 for combination 1b (slab constantly under tension), Equation 11.54:

$$\sigma_{s,max,f,Ed} - \sigma_{s,min,f,Ed} = \left(|-2224.8| \cdot \frac{0.722}{2.58 \cdot 10^{-2}} \cdot \frac{1}{10^3} + 0.2 \cdot \frac{3.2}{2.57 \cdot 0.021} \right) \cdot \left| \frac{-2224.8 - (-4861)}{-2224.8} \right| = (62.26 + 11.86) \cdot 1.18$$

$$= 87.5 \text{ MPa} = 8.75 \text{ kN/cm}^2$$

Case 2 for combination 2a (slab tension–compression), Equation 11.55:

$$\sigma_{s,max,f,Ed} - \sigma_{s,min,f,Ed} = \left(307.3 \cdot \frac{0.313}{4.47 \cdot 10^{-2}} + |-2328.9| \cdot \frac{0.722}{2.58 \cdot 10^{-2}} \right) \cdot \frac{1}{10^3} + 11.86$$

$$= 79.18 \text{ MPa} = 7.92 \text{ kN/cm}^2$$

Case 2 for combination 2b (slab tension–compression), Equation 11.55:

$$\sigma_{s,max,f,Ed} - \sigma_{s,min,f,Ed} = \left(975.6 \cdot \frac{0.313}{4.47 \cdot 10^{-2}} + |-1660.6| \cdot \frac{0.722}{2.58 \cdot 10^{-2}}\right) \cdot \frac{1}{10^3} + 11.86$$

$$= 65.16 \text{ MPa} = 6.52 \text{ kN/cm}^2$$

Subsequently, the damage equivalent factors $\lambda_{s,i}$ will be determined.

- Factor $\lambda_{s,1}$
  Span of continuous bridge, L = 25 m, Figure 11.20: $\lambda_{s,1} = 1.17$
- Factor $\lambda_{s,2}$
  For long-distance traffic: $\bar{Q} = 1$

  Equation 11.45: $\lambda_{s,2} = 1 \cdot \sqrt[9]{\frac{2 \cdot 10^6}{2 \cdot 10^6}} = 1$

- Factor $\lambda_{s,3}$
  Design life of the bridge $t_{Ld} = 100$ years.

  Equation 11.46: $\lambda_{s,3} = \left(\frac{100}{100}\right)^{1/9} = 1$
- Factor $\lambda_{s,4}$
  For two slow lanes, Equation 11.47:

  $$\lambda_{s,4} = \sqrt[9]{\frac{2 \cdot (2 \cdot 10^6)}{2 \cdot 10^6}} = 1.08$$

  $\Phi_{fat} = 1.2$ for normal roughness of the pavement layer
  Damage equivalent factor, Equation 11.44: $\lambda_s = 1.2 \cdot 1.17 \cdot 1 \cdot 1 \cdot 1.08 = 1.52$
- Damage equivalent stress ranges, Equation 11.42
  Combination 1a: $\Delta\sigma_{s,equ,glob} = 1.52 \cdot 8.45 = 12.84$ kN/cm²
  Combination 1b: $\Delta\sigma_{s,equ,glob} = 1.52 \cdot 8.75 = 13.3$ kN/cm²
  Combination 2a: $\Delta\sigma_{s,equ,glob} = 1.52 \cdot 7.92 = 12.04$ kN/cm²
  Combination 2b: $\Delta\sigma_{s,equ,glob} = 1.52 \cdot 6.52 = 9.91$ kN/cm²

Local effects

The stress range in the reinforcement calculated with FE from the passage of FLM3 is determined as $\Delta\sigma = 4.8$ kN/cm². The influence line is based on the conservative assumption that $\lambda_{s,1,loc} = \lambda_{s,1,glob} = 1.17$.

Damage equivalent factor, Equation 11.44: $\lambda_s = 1.2 \cdot 1.17 \cdot 1 \cdot 1 \cdot 1.08 = 1.52$

Damage equivalent stress ranges, Equation 11.42: $\Delta\sigma_{s,equ,loc} = 1.52 \cdot 4.8 = 7.3$ kN/cm²

Combined loading due to local and global effects

Combination 1a: $\Delta\sigma_{s,equ} = 12.84 + 7.3 = 20.14$ kN/cm²
Combination 1b: $\Delta\sigma_{s,equ} = 13.3 + 7.3 = 20.6$ kN/cm²
Combination 2a: $\Delta\sigma_{s,equ} = 12.04 + 7.3 = 19.34$ kN/cm²
Combination 2b: $\Delta\sigma_{s,equ} = 9.91 + 7.3 = 17.21$ kN/cm²

- Fatigue assessment, Equation 11.41

$$\text{Combination 1a: } 1.0 \cdot 20.14 = 20.14 > \frac{16.25}{1.15} = 14.13 \text{ kN/cm}^2 \text{ (not sufficient)}$$

$$\text{Combination 1b: } 1.0 \cdot 20.6 = 20.6 > \frac{16.25}{1.15} = 14.13 \text{ kN/cm}^2 \text{ (not sufficient)}$$

$$\text{Combination 2a: } 1.0 \cdot 19.34 = 19.34 > \frac{16.25}{1.15} = 14.13 \text{ kN/cm}^2 \text{ (not sufficient)}$$

$$\text{Combination 2b: } 1.0 \cdot 17.21 = 17.21 > \frac{16.25}{1.15} = 14.13 \text{ kN/cm}^2 \text{ (not sufficient)}$$

Fatigue assessment of concrete

More critical are the combinations leading to the largest compression stresses in concrete. These are those without temperature, creep, or shrinkage:

$$G + \left(0.75 \cdot TS + 0.40 \cdot UDL + 0.4 \cdot q_{fk}\right)$$

The noncyclic traffic load is placed in the most unfavorable position giving a positive bending moment equal to 3396 kN-m.

$$M_{perm} = 600 + 3396 = 3996 \text{ kN-m}$$

Maximum concrete stress: $\sigma_{c,max} = 3996 \cdot \dfrac{0.373}{4.47 \cdot 10^{-2} \cdot 6.18} \cdot 10^{-4} = 0.46 \text{kN/cm}^2$

Minimum concrete stress: $\sigma_{c,min} = 0 \text{ kN/cm}^2$ (conservative assumption)

Concrete age at first loading: $t_0 = 28$ days

Concrete C35/45: $f_{ck} = 35$ MPa, $f_{cd} = \dfrac{35}{1.5} = 23.3$ MPa

Cement of normal hardening rate: $s = 0.25$
Fatigue strength of concrete, Equation 11.57:

$$f_{cd,fat} = 0.85 \cdot e^{0.25 \cdot \left(1 - \sqrt{\frac{28}{28}}\right)} \cdot 23.3 \cdot \left(1 - \frac{35}{250}\right) = 17.03 \text{ MPa} = 1.7 \text{ kN/cm}^2$$

- Fatigue assessment (Table 11.11)
  $f_{ck} < 50$ MPa:

$$\frac{\sigma_{c,max}}{f_{cd,fat}} = \frac{0.46}{1.7} = 0.27 < 0.5 + 0.45 \cdot \frac{0}{1.7} = 0.5 \quad \text{and} \quad \frac{\sigma_{c,max}}{f_{cd,fat}} = 0.27 \leq 0.9 \text{ (sufficient)}$$

EXAMPLE 11.2

In the bridge of Example 11.1, crossbeams are provided at every 1/3 of the span to support the lower flange from lateral torsional buckling. The beams have an IPE 300 cross section in the span region. The flanges of the crossbeams are butt welded to the transverse frames, and their web fillet welded to them (Figure 11.24). The fatigue assessment of this connection is to be made for the beam that is placed at a distance $1/3 \cdot 25 = 8.33$ m from the internal support.

The relative vertical displacements $w_1 - w_2$ between girders 1 and 2 at the cross section under consideration have a maximum value of 6.6 mm and a minimum value of −1.8 mm when the vehicle of FLM3 crosses the bridge. The relative displacements result in end moments $M_{max,f} = 1.78$ kN-m and correspondingly $M_{min,f} = -0.492$ kN-m at the crossbeam, as well as shear forces $V_{max,f} = 0.508$ kN and $V_{min,f} = -0.141$ kN.

- Fatigue assessment of the flanges connection
  The area of the IPE 300 flanges is $A_f = 15 \cdot 1.07 = 16.05$ cm².
  The entire moment is associated to the flanges. The stress range is determined from

$$\sigma_{max,f,Ed} - \sigma_{min,f,Ed} = \left[178 - (-49.2)\right] \cdot \frac{1}{16.05 \cdot (30 - 1.07)/2} = 0.98 \text{ kN/cm}^2$$

According to Table 11.3, the critical length for cross girders is the sum of the two adjacent spans of the stiffeners → $L = 2 \cdot 8.33 = 16.66$ m.

Equation 11.19: $\lambda_1 = 2.55 - 0.7 \cdot \frac{16.66 - 10}{70} = 2.48$

Example 11.1: $\lambda_2 = 1.224$, $\lambda_3 = 1.0$, $\lambda_4 = 1.01$

Equation 11.24: $\lambda_{max} = 2.5 - 0.5 \cdot \frac{16.66 - 10}{15} = 2.278$

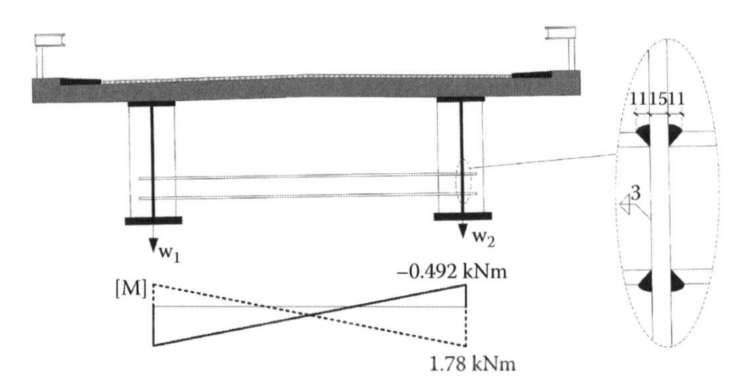

Figure 11.24 Crossbeams and their connection to the transverse frames.

Damage equivalent factor, Equation 11.18: $\lambda = 2.48 \cdot 1.224 \cdot 1 \cdot 1.01 = 3.07$
It is $\lambda > \lambda_{max}$ so that finally $\lambda = 2.278$.
Equivalent stress range, Equation 11.26: $\Delta\sigma_{E2} = 2.278 \cdot 0.98 = 2.23$ kN/cm²
Detail category
The flanges are butt welded to the cross stiffeners. From Figure 11.10,
$l = 11 \cdot 2 + 15 = 37$ mm (Figure 11.24) < 50 mm → detail category 80
Fatigue assessment, Equation 11.34
Partial safety factor for action effects $\gamma_{Ff} = 1.0$
Main girder with low failure consequences. Safe life assessment method
Partial safety factor for fatigue resistance (Table 11.8): $\gamma_{Mf} = 1.15$

$$1.0 \cdot 2.23 \leq \frac{8.0}{1.15} = 6.97\ \text{kN/cm}^2\ \text{(sufficient)}$$

- Fatigue assessment of the web connection
The entire shear force is associated to the web welds. The stress range is determined from the following relation:

$$\max\left(\tau_{max,f,Ed} - \tau_{min,f,Ed}\right) = 1.5 \cdot \frac{V_{max,f} - V_{min,f}}{A_v} = 1.5 \cdot \frac{0.508 - (-0.141)}{2 \cdot 0.3 \cdot 24} = 0.068\ \text{kN/cm}^2$$

Stress ranges, Equation 11.27: $\Delta\tau_{E2} = 2.278 \cdot 0.068 = 0.15$ kN/cm²
Detail category 80 for shear stresses, Figure 11.9

Fatigue assessment, Equation 11.35: $1.0 \cdot 0.15 < \frac{8}{1.15} = 6.96\ \text{kN/cm}^2\ \text{(sufficient)}$

11.11 RESIDUAL STRESSES AND POSTWELD TREATMENT

A welded joint contains a number of zones that differ from the base material due to metallurgic changes that develop during and near after the welding process (Figure 11.25). These zones are the following:

- The weld metal zone that consists of an alloy formed by the melted base material, the electrode material, and/or filler material.
- The fusion zone.
- The heat affected zones in which the base material does not undergo melting. However, the nonuniform distribution of the temperature due to heating and cooling lowers the strength and the ductility of the individual sections and causes residual stresses in both directions x and y.
- The zone in which the properties of the base material remain unchanged.

The residual tensile stresses exceed the yield limit of the structural steel and contribute to the transition of internal cracks from one zone to another during a dynamic excitation: brittle failure. *Grinding* of the weld toe region reduces the stress concentration and improves the fatigue life. Indeed, after taking a closer look in Figure 11.11, one can observe that ground welded connections have 25%–40% higher fatigue strength than the nongrounded ones.

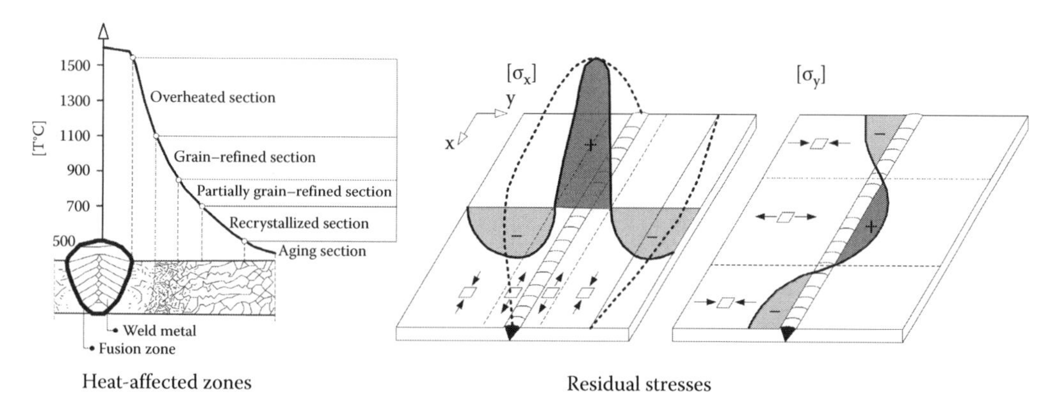

Figure 11.25 Nonuniform temperature distribution due to welding and corresponding residual stresses.

Weld grinding is only one of the many post-weld treatment techniques. Another effective method is the introduction of compressive residual stresses by *hammer peening* [11.15]. This reduces the tensile residual stresses and obviously delays the formation of cracks. EN 1993-1-9 does not provide any guidelines for the estimation of the fatigue strength improvement due to these techniques, and approval is needed.

REFERENCES

[11.1] Calgaro, J.-A., Tschumi, M., Gulvanessian, H.: *Designer's guide to Eurocode 1: Actions on Bridges*. Thomas Telford, London, U.K., 2010.

[11.2] EN 1992-1-1, CEN (European Committee for Standardization): Design of concrete structures. General rules and rules for buildings, Brussels, Belgium, 2004.

[11.3] EN 1992-2, CEN (European Committee for Standardization): Design of concrete structures. Concrete bridges. Design and detailing rules, Brussels, Belgium, 2004.

[11.4] EN 1993-1-9, CEN (European Committee for Standardization),, Eurocode 3: Design of steel structures, Part 1–9: Fatigue, Brussels, Belgium, 2005.

[11.5] EN 1993-2, CEN (European Committee for Standardization),, Eurocode 3: Design of steel structures–Part 2: Steel bridges, Brussels, Belgium, 2006.

[11.6] EN1994-2, CEN (European Committee for Standardization): Design of composite steel and concrete structures. Part 2: Rules for bridges, Brussels, Belgium, 2005.

[11.7] Hanswille, G.: *Eurocode 4 Part 2: Composite Bridges and the German National Annex*. University of Wuppertal, Institute for Steel and Composite Structures, Wuppertal, Germany.

[11.8] Hanswille, G., Porsch, M.: Fatigue behaviour of shear studs. *Stahlbau* 78(3), 2009.

[11.9] Kindmann, R., Stracke, M.: *Verbindungen im Stahl-und Verbundbau*, Ernst und Sohn, Berlin, Germany, 2003.

[11.10] Miner, M. A.: Cumulative damage in fatigue, *Journal of Applied Mechanics*, ASME 67, 159–164, 1945.

[11.11] Palmgren, A.: Die lebensdauer von kugellagern, *Verfahrenstechnik* 68, 339–341, 1924.

[11.12] Raul, J., Davaine, L.: EN 1994 Part 2. Composite bridges. Eurocodes and applications. In: *Dissemination of Information Workshop*, Brussels, Belgium, February 18–20, 2008.

[11.13] Roik, K., Hanswille, G.: Hintergrundbericht zu EC4: Nachweis des Grenzzustandes der Betriebsfestigkeit für Kopfbolzendübel, Bericht Ecc/11/90, Bochum, Germany, 1990.

[11.14] Smith, L. F. C., Hirt, M. A.: Fatigue-resistant steel bridges. *Journal of Constructional Steel Research* 197–214, 1989.

[11.15] Weich, I.: Reasons and calculation of the improved fatigue life of welds treated with high frequency peening methods. *Stahlbau* 78(8), 577–583, 2009 (in German).

Chapter 12

Shear connection

12.1 INTRODUCTION

When a concrete slab rests on a steel girder without any connection, it deflects like the girder but has its own neutral axis so that its top fibers shorten while its bottom fibers elongate. The fibers of the steel girder are also subject to similar displacements in longitudinal direction, so a differential displacement appears at the concrete–steel interface (Figure 12.1a) due to the fact that the bottom fiber of the slab elongates, while the top fiber of the girder shortens. If the differential displacements at the interface are restraint, the slab and the girder behave as a composite girder with a single neutral axis (Figure 12.1b). The restraint is provided by shear connectors, while any natural bond between concrete and steel is ignored. The shear connectors transfer a **longitudinal shear** that develops due to vertical shear. In the case of box girder bridges an additional transverse shear (Figure 9.4d) is developed and resisted by the shear connectors. If the shear connection ensures the full development of the moment resistance of the composite member, it is characterized as full, otherwise as partial. This chapter presents the rules for design of the shear connection, which must be a **full connection** for bridges designed by the rules of EN 1994-2 [1].

Shear connectors could be also subjected to direct tension. This may be the result of frame action between the slab and the girder, where the support moment M is transferred from the slab through the shear studs to the girder web (Figure 12.2). The support moments are very small due to the high flexibility of the web. However, higher moments, and therefore tension forces in the connectors, may develop at places of cross frames or transverse stiffeners. Frame action and tension in the shear connectors may be neglected if the stiffeners or the elements of the cross frame are welded by butt welds to the girder flange and if the slab is modeled and designed by introduction of hinges at its junctions with the girders or girder webs (for box sections).

Figure 2.22 shows the most usual configuration of composite box girders, where the slab is on top of the steel girder and the connectors are welded to the top flange. At internal supports of continuous bridges, the bottom rather than the top flange is in compression. It could then be advantageous to provide in the zone of negative moments concrete in the bottom flange to create a **double** composite action rather than stiffening the bottom flange (Figure 2.22b). The shear connection between the bottom plate and the concrete may be provided by studs welded on both the bottom plate and the web. The studs on the web are almost horizontal, depending on the web inclination.

Another type of double action is indicated in Figure 12.17. The top flange and its concrete are the compression flange of the girder in longitudinal direction and simultaneously act as a composite plate in transverse direction. This composite plate carries directly the traffic loads, where the steel plate is the lower reinforcement of the concrete slab.

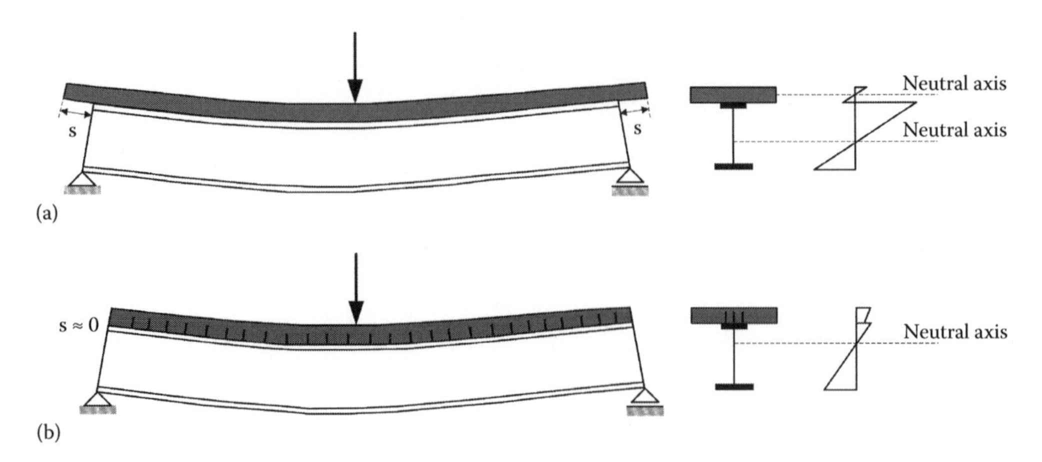

Figure 12.1 Girders in flexure (a) without and (b) with full shear connection.

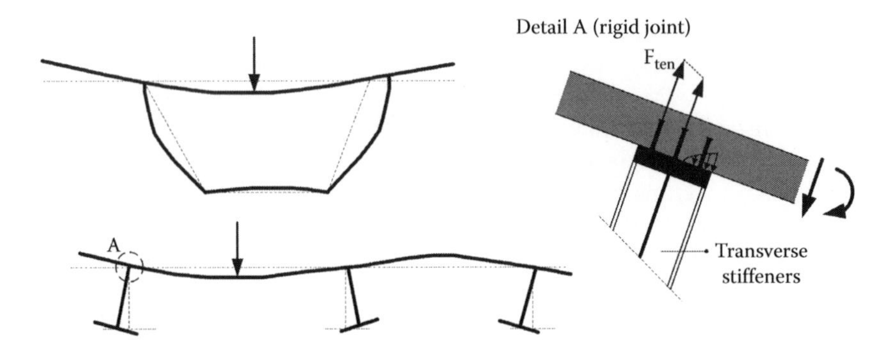

Figure 12.2 Tension forces in studs.

12.2 RESISTANCE AND DETAILING OF HEADED STUD SHEAR CONNECTORS

12.2.1 General

Among the various types of shear connectors, EN 1994-2 provides design rules only for headed studs welded to the steel girder but gives the possibility to use other types if relevant information is given in National Annexes. Headed stud shear connectors have sufficient deformation capacity, their mechanical behavior being regarded as ductile, allowing any inelastic redistribution of shear forces between them. Headed studs have a *shank*, a *head*, and a *weld collar*. The stud is supposed to be automatically welded to the steel element by appropriate machines. The diameter of the stud, d, is the diameter of the shaft; the height, h_{sc}, is the overall height, which must be not smaller than three times the diameter. By provision of minimum dimensions for the head as given in Figure 12.3, uplift separation between the slab and the girder is prevented. The diameter of the stud should not be larger than 2.5 times the thickness of the flange to which it is welded.

Verifications for the shear connection refer to

- Ultimate limit states (ULS)
- Serviceability limit states (SLS), under characteristic combinations of actions
- Fatigue limit states (FLS); see Chapter 11

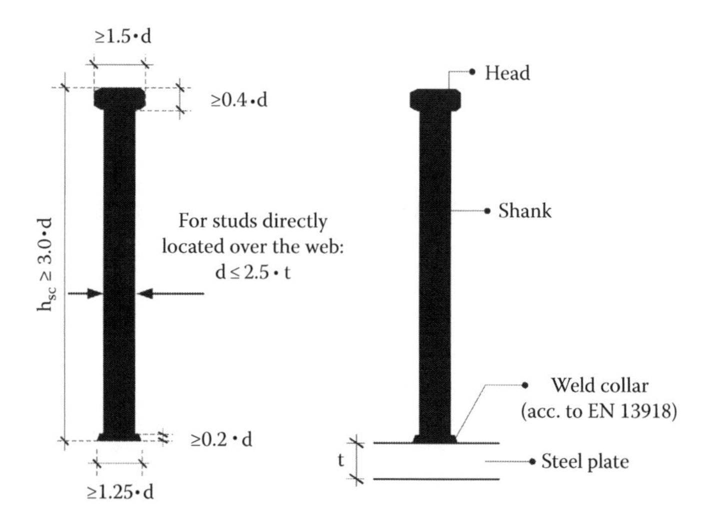

Figure 12.3 Dimensions of headed studs.

12.2.2 Shear resistance of vertical studs

The shear resistance of headed studs for the various limit states is given in Table 12.1.

Table 12.2 shows the shear resistance P_{Rd} at ULS **for solid slabs**, calculated in accordance with the analytical expressions of Table 12.1. They are valid for shear connectors with $h_{sc} \geq 4 \cdot d$, which is the most usual case for composite bridges. One can see that for shear connectors with $f_u = 450$ MPa, the stud's resistance is determined from the shear failure at stud shank toe ($P_{Rd,1}$). For $f_u = 500$ MPa and for the concrete quality C30/37, the stud's shear resistance is determined from the crushing of concrete around the shank ($P_{Rd,2}$); for all the other concrete qualities, the stud's shank toe is the weakest component. Obviously, crushing of concrete is the less ductile failure mode and this is one important reason why the lowest concrete quality for composite bridges should be C35/45.

In case of partially or fully precasted slabs, the minimum thickness of the infill (e_s in Figure 12.4) around the shear connectors should be such that concrete can be cast properly. EN 1994-2 gives no further recommendations. However, it is recommended to reduce the shear resistance P_{Rd} through the k_p-factor of Figure 12.4 for infill thicknesses lower than 25 mm [12.2]. For totally prefabricated slabs with pockets and shear connectors arranged in groups, the inclination of the side openings of the pockets should be lower than 10°, otherwise an additional reduction of the shear resistance by 20% is necessary. If there is no detailed guidance in the National Annex for the use of prefabricated slabs in composite bridges, then both static and fatigue tests may be required to demonstrate a satisfactory performance.

12.2.3 Tensile loading

In the presence of design tensile forces F_{ten}, the shear resistance is not reduced if $F_{ten} \leq 0.1 \cdot P_{Rd}$. The current design rules do not cover cases where $F_{ten} > 0.1 \cdot P_{Rd}$. Excessive tensile forces in the studs may arise in locations of bracings and plated diaphragms in closed box-girder bridges; see Figure 2.20. In order to avoid concrete's pullout failure, the strengthening solution of Figure 12.5 with welded plates and horizontal studs can be described as an appropriate one. Such a detailing ensures a safe transmission of local forces from steel to concrete by avoiding additional punching reinforcement.

Table 12.1 Shear resistance of headed studs in solid slabs for various limit states

Failure mode 1	Shear at stud shank toe	
	$P_{Rd,1} = \frac{0.8 \cdot f_u \cdot (\pi \cdot d^2/4)}{\gamma_v}$	(12.1)
Failure mode 2	Crushing of concrete around the shank	
	$P_{Rd,2} = \frac{0.29 \cdot \alpha \cdot d^2 \cdot \sqrt{f_{ck} \cdot E_{cm}}}{\gamma_v}$	(12.2)
Shear resistance at ULS	$P_{Rd} = \min\{P_{Rd,1}, P_{Rd,2}\}$	(12.3)
Shear resistance at SLS	$P_{Rd,ser} = k_s \cdot P_{Rd} = 0.75 \cdot P_{Rd}$	(12.4)
Shear resistance at FLS	See Chapter 11	
Parameters	$\alpha = 0.2 \cdot \left(\frac{h_{sc}}{d} + 1\right)$ for $3 \leq \frac{h_{sc}}{d} \leq 4$	(12.5)
	$\alpha = 1$ for $\frac{h_{sc}}{d} > 4$	(12.6)
	$\gamma_v = 1.25$ partial safety factor	

Notes:

d is the diameter of shank, but $16\text{ mm} \leq d \leq 25\text{ mm}$ and $d \leq 2.5 \cdot t_{ao}$.
t_{ao} is the thickness of the top steel flange.
f_u is the specified nominal strength of stud material but ≤ 500 MPa.
f_{ck} is the cylinder strength of concrete (Table 6.1).
h_{sc} is the height of stud, Figure 12.3.

Table 12.2 Shear resistance P_{Rd} (kN) of headed studs with $h_{sc}/d \geq 4$ in solid slabs at ULS

Shank diameter d (mm)	*Minimum h_{sc} (mm)*	*f_u = 450 MPa and C30/37 to C60/75 (Failure of shank)*	*f_u = 500 MPa and C30/37 (Concrete crushing)*	*f_u = 500 MPa and C35/45 to C60/75 (Failure of shank)*
25	100	141.30	144.27	157.00
22	88	109.42	111.73	121.58
19	76	81.61	83.33	90.68
16	64	57.88	59.09	64.31

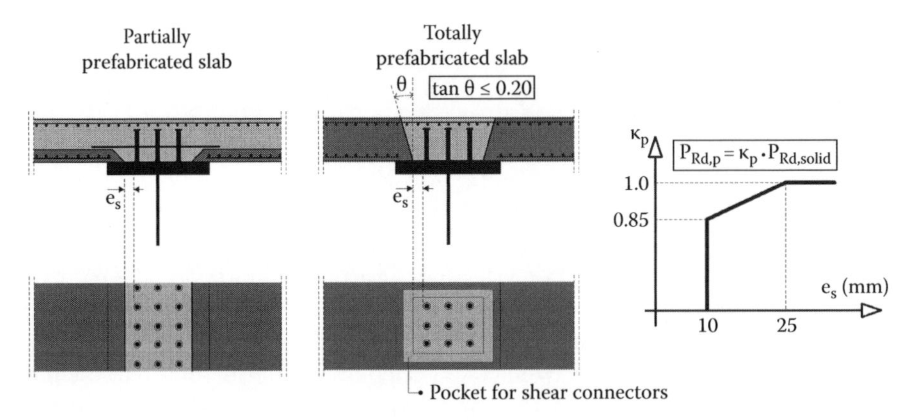

Figure 12.4 Reduction factor k_p for prefabricated deck slabs. (From Composite Bridge Design for Small and Mediums Spans, Design Guide, ECSC Steel Programme, 2002.)

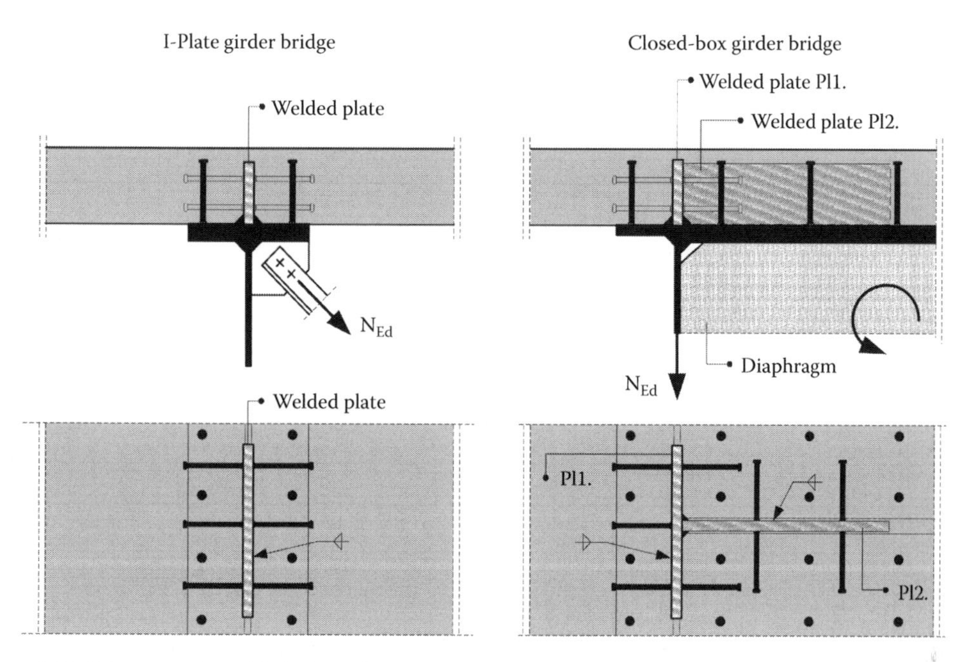

Figure 12.5 Solution with welded plates for reducing tensile forces in studs.

It is noted that for headed studs under tension and subjected to fatigue loadings, the diameter d (Figure 12.3) should not exceed 1.5 times the thickness of the flange to which it is welded.

12.2.4 Detailing of shear connectors

There exist certain detailing rules in respect to the spacing of shear connectors in longitudinal and transverse direction, the distance between transverse reinforcement and the lower side of the head of studs, and the edge distance of the stud from the steel flange. Where the slab has haunches, the sides of the haunch must lie outside a line drawn at 45° from the outside edge of the stud. In addition, transverse reinforcement must be provided at the haunch. The detailing rules are given in Table 12.3.

12.2.5 Horizontal arrangement of studs

The estimation of the studs' shear resistance according to Tables 12.1 and 12.2 is only valid when splitting forces are developed transverse to the direction of the slab thickness, in other words, for vertically arranged studs. However, in many cases, studs have to be placed in a horizontal arrangement and welded to the web(s) of the steel cross section; see Figures 2.14 and 12.6. Then adequate anchored transverse reinforcement surrounding the connectors should be provided. If the anchoring length v of the shear studs is not smaller than $14 \cdot d$, the distance e_v not smaller than $6 \cdot d$ and the spacing s of the stirrups do not exceed $18 \cdot d$ then the stud's shear resistance can be calculated from Tables 12.1 and 12.2. In addition, the stirrups should be designed against a splitting force T_d equal to $0.3 \cdot P_{Rd}$, where P_{Rd} is the design shear resistance of the connector. If the previous conditions are not fulfilled, then the shear

Table 12.3 Detailing of shear connectors

Condition	*Limitation*
Spacing in longitudinal direction (e_L)	$5 \cdot d \leq e_L \leq \min\{4 \cdot h_c, 800\ \text{mm}\}$
Spacing in transverse direction (e_T) and clear distance between edge of stud and edge of flange (e_D)	$e_D \geq 25$ mm for solid slabs $e_T \geq 2.5 \cdot d$ else $e_T \geq 4.0 \cdot d$
Studs on compression flanges that would be class 3 or 4 but are classified due to the shear connection as 1 or 2	$e_L \leq 22 \cdot \varepsilon \cdot t_{ao}$ for solid slabs and $\leq 15 \cdot \varepsilon \cdot t_{ao}$ in other cases $e_D \leq 9 \cdot \varepsilon \cdot t_{ao}$
Distance between down side of head and lower transverse slab reinforcement	≥ 30 mm for flat slabs ≥ 40 mm for slabs with haunches
Concrete cover from the side of the haunch to the connector (e_v)	$e_V \geq 50$ mm
Concrete cover for shear connectors (c)	$c \geq \max$ (20 mm, acc. to EN 1992-1-1)

If cover is not required, then a zero cover is allowed (c = 0).

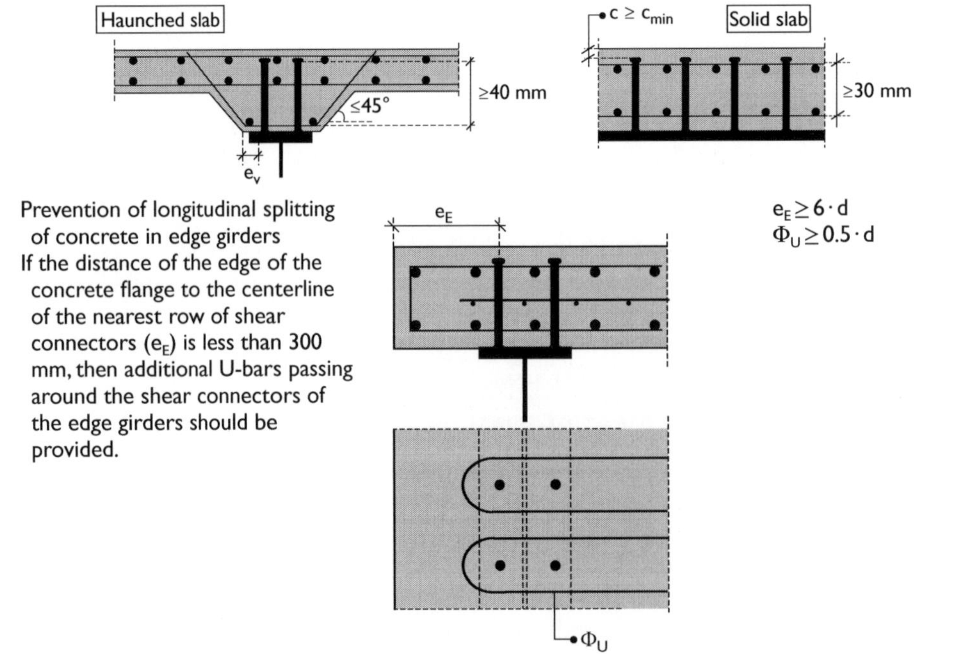

Condition	*Limitation*
Prevention of longitudinal splitting of concrete in edge girders If the distance of the edge of the concrete flange to the centerline of the nearest row of shear connectors (e_E) is less than 300 mm, then additional U-bars passing around the shear connectors of the edge girders should be provided.	$e_E \geq 6 \cdot d$ $\Phi_U \geq 0.5 \cdot d$

Note: $\varepsilon = \sqrt{f_{y,ao}/235}$ where $f_{y,ao}$ is the yield strength for structural steel of the top flange in N/mm² from Tables 6.14 and 6.15. The limitations are also valid for bottom flanges.

resistance of the horizontal headed studs becomes lower than that for the vertical ones and is calculated from the following expression; see also [12.3]:

$$P_{Rd,L} = \frac{0.0014 \cdot k_v \cdot (f_{ck} \cdot d \cdot a'_r)^{0.4} \cdot (\alpha/s)^{0.3}}{\gamma_v} \leq \min[P_{Rd,1}, P_{Rd,2}] \tag{12.7}$$

where

$a'_r = a_r - c_v - \Phi_s/2 \geq 50\,\text{mm}$ is the effective edge distance
$k_v = 1$ for connection in an edge position
$k_v = 1.14$ for connection in a middle position
$\gamma_v = 1.25$ is the partial safety factor
f_{ck} is the characteristic cylinder strength of the concrete in MPa (Table 6.1)
d is the shank diameter of the headed stud in millimeters (19 mm $\leq$ d $\leq$ 25 mm)
α is the horizontal spacing of the studs with 110 mm $\leq \alpha \leq$ 440 mm
s is the spacing of the stirrups with $\alpha/2 \leq s \leq \alpha$ and $s/a'_r \leq 3$
Φ_s is the diameter of the stirrups ($\geq$8 mm)
c_v is the vertical concrete cover in millimeters
$P_{Rd,1}$ and $P_{Rd,2}$ are the shear resistances for failure modes 1 and 2 in Table 12.1

The transverse reinforcement (stirrups of diameter Φ_s) should be designed for a tension force equal to $0.3 \cdot P_{Rd,L}$, where $P_{Rd,L}$ is the design shear resistance of the connector; Equation 12.7. The pullout failure of concrete can be avoided if the maximum value for the anchoring length v of the shear stud is greater than (Figure 12.6)

$\max\{110\,\text{mm}, 1.7 \cdot \alpha'_r, 0.85 \cdot s\}$ for uncracked concrete
$\max\{160\,\text{mm}, 2.4 \cdot \alpha'_r, 1.2 \cdot s\}$ for cracked concrete

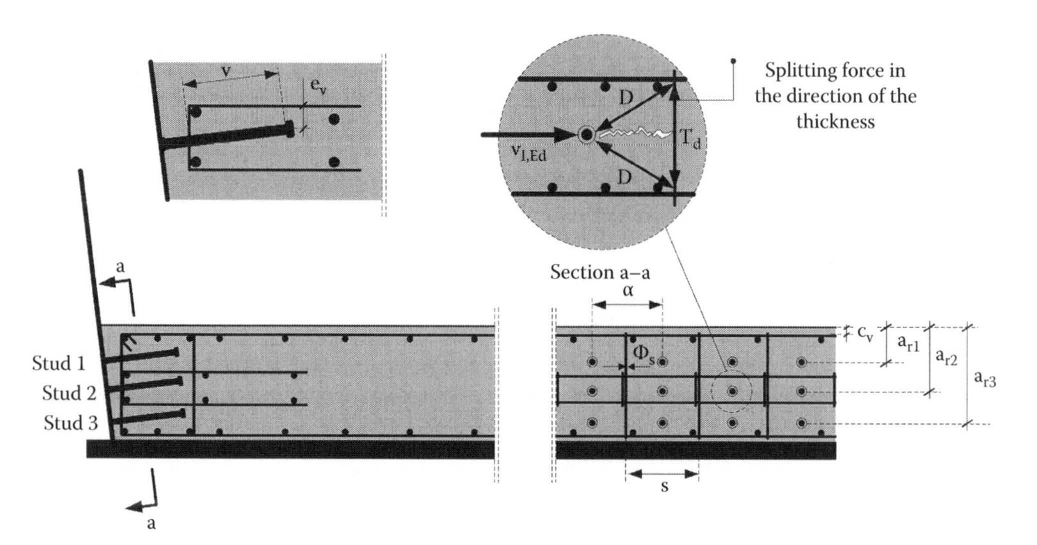

Figure 12.6 Geometric parameters of shear connections with horizontally arranged studs (bottom flange box-girder bridge with double composite action).

In case of simultaneous action of vertical and horizontal shear (e.g., solution B in Figure 2.14), the following interaction should be applied:

$$\left(\frac{F_{Ed,L}}{P_{Rd,L}}\right)^{1.2} + \left(\frac{F_{Ed,V}}{P_{Rd,V}}\right)^{1.2} \leq 1.0 \tag{12.8}$$

where
$F_{Ed,V}$ is the vertical force acting on the stud due to vertical support of the slab
$F_{Ed,L}$ is the longitudinal force acting on the stud due to bending of the main girder
$P_{Rd,L}$ is the shear resistance calculated with Equation 12.7
$P_{Rd,V}$ is estimated as follows:

$$P_{Rd,V} = \frac{0.012 \cdot (f_{ck} \cdot \Phi_L)^{0.5} \cdot (d \cdot \alpha/s)^{0.4} \cdot (\Phi_S)^{0.3} \cdot (\alpha'_{r,0})^{0.7} \cdot k_v}{\gamma_v} \leq \min[P_{Rd,1}, P_{Rd,2}] \tag{12.9}$$

where $a'_{r,0} = a_{r,0} - c_v - \Phi_s/2 \geq 50\,mm$ is the effective upper edge distance.

The nominal height of the stud connector should be at least 100 mm.

12.3 LONGITUDINAL SHEAR FOR ELASTIC BEHAVIOR

As explained in Section 12.1, the longitudinal shear at the concrete–steel interface results from vertical shear forces. The horizontal shear flow (shear force per unit length) may be determined from the well-known formula of mechanics for elastic behavior, in accordance with

$$v_{L,Ed} = \frac{V_{Ed} \cdot S}{I} \tag{12.10}$$

where
V_{Ed} is the design vertical shear force
$S = S_{1,L}$ is the first moment of area (static moment) of the concrete slab in respect to the center of gravity of the composite section for the load case and the time considered; see Figure 7.39
$I = I_{1,L}$ is the second moment of area of the composite cross section for the load case and the time considered; see Figure 7.39

The aforementioned relation is applied on the safe side also in regions of hogging moments, or generally where the concrete is considered as cracked and does not transfer direct stresses. This relation presumes elastic behavior and is applicable:

- For SLS
- For FLS
- For ULS, where the design moment is lower than the elastic moment resistance

Equation 12.10 indicates that the longitudinal shear at the concrete–steel interface follows the vertical shear forces. However, for load combinations, the longitudinal shear cannot be determined by combination of the vertical shear but has to be determined separately for each individual load case and then combined for the following reasons:

- The cross sections, and accordingly S and I of Equation 12.10, are generally not constant along the length of the bridge.
- The modular ratio of concrete and accordingly the cross-sectional properties are different for the various load cases and variable in time. The modular ratio of concrete is taken as n_0 for traffic loads, temperature, and other actions of short duration; n_P for permanent loads; and n_S for shrinkage; see Equation 6.20.
- In cross sections where concrete has still not been cast, vertical shear but not longitudinal shear exists.

The longitudinal shear is stressing the shear connectors and is always positive. However, the vertical shear has a positive or negative sign (Figure 12.7). The sign expresses the direction of the longitudinal shear and must be considered in the combination of actions, that is, longitudinal shears of the same direction are added, those of reverse direction subtracted.

Headed studs provide the longitudinal shear resistance, which may be determined at ULS and SLS from

$$v_{L,Rd} = \frac{n \cdot P_{Rd}}{e_L} \tag{12.11}$$

and

$$v_{L,Rd,ser} = \frac{n \cdot P_{Rd,ser}}{e_L} \tag{12.12}$$

where

n is the number of shear connectors at one cross section
e_L is the longitudinal spacing of connectors
P_{Rd} is the shear resistance from Equation 12.3
$P_{Rd,ser}$ is the shear resistance from Equation 12.4

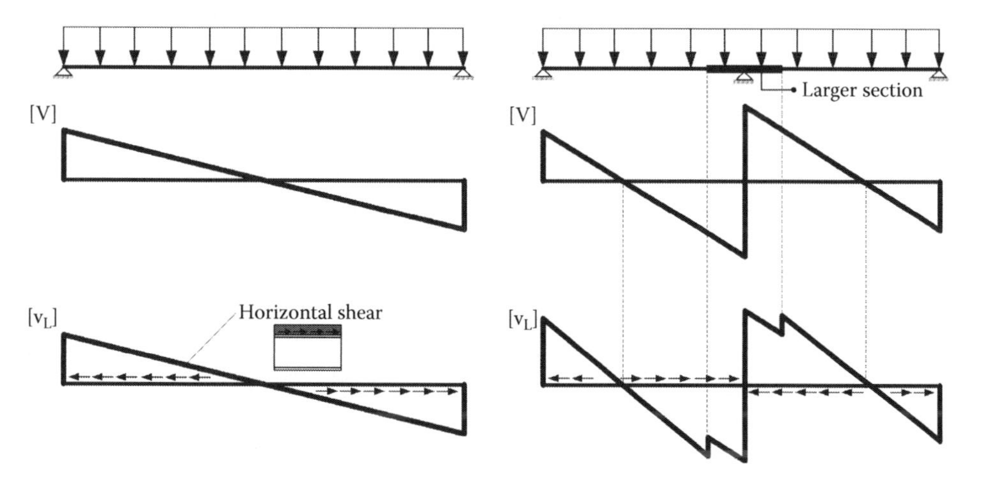

Figure 12.7 Shear forces and horizontal shear.

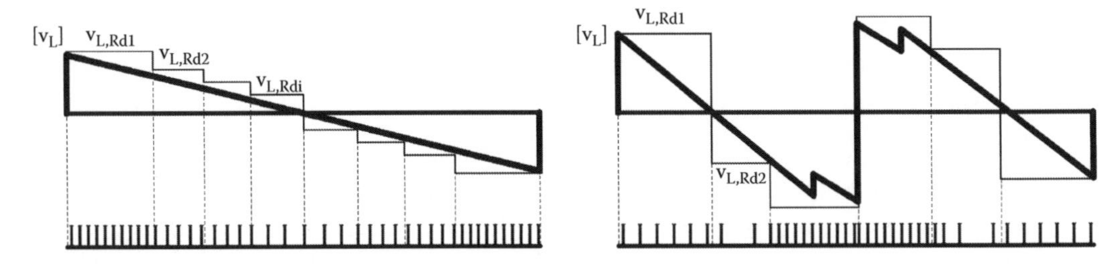

Figure 12.8 Cover of the diagram of longitudinal shear along the bridge.

The design relations at ULS and SLS may be written as

$$v_{L,Ed} \leq v_{L,Rd} \tag{12.13}$$

or

$$v_{L,Ed,ser} \leq v_{L,Rd,ser} \tag{12.14}$$

The aforementioned relations show that the diagram of longitudinal shear is covered by appropriate selection of the size of shear connectors, their number in the cross section, and their longitudinal spacing as indicatively illustrated in Figure 12.8. Usually, the number and spacing of connectors is kept constant over certain lengths to achieve constant shear resistance. The design shear $v_{L,Ed}$ at ULS may exceed the shear resistance $v_{L,Rd}$ by 10% at certain points, provided that the total resisting force in this zone is larger than the design force, that is, the area of the diagram of the shear resistance is larger than the corresponding area of the design shear (Figure 12.8).

EXAMPLE 12.1

The shear connection of the box-girder bridge in Example 9.4 will be verified. Shear forces from structural analysis are provided in Table 12.4.

SHEAR CONNECTION AT EDGE SUPPORT

The shear flow between steel and concrete will be calculated taking into account the construction sequence, the rheological behavior of concrete, and the shear lag effects both on concrete and steel (bottom) flange.

- Shear connection for short-term design at ULS

  The maximum shear force acting on the composite cross section at the edge support is calculated according to the basic combination of Equation 5.11a with leading variable traffic as follows:

$$\max V_{Ed,0} = 1.35 \cdot (537.6 + 27.0) + 1.5 \cdot 0.6 \cdot 142.6 + 1.35 \cdot 892.0 = 2094.75\ \text{kN}$$

  One can observe that the shear forces acting on the pure steel section were not taken into account because they do not produce any horizontal shear.

  The short-term static moment of the half girder is calculated from Table 9.17 and Figure 7.39:

$$S_{I,0} = \frac{30 \cdot 1{,}100/2}{6.18} \cdot (47.64 + 15) = 167{,}242.7\ \text{cm}^3$$

Table 12.4 Shear forces for the half-girder A

		At edge support	At internal support
Loading from	Figure	maxV (kN)	minV (kN)
Pure steel	*9.18*	*360.4*	*−597.2*
First concreting	*9.18*	*924.0*	*−792.0*
Second concreting + superstructures	9.19	537.6	−1140.4
Superstructures (final)	9.20	27.0	−383.0
Thermal	9.22	142.6	−139.9
Traffic (gr1a)	9.23	892.0	−1168.0
Secondary effects from second concreting	9.19	13.4	−13.4
Secondary effects from superstructures (final)	9.20	4.2	−4.2
Shrinkage secondary effects	9.21	115.6	−115.6

Note: Shear forces acting on the pure steel cross section are faced with Italic.

The horizontal shear flow is calculated from Equation 12.10:

$$v_{L,Ed,0} = \frac{2,094.75 \cdot 167,242.7}{(1.037/2) \cdot 10^8} = 6.76 \text{ kN/cm}$$

The shear resistance of a headed stud with $h_{sc} = 250$ mm, $d = 22$ mm, and $f_u = 450$ N/mm^2 is taken from Table 12.2 and is equal to $P_{Rd} = 109.42$ kN.

For four connectors per row ($n = 4$), Equations 12.11 and 12.13:

$$6.76 \leq v_{L,Rd} = \frac{4 \cdot 109.42}{e_L} \rightarrow e_L \leq 64.7 \text{ cm}$$

- Shear connection for long-term design at ULS

The shear forces act on cross sections with different cross-sectional properties. Therefore, the ULS combination refers to shear stresses and not to forces.

The loading case *2nd concreting + superstructures* refers to a cross section with an age of concrete at loading time equal to 7 days. From Table 9.17, $n_P = 18.82$, $I_{I,P} = (0.8107/2) \cdot 10^8 = 0.40535 \cdot 10^8$ cm^4, and

$$S_{I,P} = \frac{30 \cdot 1,100/2}{18.82} \cdot (84.32 + 15) = 87,076.51 \text{ cm}^3$$

The loading case *Superstructures* refers to a cross section with an age of concrete at loading time equal to 14 days. From Table 9.17, $n_P = 17.26$, $I_{I,P} = (0.8261/2) \cdot 10^8 = 0.41305 \cdot 10^8$ cm^4, and

$$S_{I,P} = \frac{30 \cdot 1,100/2}{17.26} \cdot (81.81 + 15) = 92,547.22 \text{ cm}^3$$

The loading case *Secondary effects from 2nd concreting* refers to a cross section with an age of concrete at loading time equal to 7 days. From Table 9.17, $n_{PT}=12.50$, $I_{I,PT}=(0.8878/2)\cdot 10^8=0.4439\cdot 10^8$ cm^4, and

$$S_{I,PT}=\frac{30\cdot 1{,}100/2}{12.50}\cdot(71.76+15)=114{,}523.2\text{ cm}^3$$

The loading case *Secondary effects from superstructures* (final) refers to a cross section with an age of concrete at loading time equal to 14 days. From Table 9.17, $n_{PT}=11.72$, $I_{I,PT}=(0.9008/2)\cdot 10^8=0.4504\cdot 10^8$ cm^4, and

$$S_{I,PT}=\frac{30\cdot 1{,}100/2}{11.72}\cdot(69.64+15)=119{,}160.4\text{ cm}^3$$

The loading case *Shrinkage secondary effects* refers to a cross section with an age of concrete at loading time equal to 1 day. From Table 9.17, $n_{PT}=15.22$, $I_{I,PT}=(0.8493/2)\cdot 10^8=0.4247\cdot 10^8$ cm^4, and

$$S_{I,PT}=\frac{30\cdot 1{,}100/2}{15.22}\cdot(78.02+15)=100{,}843\text{ cm}^3$$

The design horizontal shear is (see Table 12.4)

$$v_{L,Ed,\infty}=1.35\cdot\left(537.6\cdot\frac{870{,}76.51}{0.40535\cdot 10^8}+27.0\cdot\frac{92{,}547.22}{0.41305\cdot 10^8}+13.4\cdot\frac{114{,}523.2}{0.4439\cdot 10^8}\right.$$
$$\left.+4.2\cdot\frac{119{,}160.4}{0.4504\cdot 10^8}\right)+115.6\cdot\frac{100{,}843}{0.4247\cdot 10^8}+1.35\cdot\left(892.0\cdot\frac{167{,}242.7}{(1.037/2)\cdot 10^8}\right)$$
$$+1.5\cdot 0.6\cdot\left(142.6\cdot\frac{167{,}242.7}{(1.037/2)\cdot 10^8}\right)=6.27\text{ kN/cm}$$

This value is lower than for short-term design.

- Shear connection for short-term design at SLS
  The maximum shear force acting on the composite cross section at the edge support is calculated according to the characteristic combination of Equation 5.23 with leading variable traffic as follows:

$$\max V_{Ed,ser,0}=1.0\cdot(537.6+27.0)+0.6\cdot 142.6+1.0\cdot 892.0=1542.16\text{ kN}$$

The horizontal shear flow is estimated from Equation 12.10:

$$v_{L,Ed,0}=\frac{1{,}542.16\cdot 167{,}242.7}{(1.037/2)\cdot 10^8}=4.97\text{ kN/cm}$$

From Table 12.1,

$$P_{Rd,ser}=0.75\cdot 109.42=82.1\text{ kN}$$

For four connectors per row (n = 4), Equations 12.12 and 12.14:

$$4.97\le v_{L,Rd,ser,0}=\frac{4\cdot 82.1}{e_L}\to e_L\le 66\text{ cm}$$

- Shear connection for long-term design at SLS
  The shear forces act on cross sections with different cross-sectional properties. Therefore, the ULS combination refers to shear stresses and not to forces.
  The design horizontal shear is (see Table 12.4)

$$v_{L,Ed,\infty} = 1.0 \cdot \left(537.6 \cdot \frac{87{,}076.51}{0.40535 \cdot 10^8} + 27.0 \cdot \frac{92{,}547.22}{0.41305 \cdot 10^8} + 13.4 \cdot \frac{114{,}523.2}{0.4439 \cdot 10^8} \right.$$

$$\left. + 4.2 \cdot \frac{119{,}160.4}{0.4504 \cdot 10^8} \right) + 115.6 \cdot \frac{100{,}843}{0.4247 \cdot 10^8} + 1.0 \cdot \left(892.0 \cdot \frac{167{,}242.7}{(1.037/2) \cdot 10^8} \right)$$

$$+ 0.6 \cdot \left(142.6 \cdot \frac{167{,}242.7}{(1.037/2) \cdot 10^8} \right) = 4.69\ \text{kN/cm}$$

  Again, this value is lower than that for short-term design.

- Final choice
  From the aforementioned, the short-term check at ULS is the less favorable ($e_L \leq 647$ mm). It is chosen that $e_L = 640$ mm.
  According to Table 12.3,

$$5 \cdot d = 5 \cdot 22 = 110\ \text{mm} \leq e_L \leq \min\{4 \cdot 300,\ 800\ \text{mm}\}$$

$$= 800\ \text{mm (satisfied) (see Figure 12.9)}$$

SHEAR CONNECTION AT INTERNAL SUPPORT

The shear flow between steel and concrete will be calculated taking into account the construction sequence, cracking of concrete, and shear lag effects both on concrete and steel (bottom) flange. The long-term design shear force is obviously higher than the short-term one. Therefore, only the long-term design is shown.

- Shear connection for long-term design at ULS
  The design shear forces is

$$\min V_{Ed,\infty} = 1.35 \cdot (-383.0 - 4.2) - 1.0 \cdot 115.6 - 1.5 \cdot 0.6 \cdot 139.9 - 1.35 \cdot 1168$$

$$= -2341.03\ \text{kN}$$

  One can observe that the shear forces acting on the pure steel section were not taken into account because they do not produce any horizontal shear. As already mentioned, the effects of cracking of concrete cannot be taken into account accurately enough. Conservatively, the shear flow will be calculated with the properties of the uncracked composite section and with the effective[s] widths for hogging bending. For concrete, the entire deck slab is active (see Example 9.4). For the bottom flange, the shear lag effect was of negligible magnitude and the bottom plate was considered as fully active too. Therefore, the properties of the half girder's cross section can be taken from Table 9.17: $I_1 = (1.037/2) \cdot 10^8 = 0.5185 \cdot 10^8$ cm^4 and

$$S_1 = \frac{30 \cdot 1{,}100/2}{6.18} \cdot (47.64 + 15) = 167{,}242.7\ \text{cm}^3$$

The horizontal shear flow is estimated from Equation 12.10:

$$v_{L,Ed,\infty} = \frac{2,341.03 \cdot 167,242.7}{0.5185 \cdot 10^8} = 7.55 \text{ kN/cm}$$

The shear resistance of a headed stud with h_{sc}=250 mm, d=22 mm, and f_u=450 mm is taken from Table 12.2 and is equal to P_{Rd}= 109.42 kN.

For six connectors per row (n=6), Equations 12.11 and 12.13:

$$7.55 \leq v_{L,Rd} = \frac{6 \cdot 109.42}{e_L} \rightarrow e_L \leq 87.0 \text{ cm}$$

- Shear connection for long-term design at SLS
  The design shear forces is

$$\min V_{Ed,\infty} = 1.0 \cdot (-383.0 - 4.2) - 1.0 \cdot 115.6 - 0.6 \cdot 139.9 - 1.0 \cdot 1168 = -1754.74 \text{ kN}$$

The cross-sectional properties are affected from the shear lag in the bottom flange (Example 9.4) and are given in Table 10.1. For the fully cracked section,

$$A_{2,sa} = 3823.6 \text{ cm}^2, \quad \overline{z}_{2,sa} = 15 + 106.47 = 121.47 \text{ cm}, \quad I_{2,sa} = 0.46049 \cdot 10^8 \text{ cm}^4$$

For the half girder,

$$A_I = 0.5 \cdot \left(3,823.6 + \frac{30 \cdot 1,100}{6.18}\right) = 4,581.71 \text{ cm}^2, \quad \overline{z}_I = \frac{(3,823.6/2) \cdot 121.47}{4,581.71} = 50.69 \text{ cm}$$

$$S_I = \frac{30 \cdot 1,100/2}{6.18} \cdot 50.69 = 135,337.4 \text{ cm}^3, \quad I_I = \frac{0.46049 \cdot 10^8}{2} + \frac{3823.6}{2} \cdot (121.47 - 50.69)^2$$

$$+ \frac{30^3 \cdot (1,100/2)}{12 \cdot 6.18} + \frac{30 \cdot 1,100/2}{6.18} \cdot 50.69^2 = 0.3966 \cdot 10^8 \text{ cm}^4$$

From the comparison of the cross-sectional properties for the noncracked cross section in Table 9.17 with the previous values, one can easily see that the shear lag effect in the bottom flange should not be neglected.

The horizontal shear flow is estimated from Equation 12.10:

$$v_{L,Ed,ser,\infty} = \frac{1,754.74 \cdot 135,337.4}{0.3966 \cdot 10^8} = 5.99 \text{ kN/cm}$$

From Table 12.1,

$$P_{Rd,ser} = 0.75 \cdot 109.42 = 82.1 \text{ kN}$$

For six connectors per row (n=6), Equations 12.12 and 12.14:

$$5.99 \leq v_{L,Rd,ser,\infty} = \frac{6 \cdot 82.1}{e_L} \rightarrow e_L \leq 82.2 \text{ cm}$$

- Final choice
  From the aforementioned, the long-term check at SLS is the less favorable ($e_L \leq$822 mm). It is chosen that e_L=800 mm.

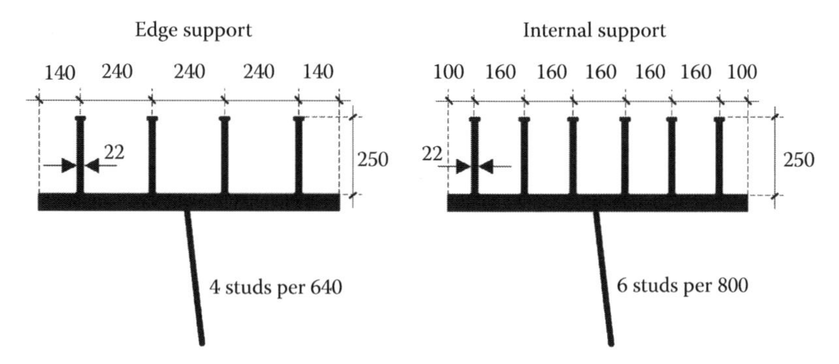

Figure 12.9 Arrangement of studs at edge and internal support (for the edge support, see also Example 12.3).

According to Table 12.3,

$$5 \cdot d = 5 \cdot 22 = 110\ \text{mm} \leq e_L \leq \min\{4 \cdot 300,\ 800\ \text{mm}\} = 800\ \text{mm (verified) (see Figure 12.9)}$$

Note: The shear forces in the headed studs due to torsion are obviously of negligible magnitude and they have not been taken into account. This would not happen in the case of a curved bridge.

LIMITATIONS ACCORDING TO TABLE 12.3

- For the edge support,

$$e_T = 240\ \text{mm} > 2.5 \cdot 22 = 55\ \text{mm}$$

$$e_D = 140 - 1.25 \cdot \frac{22}{2} = 126\ \text{mm} > 25\ \text{mm}$$

- For the internal support,

$$e_T = 160\ \text{mm} > 2.5 \cdot 22 = 55\ \text{mm}$$

$$e_D = 100 - 1.25 \cdot \frac{22}{2} = 86.25\ \text{mm} > 25\ \text{mm}$$

The distance of the edge of the concrete flange to the centerline of the nearest row of shear connectors is much greater than 300 mm; additional U-bars passing around the shear connectors of the edge girders are not necessary.

Obviously, more positions along the length of the bridge should be similarly verified and the studs should be arranged according to the magnitude of the shear flow (Figure 12.8); attention is needed in bridges with cross sections of longitudinally variable total depth.

12.4 LONGITUDINAL SHEAR FOR INELASTIC BEHAVIOR

As discussed before, the mechanics' Equation 12.10 is valid for elastic behavior. However, at ULS and for cross sections of class 1 or 2, it is possible to exploit the plastic bending resistance. In such cases, the design longitudinal shear is determined by consideration of the

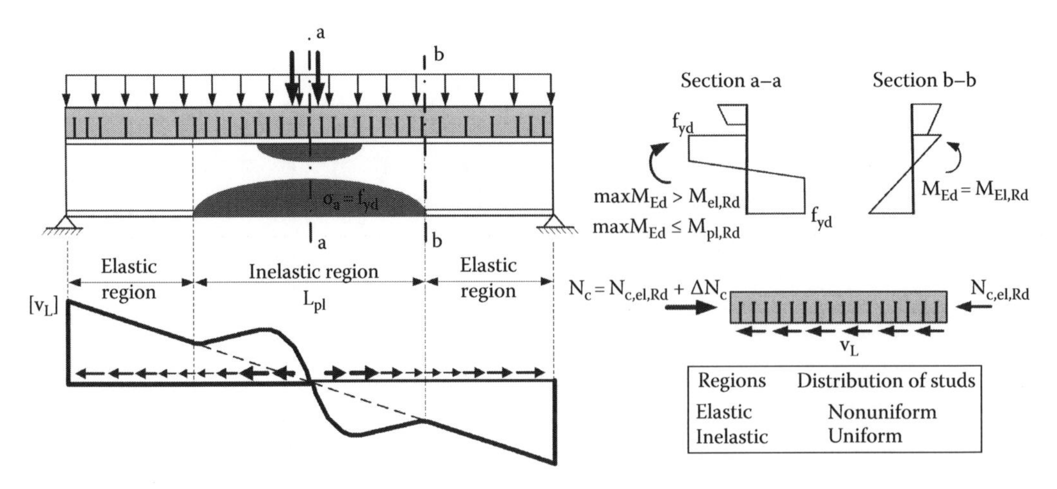

Figure 12.10 Longitudinal shear in inelastic regions.

free-body equilibrium of the concrete slab. The design shear is then determined from the difference of the design axial forces on the slab at adjacent cross sections in accordance to Figure 12.10:

$$v_{L,Ed} = \frac{N_{c,Ed,right} - N_{c,Ed,left}}{\Delta x} = \frac{\Delta N_{c,Ed}}{\Delta x} \tag{12.15}$$

The aforementioned procedure refers to positive bending. For negative bending, the design is covered by the determination of the longitudinal shear on the basis of the uncracked section as discussed earlier so that the procedure for elastic behavior is always followed.

Inelastic behavior refers to regions where the design moment is larger than the elastic moment resistance, $M_{Ed} > M_{el,Rd}$. The elastic moment is determined by consideration of the construction stages. If, for example, during construction stages the elastic moment of the steel beam during casting of concrete is reached, the elastic moment is equal to the elastic moment of the steel girder alone, $M_{a,el,Rd}$. If the steel girder is fully propped at construction stages, the elastic moment is equal to the elastic moment of the composite beam, $M_{1,el,Rd}$. Accordingly, the elastic moment resistance $M_{el,Rd}$ is not a fixed value but depends on the construction stages. $M_{el,Rd}$ is determined by adding the stresses on the relevant sections, either pure steel or composite, during construction stages. When the design stress on the steel girder, the concrete, or the reinforcement reaches its design resistance (see Table 9.11), the total moment provides $M_{el,Rd}$. The stress in reinforcement may be for simplicity omitted in this calculation. The procedure is illustrated in Table 12.5.

REMARK 12.1

- In inelastic regions, ductile connectors are mandatory so that shear forces can be transferred from one connector to the adjacent one. Connectors should be followed by a European Technical Approval (ETA) that guarantees that shear connectors posses sufficient deformation capacity, minimum characteristic capacity of 6 mm. Otherwise, an experimental verification is mandatory.

Table 12.5 Determination of elastic moment resistance $M_{el,Rd}$ for construction stages

Loading in		
Steel girder	$M_{a,Ed} = \sum_i M_{a,Ed,i}$	
Composite girder	$M_{I,Ed,0} = \sum_i M_{I,Ed,i,0}$	
Stresses in		
Steel	$\sigma_{a,Ed,0} = f_{yd} \rightarrow f_{yd} = M_{a,Ed} \cdot \frac{z_a}{I_a} + \mathbf{k}_a \cdot M_{I,Ed,0} \cdot \frac{z_{a,0}}{I_{I,0}}$	
Composite girder	$\sigma_{c,Ed,0} = f_{cd} \rightarrow f_{cd} = \mathbf{k}_c \cdot M_{I,Ed,0} \cdot \frac{z_{c,0}}{n_0 \cdot I_{I,0}}$	
Reinforcement	$\sigma_{s,Ed,0} = f_{sd} \rightarrow f_{sd} = \mathbf{k}_s \cdot M_{I,Ed,0} \cdot \frac{z_{s,0}}{I_{I,0}}$	
Coefficients k		
Concrete: $k_c = \frac{f_{cd}}{M_{I,Ed,0} \cdot (z_{c,0}/n_0 \cdot I_{I,0})}$	Steel: $k_a = \frac{f_{yd} - M_{a,Ed} \cdot (z_a/I_a)}{M_{I,Ed,0} \cdot (z_{a,0}/I_{I,0})}$	Reinforcement: $k_s = \frac{f_{sd}}{M_{I,Ed,0} \cdot (z_{s,0}/I_{I,0})}$
Critical coefficient	$k = \min(k_a, k_c, k_s)$	
Elastic bending resistance	$M_{el,Rd} = M_{a,Ed} + k \cdot M_{I,Ed,0}$	

Notes:

$M_{a,Ed,i}$ is the design bending moment acting on the pure steel cross section for the loading case i.
$M_{I,Ed,i,0}$ is the design bending moment acting on the composite cross section with short-term cross-sectional properties for the loading case i.

- In some countries, block connectors with hoops are used due to their increased resistance against uplift, in many cases, in combination with headed studs. There are no experimental investigations on the behavior of such mixed solutions and, therefore, they are not recommended.

Once the elastic bending resistance is determined, the compression force in the slab, N_c, in dependence on the design moment has to be evaluated in order to apply Equation 12.15. The compression force N_c at a section where the design moment exceeds the elastic resistance may be calculated from Figure 12.10:

$$N_c = N_{c,el,Rd} + \frac{M_{Ed} - M_{el,Rd}}{M_{pl,Rd} - M_{el,Rd}} \cdot (N_{c,f,Rd} - N_{c,el,Rd}) \tag{12.16a}$$

where

$N_{c,el,Rd}$ is the force acting in the deck slab due to $M_{el,Rd}$

$$= \left(M_{el,Rd} \cdot \frac{(z_{co} + z_{cu}) \cdot h_c \cdot b_{eff,1}}{2 \cdot n_0 \cdot I_{1,0}} \right) \tag{12.16b}$$

$N_{c,f,Rd}$ is the force acting in the deck slab due to $M_{pl,Rd}$

$$= (0.85 \cdot f_{cd} \cdot x_{pl} \cdot b_{eff,1}, x_{pl} \leq h_c) \tag{12.16c}$$

The shear flow $v_{L,Ed}$ is calculated from Equations 12.15 and 12.16 with $N_{c,Ed,right} = N_{c,el,Rd}$, $N_{c,Ed,left} = N_c$, and $\Delta x = L_{pl}/2$ (Figure 12.10) and may be written as:

$$v_{L,Ed} = \frac{2 \cdot (M_{Ed} - M_{el,Rd})}{L_{pl} \cdot (M_{pl,Rd} - M_{el,Rd})} \cdot (N_{c,f,Rd} - N_{c,el,Rd})\ [\text{kN/m}] \tag{12.17}$$

It is noted that M_{Ed} is the total design bending moment ($= M_{a,Ed} + M_{1,Ed}$). This means that **the construction sequence is not taken into account for the verification of the shear connection in the inelastic regions;** stresses due to the preloading of structural steel are absorbed due to the rotational capacity of the compact composite section (class 1 or 2).

EXAMPLE 12.2

A 25 m long simply supported plate-girder bridge is casted in one stage. Concrete casting takes place on profiled steel sheeting. The bending moment M_a acts on the pure steel and M_1 on the composite cross section; they are shown in Figure 12.11. Concrete is of quality C35/45, structural steel of quality S355, reinforcement of quality B500B. Shear studs are to be arranged for ULS.

The composite cross section is class 1 and a nonlinear behavior at midspan should be considered. The steel stresses in the bottom and the top flange are calculated according to the construction sequence (Table 12.6). One can see that yielding of the bottom flange starts at a location approximately 7 m from the supports. Therefore, the length of the nonlinear region is 11 m.

Stresses in concrete and reinforcing steel are shown in Table 12.7. Both materials remain elastic ($k > 1$).

In Table 12.8, the longitudinal variation of the elastic bending resistance is demonstrated. The lowest value arises at midspan. Seven meters away from edge supports, the design bending moment is approximately equal to the elastic resistance.

Nonuniform arrangement of studs in the elastic region

Longitudinal shear at the concrete–steel interface follows the vertical shear forces because the cross section remains constant along the bridge. Headed studs will be arranged according to the variation of the longitudinal shear; Figure 12.8.

The design shear force at the edge supports was found equal to 1483.7 kN. At a distance 3.5 m from the edge support, the shear force is reduced to $1483.7 \cdot 9/12.5 = 1068.3$ kN.

For 0–3.5 and 21.5–25 m, Equation 12.10:

$$v_{L,Ed} = \frac{1{,}483.7 \cdot (290 \cdot 30) \cdot 15.27}{2{,}303{,}750 \cdot 6.18} = 13.85\ \text{kN/cm}$$

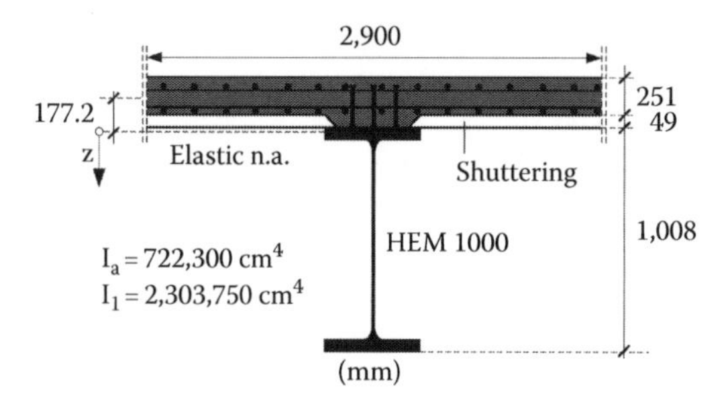

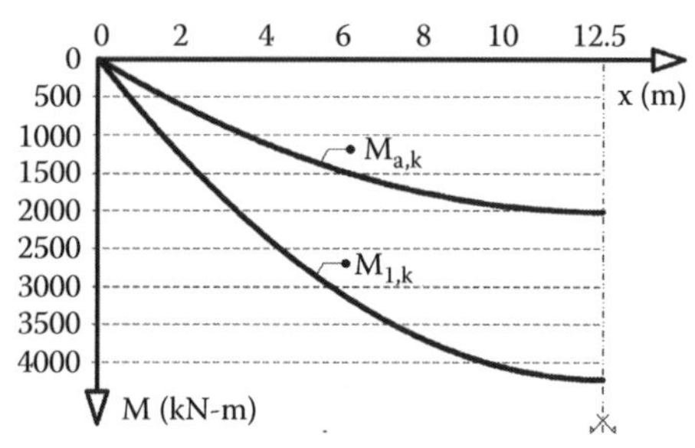

Figure 12.11 Cross section and total bending moments.

Table 12.6 Calculation of k-factors for structural steel

			Bottom flange				*Top flange*			
			$\sigma_{au,a}$	$\sigma_{au,com}$	$\sigma_{au,tot}$		$\sigma_{ao,a}$	$\sigma_{ao,com}$	$\sigma_{ao,tot}$	
x (m)	$M_{a,Ed}$ *(kN-m)*	$M_{I,Ed}$ *(kN-m)*		*(kN/cm²)*		k_{au} *Table 12.5*		*(kN/cm²)*		k_{ao} *Table 12.5*
0	0.0	0.0	0.00	0.00	0.00	∞	0.00	0.00	0.00	∞
1	415.1	874.0	2.90	3.81	6.71	8.55	2.90	0.01	2.91	3182.81
2	795.7	1675.2	5.55	7.31	12.86	4.10	5.55	0.02	5.57	1525.36
3	1141.6	2403.6	7.97	10.49	18.45	2.63	7.97	0.03	7.99	977.44
4	1453.0	3059.1	10.14	13.35	23.49	1.90	10.14	0.04	10.17	707.39
5	1729.7	3641.8	12.07	15.89	27.96	1.47	12.07	0.04	12.11	548.96
6	1971.9	4151.6	13.76	18.12	31.88	1.20	13.76	0.05	13.81	446.82
7	*2179.5*	*4588.6*	*15.21*	*20.02*	*35.23*	*≈1.00*	15.21	0.05	15.26	377.33
8	2352.4	4952.8	16.41	21.61	38.03	0.88	16.41	0.06	16.47	328.79
9	2490.8	5244.1	17.38	22.88	40.26	0.79	17.38	0.06	17.44	294.82
10	2594.6	5462.6	18.10	23.84	41.94	0.73	18.10	0.06	18.17	271.71
11	2663.8	5608.3	18.59	24.47	43.06	0.69	18.59	0.07	18.65	257.31
12	2698.4	5681.1	18.83	24.79	43.62	0.67	18.83	0.07	18.90	250.39
12.5	2702.7	5690.2	18.86	24.83	43.69	0.67	18.86	0.07	18.93	249.53

Notes:

$M_{a,Ed}$ and $M_{I,Ed}$ are the design bending moments acting on the pure steel and the composite cross section, respectively.
$\sigma_{a,a}$ and $\sigma_{a,com}$ are the steel stresses due to $M_{a,Ed}$ and $M_{I,Ed}$, respectively.

Table 12.7 Calculation of k-factors for concrete and reinforcing steel

x (m)	$M_{I,Ed}$ *(kN-m)*	σ_{co} *(kN/cm²)*	k_{co} *Table 12.5*	σ_{so} *(kN/cm²)*	k_s *Table 12.5*
0	0.0	0.00	∞	0.00	∞
1	874.0	0.19	12.56	0.96	45.35
2	1675.2	0.36	6.55	1.84	23.66
3	2403.6	0.51	4.57	2.64	16.49
4	3059.1	0.65	3.59	3.36	12.96
5	3641.8	0.77	3.01	3.99	10.88
6	4151.6	0.88	2.64	4.55	9.55
7	4588.6	0.98	2.39	5.03	8.64
8	4952.8	1.05	2.22	5.43	8.00
9	5244.1	1.11	2.09	5.75	7.56
10	5462.6	1.16	2.01	5.99	7.26
11	5608.3	1.19	1.96	6.15	7.07
12	5681.1	1.21	1.93	6.23	6.98
12.5	5690.2	1.21	1.93	6.24	6.97

Table 12.8 Determination of the length of the elastic region

x (m)	M_{Ed} *(kN-m)*	*mink Table 12.5*	$M_{el,Rd}$ *(kN-m)*
0	0.00	∞	8135.20
1	1289.15	8.55	7886.52
2	2470.88	4.10	7658.58
3	3545.17	2.63	7451.35
4	4512.04	1.90	7264.85
5	5371.47	1.47	7099.06
6	6123.48	1.20	6954.01
7	*6768.05*	*≈1.00*	*6829.67*
8	7305.20	0.88	6726.05
9	7734.91	0.79	6643.16
10	8057.20	0.73	6580.99
11	8272.50	0.69	6539.54
12	8379.48	0.67	6518.82
12.5	8392.91	0.67	6516.23

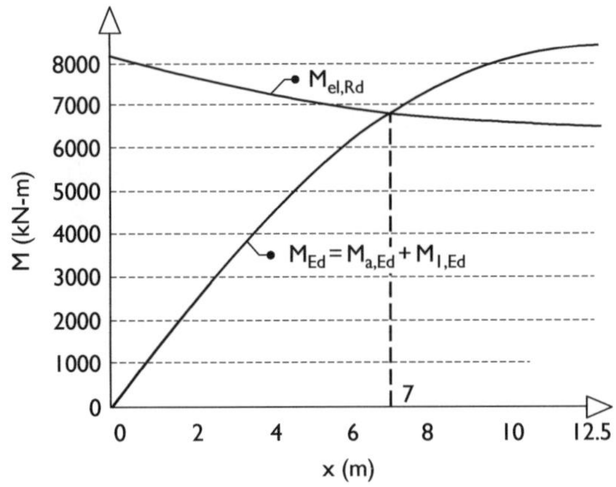

For 3.5–7.0 and 18–21.5 m, Equation 12.10:

$$v_{L,Ed} = \frac{1{,}068.3 \cdot (290 \cdot 30) \cdot 15.27}{2{,}303{,}750 \cdot 6.18} = 9.97 \text{ kN/cm}$$

Note: For the calculation of the shear flow, the safest approach is to consider the total depth of the deck slab as active.

Headed studs d = 22 mm/f_u = 500 MPa; Table 12.2 → P_{Rd} = 121.58 kN
For 0–3.5 and 21.5–25 m, Equations 12.11 and 12.13:

$$e_L \le \frac{3 \cdot 121.58}{13.85} = 26.3 \text{ cm}$$

For 3.5–7.0 and 18–21.5 m, Equations 12.11 and 12.13:

$$e_L \le \frac{3 \cdot 121.58}{9.97} = 36.6 \text{ cm}$$

For 0–3.5 and 21.5–25 m, it is chosen that there be 3 studs per 260 mm.
For 3.5–7.0 and 18–21.5 m, it is chosen that there be 3 studs per 360 mm.
According to Table 12.3,

$$5 \cdot d = 5 \cdot 22 = 110 \text{ mm} \le e_L \le \min\{4 \cdot 300, 800 \text{ mm}\} = 800 \text{ mm (both are satisfied)}$$

Studs were arranged based on the short-term cross-sectional properties because due to the presence of class 1 or 2 cross sections, creep and shrinkage are neglected.

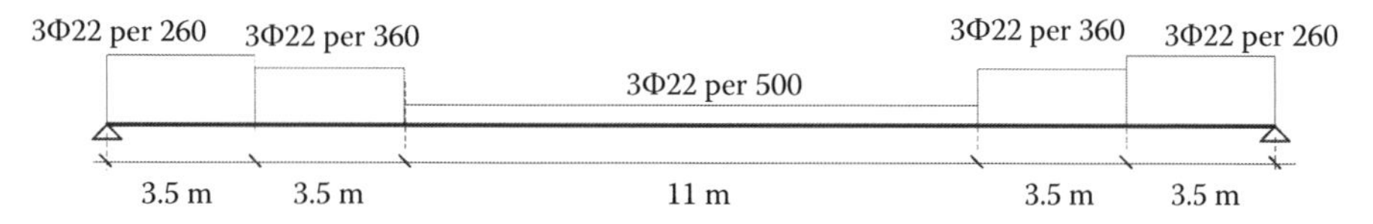

Figure 12.12 Arrangement of studs per ULS.

Uniform arrangement of studs in the inelastic region

The plastic bending resistance of the composite cross section is equal to

$M_{pl,Rd}$ = 11,089.79 kN-m with x_{pl} = 22.46 cm (plastic neutral axis in the slab).

In composite girders with ribs due to steel shuttering, it is safer to consider the whole slab under compression. Therefore, the corresponding force in the slab is equal to Equation 12.16c:

$$N_{c,f,Rd} = 0.85 \cdot \frac{3.5}{1.5} \cdot 290 \cdot 30 = 17{,}225 \text{ kN}$$

The elastic bending resistance for x = 7 and 18 m is $M_{el,Rd}$ = 6829.67 kN-m (Table 12.8)

The corresponding force in the slab is equal to Equation 12.16b:

$$N_{c,el,Rd} = \frac{682{,}967 \cdot (0.27 + 30.27) \cdot 290 \cdot 30}{2 \cdot 6.18 \cdot 2{,}303{,}750} = 6{,}372.86 \text{ kN}$$

Equation 12.17:

$$v_{L,Ed} = \frac{2 \cdot (8{,}392.91 - 6{,}829.67)}{1{,}100 \cdot (11{,}089.79 - 6{,}829.67)} \cdot (17{,}225 - 6{,}372.86) = 7.24 \text{ kN/cm}$$

Headed studs d = 22 mm/f_u = 500 MPa; Table 12.2 → P_{Rd} = 121.58 kN

For 7–18 m, Equations 12.11 and 12.13:

$$e_L \leq \frac{3 \cdot 121.58}{7.24} = 50.4 \text{ cm}$$

It is chosen that there be 3 studs per 500 mm.

According to Table 12.3,

$$5 \cdot d = 5 \cdot 22 = 110 \text{ mm} \leq e_L \leq \min\{4 \cdot 300, 800 \text{ mm}\} = 800 \text{ mm (satisfied)}$$

The arrangement of the headed studs at ULS is shown in Figure 12.12.

12.5 LONGITUDINAL SHEAR DUE TO CONCENTRATED FORCES

Longitudinal shear develops not only from vertical shear due to direct loading or secondary effects but also from concentrated longitudinal forces. Such concentrated forces are due to the primary effects of shrinkage and develop at the ends of the bridge or at the ends of

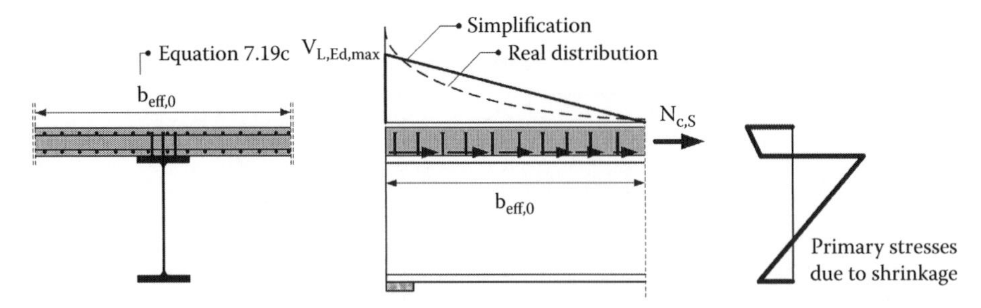

Figure 12.13 Distribution of end shear due to shrinkage at an edge support.

concrete segments at construction stages and may be determined by Equation 7.29a. The concentrated force $N_{c,S}$ is transferred to the beam by shear connectors over a length equal to the effective[s] width, $b_{eff,0}$, of the slab; see Section 7.2.2. The shear flow is approximated by a triangular distribution with a maximal value (Figure 12.13):

$$v_{L,Ed,max} = \frac{2 \cdot N_{c,S}}{b_{eff,0}} \tag{12.18}$$

Figures 12.7 and 12.13 indicate that the shear flow due to the primary effects of shrinkage is generally opposite to the corresponding one due to vertical shear. It should be mentioned that the end shear verification for shrinkage is conducted for the combination of actions of the long-term design.

The effective[s] width is determined in accordance with Section 7.2. At construction stages, the equivalent span L_e is taken as the length of the concrete segment within the span considered. Concentrated forces arise at the ends of the concrete segments due to the sudden change of the cross section (Figure 12.14). This shear flow should be added to the longitudinal shear from shrinkage, under consideration of its direction.

Concentrated longitudinal shear also develops due to the primary effects of temperature differences as presented in Figure 4.25 and has a similar distribution as for shrinkage (Figure 12.13). The direction of these forces depends on whether the slab is cooler or warmer. In the former case, the direction of the forces is the same as for shrinkage; in the latter, it is opposite.

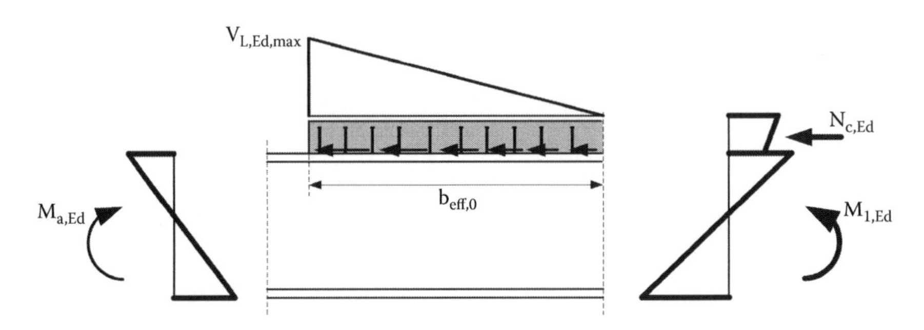

Figure 12.14 Distribution of end shear at a sudden change of cross section.

EXAMPLE 12.3

The end shear connection of the box-girder bridge in Example 9.4 at the end supports will be verified (Figure 9.16).

The shrinkage force in the composite cross section has been calculated to be equal to

$$N_{sh} = 13{,}667.46 \text{ kN}$$

The tensile force in the concrete slab is calculated according to Equation 7.29a and Table 9.17 as follows:

$$N_{c,S} = 13{,}667.46 \cdot \left(1 - \frac{33{,}000}{15.22 \cdot 6{,}540.97} - \frac{33{,}000}{15.22 \cdot 0.8493 \cdot 10^8} \cdot (108.02 - 15)^2\right)$$

$$= 6117.86 \text{ kN}$$

According to Figure 7.34, $L_e = 0.85 \cdot 60 = 51$ m
Equation 7.19c:

$$\beta_1 = 0.55 + 0.025 \cdot \frac{51}{2.0} = 1.19 > 1.0 \rightarrow \beta_1 = 1.0$$

Equation 7.19c:

$$\beta_2 = 0.55 + 0.025 \cdot \frac{51}{3.5} = 0.91 < 1.0 \rightarrow \beta_2 = 0.91$$

Equation 7.19a:

$$b_{el} = 1.0 \cdot 2.0 = 2.0 \text{ m}$$

Equation 7.19a:

$$b_{e2} = 0.91 \cdot 3.5 = 3.19 \text{ m}$$

From Figure 12.9, $b_0 = 3 \cdot 24 = 72$ cm

Note: Conservatively, it can be set $b_0 = 0$.

The effective[s] width of the whole deck slab is equal to

$$b_{eff,0} = 2 \cdot 0.72 + 2 \cdot (2.0 + 3.19) = 11.82 \text{ m} > 11.0 \text{ m} \rightarrow b_{eff,0} = 11.0 \text{ m}$$

The shear flow due to shrinkage is limited inside a length of 11 m.
Equation 12.18:

$$v_{L,Ed,max} = \frac{2 \cdot 6117.86}{11.0} \cdot \frac{1}{100} = 11.12 \text{ kN/cm}$$

Shrinkage is well developed at t = 30,000 days and, therefore, only the long-term shear flows due to vertical shear are taken into account.

The maximum shear flow due to vertical shear was calculated in Example 12.1:

$$\text{ULS: } v_{L,Ed,\infty} = 6.27 \text{ kN/cm}, \quad \text{SLS: } v_{L,Ed,ser,\infty} = 4.69 \text{ kN/cm}$$

The shear resistance of a headed stud with h_{sc}=250 mm, d=22 mm, and f_u=450 mm is taken from Table 12.2 and is equal to P_{Rd}= 109.42 kN. At SLS, this value should be reduced by 25%; see Table 12.1.

The shear flow due to shrinkage acts in an opposite direction than the shear flow due to vertical shear. Therefore, the verification of the end shear is conducted as follows:

$$\text{ULS: } v_{L,Ed,\infty} = 11.12 - 6.27 = 4.85 \text{ kN/cm} \le \frac{4 \cdot 109.42}{e_L} \rightarrow \min e_L = 90.2 \text{ cm} > 64 \text{ cm}$$

$$\text{SLS: } v_{L,Ed,\infty} = 11.12 - 4.69 = 6.43 \text{ kN/cm} \le \frac{4 \cdot (0.75 \cdot 109.42)}{e_L} \rightarrow \min e_L = 51.1 \text{ cm} < 64 \text{ cm}$$

The end shear connection at the edge supports as shown in Figure 12.9 (4 studs per 640 mm) is not sufficient since the SLS verification is not satisfied. The spacing should be changed to 4 studs per 510 mm.

Similar enhancements may be necessary at the edges of the concrete segments (construction stages); see Figure 12.14.

12.6 LONGITUDINAL SHEAR IN CONCRETE SLABS

The longitudinal shear at the steel girder–concrete flange interface is transferred from the concrete slab to the shear connectors and then to the steel girder. In order to prevent shear failure or longitudinal splitting, appropriate transverse reinforcement must be provided in the slab to allow for this transfer. The shear is then transferred by a system of compressive struts and ties in the form of transverse reinforcement and is checked at ULS. Verifications are made at various sections of potential shear failures as indicatively shown in Figure 12.15.

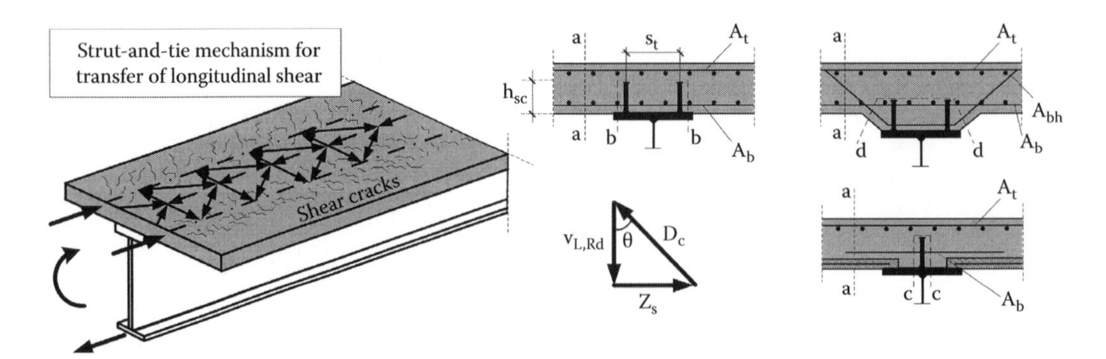

Figure 12.15 Failure mechanism and typical sections for checking shear failure.

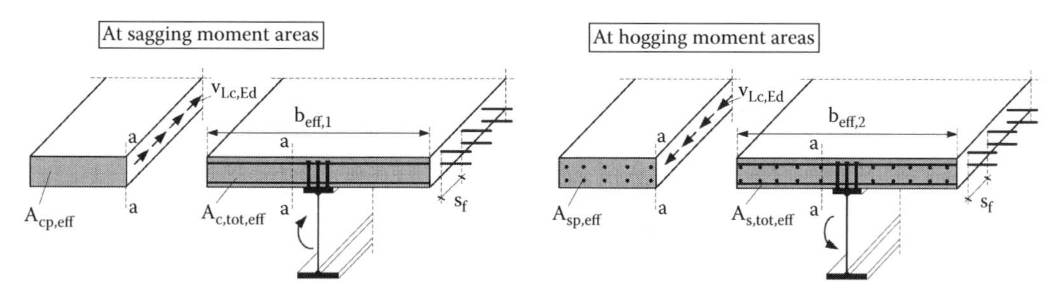

Figure 12.16 Design shear flow at section a–a.

The design shear flow may be determined in sections a–a that cut part of the concrete flange as follows (Figure 12.16):

Concrete flange in compression:
(sagging bending)

$$v_{Lc,Ed} = v_{L,Ed} \cdot \frac{A_{cp,eff}}{A_{c,tot,eff}} \quad (12.19a)$$

Concrete flange in tension:
(hogging bending)

$$v_{Lc,Ed} = v_{L,Ed} \cdot \frac{A_{sp,eff}}{A_{s,tot,eff}} \quad (12.19b)$$

where

$v_{L,Ed}$ is the design shear flow in the flange at ULS, Equation 12.10 for elastic behavior, or Equation 12.15 for inelastic behavior or alternatively, for fully covering the resistance of the connectors:

$$v_{L,Ed} = \frac{n \cdot P_{Rd}}{e_L} \quad (12.20)$$

$A_{cp,eff}$ is the partial area of the concrete flange that is cut by section a–a

$A_{c,tot,eff}$ is the total area of the concrete flange within the effective[s] width of the slab ($h_c \cdot b_{eff,1}$ for sagging bending and $h_c \cdot b_{eff,0}$ at the edge supports)

$A_{sp,eff}$ is the partial area of the longitudinal reinforcement on one side of the section a–a

$A_{s,tot,eff}$ is the total area of longitudinal reinforcement within the effective[s] width of the slab ($\rho_{s,tot} \cdot h_c \cdot b_{eff,2}$ for hogging bending and $\rho_{s,tot} \cdot h_c \cdot b_{eff,0}$ at the edge supports)

n is the number of shear connectors at one cross section

P_{Rd} is the design resistance of one connector

e_L is the longitudinal spacing of connectors

In addition, sections b–b, c–c, or d–d around the studs must be checked. The design shear in these sections is equal to the full design shear flow in the flange, $v_{L,Ed}$. The design shear flow is resisted by a strut-and-tie mechanism, where the struts are composed of concrete elements and the ties of the transverse reinforcement (Figure 12.15).

The crushing resistance of the struts may be obtained from

$$v_{c,Rd} = \nu \cdot f_{cd} \cdot L_v \cdot \sin\theta \cdot \cos\theta = \nu \cdot f_{cd} \cdot L_v \cdot \frac{1}{\cot\theta + \cot\theta^{-1}} \quad (12.21a)$$

where

L_v is the length of sections of potential shear failure as follows (Figure 12.15):

Sections a–a: L_v = thickness of the slab

Sections b–b: $L_v = 2 \cdot h_{cs} + s_t + d_{head,sc}$ (s_t transverse distance of extreme shear connectors at the cross section considered, $d_{head,sc}$ head diameter of shear connectors, h_{cs} height of the connector)

Sections c–c: $L_v = 2 \cdot h_{cs} + d_{head,sc}$ (for one shear connector in the cross section considered or for staggered connectors)

Sections d–d: Minimum length of section

θ is the angle of inclination of the strut that may be taken as

$1.0 \leq \cot\theta \leq 2.0$ for sagging bending with a recommended value $\cot\theta = 1.2$

$1.0 \leq \cot\theta \leq 1.25$ for hogging bending with a recommended value $\cot\theta = 1.0$

$$\nu = 0.6 \cdot \left(1 - \frac{f_{ck}\,[\mathrm{MPa}]}{250}\right) \quad (12.21b)$$

The resistance of the ties is determined from

$$v_{s,Rd} = \frac{A_{sf}}{s_f} \cdot f_{sd} \cdot \cot\theta \quad (12.22)$$

where

A_{sf}/s_f is the area of transverse reinforcement divided by the corresponding spacing as given in Table 12.9

f_{sd} is the design strength of the reinforcement

$\cot\theta$ takes the same values as for struts

Table 12.9 Ratios A_{sf}/s_f

Type of section for shear failure (Figure 12.15)	*a–a*	*b–b*	*c–c*	*d–d*
$\frac{A_{sf}}{s_f}$	$A_b + A_t$	$2 \cdot A_b$	$2 \cdot A_b$	$2 \cdot A_{bh}$

Note: A_b, area of bottom transverse reinforcement; A_t, area of top transverse reinforcement; A_{bh}, area of bottom transverse haunch reinforcement.

The verifications required may be written as

Concrete (compression strut):

$$v_{Lc,Ed} \le v_{c,Rd} \tag{12.23}$$

Transverse reinforcement (tie):

$$v_{Lc,Ed} \le v_{s,Rd} \tag{12.24}$$

A minimum transverse reinforcement ratio must be provided that is given by

$$\min \frac{A_{sf}}{s_f \cdot h_c} = \frac{0.08 \cdot \sqrt{f_{ck}}}{f_{sk}} \tag{12.25}$$

where

h_c is the (mean) slab thickness
f_{ck} is the characteristic concrete strength in MPa
f_{sk} is the characteristic strength of transverse reinforcement in MPa

It should be noted that the aforementioned transverse reinforcement is the main reinforcement of the slab when the slab rests on the longitudinal beams. This reinforcement is generally sufficient to resist the longitudinal shear.

EXAMPLE 12.4

The longitudinal shear in the concrete slab of Example 9.4 is to be verified.

At edge supports

According to Example 12.3, there are 4 studs d = 22 mm per 510 mm.

Equation 12.20:

$$v_{L,Ed} = \frac{4 \cdot 109.42}{51} = 8.58 \text{ kN/cm}$$

Section a–a: $L_v = 30$ cm (Figures 9.16 and 12.15)

The crushing resistance of the struts is calculated as follows:

The effective[s] width of the half girder is according to Example 12.3: $b_{eff,0} = 5.5$ m.

The most critical value of shear flow arises at the internal part of the half girder. Therefore, from Equation 12.19a:

$$v_{Lc,Ed} = 8.58 \cdot \frac{3.5 - 0.5}{5.5} = 4.68 \text{ kN/cm (Figure 9.16)}$$

Equation 12.21b:

$$\nu = 0.6 \cdot \left(1 - \frac{35}{250}\right) = 0.52$$

Equation 12.21a:

$$v_{c,Rd} = 0.52 \cdot \frac{3.5}{1.5} \cdot 30 \cdot \frac{1}{1.2 + 1.2^{-1}} = 17.9 \text{ kN/cm} > 4.68 \text{ kN/cm (sufficient)}$$

The required transverse reinforcement is obtained from Equation 12.22 (Table 12.9):

$$v_{s,Rd} = \frac{A_{sf}}{s_f} \cdot \frac{50}{1.15} \cdot 1.2 \geq 4.68 \cdot 100 \rightarrow \frac{A_b + A_t}{s_f} \geq 8.97 \text{ cm}^2/\text{m}$$

Section b–b (Figure 12.15): $L_v = 2 \cdot 25 + 3 \cdot 24 + 3.3 = 125.3$ cm (Figure 12.9)

$$v_{Lc,Ed} = 8.58 \text{ kN/cm}$$

Equation 12.21a:

$$v_{c,Rd} = 0.52 \cdot \frac{3.5}{1.5} \cdot 125.3 \cdot \frac{1}{1.2 + 1.2^{-1}} = 74.77 \text{ kN/cm} > 8.58 \text{ kN/cm (sufficient)}$$

The required transverse reinforcement is obtained from Equation 12.22 (Table 12.9):

$$v_{s,Rd} = \frac{A_{sf}}{s_f} \cdot \frac{50}{1.15} \cdot 1.2 \geq 8.58 \cdot 100 \rightarrow \frac{2 \cdot A_b}{s_f} \geq 16.45 \text{ cm}^2/\text{m} \rightarrow \frac{A_b}{s_f} \geq 8.23 \text{ cm}^2/\text{m}$$

It is chosen that there be Φ16/15 bottom + Φ10/15 top reinforcement

$$\frac{A_b}{s_f} = 13.4 \text{ cm}^2/\text{m} > 8.23 \text{ cm}^2/\text{m (satisfied)}$$

$$\frac{(A_b + A_t)}{s_f} = 13.4 + 5.2 = 18.6 \text{ cm}^2/\text{m} > 8.97 \text{ cm}^2/\text{m (satisfied)}$$

The minimum transverse reinforcement that must be provided is given by Equation 12.25:

$$\min \frac{A_{sf}}{s_f \cdot 30} = \frac{0.08 \cdot \sqrt{35}}{50} \rightarrow \min \frac{A_{sf}}{s_f} = 0.28 \text{ cm}^2/\text{m (satisfied)}$$

At midspan

From Figure 12.9, there are 4 studs d = 22 mm per 640 mm.

Equation 12.20:

$$v_{L,Ed} = \frac{4 \cdot 109.42}{64} = 6.84 \text{ kN/cm}$$

Section a–a (Figures 9.16 and 12.15): $L_v = 30$ cm

The effective[s] width of the half girder is according to Example 9.4: $b_{eff,1} = 5.5$ m.

Equation 12.19a:

$$v_{Lc,Ed} = 6.84 \cdot \frac{3.5 - 0.5}{5.5} = 3.73 \text{ kN/cm}$$

Equation 12.21a:

$$v_{c,Rd} = 0.52 \cdot \frac{3.5}{1.5} \cdot 30 \cdot \frac{1}{1.2 + 1.2^{-1}} = 17.9 \text{ kN/cm} > 3.73 \text{ kN/cm (sufficient)}$$

The required transverse reinforcement is obtained from Equation 12.22 (Table 12.9):

$$v_{s,Rd} = \frac{A_{sf}}{s_f} \cdot \frac{50}{1.15} \cdot 1.2 \geq 3.73 \cdot 100 \rightarrow \frac{A_b + A_t}{s_f} \geq 7.15\,\text{cm}^2/\text{m}$$

Section b–b (Figure 12.15): $L_v = 2 \cdot 25 + 3 \cdot 24 + 3.3 = 125.3$ cm (Figure 12.9)

$$v_{Lc,Ed} = 6.84\ \text{kN/cm}$$

Equation 12.21:

$$v_{c,Rd} = 0.52 \cdot \frac{3.5}{1.5} \cdot 125.3 \cdot \frac{1}{1.2 + 1.2^{-1}} = 74.77\ \text{kN/cm} > 6.84\ \text{kN/cm (sufficient)}$$

The required transverse reinforcement is obtained from Equation 12.22 (Table 12.9):

$$v_{s,Rd} = \frac{A_{sf}}{s_f} \cdot \frac{50}{1.15} \cdot 1.2 \geq 6.84 \cdot 100 \rightarrow \frac{2 \cdot A_b}{s_f} \geq 13.11\,\text{cm}^2/\text{m} \rightarrow \frac{A_b}{s_f} \geq 6.56\ \text{cm}^2/\text{m}$$

It is chosen that there be Φ12/15 bottom + Φ10/15 top reinforcement

$$\frac{A_b}{s_f} = 7.5\ \text{cm}^2/\text{m} > 6.56\ \text{cm}^2/\text{m (satisfied)}$$

$$\frac{(A_b + A_t)}{s_f} = 7.5 + 5.2 = 12.7\ \text{cm}^2/\text{m} > 7.15\ \text{cm}^2/\text{m (satisfied)}$$

The minimum transverse reinforcement that must be provided is given by Equation 12.25:

$$\min \frac{A_{sf}}{s_f \cdot 30} = \frac{0.08 \cdot \sqrt{35}}{50} \rightarrow \min \frac{A_{sf}}{s_f} = 0.28\ \text{cm}^2/\text{m (satisfied)}$$

At intermediate support

According to Example 12.1 (Figure 12.9), there are 6 studs d = 22 mm per 800 mm.
Equation 12.20:

$$v_{L,Ed} = \frac{6 \cdot 109.42}{80} = 8.21\ \text{kN/cm}$$

The crushing resistance of the struts is calculated as follows:

Section a–a (Figures 9.16 and 12.15): $L_v = 30$ cm

The effectives width of the half girder is according to Example 9.4: $b_{eff,2} = 5.5$ m.

The most critical value of shear flow arises at the internal part of the half girder. Therefore, from Equation 12.19b:

$$v_{Lc,Ed} = 8.21 \cdot \frac{(3.5 - 0.5) \cdot \rho_{s,tot}}{5.5 \cdot \rho_{s,tot}} = 4.48\ \text{kN/cm}$$

Equation 12.21a:

$$v_{c,Rd} = 0.52 \cdot \frac{3.5}{1.5} \cdot 30 \cdot \frac{1}{1.0 + 1.0^{-1}} = 18.2 \text{ kN/cm} > 4.48 \text{ kN/cm (sufficient)}$$

The required transverse reinforcement is obtained from Equation 12.22 (Table 12.9):

$$v_{s,Rd} = \frac{A_{sf}}{s_f} \cdot \frac{50}{1.15} \cdot 1.0 \geq 4.48 \cdot 100 \rightarrow \frac{A_b + A_t}{s_f} \geq 10.3 \text{ cm}^2/\text{m}$$

Section b–b (Figure 12.15): $L_v = 2 \cdot 25 + 5 \cdot 16 + 3.3 = 133.3$ cm (Figure 12.9)

$$v_{Lc,Ed} = 5.47 \text{ kN/cm}$$

Equation 12.21a:

$$v_{c,Rd} = 0.52 \cdot \frac{3.5}{1.5} \cdot 133.3 \cdot \frac{1}{1.0 + 1.0^{-1}} = 80.87 \text{ kN/cm} > 8.21 \text{ kN/cm (sufficient)}$$

The required transverse reinforcement is obtained from Equation 12.22 (Table 12.9):

$$v_{s,Rd} = \frac{A_{sf}}{s_f} \cdot \frac{50}{1.15} \cdot 1.0 \geq 8.21 \cdot 100 \rightarrow \frac{2 \cdot A_b}{s_f} \geq 18.88 \text{ cm}^2/\text{m} \rightarrow \frac{A_b}{s_f} \geq 9.44 \text{ cm}^2/\text{m}$$

It is chosen that there be Φ12/10 bottom + Φ10/10 top reinforcement

$$\frac{A_b}{s_f} = 11.31 \text{ cm}^2/\text{m} > 9.44 \text{ cm}^2/\text{m (satisfied)}$$

$$\frac{(A_b + A_t)}{s_f} = 11.31 + 7.85 = 19.16 \text{ cm}^2/\text{m} > 10.3 \text{ cm}^2/\text{m (satisfied)}$$

The minimum transverse reinforcement that must be provided is given by Equation 12.25:

$$\min \frac{A_{sf}}{s_f \cdot 30} = \frac{0.08 \cdot \sqrt{35}}{50} \rightarrow \min \frac{A_{sf}}{s_f} = 0.28 \text{ cm}^2/\text{m}$$

12.7 SHEAR CONNECTION OF COMPOSITE CLOSED BOX BRIDGES

In case of composite cross sections with wide flanges attached to the deck slab, the shear forces in the headed studs follow the distribution of the normal stresses due to the shear lag effect (see Figure 7.35). Therefore, an even distribution of the longitudinal shear flow $v_{L,Ed}$ according to Equations 12.11 and 12.12 in the transverse direction for SLS and

FLS would be nonrealistic if not unsafe. EN 1994-2 covers the shear lag effect on the studs' forces with the following equation:

$$P_{Ed}(y) = \frac{v_{L,Ed}}{n_{tot}} \cdot \left[\left(3.85 \cdot \left(\frac{n_w}{n_{tot}} \right)^{-0.17} - 3 \right) \cdot \left(1 - \frac{y}{b} \right)^2 + 0.15 \right] [kN] \leq P_{Rd} \tag{12.26}$$

where

$v_{L,Ed}$ is the design longitudinal shear due to global effects

n_{tot} is the total number of connectors of equal size per unit length of girder within the width b in Figure 12.17, provided that the number of connectors per unit area does not increase with y

n_w is the number of connectors per unit length placed within a distance from the web a_w equal to the larger of $10 \cdot t_{ao}$ and 200 mm, where t_{ao} is the thickness of the top flange (for these connectors, y should be taken as 0. In case of a flange projecting up to a_w outside the web, n_w may include the connectors placed on the flange)

b is equal to half the distance between adjacent webs or the distance between the web and the free edge of the top flange

P_{Rd} is the shear resistance of the headed stud according to Tables 12.1 and 12.2

For ULS, all connectors within the effective[s] width carry the same longitudinal force. This is allowed provided that the headed studs are adequately ductile to redistribute the shear forces to the adjacent connectors.

The top flange is usually a slender element and plate buckling can be avoided through an appropriate anchoring of the studs in the deck slab. For this reason, the spacing of the connectors depends on the class of the top flange. This is shown in Table 12.10:

In case of bridges with double composite action (see case B in Figure 2.22), the distribution of the shear connectors through the use of Equation 12.26 can be avoided provided that at least 50% of the total amount of the shear connectors are placed in the area a_w with $a_w = \max(20 \cdot t_{au}, 0.2 \cdot b_{ei}, 400 \text{ mm})$; for b_{ei}, see Section 7.2.2.

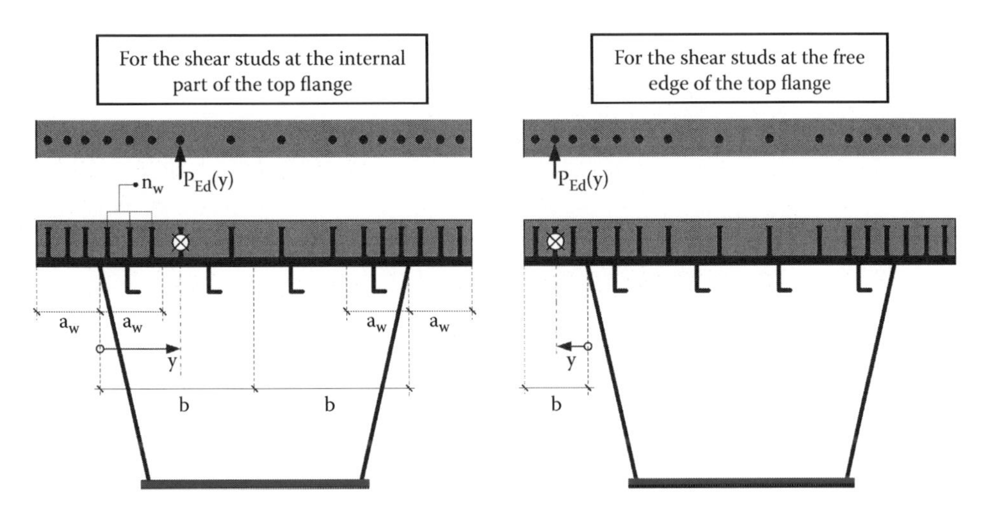

Figure 12.17 Notations for Equation 12.26.

Table 12.10 Upper limit values for spacing of connectors lying on the compression flange

		Class 2	*Class 3*
Longitudinal distribution	Outstand and interior flanges	$22 \cdot t_{ao} \cdot \varepsilon$	$25 \cdot t_{ao} \cdot \varepsilon$
Transverse distribution	Outstand flange	$14 \cdot t_{ao} \cdot \varepsilon$	$20 \cdot t_{ao} \cdot \varepsilon$
	Interior flange	$45 \cdot t_{ao} \cdot \varepsilon$	$50 \cdot t_{ao} \cdot \varepsilon$

Note: $\varepsilon = \sqrt{235/f_y}$ where f_y is the yielding strength of the compressed flange in N/mm^2.

REFERENCES

[12.1] *Composite Bridge Design for Small and Mediums Spans. Design Guide.* ECSC Steel Programme, 2002.

[12.2] EN1994-2, CEN (European Committee for Standardization): Design of composite steel and concrete structures. Part 2: Rules for bridges, Brussels, Belgium, 2005.

[12.3] Hanswille, G. and Porsch, M.: *Zur Festlegung der Tragfähigkeit von Kopfbolzendübeln in Vollbetonplatten in DIN 18800-5 und DIN EN 1994-1-1.* Festschrift Rolf Kindmann, Ruhr-Universität Bochum, Schriftenreihe des Instituts für Konstruktiven Ingenieurbau, Heft 2007-6, Shaker Verlag, Bochum, Germany, 2007.

[12.4] Körschner, K., Kuhlmann, U.: *Structural and Fatigue Behavior of Horizontally Lying Shear Studs Against Vertical and Longitudinal Shear*, Vol. 7, pp. 505–516, Stahlbau 73, Ernst und Sohn, Berlin, Germany, 2004.

Chapter 13

Structural bearings, dampers, and expansion joints

13.1 GENERAL

Bearings are structural devices that transmit loads while facilitating translations and/or rotations. In composite bridge construction, bearings transfer the support reactions and allow for displacements due to temperature, shrinkage, creep, and seismic activities, as well as rotations produced by changes in camber, traffic loads, wind loads, and misalignment of bearing seats due to construction tolerances. In general, vertical displacements are prevented, rotations are allowed to occur as freely as possible, and horizontal displacements are either accommodated or prevented. Until the middle of the twentieth century, steel bearings were used that consisted of four types: *pins*, *rollers*, *rockers*, and *metal sliding bearings*. Pins are fixed bearings allowing rotations. Rollers and rockers allow translation and rotation, while sliding bearings utilize one plane metal plate sliding against another, with polytetrafluoroethylene (PTFE), better known as teflon, as intermediate lubricant material, to accommodate translations. Figure 13.1a shows indicatively a usual layout in plan of bearings for a simply supported bridge at that time, where the bridge deck is free from rigid body translation or rotation but may expand/contract in both directions to accommodate imposed deformations due to indirect actions (Tables 4.1 and 4.2).

However, steel-only bearings (e.g., Figure 13.1b) suffered from long-term problems, such as corrosion or dust and debris collection; they were of high costs due to the need of expensive sliding surfaces and had ultimately poor performance. New trends led to the design of continuous bridges with fewer joints, widespread use of curved and skewed bridges with increased demands on bearings. This has led after 1950 to the development of modern bearings, where other materials as plastics or elastomers are used in combination with steel. Bearings are sensitive components of a bridge. As industrial products, they need certification and require manufacturing, transportation, temporary storage, and on-site installation in accordance with specifications and by qualified personnel. Such issues are covered in Europe by EN 1337 [13.5] that describes in its 11 parts the current types of bearings, gives a design methodology, and includes provisions for installation. The most common types of bearings are *reinforced elastomeric bearings*. However, the appropriate type of bearings must be chosen with due consideration of the design requirements, the initial and maintenance costs, the availability, or other parameters. EN 1337 includes in several parts provisions for some types of bearings as follows:

- Reinforced elastomeric bearings
- Roller bearings
- Pot bearings
- Rocker bearings
- Spherical and cylindrical bearings
- Guided bearings and restrained bearings

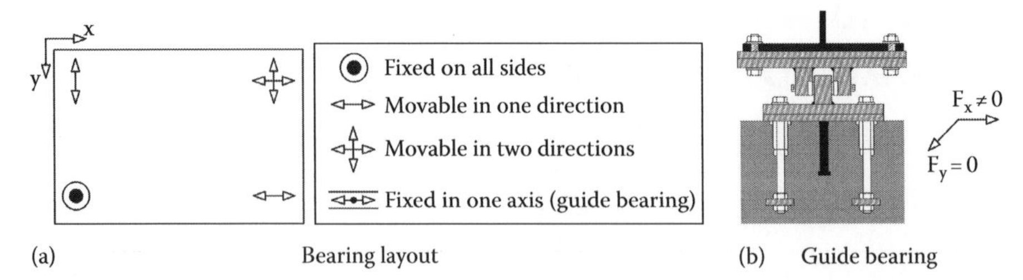

Figure 13.1 (a) Layout of bearings in plan for a simply supported bridge and (b) guide bearing.

The most usual type, reinforced elastomeric bearings, is described in Section 13.2, while spherical and pot bearings are described in Sections 13.3 and 13.4. For other types, reference is made to the code and the manufacturer's specifications (see [13.9] and [13.10]).

13.2 REINFORCED ELASTOMERIC BEARINGS

13.2.1 General

Reinforced elastomeric bearings (known also as *laminated elastomeric bearings*) consist of uniformly spaced layers of elastomer (natural or synthetic rubber) and reinforcing steel plates that obtain their bonding through the process of vulcanization [13.9]. These bearings possess durability and require low to nil maintenance. The main types of such bearings as described in EN 1337-3 are the following (Figure 13.2):

- **Type B (1)** has elastomeric top and bottom layers and is not secured against slippage. Slippage is prevented by the compression load and the friction so that a minimum pressure must always be present.
- **Type B/C (1/2)** has elastomeric top and steel bottom layer. It is secured against slippage on the bottom side by connecting the steel plate with the substructure by bolts, rods, dowels, etc. The one-sided anchorage allows easy installation and exchange so that it is the only bearing type allowed for railway bridges in some countries.
- **Type C 2** has steel top and bottom layers and may be anchored against slipping on both sides.

The bottom bearing plate is connected to the concrete of the substructure usually by headed studs or anchor bolts, while the top plate with the lower flange of the steel girder by preloaded bolts (see Figure 13.3). In most bridges, the main girder bottom flanges are not horizontal; however, the upper bearing plate needs to be set horizontal so that horizontal forces do not arise due to vertical reactions. Therefore, additional tapered bearing plates are provided in between. The previous bolted connection can be designed according to EN 1993-1-8 and additionally

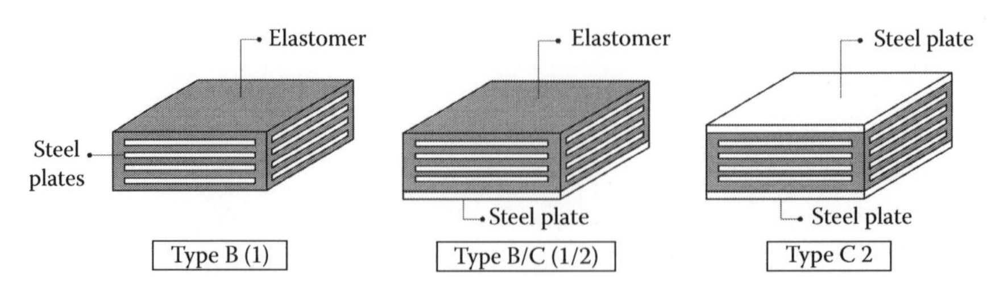

Figure 13.2 Types of reinforced elastomeric bearings.

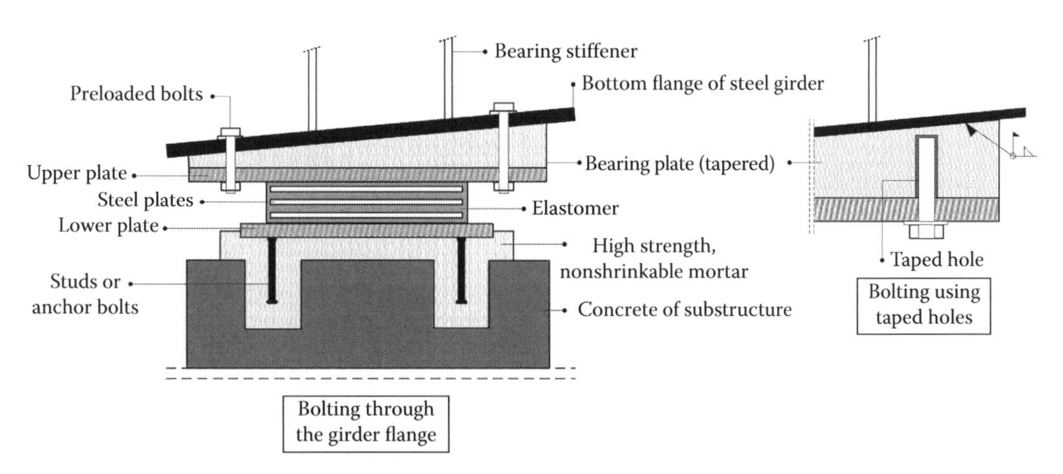

Figure 13.3 Connection of type C 2 bearings with concrete substructure and steel superstructure with bolting through the girder flange or using taped holes.

allows bearings to be replaced since elastomer has not an infinite life. A high-strength nonshrinkable mortar layer is provided between the concrete and the steel plate for leveling purposes. Figure 13.3 shows a typical anchorage of type C 2 bearings. The anchorage of type B/C (1/2) bearings to the substructure is similar with no anchorage to the superstructure.

REMARK 13.1

Bolting of the bearing plate with the lower flange of the girder is not always the best solution since clashes between bolts and bearing stiffeners may occur. Another popular method of attaching the bearing is to tap the tapered bearing plate and to weld it to the underside of the bottom flange of the girder. Bolts are tightened in the taped holes but **not tensioned** so that the threaded hole in the bearing plate is not damaged; the bolts should have higher material strength than the bearing plate. Despite the fact that this connection method is quite common among the manufacturers, some important drawbacks need to be mentioned:

- Bolting using taped holes is not covered by the design guidelines of EN 1993.
- Site welding of the bearing plates raises questions about the durability of the connection. Fatigue failure is possible to occur especially in cases of railway bridges.
- The connection using taped holes is not recommended for applications in seismic areas.

Reinforced elastomeric bearings accommodate translation and rotation (Figure 13.4) by deformation of the elastomeric layers, while the steel plates provide stiffness and resistance. Consequently, the deformation capacity increases with the total height of the elastomeric layers. The number of the steel plates depends on the total height that is between 10 and 400 mm, with a number of elastomeric layers between 1 and 16. The shape of bearings in plan is rectangular, square, or circular. The dimensions in plan depend on the loading capacity to compression, ranging between 200 and 1200 mm.

The bearings are movable in longitudinal and transverse directions of the bridge. However, by provision of *stoppers* (or *steel keep strips*), as indicatively shown in Figure 13.5, one or both displacements may be restrained. The bearings resist horizontal forces in the

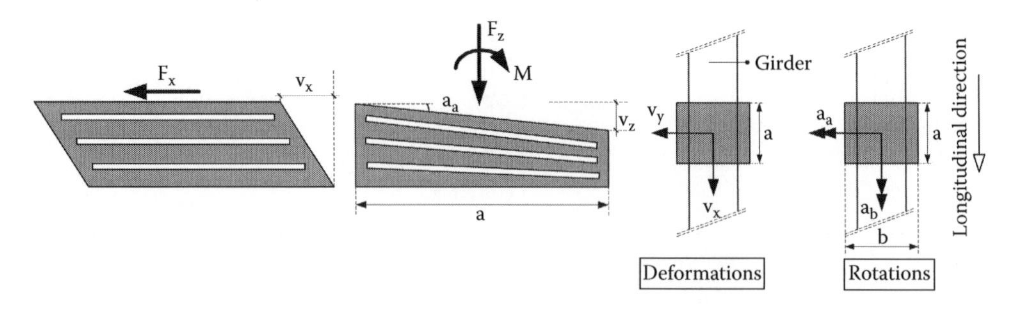

Figure 13.4 Deformations and rotations of reinforced elastomeric bearings.

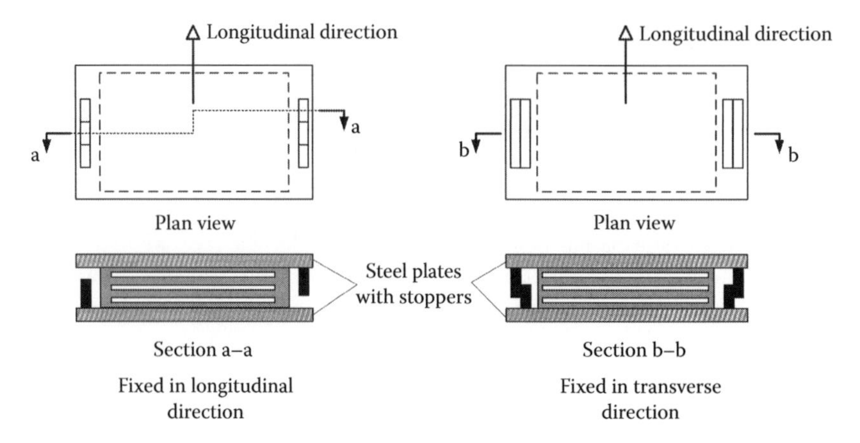

Figure 13.5 Bearings fixed in longitudinal and transverse directions.

corresponding direction and deformations are restricted to about 40 mm from the mean position. Obviously, elastomeric bearings should not be used as fixed bearings unless forces are small (i.e., footbridges or small-span road bridges).

Reinforced elastomeric bearings are not suitable to transfer tension forces. EN 1337-3 does not contain rules for such a possibility. However, for short-term loading that develops under an unfavorable load combination, tension stresses of approximately 1 MPa could be sustained. The bearing should then be appropriately detailed to give the possibility of exchange. EN 1337-3 describes the required verifications for these bearings, valid for plan dimensions up to 1200 × 1200 mm and service temperatures between –25°C and +50°C.

13.2.1.1 Check of distortion

The total design distortion must be limited in accordance with the following relation:

$$\varepsilon_{t,d} = K_L \cdot (\varepsilon_{c,d} + \varepsilon_{q,d} + \varepsilon_{a,d}) \leq 7.0 \tag{13.1}$$

where

$\varepsilon_{c,d}$ is the distortion due to compression, Equation 13.2
$\varepsilon_{q,d}$ is the distortional deformation, Equation 13.5
$\varepsilon_{a,d}$ is the distortion due to angular rotation, Equation 13.6
K_L is a factor that depends on the type of loading (=1.5 for traffic loads and 1.0 for other load types. In general, a value equal to 1.0 is recommended)

13.2.1.1.1 Distortion due to compression $\varepsilon_{c,d}$

$$\varepsilon_{c,d} = \frac{1.5 \cdot F_{z,d}}{G \cdot A_r \cdot S} \quad (13.2)$$

where

$F_{z,d}$ is the vertical design load, compression
G is the shear modulus of the elastomer (in most cases, 0.9 MPa; see Remark 13.2)
A_r is the reduced cross-sectional area of the bearing, Equation 13.3
S is the shape factor, Equation 13.4

$$A_r = A_1 \cdot \left(1 - \frac{v_{x,d}}{a'} - \frac{v_{y,d}}{b'}\right) \quad (13.3)$$

where

a′, b′ are the widths of the steel reinforcement plates (see Figure 13.6)
A_1 is the cross-sectional area of the steel reinforcement plates, possibly reduced due to holes
$A_1 = a' \cdot b'$ for rectangular bearings without holes
$v_{y,d}$ and $v_{x,d}$ are the maximal displacements parallel to side b (transverse) and correspondingly a (longitudinal) (see Figure 13.4):

$$S = \frac{a' \cdot b'}{2 \cdot t_i \cdot (a' + b')} \quad (13.4)$$

t_i is the thickness of the each elastomeric layer (see Figure 13.6)

13.2.1.1.2 Distortional deformation $\varepsilon_{q,d}$

$$\varepsilon_{q,d} = \frac{v_{xy,d}}{T_q} \leq 1.0 \quad (13.5)$$

where

$v_{xy,d}$ is the shear design deformation $\left(\sqrt{v_{x,d}^2 + v_{y,d}^2}\right)$
T_q is the nominal thickness of the shear elastomer according to Figure 13.6

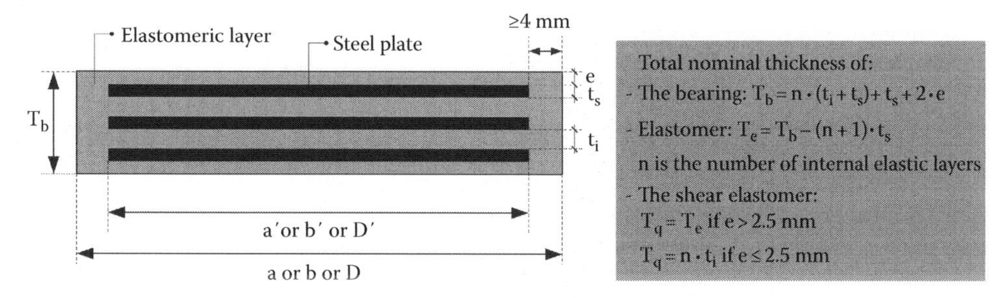

Figure 13.6 Notation for reinforced elastomeric bearings.

13.2.1.1.3 Distortion due to angular rotation $\varepsilon_{a,d}$

$$\varepsilon_{a,d} = \frac{a'^2 \cdot a_{a,d} + b'^2 \cdot a_{b,d}}{2 \cdot n \cdot t_i^2} \tag{13.6}$$

where

$a_{a,d}$ is the angle of rotation over the width a of the bearing (Figure 13.4)
$a_{b,d}$ is the angle of rotation over the width b of the bearing, if any
t_i is the thickness of the internal elastomeric layers
n is the number of internal elastomeric layers (see Figure 13.6)

13.2.1.2 Check of the tension of the steel plates

The thickness of the steel plates must be checked in order to limit their tension stresses. For constant thickness of the elastomeric layers t_i, the relevant relation is written as

$$t_s = \frac{K_p \cdot F_{z,d} \cdot 2 \cdot t_i \cdot K_h \cdot \gamma_m}{A_r \cdot f_y} \geq 2\,\text{mm} \tag{13.7}$$

where

$F_{z,d}$ is the maximum design compression force
f_y is the yield stress of the steel plates, usually 235 MPa
A_r is from Equation 13.3
$K_p = 1.3$ is a correction factor
K_h is equal to 1 for bearings without holes and 2 for bearings with holes
γ_m is a partial safety factor that may be set equal to 1.0. For seismic load combinations the recommended value is 1.15

13.2.1.3 Limitation of rotation

In order to avoid excessive unsticking, the bearing rotation must be limited in accordance to the following relations:

- *Rectangular bearings*

$$\frac{F_{z,d} \cdot n \cdot t_i}{A_1} \cdot \left(\frac{1}{5 \cdot G \cdot S^2} + \frac{1}{E_b} \right) \geq \frac{a' \cdot a_{a,d} + b' \cdot a_{b,d}}{K_{r,d}} \tag{13.8}$$

- *Round bearings*

$$\frac{F_{z,d} \cdot n \cdot t_i}{A_1} \cdot \left(\frac{1}{5 \cdot G \cdot S^2} + \frac{1}{E_b} \right) \geq \frac{D' \cdot a_{D,d}}{K_{r,d}} \tag{13.9}$$

where

$F_{z,d}$ is the vertical design load, compression
G is the shear modulus of the elastomer
A_1 is the area of the steel plates as defined in Equation 13.3
n is the number of internal elastomeric layers
$E_b = 2$ GPa is the compression modulus of the elastomer
$K_{r,d}$ is a rotation coefficient that may be taken as 3
S is the shape factor, Equation 13.4

13.2.1.4 Stability

In order to assure stability, the mean pressure must be limited as follows:

$$\frac{F_{z,d}}{A_r} < \frac{2 \cdot a' \cdot G \cdot S}{3 \cdot T_e} \tag{13.10}$$

where

T_e is sum of all elastomer layers (see Figure 13.6)
and all other symbols as explained before.

13.2.1.5 Safety against slip

To avoid slip of the bearings, the following conditions must be met:

$$F_{xy,d} \leq \mu_e \cdot F_{z,d,min} \tag{13.11}$$

and under permanent loading,

$$\frac{F_{z,Gmin}}{A_r} \geq 3 \text{ MPa} \tag{13.12}$$

where

$F_{xy,d}$ is the resulting horizontal force
$F_{z,d,min}$ is the coexisting minimum vertical design force
$F_{z,Gmin}$ is the minimum vertical design force under permanent loads
A_r is from Equation 13.3

The friction coefficient may be determined from

$$\mu_e = 0.1 + \frac{1.5 \cdot K_r}{\sigma_m} \tag{13.13}$$

where

K_r may be taken as 0.6 for concrete and 0.2 for all other materials, including resins
σ_m is the mean compression stress in [MPa] that corresponds to $F_{z,d,min}$

13.2.2 Modeling for global analysis: Provision of seismic isolation

Reinforced elastomeric bearings are usually introduced as springs in global analysis (see Figures 7.16 and 7.18). The spring stiffness in each unrestrained horizontal direction may be obtained from

$$K_x \quad \text{or} \quad K_y = \frac{A \cdot G}{T_e} \tag{13.14}$$

where

A is the plan area of the bearings ($= a \cdot b$ or $\pi \cdot D^2/4$)
T_e is the total nominal thickness of the elastomer layers
G is the shear modulus of the elastomer (=0.9 MPa for non-seismic combinations)

In vertical direction, the bearing is practically incompressible so that the vertical displacement is restrained without any use of springs. The spring forces and displacements as determined from global analysis give directly the horizontal forces on the bearings $F_{x,d}$ or $F_{y,d}$ as well as the horizontal displacements $v_{x,d}$ or $v_{y,d}$. The vertical displacement is equal to 0, while the vertical reaction is the one at the joint below the springs. The angles of rotation $a_{a,d}$ or $a_{b,d}$ are those from global analysis at the joint of the superstructure connected to the spring.

REMARK 13.2

In part of the literature, bearings are represented in the model in vertical direction also by springs (see Figure 7.16). In this case, the vertical stiffness may be estimated by the following equation [13.11], [13.12]:

$$K_z = \frac{T_e}{A} \cdot \left(\frac{1}{5 \cdot G \cdot S^2} + \frac{1}{E_b} \right) \tag{R13.1}$$

where E_b is the compression modulus of the elastomer.

Designers should be aware of the fact that the stiffness given in Equations 13.14 and R13.1 offers indicative values and changes due to creep, temperature differences, or aging phenomena may need to be taken into account. These are usually considered in global analysis by introducing in the stiffness equations varying values of the shear modulus G. According to EN 1337, $G = 0.9 \pm 0.15$ MPa. For the modulus of elasticity E_b, designers should follow the suggestions of the manufacturer.

Elastomeric bearings are important structural elements for seismic design because they provide *seismic isolation*. Their flexibility is very high compared to the flexibility of the deck, so that the deck may be considered to behave as rigid in plan (Figure 13.7). The fundamental period of the bridge analyzed by the *fundamental mode method* modeled as a single mass vibrator is obtained from Equation 13.15, while the seismic force is from Equation 13.16 (see also Section 4.9 and Table 4.19):

$$T = 2 \cdot \pi \cdot \sqrt{\frac{M}{K}} \tag{13.15}$$

$$F = M \cdot S_{a,d}(T) \tag{13.16}$$

where

M is the mass of the superstructure (see Table 4.18)

$K = \Sigma K_i$ is the stiffness of the system, equal to the sum of the stiffness of the bearings

For bridges on piers, the stiffness of piers is added to the stiffness of bearings (Figure 13.14).

$S_{a,d}(T)$ is design spectral acceleration, Equation 4.24, with $q = 1$, the behavior factor for low-damping reinforced elastomeric bearings.

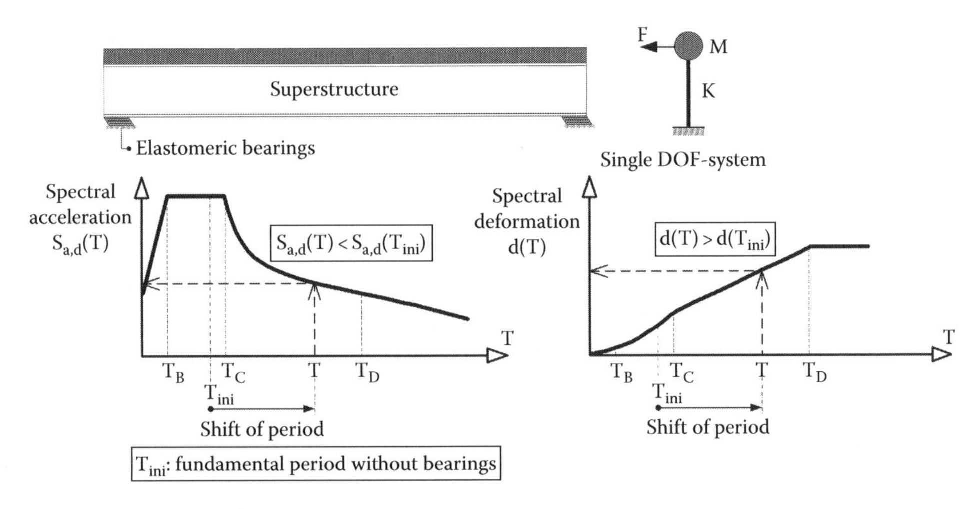

Figure 13.7 Seismic isolation by common elastomeric bearings.

The spectral displacements for 5% damping that corresponds to the damping properties of the common elastomeric bearings and elastic behavior are determined by Equation 13.17 and are also shown in Figure 13.7:

$$d_{5\%} = S_e(T) \cdot \left(\frac{T}{2 \cdot \pi} \right)^2 \tag{13.17}$$

where
 $S_e(T)$ is the elastic spectral acceleration, Equation 4.23
 T is the fundamental period of the system, Equation 13.15

The earlier mentioned relations and Figure 13.7 indicate that the introduction of common elastomeric bearings results in a shift in the fundamental period ($\Delta T = T - T_{ini}$) of the structure and accordingly a reduction of seismic forces and an increase of displacements.

The stiffness of bearings in the seismic situation is determined by Equation 13.18 and may be regulated for certain plan dimensions by the thickness of the elastomeric layers:

$$K_x \quad \text{or} \quad K_y = \frac{A \cdot G_b}{T_e} \tag{13.18}$$

where $G_b = 1.1 \cdot G$ is an increased value of the shear modulus of the elastomer in the seismic situation to account for the speed of loading and all other symbols as in Equation 13.14.

For seismic combinations, upper and lower values of the shear modulus are applied in Equation 13.18. Upper values, $G_{b,max}$, are used to determine maximum forces, and lower values, $G_{b,min}$, to determine maximum displacements. EN 1998-2 recommends to consider $G_{b,min} = 1.0 \cdot G_b$ and $G_{b,max} = 1.5 \cdot G_b$.

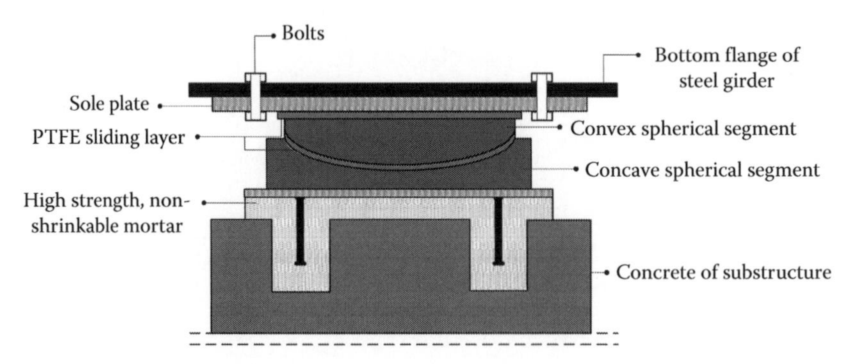

Figure 13.8 Cross section of a spherical bearing.

13.3 SPHERICAL BEARINGS

Spherical bearings allow horizontal displacements and rotations in all directions and are composed of three basic components (Figure 13.8):

- A *sole plate* that transfers the loads from the superstructure to the bearing
- A *convex spherical segment* that provides a PTFE (Teflon) sliding surface for the sole plate and a PTFE convex surface for rotation
- A *concave spherical segment* that provides a mating surface for the convex segment and transfers load to the substructure

Displacements and rotations are accommodated through two independent sliding motions, one between the sole plate and the convex segment and one between the two segments.

The plates are made of steel, plated with stainless steel, hard chrome, or similar materials. PTFE is used due to its low frictional characteristics, the chemical inertness, and high durability. Its friction coefficient varies usually between 0.04 and 0.08 depending on the pressure. Spherical bearings may be also guided to restrain displacements in one or both horizontal directions.

13.4 POT BEARINGS

Pot bearings are based on the incompressibility of natural rubber when placed in a closed steel pot where natural rubber behaves like a fluid. Pot bearings are able to transfer high compression forces in a small surface, their dimensions being primarily determined by the permissible concrete pressure of the substructure. They allow rotations around all axes, while displacements are possible by provision of a sliding material. All-round movable pot bearings are composed of the following parts (Figure 13.9):

- A top steel plate, possibly of stainless steel
- A cap with embedded sliding material (PTFE)
- A round elastomer pad
- A steel pot where the elastomer pad is placed
- Sealing rings that prevent the penetration of moisture, dust, or water in the pot

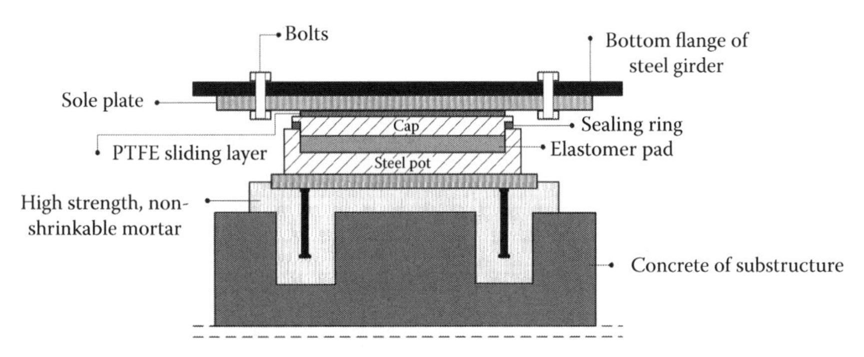

Figure 13.9 Cross section of a movable pot bearing.

13.5 SEISMIC ISOLATION

The reduction of seismic forces may be achieved by inelastic behavior of the superstructure and the piers/abutments. When the deck is rigidly connected to the abutments as in integral bridges or is fixed with the piers, energy dissipation leading to higher than unity behavior factors q may be obtained (Table 4.17). This leads to a reduction of seismic forces as indicated by Equation 4.24. However, such a reduction of seismic forces is not substantial, and in addition, damages may be caused by inelastic action. Therefore, bridges in seismic areas are commonly provided with seismic isolation systems that reduce the response to horizontal seismic action so that the deck and the piers remain elastic. Seismic isolation may be achieved by the following:

a. By introduction of low-stiffness bearings, such as reinforced elastomeric bearings, to shift the fundamental period toward higher values. As discussed in Section 13.2.2, this results in a reduction of seismic forces but increase of seismic displacements.
b. By introduction of high-damping bearings or damping devices to reduce displacements and, in most cases, forces.
c. By a combination of the two.

The most common low-stiffness seismic isolators are *low-damping elastomeric bearings*. These are bearings with an equivalent damping ratio ξ equal to 6%. They can be considered during seismic analysis as linear springs with horizontal stiffness given by Equation 13.18.

The spectral displacements for 5% damping and elastic behavior are determined by Equation 13.17. For higher damping values, the spectral displacements are determined from

$$d_{\xi eff} = \eta \cdot d_{5\%} \tag{13.19}$$

where

$$\eta = \sqrt{\frac{10}{5+\xi}} \geq 0.4$$

$\xi(\%)$ is the equivalent viscous damping ratio

The spectral acceleration is also reduced by the factor η. Figure 13.10 illustrates the effect of increased damping on spectral accelerations and displacements.

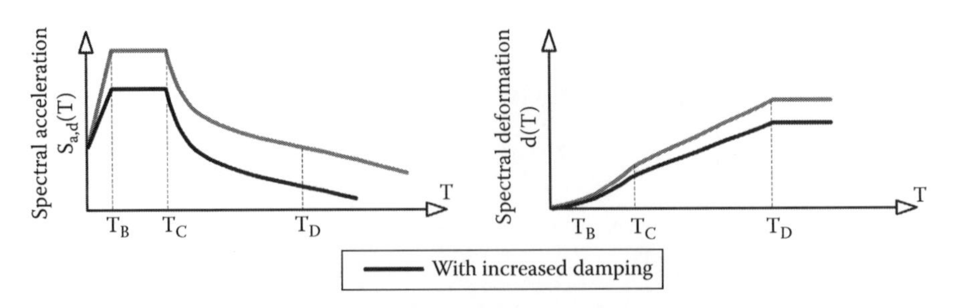

Figure 13.10 Spectral acceleration and displacements for various damping ratios.

Special devices with high-damping properties have been developed in recent times [13.10], some of which are briefly discussed in the following. Prior to use, these devices need to be tested and certified in accordance with the relevant specifications.

13.5.1 High-damping reinforced elastomeric bearings

These are bearings in which the common elastomer is substituted by high-damping elastomer. Their equivalent viscous damping ratio ξ reaches values between 10% and 20%, while common elastomeric bearings have damping ratios below 6%. The behavior of such bearings is considered as *linear hysteretic*. These bearings are modeled in global analysis like common elastomeric bearings as linear springs. The reduction in seismic forces and displacements is taken into account by the factor η. This reduction is shown in Figure 13.11. One can see the improved time-dependent behavior of a bridge with high-damping bearings.

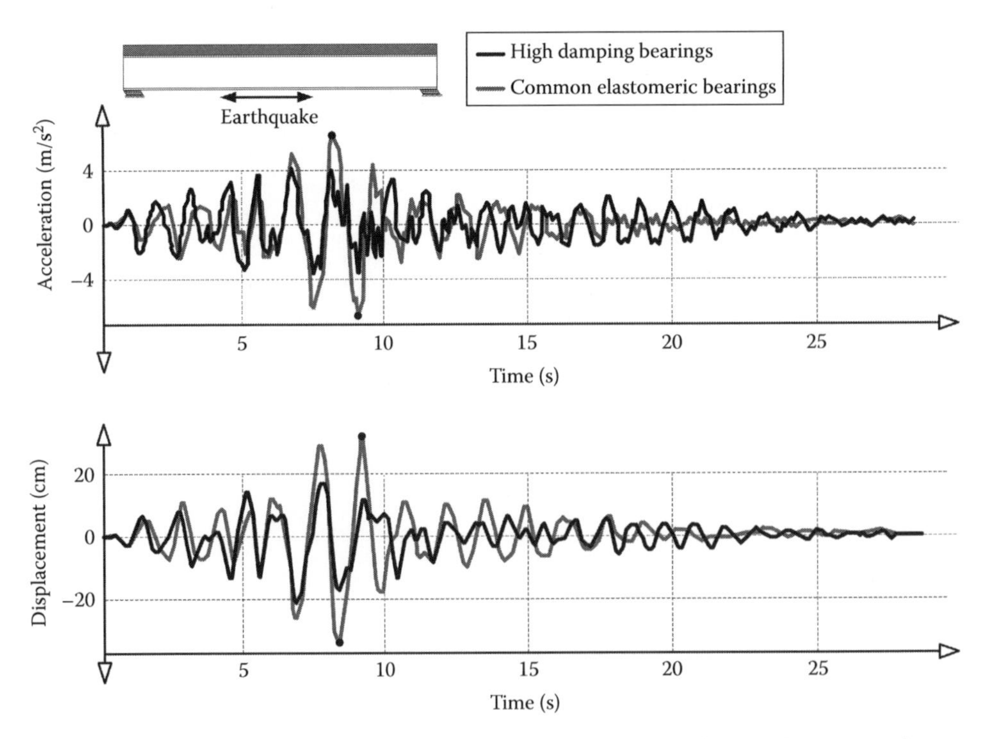

Figure 13.11 Acceleration and displacement of superstructure with high-damping bearings and common bearings.

13.5.2 Lead rubber bearings

These are common reinforced elastomeric bearings with low-damping elastomer and a cylindrical lead core that may reach damping values up to 40% (Figure 13.12).

During cyclic loading, the lead core is yielding and strain hardening so that the hysteretic response of the bearings, as illustrated in Figure 13.13, is typical for a yielding and strain-hardening material. The effective damping ratio ξ_{eff} at a certain design displacement d_{bd} may be obtained from

$$\xi_{eff} = \frac{E_D}{2 \cdot \pi \cdot F_{max} \cdot d_{bd}} \tag{13.20}$$

where

E_D is the absorbed hysteretic energy, equal to the area of the hysteresis loop
d_{bd} is the design displacement
F_{max} is the corresponding force

In analysis, the bearings may be modeled by bilinear springs with elastic stiffness K_e for displacements up to the yield displacement and post-elastic tangent stiffness K_p for larger displacements. Alternatively, they may be represented by linear springs with an effective

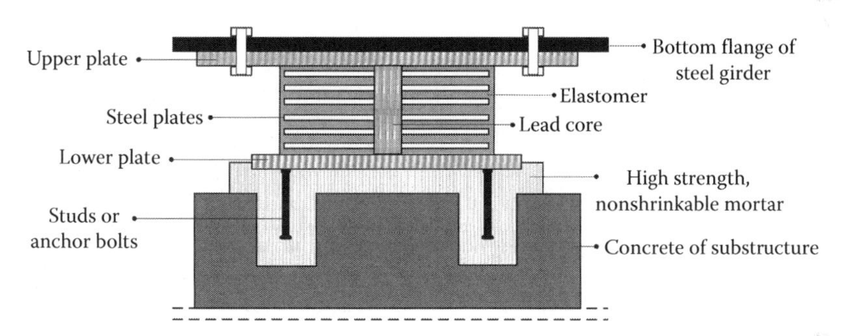

Figure 13.12 Lead rubber bearing (LRB).

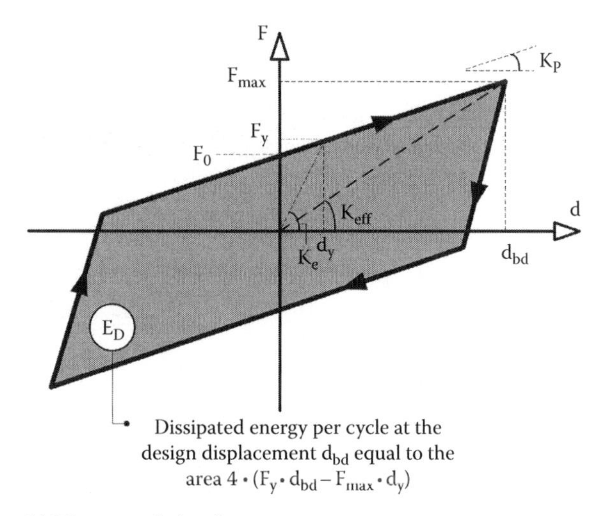

Figure 13.13 Response of LRBs to cyclic loading.

stiffness K_{eff} equal to the secant stiffness corresponding to the maximal displacement. This analysis must be iterated since the maximum displacement and accordingly the secant stiffness are not known in advance.

Elastic stiffness:

$$K_e = K_L + K_R$$

Post-elastic stiffness:

$$K_P = \frac{F_{max} - F_y}{d_{bd} - d_y} = K_R$$

Yield force:

$$F_y = F_{Ly} \cdot \left(1 + \frac{K_R}{K_L}\right)$$

where

K_R and K_L are the shear stiffnesses of the elastomeric and the lead core, respectively
F_{Ly} is the yield force of the lead core
$F_0 = F_y - K_p \cdot d_y$ is the force at zero displacement

Note: When $K_L \gg K_R$, then $K_e \approx K_L$ and $F_{Ly} \approx F_y$

REMARK 13.3

When reinforced elastomeric bearings are used as isolators, they are designed against increased displacements d_m:

$$d_m = \sum d_{Gi} + \sum d_{C,S} + 0.5 \cdot \sum d_{\Delta T} + \gamma_{IS} \cdot d_{bd} \quad \text{(R13.2)}$$

where

$\sum d_{Gi}$ are the displacements induced by the permanent actions
$\sum d_{C,S}$ are the long-term displacements due to creep and shrinkage (in most cases, $d_{C,S}$ can be neglected)
$\sum d_{\Delta T}$ are the displacements due to thermal actions
d_{bd} is the design displacement of the isolator due to earthquake that is increased by the amplification factor γ_{IS} with a recommended value equal to 1.5

Then the design verifications of EN 1337 as previously explained should be fulfilled. Attention should be paid to the following when *simple low-damping bearings* are used as isolators.

The verification in Equation 13.5 becomes $\varepsilon_{q,d} \leq 2.0$ for the design displacement of Equation R13.2.

13.6 ANCHORAGE OF BEARINGS

Anchorage of bearings should be verified at ultimate limit states (ULS) as follows:

$$V_{Ed} \leq V_{Rd} = \frac{\mu_k}{\gamma_\mu} \cdot N_{Ed} + V_{pd} \tag{13.21}$$

where

V_{Ed} is the design value of the shear force
N_{Ed} is the minimum design force acting normal to the joint in conjunction with V_{Ed}
V_{pd} is the design value of shear resistance of any fixing device in accordance with the Eurocodes (e.g., studs or anchor bolts)
μ_k is the characteristic value of the friction coefficient. Recommended values are 0.4 for steel on steel and 0.6 for steel on concrete
γ_μ is the partial safety factor for friction. Recommended values are 2.0 for steel on steel and 1.2 for steel on concrete

For railway bridges and structures subjected to earthquake, the contribution of friction in V_{Rd} is not taken into account ($\mu_k = 0$).

The reaction forces V_{Ed} for persistent design situations at fixed points are calculated for the case of **sliding bearings** as follows:

$$V_{Ed} = \gamma_Q \cdot Q_{1k} - \mu_r \cdot \left(\sum \gamma_{G,inf} \cdot G_k\right) + \mu_a \cdot \left(\sum \gamma_{G,sup} \cdot G_k\right) + \psi_1 \cdot \gamma_Q \cdot Q_{ki} + \sum \psi_{0i} \cdot \gamma_{Qi} \cdot Q_{ki} \tag{13.22}$$

where

$\sum \gamma_{G,sup} \cdot G_k$ are the unfavorable values for dead loads
$\sum \gamma_{G,inf} \cdot G_k$ are the favorable values for dead loads
$\gamma_Q \cdot Q_{1k}$ are the forces from acceleration and braking
$\psi_1 \cdot \gamma_Q \cdot Q_{ki}$ are the traffic loads
$\sum \psi_{0i} \cdot \gamma_{Qi} \cdot Q_{ki}$ are the other variable actions

μ_a, μ_r are the adverse and relieving coefficients of friction accordingly

$$\mu_a = 0.5 \cdot \mu_{max} \cdot (1 + \alpha) \tag{13.23a}$$

$$\mu_r = 0.5 \cdot \mu_{max} \cdot (1 - \alpha) \tag{13.23b}$$

For PTFE sliding bearings, $\mu_{max} = 0.03$.

α is a factor dependent on the type of bearing and the number of bearings (n_b), which are exerting either an adverse or relieving force as appropriate.

The recommended values are

$\alpha = 1$ for $n_b \leq 4$
$\alpha = (16 - n_b)/12$ if $4 < n_b \leq 10$
$\alpha = 0.5$ if $n_b > 10$

For **elastomeric bearings,**

$$V_{Ed} = \gamma_Q \cdot Q_{1k} + \psi_1 \cdot \gamma_Q \cdot Q_{ki} + \sum \psi_{0i} \cdot \gamma_{Qi} \cdot Q_{ki} + G_{sup} \cdot \sum A \cdot \varepsilon_{q,d} - G_{inf} \cdot \sum A \cdot \varepsilon_{q,d} \quad (13.24)$$

where

$\gamma_Q \cdot Q_{1k}$ are the forces from acceleration and braking
$\psi_1 \cdot \gamma_Q \cdot Q_{ki}$ are the traffic loads
$\sum \psi_{0i} \cdot \gamma_{Qi} \cdot Q_{ki}$ are the other variable actions
G_{sup}, G_{inf} are the nominal values of shear modulus; 1.05 N/mm^2 and 0.75 N/mm^2 accordingly
A is the plan area of the bearings ($= a \cdot b$ or $\pi \cdot D^2/4$)
$\varepsilon_{q,d}$ are the shear deformations of the bearings (see Equation 13.5)

For seismic design situations, the design value of the horizontal reaction force is found in EN 1998-2 (see Table 5.11). It is reminded that the variability of the shear modulus of the elastomeric bearings should be considered by conducting two analyses with $G_{b,min}$ and $G_{b,max}$.

13.7 CALCULATION OF MOVEMENTS AND SUPPORT REACTIONS

Movements and bearing forces should be calculated as accurate as possible. Wrong estimations may lead to changes in the geometry of the bridge and cause additional internal forces and deformations. Typical examples are the following cases:

- Systems in which the deformations are significant for action effects and second-order analysis needs to be conducted
- Bridges with complicated erection in which great accuracy is necessary
- Curved bridges
- Bridges with slender piers (e.g., over deep valleys)

For the aforementioned reasons, the following particularities during calculations should be considered:

- The design values for the movements of bearings are determined according to the regulations EN 1990.
- A mean value for the creep factor φ_t is used that is multiplied with 1.35 for the persistent design situations.
- A mean value for the shrinkage factor ε_{cs} is used that is multiplied with 1.60 for the persistent design situations.
- Nonuniform distribution of permanent loads is taken into account by applying $\pm 0.05 \cdot G_k$ on the influence line for uplift and for anchoring.

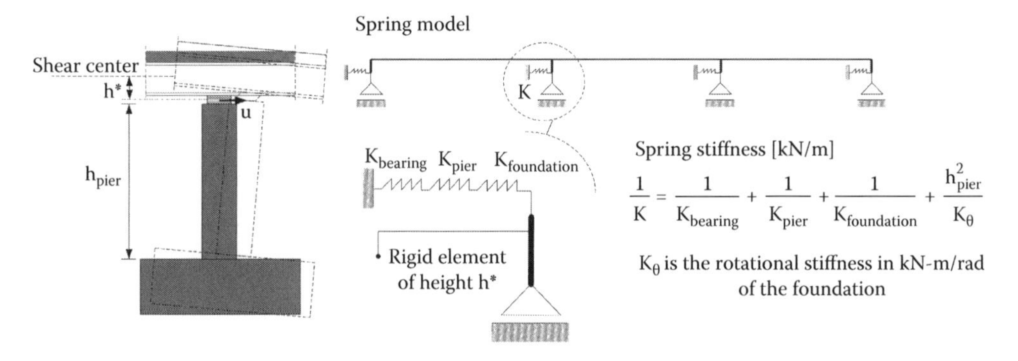

Figure 13.14 Spring model for the calculation of movements and restraints.

- The bridge should be calculated with second-order theory by taking into account deformations of the piers if required by EN 1992-1-1, 5.8.2 (6). Then equivalent geometric imperfections with only 50% of the geometric member imperfections specified in EN 1992-1-1, 5.2 are applied.
- In the case of an elastic global analysis, the elastic horizontal stiffnesses of the foundations, piers, and bearings may be modeled by individual springs (Figure 13.14). The individual springs ($K_{bearing}$, K_{pier}, and $K_{foundation}$) are combined to a global spring stiffness (K) at the bearing location for the calculation of the movements and restraints. A grillage model can be used. Moreover, the neutral displacement point shown in Figure 2.45 can be calculated reasonably accurate.

The aforementioned model can be used both for static and dynamic loadings. For concrete piers, an effective stiffness based on each cross section's moment–curvature diagram should be calculated. EN 1998-2 provides guidance for the estimation of the effective stiffness of reinforced concrete ductile members. The rotational stiffness of the foundation should take into account the soil's type.

REMARK 13.4

In bridges with piers of low height (≤10 m), the stiffness of the bearings is usually much smaller than all the other stiffness values (K_{pier}, $K_{foundation}$, etc.). In such cases, the stiffness of the springs in Figure 13.14 is practically equal to $K_{bearing}$.

The change in the position of the bearings due to climatic changes should be considered both during erection and at the final stage. Excessive changes in the position of the supports should be avoided both for structural and constructional reasons. Otherwise, resetting procedures during construction are necessary and unexpected costs may arise. For the common bridges (e.g., plate girder bridges), the most adverse combination of action effects with the uniform temperature components ΔT_N of Section 4.7.2 is sufficient for the design of bearings (see Example 13.1). However, in regions with extreme temperature variations between day and night, bearings should be calculated with the following temperature difference:

$$\Delta T_d = \Delta T_0 + \Delta T_\gamma + \Delta T_K \quad (13.25)$$

where

ΔT_0 is a temperature uncertainty for the reference temperature T_0 (T_0 refers to the final geometrical form of the bridge during installation of the bearings)
ΔT_0 is equal to:
0°C if installation takes place with measured temperature and with correction by resetting
15°C if installation takes place with estimated temperature and without correction by resetting with bridge set at $T_0 \pm 10$°C
30°C if installation takes place with estimated temperature and without correction by resetting and also one or more changes in the position of the fixed bearing
ΔT_γ is the additional safety term to allow for the temperature difference in the bridge given in the National Annex
ΔT_K is the safety term to take into account the uncertainty of the position of the bearing at the reference temperature given in the National Annex

In the absence of recommendations for ΔT_d in the National Annex, the following expression can be used:

$$\Delta T_d = T_{ed,max} - T_{ed,min} \quad (13.26)$$

with the constant temperature components being equal to

$$T_{ed,max} = T_0 + \gamma_F \cdot \Delta T_{N,exp} + \Delta T_0 \quad (13.27a)$$

$$T_{ed,min} = T_0 - \gamma_F \cdot \Delta T_{N,con} - \Delta T_0 \quad (13.27b)$$

where

γ_F is a partial safety factor (=1.35)
$\Delta T_{N,exp}$ is the extreme characteristic value of the uniform temperature component for expansion (see EN 1991-1-5 or see Section 4.7.2)
$\Delta T_{N,con}$ is the extreme characteristic value of the uniform temperature component for contraction (see EN 1991-1-5 or see Section 4.7.2)

If a second-order analysis is conducted, ΔT_d is determined incrementally.

13.8 BEARING SCHEDULES, SUPPORT PLANS, AND INSTALLATION DRAWINGS

A drawing of the *support plan* with the symbols of EN 1337-1, Table 1, should be prepared. The drawing should contain the following:

- A simplified bearing layout (see Figure 13.1)
- The type of bearing at each location
- Details at the bearing locations as recess and reinforcements
- Bedding and fixing details
- Additional information that the designer considers important for the bearings' manufacturer and the constructors, for example, indicator devices (see Remark 13.5)

REMARK 13.5

In many bridges, bearings are equipped with *indicator devices* for the measurement of translations. Such devices consist of a measuring scale and a pointer that are mounted on a well-visible area of the bearing.

The support plans are followed by the *bearings' schedules* that ensure that bearings are designed and constructed so that under the influence of all possible actions, unfavorable effects of the bearing on the structure are avoided. Therefore, a bearing schedule contains a detailed list of forces and movements of the bearings for each action. Other performance characteristics can be included. Such documents are given to the bearing producers to design the bearings according to the rules in EN 1337. EN 1993-2 gives a typical bearing schedule in Annex A similar to that of Table 13.1.

In the bearing schedule, the number of the bearing is included in order to be located in the layout of the support plan. The bearing type is also given (e.g., elastomeric C 2). One can see that the bridge's temperature reference T_0 is also provided.

Together with the bearing schedules, *bearing installation drawings* should be prepared. In these drawings, the installation procedure is explained so that a stress-free construction process is feasible.

EXAMPLE 13.1

The twin-girder bridge of Example 13.1 is supported by simple low-damping elastomeric bearings. The bridge is continuous with two spans of 25 m each (Figure 13.15). The bridge is located in a seismic area with peak ground acceleration 0.10 g. The characteristic values of the response spectrum are $T_B = 0.15$ s, $T_C = 0.50$ s, and $T_D = 2.0$ s. The importance factor of the bridge is $\gamma_I = 1.0$, and the behavior factor for elastomeric bearings is $q = 1.0$. The thickness of the surface is 50 mm.

Soil factor $S = 1.0$.

The bearing dimensions are to be specified and verified.

Actions on the bridge

The actions to be considered and the relevant load cases are described in Table 13.2. Traffic loads are represented by LM1 (see Section 4.2.2). Uniform distributed loading (UDL) and tandem system (TS) traffic loads are considered separately since they are introduced with different coefficients in the groups of traffic loads (see Table 4.7). For maximum support forces at the internal supports, traffic loads are imposed on the entire bridge deck, while for maximum forces at end supports, on one span only (Figure 13.16).

Braking forces are determined in accordance with Section 4.2.4 as follows:

- For bearings at internal supports

  Equation 4.3: $Q_{lkm} = 0.6 \cdot 1 \cdot 2 \cdot 300 + 0.1 \cdot 1 \cdot 9 \cdot 3 \cdot 50 = 495$ kN
  and $180 \text{ kN} \leq 495 \text{ kN} \leq 900$ kN

- For bearings at end supports

  Equation 4.3: $Q_{lke} = 0.6 \cdot 1 \cdot 2 \cdot 300 + 0.1 \cdot 1 \cdot 9 \cdot 3 \cdot 25 = 427.5$ kN
  and $180 \text{ kN} \leq 427.5 \text{ kN} \leq 900$ kN

Table 13.1 Typical bearing schedule

Bearing schedule no.		*Bearing no.* *Type*			*Reference temperature T_0 (°C)* *Uncertainty ΔT_0 (°C): ±*		
		Max/Min *V* *w*	*Max/Min* *Hx* *vx*	*Max/Min* *Hy* *vy*	*Max/Min* *Mz* *θz*	*Max/Min* *Mx* *θx*	*Max/Min* *My* *θy*
Permanent							
1	Self-weight	kN	kN	kN	kN-m	kN-m	kN-m
		mm	mm	mm	mrad	mrad	mrad
2	Dead load	kN	kN	kN	kN-m	kN-m	kN-m
		mm	mm	mm	mrad	mrad	mrad
3	Creep–shrinkage	kN	kN	kN	kN-m	kN-m	kN-m
		mm	mm	mm	mrad	mrad	mrad
Variable							
4	Traffic	kN	kN	kN	kN-m	kN-m	kN-m
		mm	mm	mm	mrad	mrad	mrad
5	Braking/acceleration	kN	kN	kN	kN-m	kN-m	kN-m
		mm	mm	mm	mrad	mrad	mrad
6	Centrifugal	kN	kN	kN	kN-m	kN-m	kN-m
		mm	mm	mm	mrad	mrad	mrad
7	Nosing	kN	kN	kN	kN-m	kN-m	kN-m
		mm	mm	mm	mrad	mrad	mrad
8	Footpath	kN	kN	kN	kN-m	kN-m	kN-m
		mm	mm	mm	mrad	mrad	mrad
9	Wind	kN	kN	kN	kN-m	kN-m	kN-m
		mm	mm	mm	mrad	mrad	mrad
10	Temperature	kN	kN	kN	kN-m	kN-m	kN-m
		mm	mm	mm	mrad	mrad	mrad
11	Settlement	kN	kN	kN	kN-m	kN-m	kN-m
		mm	mm	mm	mrad	mrad	mrad
Accidental							
12	Derailment	kN	kN	kN	kN-m	kN-m	kN-m
		mm	mm	mm	mrad	mrad	mrad
13	Collision	kN	kN	kN	kN-m	kN-m	kN-m
		mm	mm	mm	mrad	mrad	mrad
Seismic							
14	Earthquake (ULS)	kN	kN	kN	kN-m	kN-m	kN-m
		mm	mm	mm	mrad	mrad	mrad
15	Earthquake (SLS)	kN	kN	kN	kN-m	kN-m	kN-m
		mm	mm	mm	mrad	mrad	mrad

Table 13.1 (continued) Typical bearing schedule

	Max/Min	*Max/Min*	*Max/Min*	*Max/Min*	*Max/Min*	*Max/Min*
	V	*Hx*	*Hy*	*Mz*	*Mx*	*My*
	w	*vx*	*vy*	*θz*	*θx*	*θy*
Combinations						
1	kN	kN	kN	kN-m	kN-m	kN-m
	mm	mm	mm	mrad	mrad	mrad
2	kN	kN	kN	kN-m	kN-m	kN-m
	mm	mm	mm	mrad	mrad	mrad
3	kN	kN	kN	kN-m	kN-m	kN-m
	mm	mm	mm	mrad	mrad	mrad
4	kN	kN	kN	kN-m	kN-m	kN-m
	mm	mm	mm	mrad	mrad	mrad
5	kN	kN	kN	kN-m	kN-m	kN-m
	mm	mm	mm	mrad	mrad	mrad
6	kN	kN	kN	kN-m	kN-m	kN-m
	mm	mm	mm	mrad	mrad	mrad

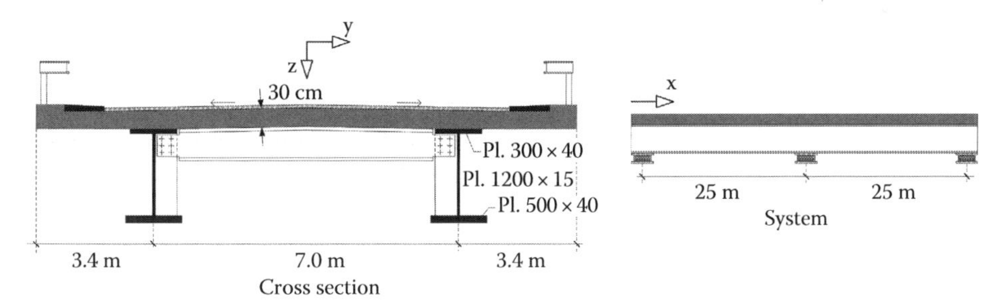

Figure 13.15 Continuous bridge of Example 13.1.

These forces are uniformly distributed along the main girders. Each main girder is assigned a longitudinal force $495/(2 \cdot 50) = 4.95$ kN/m over the entire bridge length for the first case and $427.5/(2 \cdot 25) = 8.55$ kN/m over one span for the second case.

Wind forces equal to 3.6 kN/m are considered over the entire bridge length in the transverse direction. They are shared between the two main girders.

Dimensions and mechanical properties of elastomeric bearings

Type B elastomeric bearings (Figure 13.2) are selected in accordance with preliminary calculations. The bearings do not have any holes. Their dimensions are given in Table 13.3.

The bearings are represented by means of springs acting in longitudinal and transverse directions. The spring constant is determined in accordance with Equation 13.14.

For the internal bearings, it

Total nominal thickness of the elastomer, Figure 13.6: $T_e = T_b - (n+1) \cdot t_s = 60$ mm

Plan area: $A = 45 \cdot 50 = 2250$ cm²

Spring constant, Equation 13.14: $K_x = K_y = \dfrac{0.09 \cdot 2250}{6} \cdot 100 = 3375$ kN/m

Similarly for the end bearings, it is $K_x = K_y = 2000$ kN/m

Table 13.2 Actions for dimensioning of bearings

LC	*Symbol*	*Description*
1	G	Self-weight of superstructure
2	LM1 middle	LM1, unfavorable for middle bearings
3	LM1 frequent middle	Frequent LM1, unfavorable for middle bearings
4	Q_{lkm}	Braking/acceleration force, unfavorable for middle bearings
5	LM1 end	LM1, unfavorable for end bearings
6	LM1 frequent end	frequent LM1, unfavorable for end bearings
7	Q_{lke}	Breaking/acceleration force, unfavorable for end bearings
8	$\Delta T_{N,con}$ (see Section 4.7.2)	Uniform temperature contraction for bearings 45°C
9	$\Delta T_{N,exp}$ (see Section 4.7.2)	Uniform temperature expansion for bearings 55°C
10	$\Delta T_{M,heat}$ (see Section 4.7.3)	Temperature difference (heating), top warmer 15°C
11	$\Delta T_{M,cool}$ (see Section 4.7.3)	Temperature difference (cooling), top colder 18°C
12	W	Wind in transverse direction
13	S	Shrinkage at time ∞
14	C	Creep at time ∞
15	E_x	Earthquake in x (longitudinal) direction
16	E_y	Earthquake in y (transverse) direction

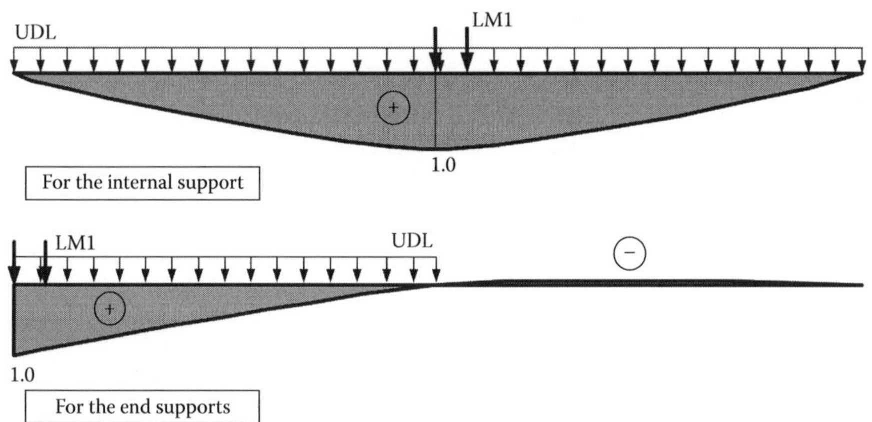

Figure 13.16 Influence lines and position of LM 1 for the support reactions at internal and end supports.

Global analysis and combination of actions

For global analysis, the bridge is represented by a grillage. The longitudinal beam elements represent the main composite girders, and the transverse beams the concrete slab.

Cracked analysis is made considering cracked cross-sectional properties at a length 15% of the span adjacent to the internal support. The bearings are introduced as springs acting in longitudinal (x) and transverse directions (y), while displacements are blocked in the vertical z direction (see Remark 13.2).

Global analysis provides reactions, displacements, and rotations at supports for all load cases considered. These are appropriately combined to form ULS combinations. For traffic loads, groups of loads are considered in accordance with Table 4.7. In the specific case, groups 1a and 2 are considered. gr1a includes the characteristic values of LM 1. gr2 includes the frequent values

Table 13.3 Dimensions of elastomeric bearings in millimeters (see Figure 13.6)

	a	*b*	T_b	*n*	t_i	t_s	e
Internal support	450	500	84	5	11	4	2.5
End support	250	400	63	5	8	3	2.5

of LM 1 set equal to 3 kN/m² over the entire deck and the characteristic values of the braking/acceleration forces. Accordingly, the following groups of traffic loads are considered:

gr1a: TS + UDL (load cases 2 or 5)
gr2: (TS + UDL) frequent + breaking/acceleration (load cases 3 + 4 or 6 + 7)

Subsequently, ULS load combinations in accordance with Table 5.6 are formed. The load combinations considered for the specific bridge are given in Table 13.4. Analysis results for the bearings at internal supports for two combinations are illustrated indicatively in Table 13.5.

Verification of bearings

The bearing verifications shall be performed for all combinations. In the following, verifications for the bearings at internal supports for the load combination 1 in accordance with Table 13.4 will be illustrated. Forces, rotations, and displacements are presented in Table 13.5, line 1.

Length of steel plates, Figure 13.6: $a' = 450 - 10 = 440$ mm
Width of steel plates, Figure 13.6: $b' = 500 - 10 = 490$ mm

Area: $A = a \cdot b = 45 \cdot 50 = 2250 \text{ cm}^2$

Area of steel plates: $A_1 = 44 \cdot 49 = 2156 \text{ cm}^2$

Table 13.4 Combinations of actions for bearing design

Line	*No. of combination*	*Combination*
1	1	$1.35 \cdot (G + C_{sec}) + S_{sec} + 1.35 \cdot gr1a + 1.5 \cdot 0.6 \cdot W$
2		$1.35 \cdot (G + C_{sec}) + S_{sec} + 1.35 \cdot gr1a + 1.5 \cdot 0.6 \cdot T$
		T is
	2	$\Delta T_{N,con}$ or
	3	$\Delta T_{N,exp}$ or
	4	$\Delta T_{M,heat}$ or
	5	$\Delta T_{M,cool}$ or
	6	$\Delta T_{M,heat} + 0.35 \cdot \Delta T_{N,exp}$ or
	7	$\Delta T_{M,cool} + 0.35 \cdot \Delta T_{N,con}$ or
	8	$0.75 \cdot \Delta T_{M,heat} + \Delta T_{N,exp}$ or
	9	$0.75 \cdot \Delta T_{M,cool} + \Delta T_{N,con}$
3	10–17	$1.35 \cdot (G + C_{sec}) + S_{sec} + 1.35 \cdot gr2 + 1.5 \cdot 0.6 \cdot T$
		For T, see line 2
4	18–25	$1.35 \cdot (G + C_{sec}) + S_{sec} + 1.35 \cdot (0.75 \cdot TS + 0.4 \cdot UDL + 0.4 \cdot q_{fk}^*) + 1.5 \cdot T$
		For T, see line 2
5	26–27	$1.35 \cdot (G + C_{sec}) + S_{sec} + 1.35 \cdot (0.75 \cdot TS + 0.4 \cdot UDL + 0.4 \cdot q_{fk}^*) \pm 1.5 \cdot W$ (loaded bridge)
6	28–29	$1.35 \cdot (G + C_{sec}) + S_{sec} \pm 1.5 \cdot W$ (unloaded bridge)

Table 13.5 Support reactions, rotations, and displacements for internal bearings

No. of combination	P_z *(kN)*	P_x *(kN)*	P_y *(kN)*	$a_{a,d}$ *(mrad)*	$a_{b,d}$ *(mrad)*	$v_{x,d}$ *(mm)*	$v_{y,d}$ *(mm)*
1	5959	≈0	37.9	4.88	0.25	≈0	9.29
26	4775	≈0	63.8	3.9	0.184	≈0	15.6

Note: For illustration purposes, values for load combination 1 and 26 are given. However, in practice, all combinations shall be examined.

Shape factor, Equation 13.4: $S = \dfrac{44 \cdot 49}{2 \cdot 1.1 \cdot (44+49)} = 10.54$

Reduced area, Equation 13.3: $A_r = 2156 \cdot \left(1 - \dfrac{0}{44} - \dfrac{0.929}{49}\right) = 2115\,\text{cm}^2$

a. **Check of distortion**

Distortion due to compression, Equation 13.2: $\varepsilon_{c,d} = \dfrac{1.5 \cdot 5959}{0.09 \cdot 2115 \cdot 10.54} = 4.46$

Shear deformation: $v_{xy,d} = \sqrt{0^2 + 0.929^2} = 0.929\,\text{cm}$

Total thickness of the elastomeric layers, Figure 13.6: $T_q = T_e = 6.0$ cm

Distortional deformation, Equation 13.5: $\varepsilon_{q,d} = \dfrac{0.929}{6.0} = 0.15$

Distortion due to angular rotation, Equation 13.6: $\varepsilon_{a,d} = \dfrac{(44^2 \cdot 4.88 + 49^2 \cdot 0.25) \cdot 10^{-3}}{2 \cdot 5 \cdot 1 \cdot 1^2} = 0.83$

Total design distortion, Equation 13.1: $\varepsilon_{t,d} = 1.0 \cdot (4.46 + 0.15 + 0.83) = 5.44 < 7$ with $K_L = 1$ and $\varepsilon_{q,d} = 0.15 < 1.0$ (sufficient)

b. **Check of the tension of the steel plates**

Equation 13.7: required $t_s = \dfrac{1.3 \cdot 5959 \cdot 2 \cdot 1.1 \cdot 1 \cdot 1}{2115 \cdot 23.5} \cdot 10 = 3.4\ \text{mm} > 2\ \text{mm}$

$t_s = 4\ \text{mm} > 3.4\ \text{mm}$ (sufficient)

c. **Limitation of rotation**

Equation 13.8: $\dfrac{5959 \cdot 5 \cdot 1.1}{2156} \cdot \left(\dfrac{1}{5 \cdot 0.09 \cdot 10.54^2} + \dfrac{1}{200}\right) = 0.38 \geq \dfrac{44 \cdot 4.88 + 49 \cdot 0.25}{3 \cdot 10^3} = 0.076$ (sufficient)

d. **Check of stability**

Equation 13.10: $\dfrac{5959}{2115} = 2.82 < \dfrac{2 \cdot 44 \cdot 0.09 \cdot 10.54}{3 \cdot 6} = 4.64$ (sufficient)

e. **Safety against slip**

$F_{z,Gmin} = F_{z,d,min} = 2231$ kN

Mean compression stress: $\sigma_m = \dfrac{2231}{2250} = 0.99\ \text{kN/cm}^2 = 9.9\ \text{MPa}$

Friction coefficient, Equation 13.13: $\mu_e = 0.1 + \dfrac{1.5 \cdot 0.6}{9.9} = 0.19$

Shear force, Equation 13.11: $F_{xy,d} = \sqrt{0^2 + 37.9^2} = 37.9\ \text{kN} \leq 0.19 \cdot 2231 = 423.89\ \text{kN}$

Equation 13.12: $\dfrac{F_{Z,Gmin}}{A_r} = \dfrac{2231}{2115} \cdot 10 = 10.5\ \text{MPa} \geq 3\ \text{MPa}$ (sufficient)

Seismic design

The masses for seismic analysis correspond to the self-weight of the bridge (G) plus $\psi_{2,1}$ = 0.2 of the $Q_{k,1} = 2 \cdot 300 = 600$ kN traffic load (see Table 4.18). The total weight in the seismic situation is accordingly equal to $7496 + 0.2 \cdot 600 = 7616$ kN. Traffic loads are placed over lane 1 to account for possible eccentricities.

For seismic analysis, minimum and maximum values $G_{b,min} = 1.0 \cdot G_b$ and $G_{b,max} = 1.5 \cdot G_b$ ($G_b = 1.1 \cdot G$) are used for the shear modulus of the elastomer. The spring constants for the internal bearings are then $K_{min} = 3712.5$ kN/m and $K_{max} = 5568.75$ kN/m, and similarly, 2200 kN/m and 3300 kN/m for the end bearings.

Multimodal response spectrum analysis is performed. The resulting fundamental modes of vibration correspond to translations in longitudinal and transverse directions. The fundamental periods are almost equal in both directions since translations are due to the flexibility of the bearings only, which is equal in both directions. The participating mass factors for both modes amount to almost 100%. Consequently, the fundamental mode method may also be applied (Table 4.19). This is done in the following for illustration purposes.

The overall stiffness of the system equals to the sum of stiffness of the bearings (four end, two middle bearings). Therefore, $K_{max,tot} = 24{,}337.5$ kN/m, and $K_{min,tot} = 16{,}225$ kN/m.

The fundamental periods are determined from Table 4.19.

For $K_{max,tot}$, it is $T = 2 \cdot \pi \cdot \sqrt{\dfrac{7616/9.81}{24337.5}} = 1.12\,\text{s}$, and similarly, T = 1.37 s for $K_{min,tot}$.

The corresponding base shears with $T_C \leq T \leq T_D$, Equation 4.24c, are

$$V_b = 7616 \cdot \left(0.1 \cdot 1.0 \cdot \frac{2.5}{1.0} \cdot \frac{0.5}{1.12}\right) = 850 \text{ kN}$$

for K_{max}, and similarly, $V_b = 694.9$ kN for K_{min}.

The deck and bearing translations, equal in the two directions, are determined from $u = (850/24337.5) \cdot 1000 = 34.9$ mm for K_{max}, and similarly, $u = 42.8$ mm for K_{min}.

It may be confirmed that upper values lead to maximum forces, and lower values to maximum displacements. The horizontal forces of bearings for K_{max} are at internal supports $F_x = F_y = 5568.75 \cdot 34.9/10^3 = 194.3$ kN and at end supports $F_x = F_y = 3300 \cdot 34.9/10^3 = 115.2$ kN.

It may be confirmed that the base shear is $V_b = 2 \cdot 194.3 + 4 \cdot 115.2 \approx 850$ kN.

Table 13.6 provides the design values of internal bearings in the seismic situation.

The bearing verification procedures for internal bearings for the seismic situation are similar as for the basic ULS combinations. The most critical are the safety against slip for K_{max} and bearing distortion for K_{min} that are illustrated subsequently.

Table 13.6 Forces, rotations, and displacements for internal bearings in the seismic situation

Calculation with	P_z *(kN)*	P_x *(kN)*	P_y *(kN)*	$a_{a,d}$ *(mrad)*	$a_{b,d}$ *(mrad)*	$v_{x,d}$ *(mm)*	$v_{y,d}$ *(mm)*
K_{max}	2331	194	194	≈0	≈0	34.9	34.9
K_{min}	2331	159	159	≈0	≈0	42.8	42.8

- Safety against slip for K_{max}

  Reduced area, Equation 13.3: $A_r = 2156 \cdot \left(1 - \frac{3.49}{44} - \frac{3.49}{49}\right) = 1831.4\ \text{cm}^2$

  Shear force: $F_{xy,d} = \sqrt{194^2 + 194^2} = 274.4\ \text{kN}$

  Mean compression stress: $\sigma_m = \frac{2331}{2250} = 1\ \text{kN/cm}^2 = 10\ \text{MPa}$

  Friction coefficient, Equation 13.13: $\mu_e = 0.1 + \frac{1.5 \cdot 0.6}{10} = 0.19$

  Equation 13.11: $F_{xy,d} = 274.4\ \text{kN} < 0.19 \cdot 2331 = 442.89\ \text{kN}$ (sufficient)

  $$\frac{F_{z,Gmin}}{A_r} = \frac{2331}{1831.4} \cdot 10 = 12.7\ \text{MPa} \geq 3\ \text{MPa (sufficient)}$$

- Bearing distortion for K_{min}

  Reduced area, Equation 13.3: $A_r = 2156 \cdot \left(1 - \frac{4.28}{44} - \frac{4.28}{49}\right) = 1758\ \text{cm}^2$

  Distortion due to compression, Equation 13.2: $\varepsilon_{c,d} = \frac{1.5 \cdot 2331}{(1.1 \cdot 0.09) \cdot 1758 \cdot 10.54} = 1.91$

  Shear distortion: $v_{xy,d} = \sqrt{v_{x,d}^2 + v_{y,d}^2} = \sqrt{4.28^2 + 4.28^2} = 6.05\ \text{cm}$

  Distortional deformation, Equation 13.5: $\varepsilon_{q,d} = \frac{1.5 \cdot 6.05}{6} = 1.51 < 2$ (sufficient, see Remark 13.3)

  Distortion due to angular rotation, Equation 13.6: $\varepsilon_{a,d} = \frac{44^2 \cdot 0 + 49^2 \cdot 0}{2 \cdot 5 \cdot 1.1^2} = 0$

  Total design distortion, Equation 13.1: $\varepsilon_{t,d} = 1.0 \cdot (1.91 + 1.51 + 0) = 3.42 < 7$ (sufficient) with $K_L = 1$.

REMARK 13.6

In Example 13.1:

- The stiffness of the piers and the foundations were not taken into account. It was assumed that $K_{pier}, K_{foundation} \gg K_{bearing}$ (see also Remark 13.4). A more detailed 3D model can be chosen.
- The verifications for the bearings were based on the results of the most adverse combination of action effects. Due to the simplicity of the bridge and the mild climate conditions, this can be described as an acceptable approach. In a different case, the increased temperature of Equation 13.25 according to the recommendations of the National Annex should be applied, alternatively Equation 13.26.

13.9 FLUID VISCOUS DAMPERS

These *dampers* behave like safety belts, developing low resistance at slow loading velocities and high resistance at high loading velocities (Figure 13.17a). Consequently, displacements are not restrained at service conditions due to temperature changes, creep, or shrinkage, but the dampers "block" and restrain deformations at higher velocities like during an

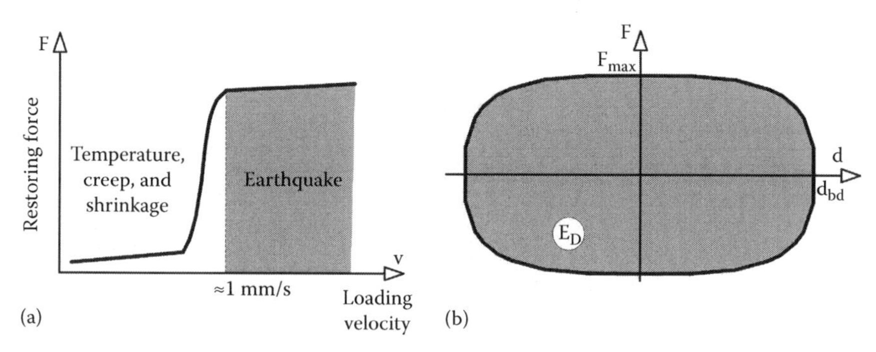

Figure 13.17 (a) Response of hydraulic viscous dampers for various loading velocities and (b) response to cyclic loading.

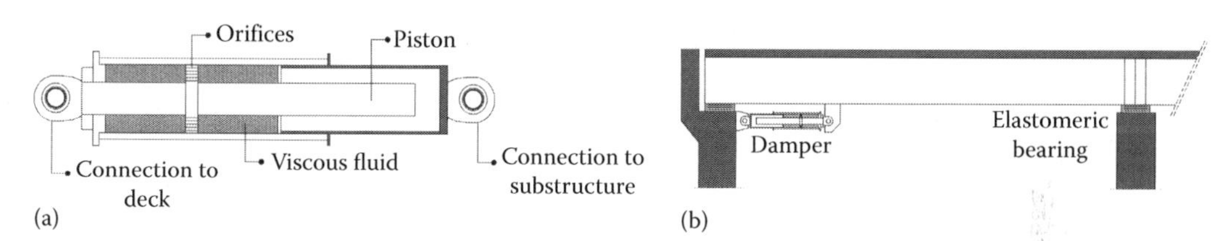

Figure 13.18 (a) Schematic representation of a fluid damper and (b) combination of dampers with elastomeric bearings.

earthquake, sudden breaking, or acceleration. Damping is produced by the displacement of a piston moving in a cylinder filled with oil, silicon, or similar materials (Figure 13.18a). Small orifices allow the flow of oil for low loading velocities, while at higher speeds, the orifices prevent the free flow and damp the movement. Dampers are called in EN 1998-2 *shock transmission units*.

The reaction (or damping) force is a function of the loading speed in accordance with Equation 13.28 and is illustrated in Figure 13.17a for a sinusoidal motion:

$$F = C \cdot v^a \tag{13.28}$$

where

C is a device-specific viscous damping coefficient [kN·s/m]
v is the loading velocity [m/s]
a is a device-specific damping exponent (usually 0.2–0.25)

Such dampers are used in combination with bearings. Their application is associated with nonlinear time history analysis since they cannot be modeled in the frame of a spectrum analysis (see Figure 13.18b).

13.10 FRICTION DEVICES

Friction devices exploit the energy dissipation with development of friction. The restoring force for flat sliding surfaces is equal to

$$F = \mu_d \cdot N_{Ed} \cdot \text{sign}(\dot{d}_{bd}) \tag{13.29}$$

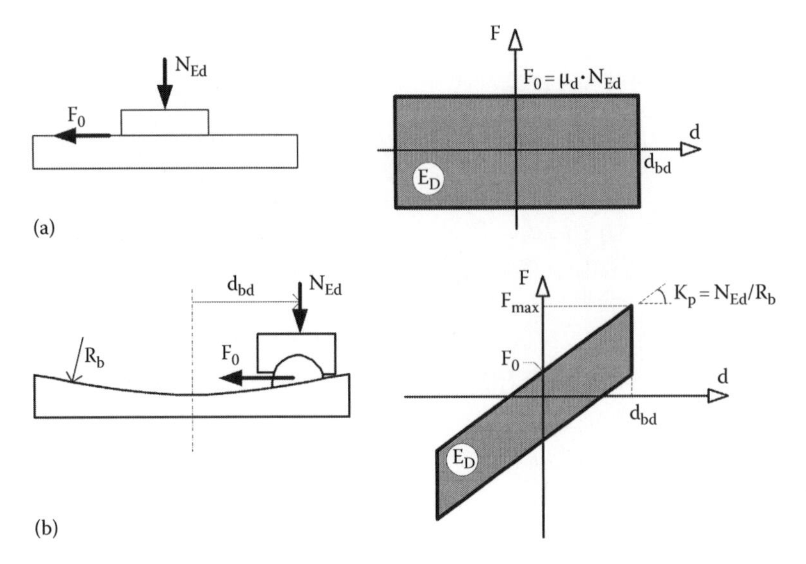

Figure 13.19 (a) Response of friction devices with flat and (b) spherical sliding surfaces.

while the dissipated energy by

$$E_d = 4 \cdot \mu_d \cdot N_{Ed} \cdot d_{bd} \tag{13.30}$$

where

μ_d is the dynamic friction coefficient
N_{Ed} is the applied vertical force
d_{bd} is the maximal design device displacement
$\text{sign}(\dot{d}_{bd})$ is the sign of the velocity vector

It may be seen that such devices (Figure 13.19) have zero stiffness and no restoring capability so that they must be complemented by additional devices. Other devices with curved sliding surfaces such as friction pendulum devices have been developed that do have restoring capabilities and improved stiffness properties (Figure 13.19b). The restoring force is given by

$$F = \frac{N_{Ed}}{R_d} \cdot d_{bd} + \mu_d \cdot N_{Ed} \cdot \text{sign}(\dot{d}_{bd}) \tag{13.31}$$

where R_b is the radius of the spherical surface and all other symbols as in Equation 13.29.

The dissipated energy is the same as for flat sliding surfaces and is given by Equation 13.30.

13.11 EXPANSION JOINTS

As already explained, horizontal deformations of the superstructure arise due to temperature, creep, shrinkage, earthquakes, traffic, etc. These deformations are associated with significant uncertainties and accurate calculations are obviously not feasible. *Expansion*

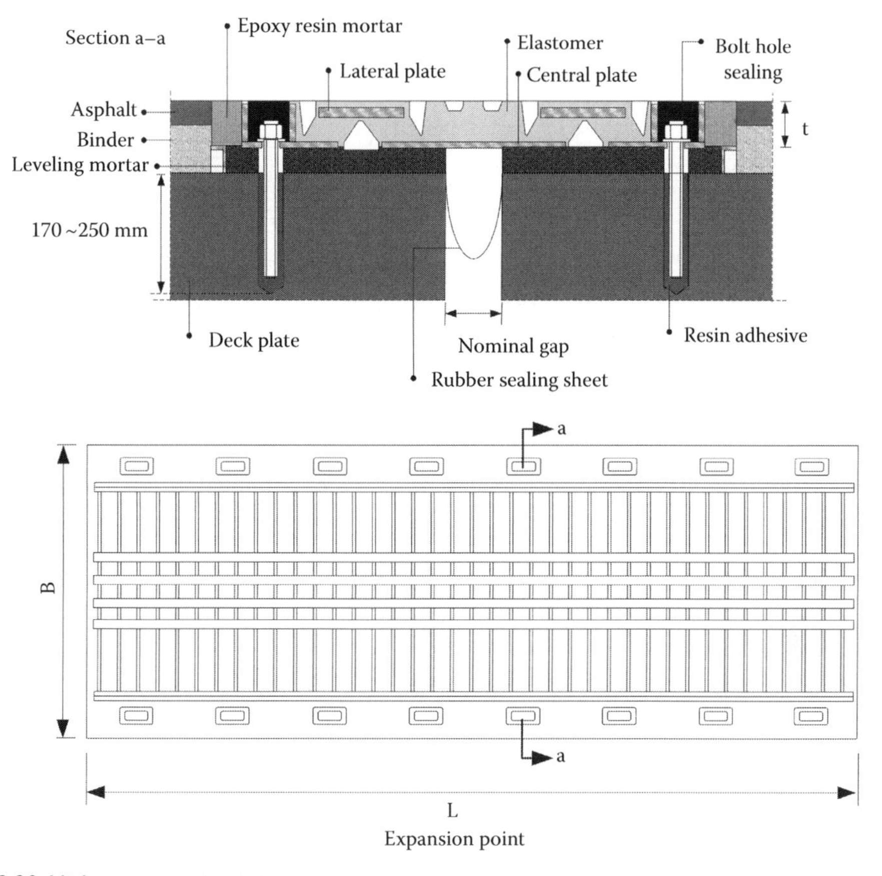

Figure 13.20 Highway expansion joint.

joints are flexible links that connect independent parts of a road bridge at piers and the abutments. They are capable of absorbing the aforementioned deformations, and in case of failure, they can be easily replaced. A typical cross section of an expansion joint is shown in Figure 13.20 [13.1]. The joint is made of natural or synthetic rubber with steel plates embedded in it (reinforced elastomer). The dimensions of the joint BxLxt depend on the required design movement. For small-span bridges, longitudinal deformations usually range between ±20 and ~40 mm with $BxLxt \approx (250 \sim 400) \times 2000 \times (30 \sim 50)$. For long-span applications, the required deformations may exceed ±300 mm and $BxLxt \geq 1000 \times 2000 \times 80$. The expansion joint in Figure 13.20 is fixed through chemical anchors in the deck plate.

Expansion joints should

- Not increase the degree of the bridge's static indeterminacy by restraining degrees of freedoms at supports
- Be waterproof
- Produce low noise when vehicles are passing over them

Expansion joints should be manufactured and designed according to the regulations of the European Technical Approval (ETA) [13.8]. Such a document specifies the design guidelines of the expansion joint that are compatible with the requirements of the Eurocodes.

Alternative types of expansion joints are described in EN 1993-2 [13.7] for the use in steel bridges and they are as follows:

- Buried expansion joint
  The surfacing is continuous over the joint gap and the expansion joint is not flush with the running surface. The joint consists of waterproofing membranes or an elastomeric pad and is formed in situ.
- Flexible expansion joint
  This is an in situ poured joint that is flush with the running surface. The joint gap is covered with steel plates that support the surfacing materials (aggregates or binder).
- Nosing expansion joint
  It has lips or edges made of concrete, resin mortar, or elastomeric. The gap between the edges is filled by a prefabricated flexible profile. The components of the joints are not flush with the running surface.
- Mat expansion joint
  The movements of the structure are absorbed by a flexible prefabricated elastic strip. The strip is fixed by bolts to the structure (see Figure 13.20). The joints' components are flush with the running surface.
- Cantilever expansion joint
  Cantilever symmetrical and nonsymmetrical elements (such as comb or sawtooth plates) are anchored on one side of the deck joint gap and interpenetrated to span the deck joint gap. The elements are flush with the running surface.
- Supported expansion joint
  This joint consists of an element that is fixed by hinges on one side and sliding supports on the other side. This element is flush with the running surface and spans the deck joint gap. Movements are allowed through sliding on the non-fixed side of the hinged element.
- Modular expansion joint
  Steel beams encased in watertight materials bridge the joint gap in a way that a moveable joint is formed. The beams are flush with the running surface.

More details are found in the "Guideline for European Technical Approval of Expansion Joints for Road Bridges" [13.8].

An *expansion joint schedule* should be prepared so that the final design is verified by the manufacturer. This schedule should contain the arrangement of the expansion joints in conjunction with the geometry of the bridge and a list of actions and imposed deformations. Moreover, the designer should describe in detail the installation procedure.

REFERENCES

[13.1] AGOM International SRL: Metal rubber engineering, Bridge Expansion Joints, Italy.

[13.2] Bridge design to Eurocodes, Worked examples, JRC scientific and technical reports, EUR 25193 EN-2012.

[13.3] Caltrans: Bridge Bearings, Memo to designers 7-1. Caltrans, Eureka, CA, 1994.

[13.4] Eggert, H., Kauschke, W.: *Lager im Bauwesen*. Ernst & Sohn, Berlin, Germany, 1995.

[13.5] EN 1337, CEN (European Committee for Standardization): Structural bearings, Parts 1 to 11 CEN, 2000 to 2006.

[13.6] EN 1998-2, CEN (European Committee for Standardization): Design of structures for earthquake resistance, Part 2 bridges, CEN 2005.

[13.7] EN 1993-2, CEN (European Committee for Standardization): Design of steel structures—Part 2: Steel bridges, 2006.
[13.8] ETAG 032: *Expansion Joints for Road Bridges*. EOTA, Albuquerque, NM.
[13.9] Gumba, Bridge bearings, 2011.
[13.10] Kawaki Core-Tech Co. Ltd.: Steel Bearings.
[13.11] Lee, D. J.: *Bridge Bearings and Expansion Joints*, 2nd edn. E & FN Spon, London, U.K., 1994.
[13.12] Setra: Laminated elastomeric bearings, Technical Guide. Setra, Neu-Ulm, Germany, 2007.
[13.13] Steel bridge bearing. Selection and design guide. AISC, Chicago, IL.
[13.14] Steel Bridge Group: Guidance notes on best practice in steel bridge construction. Attachment of bearings 2.08, 2010.
[13.15] Steel Bridge Group: Guidance notes on best practice in steel bridge construction. Bridge bearings 3.03, 2010.

Index

A

B

C

D

G

H

I

J

K

L

M

O

P

T